DEPARTMENT OF THE ARMY
U.S. Army Corps of Engineers
Washington, DC 20314-1000

EM 1110-2-1100
(Change 2)

CECW-CE

Manual
No. 1110-2-1100

1 April 2008

Engineering and Design
COASTAL ENGINEERING MANUAL

1. Purpose. The purpose of the *Coastal Engineering Manual* (CEM) is to provide a comprehensive technical coastal engineering document. It includes the basic principles of coastal processes, methods for computing coastal planning and design parameters, and guidance on how to formulate and conduct studies in support of coastal flooding, shore protection, and navigation projects. This Change 2 to EM 1110-2-1100, 1 April 2008, includes the following changes and updates:

a. Part I-1. References were checked and some were deleted (Engineer Manuals that are no longer in the USACE inventory).

b. Part I-4. Minor changes were made in the text to better reflect the contents of subsequent parts of the CEM.

c. Part II-1. Figure II-1-9 has been revised; Equations II-1-128, II-1-160, and II-1-161 have been corrected.

d. Part II-2. Equations II-2-4, II-2-5, and II-2-32 have been corrected along with other errors reported by various users.

e. Part II-5. References were checked and some were deleted (Engineer Manuals that are no longer in the USACE inventory).

f. Part II-6. The value of "e" used in Eq. II-6-28 has been corrected.

g. Part II-7. The table of contents was corrected. A new section, II-7-11, Note to Users, Vessel Buoyancy, was added at the end of the chapter.

h. Part III-3. Corrections have been made to format and spelling. Different plots were added to Figures III-3-24 and III-3-26.

i. Part IV-1. Corrections have been made to references.

j. Part V-1. Citation of an Engineer Regulation has been corrected.

k. Part V-2. Citation of references has been changed, web pages with sources of wind and wave data have been added. Some minor text changes have also been made.

l. Part V-3. Citations of unpublished reports or personal communications have been deleted, and links to other figures or parts of the CEM have been checked and corrected.

m. Part V-4. Minor text changes, corrections to references and Figure V-4-1.

n. Part V-5. Links to other parts of the CEM that were planned but never written have been deleted.

2. Applicability. This manual applies to all HQUSACE elements and all USACE commands having Civil Works and military design responsibilities.

3. Discussion. The CEM is divided into five parts in two major subdivisions: science-based and engineering-based. The first four parts of the CEM and Appendix A compose the science-based subdivision:

Part I, "Introduction"
Part II, "Coastal Hydrodynamics"
Part III, "Coastal Sediment Processes"
Part IV, "Coastal Geology"
Appendix A, "Glossary"

The engineering-based subdivision is oriented toward a project-type approach, Part V, "Coastal Project Planning and Design."

4. Distribution Statement. Approved for public release, distribution unlimited.

5. Note to Users. Revised chapters are dated 1 April 2008. Readers need to download the entire new chapters and discard earlier versions in their possession.

FOR THE COMMANDER:

STEPHEN L. HILL
Colonel, Corps of Engineers
Chief of Staff

CECW-EH

DEPARTMENT OF THE ARMY
U.S. Army Corps of Engineers
Washington, DC 20314-1000

EM 1110-2-1100
(Part V)

Manual
No. 1110-2-1100

31 July 2003

Engineering and Design
COASTAL ENGINEERING MANUAL

1. Purpose. The purpose of the *Coastal Engineering Manual* (CEM) is to provide a comprehensive technical coastal engineering document. It includes the basic principles of coastal processes, methods for computing coastal planning and design parameters, and guidance on how to formulate coastal flood studies, shore protection, and navigation projects.

2. Applicability. This manual applies to all HQUSACE elements and all USACE commands having Civil Works and military design responsibilities.

3. Discussion. The CEM is divided into six parts in two major subdivisions: science-based and engineering-based. The first four parts of the CEM and an appendix were issued in 30 April 2002 and subsequently updated by change dated 31 July 2003. These included:

Part I, "Introduction"
Part II, "Coastal Hydrodynamics"
Part III, "Coastal Sediment Processes"
Part IV, "Coastal Geology"
Appendix A, "Glossary"

The engineering-based subdivision is oriented toward a project-type approach and is divided into two parts. Part V, "Coastal Project Planning and Design," is included in this transmittal. The text and figures provide information on the design process and selection of appropriate types of solution to various coastal problems. Part VI, "Design of Coastal Project Elements," which provides engineering guidance on materials, fundamentals of design, and reliability, is in preparation.

4. Distribution Statement. Approved for public release, distribution unlimited.

FOR THE COMMANDER:

MICHAEL J. WALSH
Colonel, Corps of Engineers
Chief of Staff

Table of Contents

List of Figures

List of Tables

Page

Chapter V-1
Planning and Design Process

V-1-1. Background

a. Introduction.

(1) Purpose. This chapter provides a comprehensive description of definitions and procedures needed in the planning and design process for coastal projects. The goal of the following sections is to provide the planner/designer with sufficient engineering guidance to accomplish the level of detail necessary to produce an acceptable finished product.

(2) Applicability. Although the information on procedures presented here is directly applicable to the reconnaissance and feasibility studies of the U.S. Army Corps of Engineers (USACE), it is also applicable to all other planning and design efforts because the process is generic. The process of developing a coastal project must be iterative to ensure that the final product is optimum from both the technical and economic viewpoints and will be acceptable to the various levels of decision makers and all project partners.

b. The planning process.

(a) The solution to coastal engineering problems begins with understanding the roles of the engineer as a planner and designer. These roles may be, and often are, embodied in the same person doing sequential tasks. This chapter presents the planning and design processes, the umbrella under which the project evolves from conception through functional detailed technical design phases. Examples of detailed technical designs are found in Part VI of this manual. Although the procedures and examples are those of the USACE, the principles apply to whomever is doing the planning and designing and follow a logical engineering process regardless of project type, scale, geographic siting, or sponsor.

(b) The planning process begins with the questions "What is the problem?" and "What exactly is the project trying to accomplish?" As trite as this sounds, these are the most important and the most difficult questions to answer. To achieve a successful coastal project plan and design requires that the engineer must have a completely open mind, one without preconceived notions of the ultimate form of the project or a specific solution to advocate. An example of a pitfall of the planning process is to answer the second question with "To design a breakwater" without first clearly stating the problem. Is the problem one of navigation difficulties in the vicinity of a harbor, or potential damage to buildings near the shorefront, or water quality problems in a bay? These are symptoms of a problem, but they do not define the problem.

(c) An interdisciplinary team approach is recommended in planning coastal projects to ensure the involvement of physical, natural, and social sciences personnel. The disciplines of the coastal project planners should be appropriate to the problems and opportunities identified in the planning process, and range from coastal, geotechnical, structural, and hydraulic engineers through meteorologists, oceanographers, biologists, and geologists, to economists, urban planners, and transportation specialists. Not all disciplines are needed on all studies, but the team leader needs to be cognizant of the possible need for such talents during the project development.

(d) Interested and affected agencies, groups, coast-sharing partners, and individuals (collectively termed the public) should be provided opportunities to participate with USACE throughout the planning, design, construction, and operation of a Federal project. The purpose of public involvement is to ensure that the project is responsive to the needs and concerns of the public. The objectives of public involvement are to provide the public with information about proposed Federal activities; to make the public's desires, needs, and concerns known to the decision makers; and to consult with the public and consider the public's views before a decision is reached.

(e) In responding to the question of *problem definition*, the more detailed the engineer can make the statement, the more likely a solution can be found which will satisfy all the numerous stakeholders. A stakeholder is defined as anyone who will be affected by a problem and its potential solution, including a segment of the public. This includes the planner, designer, constructor, manager, all probable sponsors, riparian owners, project users, the political, legal, financial, and environmental entities with interests, and of course the populations that pay the taxes to support the enterprise.

(f) USACE has two levels of planning: one that looks at the problem in a relatively cursory way for a *reconnaissance* level study, and the other for a detailed study of all the factors, conditions, and alternatives for a *feasibility* level study. These terms are equivalent to conceptual design and final design as used in private practice. In the reconnaissance study, the objective is to determine if there is a Federal interest; that is, to determine if the Federal Government should invest in the solution to the problem. Responses are needed to three questions to make this determination: (a) is there a technically doable solution, (b) is it economically and politically feasible, and (c) is it environmentally sound? These are often difficult to answer, but keeping in mind all the stakeholders and the detailed problem definition, the coastal planner can proceed.

(g) Under present laws and regulations (as of the publication date of this manual), a reconnaissance level study is funded entirely by the Federal Government. To be "in the Federal interest," at least one technically adequate solution to the problem must be identified, a cost-sharing sponsor who will pay for one half of the feasibility study costs must be identified, and the economics of the potential solution must be such that the National Economic Development (NED) benefits are larger than the costs. More details of the planning process are found in ER 1105-2-100.

(h) Since reconnaissance and feasibility studies differ in their level of detail, the following discussion concentrates on feasibility requirements. The requirements of a feasibility study differ from those of a reconnaissance study in degree, not kind. A diagram is presented later in this chapter to assist the engineer planner/designer in going through all the steps needed for successful solution of a coastal engineering problem.

(i) In the case where a coastal project is planned and designed by other than USACE, Federal and state permits are usually required before construction can begin. (Requirements for USACE projects are discussed later in the chapter.) The USACE must ascertain that the proposed project would not harm the environment or adjacent property owners, that it would not interfere with navigation, and that water quality would not be impaired. Federal, state, and possibly local permits are required for construction in, across, under, or on the banks of navigable waters of the United States. Federal permits are coordinated by the applicant through the USACE District offices. Permit authorities are Section 10 of the River and Harbor Act of 1899 and Section 404 of the Clean Water Act of 1977, as amended.

Section 10 of the 1899 act requires permits for structures and dredging in navigable waters, which are those coastal waters subject to tidal action and inland waters used for interstate or foreign commerce. In tidal areas, this includes all land below the mean high-water line.

Section 404 of the Clean Water Act mandates a Federal permit for discharges of dredged or fill material in waters of the United States, which include tributaries and wetlands adjacent to navigable waters, and extends to headwaters of streams with flows greater that 0.142 m^3/sec (5 ft^3/sec). Wetlands are defined as "those areas inundated or saturated by surface or groundwater at a frequency and duration sufficient to support, and that under normal conditions do support, a prevalence of vegetation typically adapted for life in saturated soil conditions. Wetlands generally include swamps, marshes, bogs, and similar areas."

More detailed requirements for the permit process are found in Chapter 33 of the Code of Federal Regulations, paragraphs 320-330 (33 CFR 320-330).

(j) Still, the first question to be answered is: Exactly what is the problem? The more detailed the response, the easier it becomes to be sure all the important factors are considered. A more precise answer to the question posed earlier in this section might be: "The waves and currents in a harbor berthing area are of such magnitude that 30,000-ton tankers find it unsafe to berth there approximately 50 percent of the time." A statement of what the project is trying to accomplish might be "to provide safe berthing for 30,000-ton tankers at least 90 percent of the time." If this is an exact statement of the problem, then investigation is necessary to determine that the engineering, economic, and environmental conditions and constraints can all be resolved satisfactorily.

c. Six major planning steps. The planning process consists of six major steps:

(1) Specify problems and opportunities.

(2) Inventory and forecast conditions if no action is taken.

(3) Formulate alternative plans.

(4) Evaluate effects.

(5) Compare alternative plans.

(6) Select a plan.

The following paragraphs describe these steps in more detail.

(1) Specify problems and opportunities.

(a) The identification of a problem usually begins with a unit of local Government requesting, through its local congressional representative, that USACE investigate and solve a problem. The Senator or Representative then introduces a resolution, which when passed by the Senate or House of Representatives, authorizes USACE to study the problem. The problem and opportunity statements are framed in terms of Federal objectives as well as identifying state and local objectives. The statements are fashioned to ensure that a wide range of alternative solutions can be considered with identifiable levels of achievement.

(b) In the case of a small project, the Secretary of the Army, acting through the Chief of Engineers, may plan, design, and construct certain water resource improvements without specific congressional authorization. This authority is found in the six legislative authorities collectively called the Continuing Authorities Program. The per-project limit on Federal expenditures under the CAP is currently $2,000,000 for a small beach erosion control project (Section 103 of Public Law (PL) 87-874, as amended) and $4,000,000 for a small navigation project (Section 107 of PL 86-645, as amended). A complete discussion of the six legislative authorities, including that for the mitigation of shoreline erosion damage caused by Federal navigation works (Section 111 of PL 90-483, as amended), is found in ER 1105-2-100 Chapter 3.

(2) Inventory and forecast conditions if no action is taken. The next step is to identify the initial site characteristics. These include not only the physical dimension of the site and environs, but also the

biological, cultural, social, safety, political, economic, endangered species, and other environmental aspects such as auditory, olfactory, visual, etc. The complete description of these elements will help to define the positions of the various stakeholders and how they might influence the success of any proposal. If these concerns can be accommodated early on, and the parties made to feel a part of the process, then any compromises that might be called for later are easier to obtain. All alternative schemes are measured against the *without-project plan* (the conditions that would prevail in the future if no plan is constructed), in order to quantify the benefits that would accrue from the various alternative plans examined. This is an important step in the planning process since it is the foundation for evaluating potentials for alleviating the problems and realizing the opportunities. It sets a consistent base upon which alternative plans are formulated, from which all benefits are measured, against which all impacts are assessed, and from which plans are compared and selected.

(3) Formulate alternative plans. Alternative plans are formulated in a systematic manner during both the reconnaissance and feasibility studies to ensure that all reasonable alternative solutions are identified early in the planning process and are refined in subsequent iterations. Alternatives are not merely slight variations; for example, breakwater lengths of 200, 300, or 400 m, but significant ones, such as a breakwater construction, a saltwater barrier, or a harbor realignment. As the process is under way, additional alternatives may be introduced as the need or opportunity presents itself. A plan that reasonably maximizes the net NED benefits and is consistent with protecting the nation's environment, is identified as the NED Plan in the feasibility report. Other plans that reduce net NED benefits in order to further address other Federal, state, local, or international concerns should also be formulated. Each alternative plan is formulated considering the four criteria described in the Water Resources Council's Principles and Guidelines: completeness, efficiency, effectiveness, and acceptability (ER 1105-2-100). The period of analysis must be the same for each alternative plan and appropriate mitigation of adverse effects is an integral part of each alternative plan. It is usual for an economic life of 50 years to be selected for analysis. This does not imply that a coastal structure, such as a breakwater, would only last 50 years, but that analysis of benefits and costs is limited to that period. Alternative plans may include significant nonstructural components or may be completely nonstructural. It should be noted that this step usually involves many iterations in the development of the alternatives, and may also require a restatement of the problem if it is found that the functional needs are not completely met.

(4) Evaluate effects. The evaluation of effects is a comparison of the *with-* and *without-plan* conditions for each alternative plan. Differences between each with- and without-plan condition are measured or assessed, and the differences are appraised or weighted. Four accounts are established to facilitate the evaluation and display the effects of the alternative plans: NED, EQ, RED, and OSE.

NED: The *National Economic Development* account displays changes in the economic value of the national output of goods and services. A comprehensive treatment of analysis required for determining the economic benefits of coastal projects is found in U.S. Army Corps of Engineers (1991).

EQ: The *environmental quality* account displays the nonmonetary effects on ecological, cultural, and aesthetic resources.

RED: The *regional economic development* account registers the changes in regional economic activity. Regional effects are evaluated using nationally consistent projections of income, employment, output, and population.

OSE: The *other social effects* account registers plan effects from perspectives that are relevant to the planning process, but are not reflected in the other three accounts.

Display of the NED account is required. Since technical data concerning benefits and costs in the NED account are expressed in monetary units, the NED account already contains a weighting of the effects; therefore, appraisals are applicable only to the EQ, RED, and OSE account evaluations. Planners must also identify areas of risk and uncertainty in these analyses and describe them clearly so that decisions can be made with knowledge of the degree of reliability of the estimated benefits and costs, and of the effectiveness of the alternative plans (see Section V-1-3).

(5) Compare alternative plans. Plan comparison focuses on the differences among the alternative plans as determined in the preceding step. Both monetary and nonmonetary effects are compared. Again, if the functional requirements of a project are not met, it is time to go back and iterate the formulation step.

(6) Select a plan. After consideration of the various alternative plans, their effects, and public comments, the *NED Plan* is selected as the one to recommend for implementation, unless a justified exception is granted by the Assistant Secretary of the Army for Civil Works (ASA(CW)). The USACE feasibility report, along with expressions of related views by the ASA(CW) and the Office of Management and Budget, is then sent to Congress with a recommendation for authorization for construction.

d. Planning coordination requirements.

(1) Table V-1-1 lists some of the laws enacted to ensure that environmental effects of projects are fully taken into account in the planning process. These laws are briefly described below.

Table V-1-1
Major Coordination Requirements

National Environmental Policy Act, 1970, 33USC234.4

Fish and Wildlife Coordination Act, 1958, 16USC460-1(12) et seq

Coastal Zone Management Act, 1972, 16USC1451 et seq

Clean Water Act, 1977, 33USC1344 et seq

National Historic Preservation Act, 1966, 16USC470a et seq

Coastal Barrier Resources Act, 1982, 16USC3501

---

(2) The National Environmental Policy Act (NEPA) of 1970, as amended, ensures that all proposed Federal actions take full cognizance of the environmental effects of those actions. This act defines the NEPA process by which an environmental impact statement (EIS) [or an environmental impact assessment (EIA) for a lesser impact activity that does not require a statement] is prepared. (An EIA is sometimes prepared to determine whether an EIS is required.) The EIS is given wide circulation and the public comment and the Federal agency's response are incorporated in the USACE report of its proposed activity. It is through this process that any unforeseen impacts are uncovered and mitigation measures are proposed.

(3) The Fish and Wildlife Coordination Act of 1958, as amended, ensures that fish and wildlife conservation receives equal consideration with other project purposes and is coordinated with other features of any coastal development project. USACE consults with the Regional Director, U.S. Fish and Wildlife Service (FWS), the Regional Director, National Marine Fisheries Service (NMFS), and the head of the state agency responsible for fish and wildlife for the state in which the proposed work is to be performed. Funds are transferred to the FWS and NMFS to accomplish the investigations, and a Fish and Wildlife Report is made part of the USACE report.

(4) The Coastal Zone Management Act of 1972, as amended, ensures that all Federal activities are consistent with an approved coastal zone management plan for that state. If the Secretary of Commerce has approved a state's plan (all but five coastal states have approved plans), then USACE must receive state certification that its proposed activity is consistent with the state's plan.

(5) The Federal Water Pollution Control Act was enacted to regulate the discharge of pollutants. Subsequently the Clean Water Acts, as amended, required water quality certification by the Environmental Protection Agency or a designated state agency to ensure that the USACE-proposed project does not degrade the state's water quality, particularly through any dredge and fill activities.

(6) The National Historic Preservation Act of 1966, as amended, provides for the consideration of the effects of a proposed project on the cultural resources (including archeological and Indian religious and cultural sites) of the area. Studies are coordinated with the National Park Service, the Advisory Council on Historic Preservation, and the appropriate State Historic Preservation Officer.

(7) The Coastal Barrier Resources Act of 1982, as amended, requires that no Federal funds (with some exceptions) are expended in any undeveloped region designated as a unit of the coastal barrier resource system (CBRS). Most notable for coastal projects, a major exception is for beach nourishment since it mimics the natural processes; hard structures in conjunction with beach nourishment are prohibited on areas identified in this Act. Also, maintenance dredging in waters within the CBRS unit is allowed, but new work dredging is not; using sand for beach nourishment that comes from a borrow area within a CBRS unit is not generally permitted. Coordination with the Regional Director, FWS is required for all cases involving the CBRS.

e. Criteria development. The criteria needed to be established are those that must be used to determine at each step of the process whether the objectives of that step are met. This is quite site-specific, and as will be seen in the section describing the generic design chart, the questions are answered either "*yes*, it meets the objective" (go on to the next step) or "*no*, it does not meet the objective" (go back and either refine the question or pick another alternative.) The importance of adequately posing the problem in the first place is readily seen in this approach. The advantage of this method of defining the criteria is that as long as all the questions are answered, the result will be the optimum project that meets the originally stated problem.

f. The design process.

(1) Final design. To produce a final design, it is necessary to continue the iterative process begun in the planning of the project. The project is said to be at final design status when all of the objectives are met, and all the stakeholders are satisfied that an optimum design has been reached. If there are any unanswered questions, then the final design has not been reached, and another iteration is necessary. It is current USACE practice to include the final recommended design in the feasibility report, and if approved for construction, no further design work is needed. The engineer can proceed directly to plans and specifications when the project is approved for construction and funds for construction are appropriated.

(2) Plans and specifications. The documents required for a contract for construction to be awarded include the *plans and specifications* for the job. These incorporate any restrictions or other constraints on the contractor, and spell out the accuracy and tolerances appropriate to the job. The procedures for plans and specifications for construction of USACE projects are found in ER 1110-2-1200, and no further discussion is warranted here.

g. Construction and monitoring.

Although construction and monitoring are not part of the planning and design process, they must be considered before the design process is complete.

(1) Construction. In order to have a successful project, the work to be accomplished must be capable of construction, i.e., have bidability, constructibility, and operability (see ER 415-1-11). If no equipment is available that can do the work prescribed (for example, a crane capable of lifting breakwater armor units into place), then the project is not constructible. If sand for a beach nourishment job is found to be outside the limits of available dredging equipment capability, then the job is not constructible. Guidance for construction management of USACE projects is found in EP 415-1-260, and no further discussion is warranted here.

(2) Post-construction inspection and monitoring. Monitoring is an essential element of a USACE project. The degree of monitoring depends on how the data are to be used. There are two basic types of monitoring: for conformance of construction to the design and for evaluation of performance of the project. The second type is often neglected, although ER 1110-2-1407 and ER 1105-2-100 clearly specify that a monitoring plan and an operation and maintenance plan are to be prepared as part of every project for which USACE has a continuing responsibility. This is especially important in beach nourishment projects to determine when a renourishment is needed. Also, data on project performance are needed for the continuing improvement and refinement of prediction models for coastal projects.

h.   Generic design chart.

(1) Figures V-1-1 through V-1-3 show the thought processes that occur in the planning and design of a coastal project. The thought processes to successfully engineer a solution to a problem are basically the same as for the solution of any problem, but for the ease of presenting appropriate examples, will be limited here to the discussion of coastal engineering problems. The diagram is quite comprehensive and represents the steps followed by all engineers in developing successful solutions to coastal problems. It can be modified, however, to fit the needs of the planner and designer.

(2) Figure V-1-1 is generic in the sense that navigation, shore protection, coastal flooding, storm damage reduction, environmental enhancement, or any other set of problems can be treated by following the steps outlined in the diagram. As discussed earlier in this chapter, the project begins with a need, from which an accurate statement of the problem is derived. Figure V-1-1 illustrates the path (or paths) typically followed in developing a coastal project, from initiation to cost completion phases, including the reevaluation and feedback loops. There are ten major segments of the process, with several blocks in each segment. They are designated by the letters in circles, and are keyed to the following list:

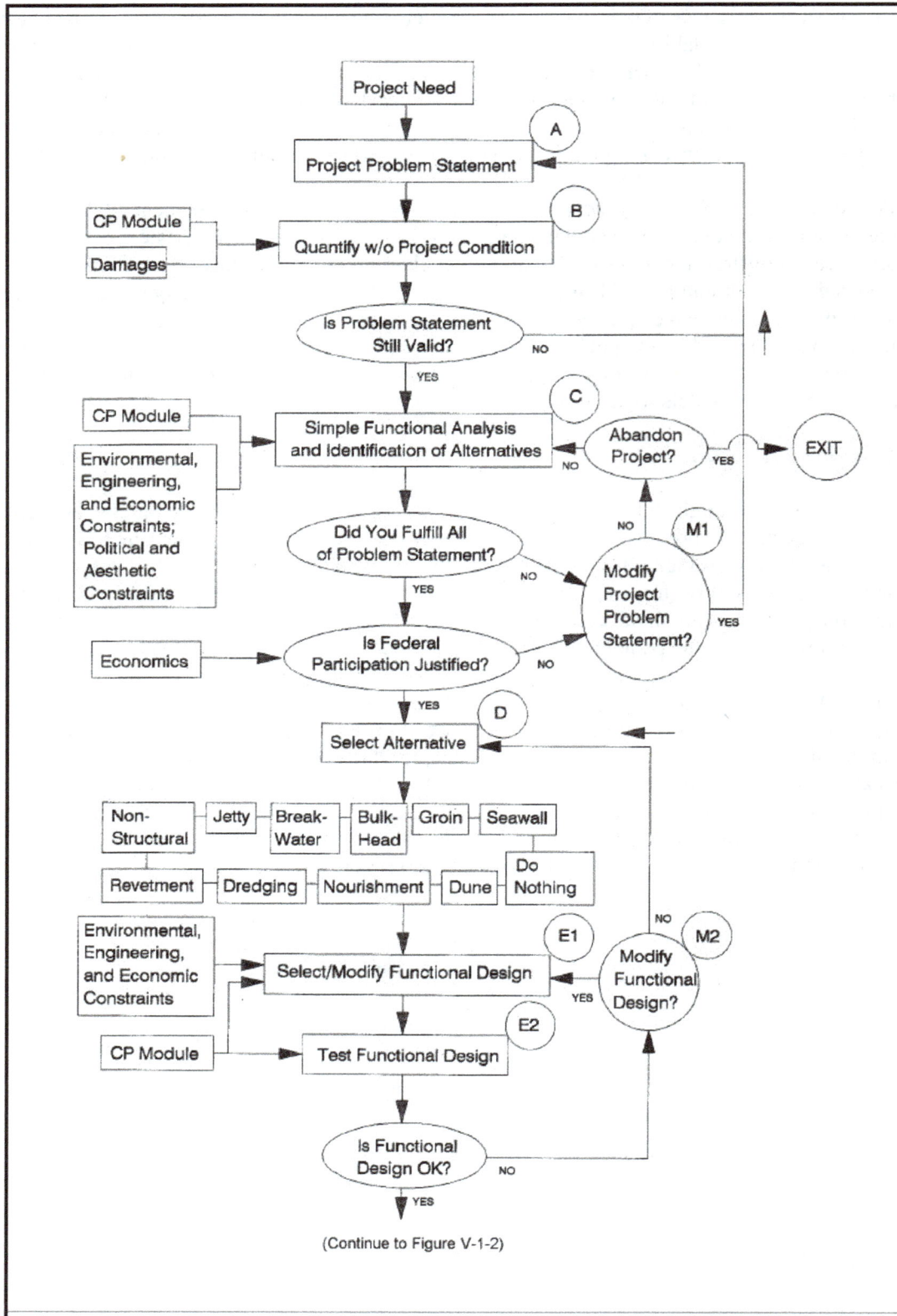

Figure V-1-1. Thought process in the planning and design of a coastal project, Part 1

(Continued from Figure V-1-1)

YES

CP Module → Test Long-Term Performance of Functional Design Including O & M and Constructibility (E3)

GO TO (M2)

Benefits → Is Functional Design OK?   NO

YES

Environmental, Engineering, and Economic Constraints; Political and Aesthetic Constraints

Add Another Component Or Try Another Alternative?   YES

NO

Develop Structural Design (F1)   YES   Modify Structural Design? (M3)   NO

CP Module →
Cost → Test Structural Design, Including O & M and Constructibility (F2)

Is Structural Design OK and Cost Acceptable?   NO

YES

CP Module → Retest Functional Design Including O & M (E4)

Is Functional Design OK?   NO

YES

CP Module → Check Overall Constructibility O & M and Life Cycle Costs (G)

Is Constuctibility, Etc., OK?   NO

YES

(Continue to Figure V-1-3)

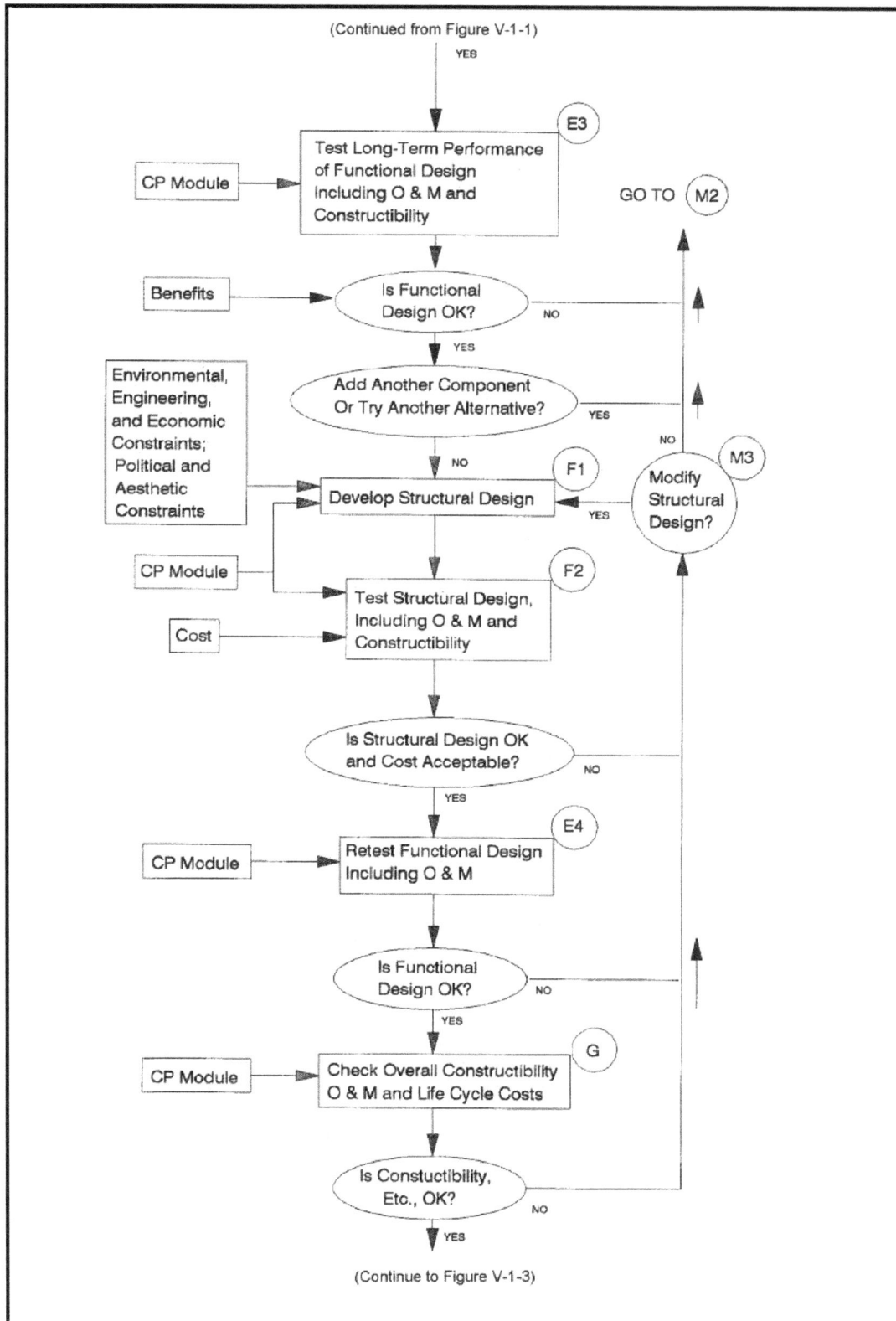

Figure V-1-2.   Thought process in the planning and design of a coastal project, Part 2

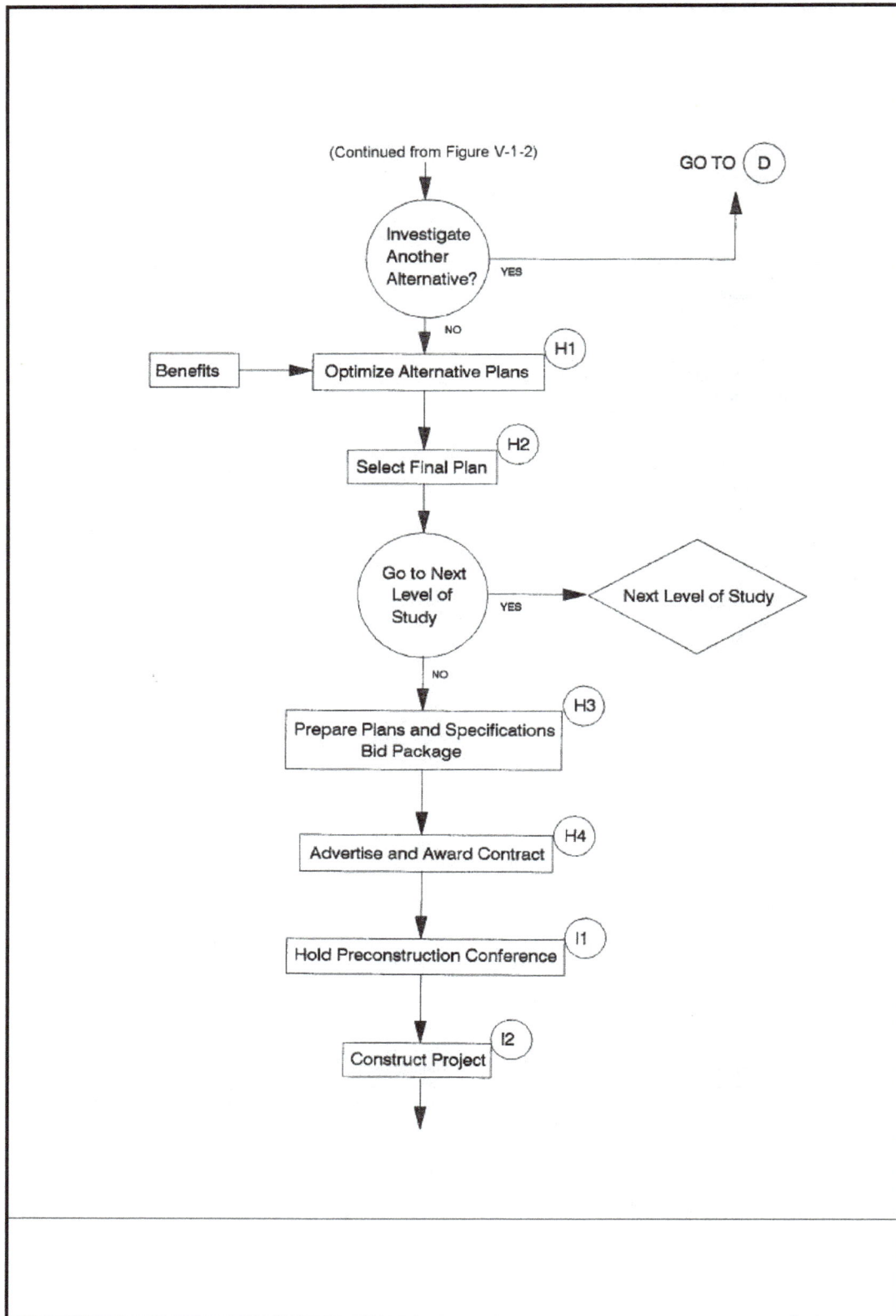

Figure V-1-3. Thought process in the planning and design of a coastal project, Part 3

(a) Clearly define project problem statement, including the project objective.

(b) Quantify existing and most likely future conditions (without project).

(c) Identify and analyze alternatives.

(d) Select alternative.

(e) Develop and test functional design.

(f) Develop and test structural design.

(g) Check for constructibility, operation and maintenance, and life-cycle costs.

(h) Select final plan, and prepare plans and specifications.

(i) Construct project.

(j) Monitor and evaluate project performance.

There are also several key segments concerning decisions about modifying various phases of the project; they are designated by the letter M.

(3) The term "functional design" refers to the effectiveness of a project at its intended function, such as the effectiveness of a breakwater at providing tranquil waters inside a harbor. "Structural design" refers to the ability of a structure to exist in the climate in which it is placed, such as the ability to withstand the effects of extreme storms without affecting its functional requirements. The term "constructibility" refers to the means, methods, and materials involved in successful project construction.

(4) In the diagram, the notation "CP Module" designates a generic concept of a Coastal Processes Module, a repository of physical data and analysis tools relevant to the coastal problem. The module includes information on wind, waves, currents, water levels, bathymetry, geomorphology, stratigraphy, sediment characteristics, sediment transport, ice processes, etc. The output from this module is needed at many points in the process. At the present time, portions of a module exist as limited databases and tools; however, a complete, integrated module, similar to the generic model, is under development, but is yet a long way from completion.

V-1-2. Design Criteria

a. Design criteria are the minimum parameters of design that are followed to ensure that the project function and structure (if any) meet the needs of a customer. A design is based on a number of design criteria that include: forcing function criteria, configuration criteria, materials criteria, geotechnical criteria, construction criteria, maintenance criteria, and economics criteria. These criteria are partly interrelated and partly independent of each other. Materials may depend on configuration, construction (including availability of equipment and manpower), and maintenance criteria. Construction may depend on configuration and materials (including availability). Maintenance depends on materials and construction.

b. Forcing function criteria must always be seen in relation to configuration, materials, and geotechnical criteria. For example, the "design storm" is an obsolete concept. Short- and long-term wave statistics are needed, and the hydrodynamics of wave interaction with its surroundings must be known in detail. The parameters of wave height, wave period, storm duration, and surge water level are quasi-independent components of a storm whose effects on a design need to be understood to adequately examine the alternatives. The frequency of occurrence (or the return period) of an event (such as a storm) is a primary component of the wave criteria. These concepts are fully discussed in Part II of this manual.

c. Criteria that need to be considered in all coastal projects are: safety, accessibility (as defined by the American Disability Act), and environmental effects. All other effects are highly site-specific, and in addition to those cited above, include project area use, materials corrosion, ice, structural unit stability, subsurface foundations, and susceptibility to seismic events (i.e., earthquakes or tsunamis).

d. In every case, the criteria must be measured against the function to be satisfied by the proposed coastal project. If the alternative does not adequately address the need, then the function is not fully satisfied, and either the problem must be restated or another alternative must be investigated. As discussed in earlier sections of this chapter, the iterative process of planning and design requires the above-cited criteria be used to evaluate each step in the process. Design criteria relative to specific project types are discussed in the remaining chapters of Part V.

V-1-3. Risk Analysis and Project Optimization

a. *Introduction.*

(1) The approach for analyzing coastal projects is undergoing some fundamental changes, shifting from the traditional deterministic basis to a probabilistic, risk-based methodology. The changes strongly impact both planning and engineering phases of project formulation and design. Concepts of risk analysis and probabilistic optimization of project design are presented in this section. The remaining chapters in Part V include further information on the subject in relation to specific project types. As of this writing, risk-based analysis is a rapidly emerging tool, and significant advances can be expected during the next few years.

(2) The changes in analysis approach, which can be expected to be distilled into a new standard for coastal practice, are driven by several progressive developments. First, general understanding of probabilistic coastal processes continues to advance, particularly due to advances in field measurement, physical modeling, and numerical modeling. Second, standard computing capabilities are increasing rapidly, facilitating lengthy probabilistic calculations which would have been impractical in the past. Third, engineers, cognizant of limitations in the traditional approach, are often eager to implement better procedures, provided that they are well-founded and clearly improve the analysis. Finally, the public is becoming more aware and concerned about coastal project performance, and expects realistic project analyses. In the United States, public involvement in coastal projects is further intensified by legislation which increases the proportion of costs borne by the client (typically state or local Government) in Federal projects.

b. *Traditional vs. Risk-based analysis.*

(1) Traditional analysis treats a coastal project in deterministic terms. The forces of nature are often represented as a design significant wave height, period, and direction, a design water level, etc. Coastal response is described as *the* response if no project is implemented, *the* response if one plan is implemented, *the* response if another plan is implemented, etc., without much formal recognition of the wide variation in possible responses.

(2) In contrast to traditional analysis, some significant developments in probabilistic treatment of coastal projects have appeared during the 1990's. Most relate to coastal structure design (Construction Industry Research and Information Association/CUR 1991, International Conference on Coastal Engineering (ICCE) 1992). Within USACE, water resource planning guidance has moved from a deterministic to a risk-based

approach, which incorporates considerations of risk and uncertainty. Similar concepts are now being adapted to USACE coastal engineering studies.

c. Reasons for risk-based analysis.

(1) Introduction. There are a number of reasons why coastal projects in the broader sense, not just structural design, may be effectively analyzed from a risk-based point of view (Table V-1-2).

Table V-1-2
Reasons for Risk-Based Analysis of Coastal Projects
(1) Forcing is probabilistic.
(2) Major uncertainties in behavior.
(3) Damage & functional performance change incrementally.
(4) Benefits and risks not fully represented in deterministic terms.
(5) Uncertain effects on adjacent areas.

(a) Coastal forcing is probabilistic. Wave characteristics vary greatly over both short-term (individual waves) and long-term (from one sea state to another). Similar considerations arise with winds, water levels, infragravity waves, and currents.

(b) Coastal engineering embodies major uncertainties. Knowledge of both the forcing processes and coastal response usually involves major uncertainties. Deterministic representations mask the uncertainties and can be misleading.

(c) Damage and functional performance change incrementally. Coastal projects rarely progress from the design condition to total failure during a single storm event. Damage usually occurs incrementally. For example, damage to a rubble-mound breakwater (when it occurs) typically begins during an unusually severe storm and progresses during subsequent severe storms until repairs are done. Similarly, beach fills erode incrementally in response to storms over a period of years. Coastal projects often continue to provide some measure of functional benefit even in a damaged state. A damaged breakwater continues to provide some protection from incident waves; a partially eroded beach fill continues to reduce coastal flooding risks.

(d) Benefits and risks not fully represented in deterministic terms. Because of the above factors, positive impacts and risks of coastal projects cannot be fully represented in deterministic terms. Some projects provide benefits beyond the design configuration, which are generally ignored in traditional practice. For example, a nearshore berm which is overbuilt to allow for progressive deterioration provides increased coastal protection during its early life. Another example is an overdredged entrance channel giving increased vessel access depths until it shoals to the design depth.

(e) Uncertain effects on adjacent areas. In addition to the uncertainties associated directly with coastal projects, the projects can introduce significant possibilities for changing adjacent areas. While projects are designed with the intent of minimizing adverse impacts on adjacent areas, it is important to recognize that uncertainties and risks can increase beyond the without-project condition. In effect, a project can transfer risk from one area or party to another. When the risks of *all* major aspects of a project are represented as best they can be determined, better-informed final decisions can be made.

(2) Professional judgement. Experienced coastal engineers are well aware of the concerns in Table V-1-2. Even with deterministic methods, they can be expected to produce project plans that include a large measure of professional judgement to ensure a technically successful project. However, the ultimate fate of a project can depend upon higher level decision makers who must weigh technical concerns against economic, environmental, aesthetic, social, and political concerns. By quantifying risks, the coastal engineer can better pass his or her experience and judgement on to other decision makers, who may not have coastal expertise.

d. Considerations for including risk-based analysis in project design

(1) Objectives. The main objectives of adopting a risk-based analysis approach rather than a traditional approach are to explicitly identify uncertainties, provide improved information for assessing tradeoffs between risks and cost, and improve decision-making for project optimization.

(2) Key variables. Although a large number of variables affect any coastal project, a small subset can usually be identified as *key variables*; that is, variables that strongly relate to project performance. The key variables will embody the main forcing mechanisms, project sizing, and project response. For example, some of the key variables for a beach nourishment project might be significant wave height (forcing), beach fill width (project size), and erosion width (response).

(3) Professional judgement. Coastal engineering requires an unusually large measure of professional judgement because of the number and complexity of processes and responses involved. Analytical and modeling tools help to represent the variability affecting coastal projects, but the judgement of an experienced engineer is a vital ingredient in risk assessment and project optimization.

(4) Resistance and functional performance vary with time. Both the resistance to damage and functional performance often vary significantly over a coastal project's design life. For example, the resistance (or structural strength) of a rubble-mound breakwater may decrease in time due to deterioration of stone such as loss of angular corners, cracking, and breaking. Resistance may also decrease due to displacement of stone and exposure of underlayers to wave attack, which would also decrease protection provided by the breakwater (functional performance). For a beach nourishment project, loss of material to storms decreases the resistance of the beach to future storms. The effectiveness of the beach as a deterrent to coastal flooding is also decreased (functional performance). In some cases, resistance *increases* with time, as in the progressive growth of protective vegetation on coastal dunes and natural cementation of beach sediments rich in calcium carbonate.

(5) Construction season and mobilization concerns. Often maintenance of coastal projects requires major mobilization efforts and is confined to a *construction season* dictated by climate and environmental factors. Therefore the risk during the *interval between construction seasons* rather than during a single storm becomes a key concern. During an unusually stormy winter (such as the winter of 1987-88 in southern California), this risk can be significantly greater than that for individual storms.

(6) Environmental, aesthetic, social, and political concerns. The role of environmental, aesthetic, social, and political factors in the ultimate planning and design of a coastal project is often at least as important as the technical engineering factors. An optimized final design includes appropriate consideration of these factors and their associated risks and uncertainty.

e. Frequency-based vs. life cycle approach.

(1) Risk-based analysis of coastal projects can be done by either of two fundamentally different approaches. The *frequency-based approach* deals with frequency-of-occurrence relationships among the key variables. By combining key forcing variables with various occurrence frequencies, information about the

frequency-of-occurrence of key project responses can be developed. For example, a traditional stage- (water level)-versus frequency curve and a stage-damage curve can be combined to generate a damage-versus-frequency curve. This approach can be applied as an add-on to traditional planning and design procedures.

(2) The *life cycle approach* deals with multiple realizations of possible evolution of the project with *time* during the span of its design life. The suite of life cycle realizations is constructed with consideration of the probabilities of key variables. For example, the realistic time variation of key forcing and response variables during a 50-year life cycle can be generated for 1,000 different possible life cycles. Probabilities and risks associated with the project are then compiled by analyzing project performance over the 1,000 life cycles.

(3) The life cycle approach appears better suited to most coastal engineering applications. Variation with time is an essential ingredient in most coastal projects, and it is directly incorporated into the life cycle approach. Time variation of resistance and functional performance, constraints imposed by construction season and mobilization, even some economic, environmental, and political factors, can be conveniently and flexibly introduced into the life cycle approach. This approach leads to a unified analysis of technical performance and many economic factors which are critical to project success. In addition to its technical and economic strengths, the life cycle approach is more easily understood by nontechnical parties involved with a project. This type of approach is evident in the Empirical Simulation Technique (Chapter II-5).

(4) The life cycle approach adapted to shore protection projects is particularly instructive (U.S. Army Corps of Engineers 1997). The life cycle embodies sequences of storms (including provisions for multiple storms of varying intensity during each year of the life cycle), erosion and post-storm recovery during each event, partial and complete property damage during each event (depending on water level, waves, extent of storm erosion, and type of building construction), cumulative property damage due to a succession of storms, optional repair or rebuilding after a suitable time lag (with conformance to any stricter building codes in effect), and periodic renourishment of the beach when needed and feasible during the life cycle. Typically, a key result from this analysis is the renourishment required during each life cycle, which can be converted to an economic present-worth dollar value. The expected cost and economic risks associated with maintaining the beach can then be realistically assessed by combining information from many different life cycle simulations.

f. Typical project elements. Risk-based analysis can be integrated into the six major planning steps (Section V-1-1b(2). Typical project elements which are especially well-suited to risk-based analysis include the following.

(1) Site characterization. Significant uncertainty can arise in documenting past and present behavior at a site. The uncertainty can be estimated based on data quality and quantity, methodologies used, observed variability, etc. (Chapter V-3).

(2) *Without-plan* alternative. Evaluation of what would happen in the future if no Federal project were built involves speculation about the natural processes and human interventions that would affect the site during the proposed project life. The impact of the *without-plan* alternative is conveniently described in probabilistic terms.

(3) Formulate, evaluate, and compare alternative plans. Risk-based analysis can be a powerful tool for formulating and comparing alternative plans. It enables decision makers to intercompare not only the expected level of performance, but also the probabilities of enhanced or reduced performance levels, which can differ greatly among alternatives. Typically, alternatives involve hard structures (such as walls, revetments, breakwaters, and jetties) and/or soft structures (such as beach nourishment projects, coastal dunes, and nearshore berms). Risk-based analysis of hard structures is discussed in Part VI. Soft structures (Chapter V-4) involve calculated risks about the movement of sediment through time and the need for future

maintenance. Uncertainties arise in forcing processes, sequencing of storms, initial state of nearshore profile when storms occur, and evolution and recovery of storm profiles (especially three-dimensional aspects). The life cycle approach to risk analysis has been shown to be a powerful tool in this type of application.

V-1-4. References

ER 415-1-11
Biddability, Constructibility, and Operability

ER 1105-2-100
Planning Guidance Notebook.

ER 1110-2-1407
Hydraulic Design for Coastal Shore Protection Projects.

ER 1110-2-1200
Plans and Specifications for Civil Works Projects.

EP 415-1-260
Resident Engineer Management Guide.

CIRIA/CUR 1991
CIRIA/CUR 1991. "Manual on the Use of Rock in Coastal and Shoreline Engineering," *CIRIA Special Publication 83*, Construction Industry Research and Information Association, London, UK: and CUR Report 154, Centre for Civil Engineering Research and Codes, Venice, Italy.

ICCE 1992
ICCE 1992. "Design and Reliability of Coastal Structures," *Proceedings of a short course attached to the 23rd International Conference on Coastal Engineering*, Venice, Italy.

Thompson, Wutkowski, and Scheffner 1996
Thompson, E. F., Wutkowski, and M., Scheffner, N. W. 1996. "Risk-Based Analysis of Coastal Projects," *Proceedings of the 25th International Conference on Coastal Engineering*, Orlando, FL.

U.S. Army Corps of Engineers 1991
U.S. Army Corps of Engineers. 1991. "National Economic Development Procedures Manual - Coastal Storm Damage and Erosion," Report 91-R-6, Institute for Water Resources, Fort Belvoir, VA.

V-1-5. Acknowledgments

Authors of Chapter V-1, "Planning and Design Process:"

John G. Housley, Headquarters, U.S. Army Corps of Engineers, Washington, DC, (retired).
Edward F. Thompson, Ph.D., Coastal and Hydraulics Laboratory (CHL), U.S. Army Engineer Research and
    Development Center, Vicksburg, Mississippi.

Reviewers:

James E. Clausner, CHL
Jeffrey A. Melby, Ph.D., CHL
Norman W. Scheffner, Ph.D., CHL (retired)
H. Lee Butler, CHL (retired)

Table of Contents

List of Figures

List of Tables

Page

Chapter V-2
Site Characterization

V-2-1. Introduction

a. Many coastal failures can be traced back to inadequate site characterization analyses. Site characterization involves identifying distinguishing qualities and features of a region that have a direct and indirect impact on the conception, design, economics, aesthetics, construction, and maintenance of a coastal project. The coastal environment varies spatially and temporally and therefore a design that is functionally, economically, and environmentally appropriate at one location may be inappropriate at another. Physical, biological, and cultural attributes need to be delineated so that an acceptable project is adopted and potential effects of the project are determined.

b. Waves bring an enormous amount of energy to the coast that is dissipated through wave breaking, generation of currents, water level changes, movement of sediment, turbulence, and heat. Incident waves vary spatially and temporally, with their properties changing with movement over the bottom. The beach is composed of sediment particles of various types, sizes, and shapes which move along and across the shore. The beach and backshore exhibit different textural properties that vary alongshore, across-shore, and with time. The coastal region supports a diverse biological community of aquatic, terrestrial, and avian organisms. It also must continue to enhance the quality of human existence by providing commercial and recreational benefits. In light of the profound diversity of the coastal zone, it is imperative that the coastal designer have a full understanding of potential impacts of engineering activities on the regional environment.

c. This chapter enumerates important physical and engineering factors that should be identified. Readers are directed to other pertinent sections of this manual for specific details on monitoring the coastal environment and for information on the physics of coastal processes. Not all factors described in this chapter pertain to all coasts. For example, ice and volcanic hazards are not applicable to a Gulf Coast project design. However, most factors should be included in all other site characterizations.

d. General Design Memorandums (GDMs) and monitoring reports (Table V-2-1) prepared by U.S. Army Corps of Engineer (USACE) Districts may be consulted for examples of site characterization studies. The Alaska Coastal Design Manual (U.S. Army Corps of Engineers 1994) also summarizes important site characterization features.

e. In summary, it is important when characterizing a site to:

(1) Include **all components** of the system.

(2) Recognize the extreme **temporal variability** in most physical and biological processes.

(3) Be cognizant of the **spatial variability** of processes, climate, land forms, underlying geology, biological habitat, and cultural resources. Process measurements made at one site may not be valid at another site only a short distance away.

f. In conclusion, the designer must think globally and engineer locally.

Table V-2-1
Some Example USACE Reports and General Design Memorandums (GDMs) Containing Comprehensive Site Characteristics

State	Author	Date	Title
Atlantic Coast			
Virginia	USAED, Norfolk	1984	Beach Erosion Control and Hurricane Protection, Virginia Beach, Virginia, Main Report Phase 1 GDM and Supplemental EIS
Virginia	USAED, Norfolk	1989	Beach Erosion Control and Hurricane Protection, Virginia Beach, Virginia, Phase II GDM
N Carolina	USAED, Wilmington	1973	Beach Erosion Control and Hurricane Protection, Brunswick County, NC, Yaupon Beach and Long Beach Segments, Phase I GDM
Florida	USAED, Jacksonville	1985	Beach Erosion Control Project, Pinellas County, Florida, Long Key Beach Nourishment and Breakwater Feature Design Memorandum
Florida	USAED, Jacksonville	1989	Sand Bypass System Canaveral Harbor, Florida, GDM
Gulf of Mexico			
Alabama	USAED, Mobile	1978	Beach Erosion Control and Hurricane Protection, Mobile County, Alabama (Including Dauphin Island), Feasibility Report
Great Lakes			
Pennsylvania	USAED, Buffalo	1980	Cooperative Beach Erosion Control Project at Presque Isle Peninsula, Erie, Pennsylvania, Phase I GDM
Ohio	USAED, Buffalo	1975	Cooperative Beach Erosion Control Project at Lakeview Park, Lorain, Ohio, Phase II GDM
New York	USAED, Buffalo	1976	Cattaraugus Harbor, New York, Phase II GDM Detailed Design (in three volumes)
Pacific Coast			
Washington	USAED, Seattle	1989	Navigation Improvement Project at Grays Harbor, Washington, GDM

V-2-2. Defining Project Area and Boundary Conditions

a. The total project area encompasses not only the physical limits of the project but also the area in which the project has an effect upon littoral processes. The physical limits of a proposed project usually are predetermined by the local sponsor for the area of interest. Often this is defined by property limits (municipal park, business district, navigation inlet, etc.) and legislative authority. Project boundaries are sometimes defined by the limits of available historic data - particularly bathymetric data. Economic analysis of the project may lead to modification of these limits if a project along the entire reach cannot be economically justified, or is not technically feasible or environmentally acceptable.

b. It is imperative that a project designer be cognizant of potential impacts of the project on the adjoining coast. Mitigation due to disruption of littoral movement is often necessary. Legal requirements for replacement of sediment loss using bypassing or introduction of an equivalent quantity downdrift vary with locality. For example, the state of Florida (1987) requires:

(1) *All construction and maintenance dredging of beach-quality sand should be placed on the downdrift beaches: or, if placed elsewhere, an equivalent quality and quantity of sand from an alternate location should be placed on the downdrift beaches.*

(2) *On an average annual basis, a quantity of sand should be placed on the downdrift beaches equal to the natural net annual longshore sediment transport.*

The lack of legal requirements does not absolve the project designer of responsibility to mitigate project impacts.

c. In order to establish the potential physical impact of the project, a sediment budget for the entire littoral cell in which the project will reside must be determined. A sediment budget quantifies the amount of sediment moving within a littoral cell. A littoral cell is a self-contained reach of coast with its own sand sources and losses or sinks. Updrift and downdrift boundaries (the location at which most, if not all, sediment enters or leaves the cell) may be formed by natural barriers such as a headland or prominent protruding structures. In developing a sediment budget, information on the sources and characteristics of materials, modes and direction of transport in the littoral zone, rates of sediment supply, and transport and loss is required. Complete understanding of the boundaries, significance of individual sources, littoral drift direction and volumes, and sinks is crucial to estimating the effect of any engineering intervention in the littoral system and helps define the project area. Part III presents a complete description of sediment transport processes.

d. Development of the sediment budget will assist in establishing the amount of mitigation required. If no mitigative measures are enacted, the more difficult question will be to determine how erosion, due to loss of sediment, is distributed downdrift as a function of time. To answer this question, the three parameters that need to be identified are (a) length of the adversely affected shore; (b) cross-sectional retreat of the erosion cut; and (c) rate of expansion of erosion, and its distribution downdrift as a function of time (Bruun 1995). It has been observed that a littoral drift barrier has a short- and long-distance influence on the downdrift shoreline as briefly discussed in Section V-2-8. The reader is directed to Bruun (1995) for a discussion on examples of downdrift shoreline impacts and the short- and long-distance influence of a littoral barrier.

V-2-3. Storm Characteristics/Meteorology

a. Storm types.

(1) A storm is an atmospheric disturbance characterized by high winds that may or may not be accompanied by precipitation. Two distinctions are made in classifying storms: a storm originating in the tropics (5 to 350 deg in both hemispheres) is a tropical storm, a storm resulting from a cold or warm front in the middle and high latitudes (30 to 60 deg) is an extratropical storm (Silvester and Hsu 1993). Both storms can generate large waves and produce abnormal rises in water level in shallow water near the edge of water bodies.

(2) A hurricane is a severe tropical storm with maximum sustained wind speeds of 120 km/hour (75 mph or 65 knots). These low pressure centers are known by different names geographically: hurricanes on the east coast of the Americas, typhoons in the western Pacific, monsoons in the Indian Ocean, and tropical cyclones in Australia (Silvester and Hsu 1993). Hurricanes, unlike less severe tropical storms, generally are well-organized and have a circular wind pattern with winds revolving around a center or eye. The eye is an area of low atmospheric pressure and light winds. The Coriolis force causes the radial inflow to rotate counterclockwise in the Northern Hemisphere and clockwise in the Southern Hemisphere. Atmospheric pressure and wind speed increase rapidly with distance outward from the eye to a zone of maximum wind speed, which may range from 7 to 110 km (4 to 70 statute miles) from the center. From the zone of maximum

wind to the periphery of the hurricane, pressure continues to increase; however, wind speed decreases. Atmospheric pressure within the eye is the best single index for estimating surge potential of a hurricane. This index is called the Central Pressure Index (CPI). Central pressures of 950-960 mb (28.0 - 28.3 in.) are common. Hurricane Ida in the Philippines in 1958 had an extremely low pressure of 877 mb (25.9 in.) (Eagleman 1983). Generally, for hurricanes of fixed size, the lower the CPI, the higher the wind speeds.

(3) Unlike tropical storms, which generally occur in summer, extratropical storms generally occur in winter. These depressions in the middle latitudes consist of warm and cold air fronts which rotate about a low pressure center. Winds are not as intense as hurricanes since their horizontal scale is greater with a lesser pressure gradient. However, extratropical storms tend to have a longer duration and their destruction effects may be felt over large areas. Extratropical storms that occur along the northern part of the east coast of the United States, when accompanied by strong winds blowing from the northeast quadrant, are called nor'easters. Nearly all destructive nor'easters have occurred between November and April. Extratropical storms produce the dominant large wave conditions in the Great Lakes and generally occur between mid-October and April. The size of a typical storm of this type is shown in Figure V-2-1 (Resio and Vincent 1976).

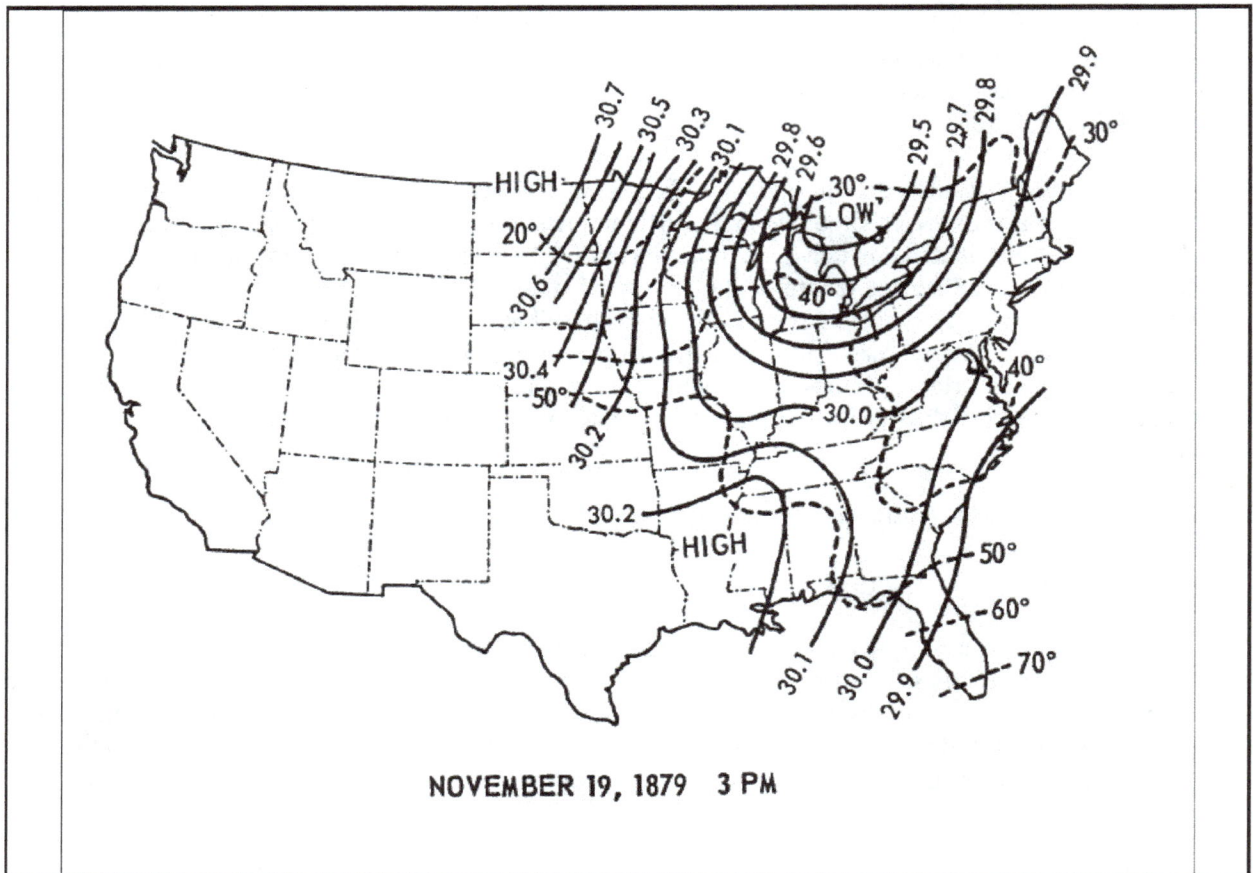

Figure V-2-1. Isobaric pattern of a typical extratropical low in the Great Lakes Region (Resio and Vincent 1976)

b.   Storm characteristics. Depending upon availability of observed hurricane data for the open ocean coast, the design analysis for coastal structures may not be based on measured water levels and waves. A statistical approach has evolved that takes into account the expected probability of occurrence of a hurricane with a specific CPI at any particular coastal location. Statistical evaluations of hurricane parameters, based on detailed analysis of many hurricanes, have been compiled for coastal zones along the Atlantic and U.S. Gulf coasts. Parameters that need to be evaluated are radius of the maximum wind, minimum central pressure of the hurricane, forward speed of the hurricane while approaching or crossing the coast, and maximum sustained wind speed at 10 m (33 ft) above the mean water level. Table V-2-2 presents extreme pressure and wind data for hurricanes recorded along the Alabama coast between 1892 and 1969 (U.S. Army Engineer District, Mobile 1978).

Table V-2-2
Extreme Pressure and Wind Data of Hurricanes Recorded along the Alabama Coast between 1892 and 1969 (U.S. Army Engineer District, Mobile 1978).

Date Hurricane Crossed Coast	Approx. no. miles and direction center passed Mobile	Lowest barometric pressure (in.)	Location of pressure measurement	Maximum wind velocity and direction (mph)	Location of wind measurement
2 Oct 1893	50 W	29.16	Mobile	80 SE	Mobile
15 Aug 1901	70 W	29.32	Mobile	61	Mobile
27 Sept 1906	20 SW	28.84	Mobile	94	Ft. Morgan
20 Sept 1909	150 SW	29.62	Mobile	52	Ft. Morgan
14 Sept 1912	20 W	29.37	Mobile	60 SE	Mobile
29 Sept 1915	100 W	29.45	Mobile	60 SE	Mobile
5 July 1916	20 W	28.38	Ft. Morgan	107 E	Mobile
18 Oct 1916	60 E	29.22	Mobile	128 E	Mobile
28 Sept 1917	100 SE	29.17	Mobile	96 NNE	Mobile
20 Sept 1926	30 S	28.20	Perdido Beach	94 N	Mobile
1 Sept 1932	25 SSW	29.03	Bayou La Batre	57 E	Mobile
19 Sept 1947	110 SW	29.54	Mobile	53 E	Mobile
4 Sept 1948	90 W	29.55	Ft. Morgan	42 S	Mobile
30 Aug 1950	20 E	28.92	Ft. Morgan	75	Ft. Morgan
24 Sept 1956	80 S	29.49	Mobile	58	Mobile
15 Sept 1960	80 W	29.48	Mobile	74	Dauphin Is.
3 Oct 1964	230 W	29.39	Alabama Port	80 NNW	Alabama Port
17 Aug 1969	90 WSW	29.44	Mobile	74	Mobile

c.   Hypothetical hurricanes. The National Weather Service and the U.S. Army Corps of Engineers have jointly established specific storm characteristics developed from statistical consideration. Since parameters characterizing these storms were statistically derived, they are known as hypothetical storms. Parameters for such storms are assumed constant during the entire surge generation period. The Standard Project Hurricane is defined as a hypothetical hurricane that is intended to represent the most severe combination of hurricane parameters that is reasonably characteristic of a region excluding rare combinations. The Probable Maximum Hurricane is defined as a hypothetical hurricane having that combination of characteristics which will make the most severe storm that is reasonably possible in the region involved if the hurricane should approach the point under study along a critical path and at an optimum rate of movement. The reader is directed to Part II, Chapter 2 for further discussions on meteorological systems and waves.

V-2-4. Hydrodynamic Processes (Design Sea State, Water Levels, Currents)

a. Design condition.

(1) In selection of design water levels and waves for a project, critical conditions must be considered. These represent critical threshold combinations of tide level, surge (or setup) level, wave conditions and local runoff, which, if surpassed, would endanger the project and/or make the project nonfunctional during their occurrence. It should be recognized that water levels have a direct impact on wave conditions in shallow water. Conversely, waves can have some impact on the water level through wave-induced setup.

(2) In selection of the design condition for the project, it is important to consider the intended structural integrity, and functional performance. Structural integrity relates to the structure's ability to withstand effects of extreme storms without sustaining significant damage. Structural integrity criteria determine the structure's life-cycle costs to the extent that a certain level of investment is necessary to prevent damages from an extreme event. Functional performance determines the incremental economic benefits of a project since it defines the structure's level of effectiveness. For example, the crest height of a shore dike to prevent flooding should be optimized for different storm frequencies with consideration of interior ponding elevations (and damages) in comparison to costs of different structure heights. In performing analyses, one must be aware of the risk and uncertainty inherent in the design process. See U.S. Army Corps of Engineers (1992) for a discussion on risk and uncertainty analysis in water resources planning.

b. Design wave height and period.

(1) Most coastal projects require an estimate of characteristics (height, period, direction and frequency of occurrence) of wind-generated gravity waves at the project site. In the Pacific basin, tsunamis also may need to be considered. Wave characteristics are determined outside the surf zone and then transferred to the project site by considering refraction, diffraction, reflection, shoaling, and breaking effects. Part II, Chapter 3 discusses methods to estimate nearshore waves with guidance for performing wave transformation studies.

(2) Wave data sets are available as summaries of visual observations, wave hindcasts and wave gauge statistics. Shipboard observations generally represent deepwater wave conditions over large areas. For coastal areas of the United States, summaries have been published in Department of the Navy (1976). The USACE's Wave Information Study (WIS) has developed wave hindcasts along U.S. coasts, the Great Lakes, and some international locations. These hindcasts generally consist of wave height, period, and direction time series and summary statistics. WIS data are available on the World Wide Web (WWW) at:
http://chl.erdc.usace.army.mil/CHL.aspx?p=s&a=DATA;1
or at:
http://frf.usace.army.mil/wis/

(3) Wave data can be measured by various forms of floating and bottom-mounted wave gauges. If available for a project site, actual measured data are preferred to hindcasts because these data are obtained by physical measurement. While the amount of wave gauge data may be spatially and temporally limited, they can be used to confirm hindcasts. If the physical measurements include significant storms (Figure V-2-2), this data can confirm hypothetical design storm size and effects. Wave gauge data collected by the Coastal and Hydraulics Laboratory (CHL) can be accessed at:
http://sandbar.wes.army.mil/public_html/pmab2web/htdocs/dataport.html

Wave data, extremal analyses, climatic summaries, buoy data, and aerial photographs, arranged geographically, can be found at the CHL web page:
http://chl.erdc.usace.army.mil/chl.aspx?p=s&a=Data;0
Wave data from the west coast is available from the Coastal Field Data Information Program:
http://cdip.ucsd.edu/
Buoy wave data and meteorological observations can be found at the National Data Buoy Center's web page:
http://www.ndbc.noaa.gov/

    c.   *Design water level.*

    (1)   Storm surge at the shoreline due to a rise in water level above the still-water level is of more interest to the design engineer. Peak surge is generally used to establish design water levels at a site. The highest water levels that occur along the Gulf Coast, the east coast of the United States from Cape Cod to south Florida, and the Hawaiian Islands are generally the result of tropical storms. At most other U.S. locations, high water levels result from extratropical storms. The state of Alaska has experienced peak storm surges over 4 m (13 ft) high. (State of Alaska 1994). Storm surge interaction with tidal elevations is discussed in Chapter II-5, along with methods to determine storm event frequency-of-occurrence relationships.

    (2)   While large inland water bodies such as the Great Lakes experience negligible tidal fluctuations, they encounter fluctuating water levels due to variations in the hydrologic cycle coupled with storm surges (wind setup) resulting from extratropical storms. Significant water level setup is possible on large shallow inland water bodies, such as Lake Okeechobee, Florida, and Lake Erie of the Great Lakes. Peak water level relationships have been determined for various locations along the Great Lakes coast (U.S. Army Engineer District, Detroit 1993a). Figure V-2-3 presents a lake elevation hydrograph for the December 2, 1985 event on Lake Erie (National Oceanic and Atmospheric Administration (NOAA) 1985).

    (3)   A complete discussion on tides, surges, and seiches is presented in Chapter II-5. This chapter should be reviewed to determine the type of data required for the study being performed. NOAA is the most comprehensive source of global tide predictions in tables of time and heights of high and low tides, tidal current tables for U.S. coasts, tidal current charts for selected harbors, and other summaries of tidal predictions for selected areas. They may be contacted at:

    NOAA National Ocean Service
    Tidal Datums and Information Section
    6001 Executive Boulevard
    Rockville, MD 20852

NOAA'S WWW site lists availability of tide data, charts, and other material.

V-2-5. Seasonal Variability

Recognizing seasonal variability of waves and currents along the coast is important to appropriately interpret historic data and predict a project's response. This variability has effects on littoral transport quantities and direction, and gross profile shape. Alternate erosion and accretion may be seasonal on some beaches; winter storm waves erode the beach and summer swell waves rebuild it. Just as insufficient data can lead to erroneous conclusions, information obtained only during a particular season can lead to similar results. Most ocean coasts experience periods during which storm or swell waves dominate, although both can occur simultaneously. As mentioned in Section V-2-3.a, tropical and extratropical storms occur only during a portion of the year. The WIS wave hindcast (Section V-2-4.b) is a source of storm and swell wave conditions along the U.S. coasts and Great Lakes.

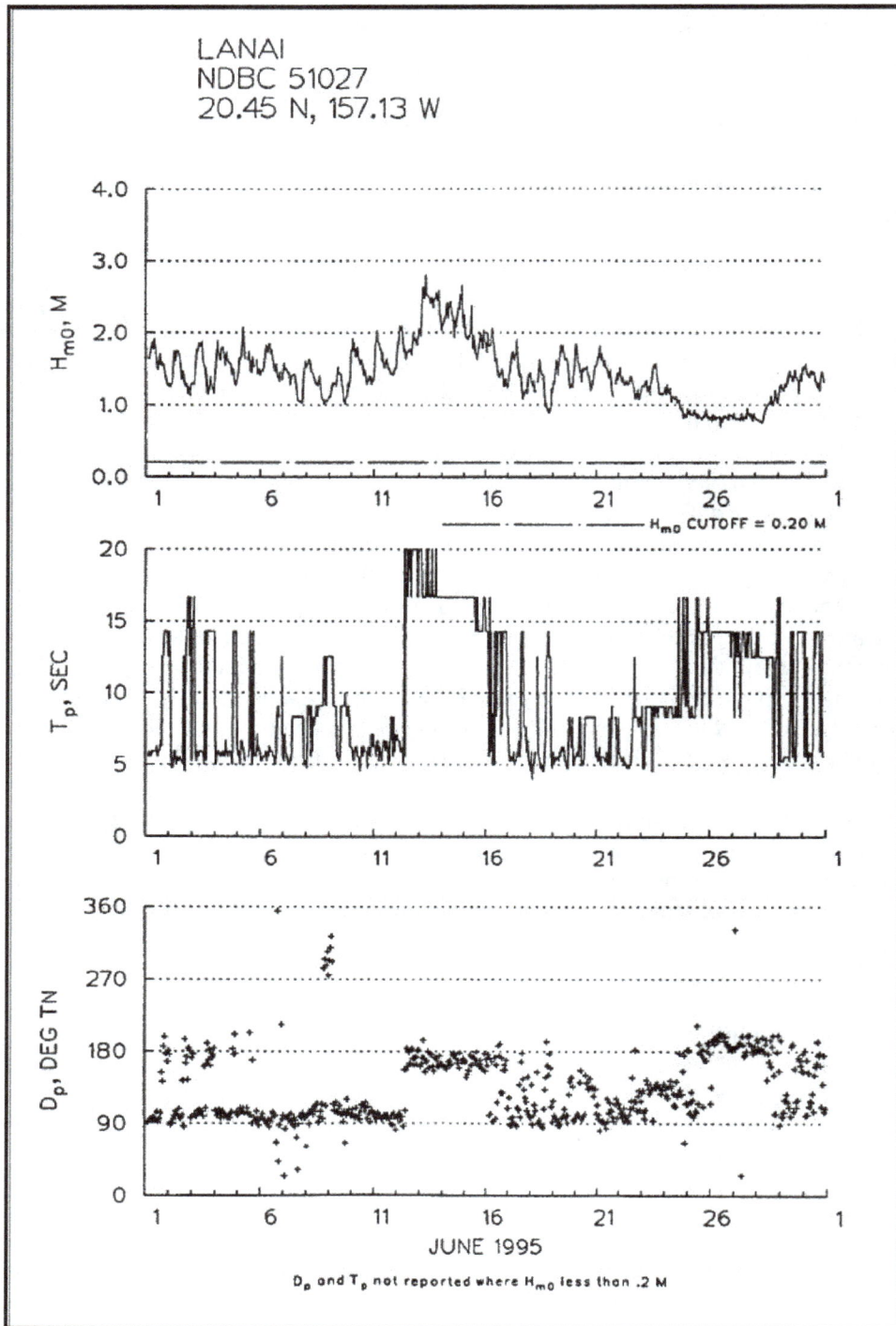

Figure V-2-2. Example of directional wave gauge data for Lanai, Hawaiian Islands
(EM 1110-2-1810)

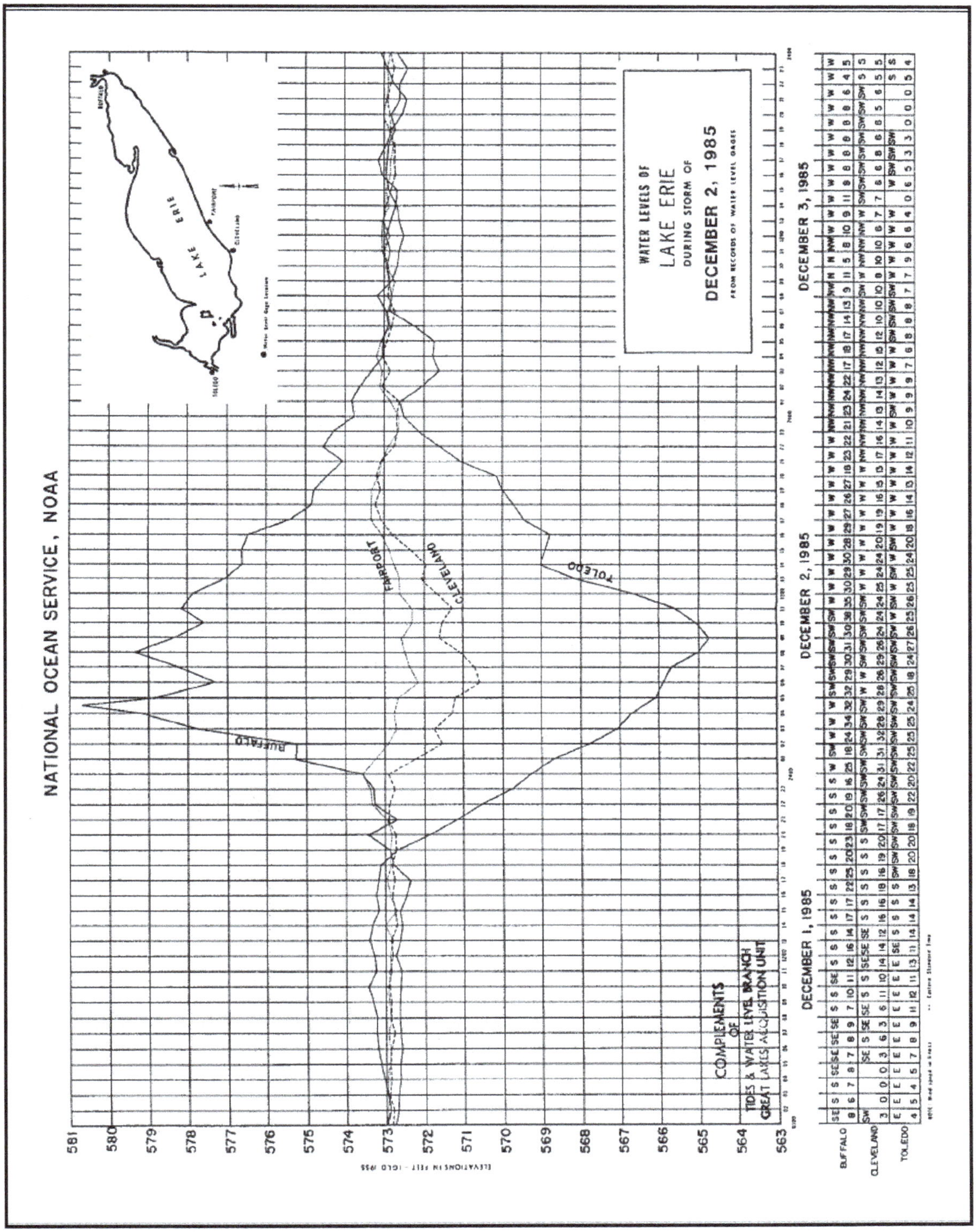

Figure V-2-3. Water levels of Lake Erie during storm of December 2, 1985 (NOAA 1985)

V-2-6. Topography and Bathymetry (Map Data)

a. *Data needs*.

(1) The amount of data needed for each project varies with the scope and type of project. Maps of the study area should include the updrift coastal zone which may affect the project and the downdrift area that will be affected by the project. Cadastral, topographic, and bathymetric information obtained from different sources should be combined into one or a series of maps. Computer Aided Drafting and Design (CADD) or Geographic Information System (GIS) programs facilitate the consolidation of map information. The advantage of a CADD or GIS drawing is that different types of information can be placed on different levels, allowing the user to access different layers for presentations, and 2- and 3-D illustrations are possible. Calculation of linear, areal, and volumetric change between elevations and distances (profiles) is possible.

(2) Recent surveys will be required to provide data in sufficient detail for cut/fill computations, sediment budgets, shoreline change, existing features, and property boundaries. Comparison with historic data will serve to illustrate shoreline and profile change and development. A review of historic maps and photographs is mandatory to fully understand the nature and processes at work, evolutionary trends, and natural ranges of variability. As windows to the past, they can present information which was at one time common knowledge, but has been lost with the passing of generations. Figure V-2-4 presents a portion of the 1836 shoreline map for Presque Isle, Pennsylvania. The location and extent of the breach along the neck (which afterward healed) give a glimpse of the shoreline position from another era. Most historic information, at least in the United States, only spans a century or less.

(3) Cadastral information relates to the showing or recording of property boundaries, subdivision lines, buildings, utilities, and related data. Topographic data presents the relief of land and the position of natural and man-made features. Often, topographic surveys with a contour interval of 0.3 m (1.0 ft) are required. Hydrographic surveys present the subsurface relief of water bodies and their shoreline position. Hydrographic survey techniques have been experiencing a renaissance. While the use of a lead line or other direct measure surveying procedures are still common, most hydrographic surveys are presently conducted using acoustical echo sounders. Other means such as the Scanning Hydrographic Operational Airborne Lidar Survey (SHOALS), a helicopter-mounted hydrographic surveying system for bathymetric measurement, are being introduced. SHOALS uses Light Detection and Ranging (LIDAR), a technology that uses a laser transmitter and receiver for water surface and water bottom detection. Because SHOALS is airborne, data can be obtained over ten times faster than shipboard echo soundings.

b. *Available sources*. Existing map data for the United States may be obtained from the following sources:

Cadastral: City, county, and state real estate records.
Topographic: State Department of Transportation, United States Geological Survey
(USGS) and the U.S. Army Corps of Engineers (USACE).
Hydrographic: NOAA and USACE.

USACE District offices prepare annual reports on their activities. Review of these historic documents presents a narrative summary of coastal activities at U.S. Government installations. Maps also may be attached to these reports. The reader is referred to Part II, Chapter 8.

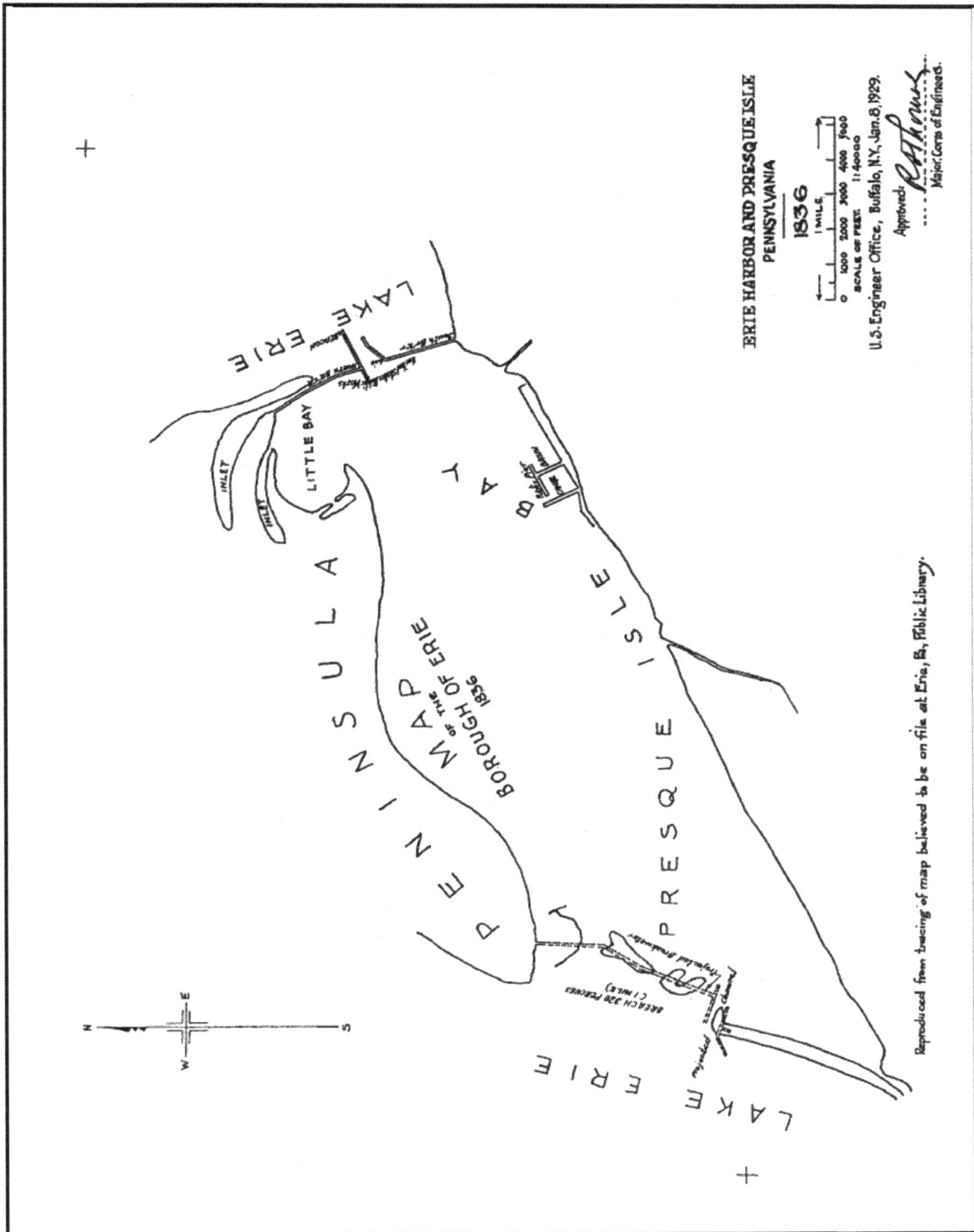

Figure V-2-4. The 1836 shoreline at Presque Isle, Pennsylvania

c.   *Reliability.*

(1)   All maps should have a north arrow (or south arrow in the Southern Hemisphere), a scale, and reference to the horizontal and vertical datum planes used. When using a historic map, one should be aware that magnetic north varies with time and one must align the map using true north. A map with a bar scale indicating the actual length is preferred. Be wary of maps that only report the scale, such as 1 cm = 1 km, unless the map is the original. The scale may be different if it is a copy, either due to slight paper shrinkage or reduction or enlargement when the copy was made. Also be aware that the datum plane can change with time or that a different datum plane may have been used. The datum used for topographic surveys may be different than that used for hydrographic work. Also, individual localities, such as cities, may have their own datum planes (Part II, Chapter 5 discusses datums). When comparing historic maps, at least two points of reference will be necessary in order to align the maps.

(2)   The accuracy of hydrographic data is affected by four types of error: sounding, spacing, closure, and error due to temporal fluctuations in the lake or sea bottom. These errors may be more significant (greater) for nearshore profiles than for beach or topographic surveys since land surveys are not affected by the latter error and measurement techniques are more precise for topographic work. The presence of errors suggests a need for caution in interpreting differences obtained from two surveys of the same profile. In other words, be sure that different profiles truly represent differences in the sea bottom and do not merely reflect survey or plotting errors. Since the nearshore seafloor can change rapidly in response to changing wave conditions, differences between successive surveys may reflect bottom differences caused by storms and seasonal wave climate changes. These fluctuations may actually be greater than long-term trends. Reviewing available historic bathymetric charts will assist in interpreting long-term trends as short-term changes are often larger than net changes.

V-2-7. Geomorphology/Geometry and Sediment Characteristics

a.   The coast is a diverse and dynamic environment. Many geologic, biological, and natural and human-made physical factors are responsible for shaping the coast and keeping it in a constant state of flux. Ancient geological events created, modified, and molded the rock and sediment bodies that form the foundation of the modern coastal zone. Over time, various physical processes have acted on this preexisting geology, subsequently eroding, shaping, and modifying the landscape.

b.   Lithology deals with the general characteristics of rock and sediment deposits and is an important factor in shaping the present coast. The most crucial lithologic parameters responsible for a rock's susceptibility to erosion or dissolution are the mineral composition and degree of consolidation. Marine processes are most effective when acting on uncemented material, which is readily sorted, redistributed, and sculpted into forms that are in a state of dynamic equilibrium with incident energy. Part IV describes coastal geomorphology/sediment characteristics.

a.   *Types of coasts/principal features.*

(1)   Upon initiation of a study in a coastal area, the investigator needs to be aware of the type of coast being examined. Coasts may generally be classified as consolidated, unconsolidated, tectonic, or volcanic. Consolidated rock consists of firm and coherent material with coasts of this type typically found in hilly or mountainous terrain such as that in Maine (Figure V-2-5). In contrast, depositional and erosional processes dominate unconsolidated coasts, which are normally found on low relief coastal plains or river deltas. The Atlantic and Gulf of Mexico coasts of the United States are mostly unconsolidated depositional environments (except locations like the rocky New England shores). Figure V-2-6 is a photo of an unconsolidated coast.

Figure V-2-5. Example of resistant rock-bound coast of Maine (Bass harbor Head Light, Maine, south of Acadia National Park)

Figure V-2-6.   Beach near entrance to North Sand Pond, Lake Ontario, NY. This is a relict dune environment at the eastern end of Lake Ontario

Forces within the earth's crust and mantle deform, destroy, and create crustal material. These tectonic activities produce large-scale uplift and subsidence of land masses. The west coast of the United States is an example of a tectonically dominated coast. Sea stacks are prevalent along the U.S. Pacific coast. Sea stacks are steep-sided rocky projections above sea level near the coast. They are formed by wave cutting back on the two sides of a promontory. With the aid of weathering and further cutting behind it, it is left as an island. Sea stacks may be located onshore (Figure V-2-7) or further offshore, are of varied sizes and locally affect the wave patterns near them. The Great Lakes were created by glacial action, and physical characteristics of the shorelands are very diverse. They vary from high bluffs of consolidated or uncon-solidated (Figure V-2-8) material to low bluffs and plains, dunes and wetlands. The eruption of lava and the growth of volcanoes may result in large masses of new crustal material, such as in the Hawaiian Islands.

(2)   Principal coastal features which have been formed by local processes give indications of the types of coast under investigation. For example, the presence of a barrier island is indicative of a coast consisting primarily of unconsolidated sediments. In contrast, the most prominent feature exhibited by a fault coast is a scarp where normal faulting has recently occurred, dropping a crustal block so that it is completely submerged and leaving a higher block standing above sea level.

b.   *Sources and sinks.* Recognition of the many sources (gains) and sinks (losses) in the coastal zone is important during development of the sediment budget for the region. In general, sand supply from rivers, cliff/bluff erosion, onshore transport from the shelf, biogenic sources (such as reefs), and alongshore sediment transport into the area constitute the major natural sources. Beach nourishment represents a human-induced gain to the budget. Natural losses can include sediment blowing inland to form dunes, offshore trans-port to deeper water, losses down submarine canyons, and the longshore transport that carries littoral

Figure V-2-7. Coast at Orick, CA. This is a pocket beach between resistant headlands

Figure V-2-8. Shore at Chimney Bluffs State Park, Lake Ontario, NY. Chimneys consist of glacial till more weather resistant than surrounding material

sediments out of the study area. Sand mining or direct removal (e.g., channel dredging with disposal outside of the littoral system) is a human-induced deficit to the budget.

   c.   *Prevailing sediment characteristics.*

   (1)   The geology of the coast and the source of littoral materials ultimately determine the prevalent shape of the shore and the composition of the beach at a specific locality. Littoral materials are classified by size, composition, shape, and other properties such as color. Littoral materials are classified by grain size in clay, silt, sand, gravel, cobble, and boulder. Several size classifications exist, with the Unified Soil Classification being the primary classification used by engineers and the Wentworth by geologists. Part III, Chapter 1 discusses sediment properties and classification.

   (2)   Littoral material is composed of materials specific to that region. While beach material is most commonly composed of quartz or feldspar particles, it also can be volcanic debris, as in the Hawaiian and Aleutian Islands, shell and coral, organics (peat), silts and clays. Littoral sands, gravel, and cobbles are usually rounded (Figure V-2-9), while departures from this shape are attributed to contributions from shell fragments and sedimentary rock such as shale. These departures from a spherical shape affect sediment motion initiation.

Figure V-2-9.   Cobble Beach along Lake Ontario, Oswego, NY. Origin of cobble is bluff and glacial drumlin erosion

Sediment color may be used to trace the source of littoral material. It also is an indication of the relative density of the material. "Light minerals" such as quartz and feldspar, which have specific gravities ranging from 2.65 to 2.76 are generally tan, cream, or transparent. The famous white sands of the Florida panhandle are a very clean, uniformly sized quartz. "Heavy minerals," such as hornblende, garnet, and magnetite, which have specific gravities ranging from 3.0 to 5.2, are dark (black, red, dark green, etc.) in color. Littoral sediments of volcanic islands are dark, often green and black (Figure V-2-10).

Figure V-2-10.   Black Sand Beach, Kalapana, HI. Sand is of volcanic origin

d.   *Sediment layering.* It is important to recognize that variation in beach material may occur with depth. Classification of the surficial material only may lead to erroneous interpretations of coastal activity. Underlying the beach material can be layers of cohesive material, peat or rock. These will have significant effects on beach profile response to storm activity and foundation conditions for proposed structures. An understanding of the geology of the area will indicate whether layering (stratigraphic variability) can be expected.

V-2-8. Littoral Drift and Sediment Transport Patterns

Littoral transport is the movement of material in the littoral zone by waves and currents. This includes movement parallel (longshore) and perpendicular (cross-shore). The littoral zone extends from the shoreline to just beyond the seawardmost breakers.

a.   *Longshore movement.* Wave-generated currents tend to dominate water movements in the nearshore zone and are important in the movement of sediments. Wave-induced currents are superimposed on the wave-induced oscillatory motion of the water. When waves break with their crests at an angle to the shoreline, a current is generated parallel to the shore that is largely confined to the nearshore between the breakers and the shoreline. The volume rate of material transport along the shore is sensitive to the breaker angle and height. The longshore gradient in breaking wave height is also a generating mechanism for longshore sand transport. This contribution is usually much smaller than that from oblique wave incidence in an open coast situation. However, in the vicinity of structures, where diffraction produces substantial change in breaking wave height, its inclusion improves transport rate prediction (Hanson and Kraus 1989). Longshore sediment processes are discussed in Part III, Chapter 2.

b.   *Cross-shore movement.*

(1)   Wave breaking generates turbulent motion and provides the necessary mechanism for suspending sediment. Sediment transport in a direction perpendicular to the beach is known as cross-shore movement. The equilibrium profile is a profile of constant shape which is approached if exposed to fixed wave and water level conditions. Waters (1939) proved that the existence of an equilibrium profile was a valid concept under laboratory conditions. The concept of the equilibrium profile is discussed in Part IV.

(2)   As a beach profile approaches an equilibrium shape, the net cross-shore transport rate decreases to approach zero at all points along the profile (Larson and Kraus 1989). The division of the profile into different transport regions has allowed empirically based relationships for the net transport rate to be formulated. The four transport zones across the profile are defined as:

Zone I: From the seaward depth of effective sand transport to the break point (prebreaking zone).

Zone II: From the break point to the plunge point (breaker transition zone).

Zone III: From the plunge point to the point of wave reformation or to the swash zone (broken wave zone).

Zone IV: From the shoreward boundary of the surf zone to the shoreward limit of runup (swash zone).

(3)   The net transport rate in zones of broken waves, where the most active transport occurs, shows good correlation with the wave energy dissipation per unit water volume. The net transport rate in the prebreaking zone decreases exponentially with distance offshore. In the foreshore zone, the net transport rate shows an approximately linear behavior decreasing in the shoreward direction. The reader is referred to

Chapter III-3 for a discussion on cross-shore transport. Larson and Kraus (1989) present an excellent literature summary of the chronological investigations on profile change and the development of cross-shore transport modelling. Dean (1991) summarizes equilibrium profile responses for differing coastal conditions. Houston (1996) applies equilibrium profile responses to beach fill design. The reader is referred to Part III, Chapter 4.

    c.   Seasonal reversals.

    (1)   Section V-2-5 discussed the seasonal variability of storms and waves. This variability results in two distinct classes of waves, storm waves and swell waves, which have completely different effects on the beach profile. In general, storm waves remove the beach berm to place it along the offshore portion of the profile while swell waves replace it back onshore (Silvester and Hsu 1993). This occurs primarily along the oceanic coasts and to a reduced degree along inland seas. Recognition of this variability is important in interpreting existing beach change and during the establishment of a monitoring program for a beach.

    (2)   The following synopsis of this action is presented eloquently in Sylvester and Hsu (1993). In describing the action of swell waves on the beach profile,

> *The broken wave swashes up the beach face with its water percolating through it, so long as there is adequate time between each wave. ...this water soaks down to the water table some distance below the face, eventually to be returned back to the sea. The downrush, when a trough arrives, is smaller than the uprush due to this percolation and therefore cannot carry much of the sediment load back down the beach face. Also, the hydraulic jump associated with this flow reversal is small. The result is that swell waves, with several seconds between each crest, will leave material on the face and hence the beach accretes.*

    (3)   In contrast to swell waves, storm waves are steeper. Silvester and Hsu (1993) wrote,

> *Now consider what happens to this swell profile when storm waves arrive. These are very steep and contain much water above the mean sea surface. They comprise waves of many periods, or constitute a wide spectrum, with heights appropriate to each. A crest arrives almost every second instead of every few seconds and hence large volumes of water are thrown over the beach face, which is quickly saturated. By this action, the water table has become almost coincident with the face itself. The downrush now nearly equals the uprush so that much sand is dragged down the face into the hydraulic jump, which is increased in size. This is one mechanism by which the berm is eroded and its contents placed into suspension.*
>
> *Another factor, of equal if not greater importance, is the flow of the excess ground water returning to the sea. At the waterline, where the hydraulic jump is also located, it is moving vertically, thus causing liquefaction or a "quicksand" effect (Longuet-Higgins 1983). This aids suspension so helping to undermine the toe of the beach, which progressively retreats landward...*

    (4)   The above briefly described the effect on the profile type (accretion/erosion) and hence a net onshore or offshore cross-shore transport due to the presence of swell or storm waves, respectively. It also is important to recognize during development of the sediment budget that the net direction of the storm and swell waves may be different, resulting in a reversal of longshore drift with season. This can have a significant effect upon the location of accretionary/erosional zones adjacent to the project. The project design needs to tolerate a range of expected conditions. One must be able to change bypassing operations during a year to accommodate these seasonal reversals.

     d.   *Long-term reversals.* Recognition of changes in the sediment budget over the long term (years to decades) is important when evaluating historic changes or predicting potential changes due to a proposed project. Modifications of the sediment gains or losses through natural causes or human intervention can shift an accretionary shore to an erosional shore. For example, the construction of large harbor structures or the stabilization of an inlet along the coast have affected the amount of sediment reaching downdrift beaches. Without sediment bypassing around the harbor, downdrift beaches may erode severely. Cyclic climatic changes such as the El Niño can cause reversals for many months. Knowledge of these type of activities is necessary to properly interpret profile change.

     e.   *Slug motion.*

     (1)   Sediment movement along the shore rarely occurs at a temporally and spatially constant rate. The presence of large beach protuberances termed "sand slugs" or "longshore sandwaves" is a phenomenon that has been reported at numerous beaches around the world. Along Long Point on the Lake Erie shore, these features have been observed with alongshore lengths of 500 to 2,500 m (1,640 - 8,200 ft) and maximum widths of 50 to 90 m (164 - 295 ft) which migrate alongshore in the direction of net drift at 150 to 300 m/year (492 - 984 ft/year) (Stewart and Davidson-Arnott 1988; Davidson-Arnott and Stewart 1987). Similarly along the Dutch coast, they have been measured with an amplitude of 30 to 500 m (98 - 1,640 ft), a celerity of 50 to 200 m/year (164 - 656 ft/year), and a period of 50 to 150 years (Verhagen 1989).

     (2) Bakker (1968) presents a mathematical theory on sandwaves and an application on the Dutch Wadden Isle of Vlieland. Stewart and Davidson-Arnott (1988) describe their formation by the onshore movement and welding of inner nearshore bars during nonstorm periods on areas of local sediment abundance. In addition, growth and downdrift extension and widening of the protuberance in the direction of net sediment transport results from attachment of an inner bar to the slug and infilling of the trough-runnel downdrift of the protuberance. The presence of sand waves results in a spatial pattern of erosion and accretion, with erosion occurring primarily in the embayments between the slugs and accretion occurring opposite the wide beach of the sand wave.

     (3)   The presence of this type of phenomenon can dwarf the effects of a relatively smaller shore project. At Presque Isle Peninsula, Pennsylvania, three prototype segmented offshore breakwaters were built in 1978. The intent was to study the effect of offshore distance, breakwater length, and gap width upon the shoreline. The movement of a sand wave through the project area a few years later totally covered the breakwaters. They began to emerge more than a decade later. Verhagen (1989) concluded that along the southern part of Holland, groin construction did not change linear long-term coastal erosion and did not change the cyclic behavior of the shore. He concluded that construction of groins on this coast did not have any effect at all.

     f.   *Hot spots.* Often at a beach nourishment project, there will be one or more areas that erode more rapidly than others. These areas are termed erosional hot spots. The location may correlate with areas which had previously experienced high erosion. Hot spots at new locations may be due to wave refraction and possibly wave focusing in response to bathymetric change from placed material (National Research Council 1995). Another cause may be a bottom composition change which affects the rate of movement. In any event, erosional hot spots may require renourishment earlier than the rest of the project. This unexpected work will place an additional financial strain on the project. Placement of a greater quantity of sand in hot spot areas may extend the time between renourishment.

V-2-9. Shoreline Change Trends

  a.   *Evidence of cyclic processes.*

  (1)   The presence of cyclic processes due to water levels and waves is observed in coastal features such as tidal inlets and seasonal bars. Inlets are the openings in coastal barriers through which water, sediments, nutrients, planktonic organisms and pollutants are exchanged between the open sea and the protected embayments behind the barriers. Inlets are not restricted to barrier environments or to shores with tides; on the west coast of the United States and in the Great Lakes, many river mouths are considered to be inlets, and in the Gulf of Mexico, the wide openings between the barriers, locally known as passes, are also inlets. Tidal inlets are analogous to river mouths but differ as they experience diurnal or semidiurnal flow reversals and they have two opposite-facing mouths (seaward/lakeward and lagoonward). Tidal inlets are characterized by large sand bodies that are deposited and shaped by tidal currents and waves. The ebb-tide shoal is a sand deposit that accumulates sea/lakeward of the inlet mouth. It is formed by ebb tidal currents and is modified by wave action. The flood-tide shoal is a sediment deposit at the landward opening that is mainly shaped by flood currents. Depending upon the size of the bay (lagoon), the flood shoal may extend into open water or may merge into a complex of meandering tributary channels, point bars, and muddy estuarine sediments.

  (2)   As indicated in Section V-2-8, seasonal variability of storms and waves results in seasonal reversals in cross-shore drift. The effect of the change from a swell-dominated to storm-dominated profile is clearly seen in the presence of seasonal bars. Storm waves remove the beach berm and place the material offshore as a bar. Swell waves attempt to reverse this action. Interpretation of long-term trends must account for these short-term changes.

  b.   *Eustatic sea level changes.* A worldwide change in sea level, referred to as eustatic sea level change, is caused by change in the relative volumes of the world's ocean basins and the total amount of ocean water (Sahagian and Holland 1991). The rise results in a slow, long-term recession of the shoreline, partly due to direct flooding and partly as a result of profile adjustment to the higher water level. Estimates of recent eustatic rise range from 15 cm/century (Hicks 1978) to 23 cm/century (Barnett 1984), although some researchers, after exhaustive studies of worldwide tide records, have not seen conclusive evidence of a continuing eustatic rise (Emery and Aubrey 1991). This topic is reviewed in greater detail in Part IV.

  c.   *Subsidence.* Subsidence is the sinking of land due to natural compaction of estuarine, lagoonal, and deltaic sediments resulting in large-scale disappearance of wetlands. This effect has been exacerbated in some areas by human intervention with groundwater and oil withdrawal. Significant subsidence occurs in and near deltas where large areas of fine-grain sediment accumulate. Land loss in the Mississippi delta has become an important issue because of loss of wetlands and rapid shoreline retreat. Along with the natural compaction of the deltaic sediments, groundwater and hydrocarbon withdrawal may have contributed to subsidence problems in southern Louisiana. The change in relative sea level in the Mississippi delta is about 15 mm/year, while the rate at New Orleans is almost 29 mm/year (data cited in Frihy (1992)).

V-2-10. Land/Shore Use

  a.   The present use of the shore and the area landward of the coast needs to be documented. The areal extent of information to be collected must be determined on a site-by-site basis. An understanding of the sediment budget will assist in defining limits. Information should be gathered not only for engineering purposes but also for environmental and economic documentation. For a project planned in the coastal zone, this information will assist in assessing environmental and economic effects of the project.

b.   Information collected on structures will include the type, ownership (residential, commercial, public, and other), density, location, and value. The elevation-versus-damage relation may be required if there is a flooding problem. The distance of structures from the top of bluffs or high-water elevation may be needed for regulatory purposes (setbacks) or expected potential damage at the present recession rate. If a shoreline erosion protection project is proposed, infrastructure data will be needed to estimate damage reduction (project benefits) due to a lower recession rate. Structures which influence sediment supply or surface/groundwater drainage patterns may also be important.

c.   In addition to the structure inventory, property boundaries have to be located. Identification of all lands, easements, and rights of way will be needed when assessing the required easements for construction limits and access to the site, as well as for future maintenance. The affected infrastructure (roads, utilities, etc.) also must be identified.

V-2-11. Potential for Project Impacts

a.   *Effects on natural tidal flushing.*

(1)   Dredging a channel through a tidal inlet usually results in increased shoaling. Channel dredging also can have a significant effect on adjacent shorelines, although the effect may be difficult to assess. A complete understanding of natural processes prior to dredging is required to discern the change due to dredging. Disposal of dredged material offshore may result in permanent removal from the littoral system if the depth is sufficient to prevent return to the nearshore littoral environment. Although the limiting depth for which material offshore will return to the beach is generally unknown, it is dependent on variables such as sediment size and wave climate. A few tests have suggested that material placed in water depths greater than 6 m (20 ft) in the ocean and about 2 m (6 ft) in the Great Lakes will not readily return to the nearshore littoral system (Harris 1954, Schwartz and Musialowski 1980).

(2)   Jetties or breakwaters often are built to stabilize inlets. These structures serve to stop the entry of littoral drift into the channel, function as training walls for tidal currents, stabilize the position of the navigation channel, may increase tidal current velocities which flushes sediment from the channel, and reduce shoaling in the channel. Despite their positive engineering effects, jetties often form a barrier to longshore sediment movement. Where there is no predominant direction of longshore transport, jetties may stabilize nearby shores, but this effect is limited only to the sand impounded at the structures. At most sites the amount of sand available to downdrift shores is reduced, at least until a new equilibrium shore is formed at the jetties. Where longshore transport predominates in one direction, an accretionary fillet will occur on the updrift side of the channel and erosion will occur on the downdrift side. Again, bypassing or nourishment will be required to mitigate this imbalance.

(3)   The increase in channel velocities also increases the potential for scour along the structure toe. This effect can be exacerbated by the presence of waves. This potential effect must be considered during project design.

b.   *Up/downdrift effects and need for bypassing.*

(1)   Construction of a coastal project that protrudes from the shore or is located nearshore and modifies the local wave climate may result in a change in sediment accretionary and erosional patterns. This perturbation of longshore drift movement will result in an accretionary zone immediately updrift or within the wave shadow of the structure and an erosion zone downdrift of the structure. The effect of the structure may be mitigated by prefilling the fillet (at a protruding structure) or salient (behind a shore-parallel structure). This preventative action may not be adequate. If the structure protrudes sufficiently to trap and deflect most of the littoral drift to deep water, bypassing of material may be required.

(2)   Prediction of erosion downdrift of a barrier is challenging. Bruun (1995) presents a discussion, series of examples, and a good reference list on this problem. He notes that downdrift erosion may in some cases be composed of short (local) as well as long-distance effects which move downdrift at various rates. The presence of a 'zero-area' or location of a temporary reduction of erosion a short distance from the barrier does not necessarily indicate the extreme limit of leeside erosion. While the short distance effect is a geomorphological feature, the long-distance effect is due to a sediment deficit. Determination of the maximum recession and length of shore affected by downdrift erosion is further complicated by the presence of natural erosionary processes.

(3)   Mitigation of shoreline damage due to the presence of an existing structure requires knowledge of recession rates (sediment budget) before and after its construction. The difference in sediment gain/loss downdrift of the structure is the minimum information required to mitigate effects of the structure. This value is used if bypassing of material is not practical or is politically infeasible (updrift owners enjoy the newly created or enlarged beach) and new material instead is introduced to the system.

   c.   *Changes in wave climate.*

(1)   Construction of a protruding or an offshore shore-parallel structure will modify the local wave and current climate near the structure. A protruding structure, such as a groin or jetty, may cause development of a rip current along the updrift face. An offshore breakwater will reduce wave activity in the lee since wave energy behind the structure occurs from a combination of diffracted waves from the ends and transmission over/through the structure. This usually will result in deposition of littoral material in the breakwater shadow. Modification of the wave pattern may limit surfing but can enhance other recreational activities such as swimming.

(2)   Reflection of incident waves from a natural or man-made structure also will affect local wave climate. The amount of wave reflection, expressed as the reflection coefficient (reflected wave height/incident wave height), is dependent upon the structure's surface roughness, structure height in relation to the wave runup (freeboard), structure slope and incident wave angle (especially as the wave direction approaches the structure orientation). Smooth, vertical, and high (no overtopping) structures (such as sheet-pile walls) reflect most incident wave energy. The resulting increase in wave height near the structure can induce additional scour at the structure toe, increase local transport rate, and result in unacceptable wave conditions within a harbor.

   d.   *Impact on benthic organisms.*

(1)   Construction of shore protection measures usually produces short-term physical disturbances which directly affect biological communities and may result in long-term impacts. Coastal structures alter bottom habitats by physical eradication and in some cases scour or deposition. However, certain hard structures may create a highly productive, artificial reef-type habitat. Beach nourishment will cover nonmotile organisms.

(2)   Species comprising marine bottom communities on most high-energy coastal beaches are adapted to periodic changes related to natural erosion and accretion cycles and storms. Burial of offshore benthic animals by nourishment material has a greater potential for adverse impacts than those in the intertidal and upper beach zone (Nagvi and Pullen 1982). Survival of vegetation and animals will depend upon the deposition depth, rate of deposition, length of burial time, season, particle size distribution and other habitat requirements.

(3) A biological survey of organisms living in and using the proposed project area must be completed before a project is initiated. This must include threatened and endangered species. This inventory and knowledge of habitat requirements will assist in defining potential impacts of the proposed project.

(4) Certain species are very sensitive to placement of beach nourishment. Moderate disturbance of a mature oyster reef can destroy it. Covering of mangrove prop roots can kill the entire plant (Odum, McIvor, and Smith 1982). Hard corals are more sensitive than soft corals to covering with fine sediments. Excessive sedimentation for a nourishment project which buries a reef results in permanent destruction or replacement by soft bottom habitat and community. Nourishment can affect sea turtles directly by burying their nests or by disturbing nest locating and digging during the spring and summer nesting season. Elimination of these adverse effects may be possible by timing of placement (to be discussed in Section V-2-12).

e. Changes in natural habitat. Construction of a project in the coastal zone can cause short- and long-term changes in the natural habitat. Placement or dragging dredge anchors and pipelines can damage environmentally sensitive habitats such as coral reefs, seagrass beds, and dunes. Short-term changes to the grain size and shape of the beach will occur depending on characteristics of the native and borrow material. Waves and currents will winnow and suspend finer sediments and deposit them in deeper water offshore. Eventually, the sediment size distribution will become comparable to the beach sediments prior to nourishment. An increase in compaction of the beach can result from beach nourishment. Burrowing animals such as crabs and sea turtles can have difficulty digging in compacted beaches. As with sediment sorting, the compaction increase will be temporary until the beach is softened by wave action, particularly storms. Construction of a hard structure can result in a permanent change locally by removal of bottom habitat. However, a rubble-mound structure may provide a different (reef-type) habitat. Scraping of the new beach fill face during its initial adjustment can also have an adverse effect on species such as sea turtles and crabs that transit the beach surface for nest building and reproduction.

V-2-12. Environmental Considerations

Selection of the best environmental and engineering solution to a coastal problem requires a thorough understanding of the complexity and diversity of the coastal zone. A clear definition and cause of the problem as well as a comprehensive review of potential solutions is required. In the previous sections, some potential impacts to be expected by a project were discussed. In this section, the principal environmental factors that should be considered in design and construction and techniques to attain environmental quality objectives are discussed briefly. The reader is directed to EM 1110-2-1204 for more detailed information.

a. Surveying the project area.

(1) An understanding of existing environmental conditions is vital to ensuring that the status quo is maintained or enhanced. An environmental survey of the project area is required in order to establish a baseline condition. As in any sampling program, the most critical aspect of data collection is identifying the proper parameters to sample and measure. The quality of information obtained will be dependent upon collecting representative samples, use of appropriate sampling techniques, protecting samples until they are analyzed, accuracy and precision of analysis, and correct interpretation of results.

(2) The purpose of collecting samples is to acquire adequate representation of the project area's characteristics. This requires that samples be taken in locations typical of ambient conditions found at the project site. The number and frequency of samples will need to be assessed. Sampling equipment should be selected based on reliability, efficiency, and the habitat to be sampled. In order to maintain the integrity of the results, preservation of samples is imperative. Preservation is intended to retard biological action and hydrolysis/oxidation of chemical constituents, and to reduce volatility of constituents.

(3) Habitat-based evaluation procedures are developed to document the quality and quantity of habitat. These procedures can be used to compare the relative value of different areas at the same time (baseline studies) or the relative value of one area with time (impact assessment). Two habitat assessment techniques available are the Habitat Evaluation Procedure (HEP) and the Benthic Resources Assessment Technique (BRAT).

(4) HEP has been computerized for use in habitat inventory, planning, management, impact assessment, and mitigation studies (U.S. Fish and Wildlife Service 1980). The method is comprised of a basic accounting procedure that computes quantitative information for each species evaluated. The inventory can pertain to all stages of a species, to a specific life stage, or to groups of species. An HEP analysis includes: scoping, development and use of Habitat Suitability Index models, baseline assessment, impact assessment, mitigation, and decision on course of action.

(5) BRAT procedures use benthic characterization information to produce semiquantitative estimates of potential trophic value of soft-bottom habitats. BRAT estimates the amount of benthos that is both vulnerable and available to target fish species that occur at a site. The utility of BRAT lies in the ability to provide meaningful information relevant to value decisions. While it does not provide an assessment of overall habitat status, it can be viewed as an in-depth assessment of a single habitat variable, that of trophic support. As such, it may contribute semiquantitative input to habitat-based assessments such as HEP.

b. Mitigation measures.

(1) During the design of a project, care must be taken to preserve and protect environmental resources, including important ecological, aesthetic and cultural values. Specific U.S. Army Corps of Engineer mitigation policy for fish and wildlife and historic and archaeological resources is included in Chapter 7 of Engineer Regulation (ER) 1105-2-100. Mitigation consists of avoiding, minimizing, rectifying, reducing, or compensating for impacts. The first elements often can be accomplished through their consideration during project design. The amount of compensation required for significant losses to important resources is quantified through documentation of the amount of actual/predicted losses. Justification must be based on significance of resource losses due to a project compared to costs necessary to carry out mitigation.

(2)   Some examples of mitigative measures for the aforementioned mitigative elements are as follows:

(a)   Avoid: Adjust the time of construction activities to avoid periods of fish migration, spawning, shorebird or turtle nesting; preserve a public access point.

(b)   Minimize: Disturb an immature reef instead of a mature one; use rough-surface facing materials on a structure.

(c)   Rectify: Replace a berm; restore flow to former wetland.

(d)   Reduce: Control erosion (sedimentation control plan); place restrictions on equipment and movement of construction and maintenance personnel.

(e)   Compensate: Use dredged material to increase beach habitat, create offshore islands, increase or develop new wetlands; construct an artificial reef.

c.   *Water quality.* Water quality impacts consist of changes to the water column's characteristics which may have short- and long-term consequences. Construction processes often are responsible for short-term increases in local turbidity levels, releases of toxicants or biostimulants from fill materials, introduction of petroleum products and/or the reduction of dissolved oxygen levels. These impacts can be minimized by construction practices, fill material selection, and in some instances, construction scheduling. These impacts are temporary unless long-term changes in hydrodynamics have occurred. It is these long-term repercussions which must be identified during the design process. The size and type of structural alternatives will result in a range of potential impacts. For example, the design of a jetty or offshore breakwater will greatly influence its impact on circulation and flushing, which affect water quality.

d.   *Disposal of materials.* Physical and chemical testing of the proposed material to be dredged is required in order to assess the appropriate disposal method. Local regulations also may dictate the manner in which material is to be disposed: open-water, upland or in a confined disposal facility. For material that is placed in open water, it may be necessary to predict long-term fate of the disposal mound. This assessment will entail determining whether the material is dispersive or nondispersive. If material is dispersive, rate of erosion and fate of the material should be computed from models or field studies.

V-2-13. Regional Considerations

a.   *Regulations.*

(1)   Construction of a shore erosion or navigation project in the coastal zone is governed by national and local regulations. Research of governing laws is an essential part of the planning process in order to determine potential limitations for constructing in the coastal zone. These stipulations may include physical structure limitations such as size, setback requirements, the need for buffer zones, restriction on hard structures (North Carolina, Massachusetts, and Maine), and ability to rebuild after sustaining damage; and environmental limitations such as season when construction can occur and required mitigation (i.e., bypassing, wetland creation, etc). These laws are the result of lessons learned from constructing along the coast (Figure V-2-11).

(2)   The Coastal Barrier Resources Act (Public Law 97-348 1982) is an example of a national regulation affecting coastal regions of the United States. The purpose of the Act is to minimize loss of human life; wasteful expenditure of Federal revenues; and damage to fish, wildlife, and other natural resources associated with the coastal barriers of the United States by restricting future Federal expenditures and financial assistance which have the effect of encouraging development of coastal barriers. The Act established the Coastal Barrier Resources System, which identified undeveloped coastal barriers on a series of maps. A coastal barrier is a depositional geologic feature (such as a bay barrier, tombolo, barrier spit, or barrier island) which consists of unconsolidated sedimentary materials, is subject to wave, tidal, and wind energies and which protects landward aquatic habitats including adjacent wetlands, marshes, estuaries, inlets, and nearshore waters. It is considered undeveloped if it contains less than one structure per 0.02 km^2 (5 acres) that is "roofed and walled" and covers at least 18.6 m^2 (200 ft^2).

(3)   Article 34 of the New York State Environmental Conservation Law (State of New York 1988) is an example of a state law that regulates the need for coastal erosion management permits and controls activities within defined structural hazard areas and natural protective feature areas. Maps of structural hazard areas are available for the entire state coastline which define a zone in which no new non-movable structures or non-movable major additions to existing structures can be built without formal approval. New public utilities must be located landward of the shore structures that they serve. Within natural protective feature areas, development is generally prohibited, only clean sand or gravel of an equivalent or slightly larger grain size may be deposited nearshore, and a permit is required for any new construction, modification, or restoration

Figure V-2-11. Structures threatened by erosion, Lake Ontario, Crescent Beach, NY

of coastal structures. Many other states and communities have similar regulations and set-back limits, which limit construction activities within a specified distance from the shore or bluff edge.

    b.   Seismic. Forces within the earth's crust and mantle deform, destroy, and create crustal blocks. These tectonic activities produce physical features such as faults and folds. Tectonic movements produce large-scale uplift and subsidence of land masses. The frequency and magnitude of seismic forces on foundations of marine structures is essential information along tectonically dominated coasts such as the west coast of North America. Rigid structures (e.g. piers and seawalls) and flexible structures (e.g. rubble-mound breakwaters and revetments) should be evaluated for seismic response, according to the same general practices used for building design (U.S. Army Corps of Engineers 1994).

    c.   Tsunami.

    (1)   Tsunamis, or seismic sea waves, are long-period waves generated by displacements of the seafloor by submarine earthquakes, volcanic eruptions, landslides and submarine slumps, and explosions. The term tsunami is derived from two Japanese words: "tsu," for harbor and "nami," meaning wave. The underwater disturbance results in uplifting the water surface over a large area, which forms a train of waves with periods exceeding 1 hr, in contrast to normally occurring wind-generated water waves which have periods less than 1 min. In the open ocean, amplitude of tsunamis is usually less than 1 m (3.3 ft) and hence may go unnoticed to passing ships. However, the wave height increases greatly as the shore is approached, resulting in potentially catastrophic flooding and damage. Tsunamis generated by volcanic activity or landslides result in the wave energy spreading along the wave crests and affect mainly the areas near their source. Those generated by tectonic uplifting may travel across an ocean basin, causing great destruction far from their source (Camfield 1980).

(2)   Tsunamis can be generated in any large water body, including inland seas and large lakes. However, the majority of seismic activity that usually generates tsunamis occurs along the boundaries of the Pacific Ocean, and some strong activity is concentrated in the Caribbean and Mediterranean Seas. While earthquakes affect the eastern United States, these usually occur inland. The only major recorded tsunami along the east coast of North America was the one which devastated the Burin Peninsula along Placentia Bay, Newfoundland, in November 1929. The tsunami was enhanced by an exceptionally high tide and high storm waves.

(3)   Because of the frequency and magnitude of tsunamis in the Pacific Ocean, a warning system has been developed. Although tsunamis travel at speeds as great as 800 km/hour (500 mph), transoceanic distances are sufficiently large to allow several hours warning prior to arrival of a distantly generated tsunami. The Tsunami Warning System (TWS) was founded in 1946 following the Aleutian tsunami of 1 April of that year that caused major damage and many casualties in the Hawaiian Islands. The TWS is a cooperative effort among Pacific Ocean nations with seismograph and tide data collected and communicated to the Pacific Tsunami Warning Center in Hawaii, which disseminates the appropriate warning.

(4)   Tsunami flood elevation - frequency information is developed using historic data and numerical models. Unfortunately, good data are available for only a limited number of tsunamis. Cox and Pararas-Carayannis (1969) and Pararas-Carayannis (1969) present a catalog of tsunamis for Alaska and the Hawaiian Islands. Where data are available, the probability of tsunami flood elevations can be determined by the same methods used for the riverine environment. If insufficient information is available, a synthetic record of tsunami activity must be generated. The geophysical and tectonic setting of the area is used to synthesize a record of tsunamigenic, tectonic deformations on the seafloor. A numeric model is used to simulate the tsunamis resulting from each deformation. The resulting data are combined with tide information to produce the combined tsunami and tide elevation - frequency relationship. Houston (1979) presents a description of tsunami modeling, elevation prediction, and structural damage. Houston (1980) and Crawford (1987) present examples of tsunami elevation predictions for the southern California coast and coast of Alaska from Kodiak Island to Ketchikan, respectively.

d.   Ice.

(1)   An understanding of ice properties and the effect of ice on the shore and marine structures is important in the Great Lakes, in coastal estuaries experiencing significant freshwater inflow which may transport ice, and along the northern ocean coasts. Ocean coasts may be subject to sea ice or pans and floes of freshwater ice originating from river discharges. Northern lakes, such as the Great Lakes, experience significant accumulations of ice during the winter. Sea ice, also known as saline or brackish ice, freezes and is most dense at -1.7 °C (29 °F), in contrast to freshwater ice at 0 °C (32 °F).

(2)   A freshwater lake freezes in two stages defined by the water temperature. Cooler water at the surface generates a circulation pattern by sinking to the bottom, exposing warmer water, which successively cools. When the lake reaches a uniform 4 °C (39 °F), that is, the lake becomes isothermal, the lake is termed to have "turned over." As the surface water cools further, it becomes less dense than the water below it until it freezes. At this point, an ice sheet grows laterally (primary ice). In turbulent water, the primary ice cover will consist of a congealed frazil sheet and ball ice. When formed in rivers, upon reaching the outlet, severe ice jams may form which can completely fill the outlet cross section. The ice cover thickens (secondary ice) with continuing freezing temperatures, which may be at a rate of 1 in. per day (Wortley 1984). While the primary ice's axis of symmetry, called the c-axis or optic axis, is oriented perpendicular to the free surface, the secondary ice's c-axis is oriented horizontally. The formation of snow ice on the surface, which is randomly oriented, insulates the underlying ice, resulting in a reduction in growth rate of the secondary ice. Ice thicknesses in the Great Lakes generally are about 0.6 m (2 ft) to 0.9 m (3 ft). However, along the shore and near coastal structures, ice can build up considerably thicker due to ridge formation and spray. Ice ridges

or hummocks 3 to 5 times higher than the flat ice thickness are common and are not only confined to along the shore. Ice ridges ranging from 3 m (10 ft) to 4.5 m (15 ft) above the normal ice surface and extending 18 m (60 ft) below have been observed on Lake Erie. The time that ice clears at spring breakup is dependent upon heat gain from the atmosphere, local snow and ice conditions, and wind and water currents (Wortley 1984).

(3)   Shoreline recession is not only caused by wave action, but also by downslope movement of material due to gravity (mass wasting). Exposure of permafrost lenses to seawater along the Arctic Alaskan coasts results in melting of these lenses. The resultant loss in strength may cause catastrophic failure of the bluff above it (U.S. Army Corps of Engineers 1994). In other areas, shore ice may protect the bluff from wave action. However, during the spring thaw, the saturated bluff will experience reduced cohesive strength, making it more vulnerable to mass wasting. This can be exacerbated by the presence of springs emerging along the bluff face, especially if the bluffs have been undercut by wave action.

(4)   In addition to recognizing the effect of ice on shore processes and the functionality of a proposed project, calculation of ice loads may be required. The effect of ice on coastal projects is summarized in Chapter VI-3-5, with guidance on calculating ice loads in Chapter VI-6-6.b.

V-2-14. Foundation/Geotechnical Requirements

a.   Every proposed coastal structure, nourishment project, or dredging operation requires knowledge of the underlying sea, lake, or river bottom materials. A geotechnical site investigation is required to assess the nature and extent of all sediments and their respective properties. The level of detail of the investigation is predicated on the study phase: reconnaissance, feasibility or detailed design, and the scope of the project. For example, only surface sediment data may be required for a beach fill, whereas a major coastal structure may require additional information on subsurface conditions. These investigations will entail researching available information from previous studies/projects in the area and obtaining new data.

b.   The investigations seek to identify the elevation, thickness, and the physical, hydraulic, and mechanical properties of the soil, depth to bedrock and its properties, and groundwater level. Knowledge of the general soil group (clay, sand, etc.) allows one to assess the general characteristics of the soil. However, each soil group includes materials with a great variety of properties, and without a proper assessment, serious consequences can result (structural failure, inability to dredge the channel, etc.). Chapter VI-3-1 and Eckert and Callender (1987) present discussions on foundation/geotechnical requirements. Terzaghi and Peck (1967) also discuss the minimum requirements for adequate soil description and present a table of data required for soil identification.

V-2-15. Availability of Materials (Sand/Stone Resources)

The specification of materials for a proposed project requires the identification of the type, location, quantity, and quality of material available. Sand is the primary choice for beach nourishment projects, although gravel and cobble may be considered in certain situations. Rock is a popular construction material for coastal structures because of the range of sizes, durability, and availability.

a.   *Sand.*

(1)   Beach nourishment projects rely on the introduction of additional sand to the littoral zone to reduce a supply imbalance. Beach sand is generally a natural material preferably derived from a borrow area close to the project area. Location of a suitable borrow area requires geotechnical investigations of sediment size, type, and quantity in addition to environmental, hazard, and regulatory restrictions. Additional processing of sand (e.g., from an offshore site, the dredge may be fitted with additional screens) may be

necessary to obtain the desired product. Since sand may not be available in unlimited amounts, other alternatives may need to be considered. Manufactured sand (rock crushed to suitable gradation) was used at Maumee Bay State Park, OH, and crushed and tumbled recycled glass was deposited at Moonlight Beach, Encinitas, CA, for an emergency repair (Finkl 1996). At Fisher Island, FL, oolithic aragonite sand imported from the Bahamas was placed due to the local scarcity and environmental sensitivity of upland and offshore sand sources, the developer's interest in creating a unique and attractive beachfront, and the relatively modest size of the beach fill required (Bodge 1992). The political and engineering issues of placing this sand on other Florida beaches is discussed in Higgins (1995) and Beachler (1995), respectively.

(2) The identification of undesirable materials in the beach sand also may be required. For example, calcareous materials in the source materials have been found to react with available water sources (precipitation, groundwater, etc.) to precipitate fine-grained carbonate cement in pore spaces of the previously unconsolidated sand (Stransky and Greene 1989). This results in undesirable cementation of the beach nourishment material. This action leads to the formation of steep beach scarps, which are susceptible to undercutting and collapse, and inhibit the access of beachgoers to the shore.

   b.   *Stone.*

(1) Coastal structures may use a large quantity of stone of various sizes. The location, quantity, cost, suitability, and quality of the stone are important aspects, which must be investigated. A firm or agency that regularly requires large quantities of stone should maintain records on quarries within its geographic area of business. The change from using cut rectangular stone to rubble-mound shot stone for the armor layer in coastal structures in the past decades has resulted in savings in time and cost. However, the importance of stone quality has only been recently recognized. Appropriate quarry inspections and quality control by an experienced geologist or inspector are essential.

(2) Stone should be durable, sound, and free from detrimental cracks (both natural and quarry-induced), seams, and other defects that tend to increase deterioration from natural causes or which cause breakage during handling and placing (EM 1110-2-2302). The stone should also be resistant to localized weathering and disintegration from environmental effects. Inspection records of potential stone suppliers are important to help ensure these conditions are met and to determine the location, sizes, and types of stone available. Table V-2-3 lists data needed to perform a quarry inspection.

(3) During annual inspection of an existing project, the presence of stone cracking should be observed, as this will affect the maintenance schedule and structural integrity. Noting characteristics, location, and number of cracks will assist in assessing whether the stone will continue to deteriorate and at what rate. Cracks are generally classified as multiple penetrating/throughgoing (MP/T) cracks and mirror image cracks (MIC). MP/T cracks are commonly blast-induced fractures and characteristically form radial fractures, which run diagonal to the shot face. They are highly destructive to the stone's integrity and longevity. These sharp penetrating cracks are often initially minute and will not enlarge until exposed to the elements. Elimination/reduction of this type of fracturing is accomplished by proper shot design at the quarry. Mirror image cracks occur when a fracture splits the stone in half and the halves split repeatedly. These cracks generally do not form until the stone has been placed in the project. MRC are generally jagged and the result of the weathering of geologic beds and/or stress relief. Reasons for MIC (Marcus 1996) are:

Table V-2-3
Quarry Inspection Checklist

Item	Comment
SOURCE INVESTIGATION	
General	Purpose of inspection, date of inspection, personnel attending.
Location	Source name, address, contacts, geographic coordinates, descriptive directions of source location.
Production/Transportation	Materials produced, complete description of lift/face development, plant equipment used for production operations, transportation modes, operating season, additional materials, sources interested in producing, summary assessment.
Geology	Stratigraphy, lithology, structure, groundwater, summary assessment. Photographs are extremely useful in documenting the geology.
Blasting Operations	Blast design (description, blast design factors, blast design relationships, analysis), blasting effects (description, fragmentation, development of radial fracturing, transverse fracturing, cratering, back break and throw), summary assessment.
Material Sampling	Number, size, location within quarry samples where taken, photographs of samples.
Summary Assessment	Summary assessment of source, quarry personnel, and quality control.
LABORATORY INVESTIGATIONS / TESTING	
General	Purpose of testing (quality, durability).
Testing Program	Petrographic examination, abrasion resistance, accelerated weathering tests (wet-dry, freeze-thaw), resistance to salt crystallization.
Testing by Others	Report/evaluate laboratory testing by others.
Summary Assessment	From laboratory testing, make assessment of suitability of material for the use intended.
SERVICE RECORD INVESTIGATIONS	
General	This investigation documents the condition of stone used on other projects. Note changes in personnel and operations which may affect evaluation of service records.
Case Histories	Project name, location, description, material types, material sources (quarry , lift, etc.), exposure conditions, dates of placement, evaluations.
Summary Assessment	Assessment of how well the stone is holding up and if deteriorating, the cause should be determined.
SOURCE EVALUATION	
General	Indicate whether quarry is now a listed source for particular material.
Source Summary Assessment	List areas/lifts within quarry which are suitable for particular products.
Conclusions of Investigating Geologist	List evaluations of the geologist performing past/present investigations.
Review/Determination of Geologist	The investigating geologist makes an overall determination of the quarry, quality control, and suitability of the stone for the intended use.

(a)   Differences in thermal expansion behavior of the component crystalline minerals (unavoidable).

(b)   Freeze-thaw expansion and contraction of interstitial pore water (avoidable by stone curing and eliminating winter quarrying in northern locations).

(c)   Wet-dry expansion and contraction of interstitial clays and clay mineral (mostly avoidable).

(d)   Stress relief (slow-term stress release).

(4)   The reader is directed to Marcus (1996), EM 1110-2-2302, and Construction Industry Research and Information Association (1991). Figures V-5-12 and V-2-13 are photographs of the progressive deterioration (after 3 to 4 years and after a maximum of 5 years, respectively) of 80- to 178-kN (9- to 20-ton) armor stone (dolomite) in the lacustrine environment (Lake Erie).

Figure V-2-12.   Cracked dolomitic limestone - Cleveland East breakwater, Ohio, 1989 (stone is dolomitic limestone)

V-2-16. Accessibility

a.   Access to the project area is required before, during, and after construction. The total project area encompasses not only the geographic limit of the actual project footprint but also extends a certain distance away due to impacts to the littoral zone. Access requirements before, during, and after construction may be very different. Prior to construction, temporary rights of entry may only be required to survey the site. Complete lands, easements, and rights of way will be required during construction. After project completion, rights of entry may only be necessary for project monitoring and inspection, but similar requirements obtained during construction may be needed for project maintenance.

Figure V-2-13. Cracked stone - - Cleveland East breakwater, Ohio, 1990. Note continuing degradation compared to 1989 (stone is dolomitic limestone)

b. Access to the site may be by land, water, or both, and may be over public or privately owned land. Individual topographic and hydrographic conditions will dictate whether the project may be constructed by water-based and/or land-based equipment. The designer will need to assess the anticipated means of construction and acquire the necessary real estate for access. Locations with difficult access can significantly increase the cost of a project. In these instances, a change in the project design or scope may be warranted.

c. The access should be designed to allow the safe and efficient movement of equipment, materials, and personnel. Access and haul roads should be of sufficient width and grade to safely accommodate large construction vehicles. Grades should be as flat as practical with the maximum allowable grade of 10 percent (EM 385-1-1). A traffic control plan must be developed. All marine work must be accomplished with certified and inspected plant and equipment. Mooring lines and cables must be clearly marked and the vessels must follow all navigation rules applicable to the waters on which the vessels will be operated. Plans to remove and secure plant and evacuate personnel to safe haven during hurricanes, storms, or floods must be considered. Safety and health requirements concerning activities and operations are discussed in EM 385-1-1.

d. Easement requirements may be temporary or permanent. Easements at a project site generally consist of temporary and permanent road easements, temporary work area easements, and permanent easements. A road easement grants possession of the land for location, construction, operation, maintenance, alteration and replacement of a road and appurtenances; together with the right to trim, cut, fell and remove therefrom all trees, underbrush, obstructions and other vegetation, structures, or obstacles within the limits of the right-of-way. A temporary work area easement grants the possession of the land as a work area which includes the right to move, store, and remove equipment and supplies, erect and remove temporary structures on the land and to perform any other work necessary and incident to the construction of the project. It may also include the right to trim, cut, fell, and remove all trees, underbrush, obstructions and other vegetation, structures, or obstacles within the limits of the right-of-way. A permanent easement is a perpetual and assignable right and easement to construct, maintain, repair, operate, patrol, and replace the project, including all appurtenances. In addition to providing the rights of entry and access to the site, it will be necessary to provide and maintain adequate utilities and install and maintain necessary connections. All necessary arrangements with local Government officials or owners must be made for use of public and private roads to the site. Recognition that other construction projects may be occurring simultaneously is essential to minimize interference and work disruptions.

e. In granting access to the site, certain restrictions and requirements will need to be identified to control environmental pollution and damage. An environmental protection plan must be created which identifies methods and procedures for environmental protection, necessary permits for waste disposal, plan of restoration upon project completion, environmental monitoring plans, traffic control plan, surface and groundwater protection methods, list of fish and wildlife which require special attention, and spill response plan. The protection of environmental resources may encompass a wide range of activities, such as:

(1) Protection of land resources and landscape.

(2) Reduction of exposure on unprotected erodible soils.

(3) Temporary protection of disturbed areas.

(4) Erosion and sedimentation control devices.

(5) Control of ground vibration.

(6) Disposal of waste materials and removal of debris.

(7) Preservation of historical, archaeological, and cultural resources.

(8) Protection of water resources.

(9) Protection of fish and wildlife resources.

(10) Protection of air resources.

(11) Protection from sound intrusions.

f. The above list summarizes of types of protection to be considered prior to granting access to the site for the construction of a project. The list is not all-inclusive and needs to be tailored to the particular project site. Certain restrictions may apply only during certain times of the year, certain days, or certain times of day.

V-2-17. Monitoring

a. Collecting data (monitoring) of coastal projects will assist in proper design of a project, and improve design procedures, construction methods, operations, and maintenance. Comparison with historic information will aid in understanding processes and changes at the site of interest. Monitoring may be classified by its purpose; operational or research monitoring (Weggel 1995). *Operational monitoring* is done to obtain data for the design, operation, and maintenance of a project. *Research monitoring* is performed to assess the performance of a project in comparison with its predicted performance or to assist in broadening the understanding of physical processes.

b. In general, there are three types of monitoring: physical processes, biological, and economic (Weggel 1995). Physical process monitoring consists of gathering data on the physical mechanisms, forces, and littoral zone responses that are characteristic of the project. Biological monitoring entails accumulating data on living organisms that may be affected by the project. Economic monitoring involves collecting information on the monetary impact of the project.

c. Data collection needs to be accomplished before, during, and after construction. Data collected prior to project construction establishes the baseline data. Data may need to be updated during the design phase. Some data such as hydrographic surveys need to be updated during the plans and specifications phase in order to have the most current information available prior to bidding the project. Information may be required such as check surveys or environmental monitoring of threatened or endangered species during project construction. Post-construction monitoring may be operational or research-oriented.

d. The success of a monitoring program depends upon creation of an extensive and implementable plan. In developing the plan, the processes and data which most affect the project should be identified. The relative importance of each element will have to be assigned with only the most pertinent selected based upon cost limitations. To be effective, monitoring should occur when changes are likely to happen and at sufficient intervals to properly assess changes. Timing of the monitoring should account for seasonal changes in order to allow differentiation from a project-induced change. Monitoring frequency also should be variable, with more data being gathered immediately after the project is in place. As project effects diminish, frequency can be reduced. It is important to recognize that the most effective monitoring plan is one that is fluid; that is, the type, amount, and frequency of data are adjusted as collected data are analyzed. Data that were important initially may be superseded by other information needs which are more critical because it is impossible to anticipate all project effects. A partial list of measurable physical properties is presented in Table V-2-4. For further information on project monitoring, the reader is directed to Weggel (1995), Morang, Larson, and Gorman (1997a, 1997b), Larson, Morang, and Gorman (1997), Gorman, Morang, and Larson (1998), and EM 1110-2-1810.

V-2-18. Data Needs and Sources

a. The intent of this chapter has been to introduce the many factors that a coastal manager or engineer should consider, measure, or collect at the beginning of a new coastal project. At the initiation of a project, it is imperative that one fully understand the problem. While this may seem obvious, one cannot assume that the request for a solution to a coastal "problem" is valid or accurately stated. Those soliciting a solution may not perceive the problem objectively due to political/personal motivations, may have an inaccurate understanding of physical processes, or may have inadvertently been the cause of the problem. It should be considered mandatory to meet with those involved and to visit the site in order to gain an appreciation of the scope of the problem and to observe the scale and relationships of the various physical/geologic features.

b. Having gained an understanding of the problem, necessary data should be listed. Table V-2-5 presents a list of data needs and sources. The problem to be confronted may entail not only balancing existing data and the need to gather more with available monetary resources, but also what to do when the project needs to be built with minimal data. The latter situation will require a more imaginative approach to assess the basic governing processes of the region. Narrative and verbal sources of information will be very helpful. These may include newspaper reports, diaries, local historians, fishermen, marina operators, harbor masters, and local residents. Long-time residents who are especially familiar with the area may give a description of

Table V-2-4
Coastal Project Monitoring Matrix

MEASURABLE PROPERTIES	Project Type			
	BEACHES/SHORES (Dunes, groins, breakwaters, sand bypass systems, submerged sills, borrow areas)	JETTIES (Disposal areas, levees, closure stations, seawalls, revetments, bulkheads)	BREAKWATERS (Piers, wharves, moles, berths, docks)	DREDGING (Inlets, channels, marinas, outfalls)
Beach profiles	▓	▓	▓	
Bathymetry	▓	▓	▓	▓
Waves	▓	▓	▓	▓
Tide height	▓	▓	▓	▓
Tidal currents		▓		▓
Tidal prism		▓		▓
Surge	▓	▓	▓	▓
Longshore currents	▓	▓	▓	
Sediment size	▓	▓		▓
Winds	▓	▓	▓	▓
Temperatures	▓	▓	▓	▓
Salinity			▓	
Ice coverage	▓	▓	▓	▓
Structural surveys	▓	▓	▓	

Note: Shaded block indicates property to be monitored. Nonshaded block indicates property usually not monitored except in unusual circumstances.

historical positions of the shore, past structures which have since deteriorated, and major storms. It is imperative that the gathered information be cross-checked with other sources to assess its validity, recognizing that proclamations with numerical statements made without actual measurements should be treated with skepticism. For example, statements such as, "The waves were 12 m (40 ft) high" or "I always had a beach at least 61 m (200 ft) wide in front of my house" should be treated with suspicion. The statement needs to be substantiated or modified by further questioning and through consideration of when the event occurred. While visual wave height measurements are generally difficult to make due to the lack of other physical objects offshore, the resulting runup may give a clue. A verbal account of a large storm(s) will narrow the time frame to search through old newspapers. The type, quantity and time of year various fish are caught will give clues to the aquatic diversity. Old town maps and property deeds will be indications of the historic shoreline/bluff position. Diaries of deceased lighthouse keepers and ship captains may give accounts of the conditions (direction and duration) and date of storms.

c. In addition to verbal and written narrations, a geologic interpretation of the present condition will assist in assessing the past or future trends. Abandoned beach lines, sea stacks, and old shore structures/foundations will suggest the position of past shorelines. Old piers and jetties may indicate the alongshore direction of the migration of an inlet. Stratigraphic information may suggest a change in depositional/erosional patterns. For instance, peat and organic deposits may indicate the presence of a barrier island in the past. The size, amount, and distribution of shoals can give clues to inlet processes. Sedimentological trends such as a change in size of offshore material may explain the presence of headlands or variations in erosion rates. The presence of a fillet at a man-made or natural protruding structure/feature will indicate the littoral transport direction and mean incident wave direction. The shape of the shoreline between headlands also will give an indication of the average incident wave direction.

Table V-2-5
Data Needs and Sources

Data Needed	Best Source(s)	Alternate Sources
Site map/real estate (Needed to locate structures, land features, utilities, roads and property boundaries.)	Recent survey	Town maps USGS quad maps Atlases Local utility companies Digitize aerial photographs Sketch of site with proposed structures located by existing features
Site topography & bathymetry (Land and underwater contours needed to transform waves to site, determine dredge and beach fill quantities, structure lengths.)	Recent USACE high-density acoustic surveys, land surveys, or SHOALS hydrographic LIDAR surveys	NOAA charts USGS quad maps Contractor surveys Emery method Visual estimation of elevations and slopes
Directional wave statistics (height, period & direction) (Needed for estimation of incident wave for structure design, time series for sediment transport evaluation, shoreline response and beach fill design, channel design (vessel effects))	Directional wave gauge deployed near site	Deepwater wave gauge with waves refracted to coast WIS hindcast Wave hindcast using wind statistics Wave hindcast using design wind
Water level • **Tidal** • **Storm setup** (Needed for incident wave computations, flood evaluations, structure toe depths)	NOAA water level gauge near site	Interpolation between gauges to site Peak stage gauge Highwater marks Numerical models
Currents • **Tidal** • **Longshore** (Needed for sediment transport evaluations, dredge disposal fate, scour potential)	Directional current meter	Nondirectional current meter Measurements using drogues/floats Seabed drifters Numerical models
Sediment characteristics • **Littoral zone (a)** • **Subsurface (b)** (Needed for sediment transport analysis, beach fill evaluation, structure foundation design)	Extensive surface samples (a) Extensive boring program (b)	Geophysical subsurface investigations (a&b) Subsurface probe to refusal (b) Regional geologic maps (b) Visual comparison of littoral sediments using sediment card (a) Interpolation between known information (a&b)
Historic shoreline positions (Needed for evaluations of natural shoreline change, effects of human activity, erosion rates, regulatory setbacks)	USACE project maps Shoreline studies from universities and state studies	Digitize from aerial photographs Historic NOAA charts Extrapolate from topographic surveys Historic town property maps and deeds Estimate from ruined structures and verbal/written accounts
Sediment budget (Needed for determination of natural sediment movement; existing and potential effects of human activity: beach fill amounts, channel dredging estimates, bypassing quantities around structures)	Direct measurement using sediment traps in conjunction with estimation from quantity of trapped material at protruding features and sediment supply from bluff & shore erosion	Numerical estimates using longshore drift equations and period of record wave hindcast Budgets in engineering reports, published literature

(Continued)

Table V-2-5 (Concluded)

Data Needed	Best Source(s)	Alternate Sources
Environmental data (Needed for evaluation of natural and existing environment, and the potential effects of human activity on flora, fauna, and water quality)	Habitat evaluation procedures	Benthic resource assessment techniques U.S. Fish and Wildlife Service, USACE and State agency reports University studies National & local environmental agencies (Audubon Soc., Sierra Club, Nature Conservancy, etc.) Sportsman's Clubs
Historic bathymetry (Needed for evaluation of shoal growth, project effect on shore, and inlet migration)	USACE condition surveys NOAA charts in digital form from National Geophysical Data Center	Historic U.S. Coast and Geodetic Survey charts Contractor surveys
Sand/stone resources • **Availability (a)** • **Quality (b)** (Needed for confirmation of size, location, and quantity of sand or stone material for project as available product can dictate design. Use of inferior quality product will result in service life reduction and possibly failure)	USACE inventories of quarries/sand deposits (a) Service records (b)	State inventories of quarries/sand deposits (a) Geotechnical search for sand deposits (a) Laboratory testing (b)

V-2-19. References

EM 385-1-1
Safety and Health Requirements

EM 1110-2-1204
Environmental Engineering for Coastal Shore Protection

EM 1110-2-1810
Coastal Geology

EM 1110-2-2302
Construction with Large Stone

ER 1105-2-100
Planning Guidance Notebook

Barnett 1984
Barnett, T. P. 1984. "The Estimation of 'Global' Sea Level: A Problem of Uniqueness," *Journal of Geophysical Research*, Vol 89, No. C5, pp 7980-7988.

Beachler 1995
Beachler, K. E. 1995. "Bahamian Aragonite: Can it be used on Florida Beaches? Engineering Issues," *Proceedings of the 1995 National Conference on Beach Preservation Technology*, Florida Shore and Beach Preservation Association, Tallahassee, FL.

Bodge 1992
Bodge, K. R. 1992. "Beach Nourishment with Aragonite and Tuned Structures," *Coastal Engineering Practice '92*, ASCE, New York, NY.

Bruun 1995
Bruun, P. 1995. "The Development of Downdrift Erosion," *Journal of Coastal Research*, Vol 11, No. 4, pp 1242-1257.

Camfield 1980
Camfield, F. E. 1980. "Tsunami Engineering," Special Report No. 6, Coastal Engineering Research Center, U.S. Army Engineer Waterways Experiment Station, Vicksburg, MS.

CIRIA 1991
CIRIA. 1991. "Manual on the Use of Rock in Coastal and Shoreline Engineering," Construction Industry Research and Information Association Special Publication 83, (Also published as Centre for Civil Engineering Research and Codes Report 154), Gouda, The Netherlands, London, U.K.

Cox and Pararas-Carayannis 1969
Cox, D. C., and Pararas-Carayannis, G. 1969. Catalog of Tsunamis in Alaska, U.S. Department of Commerce, Environmental Services Administration, Coast and Geodetic Survey.

Crawford 1987
Crawford, P. L. 1987. "Tsunami Predictions for the Coast of Alaska, Kodiak Island to Ketchikan," Technical Report CERC-87-7, U.S. Army Engineer Waterways Experiment Station, Coastal Engineering Research Center, Vicksburg, MS.

Davidson-Arnott and Stewart 1987
Davidson-Arnott, R. G. D., and Stewart, C. J. 1987. "The Effects of Longshore Sand Waves on Dune Erosion and Accretion, Long Point, Ontario," *Proceedings, Canadian Coastal Conference*, Quebec City, National Research Council, Ottawa, pp. 131-144.

Dean 1991
Dean, R. G. 1991. "Equilibrium Beach Profiles: Characteristics and Applications," *Journal of Coastal Research,* Vol 7, No. 1, pp 53-84.

Department of the Navy 1976
U.S. Naval Weather Service Command. 1976. Summary of Synoptic Meteorological Observations, prepared by the National Climatic Data Center, Asheville, NC.

Eagleman 1983
Eagle, J. R. 1983. *Severe and Unusual Weather*, VanNostrand Reinhold Company, New York, NY.

Eckert and Callender 1987
Eckert, J., and Callender, G. 1987. "Geotechnical Engineering in the Coastal Zone," Instruction Report CERC-87-1, U.S. Army Engineer Waterways Experiment Station, Vicksburg, MS.

Emery and Aubrey 1991
Emery, K. O., and Aubrey, D. G. 1991. *Sea Levels, Land Levels and Tide Gauges*, Springer-Verlag, New York, NY.

Finkl 1996
Finkl, C. W. 1996. "Beach Fill from Recycled Glass: A New Technology for Mitigation of Localized 'Hot Spots' in Florida," Department of Geology, Florida Atlantic University, Boca Raton, FL.

Frihy 1992
Frihy, O. E. 1992. "Sea-Level Rise and Shoreline Retreat of the Nile Delta Promontories, Egypt," *Natural Hazards*, Vol 5, pp 65-81.

Gorman, Morang, and Larson 1998
Gorman, L. T., Morang, A., and Larson, R. L. 1998. "Monitoring the Coastal Environment; Part IV: Mapping, Shoreline Change, and Bathymetric Analysis," *Journal of Coastal Research*, Vol 14, No. 1, pp 61-92.

Hanson and Kraus 1989
Hanson, H., and Kraus, N. C. 1989. "GENESIS: Generalized Numerical Modeling System for Simulating Shoreline Change; Report 1, Technical Reference Manual," Technical Report CERC-89-19, U.S. Army Engineer Waterways Experiment Station, Vicksburg, MS.

Harris 1954
Harris, R. L. 1954. "Restudy of Test - Shore Nourishment by Offshore Deposition of Sand, Long Branch, New Jersey," TM-62, U.S. Army Corps of Engineers, Beach Erosion Board, Washington, DC.

Hicks 1978
Hicks, S. D. 1978. "An Average Geopotential Sea Level Series for the United States," *Journal of Geophysical Research*, Vol 83, No. C3, pp 1377-1379.

Higgins 1995
Higgins, S. H. 1995. "Bahamian Aragonite: Can it be used on Florida's Beaches? Political Issues," *Proceedings of the 1995 National Conference on Beach Preservation Technology, Florida Shore and Beach Preservation Association,* Tallahassee, FL.

Houston 1979
Houston, J. R. 1979. "State-of-the-Art for Assessing Earthquake Hazards in the United States: Tsunamis, Seiches, and Landslide-Induced Water Waves," Report 15, Miscellaneous Paper S-73-1, U.S. Army Engineer Waterways Experiment Station, Vicksburg, MS.

Houston 1980
Houston, J. R. 1980. "Type 19 Flood Insurance Study: Tsunami Predictions for Southern California," Technical Report HL-80-18, U.S. Army Engineer Waterways Experiment Station, Vicksburg, MS.

Houston 1996
Houston, J. R. 1996. "Simplified Dean's Method for Beach-Fill Design," *Journal of Waterway, Port, Coastal, and Ocean Engineering*, Vol 122, No. 3, pp. 143-146.

Larson and Kraus 1989
Larson, M., and Kraus, N. C. 1989. "SBEACH: Numerical Model for Simulating Storm-induced Beach Change; Report 1: Empirical Foundation and Model Development," Technical Report CERC-89-9, U.S. Army Engineer Waterways Experiment Station, Vicksburg, MS.

Larson, Morang, and Gorman 1997
Larson, R. L., Morang, A., and Gorman, L. T. 1997. "Monitoring the Coastal Environment; Part II: Sediment Sampling and Geotechnical Methods," *Journal of Coastal Research*, Vol 13, No. 2, pp 308-330.

Longuet-Higgins 1983
Longuet-Higgins, M. S. 1983. "Wave Set-up, Percolation and Underflow in the Surf Zone. *Proc. Roy. Soc.*, Ser A 390: pp 283-291.

Marcus 1996
Marcus, D. W. 1996. "Problems and Improvements of Armor Stone Quality for Coastal Structures," *Ohio River Division Laboratory Accelerated Weathering Workshop*, Cincinnati, OH.

Morang, Larson, and Gorman 1997a
Morang, A., Larson, R. L., and Gorman, L. T. 1997a. "Monitoring the Coastal Environment; Part I: Waves and Currents," *Journal of Coastal Research*, Vol 13, No. 1, pp 111-133.

Morang, Larson, and Gorman 1997b
Morang, A., Larson, R. L., and Gorman, L. T. 1997b. "Monitoring the Coastal Environment; Part III: Geophysical and Research Methods," *Journal of Coastal Research*, Vol 13, No. 4, pp 1964-1085.

Naqvi and Pullen 1982
Naqvi, S. M., and Pullen, E. J. 1982. "Effects of Beach Nourishment and Borrowing on Marine Organisms," Miscellaneous Report 82-14, Coastal Engineering Research Center, U.S. Army Engineer Waterways Experiment Station, Vicksburg, MS.

National Oceanic and Atmospheric Administration 1985
National Oceanic and Atmospheric Administration. 1985. "Chart of Water Levels of Lake Erie During Storm of December 2, 1985," Tides and Water Level Branch, Great Lakes Acquisition Unit.

National Research Council 1995
National Research Council. 1995. *Beach Nourishment and Protection*, National Academy Press, Washington, DC.

Odum, McIvor, and Smith 1982
Odum, W. E., McIvor, C. C., and Smith, T. J. III. 1982. "The Ecology of the Mangroves of South Florida: A Community Profile," FWS/OBS-81/24, U.S. Fish and Wildlife Service, Office of Biological Services, Washington, DC.

Pararas-Carayannis 1969
Pararas-Carayannis, G. 1969. "Catalog of Tsunamis in the Hawaiian Islands," Report WDCA-T 69-2, ESSA - Coast and Geodetic Survey, Boulder, CO.

Public Law 97-348 1982
Coastal Barrier Resources Act of 1982, (16 U.S.C. 3501 Public Law 97-348).

Resio and Vincent 1976
Resio, D. T., and Vincent, C. L. 1976. "Design Wave Information for the Great Lakes, Lake Erie, Report 1," Technical Report H-76-1, U.S. Army Engineer Waterways Experiment Station, Vicksburg, MS.

Sahagian and Holland 1991
Sahagian, D. L., and Holland, S. M. 1991. Eustatic Sea-Level Curve Based on a Stable Frame of Reference: Preliminary Results, *Geology*, Vol 19, pp 1208-1212.

Schwartz and Musialowski 1980
Schwartz, R. K., and Musialowski, F. R. 1980. "Transport of Dredged Sediment Placed in the Nearshore Zone - Currituck Sand-Bypass Study (Phase I)," Technical Paper No. 80-1, Coastal Engineering Research Center, U.S. Army Engineer Waterways Experiment Station, Vicksburg, MS.

Silvester and Hsu 1993
Silvester, R., and Hsu, J. R. C. 1993. *Coastal Stabilization: Innovative Concepts*, Prentice Hall, Inc., Englewood Cliffs, NJ.

State of Alaska 1994
State of Alaska. 1994. "Alaska Coastal Design Manual," Department of Transportation and Public Facilities.

State of Florida 1987
State of Florida. 1987. Section 161.142, Declaration of Public Policy Relating to Improved Navigation Inlets (as referenced in Bruun (1995)).

State of New York 1988
Coastal Erosion Management Regulations Statutory Authority: Environmental Conservation Law Article 34, 6 NYCRR Part 505, State of New York Department of Environmental Conservation, Albany, NY.

Stewart and Davidson-Arnott 1988
Stewart, C. J., and Davidson-Arnott, R. G. D. 1988. "Morphology, Formation and Migration of Longshore Sand Waves; Long Point, Lake Erie, Canada," *Marine Geology* 81, pp 63-77.

Stransky and Green 1989
Stransky, T. E., and Green, B. H. 1989. "Presque Isle Peninsula: A Case Study in Beach Cementation," *Bulletin of the Association of Engineering Geologists*, Vol. XXVI, No. 3, pp. 352-332.

Terzaghi and Peck 1967
Terzaghi, K., and Peck, R. B. 1967. *Soil Mechanics in Engineering Practice*, 2nd ed., John Wiley, New York.

U.S. Army Corps of Engineers 1992
Institute for Water Resources, U.S. Army Corps of Engineers. 1992. "Guidelines for Risk and Uncertainty Analysis in Water Resources Planning," Volumes 1 and 2, IWR Reports 92-R-1 and 92-R-2, Washington, DC.

U.S. Army Corps of Engineers 1994
U.S. Army Engineer Waterways Experiment Station and U.S. Army Engineer District, Anchorage. 1994. "Alaska Coastal Design Manual," prepared for the Alaska Department of Transportation and Public Facilities.

U.S. Army Engineer District, Detroit 1993
U.S. Army Engineer District, Detroit. 1993. "Design Water Level Determination on the Great Lakes," Detroit, MI.

U.S. Army Engineer District, Mobile 1978
U.S. Army Engineer District, Mobile. 1978. "Beach Erosion Control and Hurricane Protection for Mobile County Alabama (Including Dauphin Island)," Feasibility Report.

U.S. Fish and Wildlife Service 1980
U.S. Fish and Wildlife Service. 1980. "Habitat Evaluation Procedures (HEP)," U.S. Fish and Wildlife Service, Ecological Services Manual 102.

Verhagen 1989
Verhagen, H. J. 1989. "Sand Waves Along the Dutch Coast," *Coastal Engineering*, 13, pp 129-147, Elsevier Science Publishers, B. V., Amsterdam.

Waters 1939
Waters, C. H. 1939. "Equilibrium Slopes of Sea Beaches," unpublished M.S. thesis, University of California, Berkeley.

Weggel 1995
Weggel, J. R. 1995. "A Primer on Monitoring Beach Nourishment Projects," *Shore and Beach*, Vol 63, No. 3, Berkeley, CA.

Wortley 1984
Wortley, C. A. 1984. "Great Lakes Small-Craft Harbor and Structure Design for Ice Conditions: An Engineering Manual," University of Wisconsin Sea Grant Institute, WIS-SG-84-426, Madison, WI.

V-2-20. Acknowledgments

Authors of Chapter V-2, "Site Characterization:"

Michael C. Mohr, U.S. Army Engineer District, Buffalo, Buffalo, New York.

Reviewers:

Andrew Morang, Ph.D., Coastal and Hydraulics Laboratory (CHL), U.S. Army Engineer Research and Development Center (ERDC), Vicksburg, Mississippi.
Joan Pope, ERDC
H. Lee Butler, CHL (retired)

Table of Contents

List of Figures

List of Tables

Chapter V-3
Shore Protection Projects

V-3-1. Introduction

The main purpose of this chapter is to summarize alternatives and their functional design for shore protection. Coastal defense and stabilization works are used to retain or rebuild natural systems (cliffs, dunes, wetlands, and beaches) or to protect man's artifacts (buildings, infrastructure, etc.) landward of the shoreline. A secondary purpose is to review the many constraints that will influence the final design.

a. Major concerns for shore protection

(1) Storm damage reduction. Coastal storms generally cause damage by two mechanisms.

(a) Coastal floodings. On the Atlantic Ocean and Gulf of Mexico coasts, tropical storms (hurricanes) produce elevated water levels, (storm surge) that inundate and damage coastal property. Extra tropical storms (northeasters) along the eastern seaboard and other coasts also create high water and flood damage. Damage from coastal flooding is arguably greater than that due to high winds on the world's coasts.

- Following the devastating flood in 1953, the Dutch people began the Delta Project to raise the dikes and construct barriers (dams) across the estuarine openings to the North Sea. The last component was the Oosterschelde (Eastern Scheldt) storm surge barrier as displayed in an aerial view in Figure V-3-1a and a photograph of the movable gates in Figure V-3-1b. It is one of the largest coastal engineering projects ever completed in the world and a major engineering achievement.

- Inland flooding also disrupts traffic, business, medical services, and normal life to produce secondary, economic and social impacts.

(b) Wave damage. Elevated water levels also bring higher wave energy inland to damage upland development. Damage is a nonlinear function of wave height. On the West Coast, the elevated ocean surface of El Nino events coupled with high storm waves causes damage to marinas, piers, and coastal infrastructure.

(2) Coastal erosion mitigation. The second major concern is coastal erosion. Storms create short-term erosional events. Natural recovery after the storm and seasonal fluctuations may not be in balance to produce long-term erosion. Shore protection projects moderate the long-term average erosion rate of shoreline change from natural or manmade causes. Reduced erosion means a wider sediment buffer zone between the land and the sea. And consequently, erosion mitigation translates into storm damage reduction from flooding and wave attack. How natural shorelines remain stable and mitigate upland damage is explicitly reviewed in Part V-3-3-a. Use of the terms flood control and erosion control are discouraged. Complete control of coastal flooding and erosion is a myth that gives a false sense of security to the client, the general public, and the media. Man cannot control nature. There is always the chance for a more powerful storm than the level of shore protection provided within the design constraints. A reduction in potential levels of flooding and erosion, i.e., mitigation means storm damage reduction benefits and the need for a risk-based, design philosophy, as discussed herein.

a. Aerial view

b. Moveable gates

Figure V-3-1. Oosterschelde storm surge barrier (courtesy Rijkswaterstaat, The Netherlands)

(3) Ecosystem restoration. A new area of concern is the restoration of lost environmental resources such as wetlands, reefs, nesting areas, etc. In 1990, the U.S. Army Corps of Engineers was directed to also consider ecosystem restoration where a Federal project has contributed to ecosystem degradation. For this chapter, Corps civil works project objectives and special design constraints (economic, environmental, institutional, etc.) have been omitted. They have all been brought together in Part V-7. Where appropriate, differences between the Corps' design approach and a general design approach are discussed here.

b. Alternatives for shore protection.

(1) Overview. Figure V-3-2 (adapted from Gilbert and Vellinga 1990) schematically displays five alternative ways to mitigate the damage of coastal storms, namely, accommodation, protection, beach nourishment, retreat and of course, the do-nothing alternative. Civilization's artifacts at the coast are here represented by the lighthouse at a fixed reference line. Storm surge and storm erosion reduce the distance between the reference line and the sea. Sea level rise and historic, coastal erosion also reduce the distance, but at slower time scales. Beach nourishment accomplishes the same objective as the retreat option (i.e., increase the distance to the sea).

(a) Pope (1997) has a similar classification system summarized in Table V-3-1.

Table V-3-1 Classes of Management and Engineering Response for Shore Protection (Pope 1997)

Type	Common Phrase
1) Armoring	Draw the line
2) Moderation	Slow down the erosion rate
3) Restoration	Fill up the beach
4) Abstention	Do nothing
5) Adaptation	Live with it

(b) The protection category in Figure V-3-2 is divided into armoring (seawalls, bulkheads, etc.) for flooding and moderation (groins, breakwaters, etc.) for erosion mitigation and shoreline stabilization. Beach nourishment or restoration is sometimes called the soft alternative to the armored or hard alternative for shore protection. Figure V-3-3 displays the shift from hard to soft, beach nourishment projects over the past 50 years by the Corps of Engineers (from Hillyer 1996).

(c) For design, consider the following six types of alternatives, namely: armoring, beach stabilization (moderation) structures, beach nourishment, adaptation and retreat, combinations (and new technologies) and the with-no-project (abstention) alternative. Table V-3-2 summarizes these alternatives for coastal hazard mitigation including the sections where full discussions are presented.

(2) Armoring. Seawalls, bulkheads, and protective revetments for cliffs and dikes are the traditional types of armored shorelines. The cost of armoring is justified when flooding and wave damage in low areas threaten substantial human investment. On historic, eroding coasts, it must be expected that erosion will continue to diminish the width of the buffer strip between armored shoreline and the sea. If a recreational beach is present, periodic beach nourishment must be anticipated. Part V-3-2 gives functional design details

Table V-3-2
Alternatives for Coastal Hazard Mitigation

Approach	Changes to the Natural, Physical System							
Class	Armoring Structures			Shoreline Stabilization Structures and Facilities				
Type	Seawall	Bulkhead	Dike/Revetment	Breakwaters	Groins	Sills	Vegetation	Groundwater Drainage
Geometry (configurations) or location	• Vertical • Curved • Gravity	• Crib • Stepped/Terraced • Cantilevered • Tie-Backed	• Sloped	• Headland • Detached • Single • System • Submerged (reef-type)	• Normal • Angled • Single • System (field) • Notched • Permeable • Adjustable • Shaped (T or L)	• Shoreline • Submerged • Perched beach • Intertidal		• Beachdrain • Bluff dewatering • Interior drainage

Materials of construction	• Concrete
	• Rock
		• Sheet-pile
		  - steel
		  - timber
		  -concrete
		  - aluminum
			• Earth
			• Rock revetment
			• Geotextiles (bags)
			• Gabions
				• Rock
				• Precast concrete units
				• Sheet-pile types
				  - steel
				  - concrete
				  - timber
				  - etc.
				• Geotextiles bags
							• wetland
							• Submerged Aquatic Vegetation
							• Mangrove
								• System of pipes and pumps with sumps

Discussion in found sections	V-3-2-a(1)	V-3-2-a(2)	V-3-2-a(1)	V-3-3-c V-3-3-d	V-3-3-e	V-3-3-f		V-3-5-b
	Part V-3-2			Part V-3-3				

(Continued)

Table V-3-2 (Concluded)

Changes to the Natural Physical System (continued)		Changes to Man's System			Changes in Both		No Change
Beach Restoration		Adaptation and Accomodation			Combinations		
Beach Nourishment	Sand Passing	Flood Proofing	Zoning	Retreat	Structural and Restoration	Structural and Restoration and Adaptation	Do Nothing

• Subaerial
• Dune
• Feeder
• Profile
• Underwater berms
• Borrow sites
  - offshore
  - land
• Dredged material
• Artificially made (crushed rock)
	• Bypassing
	• Bankpassing
	• Littoral traps
	• Smooth out hot-spots
	• Downdrift material returned updrift
		• Elevated structures
		• Raise grade
		• Sandbags
		• Flow diversion
		• Single-family homes on timber piles
			• Setbacks
			• Land use restrictions
			• Public lands (Institutional)
				• Individuals
				• Communities
				• Infrastructure
				• Move structures
					• Any combination of 1, 2, or 3 alternatives	• Any combination of all alternatives except retreat	Let nature take its course

Part V-4		V-3-4 V-3-4-b		V-3-4-c	V-3-5		V-3-6

Alternatives for
storm damage mitigation
(storm surge, sea level rise,
coastal erosion)

reference
line

Today

Adaptive responses:

Accommodation

Protection

Beach nourishment

Retreat

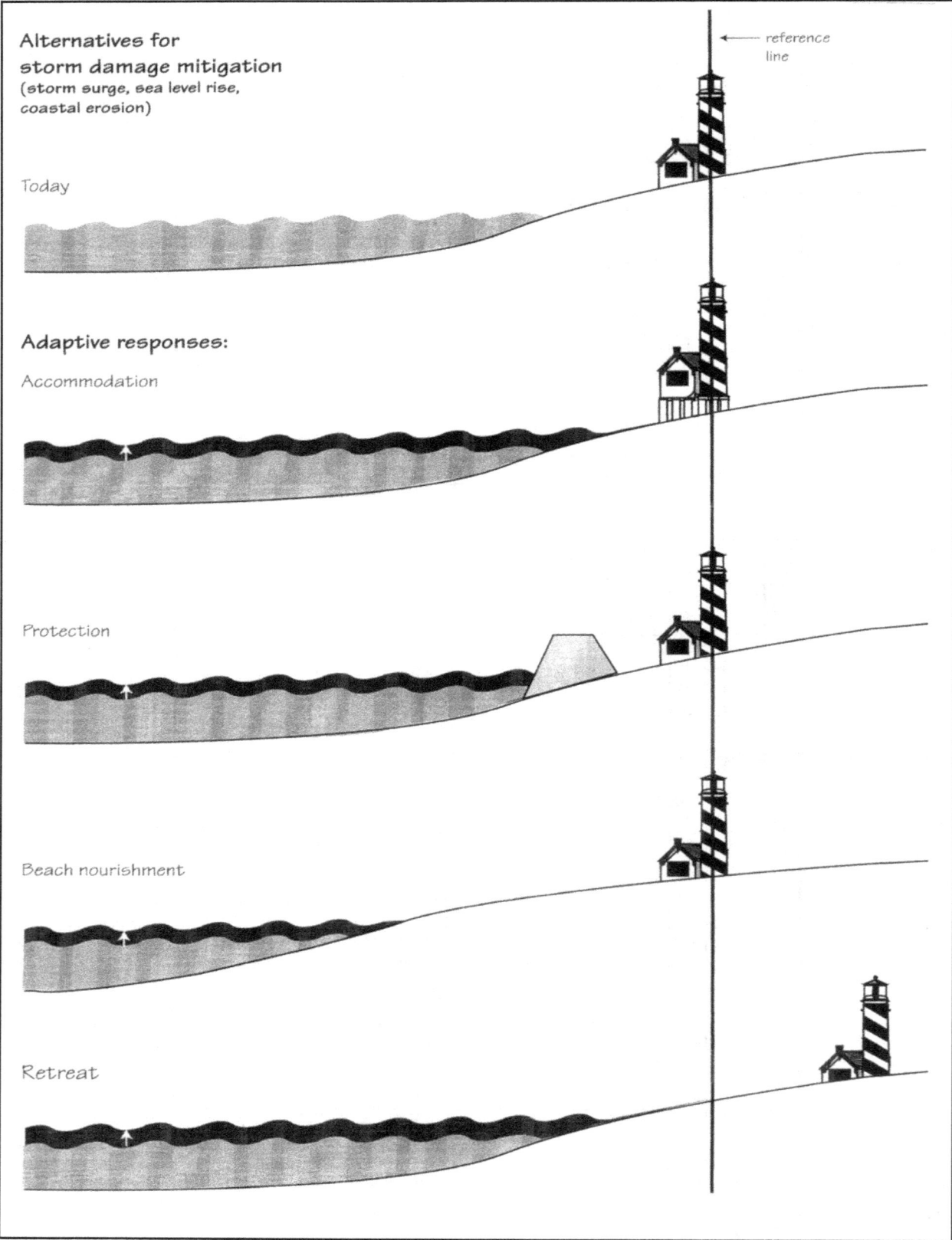

Figure V-3-2. Alternatives for shore protection

and summarizes knowledge on the interaction of armored shorelines and adjacent beaches. See also Engineer Manual 1110-2-1614, "Design of Coastal Revetments, Seawalls, and Bulkheads."

(3) Shoreline stabilization. Headland and nearshore breakwaters, groins, sills and reefs, and wetlands all moderate the coastal sediment transport processes to reduce the local erosion rate. These structures should be considered where chronic erosion is a problem due to the diminished sediment supply. They are often combined with beach nourishment to reduce downdrift impacts. Their purpose is to slow the loss of placed sand, not to trap sand from the littoral system and create more problems elsewhere. In many locations, their improper functional design, or construction without adding extra material, has produced adverse environmental impacts by starving the supply of sand to downdrift beaches. Their proper design is one of the great challenges of coastal engineering, and functional design aspects are found in Part V-3-3.

(4) Beach nourishment. Loose sediment material can be placed on the subaerial beach, as underwater mounds, across the subaqueous profile, or as dunes to rebuild the dunes. The soft alternative solution for shore protection is now the common alternative selected for a variety of reasons (constraints). Because of its importance, a separate chapter, Part V-4 contains all the details for design.

(5) Adaptation and retreat. Elevating structures, flood proofing, zoning restrictions, storm warning and evacuation planning are some of the types of coastal adaptation methods. Further details are in Part V-3-4-b. Retreat is permanent evacuation or abandonment of coastal infrastructure, and for communities subject to high erosion rates and flooding damages, this is always a possible alternative. Total costs and constraints of this alternative must include the environmental impact on the new site to where "retreat" takes place. In contrast to the engineering, decision-making process to determine the best alternative considering all the design constraints for each site, some advocate retreat as the only solution. Further discussion is in section Part V-3-4-c.

(6) Combinations and new technologies. In many locations, elevated structures combined with some type of armoring or shoreline stabilization structure together with beach nourishment are employed for shore protection. Nontraditional technologies (e.g., beach drains, geotextile bags, artificial breakwater structures, wetlands, etc.) are also being investigated in field experiments. Part V-3-5 gives more details.

(7) Do nothing. Finally, the option to allow continued erosion and storm damage with the expected, annual costs for this choice should be determined. The without project condition provides the basis for measuring the effectiveness to reduce the expected damages of each proposed alternative. Further details for estimating damage costs are in Part V-3-1-c. Part V-3-6 presents more general information on this option. See also Part VI-2-1 for more details regarding various subtypes of the armoring, shoreline stabilization, and beach nourishment alternatives. Each alternative must be considered under a wide variety of design constraints.

c. Design constraints.

One good definition of engineering is "design under constraint." Engineering is creating and designing what can be, but it is constrained by our understanding of nature, by economics (costs), by concerns of environmental impact, by institutional, social, legal issues and possibly by aesthetics. Listed are the five design constraint categories which are discussed further in the following paragraphs.

Design Constraints
Scientific and Engineering Understanding of Nature
Economics
Environmental
Institutional, Political (Social), Legal
Aesthetics

We also limit the discussion here to the general practice of coastal engineering. Part V-8 is completely devoted to special, U.S. Federal government planning requirements and design constraints.

(1) Scientific and engineering understanding of nature. The coastal setting is dynamic and influenced by land, water, and air interactions and processes. It is a regime of extremes, surprises, and constant motion as the coast responds to changing conditions.

(a) The *Coastal Engineering Manual* (CEM) demonstrates continued improvement in understanding and ability to analytically and numerically model nature. For example, Part III-3 discusses analytical methods to estimate sandy beach shoreline recession rates during storm events that will be useful later in this chapter. Part III of the CEM also introduces many new, dynamic, numerical models that simulate coastal hydrodynamics and sediment transport processes.

(b) CP Module. In V-1, the idea of a Coastal Processes Module, (CP Module) was defined as a repository of physical data and analysis tools relevant to the coastal problem. Wind, waves, currents, water levels, bathymetry, geomorphology, stratigraphy, sediment characteristics, sediment transport processes, etc. and the analysis tools (mainly numerical models) make up the CP Module that is employed many times in the design process (see Figure V-1-1, 2 and 3). However, a fully dynamic, three-dimensional, numerical model of water levels, waves and sediment transport to simulate bathymetric and shoreline change is still under development. It remains a long way from routine application for the functional design of coastal structures. An example would be the simulation of natural, sediment movement behind and through nearshore breakwaters for both normal conditions and storm events. The inability to accurately predict the short-and long-term impacts of coastal structures on the nearshore physical environment remains a design constraint in coastal engineering. Part of the difficulty is the stochastic variability of the natural environment.

(c) Empirical Simulation Technique (EST) methodology. Numerical models are deterministic tools that produce one solution for each set of boundary conditions. The EST procedure is the numerical, computer simulation of multiple, life-cycle sequences of systems such as storm events and their corresponding environmental impacts (Scheffner et al. 1997; 1999). Multiple life-cycle simulations are then used to compute frequency-of-occurrence relationships, mean value frequencies and standard error estimates of deviation about the mean. Using the EST procedure for a specific project generates risk-based frequency information that relates the effectiveness and cost of the project to the level of protection provided.

- A user's guide for application of the EST with examples is found in Scheffner et al. (1999). One example describes calculations of the frequency-of-occurrence relationship for storm-induced, horizontal recession of beaches and dunes in Brenard County, Florida. Previous references to the EST procedure are found in Part II-5 and II-8. See also Part V-1 for discussion.

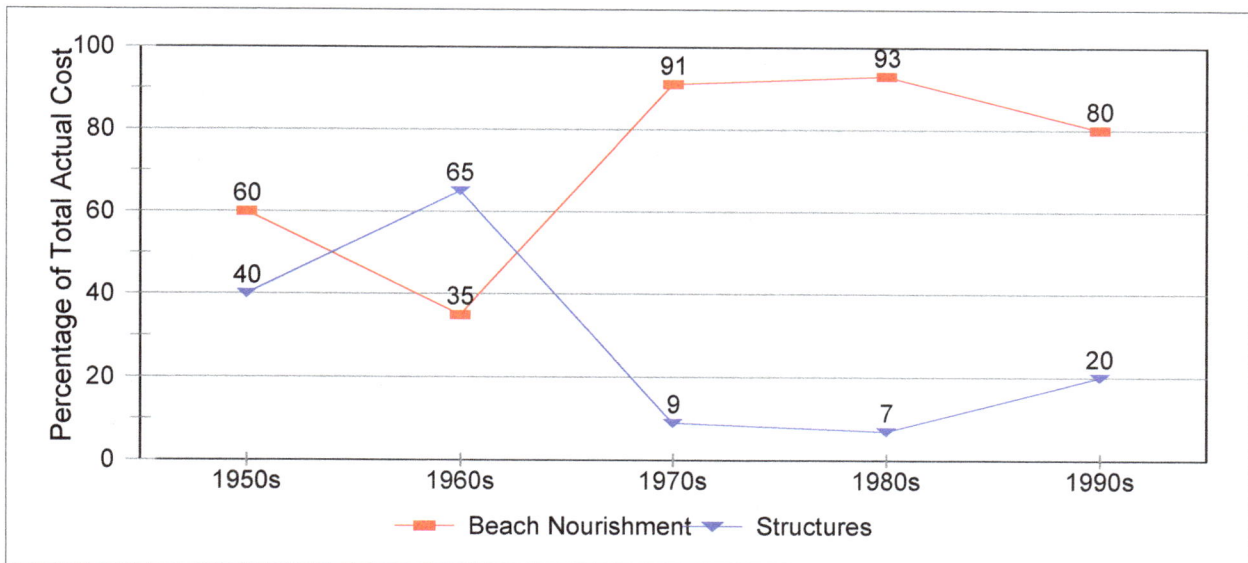

Figure V-3-3. Shift from hard (armored walls, groins, etc.) to soft (beach nourishment) alternatives by the Corps of Engineers (from Hillyer 1996)

- The Corps has specified the EST methodology as a requirement for the risk-based analysis in all shore protection studies (ER 1105-2-101; Thompson et al. 1996). The probabilistic design (functional, structural) of coastal structures remains a constraint, but recent advances such as the EST methodology are helping to produce designs that provide a realistic level of storm protection.

(2) Economics. A key constraint for each shore protection alternative is life-cycle cost. In general, the level of storm protection and, hence, costs (high, medium, low) can only be justified when the corresponding value of property and infrastructure to be protected (benefits) are comparable (high, medium, low). Costs are not subjected to any other constraints. Benefits, however, can be restricted and limited to only those that are perceived to benefit the funding authority. For example, see Part V-8 for the restriction on allowable benefits for Federal government-sponsored projects designed by the Corps.

(a) Cost. It is important to have a clear definition of all terms when discussing costs. Moreover, nothing lasts forever. Advocates for permanent, low cost, or retreat solutions for shore protection fail to understand the following and how they are applied in the professional practice of coastal engineering design economics.

- Costs: The monetary value required for a project. Without additional qualifying words, the word alone is confusing and subject to misinterpretation.

- Initial costs: The total expense for all initial construction and design costs of a project. The year of construction should be noted so that inflationary cost aspects can be estimated in the future. Distinction should be made between estimates and actual contract costs.

- Maintenance costs: The estimated annual expenses required to maintain both the functional and structural integrity of the alternatives which are altered by storm damage and natural processes. Both deterministic estimates and risk-based calculation methods can be employed. (see Part VI-7 for an example of the probabilistic method for rubble-mound breakwaters)

- Alteration/removal costs: The estimated expenses required to alter the design or completely remove the structure if there are significant, downdrift impacts. Evidence from postproject monitoring and decision criteria mechanisms are required for implementation. This cost has been ignored for most projects and could be included in the category of maintenance costs.

- Total, life-cycle cost: The combined initial, maintenance and alteration/removal costs required over the design life of the project. Annual maintenance costs are usually converted to their present worth so that they can be directly combined with the initial construction costs. The present worth (value) is determined by multiplying the annual maintenance expense by the present worth factor, PWF. The PWF is a function of the design life and the interest (discount) rate. (See any standard engineering economics text).

- Design life: An estimate of the number of years of useful life of the structure/ alternative. Usually 25 to 50 years is employed for well-designed projects. Design life selection includes structural life of materials (structural integrity), functional life (usefulness), technical life (technologically up-to-date), and aesthetics. Design life is employed for the economic analysis of the present worth of annual maintenance cost for the total, life-cycle cost comparison of all alternatives. It does not mean the length of time the project will last in the field.

- Interest rate: The second variable required to calculate the present worth of the annual maintenance costs. Often, the rate employed is related to the current, bank loan rate for construction projects. It can also be set by government policy as discussed in Part V-8 for the Federal government.

- Damage: When energy levels in storms exceed the design levels, both structural damage and some loss in functional performance may occur. Repair is possible. Damage is an expected aspect of risk-based coastal design.

- Failure: When storms below the design level cause loss of structural integrity and/or functional performance. The design has failed and a redesign is needed before repair or reconstruction. Use of the word "failure" for loss of structural integrity or performance should be avoided until it can be proven that a design failure took place under specified storm conditions.

- Balanced design: The most economical balance between the initial construction costs and maintenance (damage repair) costs so that the total cost is a minimum. Initial costs increase as the level of protection for more powerful (but rarer) storms increases, but maintenance costs decrease because damage is less frequent. The classical U-shaped, total cost curves result. (See Part VI-7 for an example with rubble-mound breakwater design)

(b) Benefits. Storm damage reduction and coastal erosion mitigation are the two major benefits of shore protection. These two along with ecosystems restoration are the only benefits allowed by the Federal government for Corps projects as discussed further in Part V-8.

- Many other benefits exist. As seen in Figure V-3-3, since the 1960s, beach nourishment has been the selected alternative for shore protection. Substantial recreation and tourism benefits have resulted for local, state, and Federal governments. Waterfront property is generally of greater value and generates higher property taxes. Innumerable secondary (ripple effect) benefits result from the coastal, beach-related travel and tourism industry. The economic value of beaches has been well documented (Houston 1995a; 1995b).

- A good example is Miami Beach, Florida, which was renourished in 1979 by a joint Corps of Engineers - City/County government project costing $52 million. The capitalized annual cost is about $4 million and the project has lasted more than 20 years without the need to renourish the beach. Attendance at the beach increased from 8 million in 1978 to 21 million in 1983 (Houston 1995a). More than 2 million foreign visitors spend over $2 billion annually at Miami Beach (Cobb 1992). The Miami Beach experience is roughly $700 return in foreign exchange for every $1 invested in beach nourishment (Houston 1995a; 1995b; 1996).

- Beach nourishment can also enhance the natural environment. Widened beaches reduce the potential for new, tidal inlet formation during storms at narrow reaches of barrier islands. The economic losses to the protected bay environment (property, recreation, farming, fishing, infrastructure, etc.) can be estimated and added to the storm damage and other benefits for the impacted barrier island. In general, however, environmental benefits of the enhanced, flora and fauna habitat are difficult to quantify monetarily.

- All benefits are site specific. Here, we briefly outline the methodology commonly employed to determine storm damage reduction benefits. A key factor, as illustrated in Figure V-3-2 is distance between the reference baseline and the sea. Steps in the methodology are:

  - Make a structure inventory (residential, commercial, public). Employ aerial, orthodigital mapping and Geographic Information System (GIS) technology where possible and adapt new technologies.

  - Obtain software to calculate the depreciated replacement cost of the structures and content value.

  - Obtain the water level, storm frequency-of-occurrence data for the site, and accompanying wave and shoreline erosion data. The EST methodology previously discussed should be employed, whenever possible.

  - Obtain and run storm damage calculation models. Long-term erosion is included to estimated damages under changing future conditions. The key variables are water level and position of each structure in relation to the shoreline. Some models only treat property structure damage and others land and infrastructure (roads, etc.) damage.

  - Apply the models for both the without project conditions and for the alternatives and subalternatives design considered for shore protection.

- The result is the average, annual damages prevented (benefits) of each alternative. Differences for each alternative to prevent or reduce storm damage are quantified by this approach. Complete details can be found in Part V-8 where names and references for some of the software and models presently employed by the Corps are presented. In general, the state of art for these damage calculation models is less well advanced than for other areas of coastal engineering design.

(c) Benefit/cost ratios. A useful indicator of economic performance of each alternative is the benefit to cost ratio (BCR). As previously noted, the total, life-cycle costs do not depend on the other constraints, therefore remain constant. However, the benefits included in the ratio can be limited by the funding authority. Consequently, the BCR calculated can be significantly different depending upon whether all the potential benefits or only limited benefits are considered. Because the Federal government limits the benefits allowable to only storm damage reduction benefits, the total or true BCR is always greater than that specified for Corps projects. In effect, two BCR's exist.

- Federal government, $(BCR)_F$

$$(BCR)_F = \frac{\text{Storm damage reduction benefits}}{\text{Total, life-cycle cost}} \qquad\qquad (V-3-1)$$

- Total, true $(BCR)_T$

$$(BCR)_T = \frac{\text{Total benefits}}{\text{Total, life-cycle cost}} \qquad\qquad (V-3-2)$$

- Further details regarding the $(BCR)_F$ for Federal-sponsored projects and other methods to measure economic performance are discussed in Part V-8. The total $(BCR)_T$ is never calculated for Corps projects. Consequently, the local sponsors, general public, and media may not understand nor appreciate the true value of shore protection projects to their community. These institutional, political (social), and legal constraints are discussed further in the following paragraphs.

(d) Sea level rise. A detailed summary of present day knowledge of mean sea level change of the world's oceans is given in Part IV-1-6. Over the last 100 years, average, relative sea level rise has been 30 cm (3 mm/year) on the East Coast and 11 cm (1.1 mm/year) along the West Coast (excluding Alaska). The Gulf of Mexico coast is highly variable ranging from 100 cm (10 mm/year) in the Mississippi Delta plain to 20 cm (2 mm/year) along Florida's west coast (National Research Council 1987). Substantial local variability exists. The question remains as to whether these average rates will increase (substantially), stay constant, or decrease in the future. Three things remain clear, however. The existing rates of mean sea level rise at specific sites have not been a severe economic constraint for the shore protection alternatives selected. At many locations, anthropogenic effects (e.g., jettied tidal inlets) causing downdrift, beach erosion are clearly much larger than those occurring due to sea level rise. And finally, long-term, relative changes in sea level can be incorporated into storm surge analysis and the economic design of coastal structures.

(3) Environmental. A third major constraint of shore protection works is their impact on the environment. The Eastern Scheldt, storm-surge barrier shown in Figure V-3-1 was the focus of much discussion in the early 1970s. Environmental scientists favored raising the dikes around the periphery to maintain the saltwater ecology of the tidal estuary. Agricultural and water boards favored a solid dam across the mouth that would create an inland, freshwater lake. A compromise was reached: a storm-surge barrier with movable gates which stay open under normal conditions but are closed at very high storm-surge events. The final design, construction methods and equipment required much research and challenged the ingenuity and technical process of Dutch coastal engineers. In the final analysis, the environmental constraint, to maintain the saltwater ecology, dictated the final design. The additional engineering and construction cost proved to not be the deciding factor.

(a) Types of environmental concerns. As in the preceding example, modification of upland habitat such as land use, resting areas for turtles and shore birds, wetlands, flora and fauna beneficial to the ecosystem, threatened and endangered species, etc. can take place. The aquatic habitat can also be important, for example, water quality, aquatic species, benthic organisms, hazardous, toxic and radiological sediment in borrow areas, increased turbidity during dredging operations and wave climate alterations by sand volume removal in borrow sites, etc.

- These potentially negative impacts must first be identified. Detailed surveys and sampling investigations are conducted to catalog the species and habitats in the project area under existing conditions. Use should be made of previous studies and summary information. The U.S. Fish and Wildlife Service (F&WS) prepares a planning aid report for large Corps projects that detail existing fish and wildlife resources and their habitats. This report also identifies threatened and endangered species and critical fish and wildlife habitats. The National Marine Fisheries Services (NMFS), state and local resource agencies, and local universities may also provide valuable information.

- The offshore, sand borrow site is the greatest environmental concern for beach nourishment projects. New, benthic sampling and collection efforts are often needed to catalog existing species and habitat. Of concern are species capable of rapid recolonization, commercially important species, or protected species. These surveys provide data on abundance and diversity together with a complete list of all species present. This knowledge can be critical in borrow site selection and hence overall cost of the project.

- In most cases, an Environmental Assessment (EA) report is sufficient to demonstrate the minor environmental impact of shore protection projects. Rarely is a full, Environmental Impact Statement (EIS) needed which is time consuming and can be expensive. Corps project needs for an EIS are discussed in Part V-8.

(b) Impact on natural sediment transport system. The negative, downdrift impact on the local and regional sediment budget can be a key environmental constraint. These concerns are addressed in detail in Part V-3 for armored (Part V-3-2) and shoreline stabilization (Part V-3-3) structures and in Part V-5 for jetties at navigation inlets. A beach nourishment project has many positive, environmental impacts by bringing new material to sand starved beaches and expanding the beach habitat. Studies in turtle nesting areas have proven that renourished beaches increase the number of turtle nests (Broadwell 1991; Nelson et al. 1987).

(c) Mitigation. Procedures, or measures which avoid, minimize and/or compensate for negative impacts are defined as mitigation. Threatened and endangered species such as the piping plover, least tern, sea turtles, and whales required special consideration during the planning and construction stages of shore protection projects. Avoidance of negative impacts is achieved by scheduling construction activities at times when the species do not normally inhabit the project area. Piping plovers and least terns are most vulnerable during the nesting/fledging period from early spring to late summer. Disturbances on the beach cause the nest to be abandoned before the eggs hatch.

- Avoidance for sea turtles and whales is not practical for the southern section of the Atlantic coast because these species inhabit the area for most of the year. Minimization of negative impacts is achieved in various ways including monitoring to document contact; using deflectors on the dragarms and collection boxes on hopper dredges; and conducting turtle relocation projects. These techniques are approved by the National Marine Fisheries Service.

- Mitigation by compensation is employed when resource loss is unavoidable. The most common example is new wetlands construction to compensate for the wetlands area lost due to project construction. Some states require more new area constructed than lost and permit wetland banks that are used to pay for planned, future wetlands loss. New and rebuilt dunes are replanted with grasses to compensate for any plants lost during construction. Mitigation by compensation methods are normally carried out and completed during project construction.

• Ecosystem restoration projects result from habitat lost due to a previous activity such as construction of jetties at a tidal inlet and the long-term, downdrift erosion of the beach. Restoration and protection of unique species habitat could also be the objective. Normally, beach nourishment projects can be designed to meet these project objectives. Methods to quantify these environmental benefits are discussed in Part V-8 as applied by the Corps.

(4) Institutional, political, legal. A fourth area that has a formidable influence on the design process are the institutional, political (social), and legal requirements for all projects.

(a) Institutional (policies and guidelines). The Federal objective of water and related land resources project planning is to contribute to the national economic development (NED) consistent with protecting the nation's environment. Applicable executive orders and other Federal government policies and guidelines as planning requirements are discussed fully in Part V-8. The Corps is responsible for shore protection designs of the Federal government. The Corps District Office engineers with the possible aid of the Coastal Hydraulics Laboratory (CHL), Engineer Research and Development Center (ERDC), do most of the design work. Some design work is performed by the private, civil engineering consulting firms. No general guidelines exist as to when and how the private sector, coastal engineering community participates in the design process for the Federal government.

• Other Federal agencies are responsible for some aspects and alternatives for shore protection. A National Flood Insurance Program (NFIP) was established in 1968 to help reduce the Federal share of costs in connection with flood losses. The NFIP is operated by the Federal Insurance Administration, a division of the Federal Emergency Management Administration (FEMA). An essential component for implementing the NFIP is the Flood Insurance Rate Map (FIRM) which delineates special flood hazard areas and insurance risk zones. These maps are prepared by FEMA's mitigation division. Places subject to flooding at the annual, one percent exceedance probability level are designated as special flood hazard areas. The associated recurrence interval is the so-called, 100-year flood. Use of the 100-year flood designation is discouraged. Many people believe that a 100-year flood happens only once every 100 years. Some areas have experienced flooding at this probability level in consecutive years. The FEMA flood hazard zone maps include special V-zones for both flooding and significant wave energy to cause structural damage. Construction standards (where applicable) and flood insurance rates are usually higher for structures located in V-zones.

• Zoning laws and improved building standards are then implemented by coastal communities based on the FIRM and designated hazard-prone areas. One common standard is the requirement that all, first-floor living quarters of new construction be set at least one foot above the mapped elevation of the one percent chance flood.

• Since 1982, all NFIP policies are actuarial, i.e., the flood insurance policy's annual rates fully reflect the buildings risk of flooding. No taxpayer subsides are required. Presently, about 23 percent of structures vulnerable to flooding damages are covered by the NFIP. And, only 3 percent of these NFIP policies are in coastal communities. These coastal communities generate more premium income that they have received in loss claims (Houston 1999).

•Beaches serving as flood protection works are eligible for disaster relief from FEMA provided the relief is assigned public beach facility status. To qualify, it must have a period renourishment program for long-term maintenance. The local government project sponsor must restore the beach to its normal design shape (template) and pay for the cost of replacing sand eroded prior to the storm. Since natural beach rebuilding begins after the storm, it is not clear what time frame is employed to define the storm erosion volume.

•   Several states have established construction setback lines to reduce damage in areas subject to coastal erosion and shoreline retreat. The setback line position is often calculated as some multiple of the annual erosion rate or a specified distance from a contour location in a particular year. In Florida, the line location is based on many factors, namely long-term erosional trends, short-term storm effects, rare water levels at the one percent chance, annual exceedance level, wave uprush, dune line position, wind forces and existing development. In Delaware, 1979 aerial photography has been employed and the restriction line set 30.48 m (100 ft) landward of specified contour elevations, dune toes, or edge of the existing boardwalk structure. These development restrictions affect the without-project calculation of storm damage benefits discussed in Part V-3-1. If homes and structures are not able to be repaired or replaced after a storm by FEMA or state policies, than this will change the without-project estimate of benefits. See also Part III-5-13 for a discussion of setback lines for cohesive shorelines.

(b) Political (social well-being). Specific national policies and laws change as administrations and public interests change. A diverse and broad range of coastal system users with varying economic, social, and environmental expectations and goals exist. Although there have been shifts in the national policy for shore protection by the Administrative Branch of the Federal government, the legislative branch controls the authorization and funding of all Corps projects. Congress has continued to approve and fund the beach nourishment alternative. Shore protection in the U.S. remains political, fragmented, and controversial for a variety of reasons that are further elaborated in Part V-8. National plans for beach management and shore protection do not exist.

•   Each project must consider many social aspects.

   -   Local, regional and state plans for the coastal zones

   -   Public health, safety, and social well-being, including possible loss of life

   -   Community cohesion

   -   Availability of public facilities and services

   -   Potential adverse effects on property values and the tax base

   -   Displacement of people, business, and livelihood

   -   Disruption of normal and anticipated community and regional growth

   -   Sufficient parking and public transport

   -   Sufficient dune crossovers

- Public access and safety during construction

- Access for people with disabilities

- Interruption of recreation

• Cultural resources must also be considered.

• Coastal project construction has the potential to severely impact important cultural resources. Project activities such as offshore sand borrowing can damage or destroy important historical sites related to the region's maritime history. Shipwrecks, native American Indian, and prehistoric sites are typically of interest. Investigations by archaeologists to identify cultural resources in the project area provide data necessary to evaluate site significance and potential project impact. Close coordination with the state Historic Preservation Office is necessary for compliance with the National Historic Preservation Act of 1966.

• The politics surrounding shoreline erosion and measures for mitigation provide a wealth of fascinating reading material. For example, the Westhampton groin field on the south shore of Long Island, New York (U.S. Army Engineer District, New York, 1958: Heikoff 1976; Kassner and Black 1983; Nersesian, Kraus, Carson 1992; Spencer and Terchunian 1997; Terchunian 1988) is a classic example of a political decision that significantly altered the original design. The groin field was built in two stages, with 11 units constructed in 1964/65 and four more in 1970/71. It was a Corps project authorized by Congress in 1960. The project area extended from Fire Island Inlet east to Montauk Point and called for beach fill and groins as needed starting at the west end since the natural, net drift of sand was from east to west. A winter storm in 1962 breached the weakened barrier island at Westhampton. Local interests including the Suffolk County government lobbied for and eventually convinced the Corps to construct the groins in reverse sequence, from east to west. In addition, the groins were not filled with sand when constructed and construction was stopped in the middle of the project for political reasons. The result was a massive sand trap along Westhampton that starved the downdrift (westerly) beaches. The interruption of natural sand transport by Shinnecock and Moriches Inlet and the Westhampton Groin Field has accelerated erosion on Fire Island at the west end of the system (Kana 1999). A lawsuit by Fire Island property owners has resulted against the Corps (see Spencer and Terchunian 1997 for more details and reference). The legal constraint has long been a factor in coastal, shore protection design.

(c) Legal (laws). Congress, through passage of the biannual Water Resources Development Acts (WRDA), authorizes studies and funds construction of Corps projects. Sections of this law also include special investigations and establish cost-sharing formulas between the Federal government, state and local interests. For example, in the 1998 WRDA, the cost-sharing law was changed to 50 percent Federal and 50 percent from local/state interest.[1] As a result, some states have passed laws and statues to provide an annual source of funding for the increased cost of participation in Federally-authorized projects.

• The Coastal Barrier Resources Act (CBRA) was passed in 1982 to minimize loss of life, damage to fish, wildlife and natural resources, and wasteful expenditures of Federal revenues on Atlantic Ocean and Gulf of Mexico barrier beaches. The goal is to restrict all Federal government expenditures and assistance that aid development on the coastal barriers. For example, the CBRA relies on the National Flood Insurance Program to discourage building by prohibiting sale of Federal flood insurance in areas covered by the act. The CBRA does permit Federal funding for shoreline

---

[1] Previously, the formula was 65 percent Federal, 35 percent state/local.

stabilization by nonstructural projects that mimic, enhance, or restore natural stabilization systems, i.e., beach nourishment projects. Federal expenditures are also allowed for the study, management, protection, and enhancement of fish and wildlife resources and habitats.

• The CBRA is one of many Federal laws designed primarily to protect environmental and cultural resources. A partial list includes the following:

- Archeological Resources Protection Act

- Clean Air Act

- Clean Water Act

- Coastal Barrier Resources Act

- Coastal Zone Management Act

- Disabilities Act

- Endangered Species Act

- Estuary Protection Act

- Federal Water Project Recreation Act

- Fish and Wildlife Coordination Act

- Land and Water Conservation Act

- Marine Protection, Research Sanctuaries Act

- National Historic Preservation Act

- National Environmental Policy Act

- Rivers and Harbors Act

- Watershed Protection and Flood Prevention Act

- Wild and Scenic River Act

• All shore protection projects must apply for and receive a permit from the USACE prior to construction. This permit is pursuant to Section 10 of Rivers and Harbors Act (1899) and Section 404 of the Clean Water Act (1977). The permit process considers and evaluates many factors, including effects on conservation, economics, aesthetics, general environmental concern, wetlands, cultural values, fish and wildlife resources, flood hazards, flood plain usage, land use, navigation, shore erosion and accretion, recreation, water supply and conservation, water quality, energy needs, safety, food and fiber production, mineral needs, and welfare of people and society. Some states have a Joint Permit Application for local boards, state agencies and the Corps of Engineers permit. This method saves considerable expense and time in that only one permit application is required to meet the needs of all three levels of government review of the proposed project.

- Some states (North Carolina, Maine) have passed laws banning the use of armored structures (seawalls, bulkheads, revetments) and shore protection on their ocean coasts. South Carolina only bans armored structures and other coastal states are considering similar laws. Florida and California have adopted sand mitigation policies and procedures to permit seawall construction but require the annual placement of sand to compensate for that trapped behind the structure. Further details on seawall and beach interactions are summarized in Part V-3-2.

- Laws for property boundaries at the land-water interface are complex and vary from state to state. A 26-article series entitled "The Law of the Sea in a Clamshell" explaining the applicability and diversity of laws pertaining to the shore has been published by the American Shore and Beach Preservation Association in the magazine *Shore and Beach* (Graber 1980). Clear and legally defendable knowledge of property ownership must be an early step in the design process for coastal protection works.

(5) Aesthetics. A final and especially challenging area in design pertains to the sense of beauty and accepted notions of good taste. Many people feel that natural shorelines (e.g., wide, sandy beaches, rocky cliffs, or vegetated marshes and trees, etc.) are more aesthetically pleasing than ones artificially manipulated for shore protection. An uninterrupted, uncluttered, and natural view of the sea is desirable for most people. Therefore, when possible, an aesthetically balanced and consistent appearance, replicating natural systems is preferred.

(a) The "do nothing" alternative may result in a destructive wake of debris from flooding and wave damage that is visually disturbing for days or weeks following a storm. Some alternatives (e.g., geotextile bags filled with sand) will not survive medium level storm events and leave a debris-strewn beach.

(b) Aesthetics played a major role in selection of the final design for the new, hurricane protection, seawall/boardwalk at Virginia Beach, Virginia. The initial design by the Corps was a massive, curved, concrete seawall patterned after the one in Galveston, Texas. The City of Virginia Beach is a popular tourist, recreation beach and the promenade (boardwalk) features a key aspect of the design. The city rejected this initial design for aesthetic and tourist-economy reasons. First floor hotel guests and restaurant patrons would not be able to see the ocean on the south end. Their view was blocked by the crest elevation of the proposed seawall. The revised design lowered the seawall elevation, modified the structural design, added interior stormwater drainage to accommodate additional overtopping, and widened the nourished beach in front to mitigate storm energy. An artist's perspective and aerial photo are found in Part V-3-2a (Figure V-3-6). The constraints that dictated the final design were aesthetics and the need to accommodate the beach-driven, tourist industry.

(c) Engineering is not applied science. Our limited understanding of nature is one constraint, but it is far from the only one, seldom the hardest one, and almost never the limiting constraint for coastal engineering design.

V-3-2. Coastal Armoring Structures

a. Types.

(1) Seawalls and dikes. The primary purpose of a seawall (and dike) is to prevent inland flooding from major storm events accompanied by large, powerful waves. The key functional element in design is the crest elevation to minimize the overtopping from storm surge and wave runup. A seawall is typically a massive, concrete structure with its weight providing stability against sliding forces and overturning moments. Dikes are typically earth structures (dams) that keep elevated water levels from flooding interior lowlands.

- Various types of seawalls and dikes are depicted in Figure V-3-4. When vertical, they are labeled nonenergy absorbing, whereas if with a sloping surface or rubble mound, they absorb some energy (Pilarczyk 1990). The front face may also be curved or stepped to deflect wave runup. Typical damage modes for seawalls include: toe scour leading to undermining; overtopping and flanking; rotational slide along a slip-surface below and shoreward of the seawall; and corrosion of any steel reinforcement. Vrijling (1990) discusses 14 damage/failure mechanisms for dikes including "stability of the protective revetment." Part VI presents details for functional design.

- Construction of the massive, concrete seawall to protect Galveston, Texas, against overflows from the sea began in 1902, in the aftermath of the major hurricane of September 1900. Over 6,000 (16 percent) of the citizens lost their lives. An original construction photo (top) and chronology of seawall and embankment (dike) cross-section development (bottom) are shown in Figure V-3-5 (from Davis 1961). Major features are the wood piles, a sheet-pile cut-off wall, riprap toe protection, and the curved face to deflect wave runup. Modification and extension occurred in 1909, 1915, 1926, and the last extension to the west completed in 1963. In the almost 100 years of existence, many lives and millions of dollars of property damage have been saved by this project (Davis 1961).

- The City of Virginia Beach has opted for a low-crest elevation, sheet-pile, concrete cap seawall that also serves as a new boardwalk. Figure V-3-6 displays an artist's perspective (top) with cross section and an aerial photo (bottom) of a recently (1998) completed section. Construction of the newly designed, interior drainage system with pumping stations for an ocean outfall and a widened sandy beach will complete the project in 2002.

- Part VI-2 also discusses many other typical cross sections and layouts of seawalls and sea dikes.

(2) Bulkheads. These are vertical retaining walls to hold or prevent soil from sliding seaward. Their main purpose is to reduce land erosion and loss to the sea, not to mitigate coastal flooding and wave damage. For eroding bluffs and cliffs, they increase stability by protecting the toe from undercutting. Bulkheads are either cantilevered or anchored sheet piles or gravity structures such as rock-filled timber cribs and gabions. Cantilever bulkheads derive their support from ground penetration; therefore, the effective embedment length must be sufficient to prevent overturning. Toe scour results in a loss of embedment length and could threaten the stability of such structures. Anchored bulkheads are similar to cantilevered bulkheads except they gain additional support from anchors embedded on the landward side or from structural piles placed at a batter on the seaward side. For anchored bulkheads, corrosion protection at the connectors is particularly important to prevent failures. Gravity structures eliminate the expense of pile driving and can often be used where subsurface conditions support their weight or bedrock is too close to the surface to allow pile driving. They require strong foundation soils to adequately support their weight, and they normally do not sufficiently penetrate the soil to develop reliable passive resisting forces on the offshore side. Therefore, they depend primarily on shearing resistance along the base of the structure to support the applied loads. Gravity bulkheads also cannot prevent rotational slides in materials where the failure surface passes beneath the structure. Typical bulkheads are shown in Figure V-3-7.

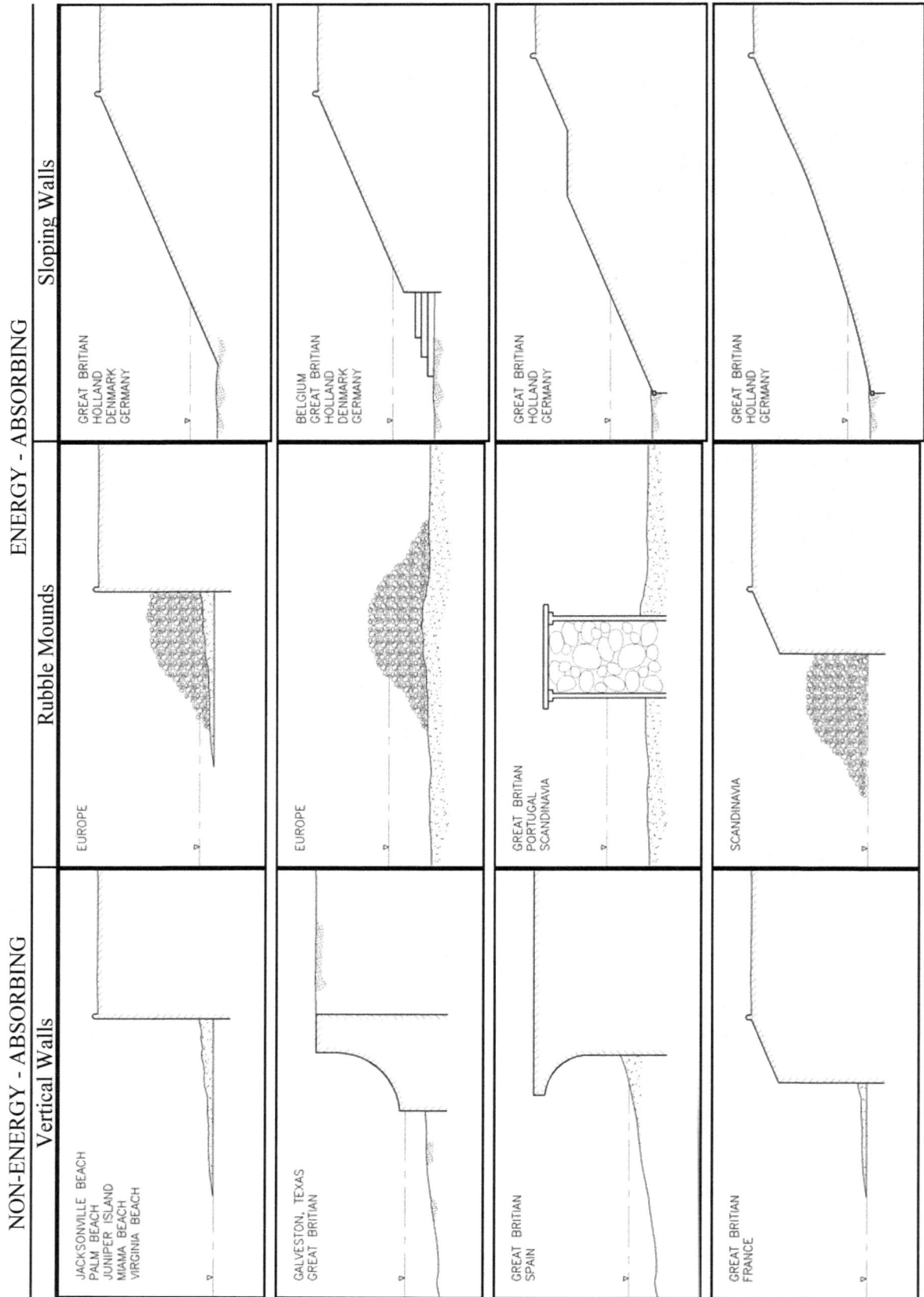

Figure V-3-4. Different types of seawalls and dikes (from Pilarczyk 1990)

a. Photograph of original construction (from USAED, Galveston)

Figure V-3-5. Galveston, Texas, seawall (Continued)

(a) The primary purpose of bulkheads is to hold land or fill in place and prevent shore side losses. A secondary purpose is to protect the land from wave attack. The strength of a bulkhead to protect against wave attack is provided almost solely by the fill, and if this material is lost, the bulkhead has no practical mechanism to adequately protect against waves. Therefore, two critical elements of a good bulkhead design that prevent or limit loss of backfill are: return walls at the alongshore ends of the structure to prevent high water from washing material away from behind the structure; and geotextiles to allow water but not fines to flow through the structure. Drainage of water through, behind, or laterally away from the structure is important to relieve pore pressure from excessive rainfall or overtopping. Drainage can be provided by drilling weep holes in the structure face to allow water to seep out.

(b) Steel and timber sheetpiling are the most commonly used bulkhead material. Steel sheet piles are individual sheets which can be interlocked and driven into hard, dense soils. The interlocking nature of individual steel sheet piles helps to limit erosion losses of backfill through the bulkhead. However, good design practice also includes installation of a geotextile or gravel filter between the bulkhead and the backfill to further prevent sediment losses. Filters are particularly important for timber bulkheads because they lack an interlocking mechanism, though overlapping timber sheets in a two-to-three layer technique can often limit the pathways for sediment to travel. An often encountered difficulty in installing geotextiles is depth of placement. For sheet piles that are driven, placing the geotextiles the full depth of the structure is practically impossible through the structure below the depth of the filter cloth. The loss of the backfill at a timber bulkhead from water (rain and/or wave overtopping) seeping and carrying sand through the bulkhead below the depth of the filter or below the depth of the timber piles is a common damage mechanism.

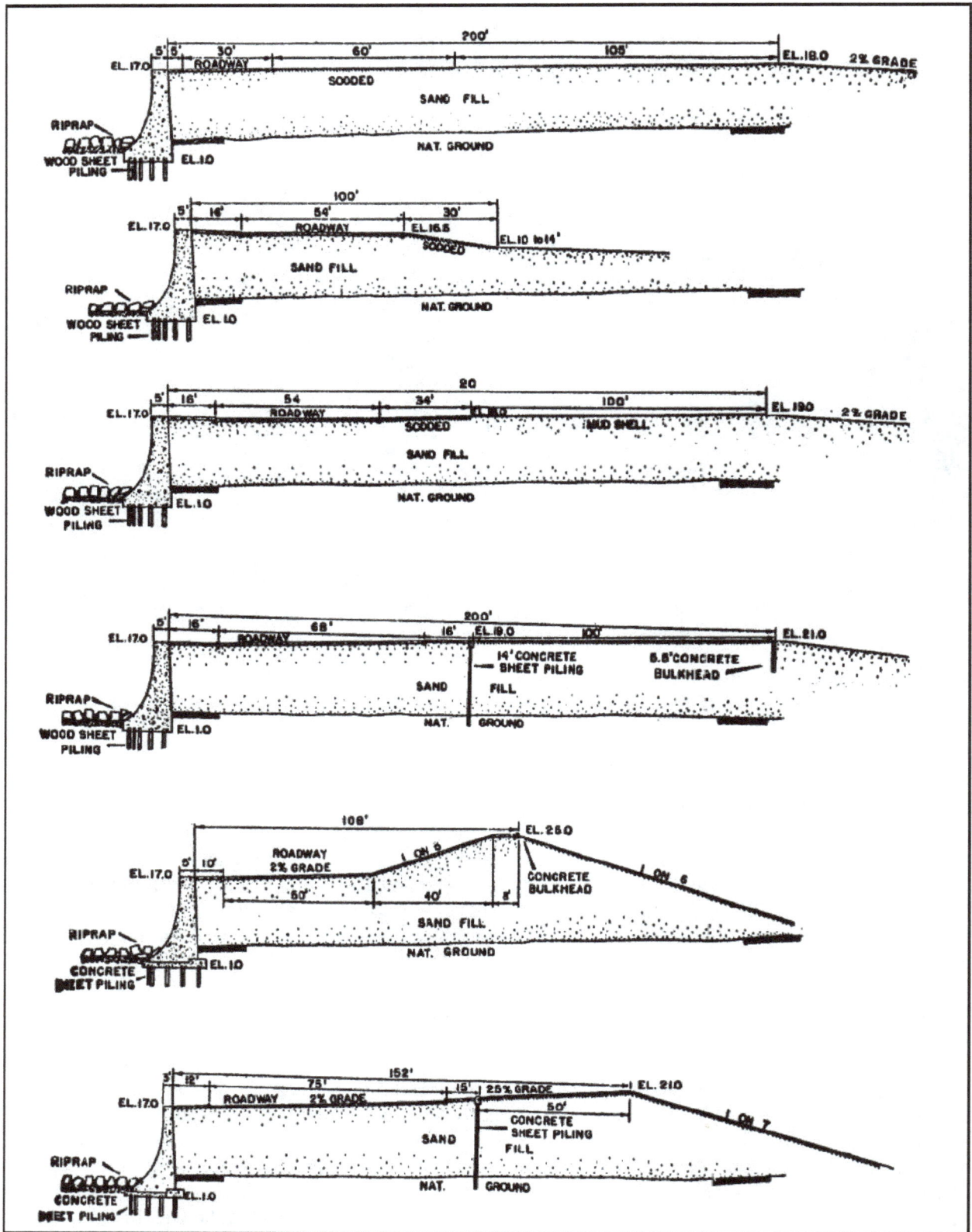

b. Chronology of development (from Weigel 1991)

Figure V-3-5.   (Concluded)

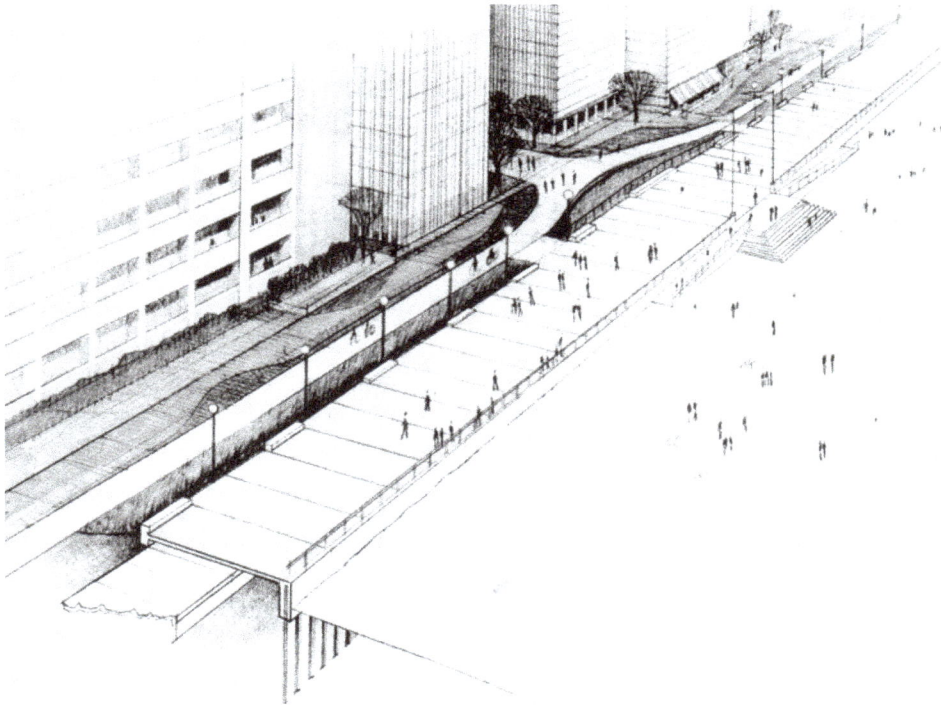

a. Artist's perspective

b. Aerial photo

Figure V-3-6. Virginia Beach seawall/boardwalk, 1997 (courtesy City of Virginia Beach, VA)

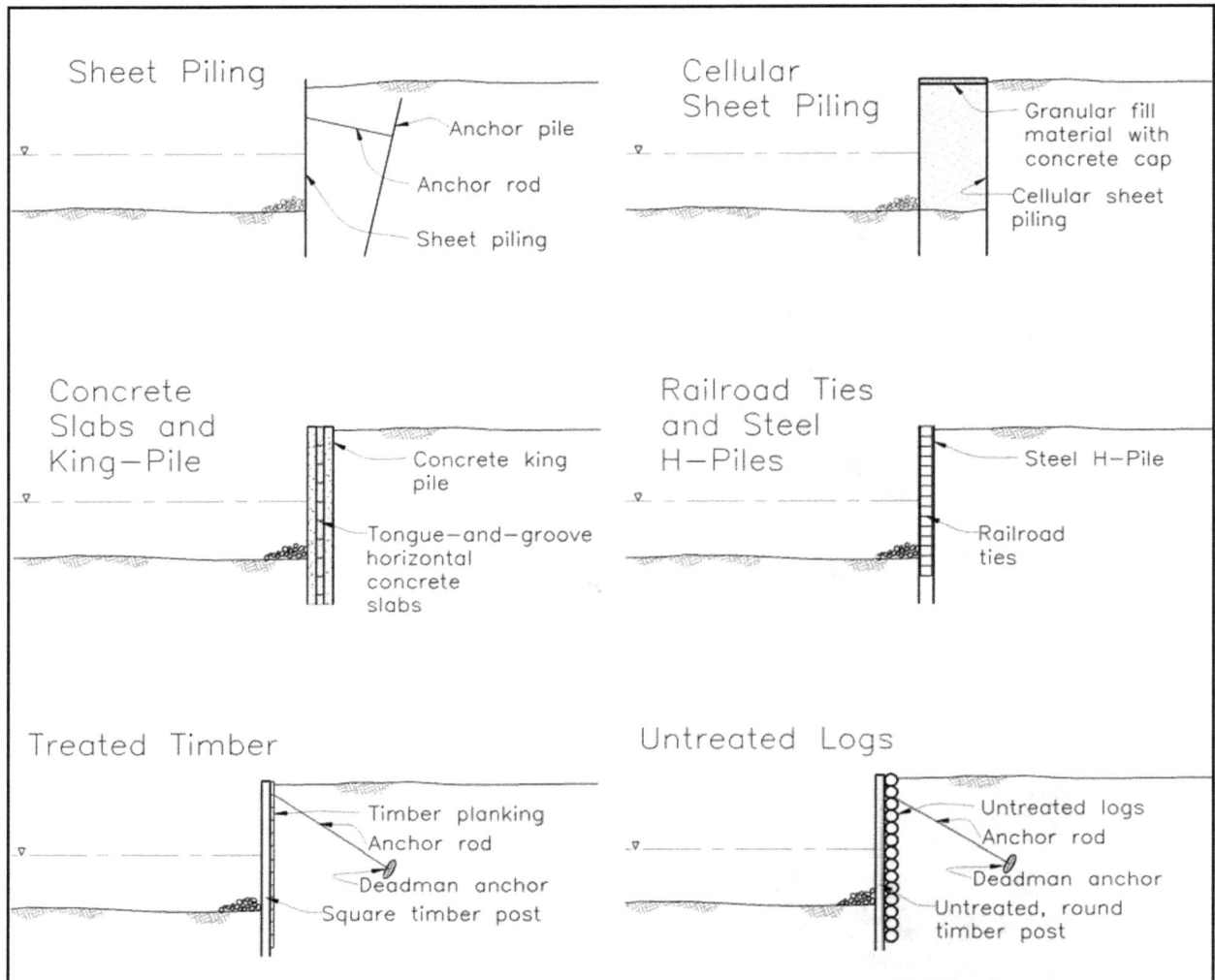

Figure V-3-7. Typical bulkhead types

(3) Revetments. Revetments are a cover or facing of erosion resistant material placed directly on an existing slope, embankment or dike to protect the area from waves and strong currents. Three major features are a stable armor layer, a filter cloth or underlayer, and toe protection. The filter and underlayer support the armor, yet allow for passage of water through the structure. Toe protection prevents undercutting and provides support for all the layer materials previously mentioned. If the toe fails, the entire revetment can unravel.

(a) Figure V-3-8 summarizes a wide range of designs and materials employed for a revetment. Armoring may be either flexible (normal) or rigid. Riprap and quarrystone designs can tolerate some movement and shifting or settling of their underlying foundation, yet remain functional. Rigid, concrete or asphalt slabs-on-grade are generally unable to accommodate any settling.

(b) Typical failure modes for revetment include:

• Armor layer damage and interior exposure.

• Overtopping and loss of foundation material.

- Toe failure and unraveling.

- Excess groundwater pressure and piping failure through the armor layer.

- Rotational sliding along the slip-circle surface

- Flanking of the end sections.

(c) Armor layer stability is discussed in detail in Part VI-5-3-a; toe stability and protection in VI-5-3-d, and filter layer design in VI-5-3-b. Shore protection by revetments can be for all levels of wave energy. For low wave energy environments in bays and rivers, relatively inexpensive and readily available stone sizes makes revetments a common choice for erosion protection by individual property owners. Vegetation and marsh grasses seaward of the revetment also diminish some wave energy to protect the revetment.

(4) Combinations and other types. Protective revetments on dikes are an example of combination coastal armoring structures. Earthen dikes, with stone revetments have been constructed to protect Texas City, Texas, and along the Lake Erie shoreline by the USACE. Due to the nature of the earthen structure, design and specifications should be evaluated by geotechnical engineers (see EM 1110-2-1913, "Design and Construction of Levees"). In the Netherlands, the sea side is heavily armored to protect the dike against the North Sea. On the land side, grazing sheep are used to continually compact the earth.

(a) As previously discussed, a storm surge barrier (see Figure V-3-1) across the opening to the sea provides alternative means of armoring for shore protection. The most notable hurricane barrier in the United States is located at New Bedford Harbor, Massachusetts. The tide (or flood) gate remains open for navigation, then closes to prevent flooding of inland areas during storms.

(b) When insufficient space or earthen dike materials are available, rigid, vertical flood walls may be constructed. Design guidance is available in EM 1110-2-2502, "Retaining and Flood Walls." Crossing the wall requires flood gates that roll, swing, or slide to close the opening EM 1110-2-2705, "Structural Design of Closure Structures for Local Flood Protection Projects." Special, interior drainage facilities may also be needed including pumping stations (EM 1110-2-3102, "General Principles of Pumping Station Design and Layout."). Grade raising, i.e., increasing ground elevation by filling with stable material is also possible. The entire city of Galveston, Texas, was elevated about 4 m to match the seawall crest elevation. (see Figure V-3-5). Sandy material was dredged from nearby Galveston Bay and pumped hydraulically to fill the island. Wiegel (1991) presents a complete history of the Galveston, Texas seawall.

b. Functional design.

(1) The functional design of coastal armoring structures involves calculations of wave runup, wave overtopping, wave transmission, and reflection. These technical factors together with economic, environmental, political (social), and aesthetic constraints all combine to determine the crest elevation of the structure.

(2) Wave runup and overtopping depend on many factors. Part VI-5-2 presents all the details. Empirically determined coefficients, formulas, tables, etc. have mainly come from laboratory scale experiments with irregular waves in large wave tanks. Independent variables include wave characteristics, water depths, slopes, roughness, degree of permeability or impermeable, wave angle, berm or continuous slope, freeboard, etc. Tables VI-6-18, 19, and 20 in Part VI-6 present partial safety factors for runup on rock-armored slopes, hollowed cubes, and dolosse armor units, respectfully.

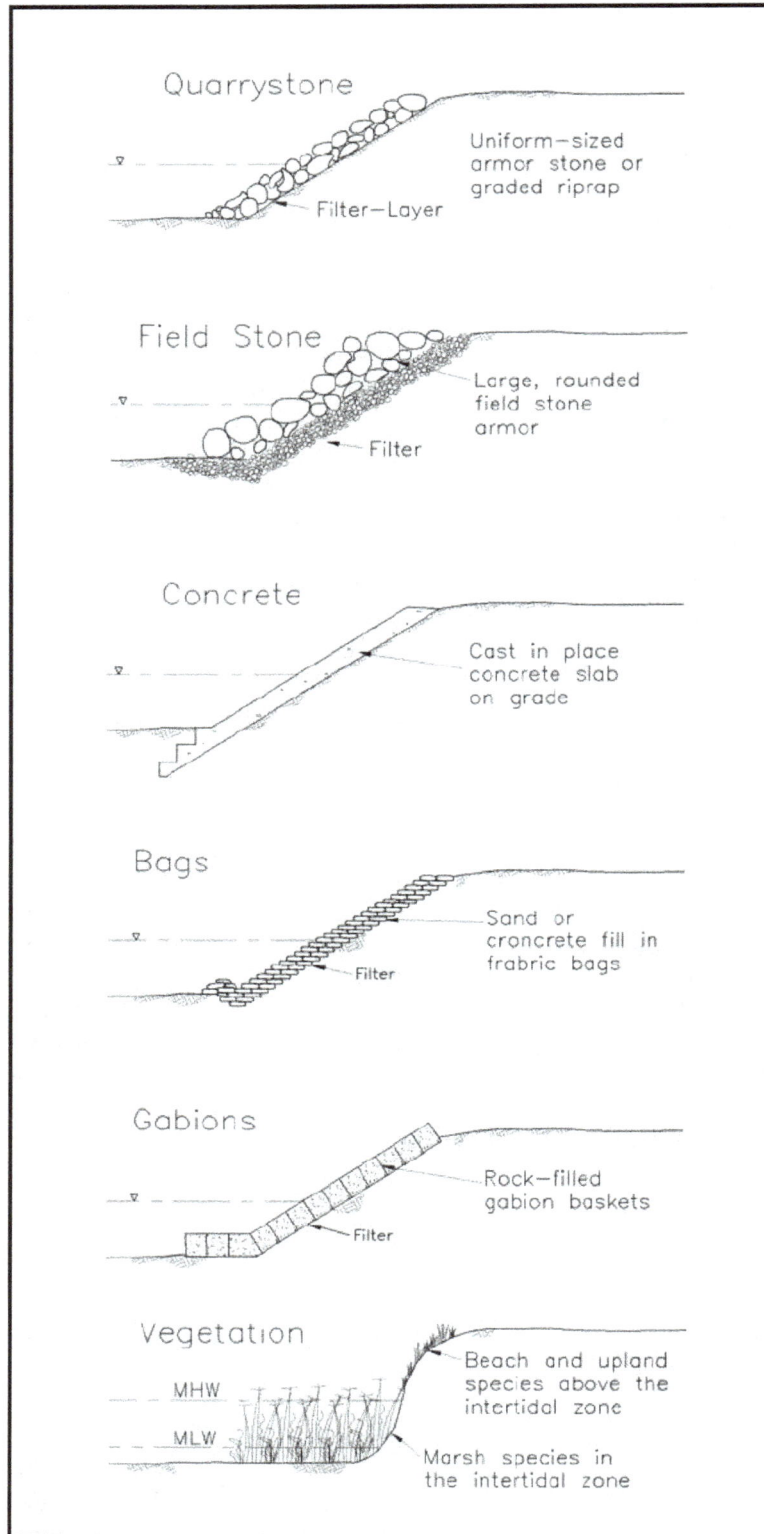

Figure V-3-8. Summary of revetment alternatives (continued)

Concrete Armor Units

Randomly- or specially-placed armor units such as tribars, dolosse, etc.

Underlayer

Filter

Concrete Revetment Blocks

concrete revetment blocks

Concrete-Filled Mattress

Concrete-filled mattress

Concrete Slabs

Concrete slabs from demolition work

Filter

Landing Mat

Landing Mat

Filter   Anchor

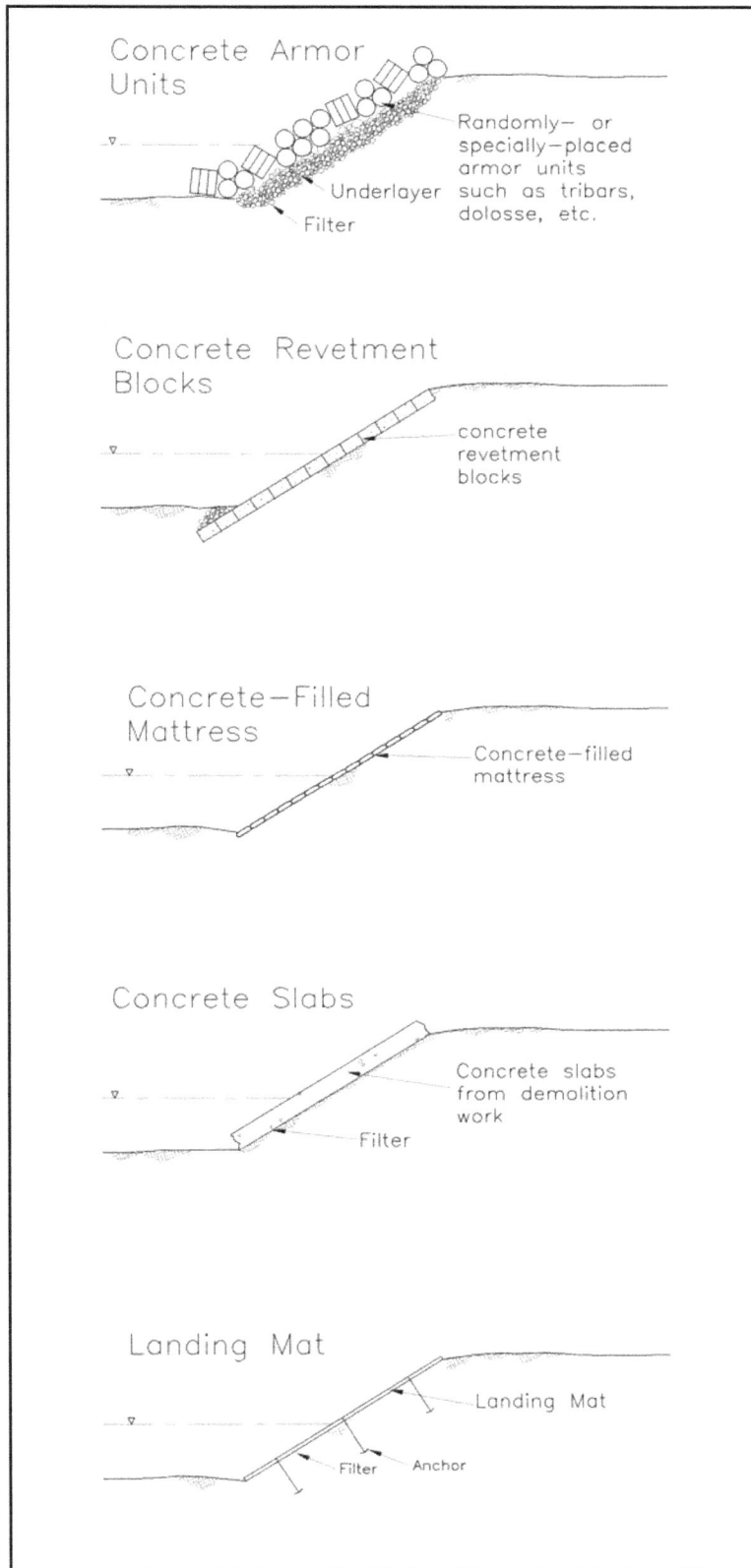

Figure V-3-8.   (Concluded)

(3) Wave reflection and transmission through and over sloping structures and beneath vertical wave screens is also found in Part VI-5-2.

c. Interaction with adjacent beaches.

There is a common perception that "… seawalls increase erosion and destroy the beach." The limited available evidence is examined in this section. The term seawall herein means any type of coastal armoring that hardens the shoreline to a fixed position, hence, also applies to bulkheads and revetments.

(1) Background. Concern with how seawalls interact with adjacent beaches can be traced to events in the 1960s and coastal geology studies on the origins and movements of barrier islands (Hoyt 1967). Barrier islands are one of the 11 types of land/water interfaces on earth (Shepard 1976). Barrier beach systems make up about 35 percent of the United States coast stretching from Maine to Texas. They protect the bays and estuaries that lie behind them from direct wave attack, but are dynamic systems with sand volumes that depend on changing ocean conditions, sand supplies, and control boundaries that define the volume.

(a) As depicted schematically in Figure V-3-9a (adapted from Dolan and Lins 1987), barrier islands are commonly perceived to migrate landward with constant volume as sea level rise continues. Storm surge with high waves produce sand overwash into the back bay. The barrier is said to roll over itself, shoreline movement is termed recession, and no volume change means no coastal erosion. Some scientific evidence disputes the rollover model. Leatherman (1988) used shoreline position data to show that tidal inlet formation processes dominate and move far greater sediment quantities over the long term. The migration model also requires the moving sand volume to overlay continuous, basal peat layer from the muds and plants in the lagoon. Stratigraphic evidence contradicts this important aspect along the East Coast of the United States (Oertel et al. 1992). Using the Bruun (1962) rule, a 1-2 mm/year rise in sea level translates to about 0.05-0.2m/year shoreline retreat rate. These are relatively small changes in shoreline position and herein labeled as those at geologic time scales. See Part IV-2-9 for a full discussion of marine depositional coasts and barriers.

(b) When man enters the picture by constructing a road on the shore, he establishes a fixed reference line. The shoreline position relative to the road decreases in time as depicted in Figure V-3-9b. Once development has been permitted, continued erosion may threaten man's artifacts (roads, buildings, bridges, etc.) and some type of shore protection may be undertaken such as seawall construction. These structures are not intended to protect the beach, but areas landward from the beach. Armoring provides a nonmoving reference point on the beach to make the existing, historic erosion more noticeable. Few argue that the road alone is "…destroying the beach", but this same logic is applied by some when a revetment or seawall is present on an eroding shoreline and the dry beach width is reduced each year in front of the hardened shoreline (Pilkey and Wright 1988). Eventually, the ocean will reach the seawall (and road) and the dry beach will be gone.

(c) As also depicted in Figure V-3-9b, a seawall traps sediment behind the structure, reduces overwash and fixes the shoreline position. Continued erosional stress over time acts to deepen the water depth at the structure that is of concern for structural design. The trapped sediment formerly in the dune, bluff or cliff) is removed from that available to contribute to subaqueous bar building during storms. This trapped material is also prevented from contributing to the longshore sediment transport processes along the coast and may alter the sediment budget. The volume trapped relative to that naturally active in the cross-shore profile will be discussed further in the following paragraphs.

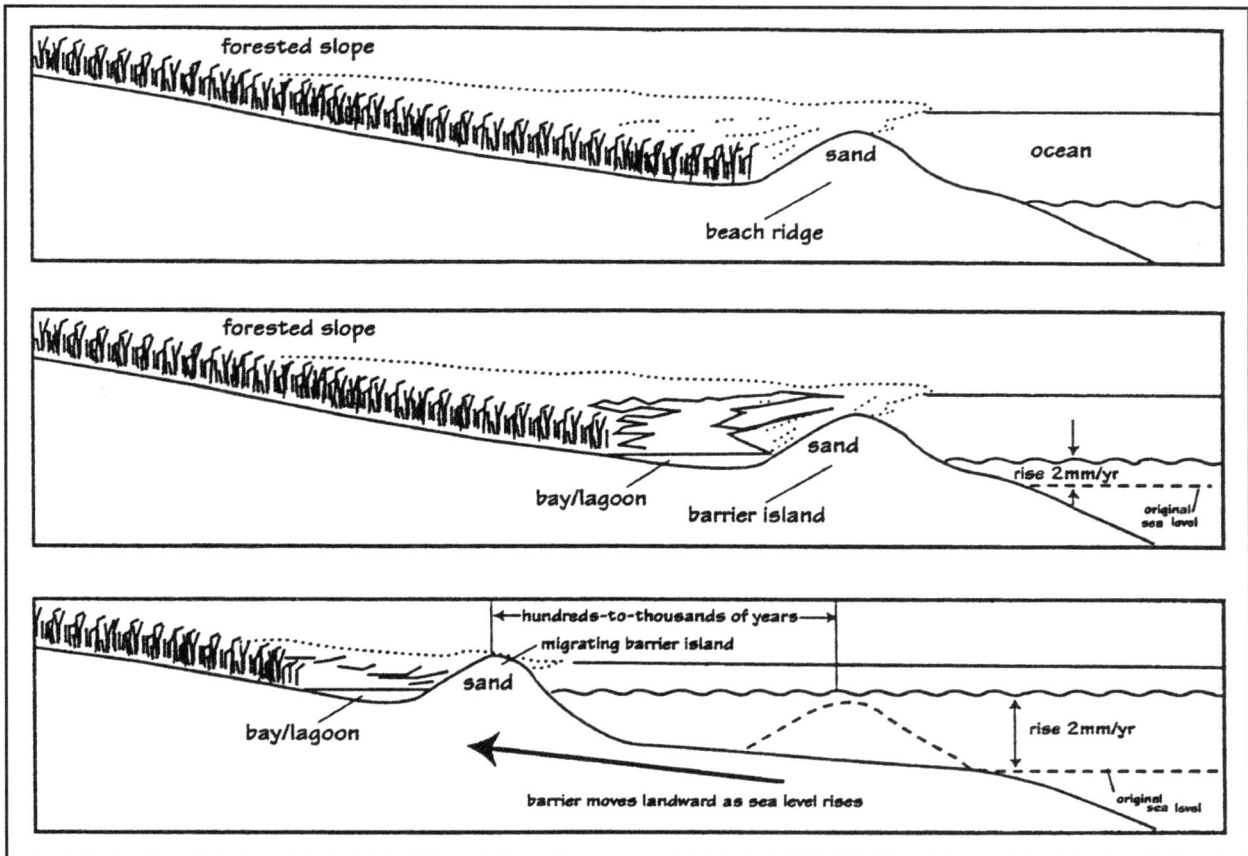

a. Barrier island migration at geologic time scales

Figure V-3-9. Time scales for shoreline movements (Continued)

(d) Natural erosion processes (wave-induced, sediment imbalances) plus anthropogenic induced erosion produce larger erosion rates than sea level rise (Komar et al. 1991). These are herein labeled as those at engineering time scales. The degree to which coastal armoring affects the adjacent beach has been the focus of some research effort.

(2) Literature review.

(a) Common concerns. Dean (1987) critically examined nine commonly expressed concerns about seawalls and adjacent beaches as summarized in Table V-3-3. Use was made of conservation of sediment mass, laboratory and field data, and the theory of sediment transport. Conclusions from this analysis were (numbers coincide with Table V-3-4) as follows:

Concerns Probably False (or Unknown)

- profile steepening (6)

- delayed beach recovery after storms (5)

- increased longshore transport (8)

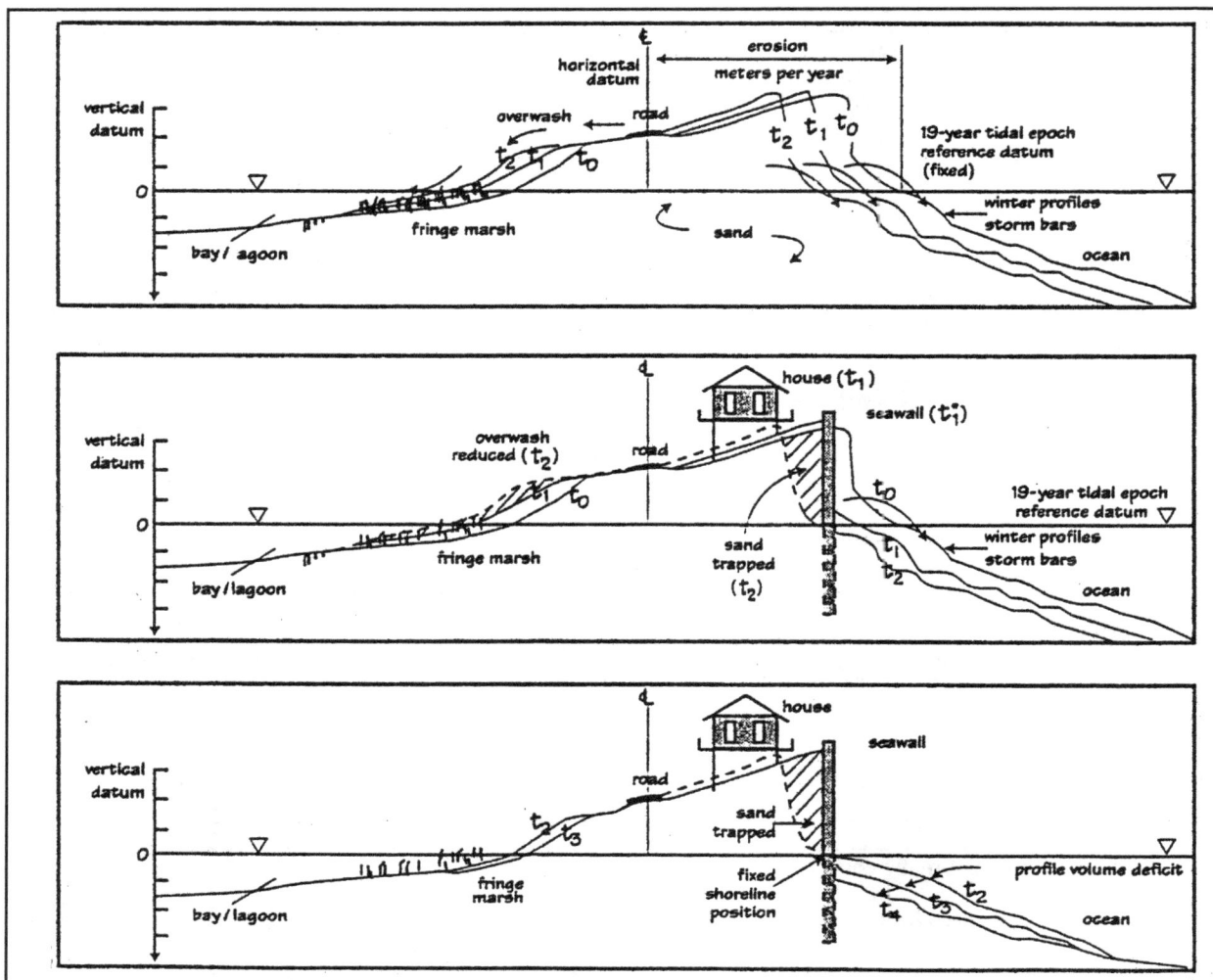

b. Coastal erosion at engineering time scales

Figure V-3-9. (Concluded)

- sand transport far offshore (9)

- increase in long-term, average erosion rate (3)

Concerns Probably True

- frontal effects (toe, scour, depth increase) (1)

- end-wall effects (flanking) (1)

- blockage of littoral drift when projecting into surf zone (groin effect) (4)

- beach width fronting armor likely to diminish (2)

Table V-3-3
Assessment of Commonly Expressed Concerns Related to Coastal Armoring (Dean 1987)

No.	Concern		Assessment
1	Coastal armoring placed in an area of existing erosional stress causes increased erosional stress on the beaches adjacent to the armoring.	True	By preventing the upland from eroding, the beaches adjacent to the armoring share a greater portion of the same total erosional stress.
2	Coastal armoring placed in an area of existing erosional stress will cause the beaches fronting the armoring to diminish.	True	Coastal armoring is designed to protect the upland, but does not prevent erosion of the beach profile waterward of the armoring. Thus, an eroding beach will continue to erode. If the armoring had not been placed, the width of the beach would have remained approximately the same, but with increasing time, would have been located progressively landward (see 2b).
2a	Beaches on eroding coastlines will diminish in front of fixed dune positions.	True	An eroding beach continues to erode relative to a fixed dune position. The width of the beach must diminish if the shoreline is eroding (Figure 1).
2b	Natural beaches on retreating barriers maintain the same beach width.	True	Relative to a retreating duneline, a shoreline eroding at the same rate results in a stable beach width.
3	Coastal armoring causes an acceleration of beach erosion seaward of the armoring.	Probably False	No known data or physical arguments support this concern.
4	An isolated coastal armoring can accelerate downdrift erosion.	True	If an isolated structure is armored on an eroding beach, the structure will eventually protude into the active beach zone and will act to some degree as a groin, interrupting longshore sediment transport and thereby causing downdrift erosion.
5	Coastal armoring results in a greatly delayed poststorm recovery.	Probably False	No known data or physical arguments support this concern.
6	Coastal armoring causes the beach profile to steepen dramatically.	Probably False	No known data or physical arguments support this concern.
6a	Coastal armoring destroys foreshore bar and trough features.	Probably False	No known data or physical arguments support this concern.
7	Coastal armoring placed well-back from a stable beach is detrimental to the beach and serves no useful purpose.	False	In order to have any substantial effects to the beaches, the armoring must be acted upon by the waves and beaches. Moreover, armoring set well-back from the normally active shore zone can provide "insurance" for upland structures against severe storms.
8	Seawalls increase the longshore sediment transport.	Unknown	No known data exists, physical arguments can support or discredit this concern. Needs research.
9	Seawalls cause sand transport a far distance offshore.	Probably False	No known data or physical arguments support this concern.
10	Other		

Kraus (1988) reviewed over 100 references (laboratory, field, theory, and conceptual studies) to make a thorough examination of the literature. This review and seven companion papers are presented in Kraus and Pilkey (eds. 1988). An updated literature review is found in Kraus and McDougal (1996) who examined 40 additional papers. In general, these extensive literature reviews agreed with Dean (1987) regarding which concerns were probably false and which many are true. The interested reader should consult these references for all the details.

(b) Definitions. The natural, background shoreline erosion rate, P_N and the rate *after* human activities P_A can define a coastal erosion ratio, R_P

$$R_p(x,t) = \frac{P_A}{P_N} \qquad\qquad\qquad\qquad \text{(V-3-3)}$$

where the subscript R_p means shoreline position is used to define R. If profile data are available, then actual, coastal erosion volume could be employed to find a volume ratio, R_V as

$$R_v(x,t) = \frac{V_A}{V_N} \qquad\qquad\qquad\qquad \text{(V-3-4)}$$

- where V_N is the natural erosion (volume loss) rate and V_A is the volume loss rate after construction of roads, seawalls, etc. at a given location. Clearly, if R_v (or R_p) is proven greater than unity under similar climatological conditions, then we may conclude that armoring has increased the natural, historical conditions at the site. The level of impact (if any) on the frontal and laterally adjacent beaches (1 percent, 5 percent, 10 percent, 50 percent, etc.) needs quantification. Pilkey and Wright (1988) use the terms passive and active erosion of the beach to distinguish between the perceived versus real natural and manmade causes, respectively.

- The volume of sediment trapped behind a seawall depends upon its position on the beach, crest elevation and length. Weggel (1988) defined six types of seawalls depending on their location on the beach and water depth at the toe. At one extreme (type 1) the wall is located landward of the limit of storm wave runup to have zero impact. At the other extreme (type 6) walls are located seaward of the normal breaker line. Types 2-5 lie in between and are said to have increasing effects on coastal sediment processes as the type number increases. Storm surges can create all six type conditions during a single storm event. Coastal erosion may also gradually alter the types.

- Dean (1987) postulated that the sediment trapped behind the wall resulted in an excess erosional stress to produce toe scour and excess erosion on unprotected adjacent property.

(c) Frontal impacts. Beach profile change, toe scour during storms and nearshore bar differences have been attributed to seawalls. Conventional wisdom has been that these impacts were due to wave reflection. Kraus and McDougal (1996) studied the field results by Griggs et al. (1997); laboratory work by Barnett and Wang (1998) and Moody and Madsen (1995) and their own research in the SUPERTANK (large scale) seawall tests (McDougal, Kraus, and Ajiwibowo 1996) to conclude that reflection is not a significant factor in profile change or toe scour. In the field, toe scour is more dependent on local, sediment transport gradients and the return of overtopping water (through permeable revetments or beneath walls) than a result of direct, cross-section wave action. Their conclusions also negate the common perception that sloping and permeable surfaces produce less effects than vertical, impermeable walls. Scour and scour protection is covered in detail in Part VI-5-6.

(d) Impacts on laterally adjacent beach. Perhaps the key environmental concern is how a seawall affects a neighbor beach with no armoring. Does the wall create end-of-wall or flanking effects, i.e., R (x,t) greater than unity? Two studies are often cited to demonstrate flanking effects. Walton and Sensabaugh (1979) provide posthurricane Eloise field observations (14 data points) of additional bluff (contour) recession adjacent to seawalls in Florida. McDougal, Sturtevant, and Komar (1987) and Komar and McDougal (1988) present small scale, equilibrium beach, laboratory measurements (nine data points) for 7-14-cm waves at 1.1-sec periods normal to a median grain-size, sandy beach. The 23 data points are then combined to demonstrate the excess flanking erosion. The extent and length of the excess erosion is related to seawall length and is explained in terms of the seawall denying sand to the littoral system (e.g., Dean 1987).

- However, other mechanisms may be responsible. If the seawall extends seaward, it may act like a groin to cause downdrift impacts. Tait and Griggs (1991) measured an area of lowered beach profile extending 150 m downcoast at Site No. 4 in California. They proved that the upcoast end of the wall produced sand impoundment or a groin effect. Toue and Wang (1990) conducted laboratory experiments with waves attacking walled and nonwalled beaches at angles and concluded that downdrift impacts were a groin effect.

- Plant (1990) and Plant and Griggs (1992) observed rip currents at interior sections and at the ends of armored sections. These rip currents were attributed to wave overtopping, return flows and elevated, beach water tables during storms. McDougal, Sturtevant, and Komar (1987) also observed rip currents in their model tests previously described and from field evidence in Oregon. They concluded that this mechanism may be more responsible for end-of-wall, flanking effects than the sand trapping theory of Dean (1987).

- Griggs et al. (1997) discuss eight full years of field monitoring including the intense winter storm of January 1995. This storm did not produce end scour on the control beach at Site No. 4. They concluded from a comparison of summer and winter beach profiles at beaches with seawalls and on adjacent, control beaches, that no significant long-term effects were revealed.

- Basco et al. (1997) summarize the results of 15 years of profile survey data with 8-9 years taken before seawall construction at Sandbridge, Virginia, on the Atlantic Ocean. The shoreline has been eroding on average 2m/year (Everts, Battley, and Gibson 1983) long before wall construction began. One part of the study used five years of monthly and poststorm profile data at 28 locations (62 percent walled; 38 percent nonwalled) of the 7,670 m study reach. They concluded that the volume erosion rate was not higher in front of seawalls. However, seasonal variability of sand volume was slightly greater in front of the walled locations. Winter waves drag more sand offshore in front of walls, but summer swell waves pile more sand up against walls in beach rebuilding. Walled sections recovered about the same time as nonwalled beaches for both seasonal transitions (winter to summer) and following erosional storm events. These results were for a weighted average of total sand volume (subaerial) in front of the walled section and seaward of a partition for the nonwalled beach sections.

- At individual profile locations adjacent to walls, using the full 15 years of data, R_v values varied considerably. The evidence for any long-term, end-of-wall effects were considered inclusive for Sandbridge beach. There was never evidence of flanking effects after storms on adjacent beaches (Basco et al. 1997). This study continues. In general, Basco et al. (1997) have confirmed all the conclusions of Dean (1987), Kraus (1988) and Kraus and McDougal (1996) except the end-wall, flanking effect.

- Natural beaches coexist in front of the rocky cliffs and naturally-hardened shorelines at many locations throughout the world. A major, comprehensive research effort is needed to quantify the effect of sand trapping on frontal and downdrift beaches.

(3) Active volume in the cross-shore profile. Successive cross-shore surveys of the beach profile to closure depth reveal spatial variations in vertical elevation at each location. In the absence of lateral transport, the eroded sections balance the accreted areas, i.e., sediment volume is conserved. The active sediment volume is defined as one-half of the total volume change between two successive surveys. The 12-year, biweekly nearshore bathymetric data set surveyed at the Corps of Engineers Field Research Facility, Duck, North Carolina, has been analyzed to quantify the total active sand volume, its spatial variation across the profile, and its relation to long-term, fair-weather, and storm periods (Ozger 2000). An empirical relationship between storm wave power and active sand volume has been developed for the Duck site. Prestorm morphology and duration of storm surge are possible factors for the scatter in the power versus active sand volume relationship. The maximum value of active sand volume was $140 m^3/m$ (350 cu yd/ft). Different tidal conditions, wave climate and hard bottoms (or reefs) limit the cross-shore movement of sediment. Determinations of the naturally active, sand volume should be made for other sites. Basco and Ozger (2001) summarize the above results and discuss various applications in coastal engineering. The seawall trap ratio, WTR can be defined as:

$$WTR = \frac{Wall\ Trap\ Volume}{Active\ Sediment\ Volume} \qquad\qquad (V\text{-}3\text{-}5)$$

to quantify the relative impact of the sand volume removed from the system. Dean (1987) failed to consider the WTR. Weggel (1988) qualitatively addresses the importance of the numerator increasing with type number but also did not consider the significance of the denominator. The spatial distribution of the WTR is also important relative to wall location. At locations with seawalls where the WTR is small, annual mitigation may be economic.

(4) Sand rights and mitigation. A few states have adopted sand mitigation polices and procedures to permit seawall construction and maintain a healthy beach. The idea is to annually replace the beach materials trapped behind the structure with a volume calculated by some formula. The methodology in Florida has both on offshore and longshore transport components which require knowledge of the annual erosion rate and net, longshore transport rate, respectively (Terchunain 1988). Sarb and Ewing (1996) present formulas for cliff and bluff erosion impacted by seawall construction in California. These formulas attempt to deterministically estimate the wall trap volume (numerator) but do not consider the active sediment volume (denominator) in Equation V-3-5. The relative volume trapped is unknown. Field research efforts to date have yet to confirm the trapping theory of Dean (1987). The reason may be that the trapped volume is only a small percentage of the total, active sand volume in the profile. The WTR is near zero so that downdrift impacts are minimal and lost in the data scatter. See also Part III-5-13 for discussion of measures to manage human influence on sediment supply. Who owns the sediments on eroding shorelines, the local property owners or downdrift interests, is a legal question facing society. The subject of sand rights (Magoon 2000) is an extremely complicated issue that may require settlement by the nation's courts.

V-3-3. Beach Stabilization Structures

a. Naturally stable shorelines. Part IV-2 classifies coasts and their morphology. Marine depositional coasts with barriers and beaches are one of the most widely distributed geomorphic forms around the world. (Part IV-2-9, 10). They form a flexible buffer zone between land and sea. The subaerial beach has two major zones. The foreshore extends from the low-water line to the limit of wave uprush at high water. At this point, the backshore extends to the normal landward limit of storm wave effects. This landward limit is usually marked by a foredune, cliff, structure, or seaward extent of permanent vegetation. The backshore is only affected during storms when surges and high waves transport backshore sediments. This exposed, subaerial beach definition is accepted by the general public. Some authors include the surf zone and bars out to closure depth, i.e., the subaqueous part of the beach, because both parts, subaerial and subaqueous, exchange sediment. Beaches may stretch for hundreds of kilometers or others, called pocket beaches, are restricted by headlands and are only tens of meters in length. Figure IV-3-31 schematically summarizes factors controlling morphodynamics along a range of coastal environments extending from rocky, to noncohesive sediments to cohesive shorelines.

(1) Many beaches are naturally stable. In general, wide beaches are exposed to more severe wave conditions at that location, but the relationship between beach width (or section volume) and storm energy for naturally stable shorelines has yet to be determined. Figure V-3-10a displays a stable, pocket beach on Bruny Island, Tasmania, Australia, where beach width increases along the more exposed section of coast (from Silvester and Hsu 1993). The protected reach behind the headland is much narrower than that receiving a direct attack by large waves during storms. The dark area is vegetation and landward limit of the backshore, subject to normal storm wave effects. If this photo were taken at high tide, a minimum beach width for a stable shoreline could be determined.

(2) This concept of a minimum beach width (or volume) is schematically illustrated in Figure V-3-10b. The volume of sediment present protects the uplands (foredune, cliff, structure, or vegetation) from damage under normal or average storm conditions. The landward boundary of the backshore is a reference baseline for shore protection. On eroding beaches, the backshore may be missing, and the high-water uprush may impinge directly on cliffs or structures. Both natural and anthropogenic agents may cause the erosion. But a minimum, beach width is still necessary for natural shore protection at the eroding site.

b. Minimum dry beach width. Professor Richard Silvester in an article on the stabilization of sedimentary coastlines (Silvester 1960) wrote:

> "...to allow for storm-cycles and the short-term reversals of drift, a sufficient width of beach should be allowed as working capital on which the sea can operate. Once the coast has been stabilized, by preventing the net movement of sediment, no long-term erosion need be anticipated and the 'active' beach width can be minimized." (p. 469)

(1) As illustrated in Figure V-3-11a,b, for both naturally open beaches and pocket beaches between headlands, the minimum, dry beach width, Y_{min} is defined as the horizontal distance between the mean highwater (mhw) shoreline and the landward boundary or base (reference) line. The mhw shoreline is employed because it is the common, land/water boundary shoreline on maps; it is more readily identified from aerial photos; and it is a more conservative, minimum width (and volume) for shore protection. It is the minimum, dry beach width required to protect the foredune, cliff, structure, or vegetation behind the baseline from normal storm conditions. The beach does the work, and it's resilience and recovery are critical for long-term shore protection.

a. Aerial view (schematic from photograph) of Cloudy Bay, Bruny Island, Tasmania, Australia, showing greater natural beach berm width along the more exposed section of coast (adapted from cover photograph and Figure 3-21 in Silvester and Hsu (1993))

b. Schematic of minimum dry beach width, Y_{min}

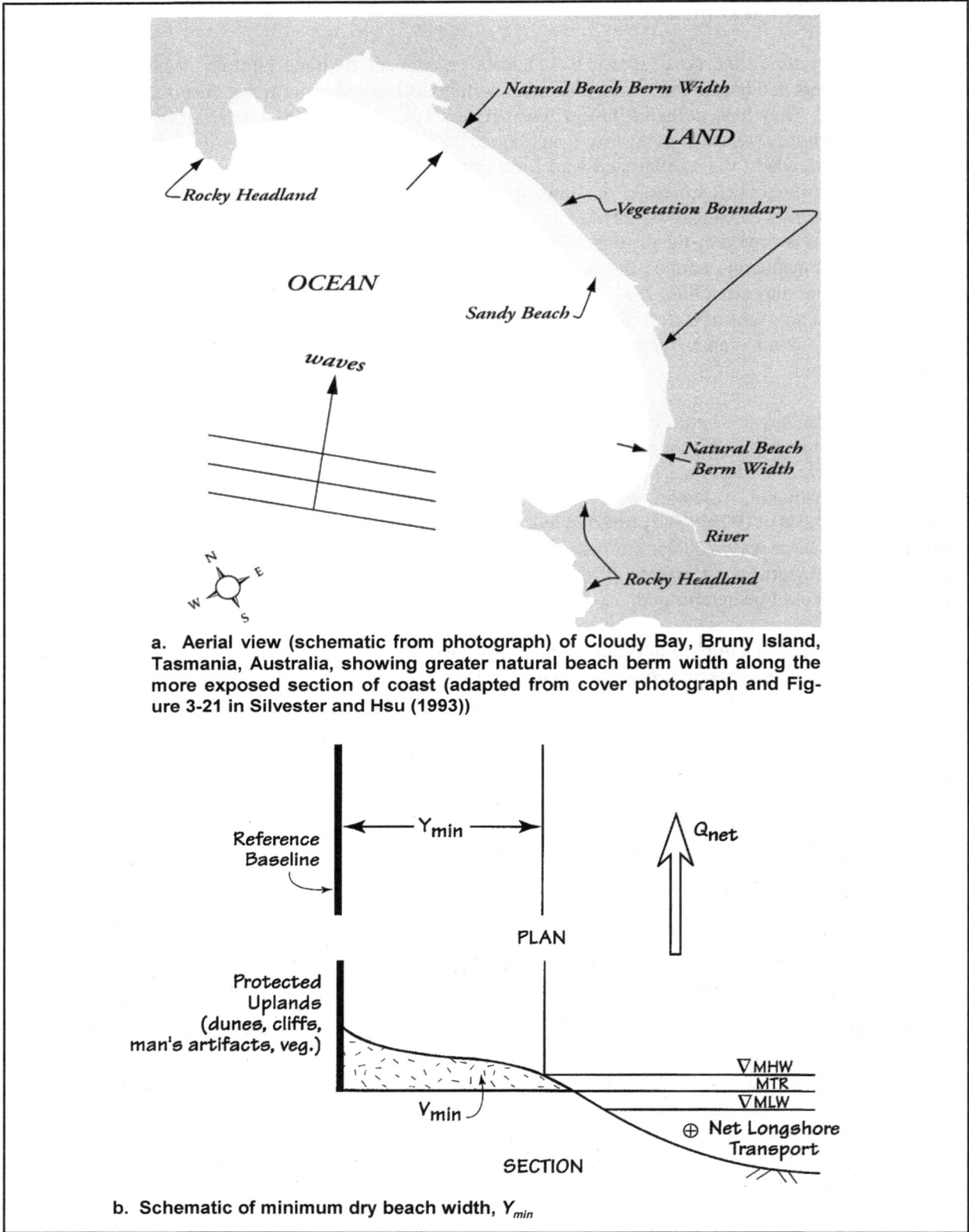

Figure V-3-10. Naturally stable shorelines with beach width dependent on stormwave energy (from Silvester and Hsu 1993)

(2) Normal storm wave conditions are expected once every two years or maybe every year. More intense, less frequent storms will reach the foredune, cliffs, structure or vegetation line. Beach stabilization structures can provide upland protection beyond the baseline for these rarer storm events. At a minimum, these structures should be designed to provide the minimum, dry beach width for shore protection.

(3) Figure V-3-11c,d,e depict the three most common beach erosion mitigation structures, namely headland breakwaters, nearshore breakwaters, and a groin field. And, each schematic displays the minimum, dry beach width, Y_{min} that is required for design. In each case, it is located in the gap area with greatest wave energy. The EST methodology discussed in V-3-1-c can be applied to determine the probability distribution of dry beach widths including the minimum for normal storm conditions. Functional design of these structures based on empirical knowledge is presented in the next sections. Two key factors are the minimum dry beach width (or volume) and the natural, sediment transport processes at the site. Explicit acknowledgment of Y_{min} as design criteria is often missing in coastal engineering design.

c. Headland breakwaters.

(1) Background and definitions. Natural sandy beaches between rocky headlands have been called a variety of names in the literature, related to the curved shape of the bay found at many coasts around the world. Silvester and Hsu (1993) summarize the literature. See also Part III-2-3-i. for a list of references. Because of their geometry, they have been called spiral beaches, crenulate-shaped bays, log-spiral and parabolic-shaped shorelines, headland bay beaches and pocket beaches. Half-Moon Bay in California is a good example as first discussed by Krumbein (1944) and shown as Figure 4.3 in Silvester and Hsu (1993). Many researchers have studied the dynamic processes of this geomorphic feature, but Silvester (1960) was the first to examine their static equilibrium and propose the creation of artificial headland breakwaters as a shore protection structure. Figure V-3-12 presents a sketch of an artificial headland system and beach plan-form (from EM 1110-2-1617, "Coastal Groins and Nearshore Breakwaters."). Normal wave conditions with a predominant swell direction produce a maximum indentation between two fixed points (breakwater structures) and a fully equilibrated, planform shape. Thus man can mimic nature by building the headland breakwaters and letting nature sculpture the beach with a limiting indentation and shoreline that is stable.

(2) Physical processes. Waves from one persistent, dominant direction, β, diffract around the upcoast headland and refract into the bay. Waves will break at angles to the shoreline causing sediment transport and shoreline shape adjustment (nonequilibrium shape) until a full equilibrium shape is reached. At this stage, waves break simultaneously around the entire periphery, no longshore currents and no littoral drifts occur within the embayment. The tangent section, adjacent to the downdrift headland is exactly parallel to the normal wave crest direction from offshore. Such a bay is said to be in static equilibrium (i.e., it is stable until there is a shift in the dominant wave direction). Minimal amounts of additional sediments enter or leave past the headland boundaries. Bidirectional, dominant, wave impact (swells and storms) and sediment bypassing the headlands are two reasons for littoral drift to continue around the bay. These bays are said to be in dynamic equilibrium and can be predicted within certain tolerances. Only the static equilibrium shapes can be related to wave input. The ability to calculate the static equilibrium shape and maximum indentation are needed for the functional design of headland breakwaters.

(3) Functional design. Early investigators employed a log-spiral curve to fit the planform shape (Yasso 1964; Silvester 1970). In practice it is difficult to apply because the center of the log-spiral does not match the point at which diffraction begins to take place.

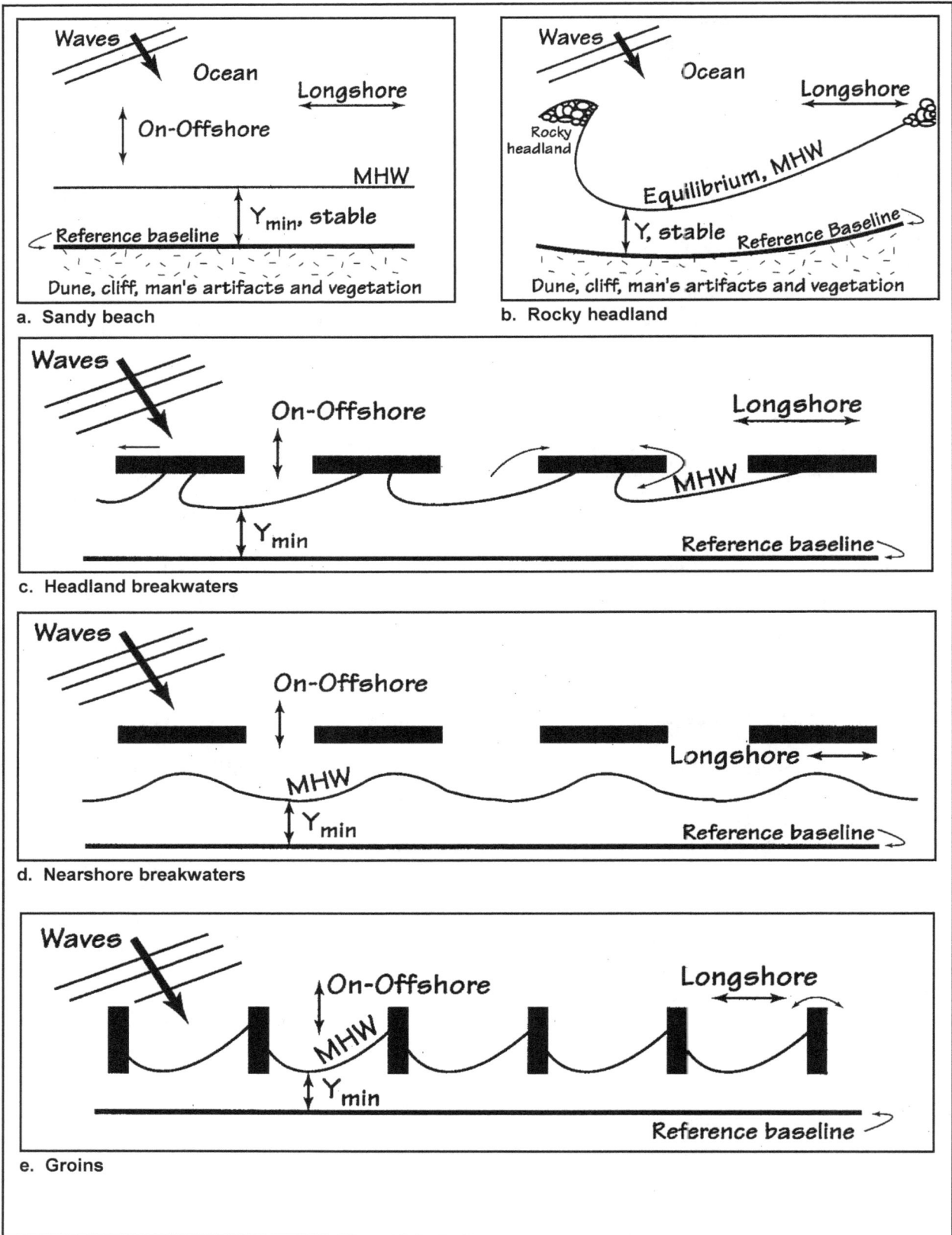

Figure V-3-11.   Natural and artificial stable shorelines with minimum dry beach width, Y_{min}

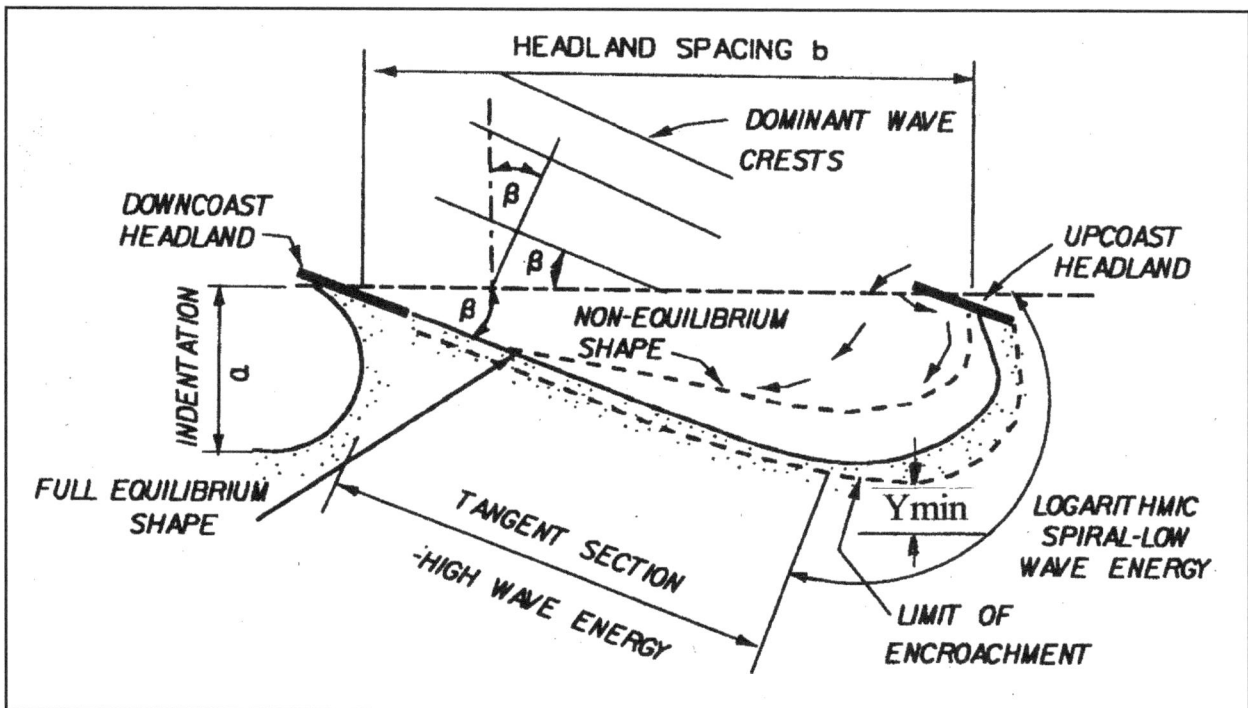

Figure V-3-12. Definition sketch of artificial headland system and beach planform (from EM 1110-2-1617)

(a) Parabolic bay shape. A new empirical approach that uses shoreline data from bays in static equilibrium and physical models has been developed by Hsu, Silvester, and Xia (1987, 1989). It is called a parabolic model because the data has been used to fit a second order polynomial. Figure V-3-13 presents a definition sketch of the four key geometric variables, R, R_o, β and θ, that form the parabolic model. The model center now exactly matches the initial diffraction point. Part III-2-3-i presents complete details including definitions of R, R_o, β and θ; the parabolic model equation (2-24); the three coefficients C_o, C_1, C_2 related to wave angle β in Figure III-2-27 and limitations of the data. An Example Problem III-2-8 is also presented to illustrate an application of the model which assumes one predominant wave direction exists at the site of interest. Many more examples and discussion is found in Silverster and Hsu 1993.

(b) Minimum width for storm protection. As illustrated in Figure V-3-12, storm waves may be from a different direction to cut back the beach and form a new limit of encroachment planform shape. A more detailed definition sketch in perspective and cross section is presented in Figure V-3-14 (adopted from Hardaway, Thomas, and Li 1991). Two nearshore breakwaters together with beach nourishment form the headland breakwater design for shore protection. The following terms are defined:

L_s Length of breakwater structure

L_g Gap distance between adjacent breakwaters

d_s Depth (average) at breakwater below mean water level

e   Erosion of shoreline (mhw) from design storm

Y_o Distance of breakwater from original shoreline

Y_g Maximum indentation under normal wave conditions

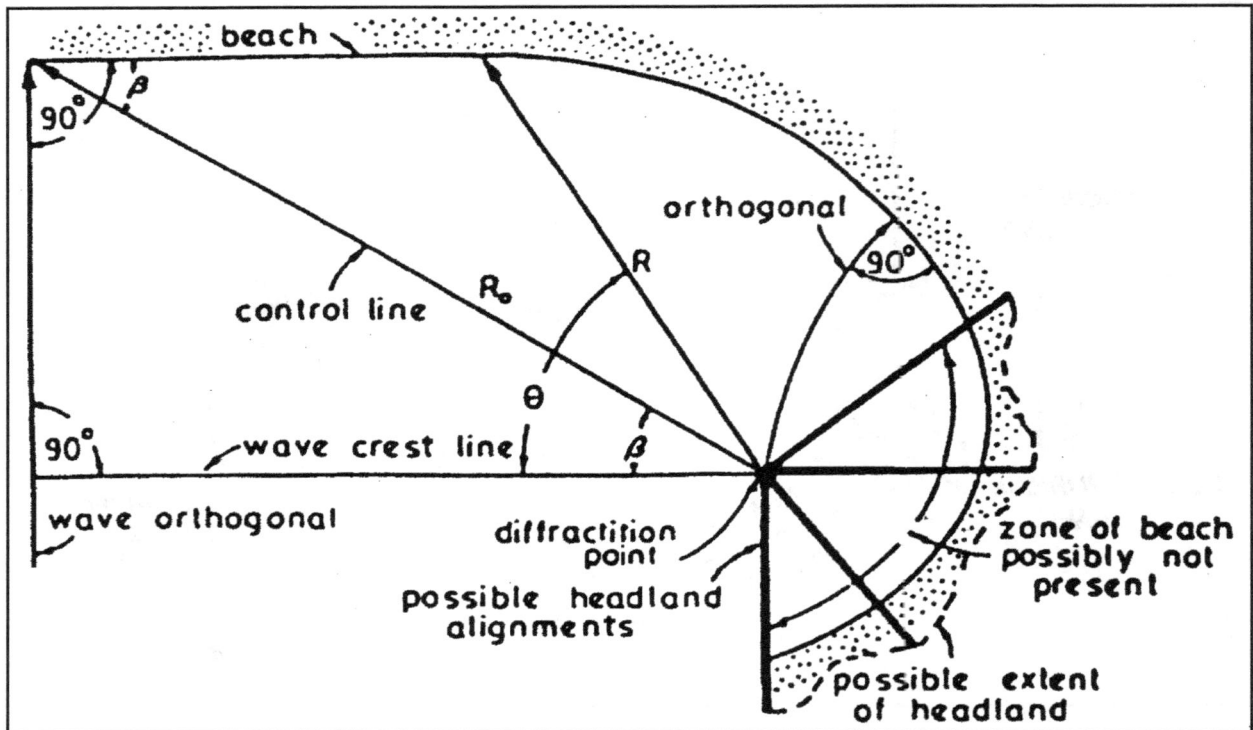

Figure V-3-13. Definition sketch of parabolic model for planform shape

Y	Distance of breakwater from nourished shoreline
Y_{min}	Minimum distance from base (reference) line to mhw shoreline after design storm event
B	Minimum beach width at mhw after nourishment
W	Width of design beach nourishment
Z_s	Backshore elevation at baseline
F_B	Breakwater freeboard, mhw to crest
Q_{net}	Net longshore sediment transport rate
Q_{gross}	Gross longshore sediment transport rate
$Q_{offshore}$	Offshore sediment transport rate for design storm

The planform shape and maximum indentation, Y_g can be estimated by the parabolic shape model previously discussed above. A hyperbolic tangent shape was developed by Moreno and Kraus (1999), which may be more convenient to apply than the parabolic shape.

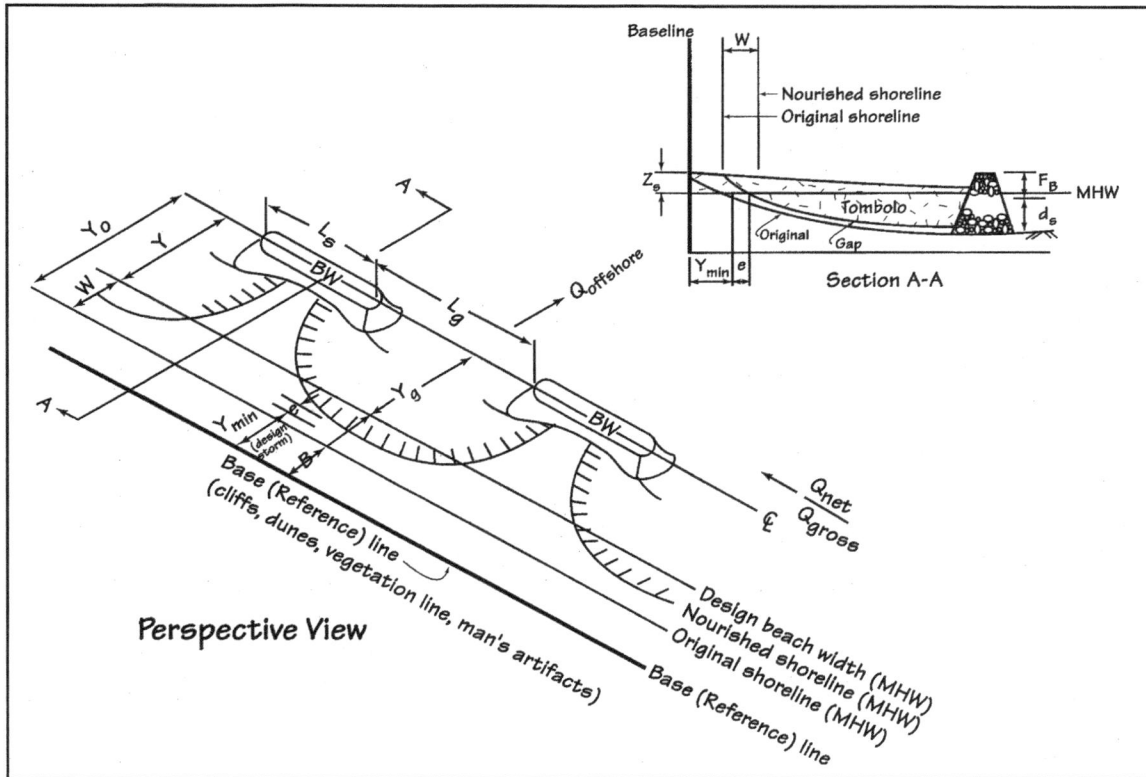

Figure V-3-14. Definition sketch, headland breakwaters

(4) Applications on Chesapeake Bay. Headland breakwater systems have been built along the shoreline of the Chesapeake Bay for shore protection and to maintain recreational beaches. Since 1985, 60 breakwaters at 19 sites have been designed, constructed, monitored and analyzed to learn about their functional performance (Hardaway, Thomas, and Li 1991; Hardaway, Gunn, and Reynolds 1995; Hardaway and Gunn 1991, 1995, 1998, 1999). The design method employs a three-step procedure that accounts for bimodal annual wave climates (annual and storm wave direction) a numerical wave transformation model for near-shore wave refraction and shoaling, and the beach planform shape model for static equilibrium (Silverster and Hsu 1993). System design also includes upland runoff, bank geology, shoreline morphology, sedimentation, and aesthetics. Potential impacts to adjacent shorelines must also be considered and minimized.

Figure V-3-15a displays before and after photos for the Van Dyke project on the James River. The dark area along the shore is vegetation after the new bank was graded to provide sand for beach nourishment as part of the construction. Figure V-3-15b displays 12 nearshore breakwaters at the Luter project site (James River) one year after construction. Note the use of Y-shaped breakwaters to refine the shape of the planform beach. Moving the breakwater ends further offshore changes the diffraction point to provide the desired planform beach shape. Short breakwaters at both ends pin the downdrift beach. Experience since 1991 indicates that on the Chesapeake Bay, the ratio of Y_g/L_g is about 1.7 for stable, equilibrium-shaped beaches. As illustrated in Figure V-3-16, at the Murphy project site (Potomac River) the present shore is still eroding and will require several years to reach the predicted embayment shoreline shape. Hardaway, Thomas, and Li 1991 present minimum design parameters for medium wave energy shorelines (average fetch 1-5 nautical

VanDyke Project
Pre-Construction
1994

VanDyke Project
Two Years After
Construction
1999

a. Van Dyke Project

Luter Project
One year after construction, May 1999
Approximate photo scale 1 inch=200 ft

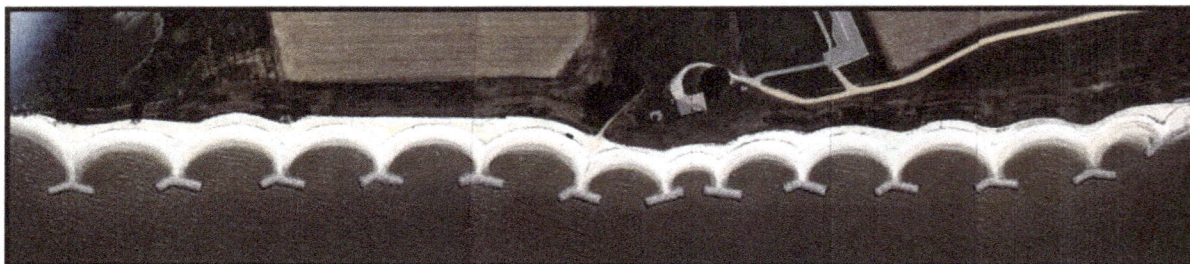

b. Luter Project

Figure V-3-15.   Headland breakwater projects on the James River estuary, Chesapeake Bay (from Hardaway and Gunn 1999)

miles) that include guidelines for distance B (Figure V-3-14) related to storm surge and wave conditions. Much more research is needed to relate Y_{min} to design storm conditions for the functional design of headland breakwater systems. The next section discusses analytical methods to estimate Y_{min} during storm conditions.

d. Nearshore breakwaters.

(1) Background and definitions. Nearshore breakwaters are detached, generally shore-parallel structures that reduce the amount of wave energy reaching a protected area. They are similar to natural bars, reefs or nearshore islands that dissipate wave energy. The reduction in wave energy slows the littoral drift, produces sediment deposition and a shoreline bulge or salient feature in the sheltered area behind the breakwater. Some longshore sediment transport may continue along the coast behind the nearshore breakwater.

(a) Figure V-3-17 displays a salient behind a single breakwater and a multiple breakwater system with both salient and a tombolo when the shoreline is attached to the breakwater. The tombolo may occur naturally or be forced during construction to produce a headland breakwater as discussed in the previous section. The tombolo blocks normal, longshore sediment transport behind the structure. Daily tidal variations may expose a tombolo at low tide while only a salient feature is visible at high tide as occurs at the Winthrop Beach, Massachusetts, nearshore breakwaters constructed in 1935 (Dally and Pope 1986). Figure V-3-18 displays the single 610 m long, rubble-mound breakwater at Santa Monica, California, and salient feature (circa 1967). Periodic dredging is needed to prevent tombolo formation. The multiple, nearshore breakwater system at Presque Isle, Pennsylvania, is shown in Figure V-3-19 (Fall 1992). Fifty-five breakwaters were built in 1989-1992 to protect 8.3 km (13.8 miles) of Lake Erie shoreline (Mohr 1994).

(b) In general, the primary objectives of a nearshore breakwater system are to:

- Increase the fill life (longevity) of a beach-fill project.

- Provide protection to upland areas from storm damage.

- Provide a wide beach for recreation.

- Create or stabilize wetland areas.

(c) In addition, adverse effects on downdrift beaches should be minimized by consideration of the impact on longshore sediment transport.

(d) Numerous variations of breakwater types exist. Here, the focus is on detached, offshore breakwaters not connected to shore by any type of sand-holding structure. They may be low-crested to permit increased wave transmission and lower construction costs. They also may be reef-type breakwaters constructed of homogeneous stone size as opposed to the traditional, multilayer, cross-section design. Headland breakwaters (natural or constructed tombolos) are discussed (Part V-3-3). Another type of shore-parallel, offshore structure is called the submerged sill or perched beach and is discussed in (Part V-3-3). Additional

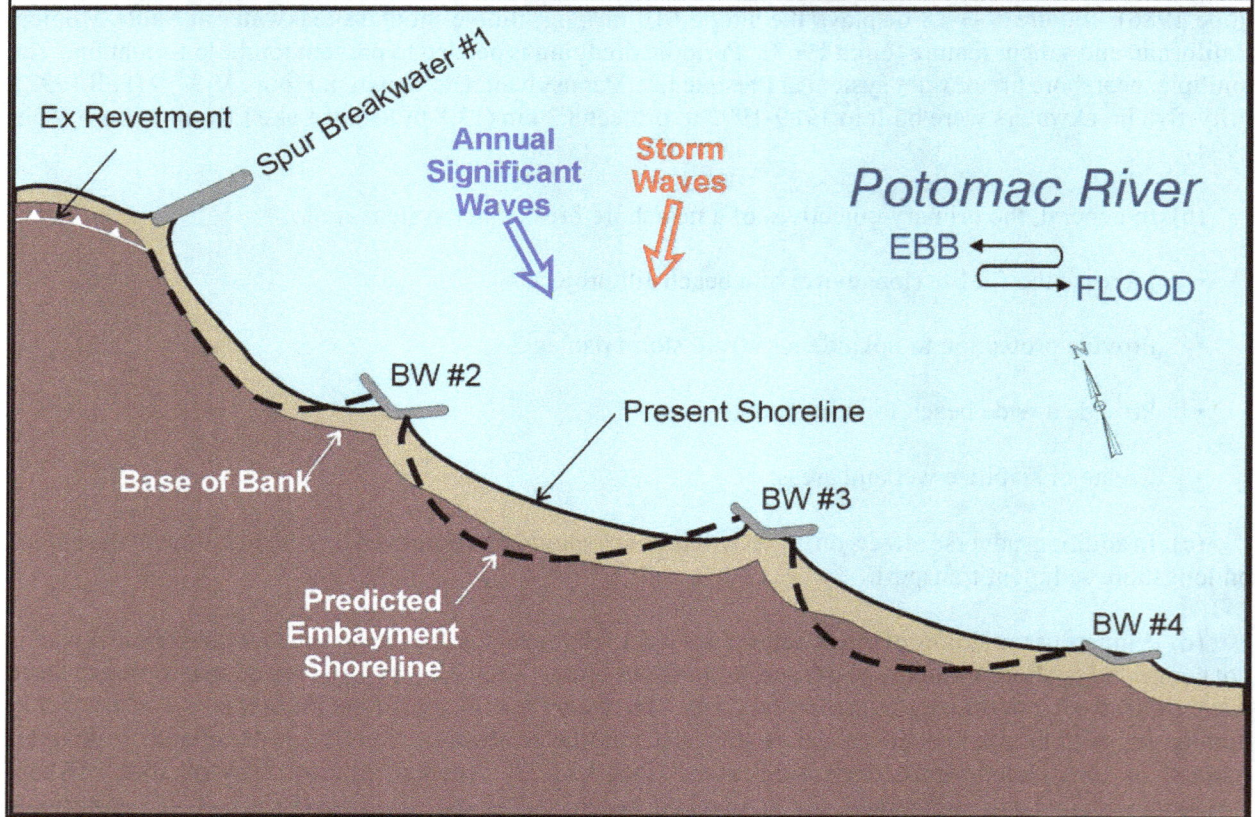

Figure V-3-16.   Headland breakwater project on the Chesapeake Bay (from Hardaway and Gunn 1999)

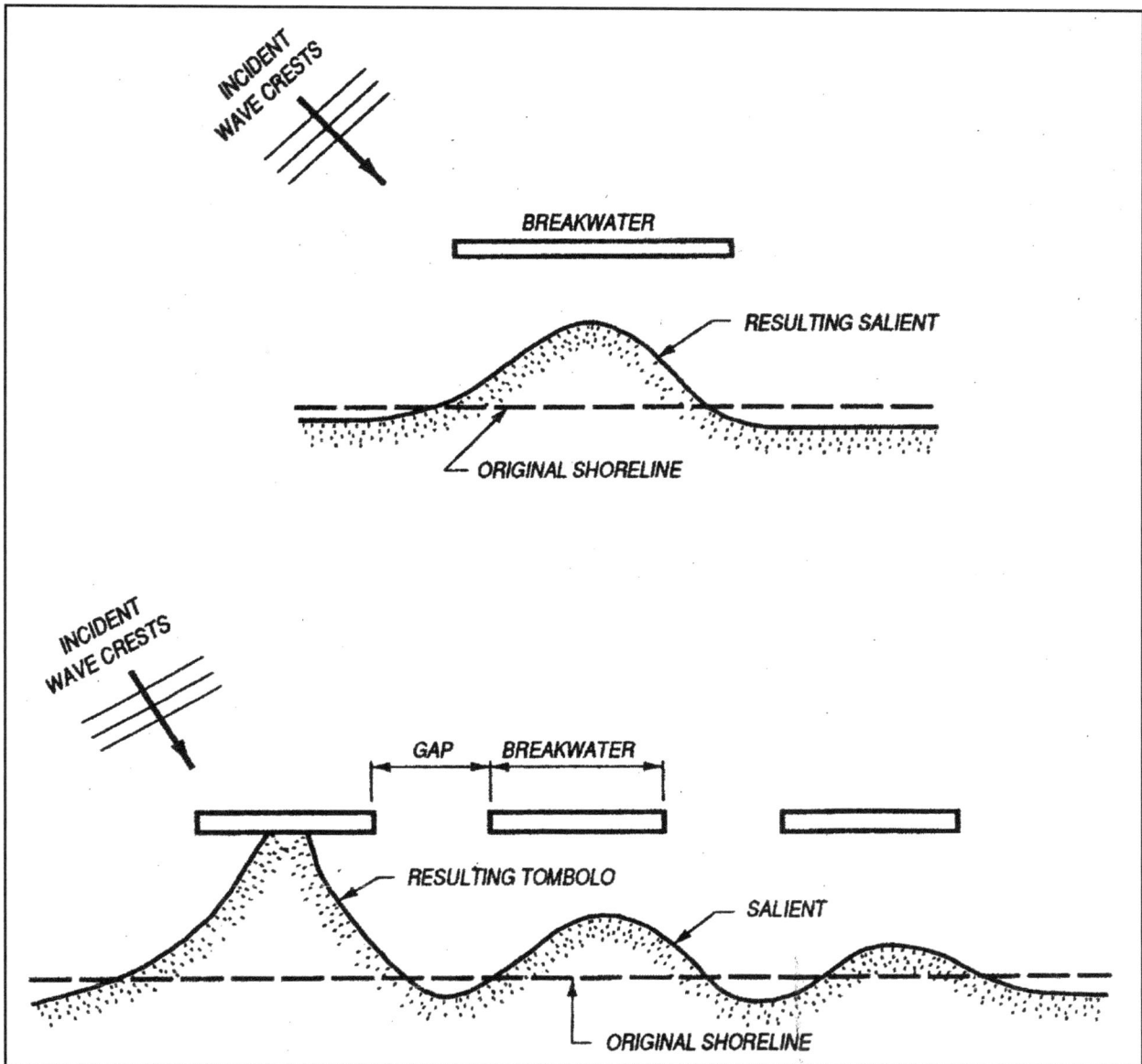

Figure V-3-17. Types of shoreline changes associated with single and multiple breakwater (from EM 1110-2-1617)

information and references on other breakwater classifications can be found in Lesnik (1979), Fulford (1985), Dally and Pope (1986), EM 1110-2-1617 and Chasten et al. (1993).

(2) Physical processes.

(a) Normal morphological responses. Waves breaking at an angle to the shore produce time-averaged, longshore (littoral) currents and longshore sediment transport. Consider the left breakwater in Figure V-3-20 with wave energy in the plus direction. Physical processes at macro-level scales in the vicinity of the breakwater for normal wave and water level conditions are as follows. The breakwater shelters the coast immediately behind the structure and adjacent areas (diffraction) from the incoming waves. Breaking wave heights are smaller in the sheltered areas. The exposed, gap areas have larger breaking wave heights. The

a. Salient with eroded downdrift (longshore drift is from bottom to top)

b. Deteriorated breakwater allowing wave transmission

Figure V-3-18. Santa Monica, California, breakwater and salient (circa 1967). Littoral drift from bottom to top
(from Chasten et al. 1993)

Shore Protection Projects

Figure V-3-19. Breakwater construction and salients; two views of Presque Isle, Pennsylvania (from Mohr 1994)

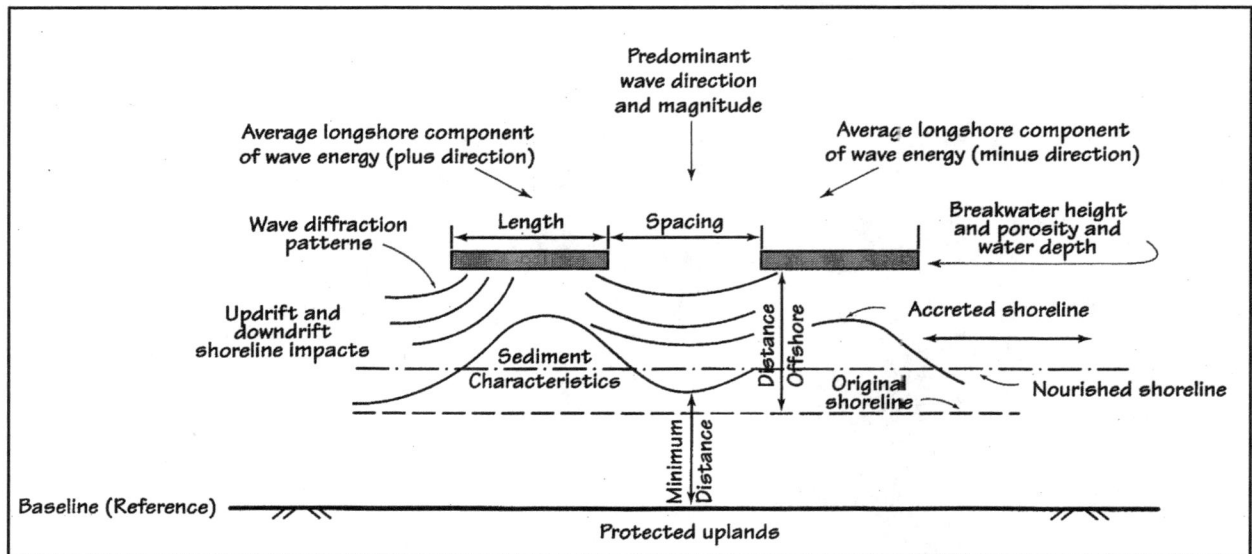

Figure V-3-20. Definition schematic for nearshore breakwaters

wave induced, mean water level change (setup) in the exposed, gap areas is larger than in the sheltered areas. Longshore variability in the wave setup produces gradients in the mean water surface. Water flows from the elevated levels in the gap area towards the lower, sheltered area to accelerate the longshore current flowing towards the sheltered area behind the structure from the left side. These gradients also change the direction of the current which is driven away from the breakwater in the region immediately downdrift of the breakwater (right side). These two current systems (littoral current and setup current) merge behind the structure to give rise to complex circulation patterns. The acceleration of the littoral current updrift causes initial erosion of the beach on the updrift side. The same occurs in the area immediately downdrift. These currents carry the eroded material towards the sheltered area, where it deposits. These mechanisms cause the patterns of deposition behind and erosion on either side that is observed in nature (see Figure V-3-20). The above physical description had been confirmed by a two-dimensional, numerical (horizontal plane) joint processes (waves, currents, sediment transport) morphological modeling system (Zyserman et al. 1998).

(b) Storm processes and response. Protection afforded by the breakwater will limit erosion of the salient during significant storms. The exposed gap area will be eroded with sediment dragged offshore during storms. Breakwater height, length, wave transmission characteristics and distance from shore contribute to its effectiveness to provide a minimum dry beach width, as discussed further in the following paragraphs.

(3) Functional design. Prototype experience for the functional design of nearshore breakwaters in the United States is generally limited to sediment-starved shores with fetch-limited wave climates on the Great Lakes, Chesapeake Bay, and Gulf of Mexico shores (Pope and Dean 1986). Table V-3-4 is a summary of U.S. projects up to 1993 (Chasten et al. 1993). Nearshore breakwaters for shore protection have also been used extensively for shore protection in Japan and Israel (Toyoshima 1982) and in Denmark, Singapore and Spain (Rosati 1990). Detailed summaries of the literature, previous projects, and design guidance are provided in a number of references (Lesnick 1979; Dally and Pope 1986; Pope and Dean 1986; Kraft and Herbich 1989; Pope 1989; Rosati 1990; Rosati and Truitt 1990; EM 1110-2-1617; and Chasten et al. 1993). One key aspect of all these efforts has been to determine under what conditions salients or tombolos will naturally form behind the breakwater.

Table V-3-4
Summary of U.S. Breakwater Projects (from Chasten et al. 1993)

Coast	Project	Location	Date of Construction	Number of Segments	Project Length	Segment Length	Gap Length	Distance Offshore Preproject	Water Depth	Fill Placed	Beach[1] Responses	Constructed By	Maintained By
Atlantic	Winthrop Beach (low tide)	Massachusetts	1935	5	625 m	91 m	30 m	Unknown	3.0 m (mlw)	No	1	State of Mass.	
Atlantic	Winthrop Beach (high tide)	Massachusetts	1935	1		100	30	305	3.0 (mhw)	No	3	State of Mass.	
Atlantic (Potomac River)	Colonial Beach (Central Beach)	Virginia	1982	4	427	61	46	64	1.2	Yes	2	USACE	USACE
Atlantic (Potomac River)	Colonial Beach (Castlewood Park)	Virginia	1982	3	335	61, 93	26, 40	46	1.2	Yes	1	USACE	USACE
Chesapeake Bay	Elm's Beach (wetland)	Maryland	1985	3	335	47	53	44	0.6-0.9	Yes	1	State of Maryland	State of Maryland
Chesapeake Bay	Elk Neck State Park (wetland)	Maryland	1989	4	107	15	15		0.6-0.9	No	2-4	USACE	USACE
Chesapeake Bay	Terrapin Beach (wetland)	Maryland	1989	4		23	15, 31, 23	38.1	0.6-0 9	Yes	5	USACE	USACE
Chesapeake Bay	Eastern Neck (wetland)	Maryland	1992-1993	26	1676	31	23		0.3-0.6	Yes		U.S. Fish and Wildlife Service, USACE	U.S. Fish and Wildlife Service
Chesapeake Bay	Bay Ridge	Maryland	1990-1991	11	686	31	31	42.7		Yes	4	Private	Private
Gulf of Mexico	Redington Shores	Florida	1985-1986	1	100	100	0	104		Yes	1	USACE	USACE
Gulf of Mexico	Holly Beach	Louisiana	1985	6	555	46, 51, 50	93, 89	78, 61	2.5	No	4	State of Louisiana	State of Louisiana
Gulf of Mexico	Holly Beach	Louisiana	1991-1993	76		46, 53	91, 84	122, 183	1.4, 1.6	Yes	3	State of Louisiana	State of Louisiana
Gulf of Mexico	Grand Isle	Louisiana		4	84	70	21	107	2	No	3	City of Grand Isle	City of Grand Isle
Lake Erie	Lakeview Park	Ohio	1977	3	403	76	49	152	3.7	Yes	4	USACE	City of Lorain
Lake Erie	Presque Isle	Pennsylvania	1978	3	440	38	61, 91	60	0.9-1.2	Yes	2	USACE	USACE
Lake Erie	Presque Isle	Pennsylvania	1989-1992	55	8300	45	107	76-107	1.5-2.4 (lwd)	Yes	3-4	USACE	USACE
Lake Erie	Lakeshore Park	Ohio	1982	3	244	38	61	120	2.1	Yes	5	USACE	City of Ashtabula
Lake Erie	East Harbor	Ohio	1983	4	550	45	90, 105, 120	170	2.3	No	5	State of Ohio	State of Ohio
Lake Erie	Meumee Bay (headland)	Ohio	1990	5	823	61	76		1.3	Yes	1	USACE	State of Ohio
Lake Erie	Sims Park (headland)	Ohio	1992	3	975	38	49		2.5	Yes	1	USACE	City of Euclid
Pacific	Venice	California	1905	1	180	180	0	370		No	5	Private	Private
Pacific	Haleiwa Beach	Hawaii	1965	1	49	49	0	90	2.1 (msl)	Yes	3	USACE/State of HI	USACE
Pacific	Sand Island	Hawaii	1991	3	110	21	23					USACE	USACE

[1] Beach response is coded as follows: 1-permanent tombolos, 2-periodic tombolos, 3-well developed salients, 4-subdued salients, 5-no sinuosity.

(a) Salients or tombolos. A salient is the preferred shoreline response for a detached breakwater system designed for the Corps as stated in EM 1110-2-1617 and in Chasten et al. (1993). This is to allow longshore sediment transport to continue to move through the project area to downdrift beaches. Salients are likely to predominate when the breakwaters are sufficiently far from shore, short relative to incident wavelength, and relatively transmissible (low crested or large gaps with low sediment input). Wave action and longshore currents tend to keep the salient from connecting to the structure.

- Sand will more likely accumulate in the structure lee and form a tombolo when the breakwater is close to shore, is long relative to incident wavelength, and is relatively impermeable (high crest and small gaps and with large sediment input). A tombolo-detached breakwater functions like a tee-shaped groin by blocking longshore transport and promoting sediment movements offshore in rip currents through the gaps. Although some longshore transport can occur seaward of the breakwater, the interruption in the littoral system may starve downdrift beaches of their normal sediment supply, causing erosion. Variable wave energy regimes may produce periodic tombolos to temporarily store and then release sediment to the downdrift region.

- Salient formation provides a recreational swimming environment and limits access for maintenance and to the public. Tombolo formation provides a recreational beach environment and allows direct access to the structure for maintenance, but public access may not be desirable.

- Figure V-3-21 presents a definition sketch for the variables that have been employed to develop empirical relationships for detached, breakwater design by many research efforts listed in Table V-3-5 dating back to the 1960s. For salient or tombolo formation, the key variables are:

Y   Distance of breakwater from nourished shoreline

L_s Length of breakwater structure

L_g Gap distance between adjacent breakwater segments

d_s Depth (average) at breakwater structure below mean water level

Three dimensionless ratios, Y/d_s, L_s/L_g and L_s/Y have emerged to separate salient and tombolo response.

- As qualitatively discussed, when the breakwater is long and/or located close to shore, conditions favor tombolo formation. As shown in Table V-3-6, "Conditions for the Formation of Tombolos," many references say $L_s/Y > 1$-2 for tombolo formation (except Gourlay 1981). Dally and Pope (1986) recommend

$$\frac{L_s}{Y} = 1.5 \text{ to } 2 \qquad \text{single breakwater} \qquad \text{(V-3-6)}$$

$$\frac{L_s}{Y} = 1.5 \qquad\qquad L \le L_g \le L_s \text{ segmented breakwater} \qquad \text{(V-3-7)}$$

where L is the wavelength at the structure.

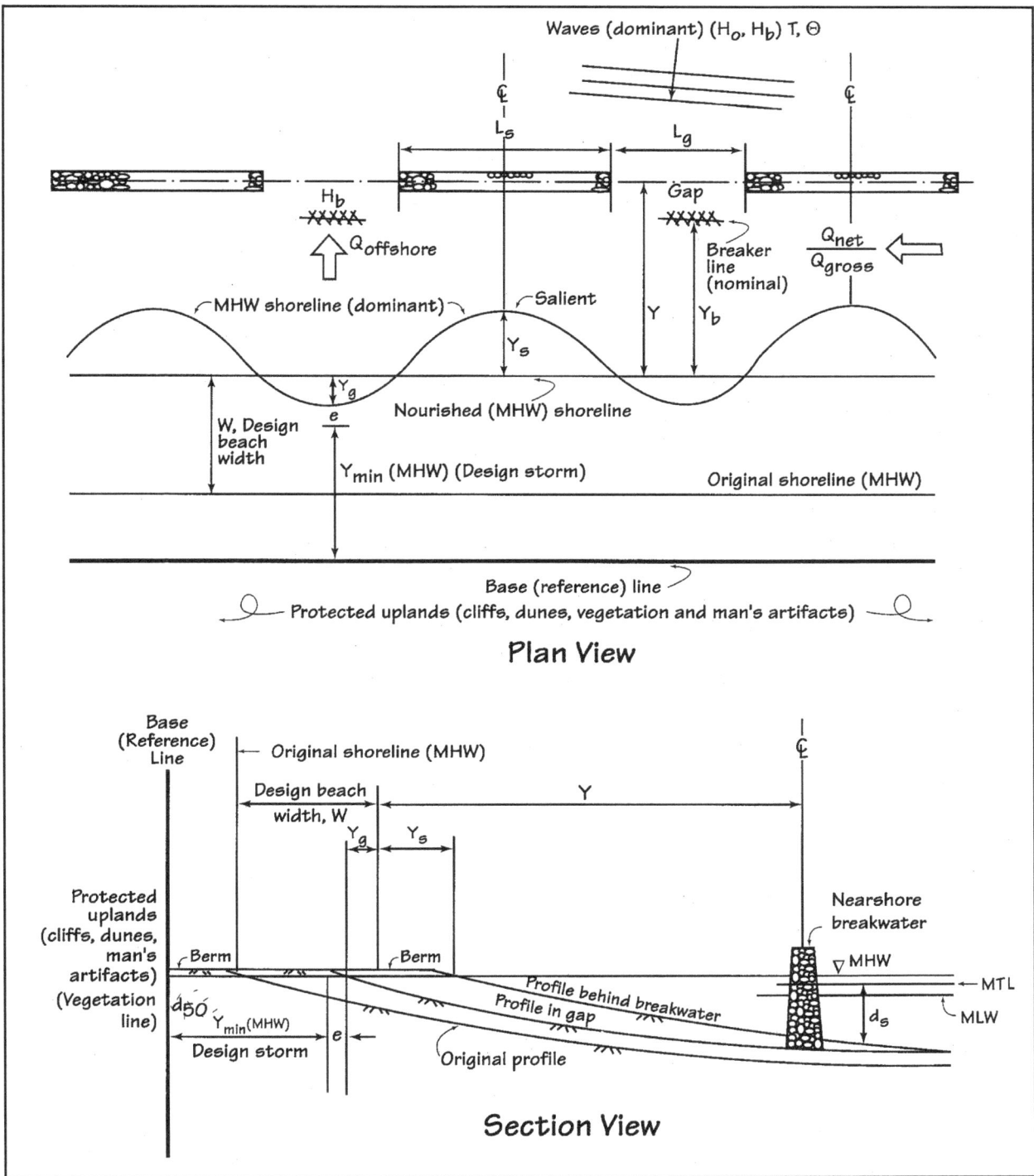

Figure V-3-21.   Definitions of key variables for nearshore breakwater

Table V-3-5
Empirical Relationships for Nearshore Breakwater Design

Inman and Frautschy (1966)	Predicts accretion condition; based on beach response at Venice in Santa Monica, CA
Toyoshima (1972, 1974)	Recommends design guidance based on prototype performance of 86 breakwater systems along the Japanese coast
Noble (1978)	Predicts coastal impact of structures in terms of offshore distance and length; based on California prototype breakwaters
Walker, Clark, and Pope (1980)	Discusses method used to design the Lakeview Park, Lorain, OH, segmented system for salient formation; develops the Diffraction Energy Method based on diffraction coefficient isolines for representative waves from predominant directions
Gourlay (1981)	Predicts beach response; based on physical model and field observations
Nir (1982)	Predicts accretion condition; based on performance of 12 Israeli breakwaters
Rosen and Vadja (1982)	Graphically presents relationships to predict equilbrium salient and tombolo size; based on physical model/prototype data
Hallermeier (1983)	Develops relationships for depth limit of sediment transport and prevention of tombolo formation; based on field/laboratory data
Noda (1984)	Evaluates physical parameters controlling development of tombolos/salients; especially due to on-offshore transport; based on laboratory experiments
Shore Protection Manual (1984)	Presents limits of tombolo formation from structure length and distance offshore; based on the pattern of diffracting wave crests in the lee of a breakwater
Dally and Pope (1986)	Recommends limits of structure-distance ratio based on type of shoreline advance desired and length of beach to be protected
Harris and Herbich (1986)	Presents relationship for average quantity of sand deposited in lee and gap areas; based on laboratory tests
Japanese Ministry of Construction (1986); Rosati and Truitt (1990)	Develops step-by-step iterative procedure, providing specific guidelines towards final design; tends to result in tombolo formation; based on Japanese breakwaters
Pope and Dean (1986)	Presents bounds of beach response based on prototype performance; response given as a function of segment length-to-gap ratio and effective distance offshore-to-depth at structure ratio; provides beach response index classification
Seiji, Uda, and Tanaka (1987)	Predicts gap erosion; based on performance of 1,500 Japanese breakwaters
Sonu and Warwar (1987)	Presents relationship for tombolo growth at the Santa Monica, CA breakwater
Suh and Dalrymple (1987)	Gives relationship for salient length given structure length and surf zone location; based on lab tests and prototype data
Berenguer and Enriquez (1988)	Presents various relationships for pocket beaches including gap erosion and maximum stable surface area (i.e., beach fill); based on projects along the Spanish coast
Ahrens and Cox (1990)	Uses Pope and Dean (1986) to develop a relationship for expected morphological response as function of segment-to-gap ratio

(from Chasten et al. 1993)

- Conversely, short breakwaters at greater distance from shore favor salient formation. The "Conditions for the Formation of Salients" in Table V-3-6 presents a wide range of conditions with generally, $L_s/Y < 1$. Dally and Pope (1986) recommend for salient formation

$$\frac{L_s}{Y} = 0.5 \text{ to } 0.67 \qquad\qquad (V\text{-}3\text{-}8)$$

for both single and segmented breakwaters. However, for very long distances, to insure that tombolos do not form, the recommended ratio for a segmented system (Dally and Pope 1986) is

$$\frac{L_s}{Y} = 0.125 \text{ (long systems)} \qquad\qquad (V\text{-}3\text{-}9)$$

- Permeable structure systems (partly submerged, large gaps) also allow sufficient wave energy to minimize the chance for tombolo formation: As shown in Table V-3-6, "Conditions for Minimal Shoreline Response," the references cited generally recommend $L_s/Y < (0.125 - 0.33)$ to produce a minimal shoreline response. Ahrens and Cox (1990) defined a beach response index, I_s

$$I_s = \exp (1.72 - 0.41 \, L_s/Y) \qquad\qquad (V\text{-}3\text{-}10)$$

where the five types of beach response (Pope and Dean 1986) give I_s value as:

Permanent tombolo formation, $I_s=1$

Periodic tombolos, $I_s=2$

Well-developed salients, $I_s=3$

Subdued salients, $I_s=4$

No sinuosity, $I_s=5$

- These results are preliminary and require verification.

- The ratio L_s/L_g is also important for salient or tombolo formation. Large gaps will let more wave energy reach the shore to promote salient formation. And, this will coincide with smaller L_s/L_g ratios. A dimensionless plot of U.S. segmented, nearshore breakwater projects by Pope and Dean (1986) using L_s/L_g is shown in Figure V-3-22 and verifies this trend. The vertical axis is Y/d_s and is the distance offshore relative to the local, mean water depth at the breakwater. The water depth is important for it is related to the nominal, surf zone width at breaking Y_b. The dimensionless ratio Y/Y_b is a measure of breaker location relative to the width of the surf zone. For a given L_s/L_g, larger Y/d_s values (or Y/Y_b) mean the breakwater is located further offshore (beyond the normal surf zone width) to foster salient formation. Obviously, breakwaters located far offshore will have less effection the shoreline.

Table V-3-6
Conditions for Shoreline Response Behind Nearshore Breakwaters (from Chasten et al. 1993)

Conditions for the Formation of Tombolos		
Condition	**Comments**	**Reference**
$L_s/Y > 2.0$		*Shore Protection Manual* (1984)
$L_s/Y > 2.0$	Double tombolo	Gourlay (1981)
$L_s/Y > 0.67$ to 1.0	Tombolo (shallow water)	Gourlay (1981)
$L_s/Y > 2.5$	Periodic tombolo	Ahrens and Cox (1990)
$L_s/Y > 1.5$ to 2.0	Tombolo	Dally and Pope (1986)
$L_s/Y > 1.5$	Tombolo (multiple breakwaters)	Dally and Pope (1986)
$L_s/Y > 1.0$	Tombolo (single breakwaters)	Suh and Dalrymple (1987)
$L_s/Y > 2\, b/L_s$	Tombolo (multiple breakwaters)	Suh and Dalrymple (1987)
Conditions for the Formation of Salients		
$L_s/Y < 1.0$	No tombolo	*Shore Protection Manual* (1984)
$L_s/Y < 0.4$ to 0.5	Salient	Gourlay (1981)
$L_s/Y = 0.5$ to 0.67	Salient	Dally and Pope (1986)
$L_s/Y < 1.0$	No tombolo (single breakwater)	Suh and Dalrymple (1987)
$L_s/Y < 2\, b/L_s$	No tombolo (multiple breakwater)	Suh and Dalrymple (1987)
$L_s/Y < 1.5$	Well-developed salient	Ahrens and Cox (1990)
$L_s/Y < 0.8$ to 1.5	Subdued salient	Ahrens and Cox (1990)
Conditions for Minimal Shoreline Response		
$L_s/Y \leq 0.17$ to 0.33	No response	Inman and Frautschy (1966)
$L_s/Y \leq 0.27$	No sinuosity	Ahrens and Cox (1990)
$L_s/Y \leq 0.5$	No deposition	Nir (1982)
$L_s/Y \leq 0.125$	Uniform protection	Dally and Pope (1986)
$L_s/Y \leq 0.17$	Minimal impact	Noble (1978)

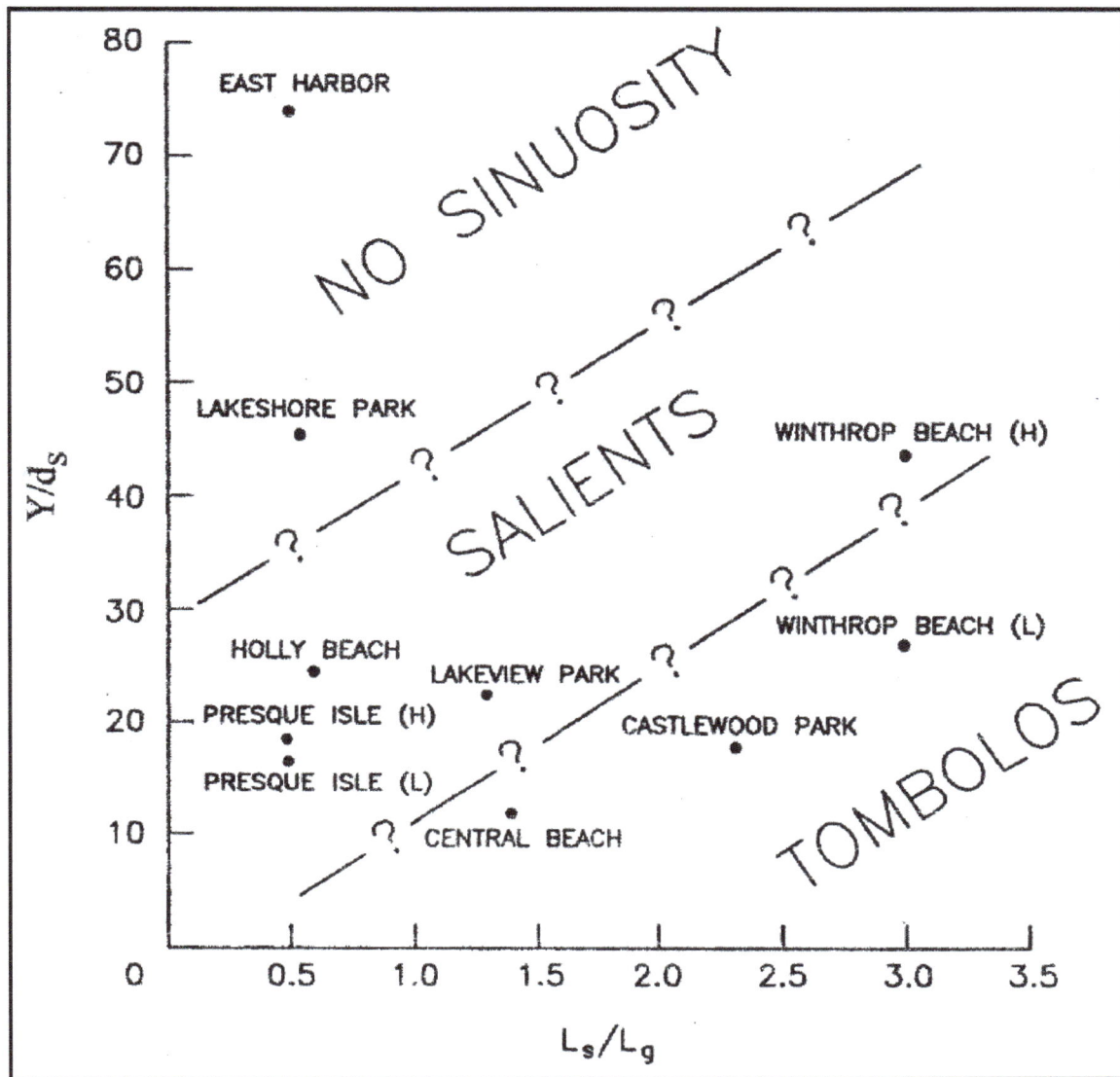

Figure V-3-22. Dimensionless plot of nearshore breakwater projects For Y/d$_s$ versus L$_s$/L$_g$ (from Pope and Dean 1986)

- Zyserman et al. (1998) used the Y/Y$_b$ ratio in their numerical model, process-oriented studies of nearshore breakwaters. The distance Y$_b$ was taken as Y$_{80}$, meaning the distance from shore where 80 percent of the undisturbed littoral transport takes place and, hence, a measure of the width of the surf zone. Their research using a numerical model confirmed the empirical formulas of Dally and Pope (1986) for tombolo and salient formation previously summarized.

(b) Planform configuration. Some research has provided insight into other variables that quantify the planform shape of the shoreline as shown in Figure V-3-21. The length of the salient, Y$_s$, increased as the L$_s$/Y ratio increased, as expected. Suh and Dalrymple (1987) developed an exponential expression involving $L_g Y L_s^2$, where L$_g$ is the gap length for prediction of Y$_s$, by combining movable-bed laboratory results and prototype data. Rosati (1990) evaluated the relation and found it over-predicted Y$_s$ for the majority of prototype data. Seiji, Uda, and Tanaka (1987) gave conditions on the L_g/Y ratio for no, possible, and certain erosion, Y_g opposite the gap. The magnitude of Y_g was not determined, but gap erosion occurred for L_g/Y greater than 0.8. Hallermeier (1981) gave an equation for the water depth to locate nearshore breakwaters when tombolo formation was to be avoided. The relation requires knowledge of wave height and period statistics at the 12 hr per year exceedance level. Rosati (1990) also evaluated Hallermeier's relation with the limited, available field data. The correlation was said to be good for seven of the nine data points tested. These relationships to quantify Y_s, Y_g and d_s can be found in Chasten et al. (1993).

- Walker, Clark, and Pope (1981) discuss a procedure to apply diffraction analysis to determine the approximate shoreline configuration behind a breakwater. Their studies indicate that if the isolines of the $K' = 0.3$ diffraction coefficients are constructed from each end of the breakwater for a range of incident wave directions (monochromatic waves) and they intersect seaward of the postproject shoreline, a tombolo will not form. This corresponds to $L_g/Y \geq 2$, where Y is after placement of the beach fill, as part of the project shown in Figure V-3-21. Waves coming around each end of the breakwater meet each other before the undiffracted, incident wave (outside the breakwater's shadow) reach the shoreline. The postproject shoreline is estimated as a smoothed crest pattern for all diffracted crests and a balance in the sediment volume.

- The Japanese Ministry of Construction presented a step-by-step interactive procedure for nearshore breakwater design (Japanese Ministry of Construction 1986). Rosati (1990) and Rosati and Truitt (1990) found that 60 percent of the designs produced tombolos and therefore the JMC method was more suitable for headland breakwater design. All are for nonpermeable, high-crested, nearshore breakwaters.

(c) Other design factors. The crest elevation and crest width, permeability, slope of front face, and type of construction are additional design factors that influence functional performance. No general guidelines presently exist.

- Generally, low crests allow more energy to penetrate into the lee of the breakwater to prevent tombolo formation or remove a tombolo by storm waves. Wide crests on low breakwaters can promote breaking to diminish wave energy penetration and encourage tombolo formation. Permeable structures can allow significant amounts of energy to propagate through them to prevent tombolo formation. Types of construction including nontraditional, patented devices are discussed in Part V-3-5.

- Waves in the lee of the breakwater are determined by three processes: diffraction around the ends, wave transmission by overtopping and wave transmission through the structure. For diffraction around single and multiple breakwaters with gaps, see Part II-7-2 for irregular waves and the *Shore Protection Manual* (1984) for many cases with monochromatic waves. Wave transmission due to overtopping and through the structure by permeability is discussed in Part VI-5-2. Wave reflection is covered in Part VI-5-2. The Automated Coastal Engineering Systems (ACES) (Leenknecht, Szuawalski, and Sherlock 1992) provides an application to determine wave transmission coefficients and transmitted wave heights for permeable breakwaters with crest elevation at or above the still-water level.

- Breakwater impact on littoral currents and creation of wave setup gradients to produce setup currents was previously discussed. If the crest elevation is low enough to permit wave overtopping, the mass carried over the structure causes a net seaward return flow of water through the gaps. Seelig and Walton (1980) present a method for estimating the strength of the seaward flowing currents. The effect of the combined littoral and setup currents on the longshore sediment transport to produce salient features with adjacent erosional areas was also previously discussed (physical processes).

• The primary function of nearshore breakwaters is to reduce the offshore sand transport during storms. Hence, these structures help retain sand on nourished beaches for longer intervals. However, overtopping can result in a net seaward flow of water and sand through the gaps between breakwater segments during storm events. The breakwater can also reduce the onshore sediment movement during normal, swell wave conditions that naturally rebuild the beach. The structure blocks the return of sediment to the beach. Following breakwater construction, a new equilibrium between onshore and offshore transport will be established.

(d) Minimum dry beach width, Y_{min}. Figure V-3-21 schematically displays the mhw shoreline for normal wave conditions. During storms, some gap erosion, e will occur to impact the minimum, dry beach width, Y_{min} required for shore protection.

• A conservative estimate of the gap erosion, e, can be made using analytical models for the dynamic response of natural beach profiles to storm effects (e.g., Kobayashi 1987; Kriebel and Dean 1993). Part III-3-3 gives a general description of these methods and also example applications of the Kriebel and Dean (1993) analytical model. The theory is for open coastal beaches so that wave diffraction in the gap area and wave overtopping to increase the return flows through the gap region are not considered, but tend to offset each other. See Example Problem V-3-1.

• Dynamic, numerical, cross-shore sediment transport models (e.g., SBEACH, Larson and Kraus 1989) could also be applied in the gap area to estimate the erosion potential, e (see Part III-3-3). These two-dimensional (vertical) models also do not consider wave diffraction and return flows in the gap area. These results would be a worst case scenario and the actual erosion can be expected to be significantly less. A general, three-dimensional, wave, current, and sediment transport model is clearly needed in this area.

• Hughes (1994) presents a complete discussion of scaling laws as applied to predicting cross-shore sediment transport in physical, hydraulic models.

(4) Nontraditional designs. Most nearshore breakwaters built in the United States and foreign countries for shore protection have been rubble-mound type structures. Availability of materials and construction equipment have made construction costs relatively inexpensive.

• Several patented, nontraditional devices have been tested in the United States. These have been precast concrete units or sand-filled geotextile tubes and bags. If constructed to the same dimensions as rubble-mound structures, they may produce similar functional performance. Their success (or failure) has been a function of structural stability of the units during storm conditions and their durability over an economic life. Their functional success (or failure) has also been dependent on maintaining the design crest elevation for wave energy reduction. Proper attention to foundation design to minimize settlement must be given for precast, concrete units.

• Some nontraditional designs reduce the bottom footprint to minimize impacts on benthic organisms. And, the costs for removal and/or adjustments to reduce downdrift impacts on adjacent shorelines can be significantly less than for traditional, rubble-mound designs. The need to reduce impact on the environment is increasing the necessity for further research and comprehensive field testing programs for nontraditional designs. (See Part V-3-5 for further details.)

EXAMPLE PROBLEM V-3-1

FIND: Functional design of a nearshore breakwater system coupled with beach nourishment for shore protection, but maintain recreational beach.

GIVEN: A sandy beach shoreline, 685 m in length with historic erosion rate of 0.6-0.9 m/year that threatens a 2.4-7.3-m bank, with road and sewer line parallel to the shoreline along the top of the bank. For an annual probability of exceedance of 2 percent (50-year recurrence interval design storm) the storm surge level is 1.83 m (mlw) and corresponding wave characteristics are H_s=2.2 m, T_p=9.7 sec at the -9.0 m (mwl) depth contour. The design, still water, storm depth is 3.05 m at this location and includes a 0.3 m, astronomical tide. Net sediment transport (sand, d_{50}= 0.6 mm) is 3,750 to 7,500 m^3/year to the north. The existing beach berm is at +0.76 m (mlw) elevation.

Solution:
1.   To maintain the longshore transport rate, the desired planform is subdued to well-developed salients. From a table of possible L_s and Y values, the ratio 0.75 is selected giving a breakwater length, L_s=30m and the offshore distance, Y=40 m. This gives the beach response index, I_s=4.1, hence, subdued salients are expected.

2.   For shore protection, waves entering the breakwater gaps and diffracting behind the structures will reach the shoreline. Sufficient beach width (Y_{min}) and berm height (Z_s) are required to dissipate this wave energy prior to reaching the toe of the banks. Analysis of nearshore, wave diffraction diagrams indicate that the 50-year design wave height of 2.2 m will be reduced to about 0.9 m at a distance of about 14 m from the bank toe when the breakwater gap is about 30 m. This gap width, L_g=30 m is considered a practical minimum width for this project. For shallow-water wave breaking, (γ_b=0.78, II-4-3), this wave will break in about 1.16 m depth of water. For a design storm surge of 1.83 m (mlw) and existing beach berm elevation of 0.76 m (mlw), these waves will break directly on the bank toe and cause significant erosion. The existing beach width and height are not sufficient to dissipate the storm wave energy at the 50-year frequency level. Two options exist. One is to further decrease the wave energy propagating through the gaps by using smaller gap widths, and resulting longer lengths of breakwater segments. The second option is to add beach fill to the shoreline area and this option is selected to provide the desired protection for the bank area.

3.   A beach-fill plan was considered that would increase the beach width to 10 m from the toe of the bank and raise the berm elevation to + 1.8 m (mlw) at a 1:8 construction slope. Wave heights are now reduced to less than 30 cm near the bank toe for the 50-year storm event. The beach fill would be expected to evolve (see Part V-4) to a stable planform with salients behind each breakwater and embayments opposite each gap. The mhw shoreline would recede about 4.5-6 m opposite the gaps as estimated from analysis of diffraction patterns. The beach shape would also evolve to a more natural and milder slope to match those in the area (1:10 to 1:15).

(Continued)

EXAMPLE PROBLEM V-3-1 (Concluded)

4. Protection of the bank toe depends on the performance of the beach berm during storm events. A worst case scenario can be evaluated first by considering the adjusted beach slope and shoreline area with storm wave conditions prior to reduction by the nearshore breakwaters. Using the analytical model of Kriebel and Dean (1993) or a numerical, storm erosion model (e.g., SBEACH, Larson and Kraus 1989) gives the results (for the worst-case, 50-year storm event scenario) that the storm berm would remain with a width of 1.5-3 m. The actual erosion would be expected to be significantly less due to reduction in the storm wave energy as a result of wave diffraction through the gaps. The gap width, L_g=30 m was selected for design.

5. Wave energy is also transmitted over the top of nearshore breakwaters during elevated water level, storm surge events. A wave transmission model (see Part VI-5-2) capable of predicting wave energy over and through submerged, reef-type breakwaters was used for analysis with crest heights of + 1.2 m, +1.5 m and 1.8 m above mlw datum. During the 50-year design storm, wave heights immediately behind the breakwaters are reduced about 60 percent, 54 percent, and 46 percent, respectively for these three breakwater crest elevations. These transmitted waves then propagate shoreward and are further dissipated by the beach salients. With the proposed beach fill in place, a breakwater crest elevation of +1.2 m (mlw) is selected to limit the transmitted, design wave height to about 1.2 m. This is the same, diffracted, design wave height opposite the gaps and is dissipated by the storm berm.

Note: Further details of this example problem are found in Appendix A of Chasten et al. 1993 and was developed by Mr. Ed Fulford of Andrews Miller and Assoc., Inc., Cambridge, Maryland. It is taken from a real project on the Chesapeake Bay for the community of Bay Ridge near Annapolis, Maryland. Construction was completed in July 1991 and postconstruction monitoring commenced soon after and at a November 1991 survey. As of 2000, the project has performed as expected with subdued salients forming behind each breakwater resulting in the overall stability of the shoreline. Numerous, significant storms occurred and the project prevented erosion of the bank area to protect the roadway and sewer pipeline. The project has been well-received by the residents of the community as a result of the stability of the shoreline and the enhancement of the recreational beach area.

e. Groins.

(1) Background and definitions. Groins are the oldest and most common shore-connected, beach stabilization structure. They are probably the most misused and improperly designed of all coastal structures. They are usually perpendicular or nearly at right angles to the shoreline and relatively short when compared to navigation jetties at tidal inlets. As illustrated schematically in Figure V-3-23, for single and multiple groins (groin field) the shoreline adjusts to the presence of the obstruction in longshore sediment transport. Over the course of some time interval, accretion causes a positive increase in beach width updrift of the groin. Conservation of sand mass therefore produces erosion and a decrease in beach width on the downdrift side of the groin. The planform pattern of shoreline adjustment over 1 year is a good indicator of the direction of the annual net longshore transport of sediment at that location.

(a) Groins are constructed to maintain a minimum, dry beach width for storm damage reduction (Figure V-3-11) or to control the amount of sand moving alongshore. Previously stated purposes such as trapping littoral drift are discouraged for this implies removal of sand from the system. Modern coastal

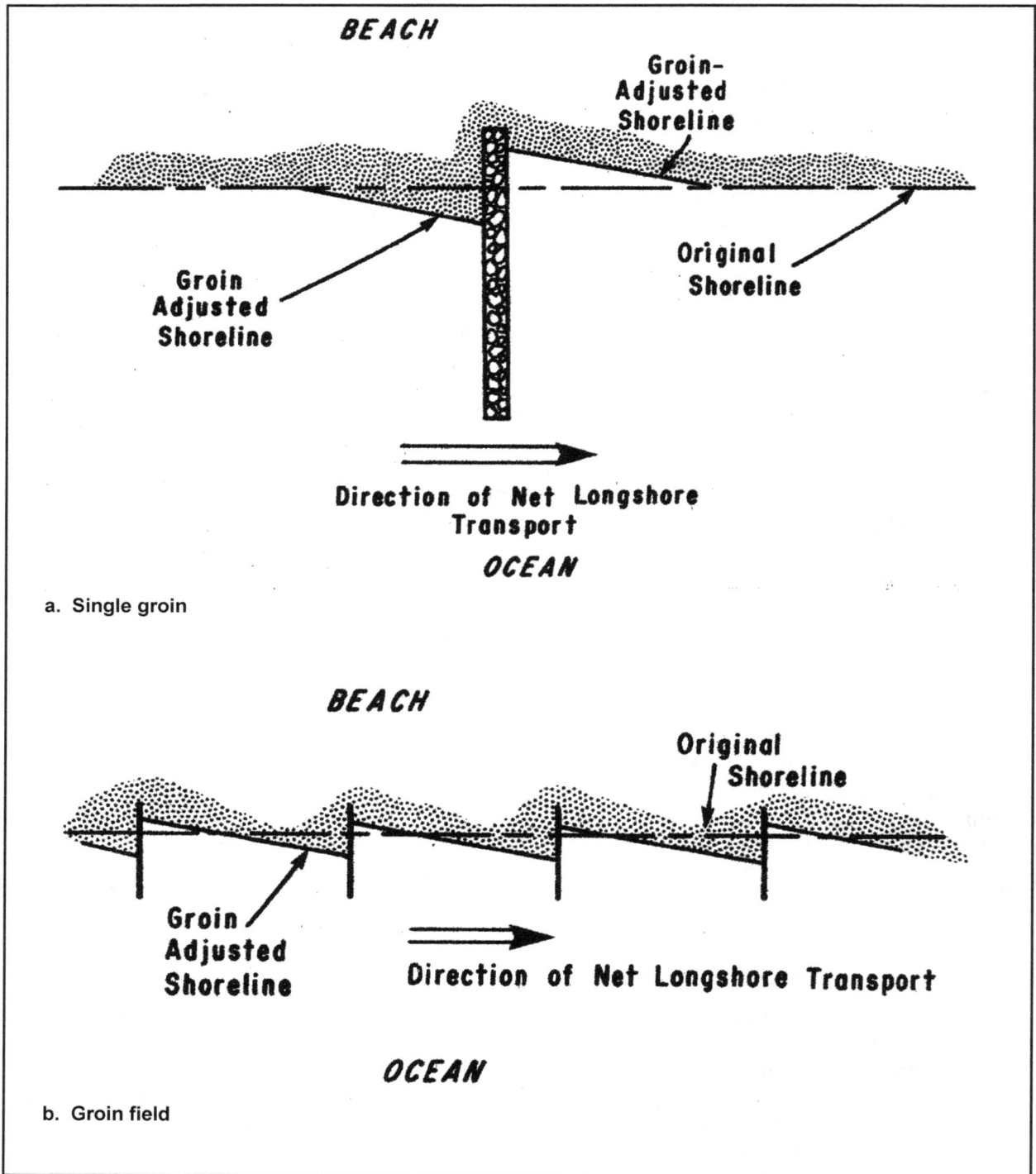

Figure V-3-23.   General shoreline adjustment for direction of net longshore transport

engineering practice is to combine beach nourishment with groin construction to permit sand to immediately begin to bypass the groin field. At the end of the sediment cell, terminal groins may be used to anchor the beach and limit the movement of sand into a navigational channel or onto an ebb-tidal shoal at tidal inlets.

(b) Figure V-3-24 shows a photo, profile and cross section of a rubble-mound groin at Westhampton Beach, New York. Sheet-pile construction with timber (Figure V-3-25) timber-steel (Figure V-3-26) prestressed-concrete (Figure V-3-27) or cellular-steel (filled) sheets (Figure V-3-28) have also been constructed in the United States.

(c) Kraus, Hanson, and Blomgren 1994 cite the following situations when the groin field alternative for shore protection and sand management should be considered.

• At divergent, nodal points for littoral drift.

• In the diffraction, shadow some of a harbor breakwater, or jetty.

• On the downdrift side of a harbor breakwater or jetty.

• At the updrift side of an inlet entrance where intruding sand is to be managed

• To reduce the loss of beach fill, but provide material to downdrift beaches in a controlled manner.

• Along the banks at inlets, where tidal currents alongshore are strong.

• Along an entire littoral cell (spit, barrier island, submarine canyon) where sand is lost without return in an engineering time frame.

(d) Groins may not function well and should not be considered under the following conditions (Kraus, Hanson, and Blomgren 1994):

• Where a large tidal range permits too much bypassing at low tide and overpassing at high tide.

• Where cross-shore sediment transport is dominant

• When constructed too long or impermeable, causing sand to be jetted seaward

• When strong rip currents are created to cause potentially dangerous swimming conditions

(e) Coastal zone management policy in many countries and the United States presently discourages the use of groins for shore protection. Many examples of poorly designed and improperly sited groins caused by lack of understanding of their functional design, or failure to implement the correct construction sequence, or failure to fill up the groin compartments with sand during construction, or improper cross-sectional shape are responsible for these restrictions. However, when properly designed, constructed and combined with beach nourishment, groins can function effectively under certain conditions, particularly for increasing the fill life (longevity) of renourished beaches.

(f) Groins are now being reevaluated as sand-retention structures (Kraus, Hanson, and Blomgren 1994; Kraus and Bocamazo 2000) by now asking the question "how much sand can be allowed to pass," while still maintaining a minimum width of beach at the groin for some level of shore protection.

Westhampton Beach, Long Island, New York, 18 Jan 1980 (courtesy USAED, New York)

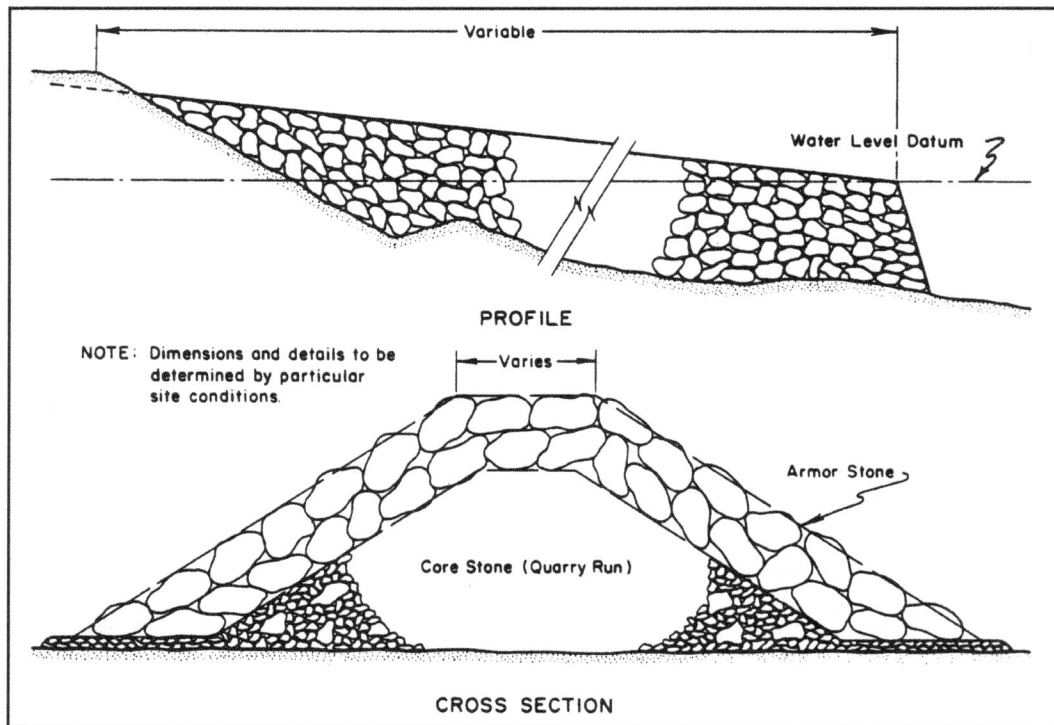

Figure V-3-24. Rubble-mound groin, Westhampton Beach, New York

Wallops Island, Virginia (1964)

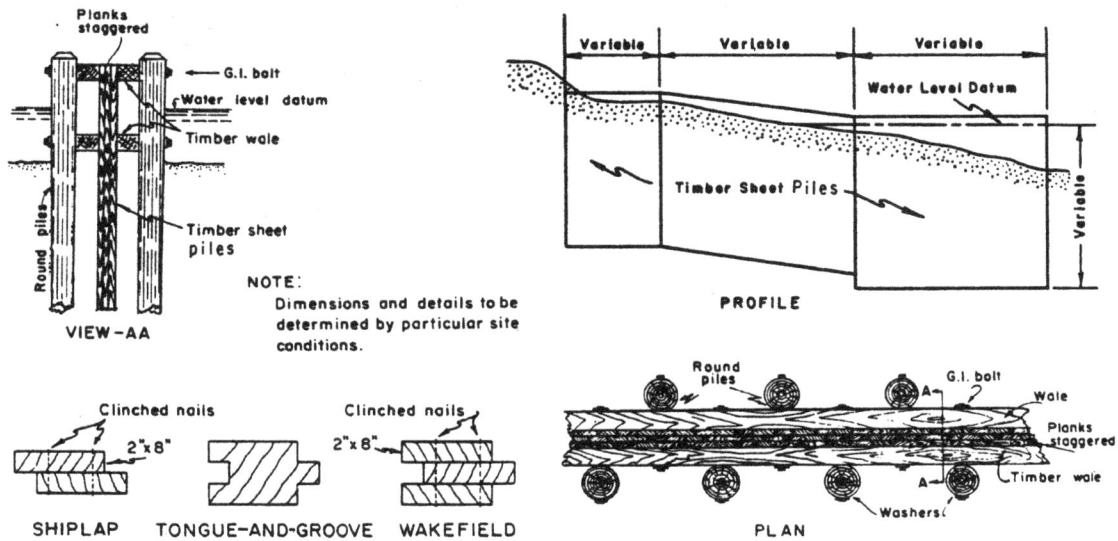

Figure V-3-25.   Timber sheet-pile groin, NASA Space Flight Center, Wallops Island, Virginia

New Jersey (September 1962)

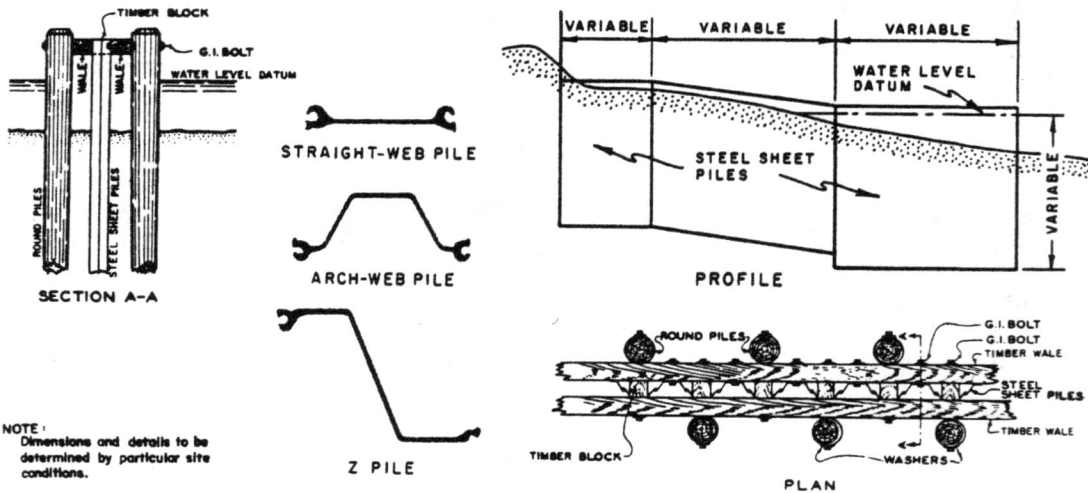

SECTION A-A

STRAIGHT-WEB PILE

ARCH-WEB PILE

Z PILE

NOTE:
Dimensions and details to be
determined by particular site
conditions.

PROFILE

PLAN

Figure V-3-26. Timber-steel sheet-pile groin, New Jersey

Doheny Beach State Park, California (October 1965)

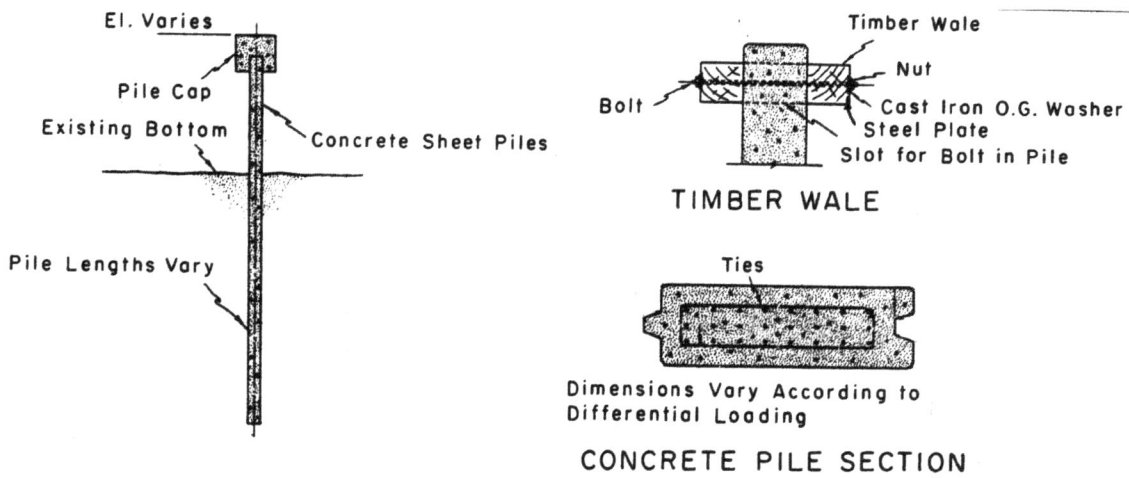

El. Varies

Pile Cap

Existing Bottom — Concrete Sheet Piles

Pile Lengths Vary

Timber Wale

Nut

Bolt

Cast Iron O.G. Washer
Steel Plate
Slot for Bolt in Pile

TIMBER WALE

Ties

Dimensions Vary According to
Differential Loading

CONCRETE PILE SECTION

Figure V-3-27. Prestressed-concrete sheet-pile groin, Dohney Beach State Park, California

Presque Isle, Pennsylvania (October 1965)

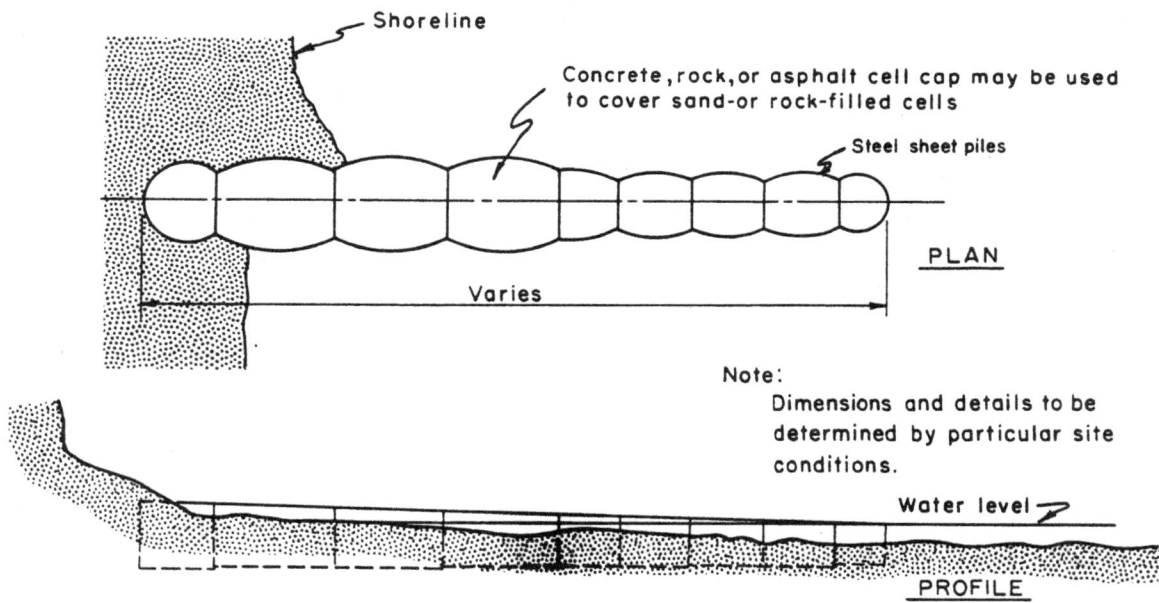

Shoreline

Concrete, rock, or asphalt cell cap may be used to cover sand- or rock-filled cells

Steel sheet piles

PLAN

Varies

Note:
Dimensions and details to be determined by particular site conditions.

Water level

PROFILE

Figure V-3-28.   Cellular-steel sheet-pile groin, Presque Isle, Pennsylvania

(2) Literature review. Although groins have been around a long time and many references exist, most only provide a few rules of thumb. No systematic methods for functional design under a wide range of structural shapes, waves, and sediment transport conditions presently exist. Reviews for the functional design of groins can be found in Bruun (1952); Bakker (1968); Balsillie and Berg (1972); Balsillie and Bruno (1972); Nayak (1976) (unpublished); Tomlinson (1980); Fleming (1990); and EM 1110-2-1617. All these reviews restate the same beliefs but fail to reference the sources that verify the concepts and conclusions from theory, model studies (laboratory or numerical), or field experiments. As stated by Kraus, Hanson, and Blomgren (1994)

"...the literature (on groins) may appear to assign validity to certain concepts and conclusions by weight of repetition (but) not by independent confirmation." (p. 1329).

(a) Laboratory investigations suffered from severe scale distortions in sediment transport to cast serious doubt and questions on their results.

(b) A fresh approach is needed that begins with a summary of over 20 parameters that govern beach response to groins as listed in Table V-3-7 (from Kraus, Hanson, and Blomgren 1994). They are grouped into three main categories (structure, beach, and hydrodynamic conditions), but the large number of variables make analysis difficult. Missing from this table is the minimum, design beach width for shore protection. Groin geometry and possible sediment size for the beach fill can be controlled in the design.

Table V-3-7
Main Parameters Governing Beach Response and Bypassing at Groins (from Kraus, Hanson, and Blomgren 1994)

Groin(s)	Beach and Sediment	Waves, Wind, and a Tide
Length	Depth at tip of groin	Wave height and variability
Elevation	Depth of closure	Wave period and variability
Porosity	Sediment availability	Wave angle and variability
Configuration (straight, T, L, etc.)	Median grain size and variability	Tidal range
Orientation to the shoreline	Sediment density	Wind speed and variability
Spacing between groins	--	Wind direction and variability
Tapering	--	Wind duration and variability

Note: Two integrated parameters governing groin functioning are the ratio of net to gross longshore sand transport, and the presence, location, and number of longshore bars.

(c) From the previously listed review papers, other references and their own experience, Kraus, Hanson, and Blomgren 1994 listed 13 functional properties attributed to groins and present a critical evaluation of each as shown in Table V-3-8. The first five are well accepted properties that have led to the general rule of the thumb to make the groin spacing to length ratio about two to four. However, this rule omits any consideration of the cross-sectional shape of the groin. The length controls water depth at the end and, hence, the amount of sediment by-passing around the tip. But the cross-sectional elevation in the swash zone controls over-passing, the length and elevation on the beach berm control shore-passing, and the structural materials control through-passing as takes place in rubble-mound and permeable groins. Tidal range, predominant wave characteristics (height, period, direction), net and gross longshore sediment transport and grain size are key hydrodynamic and sediment parameters. All these factors together produce the optimum spacing and planform configuration of the shoreline within each compartment for average climate conditions.

Table V-3-8
Functional Properties Attributed to Groins and their Critical Evaluation (from Kraus, Hanson, and Blomgren 1994)

Property	Comment
1. Wave angle and wave height are leading parameters (longshore transport).	Accepted. For fixed groin length, these parameters determine bypassing and the net and gross longshore transport rates.
2. Groin length is a leading parameter for single groins. (Length controls depth at tip of groin.)	Accepted, with groin length defined relative to surfzone width.
3. Groin length to spacing ratio is a leading parameter for groin fields.	Accepted. See previous item.
4. Groins should be permeable.	Accepted. Permeable groins allow water and sand to move alongshore, and reduce rip current formation and cell circulation.
5. Groins function best on beaches with a pre-dominant longshore transport direction.	Accepted. Groins act as rectifiers of transport. As the ratio of gross to net transport increases, the retention functioning decreases.
6. The updrift shoreline at a groin seldom reaches the seaward end of the groin. (This observation was not found in the literature review and appears to be original to the present paper.)	Accepted. Because of sand bypassing, groin permeability, and reversals in transport, the updrift shoreline cannot reach the end of a groin by longshore transport processes alone. On-shore transport is required for the shoreline to reach a groin tip, for a groin to be buried, or for a groin compartment to fill naturally.
7. Groin fields should be filled (and/or feeder beaches emplaced on the downdrift side).	Accepted. Filling promotes bypassing and mitigates downdrift erosion.
8. Groin fields should be tapered if located adjacent to an unprotected beach.	Accepted. Tapering decreases the impoundment and acts as a transition from regions of erosion to regions of stability.
9. Groin fields should be built from the downdrift to updrift direction.	Accepted, but with the caution that the construction schedule should be coordinated with expected changes in seasonal drift direction.
10. Groins cause impoundment to the farthest point of the updrift beach and erosion to the farthest point of the downdrift beach.	Accepted. Filling a groin field does not guarantee 100% sand bypassing. Sand will be impounded along the entire updrift reach, causing erosion downdrift of the groin(s).
11. Groins erode the offshore profile.	Questionable and doubtful. No clear physical mechanism has been proposed.
12. Groins erode the beach by rip current jetting of sand far offshore.	Questionable. Short groins cannot jet material far offshore, and permeable groins reduce the rip current effect. However, long impermeable jetties might produce large rips and jet material beyond the average surfzone width.
13. For beaches with a large predominant wave direction, groins should be oriented perpendicular to the breaking wave crests.	Tentatively accepted. Oblique orientation may reduce rip current generation.

(d) Kraus, Hanson, and Blomgren 1994 also discovered that the updrift shoreline rarely reached the seaward end of the groin (Table V-3-8, No. 6). On-shore sediment transport processes appear to be necessary for a groin to be filled naturally so that the shoreline reaches the tip, or the tip is buried. Longshore sand transport direction reversals, sand bypassing under water at the end, and groin permeability all normally keep the updrift shoreline well landward of the end of the groin.

(e) Properties No. 7, 8, and 9 in Table V-3-8 have long been accepted standard conditions for groin field design and construction. Even so, as noted in No. 10, filling a groin field does not guarantee that 100 percent of the original longshore transport will continue as sand bypassing. The entire, filled, groin field system will impound sand updrift to cause some erosion downdrift of the system.

(f) Properties No. 11 and 12 are often stated reasons by opponents of groins but are questionable because they cannot be supported by physical mechanisms nor principles of conservation of sand. No. 13 will be considered under innovations discussed in the following paragraphs.

(g) A critical review of the literature also shows that little, if any, previous discussion exists on how to judge success (or failure) of a groin design. As discussed further, success should be judged on two factors: to maintain a minimum, dry beach width for specified storm conditions for protection beyond a reference baseline; and to bypass an average, annual amount of sediment to minimize downdrift impacts.

(3) Physical processes.

(a) Normal morphological response. How do groins work? Waves breaking alongshore at an angle create a time-averaged, longshore current and longshore sediment transport. The cross-shore distribution of longshore sediment transport is discussed in Part III-2 (see Equation III-2-23 and Example Problem III-2-7). A key variable is the surf zone width for the theory cited (Bodge and Dean 1987) which assumes sediment mobilized in proportion to the local rate of wave energy dissipation and transported alongshore by the local, wave-induced current. The groin simply blocks a part of this normal transport of sand alongshore and causes it to accumulate in a fillet on the groin's updrift side (the side from which the sediment is coming). This accumulation reorients the shoreline and reduces the angle between the shoreline and the prevailing incident wave direction. The reorientation reduces the local rate of longshore sand transport to produce accumulation and/or redistribution of sand updrift of the groin. The amount of sand transported past the groin is greatly reduced (or eliminated) to significantly impact the downdrift area. The ratio of groin length to some statistical measure of surf zone width (or water depth at the groin tip) is a key factor in sand bypassing, as discussed further in the following paragraphs. Wave diffraction causes reduced wave energy in the lee of the groin relative to the midcompartment, mean water-level setup gradients, and setup induced currents behind the groin. These contribute to complex, current circulation patterns that move sediment alongshore and offshore along the leeside of the groin (Dean 1978). The strength of these internal current patterns depends on groin planform geometry, but also on groin cross-sectional elevation and permeability across the surf zone. Waves diffract around the groin tip, propagate over the submerged section and reflect off the body of the groin. These interactions vary with water depth changes during the tidal cycle. Consequently, sediment can also move over the top of submerged groins (over-passing), through the permeable, groin structure (through-passing) and behind the end of the structure (shore-passing). Impermeable, high crested groins created internal and external current patterns that are far different than permeable, submerged structures. Fleming (1990) discusses the results of physical model studies with current and sediment movements for both high and low groin cross sections. Complex flow patterns were produced, and it was stated that strong local currents may cause a net loss of sediment from the compartment by offshore movement during storm events.

(b) Storm response. Groins offer little or no reduction in wave energy to shore-normal waves during storms. Consequently, cross-shore sediment transport processes as discussed in Part III-3-3 for natural beaches are similar for groin field compartments. And, for near normal wave incidence, the groin system can create strong local current and rip currents which add to the offshore movement of beach material during storms.

(4) Functional design.

(a) Insight from numerical models.

- Some numerical models of shoreline change include groin field effects (e.g., GENESIS, Hanson and Kraus 1989, see Part III-2-4 for details) Boundary conditions for groins in these models give insight into how they must function (Gravens and Kraus 1989). They are as follows:

  - As the groin length increases, its impact on the shoreline regarding time evolution and equilibrium planform must increase.

  - Increasing groin permeability should decrease the impact of the structure on the shoreline. Different groin permeabilities must produce different equilibrium planforms.

  - A permeability of 100 percent should give longshore transport rates and shoreline evolution identical to that modeled with no structures.

- Groin bypassing around the seaward end is calculated at each time step in the GENESIS model. Key variables are the water depth at the tip and the breaking wave height. In GENESIS, the depth of active longshore sediment transport is taken at 1.6 times the significant breaking wave height (from Hallermeier 1981). Groin length relative to surf zone width could also be employed to calculate the bypassing factor.

- A permeability factor representing groin elevation, groin porosity and tidal range must also be estimated in the model. These three variables represent over-passing, through-passing and shore-passing respectively as sketched in Figure V-3-29. The permeability factor is assigned and must approach unity to satisfy the third criterion previously described.

- A third key factor is the ratio of net transport rate, Q_n to the gross rate, Q_g (Bodge 1992). When the Q_n/Q_g ratio is zero, a perfectly balanced transport (no net) exists to produce symmetrical fillets on both sides of a single groin. The opposite extreme is $Q_n/Q_g =1$ meaning unidirectional transport. A single fillet on one side results.

- These key factors controlling groin functioning are summarized in Table V-3-9 (Kraus and Bocamazo 2000) with symbols shown in Figure V-3-29.

- These three process factors incorporate many of the 20 or more fundamental variables. The geometry ratio of spacing, X_g to length Y_g is also the controlling factor for groin systems, as found in the literature. Note from Figure V-3-29 that Y_g represents a mean groin length measured from the average, nourished beach shoreline. Using Y_{gu} (updrift) gives a larger ratio or using Y_{gd} (downdrift) produces a smaller ratio that may account for some variability. The *Shore Protection Manual* (1984) says $X_g/Y_g=2-3$ for the proper functioning of shore-normal groins.

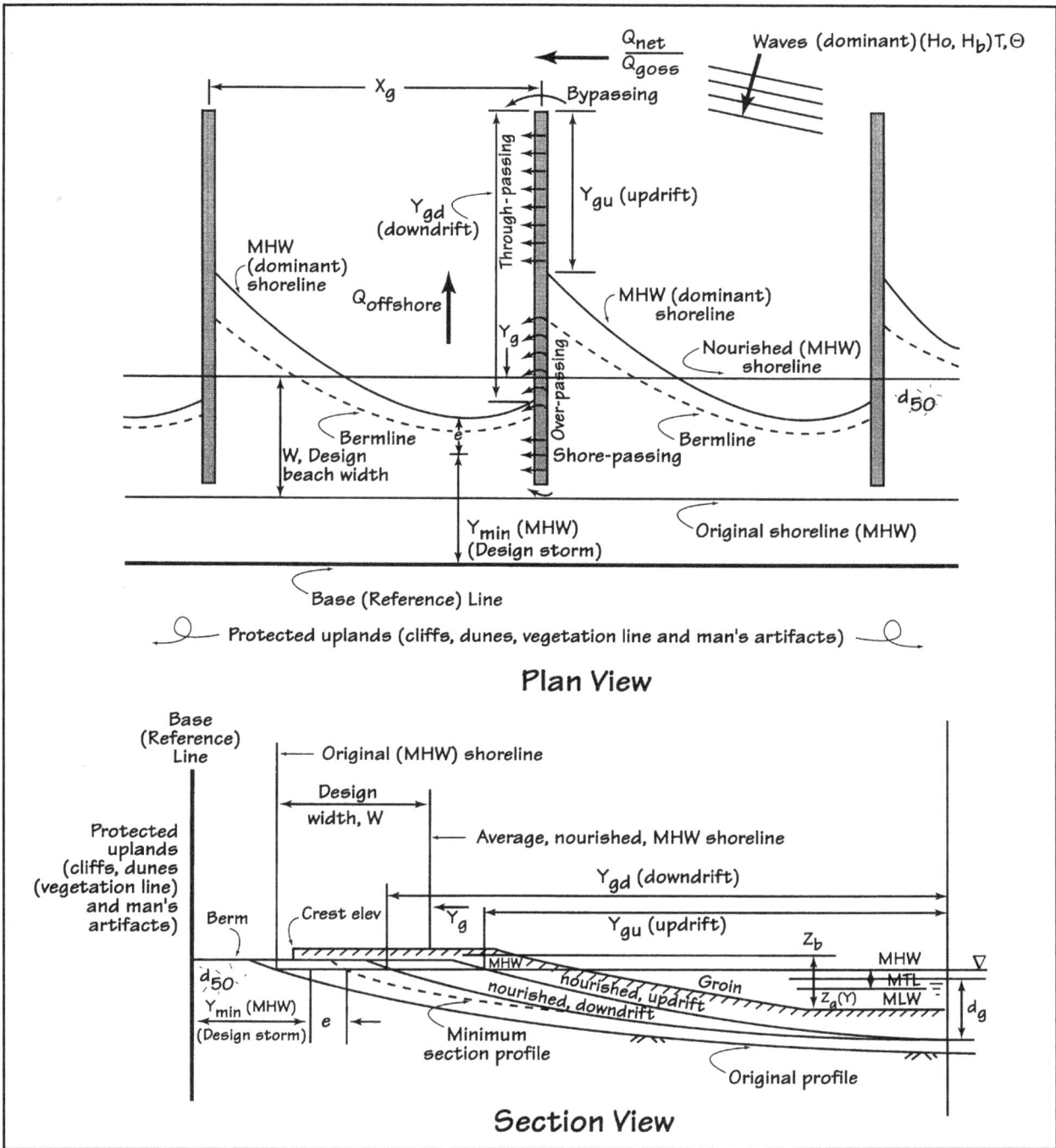

Figure V-3-29. Definition sketch of key variables in the functioning of groins

Table V-3-9
Process-Based Factors Controlling Groins

	Process	Parameter	Description
1	Bypassing	D_g/H_b	depth at groin tip/breaking wave height
2	Permeability		
	• Over-passing	$Z_g(y)$	groin elevation across profile, tidal range
	• Through-passing	$P(y)$	grain permeability across shore
	• Shore-passing	Z_b/R	berm elevation/runup elevation
3	Longshore Transport	Q_n/Q_g	net rate/gross rate

- Kraus, Hanson, and Blomgren (1994) exercised the GENESIS model for a single groin and studied the bypassing formulation. Rapid filling was followed by a gradual buildup over time that meant an increased amount of material must bypass the groin as the shoreline grows out towards the tip. Filling to capacity was only possible for $Q_n/Q_g = 1$ for the unidirectional case. As $Q_g > Q_n$, growth of the shoreline seaward decreased. These tests mean groins seldom fill to capacity by longshore transport processes alone. Cross-shore sediment transport processes must be added to understand how beach elevation and width can build beyond that capable of representation on one-line, shoreline change models. Simulations with multiple-groins and the Westhampton Beach, New York, groin field are also discussed in this paper. An aerial view looking east of the Westhampton Beach, New York, groin field with beach nourishment in 1998 is shown in Figure V-3-30.

- Realistic distributions of longshore current and sediment transport across the surf zone, beach profile shapes with bars and troughs, and other sediment transport mechanisms (wind, tide) are further complicating factors that have yet to be addressed in numerical models.

(b) Groin profile. A typical groin profile with inshore (berm) section, sloping middle section, and horizontal seaward section is shown in Figure V-3-31. In general the landward end is set at the elevation of the natural, existing beach berm, the sloping section is set at the slope of the beach face in the swash zone and the outer, seaward section at a lower elevation such as mean low water (mlw) or lower. The landward and sloping sections function as a beach template for sand to accumulate on the updrift side. The groin profile is shaped to approximately match the postproject beach profile, after nourishment is complete. The seaward end (depth) and seaward elevation are set to the planned bypassing and overpassing in the surf zone. A lower seaward section permits longshore currents to carry sediment over the structure and reduces wave reflections from the groin. A significant amount of sand is transported on the beach face in the swash zone (Weggel and Vitale 1985) and therefore overpassing also takes place in the sloped section when the groin has been filled in this area. Prevention of flanking is the main concern to locate the shoreward end. As seen in Figure V-3-29, calculations of the storm erosion distance, e, together with maximum shoreline recession are needed to establish this position. Seaward limit of the shore section is set relative to the desire, nourished beach width, W, or even further seaward to help retain the nourished beach. Seaward limit of the outer section is the groin length, Y_g, and depends on the amount of longshore sediment transport to be bypassed, as discussed.

(c) Permeability. In general, sheet-pile groins of all types are impermeable whereas rubble-mound groins permit some material to wash through the structure. Some rubble-mound design contain impermeable cores and/or are treated with sealant materials to ensure sand tightness (see EM 1110-2-1617). There are no quantitative guidelines for determining the permeability of sand for a given groin geometry of the rubble-mound type. Some patented, precast concrete groin systems are permeable, as will be discussed later.

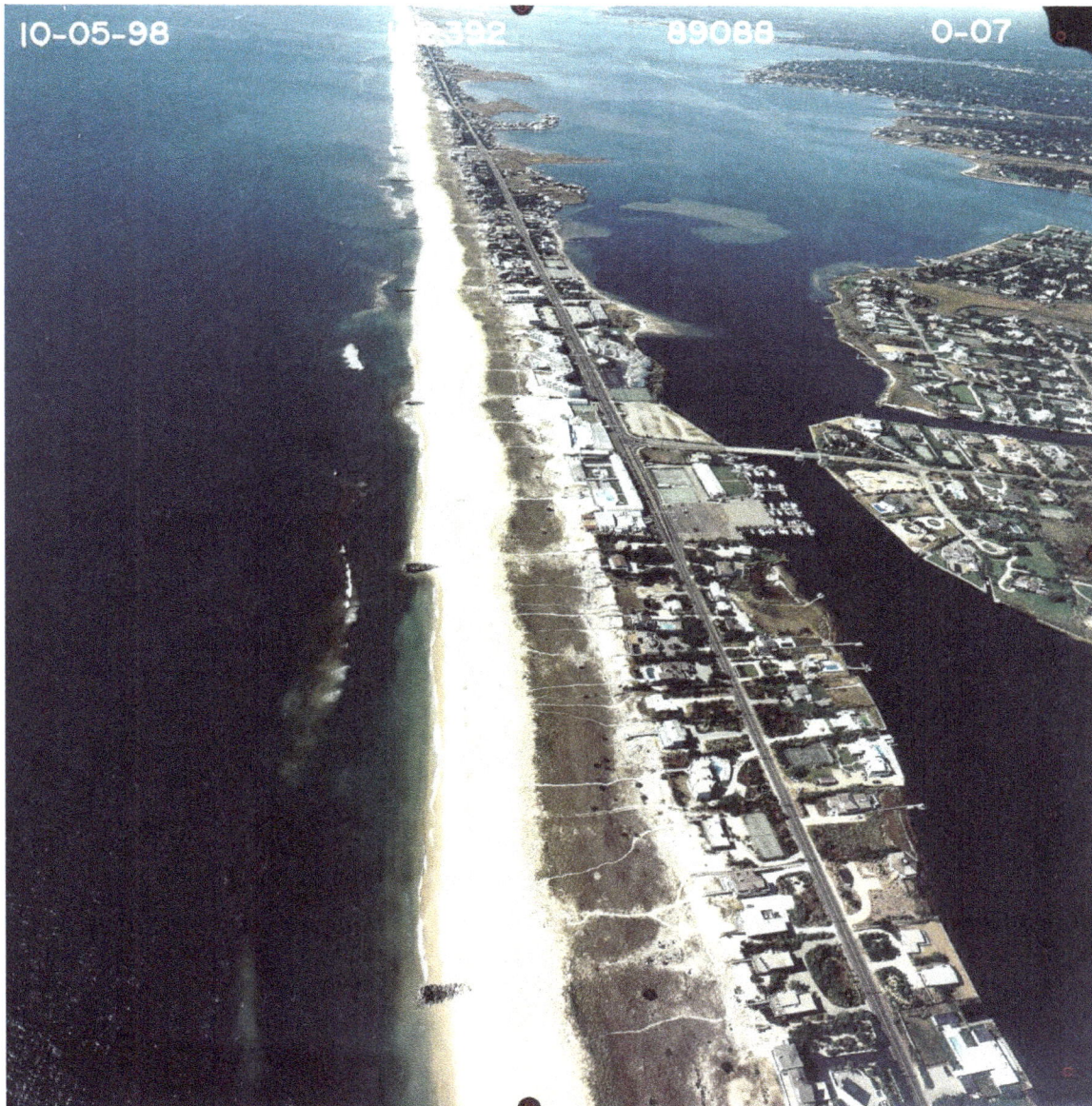

Figure V-3-30. Westhampton Beach, New York, groin field and renourished beach, 1998 (courtesy USAED, New York)

(d) Terminal groins. Groins on the updrift side of inlets can benefit nearby beach nourishment projects by controlling (or gating) the amount entering (lost) to the inlet. These terminal structures also benefit navigation projects by reducing the sediment rates within the inlet. They normally are impermeable and high and long to prevent sand from being carried through, over, or around them. Eventually, they will fill and sand bypassing around the end will be maintained. It should be noted that terminal groins are short compared with the length of navigation jetties constructed to reduce wave heights for ships entering the inlet. Consequently, the scale of interruption of normal, longshore sediment transport processes for ebb-and flood- tidal shoals are far different for navigation jetties. Terminal groins fill quickly and do not have major impacts on ebb-tidal shoals and normal, inlet, sand-passing processes. A successful terminal groin is that located on the southern bank of Oregon Inlet, North Carolina. Design and monitoring details are found in Overton et al. (1992); Dennis and Miller (1993); Miller, Dennis, and Wutkowski (1996); and Joyner,

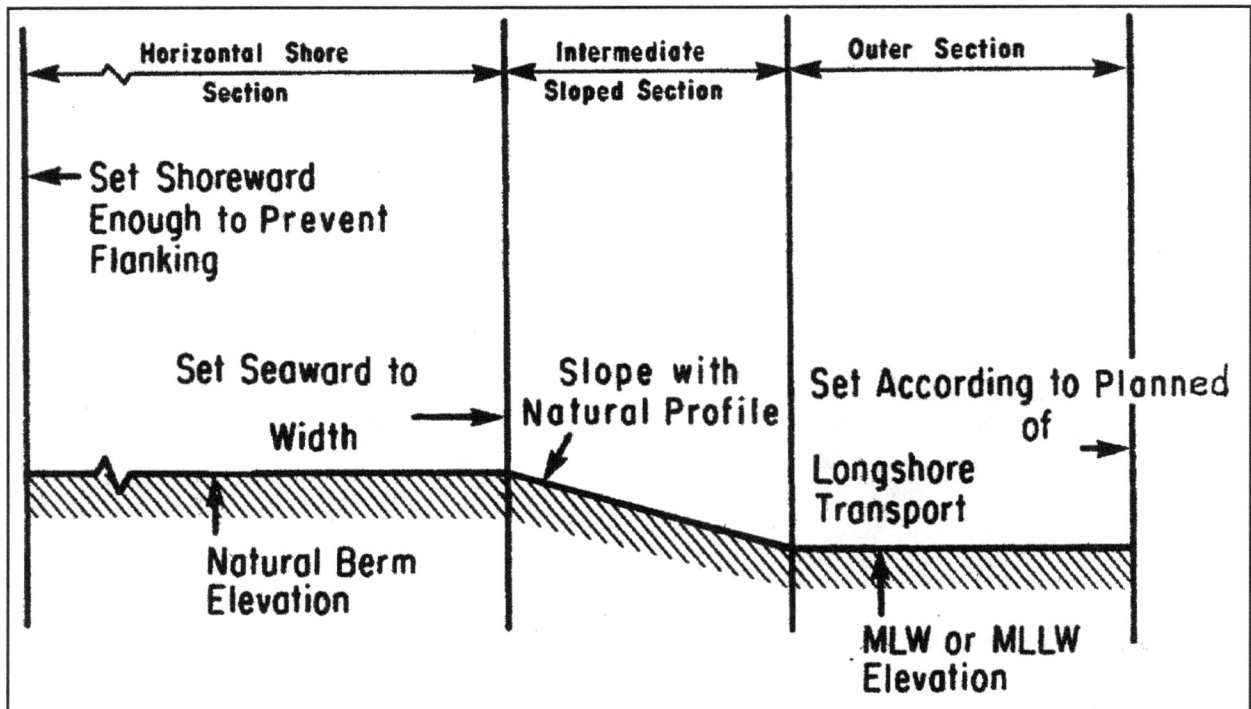

Figure V-3-31. Typical groin profile with sloping section

Overton, and Fisher (1998). See also Dean (1993; 1996) for detailed analysis of terminal structures at the ends of littoral systems.

(e) Groins system transitions. A transition reach is required from a groin field to the adjacent, natural beach. Groin lengths are gradually shortened to allow more bypassing. Generally, groin lengths are decreased along a line converging to the shoreline from the last full-length groin, making an angle of about 6 deg with the natural shoreline as depicted in Figure V-3-32 (Bruun 1952). The spacing is also reduced to maintain the same X_g/Y_g ratio (2-3) used in the design.

(f) Order of construction. The sequence in which a groin field is constructed is a practical design consideration and may not be straightforward. To minimize downdrift impacts, beach nourishment and groin construction should be concurrent. Construction of the first groin should be at the downdrift end of the project, preferably the terminal groin adjacent to an inlet. Net drift will combine with the artificial beach nourishment to fill and stabilize the first compartment. The second groin is then constructed and the process repeated. Gradually working updrift, the groin field construction is completed. This process together with tapering the ends will help to minimize the impact to adjacent, downdrift beaches. This method may increase costs, but it also may result in a more practical guide to spacing of the groins than originally designed.

(5) Nontraditional configurations. Most groins are straight structures, perpendicular to the shoreline. Figure V-3-33 illustrates other possible planform shapes. T-shaped groins are similar to nearshore breakwaters when the tee end is above the mean water level. Inclined groins may reduce rip currents along the updrift side when inclined in the direction of net sediment transport. All shapes shown have been constructed and provide some degree of shoreline stabilization. Sectional variations are also possible as listed in Table V-3-10.

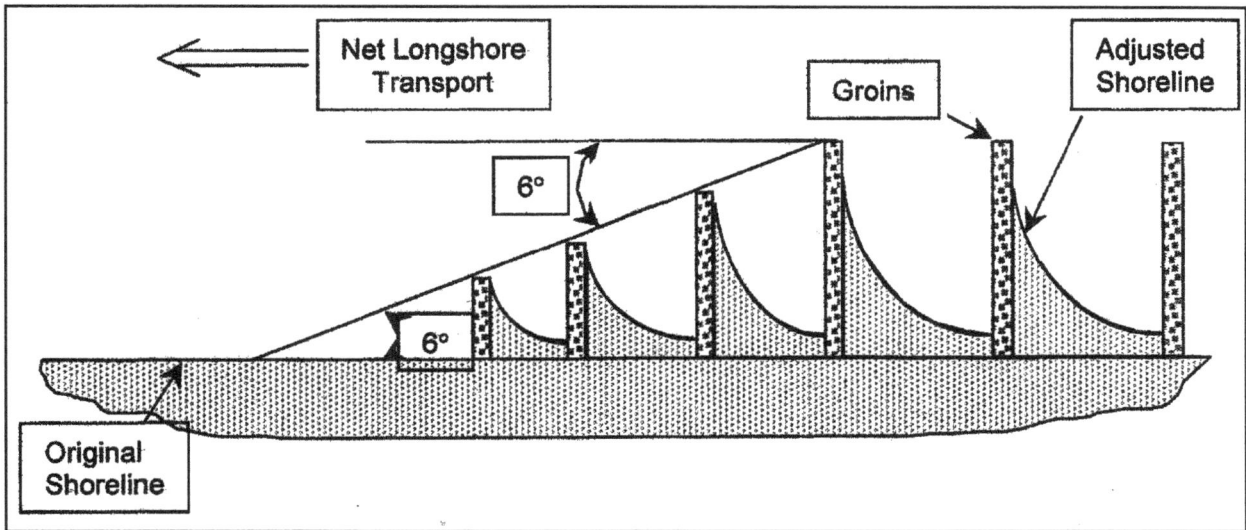

Figure V-3-32. Transition from groin field to natural beach

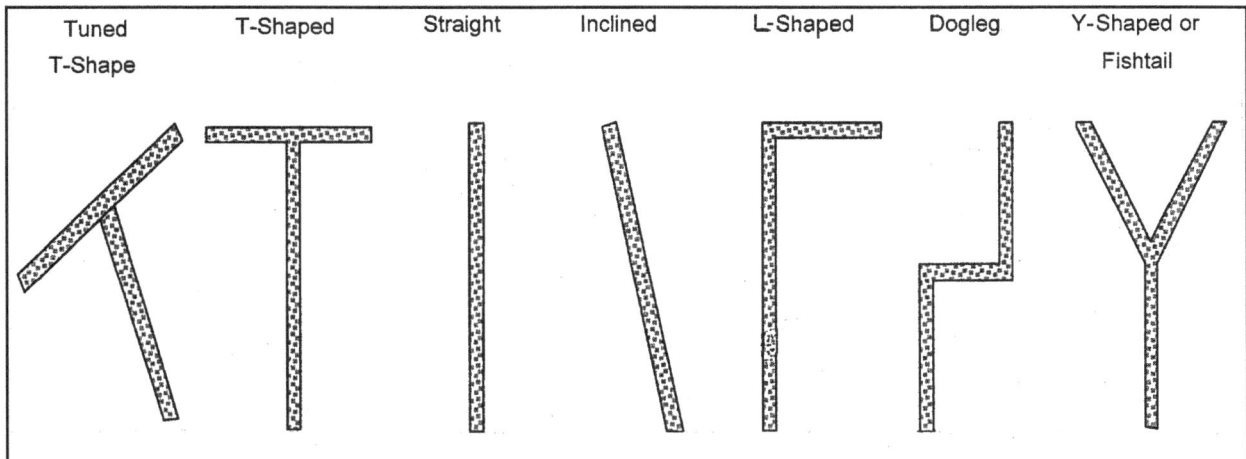

Figure V-3-33. Possible planform shapes for groins

Table V-3-10
Types of Groins-Section Views

Concepts	Comments
Flat (Horizontal)	old
Sloped	common
Notched	(swash zone, new)
Concrete-pile (permeable)	(innovative, patented)
Adjustable	(elevation, length)
Rectifying	(transport-direction sensitive)
Tuned	(elevation, T-head)
Temporary	(geotextiles, sand)
Spur	

(a) Notched groins have recently undergone laboratory and field testing by the USACE (Kraus 2000). Trial notched groins have been implemented by the U.S. Army Engineer District, New York, and the U.S. Army Engineer District, Philadelphia, along the south shore of New Jersey. The purpose of notching is to bypass beach fill and littoral sediment so that the fillet and erosive response of the shoreline are smoothed out, i.e., to straighten the shoreline adjacent to the groin. The goal is shoreline readjustment to approach a smooth, continuous shoreline through a groin field. The efficiency of notching was defined as the amount of sediment that passes the groin system, but that remains on the subaerial beach. Tentative conclusions based on both the laboratory and field experiments are: notches in the swash zone are most efficient; notches in the surf zone are less efficient and can have strong longshore currents, and be hazardous to swimmers; surf zone notches may move sediment further from shore; notching a groin in the swash zone may not be successful depending on how and at what rate sediment typically moves alongshore on the subject beach.

(b) Spur groins are short, stub groins constructed at right angles to navigation jetties and the end groins of groin fields to redirect currents and sediment transports. Sorensen (1990) describes a spur jetty constructed as an emergency measure on the leeward side of two curved jetties to prevent a breach at the landward end. Anderson, Hardaway, and Gunn (1983) document eleven locations in the Chesapeake Bay where short, shore parallel groins are attached to the terminal groin near the beach. Over 12 years of beach response at one site reveals that the spur groins cause diffraction and refraction effects to prevent the detachment of the terminal groin at its landward end.

(c) Another way to smooth the shoreline is to make the groin more permeable as illustrated in Figure V-3-34. A patented, concrete pile, permeable groin field has existed on Cayman Kai, Caribbean Sea, for over 10 years and has survived many hurricanes (from Kraus and Bocamazo 2000). Environmental restrictions against hardened shorelines in North Carolina has resulted in a groin field constructed from geotextile materials filled with sand on Bald Head Island, North Carolina (Denison 1998). In theory, these structures will provide sand for shore protection if they are damaged or fail in major storm events. Other revisited or fresh concepts for groin designs listed in Table V-3-9 have only been proposed conceptually and have yet to be field tested. These new and innovative approaches to groin design benefit from experience and modern understanding of coastal sediment processes.

(6) Basic rules for functional design of groins. Ten modern rules for groins design can be summarized as follows:

- Rule 1   If cross-shore sediment transport processes are dominant, consider nearshore breakwater systems first.

- Rule 2   Conservation of mass for transport of sediment alongshore and cross-shore means groins neither create nor destroy sediment.

- Rule 3   To avoid erosion of adjacent beaches, always include a beach fill in the design

- Rule 4   Agree on the minimum, dry beach width, Y_{min} for upland protection during storm events as a measured to judge success.

a. Aerial photo

b. Cross-section model

Figure V-3-34.   Patented concrete pile permeable groins Cayman Kai, Caribbean Sea (surviving for 10 years, including hurricanes).   Innovation is groin design to enhance sediment through-passing and smooth the shoreline variation in the groin field (from Kraus and Bocamazo 2000)

- Rule 5   Begin with X_g/Y_g=2-3 where X_g is the longshore spacing and Y_g is the effective length of the groin from its seaward tip to the design shoreline for beach fill at time of construction.

- Rule 6   Use a modern, numerical simulation model (e.g., GENESIS) to estimate shoreline change around single groins and groin fields.

- Rule 7   Use a cross-shore, sediment transport model (e.g., SBEACH) to estimate the minimum, dry beach width, Y_{min} during storm events.

- Rule 8   Bypassing, structure permeability and the balance between net and gross longshore transport rates are the three key factors in the functional design. Use the model simulation to iterate a final design to meet the Y_{min} criterion.

- Rule 9   Consider tapered ends, alternate planforms and cross sections to minimize impacts on adjacent beaches.

- Rule 10   Establish a field monitoring effort to determine if the project is successful and adjacent beach impacts.

- Rule 11   Establish a "trigger" mechanism for decisions to provide modification (or removal) if adjacent beach impacts are found nonacceptable.

f. Reefs, sills and wetlands.

(1) Background and definitions. Additional types of shore protection alternatives for both high and low wave energy coasts function by reducing the wave energy striking the shoreline. Reefs are platforms of biotic organisms built up to a strict elevation in relation to low tide. Natural reefs require high wave energy to survive. Wetlands are coastal salt or freshwater marshes that are low-lying meadows of herbaceous plants subject to periodic, water level inundations. Wetlands are fragile and only survive in low wave energy environments. See IV-2-11 and IV-2-12 for further details on wetland and reef-type coastlines, respectively. The word "sill" has evolved to take on two separate identities in coastal engineering. Both meanings imply wave attenuation in the lee of the structure. A submerged, continuous, nearshore dike to hold sand moving offshore from a nourished beach is one definition and also labeled a perched beach. Free-standing, low-profile, continuous shoreline structures to permit establishment of a marsh fringe in the lee of the structure are also called sills.

(2) Reefs.

(a) Natural types. Coral reefs are massive calcareous rock structures that slowly grow upward by secretions from simple animals living on the rock surface. They exist throughout the Florida Keys, on both Florida coasts, the Hawaiian Islands, and the U.S. Island territories and serve significantly lower the mean wave energy striking the adjacent shore. Fringing reefs border a coast, barrier reefs lie offshore enclosing a lagoon, and atolls encircle a lagoon. Under favorable growth conditions, coral reefs build upward to form wide, broad platforms that are exposed at low tide. Thus they cause waves to break and to continue breaking across the reef.

- Oyster reefs can exist in much colder, brackish water conditions of lagoons, bays and estuaries. They are found along both coasts and in the Gulf of Mexico. They are wave-resistant structures that can biologically adapt to rising sea levels. Their maximum elevation is related to the minimum time of inundation in the middle range of the intertidal zone.

- A third type of natural reef is produced by colonies of tube worms called a worm reef. The east coast of Florida has some of this type.

- With respect to natural shore protection, reefs are a biological wave damper that can accommodate rising sea levels as long as they are alive. Protection of reefs is essential.

(b) Wave attenuation. Wave transformation processes across broad, flat coral reefs include shoaling, refraction, reflection, and energy dissipation by both bottom friction and wave breaking. Wave energy is also transferred to both higher and lower frequencies in the wave spectrum as the spectral shape flattens (Hardy and Young 1991). Wave setup variations along reefs can occur due to gradients in wave breaking characteristics to produce longshore currents. For engineering purposes, depth-limited wave breaking is the dominant transformation process. A methodology to estimate random wave energy transformation across reefs is presented in USACE (1993). It is based on the breaking wave model of Dally, Dean, and Dalrymple (1985) extended to random waves following Kraus and Larson (1991). Comparison of the numerical model results with field experiments of rms, wave height attenuation on the Great Barrier Reef in Australia (Young 1989) showed good agreement with the measurements. The breaker model without bottom friction is available in the PC-based computer program NMLONG (Kraus and Larson 1991; USACE 1993).

(c) Artificial reefs. The functional design of artificial reef systems for shore protection; increasing the fill life of renourished beaches, and to enhance recreational surfing is a relatively new area of coastal engineering. No general design rules exist. Numerical and physical models have recently been employed for site specific designs in California and Australia of artificial reefs for surfing. These models aid in both wave breaker-type design (Pattiaratchi and Bancroft 2000) and to insure that the structure will not create downdrift erosion. (Turner et al. 2000).

(3) Sills.

(a) Perched beach. A beach or fillet of sand retained above the otherwise normal profile level by a submerged dike (sill) has only been used twice in the United States (National Research Council 1995). Ferrante, Franco, and Boer (1992) describe a successful, 3,000-m-long perched beach project on the coast at Lido di Ostica about 35 km from Rome, Italy, on the Tyrrhenian Sea. The rubble-mound sill was located about 150 m from shore with crest -1.5 below msl datum in -5.0 water depth. An additional 1.000-m stretch with sill closer to shore in shallower water performed better to hold a wider beach. A feasibility study of the perched beach concept in the Netherlands was reported by Ruig and Roelse (1992). Model studies and calculations of life-cycle costs demonstrated that this alternative was roughly as expensive as repeated beach nourishments with no sill construction over a 30-40-year period. Construction at the site selected, Cadzand, Tien Honderd Polder, Zeeland, The Netherlands, has yet to be implemented.

(b) Wetlands protection. In low wave-energy environments, natural, wide, fringe marshes can provide sufficient erosion protection for upland areas, as discussed later. However, for many reasons, the fringe marsh itself may be eroding and require protection, enhancement and/or to be re-established. Sills are typically low, small, continuous rock structures placed at mean low water with some sand fill in the lee to provide a substrate for marsh growth (Hardaway and Byrne 1999). Figure V-3-35b displays a curved, stone sill connecting headland breakwaters with sand fill and marsh planting on the Choptank River, Chesapeake Bay (from Hardaway and Byrne 1999). After 5 years, the sill is practically invisible as shown in Figure V-3-35c. Sills can thus be used in higher wave energy regimes to establish intertidal marsh grasses that aid in the shore protection. Periodic marsh replanting and maintenance may be required under higher wave energy conditions. Advantages and disadvantages of a wide variety of erosion mitigation structures and materials to protect wetlands can be found in the Wetlands Engineering Handbook (Olin et al. 2000).

(4) Wetlands. The final alternative in this section on shoreline stabilization structures are fringe marshes or wetlands. Part IV-2-11 considers their values, distribution classification sediment characteristics, and causes for loss of wetlands in the coastal zone. However, little is known about their importance for shoreline erosion mitigation.

(a) Tidal creeks with fetch exposures less than 0.5 nautical miles and low wave-energy environments can naturally sustain a sufficiently wide marsh fringe. Also, they generally have little or no problem with upland bank erosion because the established marsh fringe absorbs most of the wave energy before it can impact the upland area (Hardaway and Byrne 1999). On the Chesapeake Bay, Hardaway and Anderson (1980) found that low, upland banks erode almost twice as fast as marsh shorelines with similar fetch exposures and nearshore depths.

(b) Some recent field and laboratory research has focused on wave attenuation by wetland vegetation (Kobayashi, Raichle, and Asano 1993; Wallace and Cox 1997; Tschirky, Turke, and Hall 2000). Wave heights are typically reduced by 50 percent and the peak spectral period also drops as the spectrum becomes more broad banded with higher frequency components. No significant design guidance on allowable wave heights or currents for wetlands presently exists. The Wetlands Engineering Handbook (Olin, Fischenich, and Palermo 2000) provides a wealth of valuable information for the restoration and creation of wetlands.

V-3-4. Nonstructural Alternatives

a. Introduction.

Nonstructural alternatives are management strategies for coastal hazard mitigation that are not armoring (Part V-3-2) nor beach stabilization structures (Part V-3-3), nor beach nourishment (Part V-4). Society has developed ways to adapt by setting requirements for the elevation of buildings, providing insurance and planning for continual erosion with setback limits for new construction. The final nonstructural alternative is to retreat by relocation, abandonment, or demolition.

- Three Federal government agencies (the USACE, Federal Emergency Management Administration (FEMA), National Oceanographic and Atmospheric Administration (NOAA)) with different missions, planning methods, and requirements for benefits determination employ or promote the use of nonstructural methods. This section briefly summarizes the role of each and presents some examples of the retreat alternative. Full details of the Federal government's planning requirements and design constraints for all three agencies involved are presented in Part V-8.

- Coastal hazard mitigation is addressed in a piecemeal manner by multiple agencies within states, at the county, city and community level, and by businesses and individuals operating and living at the waterfront. Property ownership (Federal, state, municipal, community, and individual) and the value of scarce waterfront property strongly motivates all parties to protect their valuable investments. Part V-8 presents details of how each Federal agency (CORPS, FEMA, and NOAA) interacts with each level of responsible group, but details of approaches by these parties are beyond the scope of this chapter.

b. Adaptation.

(1) Zoning and building codes. Any structural or nonstructural change in the design, construction or alteration of a building to reduce damage caused by flooding and flood related factors (storm surges, waves, and erosion) is considered a flood proofing alternative by FEMA. The mechanism to require change in old construction practices is the National Flood Insurance Program (NFIP) administered by FEMA using Flood

a. Stone sill with marsh planting on Chester River, Kent County, MD

b. Stone sill connecting breakwaters with sand fill and marsh implantation on Choptank River, Talbot County, MD

c. Breakwater and sill project after 5 years

Figure V-3-35. Sills and breakwaters (from Hardaway and Bryne 1999)

Insurance Rate Maps (FIRM's) prepared by FEMA. A storm surge elevation at the one percent exceedance level (100-year recurrence interval) plus waves is employed to determine risk and insurance rates for individual properties located on the flood maps. Insurance rates are much lower for structures elevated above the 100-year flood level and is a requirement for all new construction in the coastal, high-hazard zone (including waves). In effect, these regulations become floodplain zoning laws applicable to individual property owners and have resulted in a reduction in Federal government expenditures for insurance claims and disaster assistance benefits (National Research Council 1990).

(2) Setback limits. A second way to adapt is to limit construction close to the shoreline. The NOAA has identified land-use planning and construction siting as the most effective means to reduce coastal storm hazards, particularly on eroding coasts. Here, the mechanism to require change in old construction practices is the Coastal Zone Management Act (CZMA) of 1972. Through the CZMA, the NOAA provides funds to individual states to help solve their own coastal hazard problems. As a result, many states have developed coastal construction setback lines and zones that include historic erosion rates at each site. The methods, definitions, widths, etc., vary from state to state as summarized in Part V-8. A key element is the historic, average erosion rate at each site. Presently, FEMA does not include delineation of erosion zones and erosion hazards on its flood maps. Methods to incorporate both coastal erosion (National Research Council 1990) and beach nourishment (National Research Council 1995) in the national, flood insurance program have been proposed but have yet to be formally adopted. Clearly, coastal erosion increases the risk and beach nourishment reduces the risk of coastal flood and wave damage.

c. Retreat.

Retreat is the final adaptation option. Relocation here also considers abandonment and demolition. To some, retreat is the only option. But practically, all constraints (economic, environmental, social, legal, etc.) must be evaluated for this alternative as well as for all others as previously discussed. This approach may be employed by the Corps as the Federal government agency designated by Congress to protect the nation's shores from the chronic effects of erosion and coastal flooding. Two examples illustrate the USACE's approach and focus on why the retreat alternative was selected.

(1) Cape Shoalwater, Washington. The northern shoreline of the inlet to Willapa Bay, Washington, has been receding at an average rate of 30 to 40 m/year for over 100 years (Terich and Levenseller 1986). Erosion of this area (Cape Shoalwater) has been faster and has lasted longer than any other site on the U.S. Pacific Ocean coast (Komar 1998). Natural, northern migration and progressive deepening of the channel inlet are the two main factors responsible for erosion of the cape. A few homes have been lost, a lighthouse destroyed, the main road moved inland and a historic, pioneer cemetery relocated in this rural area.

(a) In the 1960s and 1970s, the USACE studied the construction of a jetty to stabilize the inlet location and reduce the erosion of the cape. They concluded that a structural solution was not economically feasible to the Federal government. The USACE recommended that if Federal government funds were available at this site, they should be used to purchase land threatened by future erosion. However, no Federal project for relocation could be justified based on the relative benefits and costs to the Federal government.

(b) The rural area (low benefits) and extremely high erosion rate (high costs) were responsible for this outcome, as may have been expected. At other sites, this result is not as obvious.

(2) Baytown, Texas. The northern, upper end of Galveston Bay on the Gulf of Mexico includes Burnett, Crystal, and Scott Bays and low-lying areas that are part of Baytown, Texas, about 24 km east of Houston. Flooding occurs routinely from minor storms and is compounded by subsidence of the ground surface. Withdrawal of oil and gas and groundwater for the metropolitan area of Houston produced 2.5 m (9 ft) of subsidence between 1915 and 1975 (U.S. Army Engineer District, Galveston, 1975).

(a) Both structural (earth levee, concrete floodwall) and nonstructural (permanent relocation and evacuation) alternatives were studied in the early 1970s by the USACE to mitigate the flooding problem. The feasibility study report to Congress (U.S. Army Engineer District, Galveston, 1975) recommended the retreat alternative. Congress authorized this project under the Water Resources Development Act (WRDA 1978).

(b) The final project report (U.S. Army Engineer District, Galveston, 1979) called for the Federal government to purchase approximately 303.51 ha (750 acres), 448 homes and relocate 1,550 people within the 50-year floodplain. Total project costs were $32.131 million (1979 dollars) with the local sponsors' share to be 20 percent or $7.826 million. Annual project benefits were estimated to be $3.530 million and included:

- Reductions for insurable flood damages.

- Reductions for utility service costs.

- Reductions of temporary, emergency evacuation, public health and public relief costs.

- Value of land under new uses.

(c) Other benefits (constant anxiety from flood hazards, depressed property values, health hazards, inconvenience of repetitive, temporary evacuations, and damages to real property and personal possessions) cannot be included as net gain benefits to the national economy. Total, annual costs were $2.678 million giving a benefit to cost (B/C) ratio of 1.3 to the Federal government for the retreat alternative. In contrast, the structural alternative, B/C ratios were 0.1 - 0.3 (USAED, Galveston, 1979).

(d) In 1978, the environment assessment report to the EPA was approved. Plans called for converting the land into a natural area, with possible development of nature areas, bird sanctuaries, green belts, wildlife areas, nature walks, and other uses consistent with the high flood potential. Besides the local sponsor (City of Baytown) paying 20 percent of the final cost, they were also responsible for management of the vacated area.

(e) Unfortunately, the project floundered on local disagreement over the value of the land to be purchased and was never funded. Estimates of project costs rose from $16.980 million in 1975 to $39.131 million in 1979 (less than 5 years). People in the community who did not live in the floodplain were being asked to help buy out those who did. The community was divided over what the property was worth. Some believed their neighbors would be paid far too much for relocation. In July 1979, a bond election was held to provide the local funding for the project and it failed by a 60/40 percent margin (Pendergrass and Pendergrass 1990).

(f) In 1980, Congress was ready to authorize funding, but the local, 20 percent cost share could not be provided. The local residents decided to stay and the Corps placed the project in its inactive category. In the final analysis, the Corps' requirement for cost sharing ($7,589 per residence in 1975 and rising to $17,469 per residence in 1979) prevented an economically viable and environmentally sound project from being implemented. Because the B/C ratio was 1.3, the annual benefits to the Federal government exceeded the 20 percent, local cost share amount. In other words, the Federal government could have paid for the entire project and still realized net economic benefits to the Federal government. This result also does not consider NFIP payments then and over the last 20 years. Partial protection by the NFIP also contributed to the local residents' decisions to stay.

(g) In summary, a nonstructural solution for part of the community of Baytown, Texas, proved difficult to implement because the entire community would not share in the local cost requirement. In contrast, a flood dike and revetment to protect the Texas City – La Marque area on the lower Galveston bay has been constructed. Here, the perception that all the affected community was "protected" made the local sponsor share obtainable (Pendergrass & Pendergrass 1990).

(3) Special cases.

(a) Brighton Beach Hotel, Coney Island, New York. Komar (1998) shows etchings of the 1888 relocation of a large, beachfront hotel on Coney Island, New York. Twenty-four railway tracks were laid to span the entire hotel width, and the wooden pile-supported hotel was lifted onto freight cars on each track. Six locomotives pulled the hotel inland 150 m. No details on costs were provided for this private project which required property ownership and grading of the inland site. Its economic viability also depended on the availability of railway equipment in that era over 100 years ago. Full details are in Scientific American (1888).

(b) Cape Hatteras Lighthouse, North Carolina. Very recently, relocation of the Cape Hatteras Lighthouse has been completed by the National Park Service (NPS), U.S. Department of Commerce (see, e.g., Civil Engineering 1999). The lighthouse is on the east coast of Hatteras Island about 4.02 km (2.5 miles) north of the tip of Cape Hatteras and 64.37 km (40 miles) south of the Oregon Inlet (Figure I-2-6). It is located within Cape Hatteras National Seashore Park, administered by the NPS. The original lighthouse built at this site in 1803 was replaced in 1870 by the present structure, which is the tallest (61 m) and perhaps best-known brick lighthouse in the United States. When built in 1870, it was approximately 490 m from the shoreline. By 1935, this distance diminished to about 30 m due to landward migration of this cape feature. The mhw, average recession rate between 1852 - 1980 (128 years) has been 5.9 m/year and 3.9 m/year for the 1870 - 1980 (110 years) period (Everts, Battley, and Gibson 1983). In 1996, partly due to a wide variety of piecemeal, temporary and emergency measures (see Table V-3-11), the lighthouse stood about 50 - 90 m from the Atlantic Ocean (U.S. Army Engineer District, Wilmington, 1996) depending on the season and tidal conditions. The existing, steel, sheet-pile groin field was actually designed and constructed by the Navy to protect its installation to the north in 1970. The lighthouse, a national historic landmark, remained in operation, and the NPS decided in 1980 that a long-term solution of the erosion problem was needed.

- A 1982 conference of experts from many disciplines (engineering, geology, economy, and historic preservation) together with a 1985 study by the Wilmington District convinced the NPS to employ a structural solution. A seawall/revetment structure was selected. Also a factor in this decision was scientific and engineering opinion that the lighthouse could not be moved without suffering serious structural damage. Congress approved $5.3 million and construction was to begin during the summer of 1986.

- A Move the Lighthouse Committee Report (private) convinced the NPS to seek the advice of the National Academy of Sciences. Their final report (National Research Council 1988) recommended the retreat alternative and in December 1989, the NPS reversed its decision and announced its approval of the relocation alternative. The next 10 years were filled with further controversy and debate due to lack of congressional funding for the relocation; public sentiment in North Carolina against the move (Save the Lighthouse Committee, private), and other ad hoc committee reports with updated studies. During the period, other structural engineers and large-scale relocation experts became convinced that the lighthouse could be moved without damage.

Table V-3-11
Cape Hatteras, North Carolina, Shore Protection History

Year	Action Taken
1930s	Civilian Conservation Corps builds sand dune system along Hatteras Island
1966	Beach Nourishment – 239,000 m³ (312,000 cu yd) of sand placed along Buxton Motel area north of lighthouse
1967	Sandbags placed along 340 m of shoreline to protect former Navy Facility north of lighthouse
1970	Navy constructs three concrete & steel groins
1971	Beach Nourishment – 150,000 m³ (200,000 cu yd) of sand placed along Buxton Motel area north of lighthouse
1973	Beach Nourishment – 960,000 m³ (1,250,000 cu yd) of sand placed along Buxton Motel area north of lighthouse
1974	Repairs made to northern & southern Navy groins
1980	Emergency Repairs – 50 m landward extension of south groin
1981	Riprap and sandbags placed beyond landward extension of south groin
1982	Additional 50 m landward extension of south groin. Seven hundred sandbags placed around the base of the lighthouse
1983	Riprap scour apron placed along landward end of south groin
1992	Additional sandbags placed around base of lighthouse
1994	Additional sandbags placed around base of lighthouse
1995	Rehabilitation of landward end of south groin with 56 m of steel sheetpiling

- Also contributing to the controversy was a NPS committee study in 1992 to make recommendations for *interim* measures, as required, to ensure that the lighthouse remain protected, until the retreat solution could be implemented. An intermediate measure (15 - 25-year design life) was selected to add a fourth groin south of the existing groin field. The NPS employed the Wilmington District to make this design (USAED, Wilmington, 1996). Opponents to the move cited the costs for the fourth groin alternative ($3.5 million) in the USACE study when objecting to the retreat alternative costs (estimated as $12 million). A complete economic analysis using a 100-year design life to determine life-cycle costs of a groin field with terminal groin versus the retreat alternative has never been made by the USACE.

- The National Academy of Sciences study and report (National Research Council 1988) reveals the following NPS policies that dictated the choice of the retreat alternative.

  - Historic preservation is more important than a do nothing or "let nature take its course" policy.

  - The NPS does not have to follow a benefit/cost analysis procedure with benefits exceeding costs regarding the choice of alternative for shore protection.

  - The NPS policy for coastal shorelines is to not interfere with natural, coastal processes. This policy eliminated all structural options as discussed in Chapter 5 of the NRC 1988 report and made their analysis superficial and irrelevant. The retreat option was the only alternative given consideration.

- The NPS strategy was to use the Cape Hatteras Lighthouse relocations as an example and model of "enlightenment" (page numbers refer to NRC 1988)

  - "an exemplary response ... to the generic problem of shoreline erosion" (p. 72)
    "attract much media attention ... to educate a large national audience on the nature of coastal barriers" (p. 72)
    "use national parks as models of wise management of natural and cultural resources" (p. 71)
    "act as a signal to the country of the problems confronting the coast" (p. 39) and
    "illuminate approaches to solving the problems of living with a rising sea" (p. 39).

- The Cape Hatteras Lighthouse was moved successfully inland about 488 m (1,600 ft) from the mhw shoreline in 1999. Recognizing the engineering and construction skills required to complete the move safely, the project received the Outstanding Civil Engineering Achievement Award from the American Society of Civil Engineers in 2000.

- In the final analysis, the NPS policy against any structural interference with natural, coastal processes was the deciding factor. The decision to relocate was taken in 1989 at a time when the possible, accelerated rise in sea level was also of major concern.

(4) Impact of sea level rise. A detailed summary of present-day knowledge of sea level rise rates is presented in Part IV-1-6. Substantial variability exists when including local subsidence as previously discussed for Baytown, Texas. A National Academy of Sciences report on engineering implications (National Research Council 1987) concluded that whether to defend or to retreat depended on several factors, but mainly the future sea level rise rate and the cost of retreat which varies by site. The NRC recommended that all options should be kept open to enable the most appropriate response to be selected. Retreat is most appropriate in areas of low development. Given that a proper choice exists for each location, selecting an incorrect response alternative could be unduly expensive (National Research Council 1987).

V-3-5. Combinations and New Technologies

Pope (1997) lists the following common types of coastal erosion and flooding problems:

- Long-term, chronic land loss associated with the erosion of cohesive sediments, reduced supply of sandy sediments, and/or subsidence.

- Localized erosion impacts caused by a navigation project jetties or other coastal construction works.

- Lands and facilities impacted by storm-induced erosion.

- Flooding by a storm surge with associated wave attack damages.

- Loss of environmental resources (i.e., wetlands, oyster reefs, nesting areas, etc.).

- Need for more land.

In many cases, combinations of these problems exist that require a combination of structural measures together with nonstructural alternatives to be implemented within a comprehensive, coastal region, management plan for hazard reduction.

a. Combinations.

(1) Structural combinations.

(a) Beach stabilization structures and beach nourishment. Groins and detached breakwaters combined with beach fills are discussed in Part V-3-3e. The combination mitigates downdrift impacts and/or increases the fill life of the renourished beach. Together, their life-cycle costs and environmental impact may be less than if selectively implemented. Construction of the beach stabilization structures without fill is likely to damage adjacent beaches.

(b) Seawalls, revetments, and beach nourishment. The original design of the new seawall for hurricane protection at Virginia Beach was to be a curved, concrete-type structure as at Galveston, Texas (Figure V-3-5). To accommodate a lowered seawall crest for aesthetic reasons (see Part V-3-1-c-(5)and Figure V-3-6) a wide beach nourishment project was added to the design to reduce flooding and wave damage (USAED, Norfolk, 1994). Together with improved interior drainage and pumping equipment, this combined design provides the same hurricane flooding and wave damage protection as the original seawall design.

(c) Beach nourishment and rebuilt dune with buried seawall/revetment. The soft alternative (beach and dune with buried rock seawall/revetment) was determined to be both environmental and economically advantageous when compared against an armored revetment for storm protection against the 1 percent change storm event at Dam Neck, Virginia, on the Atlantic Ocean (Basco 1998). The final design cross section is shown in Figure V-3-36a and includes a buried rock seawall/revetment beneath the dune. In the event of a major storm causing severe dune erosion, the buried seawall will prevent storm damage if a second major storm occurs in the same season. Figure V-3-36b shows a photograph of dune construction with the buried, rock seawall (Basco 2000a). A similar approach was incorporated at Ocean City, Maryland, where a steel, sheet-pile bulkhead was incorporated as a buried backup feature in the design.

(2) Nonstructural and structural combinations.

At many locations, elevated structures combined with some type of armoring or shoreline stabilization together with beach nourishment are employed in combination for coastal hazard mitigation. Presently, 32 of 35 coastal states and territories have some type of setback requirements for new construction and existing structures found uninhabitable after a storm (Heinz Center 2000). These nonstructural, adaptive measures and structural alternatives are often combined to address the wide range of coastal problems previously cited. An example is the barrier island of Grand Isle, Louisiana, where beach nourishment, rebuilt dunes, a groin field and nearshore breakwaters are used for a community where the first floor of most residences are constructed above the 1 percent chance flood level (with waves) and the public lands on the rapidly eroding east end are restricted from any development (Pope 1997).

b. New technologies.

(1) Introduction.

Many nontraditional ways to armor, stabilize, or restore the beach including the use of patented, precast concrete units, geotextile-filled bags, and beach dewatering systems have been tried in the field. Their success depends on their stability during storm events and durability over the economic, design life. Their initial cost and cost for removal if environmental impacts warrant can be less than traditional methods, at some sites. These new technologies often involve nontraditional materials or shapes but are employed in a traditional manner, e.g., nearshore breakwaters. See Pope (1997) for more details.

a. Cross section

b. Photo during construction, Dam Neck, Virginia (from Basco 2000)

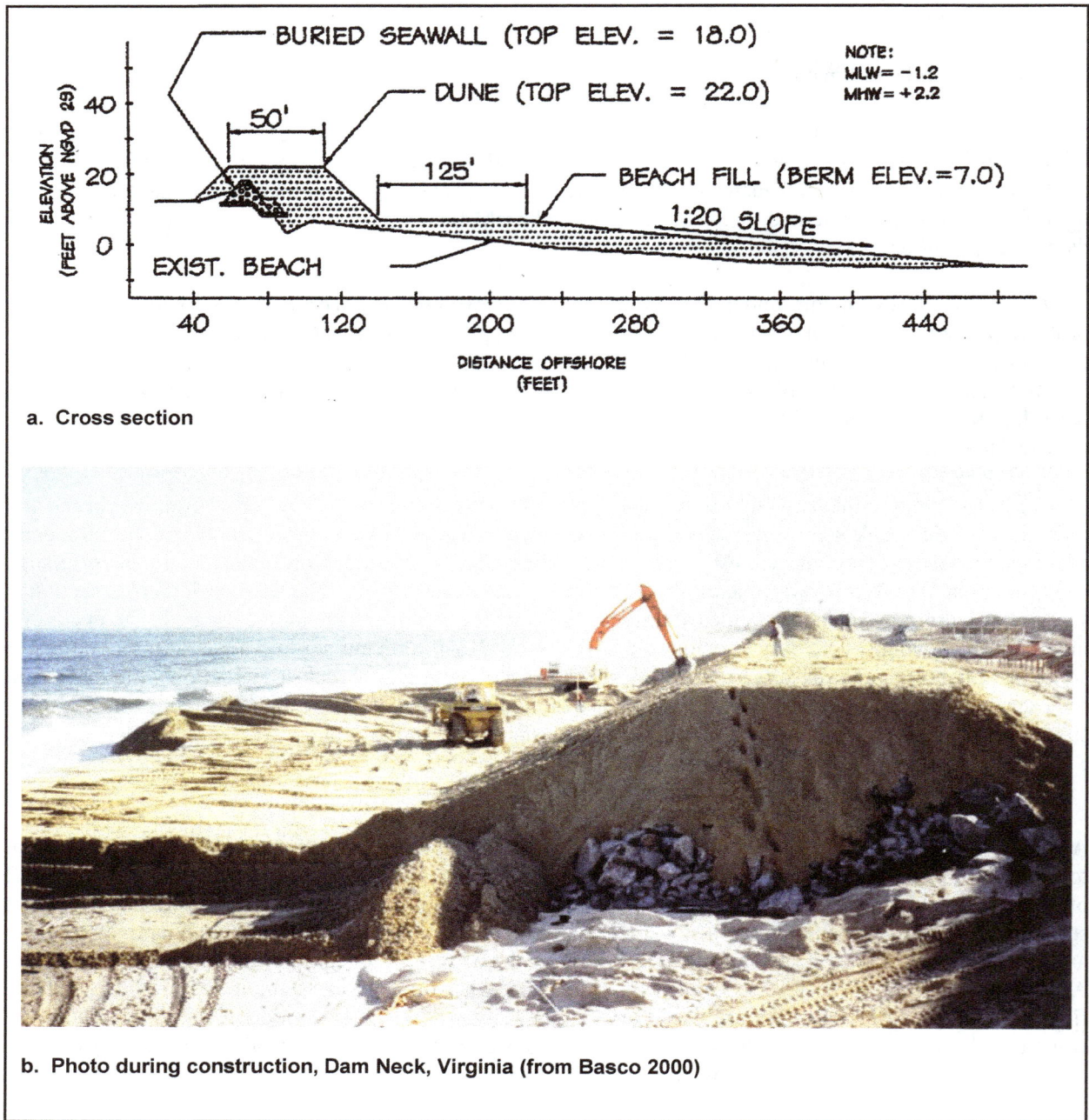

Figure V-3-36. Rebuilt dune with buried seawall (Basco 2000a)

(2) Precast, concrete units. Patented, modular, precast concrete units that can be interconnected to form one or more, nearshore breakwaters have been tested in Florida, Georgia, and New Jersey.

(a) In late 1989, two nearshore breakwater sections were installed at Sea Isle City, New Jersey, on the Atlantic Ocean. Each unit is 1.7 m long, 4.9 m wide and 2.1 m high and placed on a geotextile fabric. Each unit weighs about 12 metric tonnes and has 1.1 - 1.2 m freeboard at mean sea level (tidal range of 1.2 m). The two breakwaters were about 50 m in length with a 34-m gap and placed 37 m offshore (msl) in about 0.9-m water depth. This position was determined by the developer of the unit called the "Beachsaver" system

(Breakwaters International, Inc., Flemington, New Jersey). Independent monitoring over 18 months revealed significant structure settlement with the maximum over 1.2 m and average about 0.6 m (Sorensen & Weggel 1992). Most settlement occurred the first two months after installation. A large scour hole was found landward of one section and the sand trapping volume in two salients was only about 300 m³, before settlement (Sorensen & Weggel 1992).

(b) The Town of Palm Beach, Florida, in the early 1990s, experimented with another design labeled the "Prefabricated Erosion Protection" or P.E.P. reef system by its developer, American Coastal Engineering, Inc., West Palm Beach, Florida. Each unit is 3.7 m long, 4.6 m wide, 1.8 m high and is placed on a cloth filter fabric. Each unit weights about 22 metric tonnes with crest elevation below msl to act as a thin-crested, submerged, reef-type breakwater. One long breakwater/reef system, 1,270 m in length, placed 76 m offshore (msl) in 2.9 m National Geodetic Vertical Datum (NGVD) water depth consisting of 330 interlocking units was constructed. The submergence was 1.1 m at msl. Monitoring by the University of Florida's, Department of Coastal and Oceanographical Engineering included nearshore profiles and wave gauges landward and seaward of the structure (Dean and Chen 1996). About 17 percent of the units settled 0.8 m with the remainder settling 0.5 m. Wave transmission coefficients ranged from 0.65 for normal conditions (tide range 1.0 m) to 0.85 for storm conditions. A detailed analysis of the leeward project area revealed that wave overtopping, wave setup, pounding, excess longshore currents and sediment transport (both directions) resulted in an increased erosion trend relative to that present before construction (Dean and Chen 1996; Martin and Smith 1997). The conclusion, that the reef breakwater provided little benefit, resulted in its removal. The units have since been salvaged and used to construct a groin field with beach nourishment at the site (Erickson 1998).

(c) These negative results at West Palm Beach, Florida, resulted in a staggered and gapped, planform configuration being the modified plan for the next P.E.P. reef installation at Vero Beach, Florida, in the late 1990s. Monitoring is being conducted by the USACE (Stauble and Smith 1999) with the latest results reported by Wooduff and Dean (2000). Analysis of nearshore profiles over 25 months (August 1996 - June 1999) when compared with historic volume change (1972 - 1986) revealed that erosion has increased landward of the reef. And, background erosion decreased north but remained the same south in control areas outside the reef area (Stauble and Smith 1999).

(d) The Beachsaver breakwater unit and P.E.P. reef unit shapes are similar, namely triangular with flatter slope on the seaward side than on the landward end. The have been called pyramid-shaped. Both have a thin crest width. In contrast, a vertical, cylindrical shape (circular, concrete pipe) combined with a concrete base forms the Hollow Core Reef System (HCR) as patented by Hollow Core Reef Enterprises, Inc., Newport News, Virginia. Each unit is typically 3.6 m long at its base and 1.7 m wide, but the height varies (pipe length) and is specifically designed for intertidal regions with 1-3-m tidal range. Since 1996, a 116-m-long, nearshore breakwater consisting of 46 units, 2.6 m in height has been installed at Sea Island, Georgia, on the Atlantic Ocean (tide range 1.8 - 2.7 m). Each unit weighs about 22 metric tonnes. Freeboard above msl is unknown. The foundation is timber cross-members to form a subbase and initial settlement was measured at less than 0.5 m. The units have been stable since construction and remain in place after several hurricanes along the south Georgia coast.

(e) In summary, crest width, planform configuration, siting and foundation design appear to be the weakest aspect as revealed by field monitoring of precast, concrete, modules for breakwaters and submerged reefs.

(3) Geotextile filled bags.

(a) Geotextile materials or filter fabrics have a long history for foundation mats beneath rubble-mound structures (See Part VI-5-3); and, they have been used as silt curtains to contain dredged materials in the water column. They have also been formed into bags and long, sausage-shaped cylinders (called Longard Tubes) and filled with sand. They have been deployed as revetments for dune protection, as nearshore breakwaters, and as groins. In the 1980s and 1990s, there has been significant improvement in the quality and durability of geotextile fabrics, making them suitable for a variety of coastal applications.

(b) The design life of a geotextile filled bag depends on many factors. It is generally less than properly designed rock structures serving the same function. However, if found to cause negative impacts to adjacent shorelines, the bags can be cut open and removed with the filled sand remaining on the beach. It is for this reason that a soft groin field was permitted by the North Carolina Coastal Resource Commission (CRC) together with a beach nourishment project for South Beach on the western end of Bald Head Island, North Carolina (Denison 1998). The North Carolina CRC prohibits hardened structures on its ocean coast. Fourteen 2.75-m diameter and 100-m-long geotextile bags filled with sand were constructed to grade out to - 1.2 to - 1.5-m (mlw) depths at 120-m spacing to form the groin field. The beach was also prefilled with 496,960.7 cu m (650,000 cu yd) of sand to + 1.8 m (NGVD) berm elevation and 25 - 30-m berm width over 3.65 km. Two hurricanes in 1996 removed about 76,455.49 cu m (100,000 cu yd) from the beach, but caused no damage to the groin tubes (Denison 1998). The long-term survivability of this system has yet to be determined.

(4) Beach drains.

(a) Beach face dewatering by lowering the groundwater table along the coastline began in Denmark in the early 1980s, by accident. After installation of a filtered, seawater system for a seaside aquarium, it was discovered that the sandy beach width increased where the beach parallel, longitudinal pipe intake was buried beneath the surface (Lenz 1994). Patents were obtained by the Danish Geotechnical Institute (DGI) in many countries including the United States where the system is called Stabeach by the licensee, Coastal Stabilization, Inc., Rockaway, NJ.

(b) Lowering the groundwater table is accomplished by draining water from buried, almost horizontal, filter pipes running parallel to the coastline. The pipes are connected to a collector sump and pumping station further inland. Gravity drains the groundwater beneath the beach and through the pipes to the sump and then the water is pumped from the sump. The sand-filtered seawater can be returned to the sea or used for other purposes.

(c) Lenz (1994) describes laboratory experiments (no reference) and field tests at Stuart, Florida, and Englewood Beach, Florida. The patent belongs to Hans Vesterby, DGI, Denmark. Vesterby (1994) reviews the theory and design elements and describes field tests at three sites on the west coast of Denmark in the North Sea.

(d) Long-term, independent field monitoring is needed to learn more about the functional performance of dewatering systems. The system by itself does not produce new sand, so that its greatest contribution may be in increasing the fill life of renourished beaches.

(5) Innovative technology demonstration program.

(a) Many other ideas and devices have been deployed and/or proposed including beach cones (Davis and Law 1994); ultra-low profile, geotextiles injected with concrete (Janis and Holmberg 1994) and fishnets, stabilizers and artificial seaweed (Stephen 1994) for erosion mitigation. Most do not satisfactorily address nor answer the questions listed by Pope (1997). Alternative technologies for beach preservation was the theme for the 7th National Beach Technology Conference (Tait 1994, ed.).

(b) Section 227 of the Water Resources Development Act of 1996 authorized a National Shoreline Erosion Control Development and Demonstration (NSECDD) program. Emphasis is on "... the development and demonstration of innovative technologies" to advance the state of the art in coastal shoreline protection. Funding the 6-year effort began in fiscal year 2000. A minimum of seven projects on the Atlantic, Pacific, Gulf of Mexico, and the Great Lakes is mandated by this legislation. Pope (1997) discusses many issues surrounding the application of nontraditional and innovative technologies. The ability to perform their promised function, survive for a predictable life, impact on the environment, and total costs (initial and long-term maintenance) must be carefully examined. Pope (1997) then raises many questions that should be addressed, when considering innovative, alternative, shore protection approaches including:

- Is it heavy enough, particularly considering storm waves?

- Will it be properly anchored so that it doesn't fall apart?

- If the structure does fail, could the loose components become an environmental or public safety hazard?

- Will the installation be tolerant of erosion or scour effects around its base?

- Will the material from which it is being constructed last?

- What are the design criteria in terms of events for longevity?

- How will it perform and will it do what we want it to do?

- If it does perform as promised, could there be adverse impacts to adjacent areas?

- How much will constructing the nontraditional or innovative system cost compared to more traditional methods?

- What will it cost to maintain and can it be repaired when damaged?

- What is its effective (functional and economic) life?

- What will it cost to remove the system, if necessary?

(c) All of these questions and more will be addressed for the nontraditional and innovative technologies deployed in the National Demonstration Program. The search for new and innovative approaches has primarily been driven by the shift from hard to soft alternatives for shore protection (Figure V-3-3) and the need to reduce long-term costs of beach nourishment projects by increasing the time interval between renourishment events. As an example, precast, concrete modules, serving as nearshore breakwaters could replace conventional, rubble-mound breakwaters at some locations, and serve the same function of increasing the fill life of the renourished beach. If properly sited (see Part V-3-3) and set on a proper foundation, these units are attractive to permitting agencies because they have a smaller footprint on the bottom and can be adjusted or removed easily, if downdrift impacts are detected, by monitoring. Nontraditional and innovative technologies are subject to all the same design constraints previously discussed (Part V-3-1) plus the extra need to overcome previous shortcomings and prove that they work. The National Demonstration Program will greatly aid in this effort.

V-3-6. Do-Nothing

a. *Introduction.*

One final alternative that must always be evaluated is the do-nothing or no-project case. The risk of flooding and wave damage continues or increases if historic erosion is also present at the site. When this response is appropriate, what happens to the area, and what government programs are available to help are briefly discussed in the following paragraphs. Further details of the Federal government's response are found in Part V-8.

b. *Appropriate response.*

Whenever all structural and nonstructural alternatives considered are too costly, then no economically viable solution exists. If the life-cycle costs for protection or relocation exceed the value of the investment, then do-nothing is the appropriate response. This standard for economic feasibility is adopted automatically in Federal project studies of the Corps. If the benefit to cost ratio exceeds unity but social and environmental constraints govern, then the no-action alternative plan can become the Federally recommended plan. When the natural, coastal sediment transport processes (erosion and accretion) are the most important aspect (character, attractiveness, aesthetics, etc.) of the system, then do-nothing may also be the appropriate response. Many examples exist of highly dynamic barrier island systems that are best left alone. An example is the National Park Service policy. The exception for Cape Hatteras was the historic importance of the lighthouse as previously discussed. Individuals may also explicitly decide to take no action (flood proofing or retreat). The homeowner is willing to take the risk when the potential rental income from the property is high. If the house is eventually damaged or destroyed, it would still be covered by the NFIP, if a policy is in effect for the residence. The problem is when this no-action policy is taken, no NFIP policy exists, flooding damage takes place, and the Federal government declares the damaged region eligible for emergency financial assistance.

c. *After the flood.*

What happens when the do nothing alternative is selected by economics policy decisions? It is clear that the flood and wave damage potential remains, and the risk increases where erosion exists. It is almost certain that the area impacted will decline economically. Social and economic stresses will continue. Examples include: social stresses from apprehension and helplessness; economic stresses of depressed property values; personal property losses continue; cost and inconvenience of restoration after repetitive flooding continues; reduced recreational opportunities for citizens; reduced tourism benefits; reduced employment opportunities for tourism; property values decrease and related property taxes diminish. The no-action alternative may perpetuate a more costly Federal commitment than would be realized otherwise, because other Federal assistance programs exist (see Part V-8 for further details).

d. *Government programs available.*

When the no action plan remains, the Federal government relies on three methods to mitigate coastal damages and the possibility of loss of human life. As previously discussed (Part V-3-4), the NFIP provides compensation for flooding and wave damages, but it does not protect property from flooding. The NFIP also only encourages adaptation measures; but they are not mandatory, except for new construction where local authorities have adopted flood zoning ordinances. The Federal government also participates in measures for emergency, evacuation route planning. As discussed in Part V-8, because many coastal communities lack sufficient bridge and highway capacities, emergency evacuation is not a dependable means of hazard mitigation. The third program is Emergency Assistance from FEMA when the President declares the region eligible for this financial package after a major, coastal storm event.

e. National coastal hazard mitigation plan.

The need for a national plan of shore protection and coastal hazard mitigation is addressed in Part V-8.

V-3-7. References

EM 1110-2-1614
"Design of Coastal Revetments, Seawalls, and Bulkheads."

EM 1110-2-1617
"Coastal Groins and Nearshore Breakwaters."

EM 1110-2-1913
"Design and Construction of Levees."

EM 1110-2-2502
"Retaining and Flood Walls."

EM 1110-2-2705
"Structural Design of Closure Structures for Local Flood Protection Projects."

EM 1110-2-3102
"General Principles of Pumping Station Design and Layout."

ER 1105-2-101
"Planning Risk-Based Analysis for Evaluation of Hydrology/Hydraulics, Geotechnical Stability, and Economics in Flood Damage Reduction Studies."

Anderson, Hardaway, and Gunn 1983
Anderson, G. L., Hardaway, C. S., and Gunn, J. R. 1983. "Beach Response to Spurs and Groins," *Proceedings, Coastal Structures '83,* American Society of Civil Engineers, NY, pp 727-739.

Ahrens and Cox 1990
Ahrens, J. P., and Cox, J. 1990. "Design and Performance of Reef Breakwaters," *Journal of Coastal Research,*" Vol 6, No.1 pp 61-75

Bakker 1968
Bakker, W. T. 1968. "The Dynamics of a Coast with a Groyne System," *Proceedings, 11th International Coastal Engineering Conference,* ASCE, NY, pp 492-517.

Barnett and Wang 1998
Barnett, M. R., and Wang, H. 1998. "Effects of a Vertical Seawall on Profile Response," *Proceedings, 21st International Conference on Coastal Engineering,* ASCE, NY, pp 1493-1507.

Basco 1998
Basco, D. R. 1998. "The Economic Analysis of 'Soft' Versus 'Hard' Solutions for Shore Protection: An Example," *Proceedings, 26th International Conference on Coastal Engineering,* ASCE, NY, pp 1449-1460.

Basco 2000a
Basco, D. R. 2000a. "Beach Monitoring Results and Management Plan FCTCLANT, Dam Neck, Virginia Beach, VA," Final Report, Beach Consultants, Inc, Norfolk, VA.

Basco 2000b
Basco, D. R. 2000b. "Seawalls, Bulkheads and Revetments," *Coastal Engineering: Beyond It's Golden Anniversary*, Japan Press, Tokyo, Japan.

Basco and Ozger 2001
Basco, D. R., and Ozger, S. S. 2001. "Toward Quantification of the Active Sand Volume in the Surf Zone," Abstract, 28th International Conference on Coastal Engineering, (Cardiff), American Society of Civil Engineers, NY.

Basco et al. 1997
Basco, D. R., Bellomo, D. A., Hazelton, J. M., and Jones, B. N. 1997. "The Influence of Seawalls on Subaerial Beach Volumes With Receding Shorelines," *Coastal Engineering*, Vol 30, pp 203-233.

Balsillie and Berg 1972
Balsillie, J. H., and Berg, D. W. 1972. "State of Groin Design and Effectiveness," *Proceedings, 13th International Conference on Coastal Engineering*, ASCE, NY, pp 1367-1383.

Balsillie and Bruno 1972
Balsillie, J. H., and Bruno, R. O. 1972. "Groynes: An Annotated Bibliography," Miscellaneous Paper 1-72, U.S. Army Corps of Engineers, Coastal Engineering Research Center, Springfield, VA.

Berenguer and Enriquez 1988
Berenguer, J., and Enriquez, J. 1988. "Design of Pocket Beaches: The Spanish Case." *Proceedings, 21st International Conference on Coastal Engineering.* Malaga, Spain, pp 1411-25.

Bernd-Cohen and Gordon 1998
Bernd-Cohen, T., and Gordon, M. 1998. "State Coastal Program Effectiveness in Protecting Beaches, Dunes, Bluffs and Rocky Shores: A National Overview," *Final Report*, NOAA, National Ocean Service, Washington DC.

Bodge 1992
Bodge, K. R. 1992. "Gross Transport Effects at Inlets," *Proceedings, 6th National Conference on Beach Preservation Technology*, Florida Shore and Beach Preservation Assoc., Tallahassee, FL, pp 112-127.

Bodge and Dean 1987
Bodge, K. R., and Dean, R. G. 1987. "Short-Term Impoundment of Longshore Transport," *Proceedings Coastal Sediments '87*, ASCE, NY, pp 468-483.

Broadwell 1991
Broadwell, A. L. 1991. "Effects of Beach Renourishment on the Survivial of Loggerhead Sea Turtle," M.S. thesis, Florida Atlantic University, Boca Raton, FL.

Bruun 1952
Bruun, P. 1952. "Measures Against Erosion at Groins and Jetties," *Proceedings, 3rd International Conference on Coastal Engineering*, ASCE, NY.

Bruun 1962
Bruun, P. 1962. "Sea Level Rise as a Cause of Shore Erosion," *Journal, Waterways and Harbors Division*, ASCE, Vol 88, pp 117-130.

Chasten et al. 1993
Chasten, M. A., Rosati, J. D., McCormick, J. W., and Randall, R. E. 1993. "Engineering Design Guidance for Detached Breakwaters as Shoreline Stabilization Structure," Technical Report CERC-93-19, U.S. Army Engineer Waterways Experiment Station, Vicksburg, MS.

Civil Engineering 1999
Civil Engineering. 1999. "Back from the Brink," *Civil Engineering*, ASCE, NY, October, pp 52-57.

Cobb 1992
Cobb, Charles E., Jr. 1992. "Miami," *National Geographic*, Vol 181, No.1, pp 86-113.

Dally, Dean and Dalrymple 1985
Dally, W. R., Dean, R. G., and Dalrymple, R. A. 1985. "Wave Height Variation Across Beaches of Arbitrary Profile," *Journal of Geophysical Research,* Vol 90, No. C6, pp 11917-11927.

Dally and Pope 1986
Dally, W. R., and Pope, J. 1986. "Detached Breakwaters for Shore Protection," Technical Report CERC-86-1, U.S. Army Engineer Waterways Experiment Station, Vicksburg, MS.

Davis 1961
Davis, A. B. 1961. "Galveston's Bulwark Against The Sea: History of the Galveston Seawall," Technical Report, U.S. Army Engineer District, Galveston, Galveston, TX.

Davis and Law 1994
Davis, E. W., and Law, V. J. 1994. "Field Studies of Beach Cones as Coastal Erosion Control/Reversal Devices," *Proceedings, 7th National Conference on Beach Presentation Technology,* Florida Shore and Beach Preservation Assoc. Tallahassee, FL.

Dean 1978
Dean, R. G. 1978. "Coastal Structures and their Interaction with the Shoreline." *Application of Stochastic Processes in Sediment Transport.* H. W. Shen and H. Kikkaua, eds., Water Resources Publications, Littleton, CO, pp 18-1 to 18-46.

Dean 1987
Dean, R. G. 1987. "Coastal Armormy: Effects, Principles, and Mitigation," *Proceedings, 20th International Conference on Coastal Engineering,* ASCE, NY, pp 1843-1857.

Dean 1993
Dean, R. G. 1993. "Terminal Structures at the End of Littoral Systems," *Journal of Coastal Research*, Special Issue No. 18, pp 195-210.

Dean 1996
Dean, R. G. 1996. "Interaction of Littoral Barriers and Adjacent Beaches: Effects on Profile Shape and Shoreline Change," *Journal of Coastal Research*, Vol 23, pp 103-112.

Dean and Chen 1996
Dean, R. G., and Chen, R. 1996. "Performance of the Midtown Palm Beach PEP Reef Installation," Final Report, Department of Coastal and Oceanographic Engineering, University of Florida, Gaineville, FL.

Denison 1998
Denison, P. S. 1998. "Beach Nourishment/Groin Field Construction Project: Bald Head Island, North Carolina," *Shore and Beach*, Vol 66, No. 1, pp 2-9.

Dennis and Miller 1993
Dennis, W. A., and Miller, H. C. 1993. "Shoreline Response: Oregon Inlet Terminal Groin Construction," *Proceedings, International Coastal Engineering Symposium,* Hitton Head, SC, Vol I, pp 324-332.

Dolan and Lins 1987
Dolan R., and Lins, H. 1987. "Beaches and Barrier Islands," *Scientific American*, Vol 255, No. 7, pp 68-77.

Erickson 1998
Erickson, K. 1998. "Conversion of the P.E.P. Reef Breakwater to a Groin Field for West Palm Beach, FL," *Proceedings, Eighth Beach Technology Conference*, ASBPA, Tallahassee, FL.

Everts, Battley, and Gibson 1983
Everts, C. H., Battley, J. P., Jr., and Gibson, P. N. 1983. "Shoreline Movements: Cape Henry, Virginia to Cape Hatteras, North Carolina, 1849-1980," Technical Report CERC-83-1, U.S. Army Engineer Waterways Experiment Station, Vicksburg, MS.

Ferrante, Franco, and Boer 1992
Ferrante, A., Franco, L., and Boer, S. 1992. "Modeling and Monitoring of a Perched Beach at Lido di Ostia (Rome)," *Proceedings, 23rd International Conference on Coastal Engineering,* Vol 3, ASCE, NY, pp 3305-3318.

Fleming 1990
Fleming, C. A. 1990. "Guide on the Users of Groynes in Coastal Engineering," Report 119, Construction Industry Research and Information Assoc. (CIRIA), London, England.

Fulford 1985
Fulford, E. T. 1985. "Reef Type Breakwaters for Shoreline Stabilization," *Proceedings, Coastal Zone 85,* ASCE, NY.

Gilbert and Vellinga 1990
Gilbert, J., and Vellinga, P. 1990. "Strategies for Adaption to Sea Level Rise," Report of the Coastal Zone Management Subgroup, Intergovernmental Panel on Climate Change, World Meteorological Organization and U.N. Environmental Programme, The Netherlands.

Gourlay 1981
Gourlay, M. R. 1981. "Beach Processes in the Vicinity of Offshore Breakwaters," *Proceedings, 5th Australian Conference on Coastal and Ocean Engineering*, pp 129-134.

Graber 1980
Graber, P. H. F. 1980. "The Law of The Coast in a Clamshell, Part I: Overview of an Interdisciplinary Approach," *Shore and Beach*, Vol 48, No.4, pp 14-20.

Gravens and Kraus 1989
Gravens, M. B., and Kraus, N. C. 1989. "Representation of Groins in Numerical Models of Shoreline Response," *Proceedings, 23rd International Conference on Coastal Engineering*, ASCE, NY, pp 515-522.

Griggs et al. 1997
Griggs, G. B., Tait, J. F., Moore, L. J., Scott, K., Corona, W., and Pembrook, D. 1997. "Interaction of Seawalls and Beaches: Eight Years of Field Monitoring, Monterey Bay, California," Technical Report, CHL-97-1, U.S. Army Engineer Waterways Experiment Station, Vicksburg, MS.

Hall, Tschirky, and Turcke 1998
Hall, K. R., Tschirky, P., and Turcke, D. J. 1998. "Coastal Wetland Stability and Shore Protection," *Journal of Coastal Research*, Special Issue No.26, pp 96-101.

Hallermeier 1981
Hallermeier, R. J. 1981. "A Profile Zonation for Seasonal Sand Beaches from Wave Climate," *Coastal Engineering*, Vol 4, pp 253-277.

Hallermeier 1983
Hallermeier, R. J. 1983. "Sand Transport Limits in Coastal Structure Design." *Proceedings, Coastal Structures '83*. March 9-11, Arlington, VA, American Society of Civil Engineers, pp 703-16.

Hanson and Kraus 1989
Hanson, H., and Kraus, N. C. 1989. "GENESIS: Generalized Model for Stimulating Shoreline Change," Technical Report, CERC-89-19, U.S. Army Engineer Waterways Experiment Station, Vicksburg, MS.

Hardaway and Anderson 1980
Hardaway, C. S., and Anderson, G. L. 1980. "Shoreline Erosion in Virginia," *Sea Grant Program*, Virginia Institute of Marine Science, Gloucester Point, VA.

Hardaway and Byrne 1999
Hardaway, C. C., and Byrne, R. J. 1999. "Shoreline Management in Chesapeake Bay," *Virginia Sea Grant Publ.* VSG-99-11, Virginia Institute of Marine Science, Gloucester Point, VA.

Hardaway, Thomas, and Li 1991
Hardaway, C. S., Thomas, G. R., and Li, J. -H. 1991. "Chesapeake Bay Shoreline Study: Headland Breakwaters and Pocket Beaches for Shoreline Erosion Control," Final Report No.313, Virginia Institute of Marine Science, Gloucester Point, VA.

Hardaway and Gunn 1991
Hardaway, C. S., and Gunn, J. R. 1991. "Headland Breakwaters in the Chesapeake Bay," *Proceedings, Coastal Zone '91*, ASCE, NY, pp 1267-1281.

Hardaway and Gunn 1995
Hardaway, C. S., and Gunn, J. R. 1995. "Headland Breakwater Performance in Chesapeake Bay," *Proceeding 1995 National Conference on Beach Preservation Technology*, Florida Shore and Beach Preservation Assoc., Tallahassee, FL, pp 365-383.

Hardaway and Gunn 1998
Hardaway, C. S., and Gunn, J. R. 1998. "Chesapeake Bay: Design, Installation and Early Performance of Four New Headland Breakwater Composite Systems," *Proceedings 1998 National Conference on Beach Preservation Technology*, Florida Shore and Beach Preservation Assoc., Tallahassee, FL.

Hardaway and Gunn 1999
Hardaway, C. S., and Gunn, J. R. 1999. "Chesapeake Bay: Design and Early Performance of Four Headland Breakwater Systems," *Proceedings, Coastal Sediments 99*, ASCE, NY, pp 828-843.

Hardaway, Gunn, and Reynolds 1995
Hardaway, C. S., Jr., Gunn, J. R., and Reynolds, R. N. 1995. "Headland Breakwater Performance in Chesapeake Bay," *Proceedings, 1995 National Conference on Beach Preservation Technology, St. Petersburg, FL.*

Hardy and Young 1991
Hardy, T. A., and Young, I. R. 1991. "Modeling Spectral Wave Transformation on a Coral Reef Flat," *Proceedings, Australasian Conference on Coastal and Ocean Engineering,*" pp 345-350.

Harris and Herbich 1986
Harris, M. M., and Herbich, J. B. 1986. "Effects of Breakwater Spacing on Sand Entrapment," *Journal of Hydraulic Research* Vol 24, No. 5, pp 347-57.

Heikoff 1976
Heikoff, J. M. 1976. *Politics of Shore Erosion-Westhampton Beach, NY.* Ann Arbor Press, Ann Arbor, MI.

Heinz Center 2000
Heinz Center. 2000. "Evaluation of Erosion Hazards," The H. John Heinz III Center for Science, Economics and the Environment, Washington, DC, pp 203.

Hillyer 1996
Hillyer, T. M. N. 1996. "An Analysis of United States Army Corps of Engineers Shore Protection Program," Final Report, IWR 96-PS-1, Institute of Water Resources, U.S. Army Corps of Engineers, Alexandria, VA.

Houston 1995a
Houston, J. R. 1995. "Beach Nourishment," Coastal Forum, *Shore and Beach*, Vol 64, No.1, pp 21-24.

Houston 1995b
Houston, J. R. 1995. "The Economic Value of Beaches," *The CERCular*, Vol CERC-95-4, U.S. Army Engineer Waterways Experiment Station, Vicksburg, MS.

Houston 1996
Houston, J. R. 1996. "International Tourism and U.S. Beaches," *Shore and Beach*, Vol 165 No. 2 pp 3-4.

Houston 1999
Houston, J. R. 1999. "The Myth of the Subsidized Beach Resident," *Shore and Beach*, Vol 67, No. 2&3, pp 2-3.

Hoyt 1967
Hoyt, J. H. 1967. "Barrier Island Formation," *Geological Society America Bulletin*, Vol 78, pp 1125-1136.

Hsu, Silvester, and Xia 1987
Hsu, J. R. C., Silvester, R., and Xia, Y. M. 1989. "Static Equilibrium Bays: New Relationships," *Journal Waterways, Port, Coastal and Ocean Engineering*, ASCE, Vol 115 (No.3), pp 285-298.

Hsu, Silvester and Xia 1989
Hsu, J. R. C., Silvester, R., and Xia, Y. M. 1989. "Applications of Headland Control," *Journal Waterway, Port, Coastal and Ocean Engineering*, Vol 115, No. 3, ASCE, NY, pp 299-310.

Hughes 1994
Hughes, S. A. 1994. *Physical Models and Laboratory Techniques in Coastal Engineering*. World Scientific Publishing Co., Singapore.

Inman and Frautschy 1966
"Littoral Processes and the Development of Shorelines." *Proceedings, Coastal Engineering*. Santa Barbara, CA, pp 511-36.

Janis and Holmberg 1994
Janis, W. A., and Holmberg, D. L. 1994. "The Accretion Technology System," *Proceedings, 7th National Conference on Beach Preservation Technology*, Florida Shore and Beach Preservation Assoc., Tallahassee, FL.

Japanese Ministry of Construction 1986
Japanese Ministry of Construction. 1986. "Handbook of Offshore Breakwater Design," River Bureau of the Ministry of Construction, Japanese Government (Translated to English) U.S. Army Engineer Waterways Experiment Station, Vicksburg, MS.

Joyner, Overton, and Fisher 1998
Joyner, B. P., Overton, M. F., and Fisher, J. S. 1998. "Analysis of Morphology of Oregon Inlet, NC, Since Construction of a Terminal Groin," *Proceedings, 26th International Conference on Coastal Engineering*, ASCE, NY.

Kana 1999
Kana, T. W. 1999. "Long Island's South Shore Beaches: A Century of Dynamic Sediment Management," *Proceedings, Coastal Sediments '99*, ASCE, NY, pp 1584-1596.

Kassner and Black 1983
Kassner, J., and Black, J. A. 1983. "Inlets and Barrier Beach Dynamics: A Case Study of Shinnecock Inlet, NY," *Shore and Beach*, Vol 51, No.2, pp 22-26.

Kobayashi 1987
Kobayasi, N. 1987. "Analytical Solutions for Dune Erosion By Storms," *Journal of the Waterway Coastal and Ocean Engineering Division*, ASCE, Vol 113, No. 4, pp 401-418.

Kobayashi, Raichle and Asano 1993
Kobayashi, N., Raichle, A. W., and Asano, T. 1993. "Wave Attenvation by Vegetation," *Journal of Waterway, Port, Coastal and Ocean Engineering*, ASCE Vol 199, No. 4, pp 30-48.

Komar 1998
Komar, P. D. 1998. *Beach Processes and Sedimentation*. Second Edition. Prentice Hall, N.J.

Komar et al. 1991
Komar, P. D., Lanfredi, N., Baba, M., Dean, R. G., Dyer, K., Healy, T. Ibe, A. C., Terwindt, J. H. J., Thom, B. G. 1991. "The Response of Beaches to Sea-Level Changes: A Review of Predictive Models," *Journal of Coastal Research*, Vol 7, pp 895-921.

Kraft and Herbich 1989.
Kraft, K., and Herbich, J. B. 1989. "Literature Review and Evaluation of Offshore Detached Breakwaters," Report No. COE-297, Texas A&M University, Ocean Engineering Program, College Station, TX.

Kraus 1988
Kraus, N. C. 1988. "The Effects of Seawalls on the Beach: An Extended Literature Review," *Journal of Coastal Research*, Special Issue No.4, pp 1-29.

Kraus 2000
Kraus, N. C. 2000. "Coastal Forum-Groin Notching," *Shore and Beach*, Vol 68, No. 2, pp 18.

Kraus and Bocamazo 2000
Kraus, N. C., and Bocamazo, L. M. 2000. "State of Understanding of Groin Functioning and Recent Promising Innovations," *Proceedings, 3rd Annual Conference*, Northeast Shore and Beach Preservation Assoc.,

Kraus, Hanson, and Blomgren 1994
Kraus, N. C., Hanson, H., and Blomgren, S. H. 1994. "Modern Functional Design of Groin System," *Proceedings, 24th International Conference on Coastal Engineering*, ASCE, NY, pp 1327-1342.

Kraus and Larson 1991.
Kraus, N. C., and Larson, M. 1991. "NMLONG: Numerical Model for Simulating the Longshore Current," Technical Report DRP-91-1, U.S. Army Engineer Waterways Experiment Station, Vicksburg, MS.

Komar and McDougal 1988
Komar, P. D., and McDougal, W. G. 1998. "Coastal Erosion and Engineering Structures: The Oregon Experience," *Journal of Coastal Research*, Special Issue No.4, pp 79-94.

Kraus and McDougal 1996
Kraus, N. C., and McDougal, W. G. 1996. "The Effects of Seawalls on The Beach: Part I, An Updated Literature Review," *Journal of Coastal Research*, Vol 12, pp 691-701.

Kraus and Pilkey (eds.) 1988
Kraus, N. C., and Pilkey, O. H., (eds.) 1988. "Seawalls and Beaches," *Journal of Coastal Research*, Special Issue No. 4, Gainesville, FL.

Kriebel and Dean 1993
Kriebel, D. L., and Dean, R. G. 1993. "Convolution Method for Time-Dependent Beach-Profile Response," *Journal of Waterway, Port, Coastal and Ocean Engineering*, ASCE, Vol 119, No. 2, pp 204-227.

Krumbein 1944
Krumbein, W. C. 1944. "Shore Processes and Beach Characteristics," Technical Memorandum No.3, Beach Erosion Board, U.S. Army Corps of Engineers, Washington, D.C.

Larson and Kraus 1989
Larson, M., and Kraus, N. C. 1989. "SBEACH: Numerical Model for Simulating Storm-Induced BeachChange; Report 4, Empirical Foundation and Model Development," Technical Report CERC-89-9, U.S. Army Engineer Waterways Experiment Station, Vicksburg, MS.

Leatherman 1988
Leatherman, S. P. 1988. *Barrier Island Handbook.* Coastal Publication Series, University of Maryland, College Park, MD.

Leenknecht Szuawalski, and Sherlock 1992
Leenknecht, D. A., Szuwalski, A., and Sherlock, A. R. 1992. "Automated Coastal Engineering System (ACES) User Guide and Technical Reference, Version 1.07," U.S. Army Engineer Waterways Experiment Station, Vicskburg, MS.

Lenz 1994
Lenz, R. G. 1994. "Beachface Drainage-A Tool for Coastal Stabilization," *Proceedings, 7th National Conference on Beach Preservation Technology*, Florida Shore and Beach Preservation Assoc., Tallahassee, FL.

Lesnick 1979
Lesnick, J. R. 1979. "An Annotated Bibliography on Detached Breakwaters and Artificial Headlands," Miscellaneous Report 79-1, Coastal Engineering Research Center, U.S. Army Engineer Waterways Experiment Station, Vicksburg, MS.

Magoon 2000
Magoon, O. 2000. "Sand Rights," *Proceedings, 3rd Annual Conference*, Northeast Shore and Beach Preservation Association, Voorhees, NJ.

Martin and Smith 1997
Martin, T. R., and Smith, J. B. 1997. "Analysis of the Performance of the Prefabricated Erosion Prevention (PEP) Reef System, Town of Palm Beach, Florida," Coastal Engineering Technical Note, CETN-II-36, U.S. Army Engineer Waterways Experiment Station, Vicksburg, MS.

McDougal, Kraus and Ajiwibowo 1996
McDougal, W. G., Kraus, N. C., and Ajiwibowo, H. 1996. "The Effects of Seawalls on the Beach: Part II, Numerical Modeling of SUPERTANK Seawall Tests," *Coastal Sediments '87*, ASCE, NY, pp 961-973.

McDougal, Sturtevant and Komar 1987
McDougal, W. G., Sturtevant, M. A., and Komar, P. D. 1987. "Laboratory and Field Investigations of Shoreline Stabilization Structures on Adjacent Property," *Coastal Sediments '87*, ASCE, NY, pp 961-973.

Miller, Dennis, and Wutkowski 1996
Miller, H. C., Dennis, W. A., and Wutkowski, M. J. 1996. "A Unique Look at Oregon Inlet, NC, USA," *Proceedings, 25th International Conference on Coastal Engineering,* ASCE, NY, pp 4517-4529.

Mohr 1994
Mohr, M. C. 1994. "Presque Isle Shoreline Erosion Control Project," *Shore and Beach*, Vol 62, No. 2, pp 23-28.

Moody and Madsen 1995
Moody, P. M., and Madsen, O. S. 1995. "Laboratory Study of the Effect of Seawalls on Beach Erosion," Technical Report, MITSG 95-3, Massachusetts Institute of Technology, Cambridge, MA.

Moreno and Kraus 1999
Moreno, L. J., and Kraus, N. C. 1999. "Equilibrium Shape of Headland-Bay Beaches for Engineering Design," Proceedings, *Coastal Sediments '99*, ASCE, NY, pp 860-875.

National Research Council 1987
National Research Council. 1987. "Responding to Changes in Sea Level: Engineering Implications," Final Report, Committee on Engineering Implications of Changes in Relative Mean Sea Level, National Academy Press, Washington, DC.

National Research Council 1988
National Research Council. 1988. "Saving Cape Hatteras Lighthouse from the Sea," Final Report, Committee on Options for Preserving Cape Hatteras Lighthouse, National Academy Press, Washington, DC.

National Research Council 1990
National Research Council. 1990. "Managing Coastal Erosion," Final Report, Committee on Coastal Erosion Zone Management Water Science and Technology Board, National Academy Press, Washington, DC.

National Research Council 1995
National Research Council. 1995. "Beach Nourishment and Protection, Final Report, Committee on Beach Nourishment and Protection, Marine Board," National Academy Press, Washington, DC.

Nayak 1976
Nayak, U. B. 1976. "On the Functional Design and Effectiveness of Groins in Coastal Protection," Ph.D. diss., University of Hawaii, Honolulu, HI.

Nelson et al. 1987
Nelson, D. A., Mauck, K. A., and Fletemeyer, J. 1987. "Physical Effects of Beach Nourishment on Sea Turtle Nesting, Delray Beach, Florida," Technical Report, TR-87-15, U.S. Army Engineer Waterways Experiment Station, Vicksburg, MS.

Nersesian, Kraus, and Carson 1992
Nersesian, G. K., Kraus, N. C., and Carson, F. C. 1992. "Functioning of Groins at Westhampton Beach Long Island, New York," *Proceedings, 23rd International Conference on Coastal Engineering*, ASCE, NY, pp 3357-3370.

Nir 1982
Nir, Y. 1982. "Offshore Artificial Structures and Their Influence on the Israel and Sinai Mediterranean Beaches." *Proceedings, 18th International Conference on Coastal Engineering*. American Society of Civil Engineers, pp 1837-56.

Noble 1978
Noble, R. M. 1978. "Coastal Structures' Effects on Shorelines." *Proceedings, 17th International Conference on Coastal Engineering*. Sydney, Australia, American Society of Civil Engineers, pp 2069-85.

Noda 1984
Noda, H. 1984. "Depositional Effects of Offshore Breakwater Due to Onshore-Offshore Sediment Movement." *Proceedings, 19th International Conference on Coastal Engineering.* American Society of Civil Engineers, Houston, TX, pp 2009-25.

Oertel et al. 1992
Oertel, G. F., Kraft, J. C., Kearney, M. S., and Woo, H. J. 1992. "A Rational Theory for Barrier-Lagoon Development," *Society for Sedimentary Geology,* Special Publication No.48, pp 77-87.

Olin, Fischenich, Palermo 2000
Olin, T. J., Fischenich, J. C. and Palermo, M. R. 2000. "Wetlands Engineering Handbook," Technical Report, WRP-RE-21, U.S. Army Engineer Research and Development Center, Vicksburg, MS.

Overton et al. 1992
Overton, M. F., Fiser, J. S., Dennis, W. A., and Miller, H. C. 1992. "Shoreline Change At Oregon Inlet," *Proceedings, 23rd International Conference on Coastal Engineering,* ASCE, NY, pp 2332-2343.

Ozger 2000
Ozger, S. S. 2000. "Towards Quantification of Active Sand Volume in the Nearshore Profile," M.S. thesis, Civil Engineering Department, Old Dominion University, Norfolk, VA.

Pattiaratchi and Bancroft 2000
Pattiaratchi, C., and Bancroft, S. 2000. "Design Studies and Performance Monitoring of an Artificial Surfing Reef: Cable Station, Western Australia," *Abstracts, 27th International Conference on Coastal Engineering,* ASCE, NY, Paper No. 123.

Pendergrass and Pendergrass 1990
Pendergrass, B. B., and Pendergrass, L. F. 1990. "In the Era of Limits: A Galveston District History Update 1976-1986," U.S. Army Engineer District, Galveston, Galveston, TX.

Pilarczyk 1990
Pilarczyk, K. W., ed. 1990. *Coastal Protection.* A.A. Balkema, Rotterdam, The Netherlands.

Pilkey and Wright 1988
Pilkey, O. H., and Wright, H. L. 1988. "Seawalls Versus Beaches," *Journal of Coastal Research,* Special Issue No.4, pp 41-67.

Plant 1990
Plant, N. 1990. "The Effects of Seawalls on Beach Morphology and Dynamic Processes," unpublished M.S. thesis, University of California, Santa Cruz, CA.

Plant and Griggs 1992
Plant, N., and Griggs, G. D. 1992. "Interactions Between Nearshore Processes and Beach Morphology Near a Seawall," *Journal of Coastal Research,* Vol 8, No. 9, pp 183-200.

Pope 1989
Pope, J. 1989. "Role of Breakwaters in Beach Erosion Control," *Proceedings, Beach Preservation Technology '89,* Florida Shore and Beach Preservation Association, Tallahassee, FL, pp 167-176.

Pope 1997
Pope, J. 1997. "Responding to Coastal Erosion and Flooding Damages," *Journal of Coastal Research* Vol 13, No. 3, pp 704-710.

Pope and Dean 1986
Pope, J., and Dean, J. L. 1986. "Development of Design Criteria for Segmented Breakwaters," *Proceedings, 20th International Conference Coastal Engineering*, ASCE, NY, pp 2144-2158.

Rosati 1990
Rosati, J. D. 1990. "Functional Design of Breakwaters for Shore Protection: Empirical Methods," Technical Report, CERC-90-15, U.S. Army Engineer Waterways Experiment Station, Vicksburg, MS.

Rosati and Truitt 1990
Rosati, J. D., and Truitt, C. L. 1990. "Am Alternative Design Approach for Detached Breakwater Projects," Miscellaneous Paper, CERC-90-7, U.S. Army Engineer Waterways Experiment Station, Vicksburg, MS.

Rosen and Vadja 1982
Rosen, D. S., and Vadja, M. 1982. "Sedimentological Influence of Detached Breakwater." *Proceedings, 18th International Conference on Coastal Engineering*. American Society of Civil Engineers, pp 1930-1949.

Ruig and Roelse 1992
Ruig, J. H. M., and Roelse, P. 1992. "A Feasibilty Study of a Perched Beach Concept in the Netherlands," *Proceedings, 23rd International Conference on Coastal Engineering*, ASCE, NY, pp 2581-2598.

Sarb and Ewing 1996
Sarb, S., and Ewing, L. 1996. "Mitigation for Impacts of Seawalls on Sand Supply, Coastal Development Permit Condition and Findings of Support With Methodology and Graphics," Internal Memorandum, California Coastal Commission.

Scheffner et al. 1997
Scheffner, N. W., Borgman, L. E., Clasusner, J. E., Edge, B. L., Grace, P. J., Militello, A., and Wise, R. A. 1997. "Users Guide to the Use and Application of the Empirical Simulation Technique," Technical Report CHL-97-00, U.S. Army Engineer Waterways Experiment Station, Vicksburg, MS.

Scheffner et al. 1999
Scheffner, N. W., Clausner, J. E., Militello, A.,Borgman, L. E., Edge, B. L., and Grace, P. J. 1999. "Use and Application of the Empirical Simulation Technique: User's Guide," Technical Report CHL-99-21, Engineer Research and Development Center, Corps of Engineers, Vicksburg, MS.

Scientific American 1888
Scientific American. 1888. "Relocation of Brighton Beach Hotel on Coney Island, NY," *Scientific American* Vol VIII, No.15, p 230.

Seelig and Walton 1980
Seelig, W. N., and Walton, T. L. 1980. "Estimation of Flow through Offshore Breakwater Gaps Generated By Wave Overtopping," *Coastal Engineering Technical Aid* (CETA) 80-8, Coastal Engineering Research Center, Fort Belvoir, VA.

Seiji, Uda, and Tanaka 1987
Seiji, M., Uda, T., and Tanaka, S. 1987. "Statistical Study on the Effect and Stability of Detached Breakwaters," *Coastal Engineering in Japan*, Vol 30, No. 9, pp 131-141.

Shepard 1976
Shepard, F. P. 1976. "Coastal Classification and Changing Coastlines," *Geoscience and Man*, Vol 14, pp 53-64.

Shore Protection Manual 1984
Shore Protection Manual. 1984. 4th Edition Vol. 1 & 2, U.S. Army Engineer Waterways Experiment Station, U.S. Government Printing Office, Washington, DC.

Silvester 1960
Silvester, R. 1960. "Stabilization of Sedimentary Coastlines," *Nature*, Vol 188, Paper 4744, pp 467-469.

Silvester 1970
Silvester, R. 1970. "Development of Crenulate-Shaped Bays to Equilibrium," *Journal Waterways and Harbors Division*, ASCE, Vol 96 (WWZ) pp 275-287.

Silvester and Hsu 1993
Silvester, R., and Hsu, J. R. C. 1993. *Coastal Stabilization: Innovative Concepts.* PRT Prentice-Hall, Englewood Cliffs, NJ.

Sonu and Warwar 1987
Sonu, C. J., and Warwar, J. F. 1987. "Evolution of Sediment Budgets in the Lee of a Detached Breakwater." *Proceedings, Coastal Sediments '87.* American Society of Civil Engineers, New Orleans, LA, pp 1361-68.

Sorensen 1990
Sorensen, T. 1990. "Review of Coastal Engineering Works at the Pengkalan Datu Seaworks," Final Report Danish Hydraulic Institute, Horsholm, Denmark.

Sorensen and Weggel 1992
Sorensen, R. M., and Weggel, J. R. 1992. "Field Monitoring of a Modular Detached Breakwater System," *Proceedings, Coastal Engineering Practice 1992*, ASCE, New York, pp 189-203.

Spencer and Terchunian 1997
Spencer, R., and Terchunian, A.V. 1997. "The Sand Thieves of Long Island's South Shore," *Shore and Beach*, Vol 65, No.3, pp 4-12.

Stauble and Smith 1999
Stauble, D. K., and Smith, J. B. 1999. "Performance of the PEP Reef Submerged Breakwater Project, Vero Beach, Florida," Twenty-Five Months Report, Coastal Engineering Research Center, U.S. Army Engineer and Research Development Center, Vicksburg, MS.

Stephen 1994
Stephen, M. F. 1994. "Assessment of Alternative Erosion Control Technology in Southwest Florida: A Saga of Fishnets, Stabilizers and Seaweed," *Proceedings, 7th National Conference on Beach Preservation Technology*, Florida Shore and Beach Preservation Assoc., Tallahassee, FL.

Suh and Dalrymple 1987
Suh, K., and Dalrymple, R. A. 1987. "Offshore Breakwaters in Laboratory and Field," *Journal, Waterway, Port, Coastal, and Ocean Engineering*, ASCE, Vol 113, No. 2, pp 105-121.

Tait 1994
Tait, L. S. ed. 1994. "Alternative Technologies in Beach Preservation," *Proceedings, 7th National Conference on Beach Preservation Technology*, Florida Shore and Beach Preservation Assoc., Tallahassee, FL.

Tait and Griggs 1991
Tait, J. F., and Griggs, G. B. 1991. "Beach Response to the Presence of a Seawall: A Comparison of Field Observations," *Shore and Beach*, Vol 58, No.2, pp 11-28.

Terchunian 1988
Terchunian, A. V. 1988. "Permitting Coastal Armoring Structures: Can Seawalls and Beaches Coexist?," Special issue No.4, *Journal of Coastal Research*, pp 65-75.

Terich and Levenseller 1986
Tercih, T., and Levenseller, T. 1986. "The Severe Erosion of Cape Shoalwater, Washington," *Journal of Coastal Research*, Vol 2, pp 465-477.

Thompson et al. 1996
Thompson, E. F., Wutkowski, M., and Scheffner, N. W. 1996. "Risk-Based Analysis of Coastal Projects," *Proceedings, 25th International Conference on Coastal Engineering*, ASCE, NY, pp 4440-4450.

Tomlinson 1980
Tomlinson, J. H. 1980. "Groynes in Coastal Engineering: A Literature Survey and Summary of Recommended Practice," Report No. IT 199, Hydraulics Research Station, Wallingford, England.

Toue and Wang 1990
Tove, T., and, Wnag, H. 1990. "Three Dimensional Effects of Seawall on The Adjacent Beach," *Proceedings, 22nd International Conference on Coastal Engineering*, ASCE, NY, pp 2785-2795.

Toyoshima 1972
Toyoshima, O. 1972. *Coastal Engineering for Practicing Engineers-Beach Erosion.* Morikita Publishing Co., Tokyo, Japan, pp 227-317. (English translation of Chapter 8, "Offshore Breakwaters," available through the Coastal Engineering Research Center, U.S. Army Engineer Waterways Experiment Station, Vicksburg, MS.)

Toyoshima 1974
Toyoshima, O. 1974. "Design of a Detached Breakwater System." *Proceedings, 14th International Conference on Coastal Engineering.* American Society of Civil Engineers, Copenhagen, Denmark, pp 1419-31.

Toyoshima 1982
Toyoshima, O. 1982. "Variation of Foreshore Due to Detached Breakwaters," *Proceedings, 18th International Conference on Coastal Engineering*, ASCE, NY, pp 1873-1892.

Tschirky, Turke, and Hall 2000
Tschirky, P., Turke, D., and Hall, K. 2000. "Wave Attenuation by Emergent Wetland Vegetation," Abstracts, *27th International Conference on Coastal Engineering*, ASCE, NY, paper no. 224.

Turner et al. 2000
Turner, I., Leyden, V., Symonds, G., McGrath, J., Jackson, A., Jancar, T., Aarninkhof, S., and Elshoff, I. 2000. "Comparison of Observed and Predicted Coastline Changes at the Gold Coast Artificial (Surfing) Reef," *Abstracts, 27th International Conference on Coastal Engineering*, ASCE, NY, Paper No. 125.

U.S. Army Corps of Engineers 1993
U.S. Army Corps of Engineers. 1993. "Wave Attenuation Over Reefs," Coastal Engineering Technical Note, CETNI-56, U.S. Army Engineer, Waterways Experiment Station, Vicksburg, MS.

U.S. Army Engineer District, Galveston 1975
U.S. Army Engineer District, Galveston. 1975. Burnett, Crystal, and Scott Bays and Vicinity, Baytown, Texas," Feasibility Report, Galveston, TX.

U.S. Army Engineer District, Galveston 1979
U.S. Army Engineer District, Galveston. 1979. "Baytown, Texas," Phase I, Phase II, Design Memorandum, Galveston, TX.

U.S. Army Engineer District, New York 1958
U.S. Army Engineer District, New York. 1958. "Atlantic Coast of Long Island, NY, Fire Island Inlet to Montauk Pout: Beach Erosion Control and Interim Hurricane Protection Study," Survey Report, New York.

U.S. Army Engineer District, Norfolk 1994
U.S. Army Engineer District, Norfolk. 1994. "Hurricane Protection Project for Virginia Beach, VA," *General Design Memorandum*-Revised, Norfolk, VA.

U.S. Army Engineer District, Wilmington 1996
U.S. Army Engineer District, Wilmington. 1996. "Cape Hatteras Lighthouse, North Carolina Fourth Groin Alternative," *Design Report and Environment Assessment*, Wilmington District Office, Wilmington, NC.

Vesterby 1994
Vesterby, H. 1994. "Beach Face Dewatering-The European Experience," *Proceedings, 7th National Conference on Beach Preservation Technology*, Florida Shore and Beach Preservation Assoc., Tallahassee, FL.

Vrijling 1990
Vrijling, J. K. 1990. "Probabilistic Design of Flood Defenses," *Coastal Protection*. K. W. Pilarczyk, ed., A.A. Balkema, Rotterdam, The Netherlands, pp 39-97.

Wallace and Cox 1997
Wallace, S., and Cox, R. 1997. "Seagrass and Wave Hydrodynamics," *Proceedings of Combined Australian Coastal Engineering and Ports Conference*, Christchurch, NZ, Vol I, pp 69-73.

Walker, Clark, Pope 1980
Walker, J. R., Clark, D., and Pope, J. 1980. "A Detached Breakwater System for Shore Protection." *Proceedings, 17th International Conference on Coastal Engineering.* American Society of Civil Engineers, Sydney, Australia, pp 1968-87.

Walker, Clark, Pope 1981
Walker, J. R., Clark, D., and Pope, J. 1981. "A Detached Breakwater System for Beach Protection," *Proceedings, 17th International Conference on Coastal Engineering*, ASCE, NY, pp 1968-1987.

Walton and Sensabaugh 1979
Walton, T. L., and Sensabaugh, W. 1979. "Seawall Design on the Open Coast," Florida Sea Grant College Rept. No.29, University of Florida, Gainesville, FL.

Weggel 1988
Weggel, R. 1988. "Seawalls: The Need for Research, Dimensional Consideration and a Suggested Classification," Special Issue No.4, *Journal of Coastal Research*, pp 29-40.

Weggel and Vitale 1985
Weggel, J. R., and Vitale, P. 1985. "Sand Transport Over Weir Jetties and Low Groins," *Physical Modeling in Coastal Engineering*. A.A. Balkema, Boston, MA.

Weigel 1991
Weigel, R. L. 1991. "Protection of Galveston, Texas From Overflows by Gulf Storms: Grade Raising, Seawall and Embankment," *Shore and Beach*, Vol 59, No. 7, pp 4-10.

Woodruff and Dean 2000
Woodruff, P. E., and Dean, R. G. 2000 "Innovative Erosion Control Technology in Florida," *Book of Abstracts 27th International Conference on Coastal Engineering*, Vol 2, ASCE, NY.

WRDA 1996
WRDA. 1996. "National Shoreline Erosion Control Development and Demonstration Program," Section 227 Water Resources Development Act (WRDA) of 1996, *Congressional Record*, Washington, DC.

Yasso 1964
Yasso, W. E. 1964. "Plan Geometry of Headland-Bay Beaches," Final Report, Department of Geology, Columbia University, New York.

Young 1989
Young, I. R. 1989. "Wave Transformation Over Coral Reefs," *Journal of Geophysical Research*, Vol 94, No C7, pp 9779-9789.

Zyserman et al. 1998
Zyserman, J. A., Broker, I., Johnson, H. K., Mangor, K., and Jorgensen, K. 1998. "On The Design of Shore-Parallel Breakwaters," *Proceedings, 26th International Conference Coastal Engineering*, ASCE, NY, pp 1693-1705.

V-3-8. Definition of Symbols

γ_b	Breaker depth index (Equation II-4-3) [dimensionless]
B	Minimum beach width at mhw after nourishment [length]
d_{50}	Size of the 50th percentile of the sediment sample [length]
d_S	Average depth at breakwater below mean water level [length]
e	Erosion of shoreline (mhw) from design storm [length]
F_B	Breakwater freeboard, mhw to crest [length]
I_S	Beach response factor (Equation V-3-10) [dimensionless]
L_g	Gap distance between adjacent breakwaters [length]
L_S	Length of breakwater structure [length]
P_A	Erosion rate after human activities [length³/time]
P_N	Natural background erosion rate [length³/time]
Q_{gross}	Gross longshore sediment transport rate [length³/time]
Q_{net}	Net longshore sediment transport rate [length³/time]
$Q_{offshore}$	Offshore sediment transport rate for design storm [length³/time]
R_p	Coastal erosion ratio [dimensionless]
R_v	Coastal erosion volume ratio [dimensionless]
V_A	Volume loss rate after construction of roads, seawalls, etc. [length³/time]
V_N	Natural erosion volume loss rate [length³/time]
W	Width of design beach nourishment [length]
WTR	Seawall trap ratio [dimensionless]
X_g	Longshore groin spacing [length]
Y	Distance of breakwater from nourished shoreline [length]
Y_g	Maximum indentation under normal wave conditions [length]
Y_g	Effective groin length measured from its seaward tip to the design shoreline for beach fill at the time of construction [length]
Y_{min}	Minimum dry beach width [length]
Y_{min}	Minimum distance from base (reference) line to mhw shoreline after design storm event [length]
Y_o	Distance of breakwater from original shoreline [length]
Z_S	Backshore elevation at baseline [length]

V-3-9. Acknowledgments

Author of Chapter V-3, "Shore Protection Projects:"

David R. Basco, Ph.D., Department of Civil Engineering, Old Dominion University, Norfolk, Virginia.

Reviewers:

Andrew Morang, Ph.D., Coastal and Hydraulics Laboratory (CHL), U.S. Army Engineer Research and Development Center (ERDC), Vicksburg, Mississippi.
Joan Pope, CHL

Table of Contents

List of Figures

List of Tables

Chapter V-4
Beach Fill Design

V-4-1. Engineering Aspects of Beach-Fill Design

a. Project objectives.

(1) The primary function of a Federal beach nourishment project is to provide improved protection to upland structures and infrastructure from the effects of storms. Figure V-4-1 shows how storms damage upland property. The top panel shows the beach under normal wave and water level conditions. The letters mhw and mlw denote the mean high and low water lines, respectively, which represent the normal range of tidally-induced water level fluctuations. The elevation of the natural beach berm is above the normal high tide elevation. Under nonstorm conditions, breaking waves are confined to the seaward face of the berm. The berm and dune act as a protective buffer between upland structures and the water and waves. Sites with little or no sand buffer are candidates for shore protection projects. The middle panel shows the beach during a storm. The water level is elevated above the normal range. This exposes higher beach elevations to the action of breaking waves, which erode the berm and transport the sand offshore and along the beach. Sand moved offshore during the storm continues to aid in dissipating wave energy; it often forms a shore-parallel bar system. In this panel, a scarp forms on the seaward face of the dune, but the dune remains relatively intact and protects the structures behind it. Had there been no dune, or a much narrower berm, the structure may have been damaged as shown in the bottom panel. The lower panel shows a more severe condition, in which even higher water levels and wave action have completely eroded the dune. Some of the dune material is transported offshore. Some is transported onshore and deposited as a result of wave runup and overwash processes. The exposed structure is subject to damage by undermining, flooding, and by waves breaking directly on it.

(2) A beach nourishment project typically involves constructing a wider beach and/or more substantial dune to reduce storm damage relative to the level of damage that would have resulted without the project. The level of storm protection provided by a nourishment project is not an absolute measure due to the uncertainties in the frequency of high intensity storms. There is always some chance, or risk, that a storm will cause property damage even with the project in place. The level of protection, reduced in the aftermath of a major storm, will remain compromised if proper poststorm maintenance is not performed. The level of protection will also be compromised if scheduled periodic renourishment, which is usually a key element of the design, is not performed when needed.

(3) The wider beach created through construction of a nourishment project also provides recreational benefits. Enhanced recreational opportunity can also be a project objective.

b. Project features. Beach nourishment projects typically involve construction of one or several of the following features: berm, dune, feeder beach, nearshore berm, dune stabilization (i.e., sand fences or vegetation), or structural stabilization (i.e., groins). There are also several aspects of a beach nourishment project that specifically address the future integrity of the dune/berm. These include: periodic renourishment, advance nourishment, and emergency maintenance. These project features are discussed in more detail in the following paragraphs.

(1) Beach berm.

(a) The beach berm is the primary feature of most beach nourishment projects. Most beaches have a natural berm or berms. The lowest berm closest to the water is formed by the uprush of wave action during the ordinary range of water-level fluctuations. Sometimes, several berms will be noticeable at slightly higher

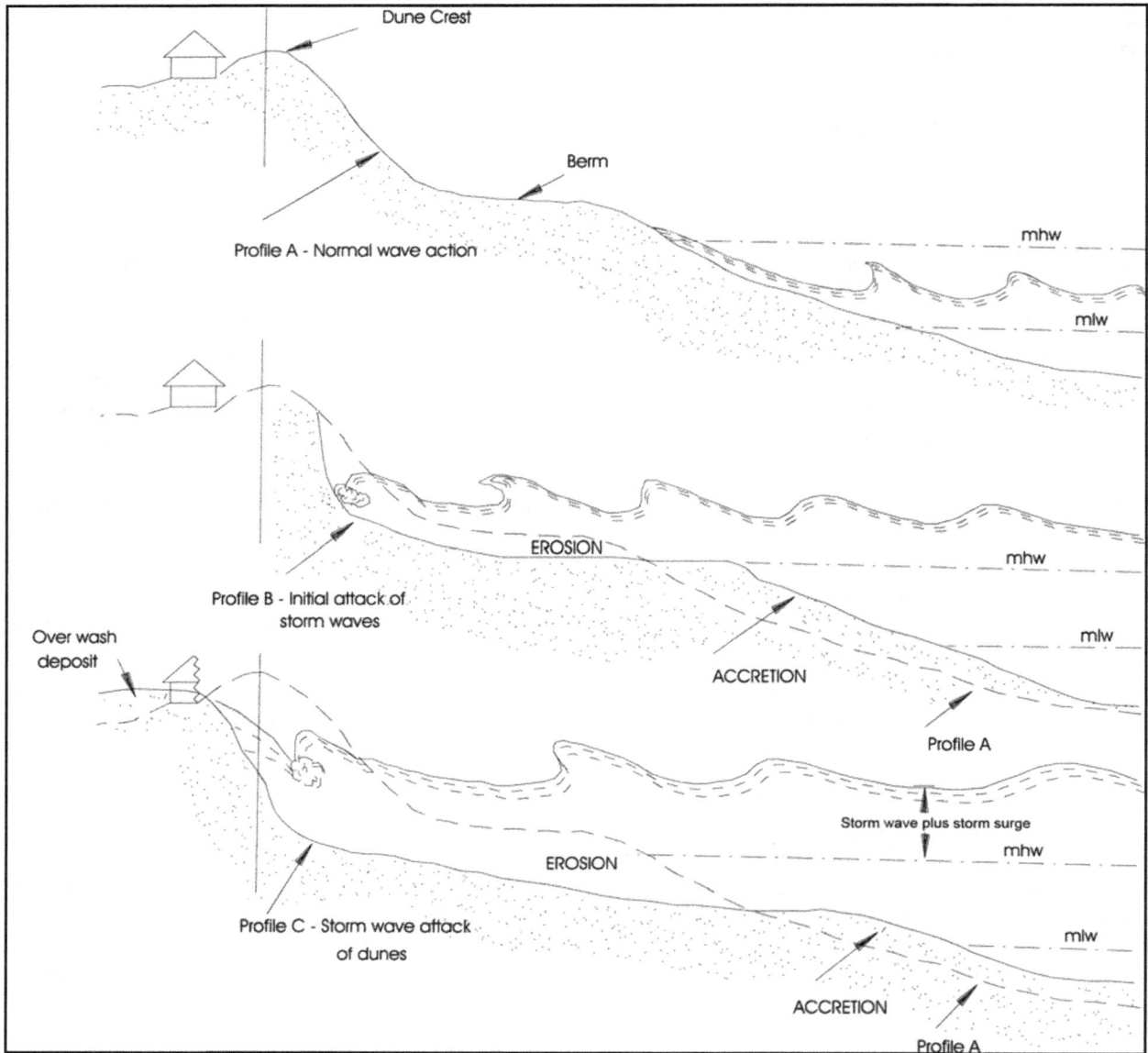

Figure V-4-1. Schematic diagram of storm wave attack on a beach, dune, and upland structures

elevations on the beach. These were formed during previous storms and are either remnant scarps left behind as a result of erosion of the lower berm or as a result of deposits left from wave uprush during higher-than-normal water levels associated with storms. Beaches that are in a severely eroded condition might have little or no berm at high tide.

(b) A nourishment project usually involves widening the beach (i.e., translating it seaward) to create a wider sand buffer for dissipating storm wave energy. The amount of additional width is determined based on the desired level of storm protection, the persistent long-term erosional trends that characterize the project area, and the target renourishment interval. The design berm width is determined through an iterative process that evaluates economic benefits as a function of width. The elevation of the constructed berm is usually set at the same elevation as the natural berm, or slightly higher. If the berm is constructed at an elevation lower than the natural berm crest, a ridge will form along the seaward edge of the fill. Wave uprush during higher water will overtop the ridge, causing temporary undesirable ponding on the beach until the berm elevation increases naturally. If the berm is constructed to an elevation that is much higher than the natural berm crest, an undesirable persistent scarp may form along the shoreline.

(c) For practical and economic reasons, during the construction of the beach berm, the total fill volume required to advance the berm to the desired width is placed on the visible portion of the beach. This construction method, sometimes referred to as the "over-building" method, enables the economic use of standard earth-moving equipment for the distribution of the fill and minimizes relocation of the discharge point. The method also enables effective verification that the sectional fill volume (design fill volume per unit length of shoreline) has been placed on the beach by the contractor using standard land-based survey techniques. The result of this construction technique is a beach berm that is initially considerably wider than the target design width. Postconstruction berm widths are often two to three times wider than the target design width. Design specifications required a 30-m (100-ft) design berm width in the northern New Jersey beach nourishment project while postconstruction surveys indicated constructed berm widths of 60 to 90 m (200 to 300 ft) and along some survey lines berm widths approaching 120 m (400 ft) were measured. While recognized by project designers, lay persons are often unaware that the initial overbuilt berm is a temporary condition. Consequently, they incorrectly judge the project as a failure when the beach berm adjusts dramatically landward immediately after construction, especially during the first storm season. For this reason, it is important that public outreach programs include easy-to-understand information concerning the expected remolding of the fill material after initial construction. Figure V-4-2 provides a schematic illustration of the preproject beach profile, the post-construction over-built beach berm, and the expected design berm after cross-shore equilibration.

(2) Dune.

(a) Sand dunes are an important protective feature. Naturally occurring dune ridges along the coast prevent storm tides, wave runup, and overtopping from directly damaging oceanfront structures and flooding interior areas. Dune features of beach nourishment projects are intended to function in the same way. A beach nourishment project may involve either reinforcing an existing natural dune by adding elevation and/or cross-sectional area, or constructing a dune where none existed beforehand. Dunes also provide a small reservoir of sand for nourishing the beach during severe storms. However, after the storm, maintenance is required to rebuild the dune because natural dune rebuilding occurs at a much slower rate than natural berm rebuilding.

(b) During hurricanes and very severe northeasters, substantial sections of dune can disappear. This is caused by offshore transport of dune material into the surf zone and by beach and dune sediments being swept landward by wave uprush and overtopping. In the case of overtopped barrier islands, flooding from ocean-side storm surges and waves and/or return flow of water from flooded bays can erode enough sand to cut shallow channels, or breaches, through the island. Occasionally, the channels will evolve into new inlets. Areas most prone to breaching are those where the barrier island is narrow and the dunes are lowest or nonexistent. The crest elevations of natural dunes often varies considerably, and nature tends to erode the low spots first. Dunes and berms built as part of a nourishment project can reduce the potential for barrier island breaching because a relatively uniform dune elevation eliminates the low spots where breaches are most likely to form.

(c) Dune growth can be promoted and the dune structure can be made more resistant to erosion if suitable vegetation can be grown on the dunes for an adequate length of time to establish an extensive root system. It generally takes 2 to 5 years for beach grass to establish a healthy root system, and up to 10 years before the maximum resistance to erosion and breaching is obtained. An active grass fertilization and

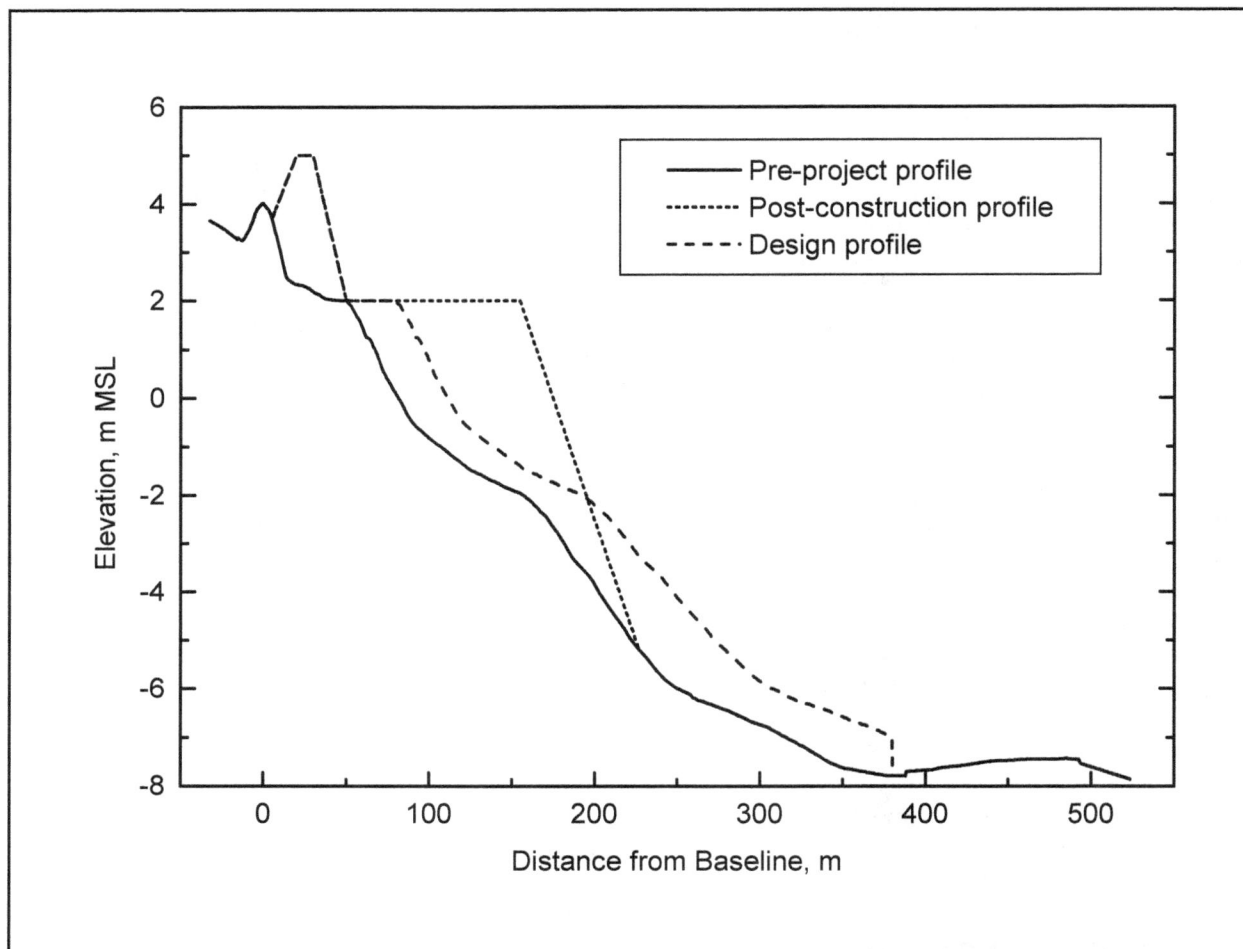

Figure V-4-2. Schematic illustration of pre-project, post-construction and design profile

maintenance program can greatly enhance the survival and effectiveness of beach grass. Part V-4-1-i provides more detailed information on dune stabilization.

(d) Preliminary estimates of dune height, width, and side slope for a beach nourishment project can be made based on characteristics of natural dunes in the vicinity of the project. To be most effective, the crest height of the dune should be above the limit of wave runup for the types of storms for which protection is sought, and the beach berm in front of the dune must be of sufficient width to withstand the erosion associated with these types of storms. If the berm in front of the dune is too narrow, the dune can quickly erode, even for relatively frequent storms, and the benefits of the higher dune elevations will be lost. The design dune height and width (along with the width of the berm) are usually selected based on results of an iterative process in which the benefits are compared with the cost of each configuration. Part V-4-1-f provides more detail on the dune/berm design process. Sometimes other factors such as real estate acquisition issues or aesthetics are factored into the selection of dune crest elevation and width.

(3) Nearshore berm. Beach-fill projects are usually constructed via direct placement of sand on the beach. Sometimes, in an effort to reduce cost, or because of limits in available dredge equipment, or in response to concerns with direct placement of fill material on the beach, material is placed offshore in an underwater berm (i.e., bar). Nearshore berm creation is intended to simulate storm bar formation by creating an artificial shore-parallel bar to dissipate storm wave energy before impacting the beach. If the berm is

placed shallower than the depth of closure (see Part III-3-3-b), it is expected that with time, the material will move onshore, eventually becoming part of the beach berm and beach face system. The material placed in the nearshore berm, however, does not provide direct shore protection to the upland, and prototype experience with the practice has met with mixed results and uncertain net benefit.

(4) Feeder beach. Beach-fill projects usually involve placement of a berm along a finite length of shoreline. Sometimes, beach nourishment projects include the creation of a feeder beach, in which fill material is introduced at the updrift end of the area intended to receive the fill. Then, longshore transport distributes the fill to the rest of the project area. Feeder beaches work best in areas that serve as a source of littoral material for downdrift beaches, in areas that are presently experiencing a deficit in the supply of littoral material and have unusually high loss rates, and in areas where the net longshore transport direction is predictable and the net transport rate is "strong" (i.e., longshore sand transport in one direction greatly exceeds the transport in the other direction). Candidate sites for feeder beaches include areas immediately downdrift from inlets or other manmade structures that form a littoral barrier or in areas that have been identified as erosional "hot spots." As the feeder material spreads under the influence of waves, the orientation of the feeder beach shoreline approaches that of the adjacent beach, resulting in longshore transport out of the feeder area equal to the transport along the adjacent area. Eventually, the shoreline orientation in the feeder beach area will return to its original configuration. Protection provided by feeder beaches will not have the same degree of alongshore uniformity as that provided by placing fill in a prescribed manner throughout the project area.

(5) Structures in conjunction with beach nourishment. Structures can enhance the performance of a beach nourishment project. Figure V-4-3 shows several such examples. When the project is relatively short in length, or significantly affected by an inlet, it may be desirable to limit alongshore losses through the use of a terminal structure or structures (Figures V-4-3a and V-4-3b show examples). Another use of structures is to place them in the interior of the nourishment project, with the intent of increasing project longevity (Figure V-4-3c) by reducing the longshore sand transport rate and minimizing end losses. Structures can also be used locally within a project to maintain the desired level of protection. For example, structures may be used to compartmentalize and stabilize a beach in anticipation of, or in response to, an area of unusually high volume losses (i.e., presence of a hotspot). Whenever structures are used, their potential updrift and downdrift impacts should be assessed. It is important to note that structures do not create sand, they only control its movement. If structures are built without adding beach fill, then sand may accumulate at one location at the expense of erosion at another area. As a general rule, compartments between structures, and the beach immediately updrift and/or downdrift of the structures, should be filled with sand to minimize adverse effects on adjacent beaches. Potential adverse effects of a groin field can be minimized by tapering the lengths of groins at the end of the groin field (see Figure V-4-3c), and adding sand on the downdrift side of the project. Part V-4-1-i provides additional information about the functional design of structures used in conjunction with beach fill projects. Part VI presents information on the structural design aspects of coastal structures.

    c. Define regional setting and site history. To maximize a beach nourishment project's effectiveness, it is important to understand the project's physical setting. The term "setting" encompasses local- and regional-scale coastal processes, the geology, and infrastructure that characterize the site and surrounding area. Part V-3 provides a general discussion of site characterization for all coastal projects. The following sections focus on those aspects that are most pertinent to engineering design of beach nourishment projects. A beach fill project can be a significant perturbation to the coastal system, and project performance is directly related to its interaction with its surroundings. The following are a few questions regarding project setting that should be answered at the beginning of the design process: In what type of littoral system, or littoral cell, will the project be constructed?

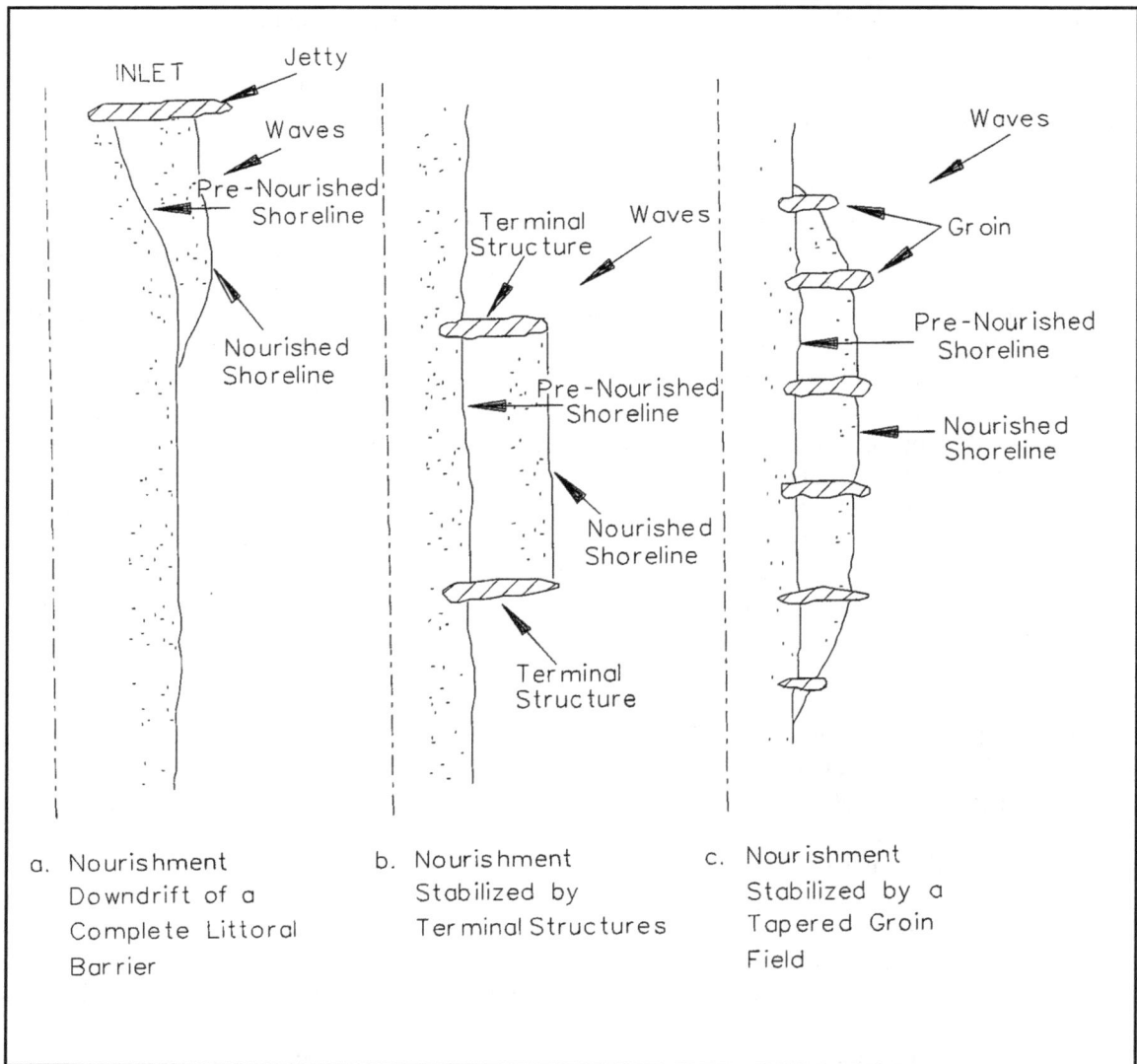

Figure V-4-3. Use of groin structures to enhance beach nourishment projects

What is the extent of the littoral cell, and where within the cell will the project reside? What are the important coastal processes that characterize the littoral cell and the project site? The answers to these questions will influence how a project is designed, how it performs, and what if any impacts it may have on adjacent beaches.

(1) Site location within littoral cell. The location of the project within a littoral cell has great influence on all aspects of a beach nourishment project, from understanding the underlying coastal processes at work, to determining the design concept that might be most effective, to the choice of methods and tools used in the design process. For example, will the project be constructed well within the interior of a littoral cell, miles from the inlets that serve as boundaries of the cell, or will it be built immediately downdrift of an inlet stabilization structure, or along the throat of an unstructured inlet? Sand transport by tidal currents, wave refraction over complex bathymetry, and wave-current interaction may not be important design issues in the former case whereas these processes may be critical in designing an effective project in the latter. In the former case, the project may experience beach recovery following a storm, where much of the sand moved offshore into a storm bar, migrates back onto the shore face with little net loss of sand. In the latter case, sand

moved offshore can be swept into tidal channels, carried away by currents into the inlet's shoals, and not return to the beach face. This will result in a significant permanent loss of sand. Different design concepts may be utilized in each of these two cases. In the former case, beach fill alone may meet the design requirements. In the latter, structures such as T-groins, may be needed to improve beach-fill retention. Different design methods may be utilized in each case. The former case may lend itself to simple analytical design tools, whereas the latter case may require more sophisticated physical or numerical modeling to aid in the design.

(2) Pathways of sediment movement.

(a) Because beach nourishment projects involve placing additional sand into the littoral system, project design is greatly aided by knowing how sand presently moves within the project domain and the littoral cell that is the setting for the project. Specifically, the following pieces of information are valuable: knowledge of the quantities of sand that presently enter and leave the littoral cell, where the sand enters and leaves, and what quantity moves through the project site. The more information known about the movement of sand the better, although this "picture" can be difficult to develop, especially for project sites with sparse data. Estimates of these quantities are often developed through formulation of a sand budget for the littoral cell and project domain. Parts IV and VI discuss development of a sediment budget. The following are types of questions regarding pathways and quantities of sand movement that might be important to the design of the project. If a project is to be constructed within a littoral cell flanked by inlets, are the inlets bypassing sand into the cell? If so, how much is bypassed? Are structures present that might be blocking (partially or completely) the flux of sand into the site from adjacent beaches? Is the project site located on a convex stretch of the barrier island that may be undergoing persistent erosion, or is the site in an area that experiences intermittent periods of erosion and accretion? Is the project to be located in a pocket beach flanked by headlands, and are those headlands blocking longshore sand transport? Are there sources of sand within the littoral cell, such as a river or lagoon, that might periodically discharge sediments to the beach system, or bluffs behind the beach that might serve as a sediment source? Are there sinks within the littoral cell, such as offshore canyons, impoundment fillets updrift of coastal structures, or loss of sand into an adjacent inlet and its shoal system? The closer to inlets, the greater the chance that tidal currents will have an important role in defining sediment pathways. It is important to gain an understanding of the littoral processes at work in the region, including the magnitude and direction of longshore sand transport (net and gross transport rates), sand sources and sinks, and the effects of existing coastal structures on the movement of sand. Parts III-2 and III-3 address the subject of longshore and cross-shore sediment transport processes, respectively, in greater detail.

(b) Historical and current charts, maps, and aerial photographs provide valuable information about the regional setting for a project. They can be valuable data sets for characterizing littoral processes at a project site, and aid in developing a sand budget. A persistent signature of impoundment at coastal structures over several years provides evidence of predominant wave direction and net longshore sand transport direction. Formation and evolution of spits, or migration paths of submerged relic ebb tidal shoals, can provide the same information. Noticeable changes in shoreline orientation and curvature, or persistent changes in bathymetric contour orientation may indicate gradients in longshore sand transport rates or a change in net transport direction. Shoreline positions that are accurately digitized from properly rectified aerial photos and/or charts provide information for identifying current and past erosional and accretional areas, and for calculating shoreline change rates. Calculated change rates can be used to estimate changes in sand volumes along different portions of the beach. Nautical charts and bathymetric surveys show the presence of canyons and proximity of the canyons to the shore, tidal channels, shoals, other morphologic features, and changes in these features through time. Controlled bathymetric surveys (relative to common horizontal and vertical datums) can be analyzed to determine volume changes for use in formulating a sand budget.

(3) Beach topography.

(a) The shape of the beach, above and below water, sheds light on the coastal processes that are at work at the project site. Beach shape also is an important factor in determining the quantity of beach nourishment material needed. The existing beach profile shape seaward of the natural berm crest can be a good indicator of the expected postnourishment beach shape, provided the sand to be placed has similar grain size characteristics as the native beach and there are no coastal structures or other features that are controlling the shape of the beach. One situation where the present profile shape may not be a good indicator of the postconstruction shape would be a beach that is heavily seawalled or reveted, with little to no dry beach in front of the structure at high tide. In this instance, the present beach shape might be unnatural (overly steep) due to the loss of a sand supply at the shoreline (i.e., the beach profile is very much starved of sand). Proximity to a tidal channel or a coastal structure may also produce an "unnatural" beach profile shape, compared to the shape that might exist for the same sediment characteristics but located away from the inlet or coastal structure. Also, if the material to be placed has grain size characteristics that are much different than the native beach, then the present beach profile shape may not be a good indicator of the shape following nourishment. Use of sand finer than the native sand will produce a beach with gentler slopes; use of coarser material will result in a beach with steeper slopes.

(b) Dunes are also an important aspect of project setting. Dune elevation, continuity of the dune "line," position of the dunes relative to the shoreline, and volume of sand in the dune above the natural berm crest elevation are important factors in determining the existing level of protection to property and infrastructure. Well-established vegetated dunes are usually a sign of a healthy beach system. A scarped dune, evidence of regular overwash, or no dune at all indicate an area vulnerable to storm damage. Beach profile surveys, or fully 3-D bathymetric and topographic surveys acquired by lidar systems such as systems such as SHOALS, can effectively characterize the beach/dune system as well as submerged morphologic features. Analysis of well-controlled beach profiles (relative to consistent horizontal and vertical datums) can provide volume change information for use in developing a sand budget, selecting design profiles for the nourishment project, and estimating the required amount of nourishment material.

(4) Sediment characteristics.

(a) Information about the grain size characteristics of the native beach material can shed light on the coastal processes at work. Systematic variations in median grain diameter along the beach, or evidence of natural tracers in the sand, may suggest the direction of net longshore transport. Grain size characteristics are a critical design parameter. Most often, sand with grain size characteristics similar to those of the native beach is sought as beach-fill. This is done to maximize compatibility with the existing beach system. Indirectly, selecting compatible material also maximizes the accuracy of predictions of future project performance, which is often based on past observations of the native beach response. Occasionally, fills are designed using material with different size properties because of limitations on sand availability and the cost to transport it to the project site. Sometimes the choice of a nourishment material with different characteristics is made to satisfy a particular design objective, such as use of coarser-grained fill material to improve resistance to erosion.

(b) Grain size characteristics are quantified based on a sieve analyses of samples which are collected throughout the project domain. Those samples acquired on the profile between the berm crest (or mean high water line) and a water depth corresponding to the position of the typical storm bar should be used to characterize native beach sand for the purpose of assessing the compatibility of sand from potential borrow sources. Compatibility of borrow and native beach material is primarily based on grain size characteristics, and to a lesser extent on color. Part V-4-1-e discusses sediment characterization and compatibility of fill in more detail.

(5) Wave and water level climate.

(a) The wave and water level conditions at the project site represent the major forces that shape the beach, and determine both the longer-term lateral spreading of material comprising the beach nourishment project and the short-term response of the project to storms. Exposure of the project site to wave energy from various directions determines the predominant longshore sand transport rate and direction. Offshore islands or coastal structures, peninsulas, or adjacent land masses may partially or completely shelter the project site from certain waves. The presence of these features modifies the energy, frequency, and directional characteristics of the incident waves that approach the site from deeper water. The presence of submerged offshore shoals, reef outcroppings, shore-attached shoals, or depressions (in general, any irregular bathymetry), also can have a significant persistent influence on local wave transformation and the longshore sand transport regime created by the incident breaking waves. It is important to assess the degree in which the wave climate varies from one end of the project to the other. Persistent variations in wave conditions of 5 to 10 percent can create significant local differences in project performance.

(b) There are several time scales of importance with regard to wave climate. The design life of a beach nourishment project is usually tens of years. Periodic renourishment is typically done every 3 to 5 years. There is evidence of cyclical patterns in weather (e.g., El Niño) that vary on the order of years, which would tend to produce wave climate patterns having similar multiple-year cycles. Annual changes in weather, and annual variations in the frequency and intensity of storm activity create a longshore sand transport regime that can vary considerably from year to year. Definition of the wave climate at these different time scales helps in assessing long-term beach nourishment project performance, potential variation in performance from one renourishment cycle to the next, and from year to year.

(c) Wave and water level conditions that accompany extratropical and tropical storms are also an important aspect of project setting and project design. The time scale of these events is on the order of days. Since most beach nourishment projects are justified based on storm protection benefits, it is important to quantify the frequency and severity of storms that can impact the project site during its design life. Lower-intensity storms typically erode the beach berm. More severe storms, particularly those with higher water levels, can inundate the berm and focus direct wave attack on dunes and exposed upland property. The most important storm parameter influencing beach and dune erosion is maximum water level, followed by wave energy and storm duration.

(d) Hindcast wave and water level information, or data measured nearby, can be used to characterize the wave and water level climate near the project site at the time scales of importance. Sometimes wave information is available in deep water, and other methods must be used to "transform" the information to the project site (e.g., to transform deepwater wave information past islands or very irregular bathymetry). Part II-2 discusses methods for developing wave climate information, including wave hindcasting. Part II-3 discusses techniques for estimating nearshore waves, including wave transformation methods. Part II-5 discusses methods for estimating water levels due to storms.

(6) Existing structures and infrastructure.

(a) How threatened are commercial and private structures, and infrastructure such as roads and utilities, by storm waves and water levels? Formulation of a beach nourishment project often involves compiling an inventory and description of the location of existing infrastructure and commercial and private properties to assess their vulnerability. The properties are valued to aid in quantifying the benefits that accrue from construction of a beach nourishment project. Are there structures that protrude beyond the predominant shoreline position? If so, it may be difficult and not cost-effective to provide a lasting beach having the design width in front of those structures. Attempting to do so may result in a persistent apparent erosional "hot-spot" in these locations. Aerial photos provide a good source of information for describing the characteristics of infrastructure and property, and the positions of structures relative to the present shoreline. Historical surveys or charts are useful in determining the degree to which existing infrastructure encroaches upon, or is seaward of, the historical beach location. Ground inspections and photos also can be used to characterize the condition and value of structures.

(b) The presence of existing coastal structures, and their characteristics, are also important parameters. What measures are already in place to provide protection to structures (e.g., seawalls and revetments), and what is the condition and effectiveness of those structures? Structures that alter or block the alongshore movement of sand, such as groins, detached breakwaters, or artificial headlands influence the pathways of sand movement at the site. Crest elevation, composition, and condition of groins, revetments, and seawalls determine the structures' functional effectiveness. The "signature" of impoundment adjacent to groins, or lack thereof, can be an indicator of groin functionality. Aerial and ground photos can be used to characterize the condition and effectiveness of existing coastal structures. Engineering drawings that show the subsurface characteristics of the structures, including any toe protection, in concert with present beach profile surveys, can be used to characterize the vulnerability of revetments and seawalls to damage caused by recession of the beach during storms.

(7) Prior engineering activities.

(a) Usually, areas being considered for a beach nourishment project have experienced problematic erosion for some time. Often, there is a record of previous studies and perhaps a record of past engineering activities at the site. This information can shed light on what may or may not work in the future, and why; and aid in designing a nourishment project. For example, records of past beach-fill activities can provide information about expected fill longevity and net longshore transport rates. Historical records documenting impoundment at groins can provide information about net longshore sand transport rates. Dredging and placement records are vital information in the development of a sand budget. The record of past engineering activities may also help explain a particular beach response that has been observed. Compiling a complete chronology and record of past engineering activities can prove to be a very valuable design aid.

(b) This section briefly highlighted some of the more important aspects of project setting and site characterization and history, and how they relate to design of beach nourishment projects. Subsequent sections provide more detailed information about how these factors enter into the design process.

d. Reach delineation.

(1) In addition to the setting for a beach nourishment project and its design features, an equally important design issue is delineation of sections of coast along which a project is to be constructed. Project economics is often a controlling factor in the process of reach delineation. The values of property and infrastructure assets fronting the beach and benefits gained by providing storm-damage reduction enter in assigning bounds of a project reach or subreaches. Project boundaries may be determined by limits of political jurisdiction such as municipal or state boundaries, or may coincide with physical features such as inlets or headlands. Environmental considerations and local preferences may also influence project boundaries.

(2) From an engineering perspective, reach delineation should be evaluated based on physical processes controlling project response. For example, location and characteristics of project terminal boundaries may be evaluated on the basis of fill retention within project bounds and project impacts on adjacent shorelines. Where reaches terminate along the open coast, fill transitions may be used to reduce the rate of spreading losses from project bounds. Transitions may be placed either within or outside project bounds based on design objectives and/or constraints. A more detailed discussion of beach fill transitions is provided in Part V-4-1-h. In cases where reaches terminate at shore-normal structures (e.g., groins and jetties), effects of interrupting littoral drift on downdrift beaches should be assessed. Whether boundaries occur at structures

or on the open coast, project reaches should be of sufficient length to minimize the effect of end losses on the central portion of the project. Part V-4-1-g discusses design parameters and environmental processes that affect project longevity. Economic analyses may identify several discontinuous subreaches within the project main reach (e.g., where areas of development are separated by areas of undeveloped coast). In such cases, a project may function more effectively by designing a single reach spanning both developed and undeveloped sections to avoid multiple areas of end losses.

(3) Beach-fill projects have often employed a uniform design template, with constant berm width and dune elevation, along the entire project. Under most circumstances, however, improved performance can be achieved by modifying the design template along specific subreaches where longshore nonuniformity exists in the without-project condition. For example, consider an existing condition where a particular shore-fronting structure is positioned closer to the shoreline than adjacent structures. A design calling for a uniform width of beach in front of all structures along the entire project reach will produce a planform perturbation at the protruding structure, and will lead to an ongoing problem of accelerated recession in front of the structure. In this case, a viable alternative may be to design a narrower berm in the area of the protrusion with additional storm damage protection provided by higher dunes or protective structures such as seawalls. Other cases where nonuniform design templates may be appropriate include presence of erosional hot spots, changes in shoreline orientation, or nonuniform placement of shore protection structures along the project reach. In design, a practical goal is to distribute the sand fill volume alongshore so as to yield a more or less uniform shoreline location after initial equilibration of the placed fill. Care must be taken in using a variable design template to avoid compromising project performance. For example, dunes placed in front of a developed subreach to prevent overtopping and flooding may not be functional if lower dunes or no dunes are placed on adjacent subreaches of undeveloped shore. In this case, storms could erode and overtop the adjacent low dunes and flank the high dunes, resulting in flooding of the developed area.

   e. Evaluate sediment sources.

(1) Borrow source types. Borrow sources for beach fill can be divided into four general categories: terrestrial, backbarrier, offshore, and navigation channels. Each category has favorable and unfavorable aspects; however, selection of an optimum borrow source depends more on individual site characteristics relative to project requirements than type of source. The single most important borrow material characteristic is the sediment grain size.

(a) Terrestrial sources. Terrestrial sources of beach-fill material can be found in many coastal areas. Ancient fluvial and marine terrace and channel deposits, and certain glacial features such as eskers and outwash plains often contain usable material. Because of their potential economic value, information on sand and gravel deposits is often collected by state geological surveys. With this information, field investigations can focus on a few likely sources, thus eliminating need for more general exploration. In some places, commercial sand and gravel mining operations may provide suitable material for direct purchase. In their absence, it would be necessary to locate a suitable deposit and set up a borrow operation specifically for the project. Use of terrestrial borrow sites usually involves lower costs for mobilization-demobilization operations and plant rental, and less weather-related downtime than the use of a submerged borrow source. However, the production capacity of terrestrial borrow operations is comparatively low, and haul distances may be long. Thus, costs per unit volume of placed material may exceed those from alternate submerged sites. In general, terrestrial borrow sources are most advantageous for projects where exploration and mobilization-demobilization costs, relative to the cost of fill, are a large part of overall expense of the operation. Unfavorable aspects of terrestrial borrow sources are typically related to adverse secondary impacts caused by mining and overland transport. Compared to hydraulic placement, mechanical (dump truck) placement of fill additionally results in practical limits in fill volume, and fill placement is mostly limited to the dry and intertidal beach. Consequently, more rapid equilibration and recession of the placed fill is experienced.

(b) Backbarrier sources. Sediment deposits in the backbarrier marsh, tidal creek, bay, estuary, and lagoon environments behind barrier islands and spits have been used in the past for beach fill. They are an attractive source because they are protected from ocean waves and are often close enough to the project beach to allow direct transfer of the material by pipeline. This eliminates the need for separate transport and transfer operations. However, most backbarrier sediments are too fine-grained to use as beach fill. In addition, many backbarrier areas are highly important elements in the coastal ecosystem and are sensitive to disturbance and alteration by dredging. Material in backbarrier sediments coarse enough for beach fill is generally confined to overwash deposits and flood tidal shoals associated with active or relic inlets. Overwash deposits occur on the landward margin of the barrier where storm waves have carried beach and dune sediments across the island or spit. Flood tidal shoals occur inshore of tidal inlets and consist of sediment transported by tidal currents flowing in and out of the inlet. These sediments are usually derived from littoral drift from adjacent beaches. Overwash deposits and relict flood tidal shoals may be ecologically important because they may provide suitable substrate for marsh growth. In addition, on retreating barriers, they may comprise a reserve of sand that will be recycled into the active beach deposits as retreat progresses. Flood tidal shoals at an active inlet may be suitable for borrow sites because the material removed is likely to be replaced by ongoing inlet processes. However, dredging material from active flood tidal shoals can adversely alter both the hydraulic conditions in the inlet and wave action on adjacent shores. A study of the hydraulic effects should be made prior to dredging flood-tidal shoals.

(c) Harbors, navigation channels, and waterways. Creation of harbors, navigation channels, and waterways, and deepening or maintenance dredging of existing navigation projects often requires the excavation and disposal of large volumes of sediment. In some cases, where the dredged sediment is of suitable quality, it can be used as fill on nearby beaches rather than placing it in offshore, upland, or contained disposal sites. Operations of this type are economically attractive because dual benefits are realized at considerably less cost than if both operations were carried out separately.

- Maintenance dredging of projects in low energy environments such as estuaries or protected bays is least likely to produce suitable beach-fill material. In such areas, the dredged material often consists of clay, silt, and very fine sand. However, when dredging new harbors, channels or waterways, or deepening existing channels in low energy areas, the dredge may cut into previously undisturbed material of suitable characteristics.

- Dredged material from higher energy areas, such as rivers above tidewater and open coast inlet shoals, is often more acceptable for beach fill. On barrier coasts, inlets trap beach sediment that has been carried to the inlet by littoral drift. Therefore, material dredged from inlets is typically similar to the native material on the project beach. However, sediment compatibility tests should be performed to determine its suitability for use as beach fill.

(d) Offshore sources. Investigations of potential offshore sources of beach-fill material under the Coastal Engineering Research Center's (CERC's), Inner Continental Shelf Sediments Study (ICONS), by the USACE Districts and others (i.e., Bodge and Rosen (1988)), indicate that large deposits of suitable material often occur in offshore deposits. Data from the Atlantic coast show that the most common suitable sources are in ebb tidal shoals off inlets, and in linear and cape-associated shoals on the inner continental shelf, such as those shown in Figure V-4-4. Potential sources on the inner shelf have also been identified in submerged glaciofluvial features, relic stream channels, and featureless sheet-type deposits.

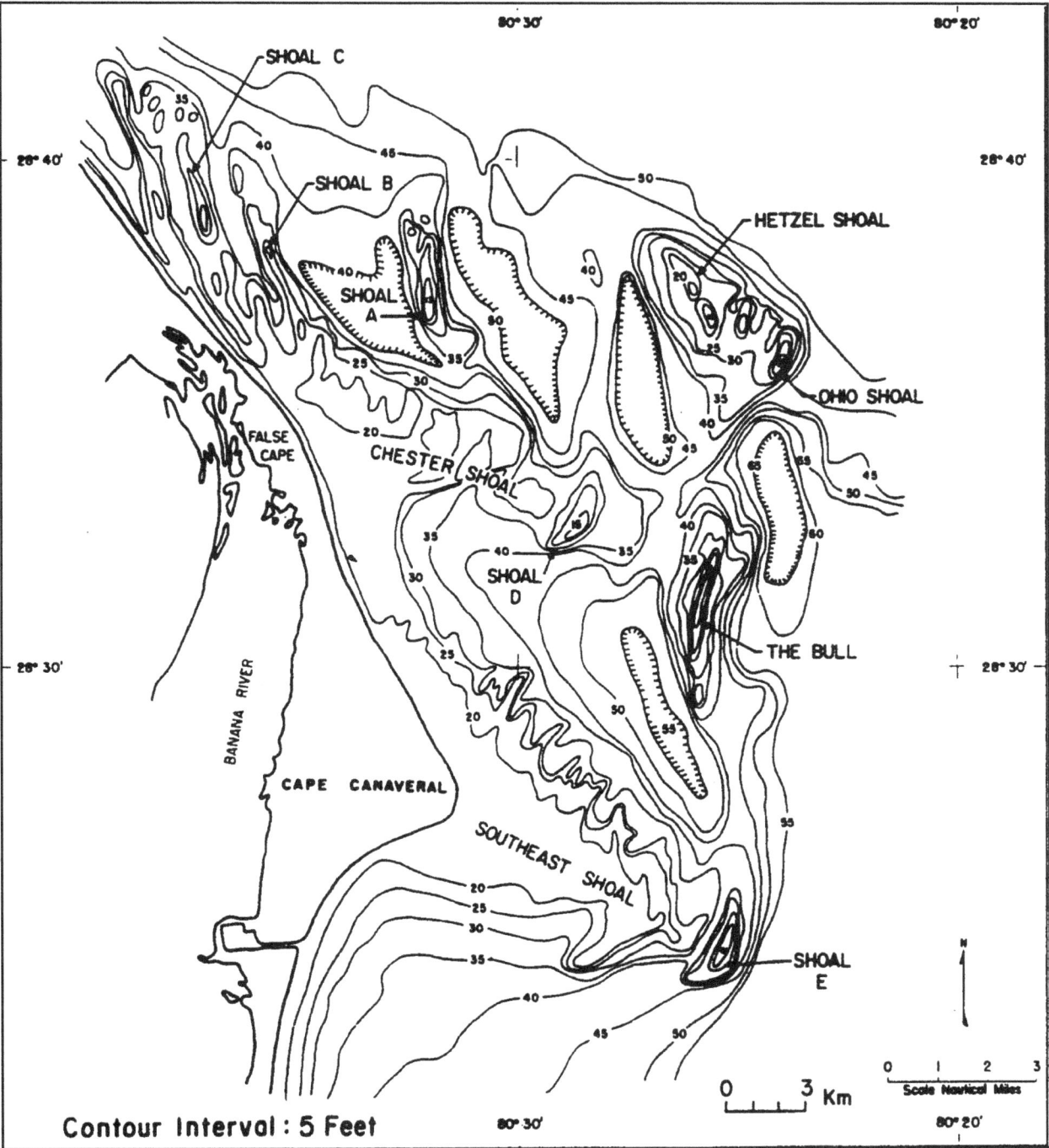

Figure V-4-4. Cape-associated inner continental shelf shoals off Cape Canaveral, Florida (Field and Duane 1974)

- Offshore deposits can be excavated by dredges designed to operate in the open sea. When borrow material is obtained by dredges, it is typically either pumped directly to the beach via pipelines or, in the case of self-propelled hopper dredges with pumpout capability, transported to the shore and pumped onto the beach. Hopper dredges are typically more cost-effective for borrow areas located more than a few kilometers from the project site. These dredges practically require a borrow area with at least one long dimension that allows the hopper vessel to transit the site; i.e., on the order of at least 1.5 km (5,000 ft). In some cases, the dredged material is taken to a rehandle site and offloaded, then transferred to the beach by hydraulic pipeline or truck haul.

• An alternate placement method is to dump material in a nearshore berm as close as possible to the project beach, in water depths shallower than the depth of closure, where it will possibly be moved ashore by wave action. Experiments in offshore dumping near New River Inlet, NC, in 1.82 to 3.65-m- (6- to 12-ft) depth, resulted in a general onshore and lateral migration of fill material (Schwartz and Musialowski 1980). Placing material in water this shallow requires special equipment such as split hull barges, dredges, or other equipment to cast the material shoreward.

• Offshore borrow sources have several favorable features. Suitable deposits can often be located close to the project area. Offshore deposits, particularly linear and cape-associated shoals usually contain large volumes of sediment with uniform characteristics and little or no silt or clay. Large dredges with high production rates can be used. Environmental effects can be kept at acceptable levels with proper planning.

• An unfavorable aspect of offshore borrow operations is the necessity of operating under open sea conditions. Restrictions on the placement of fill material on beaches during the sea turtle nesting season often requires dredging during the winter, when wave energy is highest. In more protected places, such as backbarrier or otherwise sheltered sources, less seaworthy dredging plant can be used. Dredges capable of working in open sea conditions generally have higher rental and operating costs, although this may be offset by greater production capacity.

• Evaluation of offshore sources should also consider the possible effect of dredging a borrow area on littoral processes along and adjacent to the project area. This analysis should include the use of a numerical wave transformation model and the calculation of longshore sand transport rates and transport rate gradients. Nearshore transformation of a project area's principal incident wave conditions over the pre- and postdredging bathymetry should be simulated. The incipient breaking wave conditions and littoral transport potential alongshore, leeward of the borrow area, should be compared between conditions. The proposed limits or geometry of the borrow area may require alteration to avoid unintended concentrations of wave energy, or alongshore transport gradients, produced by the excavated topography. Additionally, borrow areas near the shoreline or inlet shoals may result in accelerated transport of sediment from the beach to the dredged borrow area. In general, where practicable, borrow areas should be sited in water depths greater than the estimated depth of closure (a rough rule of thumb would be twice as deep).

• In some cases, the original relief maybe restored by natural processes over time. This is more likely to occur in active features such as inlet shoals than in relic features, or on ones that are active only during intense storms. Because the depth of closure is well inshore of offshore relic sand borrow sites, these borrow pits usually fill in with fine-grained material that is not suitable for beach fill.

(e) Environmental factors. In general, environmental effects of borrow operations can be made acceptable by careful site selection, and choice of equipment, technique and scheduling of operations. Restoration of flora and fauna often takes place in a short time after operations (Stauble and Nelson 1984). Alterations in physical features (the pit left behind after excavation) may, in some circumstances, be restored by natural processes.

- One effect of borrow operations is direct mortality of organisms due to the operation itself, and destruction or modification of the natural habitat. Direct mortality of motile fauna, such as fish, is usually not great because they move to other areas during the borrow operation. Sessile flora and fauna cannot vacate the area; mortality of these organisms is therefore higher. However, they usually are replaced by the reproduction of survivors or stocks in unaffected peripheral areas (Nelson 1985; Johnson and Nelson 1985).

- Another consideration is the destruction or modification of the habitat needed for survival of native species. A common alteration is the exposure of a substrate that differs from the natural substrate as a result of excavating overlying material. Many marine benthic, and some pelagic, organisms are adapted to specific substrate conditions. Even though larvae of the native species reach the affected area, they may not survive.

- In comparing borrow sites, it is necessary to consider whether or not natural substrate will be modified by the planned operation. This depends on the thickness of the surficial layer and the depth of excavation needed to produce sufficient fill. In many instances, where the layer of suitable fill material is thin, an increase in the areal extent of the borrow area will allow excavation of sufficient material without altering substrate conditions. While this alternative increases direct mortality, it will preserve favorable conditions for repopulation of native organisms.

- In subaqueous areas, detrimental effects on native organisms, both within and near the borrow site, may occur due to suspending silt and clay size material in the water column as a result of the dredging operation. Deposits containing more than a small amount (generally taken as less than 10 percent by volume) of silt and clay are thus less desirable sources of fill from an environmental standpoint. In addition, the fine fraction will be unstable in the beach environment.

(f) Utilization of bypassing/backpassing material. Consideration should be given to bypassing sand across tidal inlets from accreted areas at updrift jetties and from ebb and flood deltas at inlets. Likewise, back-passing of sand from a terminal downdrift jetty to an updrift beach-fill project should be evaluated as a possible cost-effective sand recycling measure. Different types of sand transfer systems are discussed in Richardson (1977). The effect of these measures on adjacent beaches must be evaluated.

(2) Borrow site exploration. A field exploration program to locate and characterize potential borrow sources is usually necessary for offshore and backbarrier environments. For a detailed discussion of procedures, see Prins (1980) and Meisburger (1990). In terrestrial areas, information on deposits is usually available from state geological surveys. There may be existing commercial sand and gravel mining operations. For existing navigation projects, or planned improvements, information on the dimensions and characteristics of material to be dredged is usually available. Field exploration programs involve four phases: a preliminary office study, general field exploration, detailed survey of the site, and characterization of potential sites. The geographical area covered by these investigations is limited by the distance from the project site that is within an economically feasible range for transportation of fill material. Borrow sources within a few miles of the site should be considered initially. Sources farther away should be considered only if no suitable sources are within this range.

(a) Office study. The first phase of the exploration program is an office study of maps, charts, aerial photographs, and literature sources concerning the survey area (Morang, Mossa, and Larson 1993). A study of these materials provides general information on the geomorphology and geology of the area, and helps to identify features that might contain potential fill material. The office study also involves laying out a plan for the next phase, general field exploration of potential sources. Such a plan would include specification of an exploratory field data collection program and definition of the equipment needed to execute the program.

- The most important equipment used for general field exploration and detailed site surveys are: seismic reflection equipment, vibracore apparatus, navigation positioning system, and vessels. Grab samples of surface sediments and side-scan sonar records are also valuable components of the general exploration phase, and can usually be conducted for a relatively small additional cost. Prins (1980) provides a detailed list of equipment and equipment capabilities recommended for use in general field exploration operations. Seismic reflection equipment should provide the highest resolution possible, consistent with achieving a subbottom penetration of at least 15 m (50 ft). High powered seismic reflection systems used for many deep penetration studies are not suitable because of their relatively poor resolution of closely spaced reflectors. Obtaining sediment cores using vibratory coring equipment is more economical than standard soil boring methods, which require more expensive support equipment. Vibratory coring equipment having 3-, 6-, and 12-m (10-, 20-, and 40-ft) penetration capability is available. A 6-m coring device is necessary; a 12-m capability is desirable. Navigation control should be established using an electronic navigation positioning system having an accuracy of about 3 m (10 ft) at the maximum range anticipated for survey and coring operations. Global Positioning Satellites (GPS) technology provides this type of accurate positioning.

- An important task of an office study is to lay out trackline plots, similar to those shown in Figure V-4-5, that are to be followed by the survey vessel in collecting seismic reflection data during the general reconnaissance phase. A grid pattern as illustrated in Figure V-4-6, having lines approximately (0.8 km (0.5 miles) apart, should be employed for areas that are judged to be the most viable either because they are located near the project site or give promise of containing deposits of usable fill material (Meisburger 1990). Zigzag lines are used to cover areas between grids. The detail of coverage is determined by trackline spacing. More complex or promising areas may call for closer spacing.

- Core sites can be tentatively selected during the office study. However, final locations should be determined based on analysis of the seismic reflection records.

(b) General field exploration. During the general field exploration program, data are collected throughout the survey area to locate and obtain information on potential borrow sources and shallow subbottom stratigraphy. This phase involves collection of a comprehensive set of seismic reflection profiles to identify sediment bodies, and a small number of cores to identify and test potential borrow sources.

- The initial part of the general exploration phase is the collection of echosounder and seismic reflection records along predetermined tracklines. The basic survey procedure is for the survey vessel to proceed along each trackline, collecting data while its position is being monitored by an electronic positioning system with fixes recorded at a minimum of 2-min intervals. Fixes are keyed to the records by means of an event marker and identified by a serial fix number. Because seismic reflection records tend to deteriorate in quality with increasing boat speed, the survey vessel's speed should be slow enough to avoid significant reduction in record quality. In general, a suitable boat speed is likely to be less than 4 or 5 knots. The records should be continuously monitored as they become available. Changes in trackline patterns, if considered desirable, can be made as work progresses.

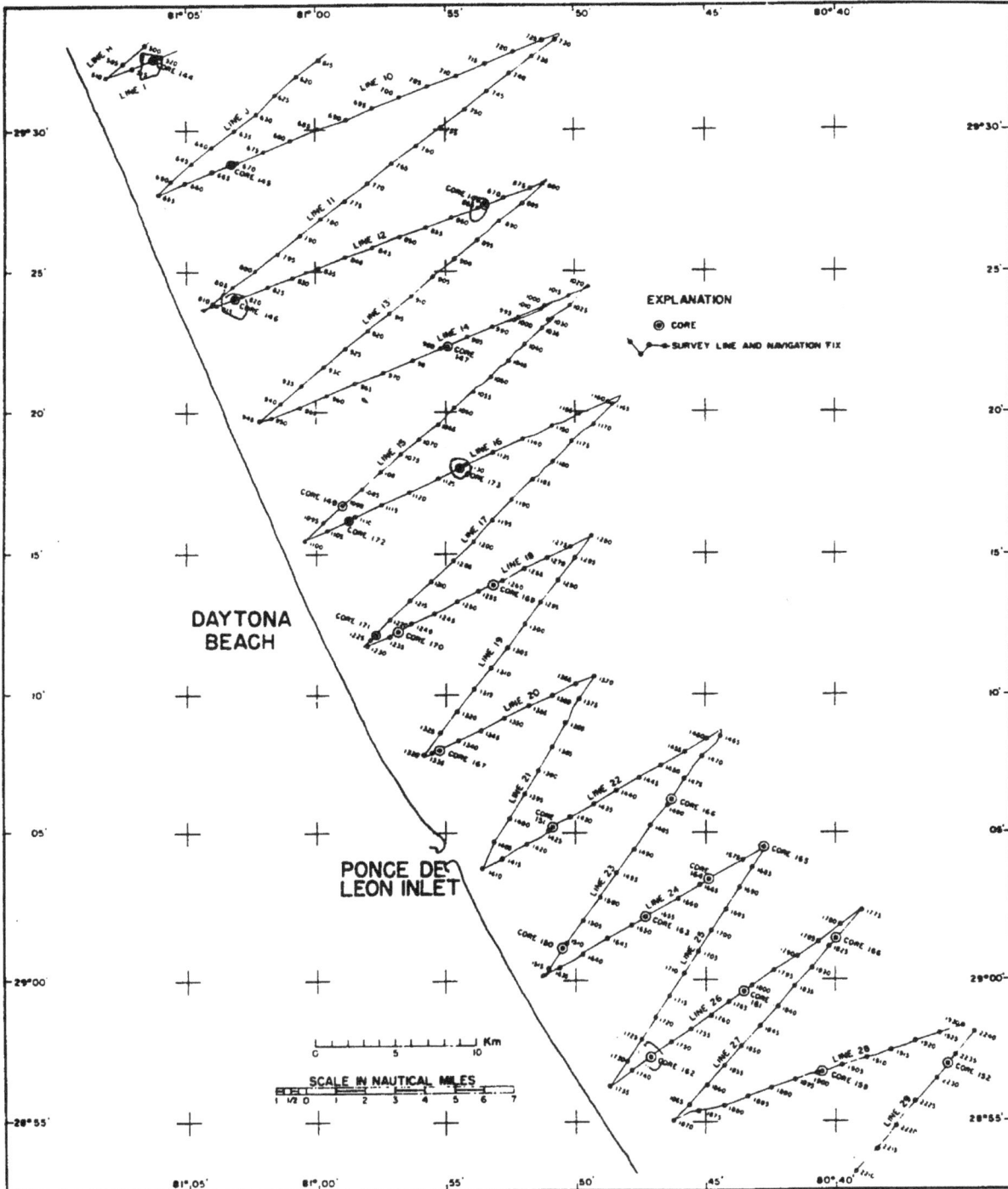

Figure V-4-5. Reconnaissance zigzag line plot from the north Florida coast (from Meisburger and Field 1975)

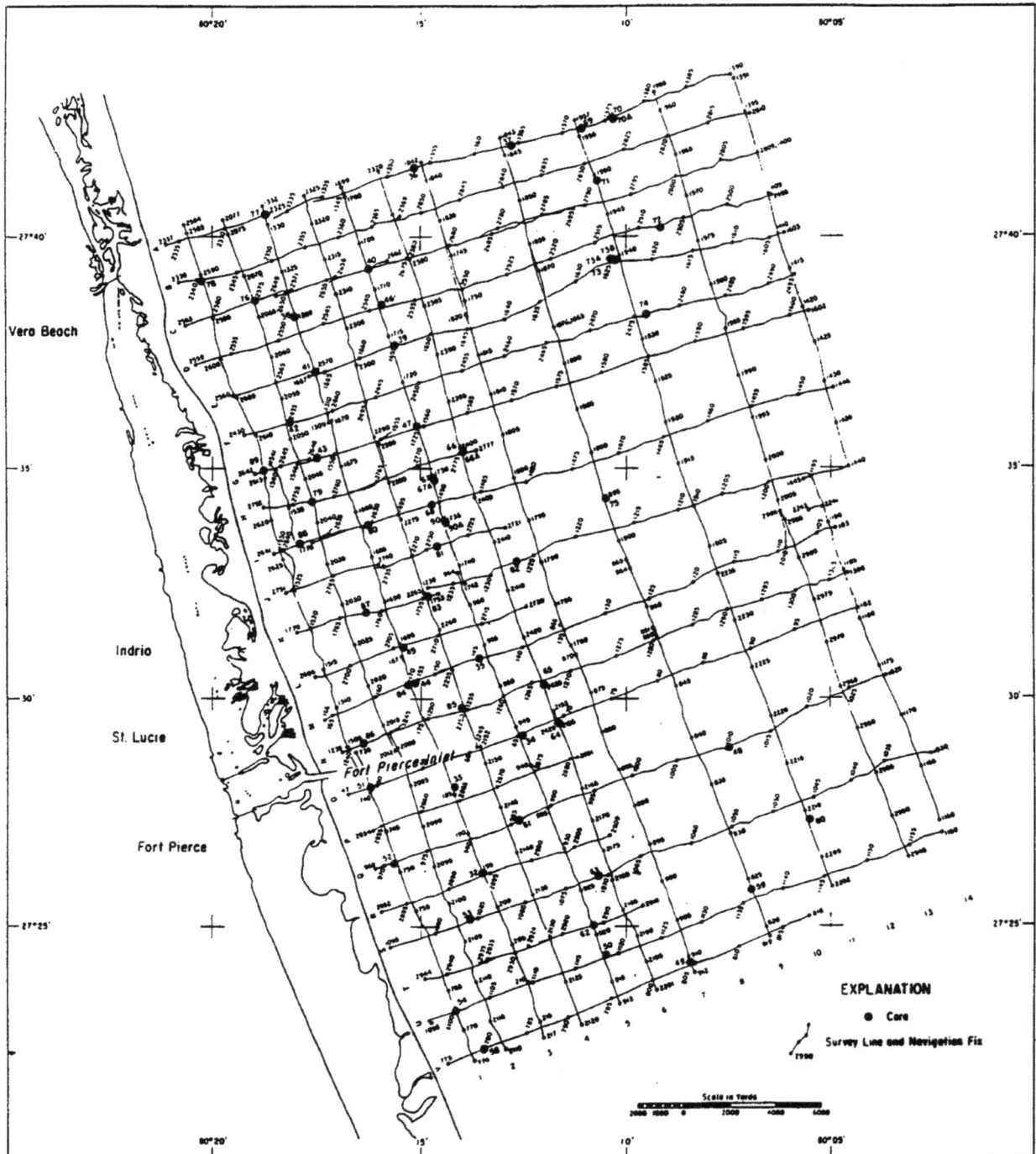

Figure V-4-6.   Grid lines covering a detailed survey area off Fort Pierce, Florida (from Meisburger and Duane 1971)

- Sediment core sites are usually selected after the seismic reflection survey, to allow time for preliminary analysis of the records to determine the most prudent and informative core locations.

- Cores should be examined as they are taken. Inspection of cores is often hampered by the presence of silt and by scratching of the acrylic core liners. However, the top and bottom sediments can be directly viewed before the core is capped. As work progresses, changes can be made in core locations.

(c) Detailed site survey. The third phase of borrow site exploration and investigation consists of a detailed investigation of potential sites, which are selected on the basis of data collected during the general exploration survey. If sufficient seismic reflection data were collected at potential sites during the general exploration phase, the detailed site study may only require the collection of additional, more densely-spaced, cores. Since few large sand bodies have uniform size characteristics, it is important to obtain a sufficient number of cores and borings to accurately reflect the variations in size characteristics throughout the deposit. The number of cores and spacing between cores should be determined based on a review of survey and seismic data as well as other geological studies of the area. These values will vary throughout a borrow site, and from one borrow site to another. But in most cases, collection of cores at less than 300-m (1000-ft) nominal spacing is recommended for purposes of ultimately defining the borrow site(s) to be used in construction. It is important to have adequate data for reliably defining the borrow site. Additional seismic reflection data, if needed, should be collected at this time.

- Sometimes the general and detailed field surveys are made in succeeding years so that ample time is available to study results of the seismic reflection survey before coring is undertaken. However, it is possible to complete the operation entirely in one season. This can be done by mobilizing both geophysical and coring equipment early in the most favorable season, and allowing sufficient lag time between the seismic work and the bulk of the coring work for thorough data analyses and selection of core sites.

- Additional data collection is typically required to fully examine the selected borrow area or areas. These additional data include: magnetometer survey to detect cultural resources of historical significance (shipwrecks, etc.) and obstructions to dredging; archaeological-diver survey to investigate magnetic anomalies; sidescan sonar survey to detect obstructions, hard-bottom, or other environmental resources; detailed bathymetric survey suitable for preparation of construction drawings; and benthic or other biologic surveys, if required, to assess the existing biotic resource.

(3) Borrow site characterization. Any beach erosion or shore protection study in which beach fill is considered should include examination of all potential borrow sources and a comparative evaluation of their suitability. The characteristics of potential borrow sources that are most important in evaluating suitability are: location, accessibility, site morphology, stratigraphy, volume of material available, sediment characteristics, geological history, environmental factors, and economic factors.

(a) Location. The distance that the material must be transported and the feasible means of transport have a large influence on project costs and may be decisive in selecting the most suitable source. Location is also important in terms of the surroundings. Use of terrestrial sources located in developed areas may have a direct impact on the population by creating undesirable noise, traffic congestion, and spillage. Offshore sources may involve questions of jurisdiction and be situated in areas where the dredging and transport activities impede or endanger navigation.

(b) Accessibility. In order to be usable, a borrow source must be accessible or made accessible for the equipment needed to excavate and transport the material. Access to terrestrial deposits may involve road construction or improvement of existing routes. Onsite reconnaissance is the best method of determining the adequacy of access and any necessary improvements. A cost estimate of work needed to create accessibility should be prepared and included in the economic analysis.

- In evaluating subaqueous deposits, one of the principal factors is water depth. To be accessible, the deposit must lie in the depth range between the maximum depth to which the dredge can excavate material, and the minimum depth to keep the dredge afloat when laden with fuel and/or sediment. Subaqueous borrow sites should be located sufficiently far offshore and in deeper water so that excavation does not induce adverse shoreline impacts by altering to the incident wave climate.

- Another aspect of accessibility is the presence of incompatible overburden above the usable sediments. The composition, areal extent, and thickness of any overburden should be determined and considered in the economic analysis (i.e., the cost to remove and dispose of it).

(c) Site morphology. Information on borrow site morphology is valuable in defining and evaluating site characteristics. In many cases, the source deposit has surface morphological features that can be used to delineate boundaries and to assist in interpolating between seismic and coring data points. In addition, site morphology may provide indications of the origin and history of the deposit. Subsurface deposits such as filled stream channels are more difficult to delineate because the only sources of data are seismic reflection records, cores, and borings.

- Description of borrow site morphology should contain information on dimensions, relief, configuration, and boundaries, and be illustrated by large-scale maps or charts. Information for compiling the reports can usually be found in hydrographic survey data available from the National Ocean Survey (NOS), National Oceanic and Atmospheric Administration (NOAA), for submerged deposits, and in published U.S. Geological Survey (USGS) topographic maps for terrestrial sources. Fathometer records, which should be made concurrently with the seismic reflection profiles, are valuable for supplementing and updating other sources.

- In cases where the existing information is inadequate, a special detailed bathymetric survey of the site should be made before the main field collection effort is undertaken.

(d) Site stratigraphy. The stratigraphic relationships within and peripheral to the site deposits should be developed from the existing sources and the seismic and coring records to define: limits of the deposit; thickness of usable material; thickness of any overburden; sedimentary structures, and sediment characteristics of each definable bed. The detail and reliability of the stratigraphic analysis depends on the complexity of the deposit, the number of outcrops, number of cores or borings available, and the degree to which stratigraphic features are revealed by seismic reflection profiles.

- In terrestrial areas, outcrops of potentially useful materials may or may not be present. In many cases, such deposits have no topographic expression and must be defined solely on the basis of borings. Seismic refraction surveys in such situations are valuable in defining the areas between data points. Seismic refraction techniques for subsurface exploration are covered in detail in Engineer Manual 1110-1-1802, "Geophysical Exploration for Engineering and Environmental Investigations."

- In submerged areas, site characteristics must be determined by a combination of bathymetric survey, seismic reflection profiling, and sediment coring. It is important, in both seismic reflection and refraction surveys, to collect enough cores or boring samples to identify and correlate the reflectors with reliable data on sediment properties, and to show significant boundaries that may not have been recorded by the seismic systems.

(e) Volume available. Most beach-fill projects require thousands or millions of cubic meters of suitable fill material. The volume in each potential source must be calculated to determine if a sufficient amount is available to construct and maintain the project for its entire economic life (including initial construction, all subsequent renourishment, and emergency maintenance). In order to do this, it is necessary to delineate the lateral extent and thickness of the deposit. Boundaries may be defined by physical criteria or, in large deposits, arbitrarily set to encompass ample material for the projected fill operation. The thickness of the usable material can be determined from an analysis of site stratigraphy.

- If deposits have a uniform thickness throughout, the available volume can be calculated by multiplying their areas by their thicknesses. Many deposits such as shoals and filled stream channels have more complex shapes, including sloping boundaries and variable thickness. To determine the volume of these deposits, an isopach map of the deposit must be created. An isopach map is a contour map showing the thickness of a deposit between two physical or arbitrary boundaries. Figure V-4-7 shows an isopach map of a borrow area used at Ocean City, Maryland. In this case, the upper boundary of the deposit is defined by the surface of the shoal and can be delineated by bathymetric data. The lower boundary was fixed at a level seismic reflection horizon passing beneath the shoal. Contours, at 1.52-m (5-ft) intervals, were drawn for all the shoal area above the base reflector. Measurements from this type of map can be used to calculate the volume. Commercially available Geographical Information System (GIS) software with Digital Terrain Modeling (DTM) capabilities is now routinely used to generate isopach maps and calculate available sediment volumes.

- Computation of the source's available volume must account for practical limitations of excavation. Particularly for hydraulic dredging (excepting small suction dredge systems), sediment deposits less than about 1-m (3-ft) thickness are impractical to specify. Buffers must be delineated between suitable and nonsuitable sediments, which cannot be included in the source's available volume. These buffers vary with the site and the nature of the sediment strata, but they typically have a minimum thickness of 0.3 m (1 ft) to 0.6 m (2 ft) in subaqueous sources. Buffer areas around sensitive environmental or cultural resources, or around known obstructions, must also be excluded. The size of these radial buffers depends upon the resource or obstruction to be avoided, but a typical radius is 45 to 90 m (150 to 300 ft). Computation of the source's volume must also be limited to those areas or strata in which the sediment is known to be beach-compatible.

(f) Sediment composition. The physical properties of a sediment sample that are most important for determination of suitability for fill on a project beach are composition and grain size distribution. The desirable physical properties of beach-fill material are mechanical strength, resistance to abrasion, and chemical stability.

- In most places, sand-sized sediment is predominantly composed of quartz particles with lesser amounts of other minerals such as feldspar. Quartz has good mechanical strength, resistance to abrasion, and chemical stability. In some deposits, particularly those of marine origin, there is a large and sometimes dominant amount of calcium carbonate that is in most cases of organic origin (biogenic). Calcium carbonate is more susceptible than quartz to breakage, abrasion, and chemical dissolution; but, if it is not highly porous or hollow, it will make serviceable beach fill.

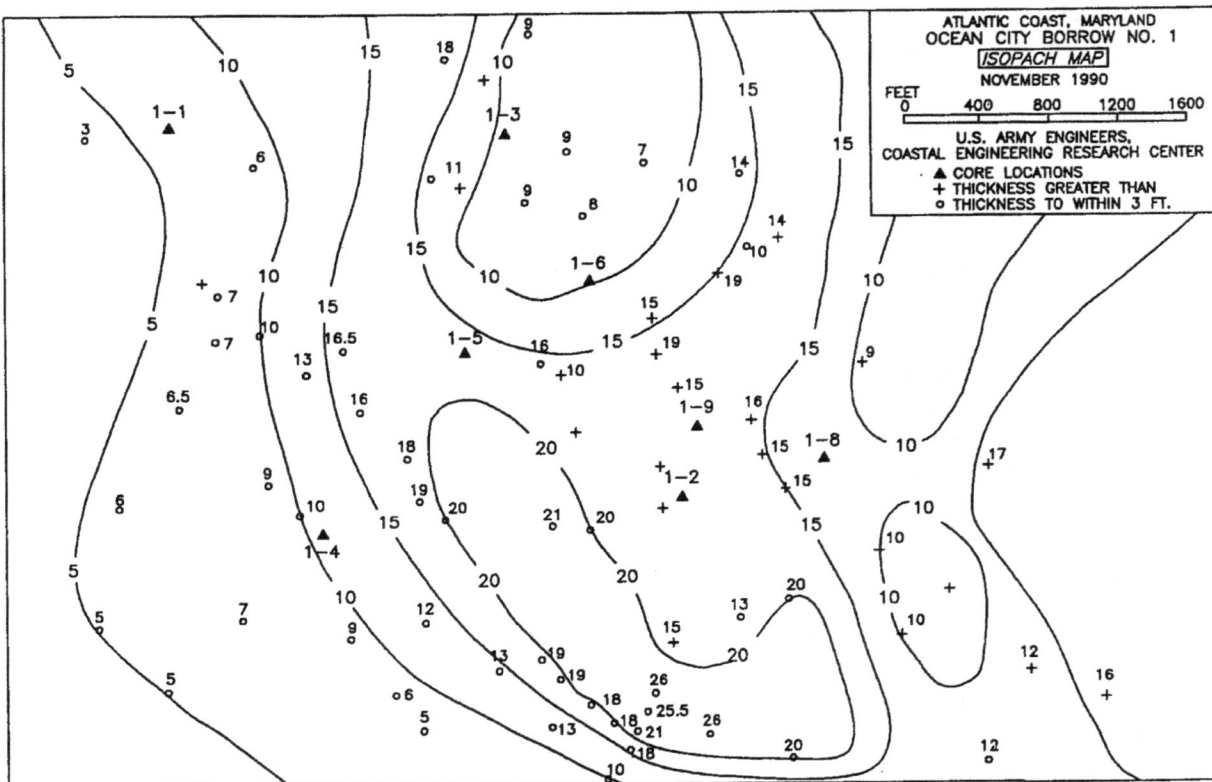

Figure V-4-7. Isopach map of a borrow area used at Ocean City, Maryland

- Sediment composition can be determined by examining representative samples under a binocular microscope. Samples should be prepared by thorough washing to remove fines and clean the surface of the particles. If the material is not well sorted, it should be subdivided into sieve fractions for analysis. A subdivision into the Wentworth classes (see Part III-1) for sand-sized and coarser material is convenient for this purpose. The percentage of carbonate in the material can be estimated by dissolving the carbonate fraction in multiple baths of hydrochloric acid with a subsequent sodium hydroxide wash.

(g) Sediment size characteristics. Generally, suitable material will have grain sizes predominantly in the fine to very coarse sand size range. The presence of very fine sand, silt, and clay in small amounts (generally less that 10 percent) is acceptable, but sources having a substantial amount of fines should be avoided if other more suitable sources are available. When using a borrow area with higher silt or clay content, a large amount of material must be handled to obtain the usable portion, thereby increasing costs. Also, the creation of turbidity during excavation and placement on the beach is environmentally undesirable. However, in the future, as sand resources become more scarce, sand separation may prove to be economically justified depending on the volume of material required and the relative silt and clay content of the borrow site. Borrow material presently discounted because of a comparatively high percentage of fines may become economically viable sand sources in the future.

- One of the main considerations in selecting a borrow source is the similarity between the grain size distributions of the borrow material and the native beach, i.e., the borrow material's compatibility with the native material. To make this comparison, it is necessary to determine, for both native beach and each potential borrow source, a composite grain size distribution representative of overall textural properties. The method used to determine grain size characteristics for both the fill and borrow sites should be the same. Beach and borrow material should be analyzed with standard sieves as described in Part III-1.

- Native beach composite sediment statistics should be computed based on sieve analyses of samples collected along cross-shore transects through the most active portion of the beach profile (see Part III-1 for information concerning sediment composite statistics). The most active portion is located between the crest of the natural berm (immediately landward of the mean high waterline) and the depth corresponding to the position of the typical storm bar. The composite statistics should be developed for a number of cross-shore transects throughout the project domain. Grain size statistics calculated from such a sampling scheme will account for most of the natural variability on the profile. If cross-shore composites exhibit a wide range of median grain size and sorting values, an alongshore composite should be calculated for the entire project domain to reduce the variability.

- Borrow area composite statistics should be determined using grain size distributions computed for samples taken from several cores within the potential borrow site. For general uniform cores, samples should be collected from the top, middle, and bottom of useable sand within the core. The composite characteristics of the borrow material should be weighted based on the estimated volume of each type of material present in the deposit.

- Figure V-4-8 shows a comparison between the native beach and borrow material used for nourishment at Ocean City, Maryland. The shaded area represents common characteristics between the native beach and fill material.

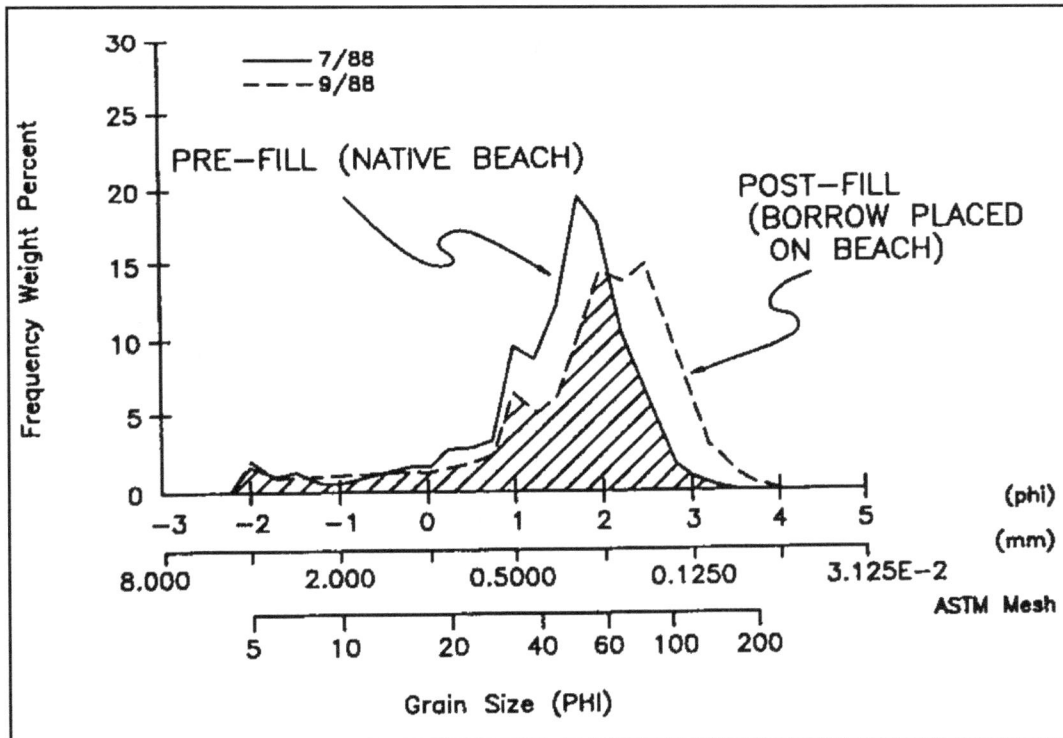

Figure V-4-8. Comparison of composite grain size analysis between the native beach and the borrow material used at Ocean City, Maryland

(h) Sediment suitability. The grain-size distribution of the borrow material will affect the cross-shore shape of the nourished beach profile, the rate at which fill material is eroded from the project, and how the beach will respond to storms. Typically, borrow material will not exactly match the native beach (except perhaps in some bypassing projects). An analysis is required to assess the compatibility of the borrow material with the native beach, from a functional perspective. A comparative analysis of sand suitability is also required to economically evaluate alternative borrow areas for a given project.

- Early research into compatibility of borrow area material by Krumbein (1957), Krumbein and James (1965), James (1974, 1975), and Dean (1974) addressed this issue by various comparative analysis techniques that utilize the sand-size distributions of the natural beach in the fill area and the borrow material in the candidate borrow sites. These approaches develop a factor, or parameter, indicating how much fill material is required in light of the different sediment characteristics between the borrow and native beach materials. They assume that borrow material placed on the beach will undergo sorting as a result of the coastal processes; and given enough time, will approach the native grain-size distribution. The portion of borrow material that does not match the native sediment grain-size distribution is assumed to be lost to the offshore. James (1975) developed this concept into a method to calculate an overfill factor, R_A, and a renourishment factor, R_J. Conceptually, the overfill factor is the volume of borrow material required to produce a stable unit of usable fill material with the same grain size characteristics as the native beach sand. The renourishment factor addresses the higher alongshore transportability of the finer grain sizes in the borrow sands and provides an estimate of renourishment needs. Use of the renourishment factor is no longer recommended in beach-fill design calculations; however, details concerning the renourishment factor and it's calculation may be obtained from the *Shore Protection Manual* (1984).

- Recent research and beach nourishment experiences have questioned the continued use of these grain-size-based factors, R_A and R_J, to estimate beach-fill performance (Dean 2000). Present guidance recommends that design be based on equilibrium beach profile concepts, an assessment of storm-induced erosion, and an assessment of wave-driven longshore transport losses; and that these methods be used to replace or complement the overfill and renourishment factor approaches (National Research Council (NRC) 1995). In practice, these recommended methods treat sediment characteristics using a single grain size parameter, the median grain diameter. They do not consider natural variations in grain size that occur on natural and nourished beaches. However, they have the advantage of incorporating more of the physics of coastal processes into the design, much more so than use of the overfill and renourishment factors. The overfill factor attempts to consider the distribution of grain sizes. Therefore, it does provide an additional piece of information on the amount of borrow material that might be needed to construct a beach nourishment project in more difficult design cases where the grain size characteristics of the borrow material differ significantly from those of the native beach material, especially the case where the borrow sediments are finer than the native sediments. The overfill factor is discussed in more detail in the next section.

- As a general recommendation, a nourishment project should use fill material with a composite median grain diameter equal to that of the native beach material, and with an overfill factor within the range of 1.00 to 1.05. This is the optimal level of sediment compatibility. However, obtaining this level of compatibility is not always possible due to limitations in available borrow sites. Both the overfill factor and equilibrium beach profile concepts indicate that sediment compatibility is sensitive to the native composite median grain diameter. As such, the compatibility range varies depending on the characteristics of the native beach material, with coarse material being less sensitive to small variations between the native and borrow sediments than fine material. As a rule of thumb, for native beach material with a composite median grain diameter exceeding 0.2 mm, borrow material with a composite median diameter within plus or minus 0.02 mm of the native median grain diameter is considered compatible. For native beach material with composite median diameter between 0.15

and 0.2 mm, borrow material can be considered compatible if its composite median diameter is within plus or minus 0.01 mm of the native diameter. For native beach material with a composite median diameter less than 0.15 mm, use of material at least as coarse as the native beach is recommended. Even though material is deemed compatible based on these rules, the designer should factor grain-size differences into estimates of required fill volume through use of equilibrium beach profile methods, or the overfill factor, or both. Methods for computing beach-fill volumes are discussed in Part V-4-1-f. These guidelines are based on composite median diameters established for the entire project and borrow site. Typically, composites for individual profiles, or subsections of the borrow site, will have variations in median diameter which may exceed the compatibility ranges previously discussed.

- Materials that are not compatible according to these guidelines may still be suitable. Borrow material that is coarser than the native material will produce a beach which is at least as stable as a fill comprised of native material. Fills with coarser material provide improved resistance to storm-induced erosion. A lesser volume of coarser fill will be required to create a beach of a given width, compared to the volume of native beach sand that would be needed. If the median diameter of the borrow material exceeds the median diameter of the native material by more than 0.02 mm, a noticeably steeper beach may form. A steeper beach may become a design issue, along with the different texture of the coarser fill. Use of material finer than the native material should be avoided, if possible, but such material still may be suitable. A much greater volume of material will be required to form a beach of a given width, compared to the volume of native sand that would be required. Use of finer sand will produce a beach with flatter slopes, which could be a design issue too. For example, it may be problematic to construct a more gentle beach adjacent to an existing groin or jetty that is intended to block the longshore movement of sand. Sand transport around the structure and into a navigation channel may increase.

(i) Overfill factor. The overfill factor, R_A, is determined by comparing mean sediment diameter and sorting values of the native beach and borrow sediments (in phi, ϕ, units). The phi, ϕ, scale of sediment diameter is defined and discussed in Part III-1, and Equations III-1-1a and III-1-1b enable conversion between sediment grain size diameter in millimeters and the ϕ scale and vice versa. The overfill factor is computed using the following relationships between the borrow and native beach material:

$$\frac{\sigma_{\phi b}}{\sigma_{\phi n}} = \frac{\left[\dfrac{(\phi_{84}-\phi_{16})}{4} + \dfrac{(\phi_{95}-\phi_{5})}{6}\right]_b}{\left[\dfrac{(\phi_{84}-\phi_{16})}{4} + \dfrac{(\phi_{95}-\phi_{5})}{6}\right]_n} \qquad \text{(V-4-3)}$$

and

$$\frac{M_{\phi b} - M_{\phi n}}{\sigma_{\phi n}} = \frac{\left[\dfrac{(\phi_{16}+\phi_{50}+\phi_{84})}{3}\right]_b - \left[\dfrac{(\phi_{16}+\phi_{50}+\phi_{84})}{3}\right]_n}{\left[\dfrac{(\phi_{84}-\phi_{16})}{4} + \dfrac{(\phi_{95}-\phi_{5})}{6}\right]_n} \qquad \text{(V-4-4)}$$

where:

$\sigma_{\phi b}$ = standard deviation or measure of sorting for borrow material (Equation III-1-3)

$\sigma_{\phi n}$ = standard deviation or measure of sorting for native material (Equation III-1-3)

$M_{\phi n}$ = mean sediment diameter for native material (Equation III-1-2)

$M_{\phi b}$ = mean sediment diameter for borrow material (Equation III-1-2)

- Values obtained using the relationships in Equations V-4-3 and V-4-4 are then plotted on the graph presented in Figure V-4-9. The value of R_A can be obtained by interpolating between the values represented by the isolines. Values of the overfill factor greater than 1.0 indicate that more than one unit of borrow material will be needed to produce one unit of fill material. Example V-4-1 illustrates computation of the overfill factor.

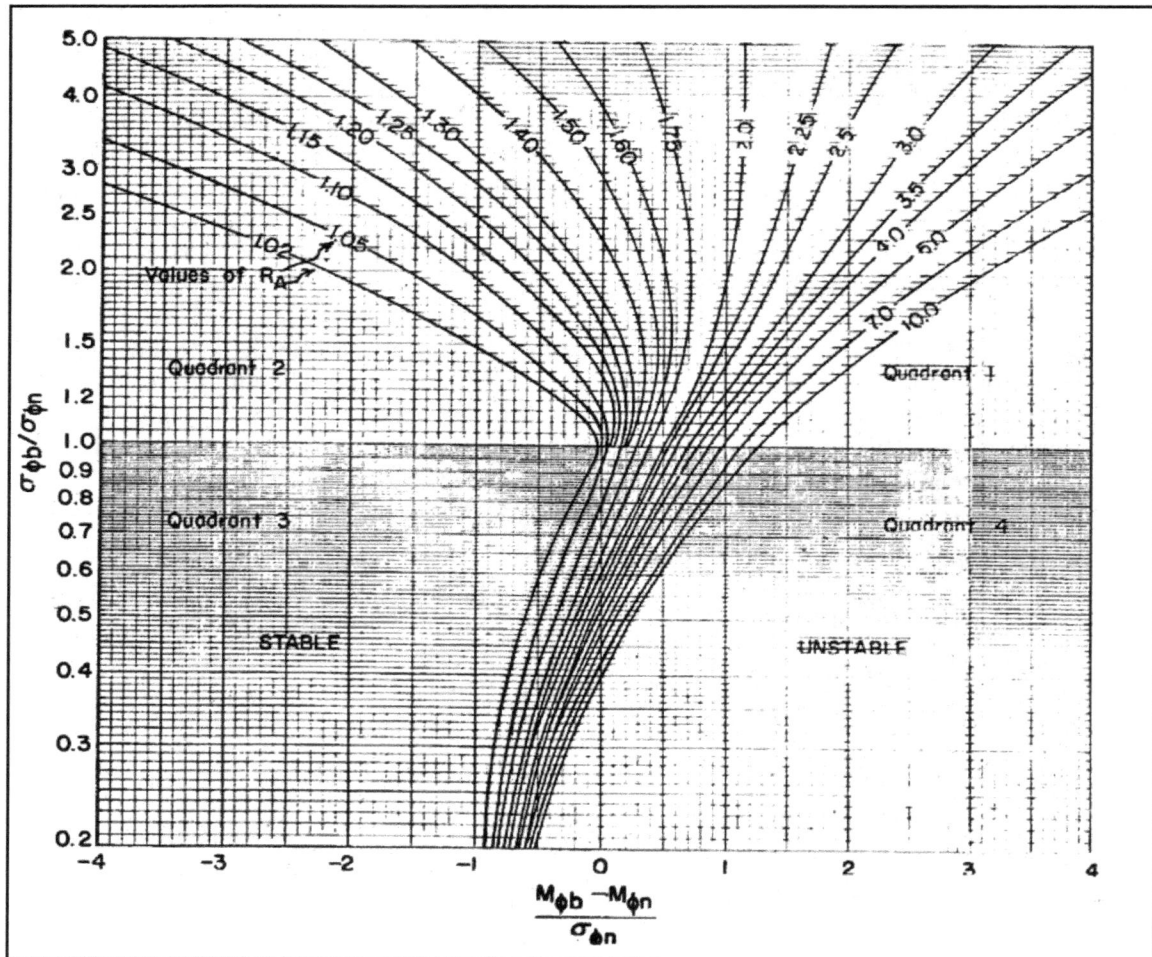

Figure V-4-9. Isolines of the adjusted overfill ratio (R_A) for values of ϕ mean difference and ϕ sorting ratio (*Shore Protection Manual* 1984)

EXAMPLE PROBLEM V-4-1

FIND:

The overfill factor, R_A.

GIVEN:

The native and borrow area phi values are:

native beach: $\phi_{05}=0.95$, $\phi_{16}=1.31$, $\phi_{50}=1.91$, $\phi_{84}=2.66$, $\phi_{95}=2.90$
borrow area: $\phi_{05}=1.42$, $\phi_{16}=1.63$, $\phi_{50}=2.49$, $\phi_{84}=3.08$, $\phi_{95}=3.55$

The median grain diameter for native and borrow material is 0.27 and 0.18 mm, respectively.

SOLUTION:

The mean sediment diameter in phi units is given in Part III-1 as

$$M_\phi = (\phi_{16} + \phi_{50} + \phi_{84}) / 3 \qquad \text{(III-1-2)}$$

$M_{\phi n} = (1.31 + 1.91 + 2.66) / 3 = 1.96$
$M_{\phi b} = (1.63 + 2.49 + 3.08) / 3 = 2.40$

The standard deviation in phi units is given in Part III-1 as

$$\sigma_\phi = (\phi_{84} - \phi_{16}) / 4 + (\phi_{95} - \phi_{05}) / 6 \qquad \text{(III-1-3)}$$

$\sigma_{\phi n} = (2.66 - 1.31) / 4 + (2.90 - 0.95) / 6 = 0.66$
$\sigma_{\phi b} = (3.08 - 1.63) / 4 + (3.55 - 1.42) / 6 = 0.72$

The sorting ratio from Equation V-4-3 is

$\sigma_{\phi b} / \sigma_{\phi n} = 0.72 / 0.66 = 1.09$

The phi mean differences ratio from Equation V-4-4 is

$(M_{\phi b} - M_{\phi n}) / \sigma_{\phi n} = (2.40 - 1.96) / 0.66 = 0.67$

Entering Figure V-4-9 with x = 0.67 and y = 1.09 results in an overfill ratio (R_A) equal to 2.5. The finer borrow material may be suitable for use, but it is quite incompatible with the native beach sand. The value of the overfill ratio suggests that 2.5 units of borrow material will be required to create 1.0 unit of stable native beach material.

• The overfill method previously described is the Krumbein-James technique (Krumbein and James 1965). Dean (1974) presents an alternative method for computing the overfill factor, not shown, which generally yields less conservative (lower) estimates of the overfill factor.

f. Beach-fill cross-section design. The design of Federal beach-fill projects is based on the optimization of net annual benefits defined as the difference between average annual costs and average annual benefits. This optimization procedure produces a plan known as the National Economic Development (NED) plan. The NED plan considers the storm damage reduction potential of various beach fill design alternatives and the averaged annual cost. Primary design parameters of each alternative include the physical dimensions of the cross-sectional design profile and the volume of sand required to obtain the design profile. Beach-fill design alternatives typically include combinations of beach berms of varying width and dunes of varying height (see Part V-4-1-b for a description of the characteristics and functions of beach berms and dunes). Design berms are characterized by berm crest elevation and berm width. Dune design dimensions include crest elevation, crest width, and side slopes.

(1) Berm elevation. The elevation of the design berm should generally correspond to the natural berm crest elevation. If the design berm is lower than the natural one, a ridge will form along the crest, which, when overtopped by high water will produce flooding and ponding on the berm. A design berm higher than the natural berm will produce a beach face slope steeper than the natural beach and may result in formation of scarps that interfere with sea turtle nesting and recreational beach use. Many healthy, natural beaches exhibit a gentle downward slope from the toe of the dune to the seaward limit of the berm. Therefore, a gentle berm slope can be specified as an element of the design profile. The berm slope is most appropriately estimated from profiles that represent a nearby, healthy beach. Or the slope can be estimated to fall in the range from 1:100 to 1:150. Adding a gentle slope to the berm also helps prevent overtopping and ponding.

(a) The natural berm elevation can be determined by examining beach profile surveys of existing and historical conditions at the project site. Because beach berms form naturally under low-energy waves, they are typically most well-developed in form at the end of the summer season. Seasonal profile surveys can be used to examine temporal changes in berm shape and to identify well developed berm features from which to estimate the natural berm height. When survey data indicate alongshore variations in the natural berm height, a representative berm height may be determined either by visual inspection of plots showing the alongshore variations or by computing an average profile shape. The Beach Morphology Analysis Package (BMAP) provides automated calculation and visualization tools for performing such analyses. Sommerfeld et al. (1994) provide an overview of the capabilities and a user's guide for operation of the BMAP software.

(b) Figure V-4-10 shows an example of seasonal variation in berm shape measured over two consecutive years at a given profile station. The fall surveys show that the beach berm is widest following the calmer summer waves, whereas the spring surveys show the berm to be in a more eroded condition following winter waves. Based on visual inspection of Figure V-4-10, a natural berm height for this profile can be approximated by the horizontal dotted line, corresponding to an elevation of 3.6 m. Figure V-4-10 shows that it would be difficult to identify the natural berm height based solely on the first spring survey, and illustrates the advantage of using multiple surveys for profile characterization.

(c) Figure V-4-11a shows beach profiles measured during a single fall survey at five different profile stations along the beach. The berm is seen to vary alongshore in height and width. To determine a representative berm height, the beach profiles are horizontally aligned on the seaward face of the dune to superimpose the berm profiles at the base of the dune, as shown in Figure V-4-11b. An average profile is computed by averaging the elevations of the aligned profiles at 1-m increments in distance offshore. Figure V-4-11c shows the average profile, from which a representative berm elevation of 3.5 m is obtained by inspecting the horizontal portion of the profile between the offshore distances of 75 and 100 m.

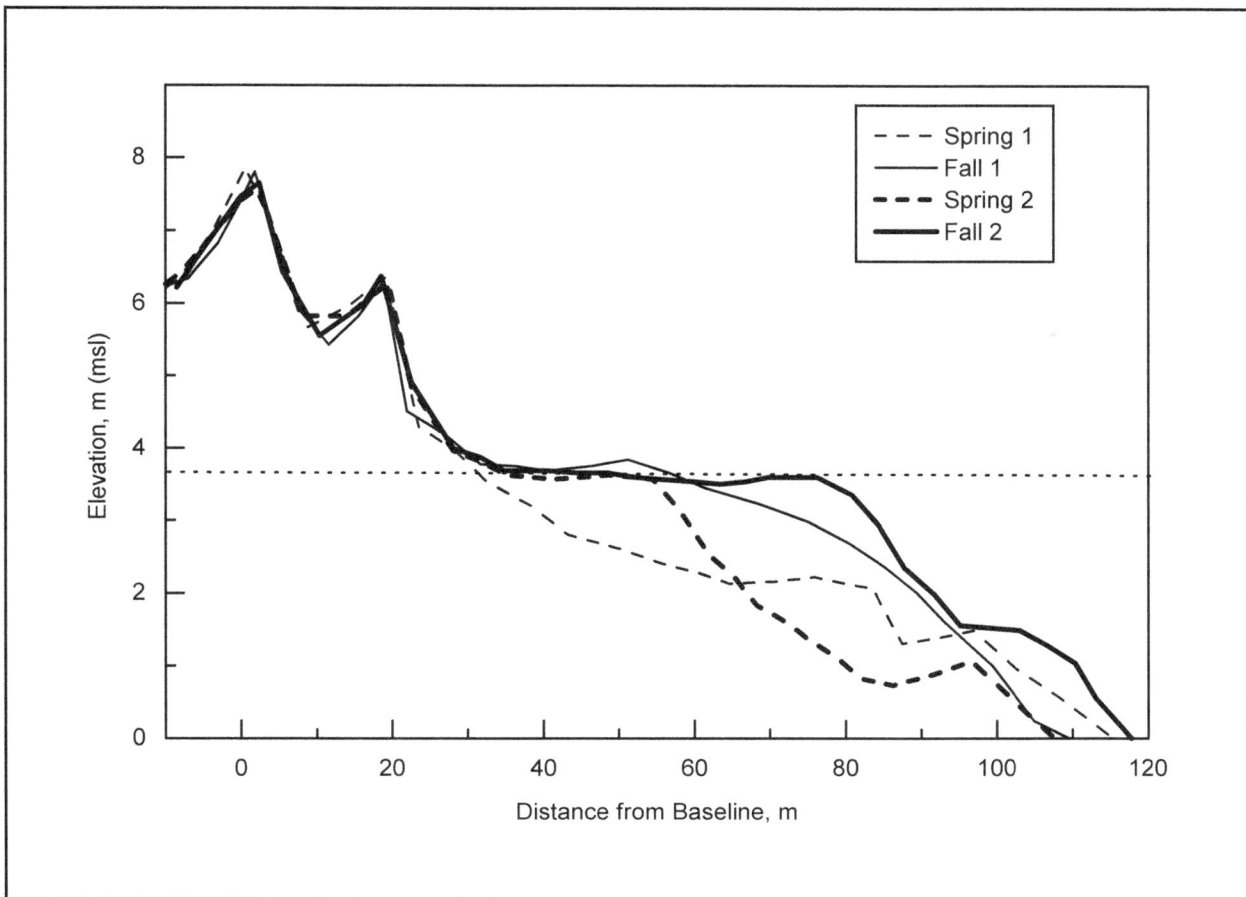

Figure V-4-10. Seasonal variation of a beach berm

(d) In cases where no dry beach exists at the project site, or where the existing beach has a deficit of sand due to substantial reduction or elimination of a critical sediment source, such as in front of a seawall or downdrift of a shore-perpendicular structure, the natural berm elevation should be estimated using profile data from adjacent beaches which are healthier in terms of sand availability but are exposed to similar waves and water levels. Priority should be given to identifying a natural berm elevation using beach measurements from the project site or a similar site. As a last resort, when no suitable beach profile data are available to determine a natural berm height, the limit of wave runup under average (nonstorm) wave and tide conditions at the site can be estimated to establish a design berm height (see Part II-4-4 for calculation of wave runup on beaches).

(2) Berm width.

(a) Selection of the design berm width depends on the purpose of the project and is often constrained by factors such as project economics, environmental issues, or local sponsor preferences. For Federal beach nourishment projects, the berm width is determined through a process of optimization based on storm damage reduction. The design beach width is optimized by computing costs and benefits of various design alternatives and selecting the alternative that maximizes net benefits (USACE 1991). Numerical models of beach profile change such as Storm-Induced BEAch CHange (Larson and Kraus 1989) provide a means of evaluating beach response as a function of berm width. Figure V-4-12 illustrates how berm width influences the landward extent of erosion during a storm. Figure V-4-12a shows four beach profiles with identical dune

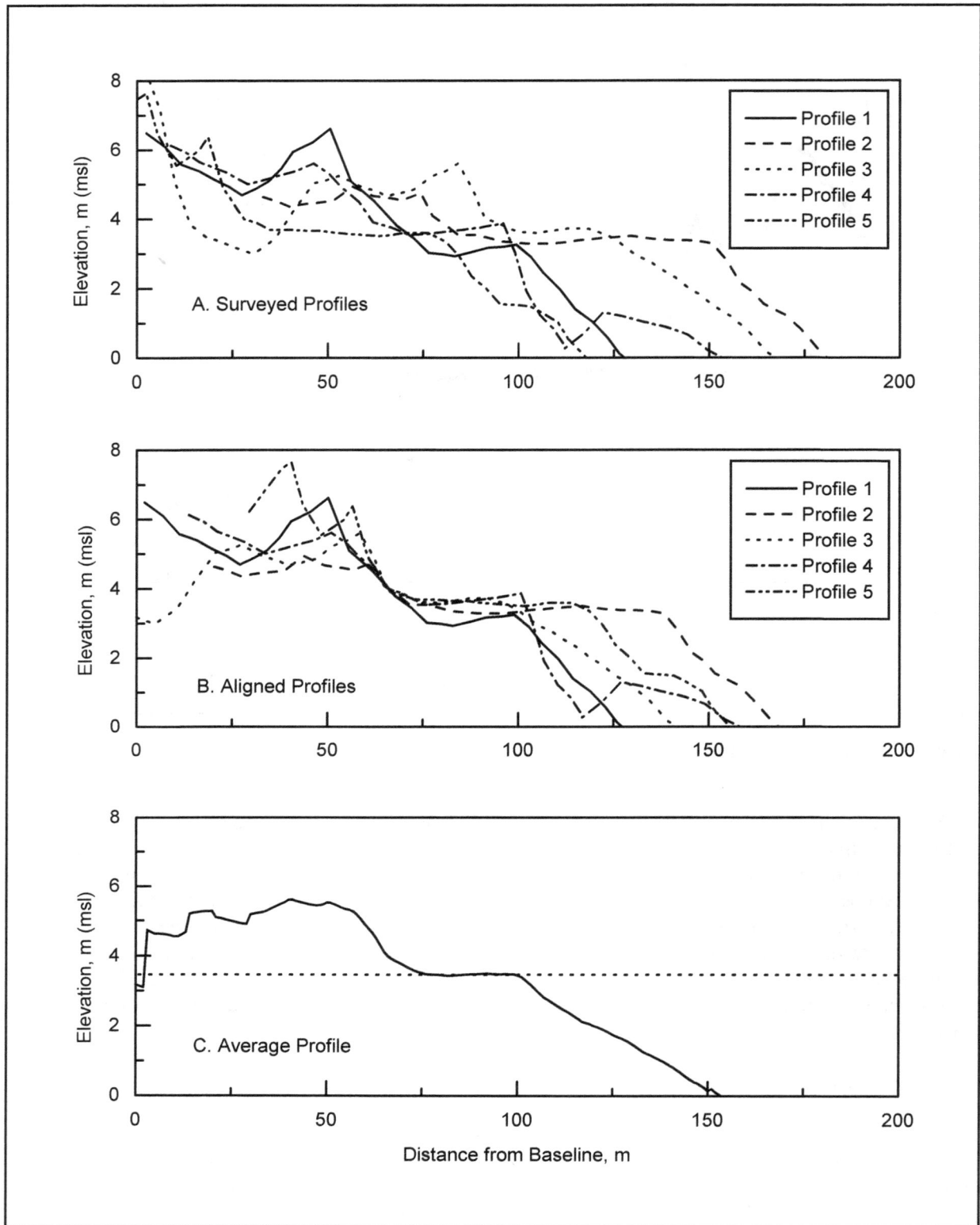

Figure V-4-11. Berm elevation determined from beach profiles measured at various alongshore stations

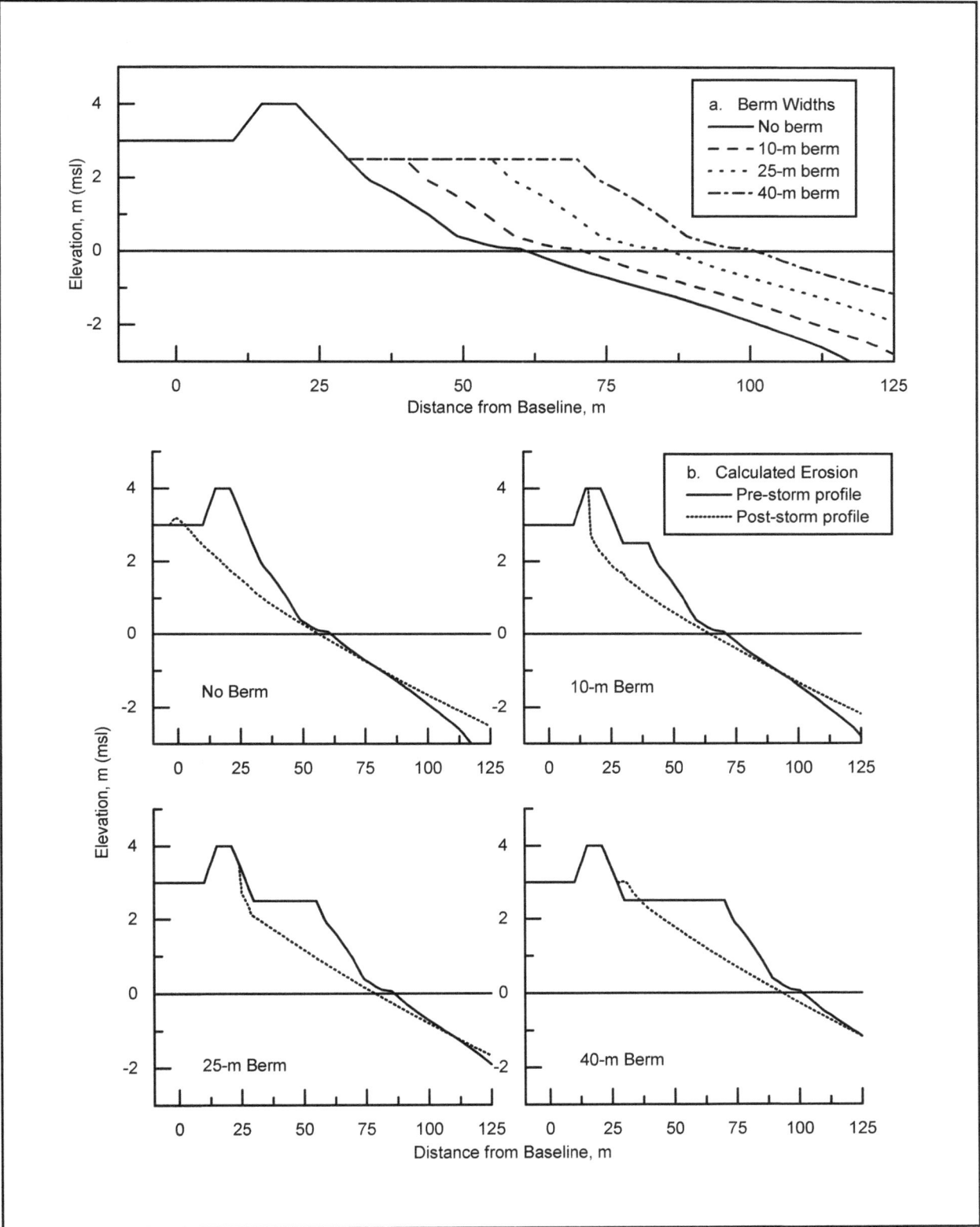

Figure V-4-12. Example of storm-induced beach erosion as a function of berm width

cross sections and varying berm widths. The SBEACH model was used to simulate profile change, and each profile was subjected to a constant wave height of 3 m, wave period of 10 sec, and water level of 1.5 m msl over a duration of 24 hr. Figure V-4-12b shows the calculated results for each profile. The profile with no berm experienced complete erosion and overtopping of the dune. Most of the dune was eroded on the 10-m berm profile, whereas the 25-m berm profile experienced only minor dune erosion. The 40-m berm provided full protection against erosion of the dune and backbeach, and some sand was pushed up against the base of the dune due to overwash across the wide berm. In this example, a shorefront property located immediately landward of the dune would experience varying degrees of damage and/or vulnerability to future storm erosion and flooding as a function of the beach berm width.

(b) Factors other than storm erosion that influence beach width include the rate and variability of long-term shoreline recession, planform spreading losses, and presence of erosional hot spots. These factors typically do not enter in the optimization of the design berm width for storm damage reduction projects, but should be accounted for in optimization of the advanced fill section and renourishment interval as discussed in Part V-4-1-g.

(c) Storm berms may be used in conjunction with a natural berm to provide added protection against damage during storms. Storm berms are constructed at an elevation higher than the natural berm and are set back, landward, from the crest of the natural berm. Storm berms are built to reduce the chance for wave action and erosion from reaching the dune during higher water levels associated with a specified degree of storm intensity (usually, the type of storm that can be expected once every few years). The crest elevation of a storm berm should be set based on the water level and runup elevation associated with the type of storm(s) against which protection is sought. If a storm berm is included in the design, the width of the storm berm can be optimized to maximize net benefits. The seaward extent of the storm berm should also consider the possibility for undesirable, persistent scarp formation.

(3) Dune dimensions.

(a) Dunes protect upland property against wave attack, erosion, and flooding during extreme storm events which overtop or severely erode the beach berm. Design parameters of a dune include the crest elevation, crest width, and side slopes. The design dune crest elevation is typically determined through economic optimization. The dune crest width may also be optimized but is typically fixed at a selected width for all design alternatives. In selecting dune crest width and side slopes, constructibility constraints and angle of repose of the fill material grain size should be considered. A typical dune design may have dimensions on the order of 5-m crest elevation above msl, 10-m dune crest width and one on five side slopes. Planting beach grasses on the constructed dune helps to maintain and build dune volume over time by trapping wind-blown sand.

(b) To illustrate the influence of dune height on storm-induced beach profile change, the no-berm profile shown in Figure V-4-12 was modified to increase the dune elevation from 4 to 4.5 m while maintaining the same crest width and side slopes. The original and modified dune configuration are shown in Figure V-4-13a. The increase in dune height translates to an added volume of 10 cu m/m for this profile. Figure V-4-13b shows profile erosion modeled by SBEACH for the same storm conditions used in the previous berm erosion example. The calculated results in Figure V-4-13b indicate the added dune height prevented dune overtopping and back-beach erosion for these particular wave and water level conditions.

(4) Design profile shape. The shape of the design profile is needed to compute cross-sectional fill volume requirements and as input to storm-induced erosion modeling that is done to optimize berm and dune dimensions. In order to obtain the design dune and berm template on the beach, sufficient sand must be placed to nourish the entire profile out to the depth of closure (see Figure V-4-14 and Part III-3-3-b for details

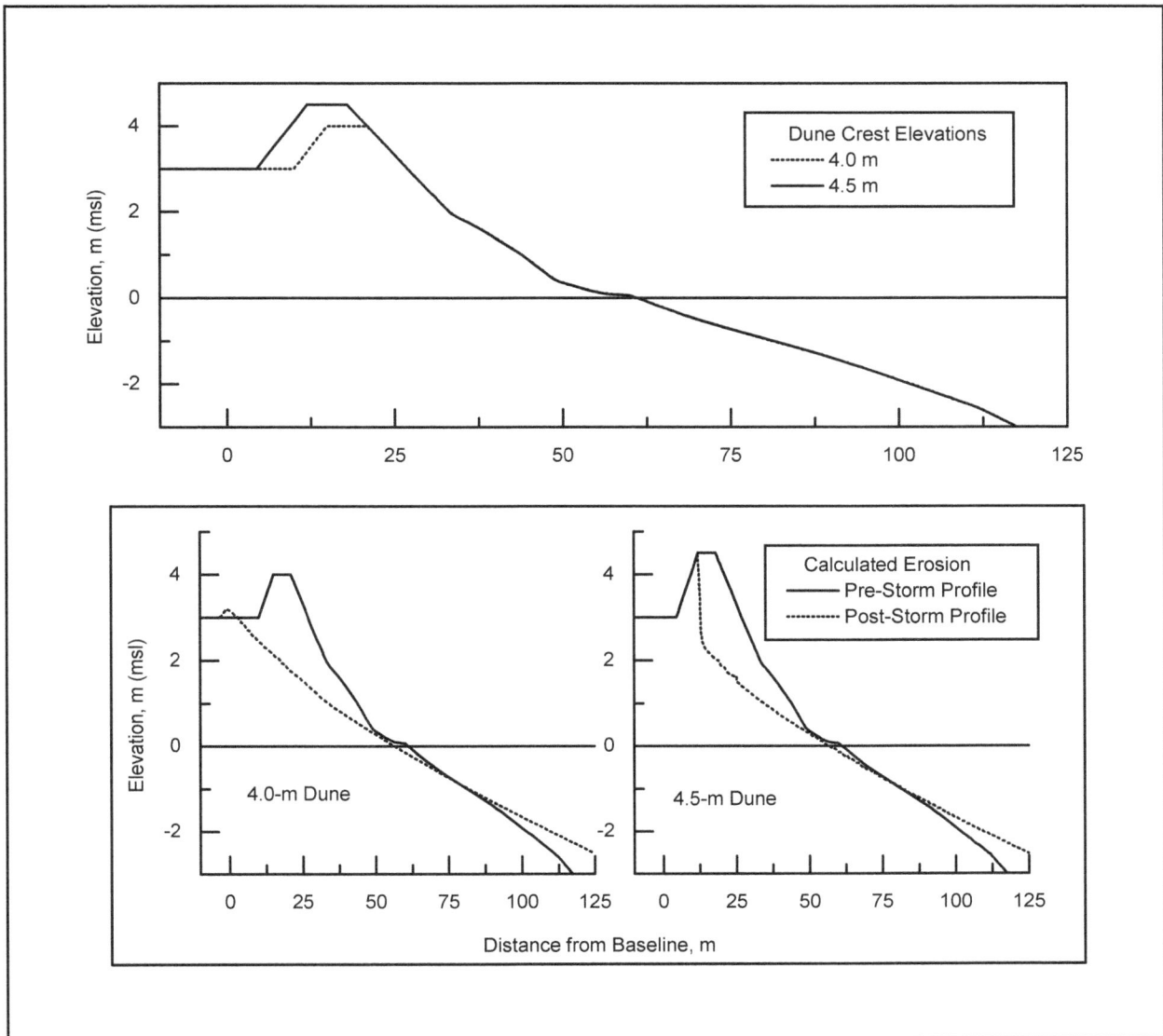

Figure V-4-13. Example of storm-induced beach erosion as a function of dune crest elevation

on depth of closure). Whereas dune and berm dimensions are determined through optimization, the shape of the design profile below the beach berm is a function of the local morphology and grain size of the fill. Local beach morphology often includes a nearshore bar system, which may be absent in erosion-stressed preproject beaches. In such cases, a berm also might be absent from the profile, or may be unnaturally low in elevation; or the preproject profile may reflect an overly steep beach face. A key aspect of defining the design profile shape is to recognize whether or not the preproject beach reflects an unnatural, sediment-starved condition, in which the preproject shape is different from that which will evolve once the fill is placed. For example, a severely eroded beach may lack the commonly observed nearshore bar system, but the bar system will likely form under sediment-rich conditions that follow nourishment. Consequently, the sectional fill volume should include an estimate of bar volume. In these situations, design profile shape can be defined by examining nearby beaches that are healthier in terms of available sediment supply (from the upland beach and from longshore sources). Profile data from adjacent beaches within the project domain, or data from a nearby site that is exposed to similar wave and

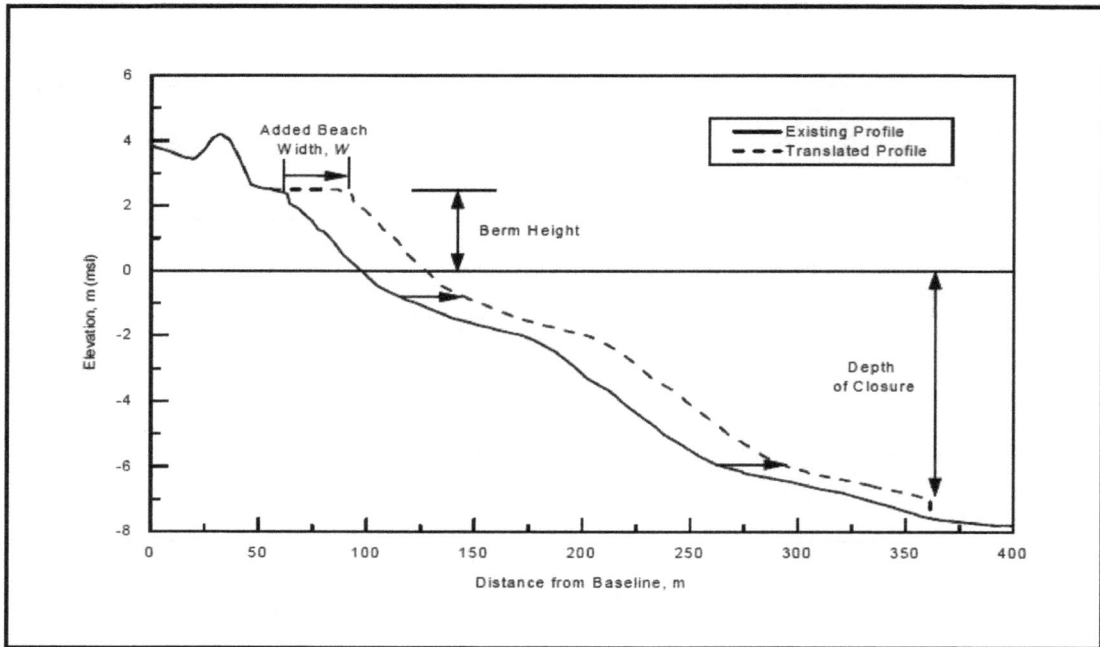

Figure V-4-14. Beach-fill design profile shape determined by profile translation

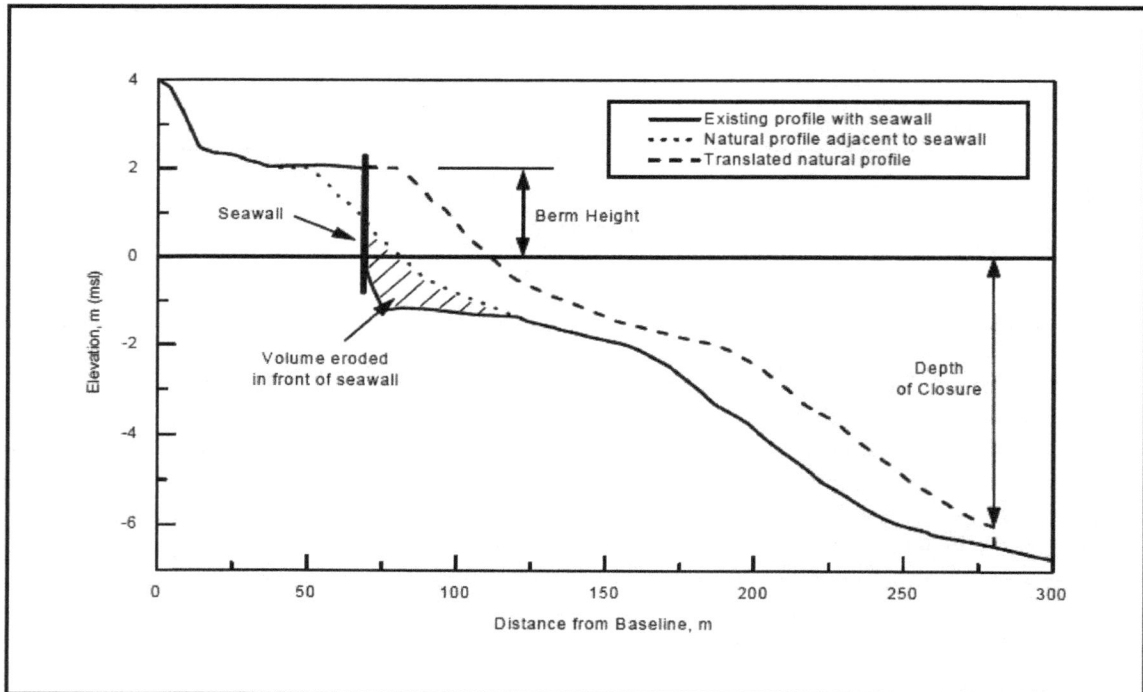

Figure V-4-15. Design profile in front of a seawall

tide conditions and has similar grain size characteristics, can be used to estimate the healthy beach profile shape for reaches where in situ profiles might be misleading. Figure V-4-15 shows an eroded profile in front of an exposed seawall. Translating the existing profile would incorrectly estimate the shape of the nourished beach and would result in an underestimate of the volume required to obtain the design beach width in front of the seawall. In Figure V-4-15, the design profile is determined by translating the natural (healthy) profile shape seaward to obtain the design berm width in front of the seawall. Grain-size differences between the native beach and the fill material also must be considered in defining the design profile shape. If the median grain size of the fill is the same as that of the native beach, the design profile shape should be obtained by translating an average profile shape that represents locally healthy (sediment-rich) beach conditions. For example, given the same composite median grain sizes, the design profile for a beach with 30 m of added berm width is determined by translating the existing profile 30 m seaward between the elevation of the berm crest and the depth of closure, as shown in Figure V-4-14. When applying the profile translation method, the existing beach shape should be determined based on an average of multiple surveys to account for seasonal and/or alongshore variability in profile shape and to avoid including anomalous profile features in the design profile shape.

(a) When fill material is finer or coarser than the native sediment, the design beach profile shape should be estimated based on equilibrium profile concepts (see Part III-3-3-c for details on equilibrium beach profiles). According to equilibrium profile theory, coarser sand will produce a steeper design profile whereas finer sand will produce a profile with a gentler slope as illustrated in Figure V-4-16. Dean (1991) provides additional discussion of equilibrium beach profile concepts and their application. To estimate the design profile shape using equilibrium profile concepts, the average profile shape that represents locally healthy (sediment-rich) beach conditions should first be translated seaward a distance equal to the added berm width (see Figure V-4-14 and Figure V-4-15). To account for the difference in profile shape due to different composite sand sizes, the profile is translated an additional distance as a function of depth between the still-water level and depth of closure, based on differences in the theoretical equilibrium profile shapes as shown in Figure V-4-17. The added distance of translation W_{add} as a function of depth y is given by

$$W_{add}(y) = y^{3/2}\left[\left(\frac{1}{A_F}\right)^{3/2} - \left(\frac{1}{A_N}\right)^{3/2}\right]$$
(V-4-5)

where A_N is the A parameter for native sand and A_F is the A parameter for fill sand (see Table III-3-3 for values of the A parameter for different sand sizes). In Equation V-4-5, when the fill material is finer than the native sand, W_{add} is positive, which produces a design profile that is gentler in slope than the native profile. Conversely, for fill that is coarser than the native beach, W_{add} is negative which produces a steeper design profile. If the representative sediment-rich beach profile includes a bar, some smoothing in the bar region may be necessary. The BMAP software provides automated methods for estimating the design profile shape for different native and fill sand sizes based on Equation V-4-5, and for the same sand sizes using the uniform profile translation technique.

(5) Optimization of design profile. Optimization of the design profile involves selecting a range of design alternatives with different dune and berm dimensions, and evaluating the design alternatives together with the existing condition to determine the alternative that provides maximum net economic benefits. Storm-induced beach erosion modeling is the primary engineering analysis performed as part of the optimization process. Model simulations of profile response to a suite of historical or characteristic storms are performed for each alternative to assess erosion, flooding, and wave attack damages to shorefront property and infrastructure.

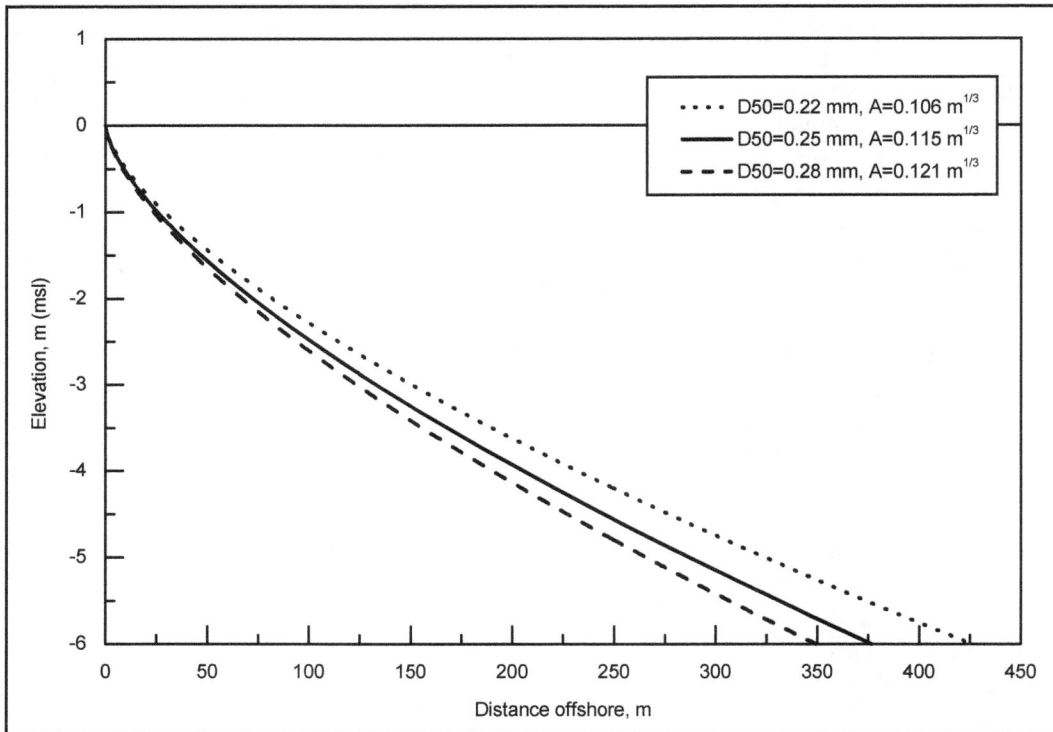

Figure V-4-16.   Theoretical equilibrium profile shapes for different sand grain sizes

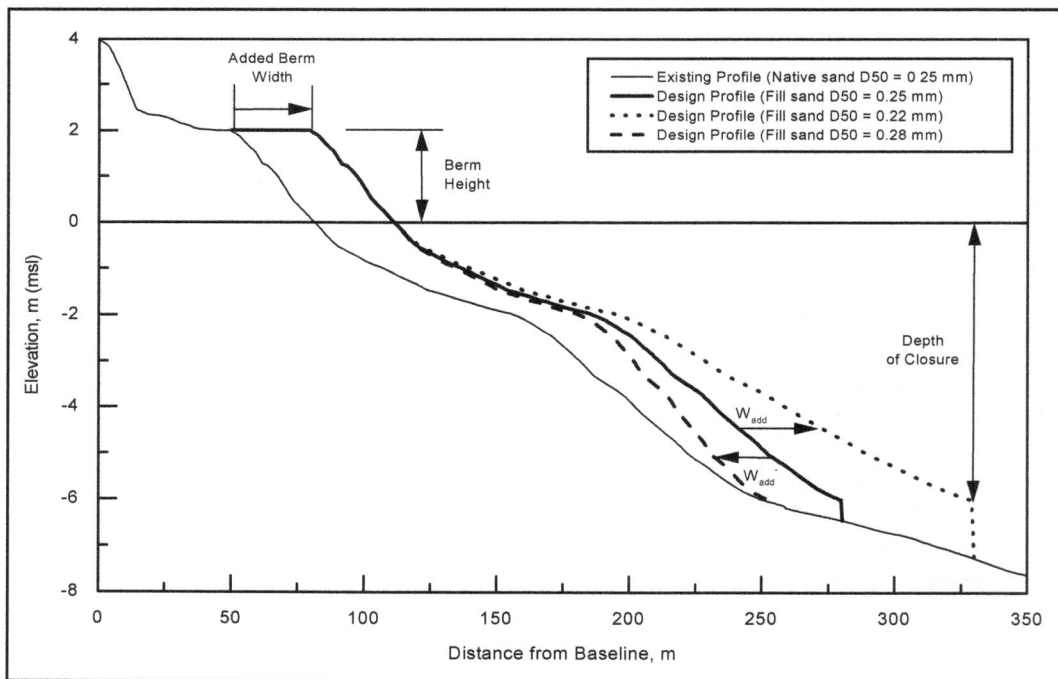

Figure V-4-17.   Design profiles for different fill sand sizes

(a) Without-project condition.

• The existing or without-project condition is included in the optimization process to determine baseline damages. Morphologic features of the existing beach, such as dune height, berm width, and offshore profile shape, typically vary along the project study domain. To accurately estimate storm erosion response for the existing condition, a set of representative morphologic reaches should be developed to describe variations in profile shape along the project domain. The BMAP software can be used to define morphologic reaches by analyzing profiles, grouping similar profiles, and calculating an average representative profile for each reach.

• Profile characteristics that should be considered when developing morphologic reaches include dune height and width, berm width, nearshore and offshore profile slopes, sand grain size, presence of seawalls or other structures, and proximity to inlets. As part of the economic analysis to evaluate damages and benefits, the project domain is divided into a series of economic reaches based on value of property and infrastructure (see Part V-4-1-d). Boundaries of economic domains should also be considered in morphologic reach delineation to ensure that storm erosion modeling is consistent with the economic analysis.

(b) Storm selection.

• Evaluation of potential storm damages requires selection of a set of storms representative of future events that may impact the project area. The set of storms should reflect a range of intensities and frequencies of events consistent with the historical record. In developing the storm set, tropical and extratropical events should be treated distinctly because of differences in storm characteristics and frequencies of occurrence.

• One approach to storm selection has been to develop a set of storms characterized by peak surge return period ranging from frequent events (5-year return period or less) to extremely rare events (100-year return period). Peak surges for selected return periods are obtained from available stage-frequency information, and representative storm surge hydrographs are developed using assumed hydrograph shapes and durations. Differences in hydrograph shape between tropical and extratropical storms should be considered; tropical storms typically have much shorter durations. The storm surge hydrographs are combined with corresponding wave height and wave period time histories to fully describe the storms. Using this approach, the frequency of modeled responses are assumed to correspond to return periods of the input storm surges (e.g., a 50-year storm surge produces a 50-year erosion response). This assumption simplifies the analysis but is not strictly accurate because, in addition to peak storm surge, other factors influence the magnitude of erosion (storm duration, hydrograph shape, and wave characteristics). Inconsistencies may arise with this approach related to characterizing the storm based on peak surge alone. For example, a 20-year storm (where frequency is defined solely on maximum surge) may produce more erosion than a 50-year storm, if the 20-year storm has higher waves or a longer duration.

• An alternate and preferred approach is to develop a "training set" of storms by selecting events from historical records and/or hindcasts. A sufficiently long historical period is identified, such as 20 years for extratropical storms or 100 years for tropical storms. All historical events within the period exceeding a selected threshold are included in the training set. For example, all events which have a peak surge exceeding 0.3 m may be considered significant from a storm erosion standpoint and included in the training set. No return periods are assigned to storms in the training set a priori. Each storm is modeled for the existing condition and each project alternative to calculate corresponding erosion responses. Using the training set of storms and modeled responses, life cycle analyses are performed by employing the Empirical Simulation Technique (EST) to generate frequency-of-occurrence relationships, whereby return periods are associated with storm responses rather than storm input. Scheffner and Borgman (1999) provide detailed guidance in applying the EST to coastal studies. Advantages of this approach are that it involves no arbitrary assignment of recurrence relationships and it utilizes historical rather than representative or hypothetical events to determine frequency-of-occurrence relationships. Part II discusses methods for estimated storm waves and water levels.

- Because water level is a primary factor controlling beach erosion, tide variations should be considered when developing the input storm set. Tide variations can be accounted for by combining storm surges with different tidal phases and ranges. For example, peak storm surges may be aligned with high tide, low tide, midtide preceding high tide, and midtide preceding low tide to generate four different but equally probable events derived from each base storm. Variations in tide ranges (neap, spring, and average) may also be considered in developing a full set of storms.

(c) Storm erosion modeling.

- Upon selection of design alternatives, characterization of the without-project condition, and selection of the storm set, storm-induced beach erosion modeling is performed to calculate parameters required for assessing economic damages based on beach erosion, flooding, and wave attack. The SBEACH model computes relevant storm response parameters such as recession distance, maximum water level and wave height at the shoreline, and wave runup. Required input to the model includes beach profiles describing the design alternatives and without-project condition, time-histories of storm water level, wave height and wave period, median sand grain size, and calibration parameters.
- Detailed examples and guidance for applying the SBEACH model to predict storm erosion are provided in the SBEACH technical report series (Larson and Kraus 1989; Larson, Kraus, and Byrnes 1990; Rosati et al. 1993; Wise, Smith, and Larson 1996; and Larson and Kraus 1998).

(d) Storm damage recurrence relationships and risk analysis. Modeled erosion responses are expressed in terms of frequency of occurrence for input to economic damage models. Typically, tropical and extra-tropical erosion responses are joined to generate a combined frequency curve spanning the recurrence intervals of interest. Risk and uncertainty of storm damage parameters can be addressed by developing mean-value frequency-of-occurrence curves together with confidence bands that indicate the variability or uncertainty associated with the calculated storm responses.

- The EST and supporting analysis tools can be used to calculate mean-value frequency curves and confidence bands. Figure V-4-18 shows an example of frequency-of-occurence relationships generated by the EST technique. The solid line represents the mean or expected value of beach recession as a function of return period, and the dashed lines show the 90 percent confidence band, indicating that 90 percent of variability in beach recession for a given return period falls within these limits.

(6) Cross-sectional fill volume requirements.

(a) A key quantity in beach-fill design is the volume of sand required to produce the desired beach cross-section. The design profile is determined using methods presented in Part V-4-1-f-(4) and results from the optimization process outlined in Part V-4-1-f-(5). The berm width is then increased to reflect the amount of advance nourishment needed to maintain the design profile prior to the first renourishment. The modified design profile shape, which includes advance nourishment, is then estimated

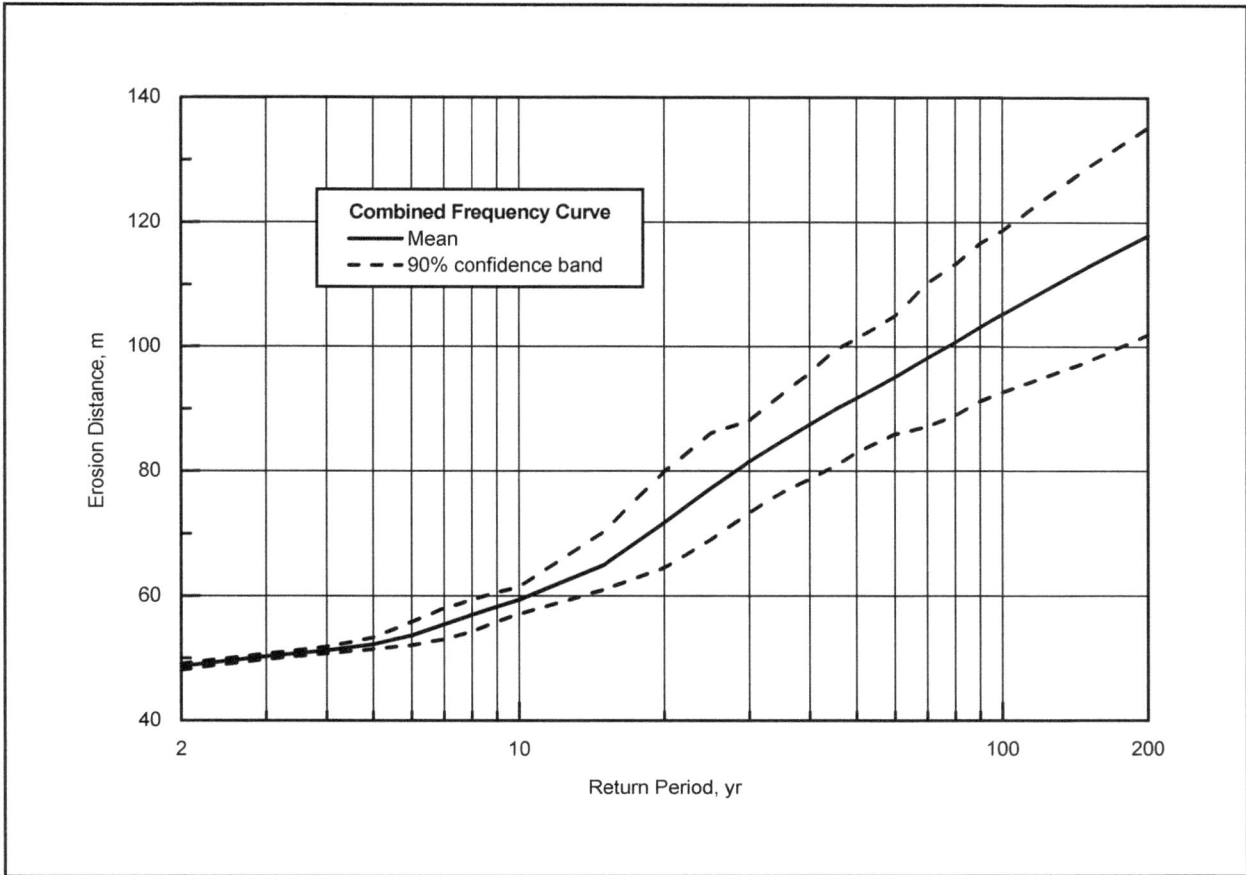

Figure V-4-18. Combined tropical and extratropical erosion-frequency curve

by translating the design profile at elevations between the design berm elevation and the depth of closure (see Figure V-4-14) by an amount equal to the advance nourishment berm width. The sectional fill volume required for initial construction (volume per unit length of shoreline) is calculated as the difference in cross-sectional area between the preproject profile and the modified design profile shape. Example V-4-2 illustrates calculation of sectional fill volume for a case where the preproject profile is in a severely eroded, unnatural, condition.

(b) Advance nourishment beach width, and design berm width, might not have uniform values from reach to reach within the project domain. Variations in the desired level of protection, assessment of long-term fill evolution and renourishment requirements, and consideration of the potential for hot spot formation, will most likely lead to alongshore differences in desired beach width. Therefore, the required sectional fill volume also will vary by reach. Volume calculations for the entire project domain are made by summing results on a reach-by-reach basis, where the volume requirement in a particular reach is calculated as the product of the cross-sectional volume requirement and the length of shoreline in that reach.

(c) For the case of a healthy (sediment-rich), preproject beach profile and a fill material that has a median grain size equal to that of the native beach sand, the volume V per unit length of shoreline required to produce a beach width W can be estimated by

$$V = W\left(B + D_C\right) \tag{V-4-6}$$

EXAMPLE PROBLEM V-4-2

FIND:

Volume per unit length of shoreline required to widen the dry beach by 30 m (at the msl datum), includes both the design berm and advance nourishment. Plot the existing profile and the final design profile.

GIVEN:

Preproject beach profile is not representative of healthy (sediment-rich) conditions.
Existing condition beach profile (Example Figure A).
Representative design beach profile reflecting healthy (sediment-rich) conditions (Example Figure B).
Berm height of 1.5 m and depth of closure of 5.5 m.
Native and fill sand median grain size are the same.

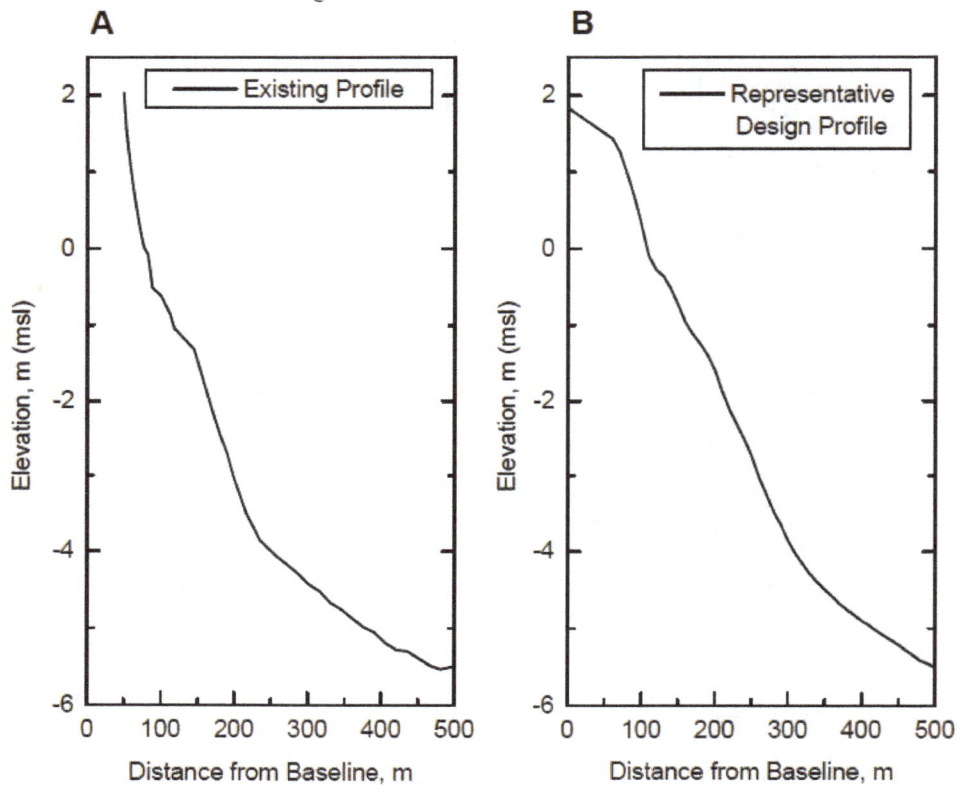

A

B

(Continued)

EXAMPLE PROBLEM V-4-2 (Continued)

SOLUTION:

Step 1: Compute sediment deficit volume in existing profile.

Align existing and representative design profile at MSL datum. Compute volume difference between existing and representative design profiles. This volume represents the sediment deficit between the existing condition and the healthy or sediment-rich condition expected to occur after nourishment. The elevation at which the existing and design profiles are aligned will influence the magnitude of the computed profile sediment deficit. In the present example the MSL datum is selected because here beach width is defined relative to this datum. The sediment deficit in the pre-project profile can vary significantly along the project reach. The computed pre-project sediment deficit in the beach profile is 100 m^3/m in this example and is illustrated in Example Figure C. The data manipulations and calculations discussed in this example are automated within the BMAP software (Sommerfeld et al. 1994).

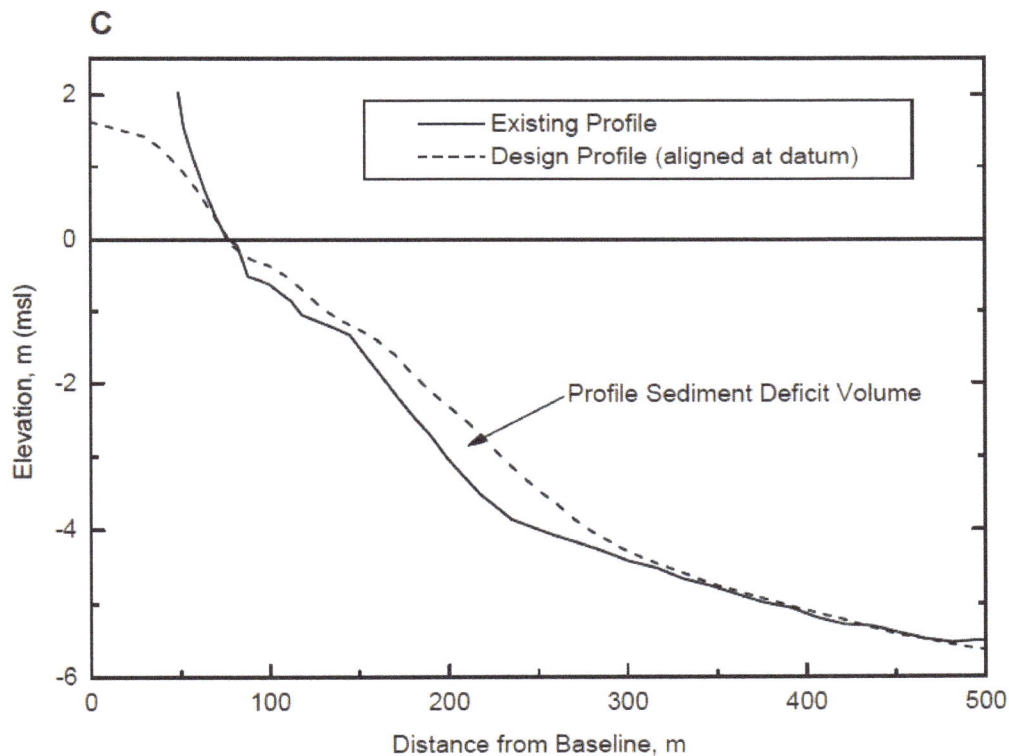

(Continued)

EXAMPLE PROBLEM V-4-2 (Concluded))

Step 2: Compute required sectional fill volume.

Translate the aligned design profile (from Step 1) 30 m seaward to obtain the final design profile. Compute volume difference between the existing and final design profile between the 1.5-m berm elevation and the 5.5-m depth of closure. This volume represents the total sectional fill volume required to advance the dry beach 30 meters seaward at the msl datum. The total sectional fill volume can vary significantly from reach to reach. In the present example, the total sectional fill volume is 298 m³/m and is illustrated in Example Figure D. Approximately one-third of the total sectional volume was required to offset the sediment deficit associated with the over-steepened erosion-stressed condition of the existing profile.

D

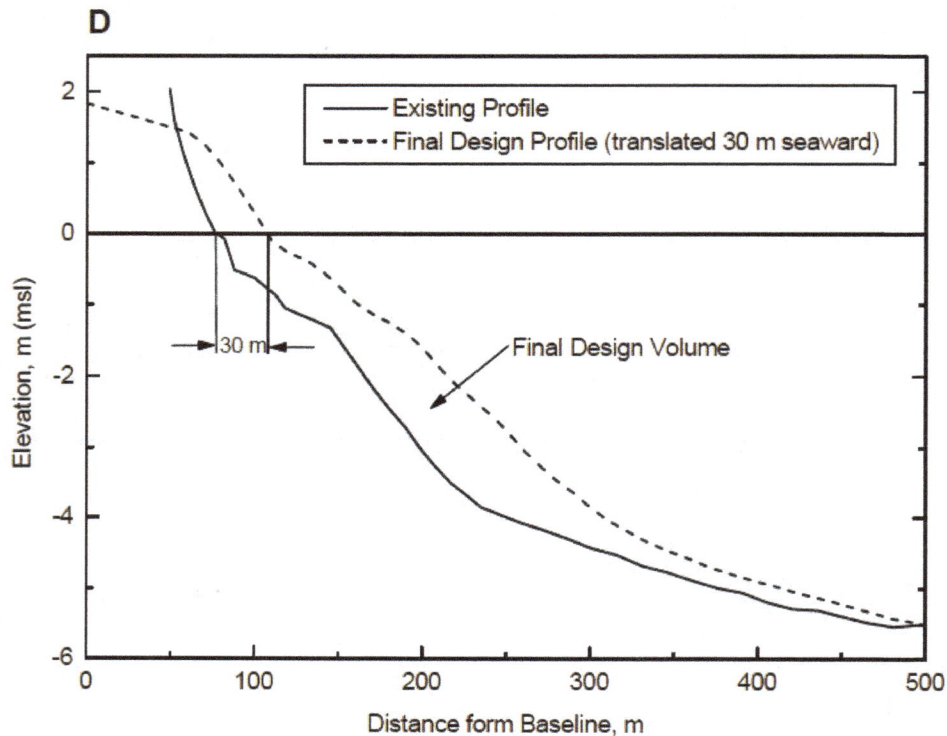

where

B = the design berm elevation

D_c = the depth of closure

Equation V-4-6 is derived by computing the area of the parallelogram formed by translating the existing profile a distance W as shown in Figure V-4-14. Example V-4-3 illustrates this calculation.

EXAMPLE PROBLEM V-4-3

FIND:

Sectional fill volume (fill volume per unit length of shoreline) required to widen the dry beach by 30 m.

GIVEN:

Preproject beach profiles are representative of healthy (sediment-rich) conditions.
Berm height of 2.5 m and depth of closure of 6 m.
Fill material (composite median diameter) same as native beach sand.

SOLUTION:

Equation V-4-6 gives

$$V = 30(2.5 + 6) = 255 \ m^3/m$$

(d) Sectional volume computations in situations when the fill and native sediments differ should be made by considering differences between the existing profile and the design profile shape as outlined in Part V-4-1-f-(4) and illustrated in Figure V-4-17. The BMAP software provides capabilities for calculating sectional fill volume in cases where native and fill sediments differ in median grain size.

(e) Equilibrium profile concepts also can be used directly to make preliminary estimates of required fill volume, when the native and fill sediments have differing composite median grain sizes. While not recommended for final fill volume computations, these methods provide valuable insight regarding the implications of using fill material with different grain size characteristics. Dean (1991) defines three basic types of nourished profiles. Figure V-4-19 shows an intersecting profile, where the profile after nourishment intersects the native profile at a depth shallower than the depth of closure; a nonintersecting profile, where the nourished profile does not intersect the native profile before closure depth; and a submerged profile, where after equilibrium there is no dry beach. A submerged profile is a special case of a nonintersecting profile which occurs when insufficient volume is placed to fully develop the underwater equilibrium profile. Dean (1991) shows that whether a profile is intersecting or nonintersecting is determined by the following inequality:

$$W\left(\frac{A_N}{D_C}\right)^{3/2} + \left(\frac{A_N}{A_F}\right)^{3/2} \quad <1, \quad \text{Intersecting profile} \tag{V-4-7}$$

$$>1, \quad \text{Nonintersecting profile}$$

- For fill material different from the native beach sand, cross-sectional volume requirements should be estimated with consideration given to the differences in profile slope given by equilibrium profile concepts. Based on theoretical equilibrium beach profile shapes, where A_N and A_F are the A parameters for native and fill sands, respectively (see Table III-3-3 for values of the A parameter for different sand sizes), fill sand that is finer than the native material will always produce a nonintersecting profile according to Equation V-4-7. Fill sand that is coarser than native sand may

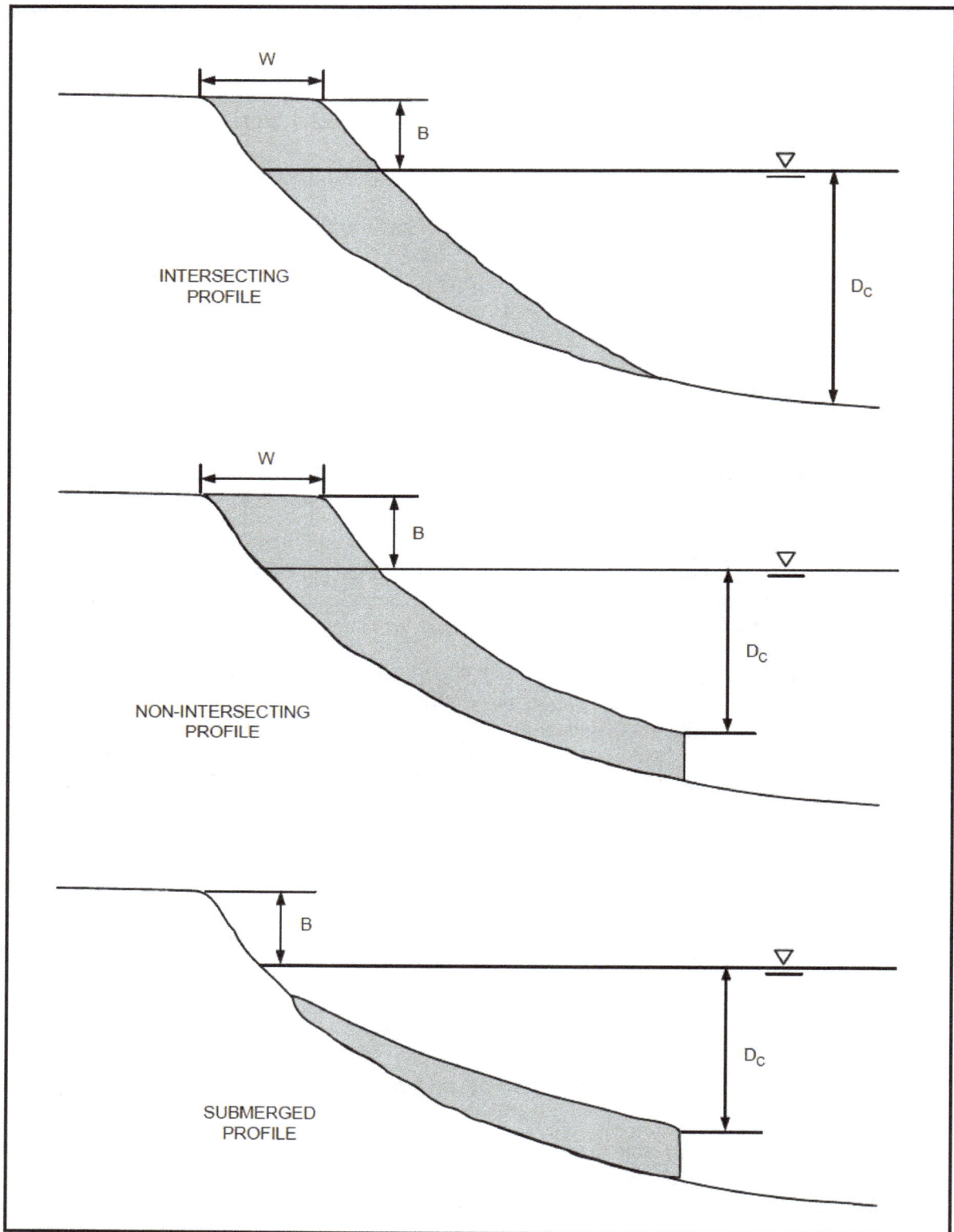

Figure V-4-19. Three basic types of nourishment profiles (adapted from Dean 1991)

produce either an intersecting or a nonintersecting profile. For a nonintersecting profile, the volume of sand per unit length of shoreline that must be placed before there is any dry beach after equilibrium is estimated as

$$V = \frac{3}{5}\left(\frac{D_C}{A_F}\right)^{5/2}\left(A_N - A_F\right)$$

(V-4-8)

- If the volume placed is less than that given by Equation V-4-8, a submerged profile is produced after equilibration. Example V-4-4 illustrates volume calculations using Equation V-4-8.

- For nonintersecting profiles with a dry beach after equilibrium (i.e., volume placed is equal to or exceeds that in Equation V-4-8) the volume per unit length of beach required to produce a dry beach of width W may be estimated as

$$V = WB + \frac{3}{5}\left(\frac{D_C}{A_F}\right)^{5/2}\left(A_N\left[1 + W\left(\frac{A_F}{D_C}\right)^{3/2}\right]^{5/3} - A_F\right)$$

(V-4-9)

Example V-4-5 illustrates volume calculations using Equation V-4-9.

For intersecting profiles, the volume per unit length of beach required to advance the beach a distance W after equilibriation can be estimated by

$$V = WB + \frac{\frac{3}{5}W^{5/3}A_N A_F}{\left(A_F^{3/2} - A_N^{3/2}\right)^{2/3}}$$

(V-4-10)

- It is noted that the depth of closure does not enter Equation V-4-10, because by definition, the nourished profile intersects the native profile landward of the depth of closure (see Figure V-4-19). Example V-4-6 illustrates volume calculations using Equation V-4-10.

- Equilibrium profile methods do not account for a sediment deficit in the preproject beach profile, which is common along erosion-stressed shorelines where beach nourishment projects are typically considered. The methods also only account for volume below the berm elevation. Volume contained in the dune must be added to the estimate. These methods are recommended for quick calculations, and to compliment calculations based on differences between preproject profile shapes and design profiles. They are not recommended for use in computing final sectional fill volume estimates.

- A third method for computing sectional fill volume when fill and native sediments have different grain size characteristics is to translate the healthy, sediment-rich, design profile as shown in Figures V-4-14 and 15, calculate the volume using Equation V-4-6 , and then apply the overfill factor to the volume determined from profile translation (see Part V-4-1-e-(3i) and Example V-4-1). The overfill factor would also be applied to any advance renourishment volume. Dune volume must be

---

EXAMPLE PROBLEM V-4-4

FIND:

      Sectional fill volume that must be placed before any dry beach width is added after equilibrium. Disregard any volume necessary to makeup for a preproject sediment deficit in the beach profile.

GIVEN:

      Berm height of 2.5 m and depth of closure of 6 m. Native sand median grain size of 0.26 mm and fill sand median grain size of 0.19 mm.

SOLUTION:

Values of the A parameter for native and fill sand are read from Table III-3-3.

    $A_N = 0.117 \text{ m}^{1/3}$, $A_F = 0.097 \text{ m}^{1/3}$

Profile is nonintersecting because $A_F < A_N$, therefore Equation V-4-8 is applicable.

Equation V-4-8 gives

$$V = \frac{3}{5} \cdot \left(\frac{6}{0.097} \right)^{5/2} \cdot (0.117 - 0.097) = 361 \text{ m}^3/\text{m}$$

This illustrates that when filling with sand finer than the native beach material, a significant amount of sand must be placed before any dry beach width is produced after equilibrium. Note that the sectional fill volume is 140 percent of the sectional fill volume computed in Example V-4-3, and the present example will produce no additional dry beach width whereas 30 m of additional dry beach width is obtained in Example V-4-3.

---

added. The overfill method can compliment calculations made using the other methods, but it is not recommended for final fill volume computations.

   g. Evaluating project longevity

The longevity of a beach nourishment project is primarily determined by the degree to which the placed sand volume addresses any preproject profile volume deficit, and the rate at which fill material is transported out of the project domain in the alongshore direction, i.e, lateral spreading losses. Wave-driven longshore sand transport processes are the major cause of lateral spreading. Projects tend to be built in erosional areas where waves act to move sand out of the project domain, in a long-term, net sense. In addition to the wave climate and its interaction with the local morphology, there is another important aspect of lateral spreading. The project itself creates a perturbation in shoreline and beach orientation, particularly where the fill transitions into the adjacent beaches. At these transitions, local wave transformation patterns and consequently the longshore sand transport regime are altered, which can lead to high rates of fill loss from the ends of the project (often called end losses). Coastal structures that exist within the project domain, or are built as part of the project, also can impede alongshore sand movement and influence the rate of sand loss. Grain size characteristics of the fill material may be a factor in determining beach-fill longevity. Part III discusses dependency of longshore sand transport on grain size. However, in practice, the role of grain size is not usually considered in evaluating lateral spreading losses.

EXAMPLE PROBLEM V-4-5

FIND:

Volume per unit length of shoreline required to widen the dry beach by 30 m. Disregard any volume necessary to makeup for a preproject sediment deficit in the beach profile.

GIVEN:

Berm height of 2.5 m and depth of closure of 6 m. Native sand median grain size of 0.27 mm and fill sand median grain size of 0.18 mm.

SOLUTION:

Values of the A parameter for native and fill sand are read from Table III-3-3.

$$A_N = 0.119 \text{ m}^{1/3}, A_F = 0.094 \text{ m}^{1/3}$$

Profile is nonintersecting because $A_F < A_N$, therefore Equation V-4-9 is applicable.

Equation V-4-9 gives

$$V = 30 \cdot 2.5 + \frac{3}{5} \cdot \left(\frac{6}{0.094} \right)^{5/2} \cdot \left(0.119 \cdot \left[1 + 30 \cdot \left(\frac{0.094}{6} \right)^{3/2} \right]^{5/3} - 0.094 \right) = 796 \text{ m}^3/\text{m}$$

By comparing with Example V-4-3, it is seen that using the finer material specified in this example requires 3.1 times the volume required using compatible material to generate the same equilibrium beach width. This example illustrates that much higher fill volumes are required when using finer-than-native sand to obtain a given beach width, which is consistent with sand compatibility calculations performed using the overfill ratio (see Example V-4-1).

Although alongshore spreading of fill material represents a loss to the project area, this material is in most cases transported to adjacent beaches. Nearby beaches, particularly those downdrift of the project, realize protective benefits from the neighboring project. Over the life of a beach nourishment project, adjacent beach benefits can be quite significant. If the project is built adjacent to an inlet with navigation channels, the impact of the nourishment project on channel operation and maintenance should be assessed.

(1) Periodic renourishment. Beginning immediately after construction, fill material will be lost from the project due to lateral spreading. Periodic renourishment will be required to maintain the desired beach cross section. It should be recognized by the designer that year-to-year loss rates can deviate from the long-term average erosion rates. In addition to end effects at transitions, losses are significantly influenced by the occurrence of major storms. Annual losses will likely vary from year to year because of the dependency on storm activity. Therefore, while an average renourishment interval and quantity can be estimated, the actual

EXAMPLE PROBLEM V-4-6

FIND:

Volume per unit length of shoreline required to widen the dry beach by 30 m. Disregard any volume necessary to makeup for a preproject sediment deficit in the beach profile.

GIVEN:

Berm height of 2.5 m and depth of closure of 6 m. Native sand median grain size of 0.27 mm and fill sand median grain size of 0.36 mm.

SOLUTION:

Values of the A parameter for native and fill sand are read from Table III-3-3.

$A_N = 0.119 \text{ m}^{1/3}, A_F = 0.137 \text{ m}^{1/3}$

Determine whether profile is intersecting or nonintersecting.

Equation V-4-7 gives

$$30 \cdot \left(\frac{0.119}{6}\right)^{3/2} + \left(\frac{0.119}{0.137}\right)^{3/2} = 0.893 \ < 1$$

therefore, the profile is intersecting and Equation V-4-10 is applicable.

Equation V-4-10 gives

$$V = 30 \cdot 2.5 + \frac{\frac{3}{5} \cdot 30^{5/3} \cdot 0.119 \cdot 0.137}{\left(0.137^{3/2} - 0.119^{3/2}\right)^{2/3}} = 137 \text{ m}^3/\text{m}$$

By comparing with Example V-4-3, it is seen that using the coarser fill material specified in this example requires approximately 45 percent less volume than that required using compatible material, to generate the same equilibrium beach width.

required interval/quantity will vary depending on the climatic conditions that occur. Ideally, the need for renourishment will be determined by monitoring performance of the fill. Some level of renourishment, or maintenance such as redistribution of sand within the project domain, is needed when the project design cross section is no longer in place. In this situation, the desired level of protection is compromised. However, the schedule for periodic renourishment may be more fixed, in the sense that budgeting for it may have been set at the outset of the project. If renourishment is needed before the scheduled time, it can be handled as an emergency maintenance action (see Part V-4-1-m). Should permanent changes to the periodic renourishment cycle (volume and/or frequency) be necessary, a reformulation of the project may be needed. Having an adequate project monitoring plan in place is very important. Monitoring data are particularly valuable if the project does not perform as designed. The data can be analyzed to evaluate the nature of the conditions that prompted the need for unexpected renourishment and assess their frequency of occurrence. Analysis of the monitoring data also can shed light on a design deficiency. Monitoring and analysis of data are discussed in more detail in Part V-4-1-l.

(2) Advance nourishment. Advanced nourishment is the volume of sand that is placed for "sacrificial" purposes during initial construction to maintain the design fill section during the initial renourishment interval (i.e., the time from project completion to the first scheduled renourishment). The magnitude of the advance nourishment should be determined based on results from work done to assess lateral spreading losses and volumetric losses due to long-term shoreline recession, i.e., the historic or background erosion rate. The postproject shoreline erosion rate will be greater than the preproject historic or background erosion rate in those cases where the preproject beach featured a sediment deficit or was otherwise sediment starved. For example, project reaches that feature an armored shoreline may historically exhibit little or no erosion, but can exhibit significant background erosion when replenished with sand fill. Advanced nourishment quantities are included in the initial total construction volume.

(3) Fill parameters affecting lateral spreading. Both simple and detailed methods are available for estimating the rate of alongshore spreading, and identifying renourishment requirements (both volume and interval). Simple methods treat the incident wave climate in a more approximate manner, through use of a representative wave height and neglecting wave direction. They consider the background erosion rate as an input parameter, and they assume the erosion rate is uniform over the project domain. Simple methods are generally most applicable to cases that do not involve coastal structures. On the other hand, detailed methods treat the effects of coastal structures and wave climate more rigorously. They address the issue of alongshore variation in wave conditions that produce the background erosion rates, as well as alongshore variations in erosion rates within the project bounds. Detailed methods treat the directionality of the wave climate. Dominant wave directions become important for projects constructed in the vicinity of engineered structures, littoral barriers, or sediment sinks, such as inlets. Detailed methods consider the actual planform layout of the shoreline and structures, whereas simple methods represent them in an idealized manner. In this section, analytical solutions to the one-line theory of shoreline evolution are examined to reveal the importance of the following beach nourishment design parameters on lateral spreading losses: length of the nourishment project, incident wave climate, and ambient background erosion. The analytical approach is extremely useful in preliminary design and to gain an understanding of the relative importance of these parameters. Detailed methods are also presented later, which rely on the use of numerical models to evaluate project longevity.

(a) Effect of fill length.

• In this section, the effect of beach-fill length (the alongshore extent of the fill) on project longevity will be examined. The influence of length is best illustrated by considering the most simple case of an initially rectangular beach fill constructed on a long straight beach with no background erosion. This situation was first introduced in Part III-2 of this manual where the linearized equation of longshore sediment transport was combined with the equation of continuity to develop the one-line theory of shoreline evolution (see Equations III-2-25 and III-2-26). A number of analytical solutions to this equation were presented. In this section, Equations III-2-31 and III-2-32 will be examined further to extract additional information pertinent to beach nourishment design. Upon close examination of Equation III-2-31 it is seen that the important parameter is

$$\frac{a}{2\sqrt{\epsilon t}} \qquad\qquad\qquad\qquad \text{(V-4-11)}$$

where a is one-half the length of the rectangular project, ϵ is the "shoreline diffusivity" parameter defined in Equation III-2-26, and t is time. Here it is seen that if the quantity in Equation V-4-11 is the same for two different projects their planform evolution would be the same. However, if two projects were exposed to the same wave climate but had different alongshore lengths, then the project with the greatest length would be predicted to last longer (with all other factors being the same). In fact, according to Equation III-2-32 the longevity of a project varies as the square of its length. If more than 50 percent of the placed beach-fill volume remains within the placement area (0.5<p(t)<1.0), Equation III-2-32 can be approximated using the following relationship (with an accuracy of ±15 percent).

$$p(t) = 1 - \frac{\sqrt{\epsilon t}}{a\sqrt{\pi}}$$

(V-4-12)

- Example Problem V-4-7 illustrates the importance of project length on project longevity. In this example, a fill with twice the length will last four times as long. The effect of project length on fill longevity is critical for short fills. It is also important in long fills which may be built in stages. For example, construction may be limited to a particular season to avoid turtle nesting season or the tourist season. Therefore it may take 2 or 3 years to complete the work. Projects built in stages will temporarily perform as short fills until the other portions of the project are completed. Actual loss rates from the constructed subreaches will likely exceed losses predicted for the completed as designed project. Any short-term accelerated losses due to construction of the project in stages should be factored into the advance nourishment quantity.

(b) Effect of wave environment.

- The rate of alongshore spreading losses is also a function of the incident wave climate. In Equation III-2-26 it is seen that the shoreline diffusivity term (ϵ) varies inversely with the breaking wave height raised to the 5/2 power. Dean and Yoo (1992) present a method for calculating a representative wave height and period based on assumptions of Rayleigh-distributed wave height, shallow-water linear-wave theory, simplified and linearized wave refraction and shoaling relations, and a constant proportionality between breaking wave height and corresponding water depth. Use of an effective wave height is recommended in the calculation of the shoreline diffusivity term (ϵ). Dean and Yoo (1992) defined the effective wave as one that produces the same spreading of the beach nourishment material as the actual time-varying wave conditions (expressed as pairs of height and period). They provided the following equation to calculate the effective wave height, H_{eff},

$$H_{eff} = \left(\frac{\frac{1}{N}\sum_{n=1}^{N}(K_s H_s^n)^{2.4}\frac{\left(C_{go}^n\right)^{1.2}}{C_*^n}}{\frac{1}{N}\sum_{n=1}^{N}\frac{\left(C_{go}^n\right)^{1.2}}{C_*^n}} \right)^{\frac{1}{2.4}}$$

(V-4-13)

where K_s is the proportionality factor between significant deepwater wave height and effective deepwater wave height and is equal to 0.735 (Dean and Yoo 1992), H_s^n is the significant wave height of the nth record in the time series of N wave records, C_{go}^n is the deepwater wave group speed of the nth record, and C_*^n is the wave celerity at breaking of the nth record. The effective wave period T_{eff} is defined as the period corresponding to the expression in the denominator.

EXAMPLE PROBLEM V-4-7

FIND:

The "half-life" of the specified beach fills (time at which 50 percent of the beach-fill material remains within the placement area).

GIVEN:

Both projects have a rectangular planform and differ only in alongshore length. Beach fill A has an alongshore length of 3 km, whereas beach fill B has an alongshore length of 6 km. Both projects are subjected to the same wave environment.

Assume:
$K = 0.77$
$H_b = 0.95$ m
$C_{gb} = (g\,h_b)^{1/2}$
$h_b = H_b/0.78$
$g = 9.81$ m/s^2

$\rho_s/\rho = 2.65$
$n = 0.4$
$d_b = 2.5$ m
$d_c = 6.0$ m

SOLUTION:

Equation III-2-26 gives

$$\epsilon = \frac{0.77(0.95)^{2.5}\sqrt{9.81/0.78}}{8} \cdot \frac{1}{(2.65-1)} \cdot \frac{1}{(1-0.4)} \cdot \frac{1}{(2.5+6.0)} = 0.03568\ \frac{m^2}{sec}$$

Solving Equation V-4-12 for t and $p(t) = 0.5$ gives

$$t_{50\%} = \frac{a^2\,\pi}{4\,\epsilon}$$

Half-life of beach fill A

$$t_{50\%} = \frac{(1500)^2\,(3.14)}{(4)\,(0.03568)} = 49.526 \times 10^6\ sec \approx 1.57\ years$$

Half-life of beach fill B

$$t_{50\%} = \frac{(3000)^2\,(3.14)}{(4)\,(0.03568)} = 198.102 \times 10^6\ sec \approx 6.28\ years$$

- Example Problem V-4-8 illustrates the relative importance of the breaking wave height on the expected longevity of a beach-fill project, with all other factors being equal. In this example, a 19 percent increase in breaking wave height resulted in a 35 percent decrease in the project's half-life, which is a measure of fill longevity.

- Differences in longevity are even more pronounced if greater differences in breaking wave height are considered. For example, if the average summer breaking wave height at the project site is 0.6 m, and the average winter breaking wave height is 1.2 m, one would expect that loss rates to be much higher in the winter compared to those in the summer (because of the diffusivity parameter dependence on breaking wave height raised to the 2.5 power). Many beach-fill projects are built during the winter season. The strong dependence of longevity on wave environment helps explain the high rates of lateral spreading loss that can occur during the winter season.

(c) Effect of background shoreline recession.

- The effect of project length and the incident wave environment have been shown to have a significant influence on expected project longevity. The analysis presented to this point has not considered losses due to ambient coastal processes, such as a gradient in the longshore sand transport rate, which tends to produce background erosion at the project site. Equation V-4-12 can be modified to include the effect of a uniform background shoreline recession rate, E, as follows:

$$p(t) = 1 - \left(\frac{\sqrt{\epsilon t}}{a\sqrt{\pi}} + \frac{Et}{\Delta y_o} \right) \qquad\qquad 0.5 < p(t) < 1.0 \qquad\qquad \text{(V-4-14)}$$

- Solving Equation V-4-14 for time t yields an expression that predicts the time required for a fraction $(1-p)$ of the material placed to be lost from the project area (or equivalently the time at which a fraction p of the material placed remains in the project area). This expression is provided as

$$t_p = \frac{-m - \sqrt{m^2 - 4ln}}{2l} \qquad\qquad \text{for } \sqrt{\epsilon}t/a < 1.0 \qquad\qquad \text{(V-4-15)}$$

in which

$$l = \left(\frac{E}{\Delta y_o} \right)^2$$

$$m = \frac{2E(p-1)}{\Delta y_o} - \frac{\epsilon}{\pi a^2}$$

and

$$n = (1 - p)^2$$

EXAMPLE PROBLEM V-4-8

FIND:

The "half-life" of the specified beach fills (time at which 50 percent of the beach-fill material remains within the placement area).

GIVEN:

Both projects have a rectangular planform with an alongshore length of 6 km. The effective breaking wave height at beach fill A is 0.80 m whereas the effective breaking wave height at beach fill B is 0.95 m.

$K = 0.77$ $\rho_s/\rho = 2.65$ $C_{gb} = (g\,h_b)^{1/2}$ $n = 0.4$
$h_b = H_b/0.78$ $g = 9.81 \text{ m/s}^2$ $d_b = 2.5 \text{ m}$ $d_c = 6.0 \text{ m}$

SOLUTION:
Equation III-2-26 gives: (for beach fill A)

$$\epsilon = \frac{0.77(0.80)^{2.5}\sqrt{9.81/0.78}}{8} \cdot \frac{1}{(2.65-1)} \cdot \frac{1}{(1-0.4)} \cdot \frac{1}{(2.5+6.0)} = 0.02322\ \frac{\text{m}^2}{\text{sec}}$$

(for beach fill B)

$$\epsilon = 0.03568 \text{ m}^2/\text{sec} \quad \text{(see EXAMPLE PROBLEM V-4-7)}$$

Solving Equation V-4-12 for t and $p(t) = 0.5$ gives

$$t_{50\%} = \frac{a^2\,\pi}{4\,\epsilon}$$

Half-life of beach fill A

$$t_{50\%} = \frac{(3000)^2\,(3.14)}{(4)\,(0.02322)} = 304.420 \times 10^6 \text{ sec} \approx 9.65 \text{ years}$$

Half-life of beach fill B

$$t_{50\%} = \frac{(3000)^2\,(3.14)}{(4)\,(0.03568)} = 198.102 \times 10^6 \text{ sec} \approx 6.28 \text{ years}$$

where $\Delta\,y_o$ is the initial dry beach width (after cross-shore equilibration), E is the historical shoreline recession rate and a is the beach-fill half length. Example Problem V-4-9 illustrates the effect of background erosion rate. Comparison of results from this example with results from Example V-4-7 show that the specified background erosion rate decreased the half-life of the fill by about 20 percent. Note that the historical shoreline erosion rate E, may underestimate the postnourishment erosion rate if the preproject beach is armored or otherwise features a deficit in sand volume or sand supply.

EXAMPLE PROBLEM V-4-9

FIND:

The approximate time required for 50 percent of the placed beach fill volume to be lost from the placement area due to alongshore spreading and a background shoreline recession rate of 0.5 m/year.

GIVEN:

Assume the planform of the beach fill is initially rectangular with an alongshore length of 6 km and an estimated offshore width of 50 m (after equilibration).

Assume:

$K = 0.77$ $\qquad$ $h_b = H_b/0.78$ $\qquad$ $n = 0.4$

$H_b = 0.95$ m $\qquad$ $g = 9.81$ m/s^2 $\qquad$ $d_b = 1.5$ m

$C_{gb} = (g\,h_b)^{1/2}$ $\qquad$ $\rho_s/\rho = 2.65$ $\qquad$ $d_c = 6.0$ m

SOLUTION:

Equation III-2-26 gives $\epsilon = 0.03568$ m^2/sec (see EXAMPLE PROBLEM V-4-7)
Equation V-4-15 gives

$$l = \left(\frac{E}{\Delta y_o}\right)^2 = \left(\frac{0.5}{50}\right)^2 = 0.0001 \text{ years}^{-2}$$

$$m = \frac{2E(p-1)}{\Delta y_o} - \frac{\epsilon}{\pi a^2} = \frac{(2)(0.5)(.5-1)}{50} - \frac{0.03568}{(3.14)(3000)^2}(31.536\text{X}10^6) = -0.04980 \text{ years}^{-1}$$

$$n = (1-p) = (1-.5)^2 = 0.2500$$

and

$$t_{50\%} = \frac{-m - \sqrt{m^2 - 4ln}}{2l} = \frac{0.04980 - \sqrt{(0.04980)^2 - (4)(0.0001)(0.2500)}}{(2)(0.0001)} \approx 5.07 \text{ years}$$

(4) Shoreline change modeling to estimate advance fill and renourishment requirements. In detailed design, one-line numerical models of shoreline evolution such as the GENEralized model for SImulating Shoreline Change model (Hanson 1987; Hanson and Kraus 1989) are typically used to provide more realistic estimates of project longevity and renourishment requirements than may be possible using the analytical approach discussed previously. Numerical shoreline change models provide the designer with an objective tool for evaluating a variety of potential project design alternatives, which may involve coastal structures in addition to beach fill. The use of numerical models of shoreline evolution for the design, optimization, and comparative evaluation of competing project alternatives has many advantages over the use of the analytical approach. For example, the GENESIS model allows examination of multiple renourishment cycles leading to a complete project life cycle, the effects of beach-fill stabilization structures, the effects of many possible future wave conditions (as defined by the Wave Information Study (WIS) wave hindcast) leading to a suite of possible future shoreline conditions. The numerical approach allows for the evaluation of the project design and its performance within the context of a realistic sediment budget developed for the project reach. Employing a shoreline change model like GENESIS allows the designer to examine project performance under conditions much more representative of the actual project setting than is possible with the analytical approach, and is therefore the recommended approach for final design of major projects.

(a) Simple model application.

- In this section, the GENESIS model is applied in a simplified manner to simulate the evolution of an idealized beach-fill project, throughout an anticipated 50-year project life, to investigate the design issues of advance fill and renourishment requirements. The hypothetical project is 6 km (3.7 miles) long with a design berm width of 20 m (65 ft). Two series of model simulations were performed for each of four renourishment intervals, 3, 5, 7, and 9 years. The first series of simulations were for a hypothetical condition where the project shoreline was subjected to no ambient or background erosion. The second series of simulations were for a project shoreline experiencing a background erosion rate of 0.5 m/year. All simulations were made for the entire 50-year project life. Note that the 3-year renourishment cycle project was simulated as a series of fifteen 3-year cycles plus one 5-year cycle beginning in project year 45. Likewise, the 7-year renourishment cycle project was simulated as a series of six 7-year cycles plus one 8-year cycle, and the 9-year renourishment cycle project was simulated as a series of five 9-year cycles plus one 5-year cycle.

- In the execution of the simulations, the computed shoreline position at the end of each renourishment cycle was used to define the initial shoreline position at the beginning of the next renourishment cycle, except of course in the project area where the required renourishment fill volume was placed with tapered transitions to the existing shoreline adjacent to the project. The incident wave climate was specified as an effective wave condition that approached the project normal to the shoreline and did not change throughout the simulation. The average berm height and depth of closure were specified to be 1.5 and 6 m, respectively, for a total active depth of 7.5 m. The preproject shoreline was straight. The initial construction planform included 2-km-long transitions to the preproject shoreline position adjacent to the project limits, which were added to reduce the rate of end losses from the project. The model was then run for the first renourishment interval, an appropriate nourishment volume was added, the model was run for another renourishment interval, and so on, until the 50-year life was reached. Each prenourishment shoreline within the project limits was advanced seaward a sufficient distance so that the predicted shoreline position at the end of the renourishment cycle was 20 m seaward of the preproject shoreline at all locations within the project, except within 400 m in from both ends of the 6-km-long project. That is, the design width was allowed to be violated only at the lateral limits of the project, and for an alongshore distance of only 400 m. The results of these model simulations bring to light a number of important considerations with respect to beach-fill design.

Plots of the placed advanced fill and renourishment volume throughout the life of each of the projects simulated are shown in Figure V-4-20. Volumes shown in this plot exclude the design fill volume of 960,000 cu m contained within the project limits and small 400-m transitions at the ends of the project. The volume values include the volume contained in the transition zones that are constructed initially, and advance nourishment placed within the project limits needed to see that the design width is not compromised during the first renourishment cycle. As can be seen in Figure V-4-20a, in the absence of background erosion, the required renourishment volume is expected to decrease slightly over the life of the project. With a background erosion rate of 0.5 m/year the required renourishment volume is seen to remain nearly constant throughout the project life. Most project sites experience background erosion.

- Results indicate that the advance fill volumetric requirement is nearly twice the volumetric requirement of subsequent renourishments, and nearly three times as much for the shortest renourishment interval, or 3 years. The advance fill volume includes the volume initially placed in the transition sections. The explanation for this result is that a substantial volume of material is required to provide a natural transition from the adjacent shorelines to the more seaward-advanced project shoreline. This volume of material is required at the time of initial construction, as part of the advanced fill. Without the transition sections, high end losses at the project transitions would have compromised the design berm width over a significant length of the lateral portions of the project. Subsequent renourishments do not require this additional volume because the adjacent beaches are already prograded toward the design beach width. Part V-4-1-h examines the subject of fill transitions in more detail.

- Figure V-4-21 provides plots of the cumulative volume placed for each of the beach nourishment projects simulated. Results indicate that the cumulative beach nourishment volume requirement for a 50-year project life is nearly the same regardless of the renourishment cycle (at least for the renourishment intervals tested, which span the typical range of most beach-fill projects). A detailed examination of the model output indicates that the average annual rate of loss of sand from the project area increases only slightly as the renourishment cycle increases (3 or 4 percent difference between annual loss rates for 3- and 9-year renourishment cycles). Selection of a renourishment cycle for Federal projects is made through an optimization process that optimizes the amortized initial construction costs and the annual cost of periodic renourishment to minimize the total average annual equivalent cost. Selection of renourishment volumes/intervals is typically done based on average expected losses over the life of the project. Selecting a longer renourishment interval, with strict adherence to amortized life-cycle-cost based selection, may reduce flexibility in dealing with annual variations in loss rates, which might be caused by several years of higher than normal storm activity, or design deficiencies discovered within the first few years following construction. From both engineering/design and monitoring/maintenance perspectives, 3- to 4-year renourishment cycles are desirable. If longer renourishment intervals are desired from an economic perspective, it is prudent to plan a short first renourishment interval (about half the optimized interval, or 3 to 4 years, whichever is smaller).

- If the volume remaining within the project area is subtracted from the cumulative placed volume, one can estimate the cumulative spreading losses. In the simulations without background erosion, the cumulative spreading losses range between 2.75 to 2.83 million cu m of sand, depending on the renourishment cycle. However, the simulations that included the effect of a 0.5 m/year background erosion rate indicate substantially higher cumulative losses, between 6.36 to 6.68 million cu m of sand, depending on the renourishment cycle. Some past projects have underestimated renourishment requirements by assuming that a historically based volumetric rate of erosion within the project domain can be used to estimate the required renourishment. This assumption incorrectly neglects the effect of background erosion outside the project on project end losses. For example, the

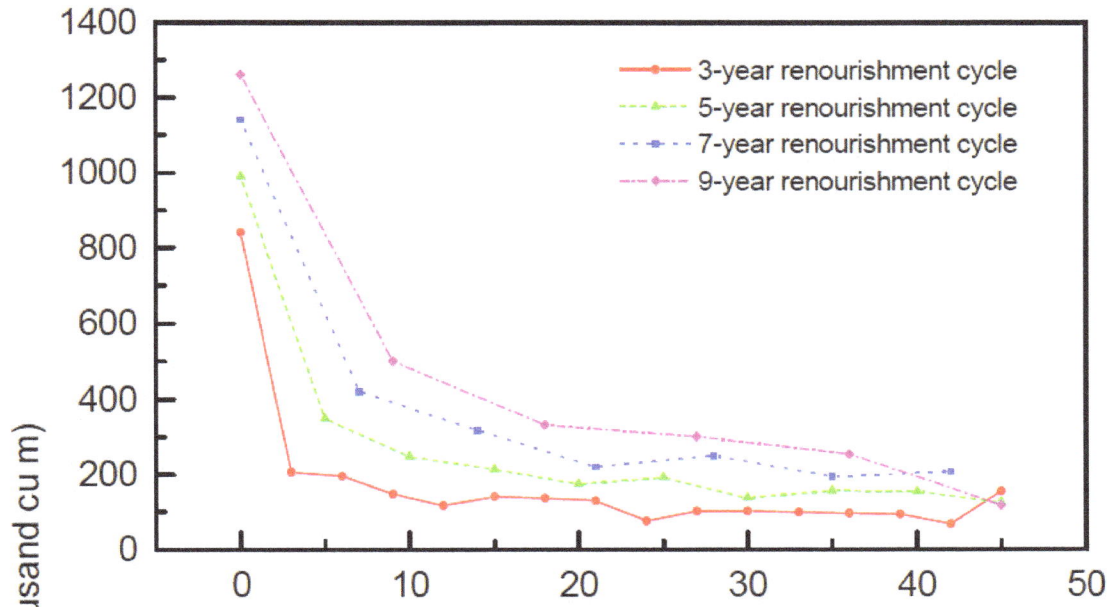

a. No background erosion

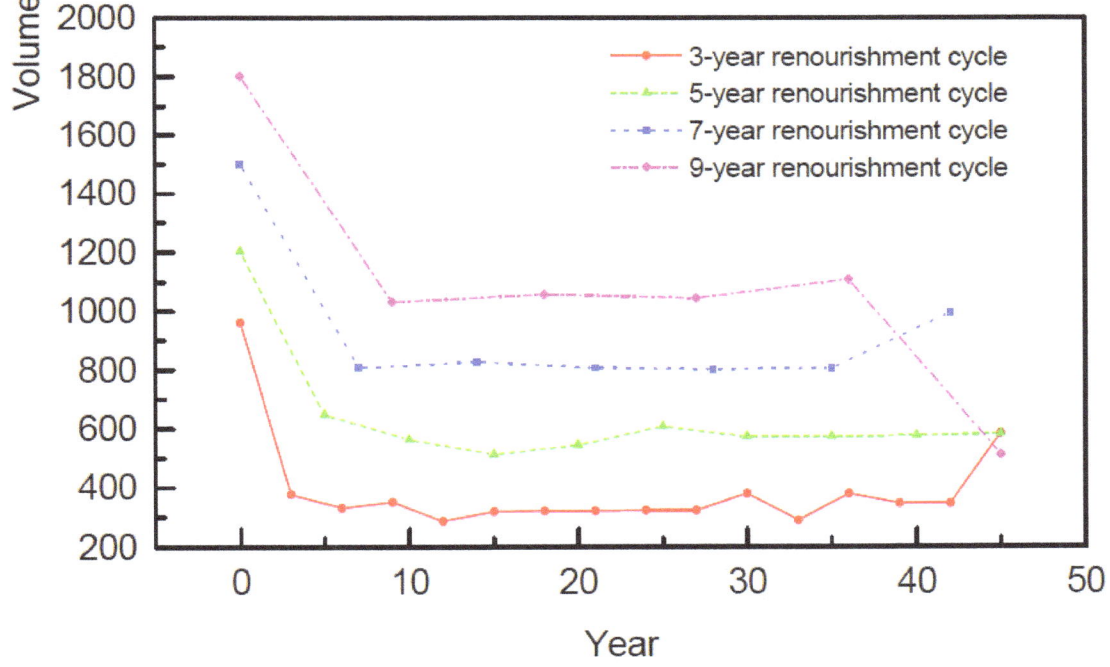

b. 0.5 m/year background erosion

Figure V-4-20. Advanced fill and renourishment volumes

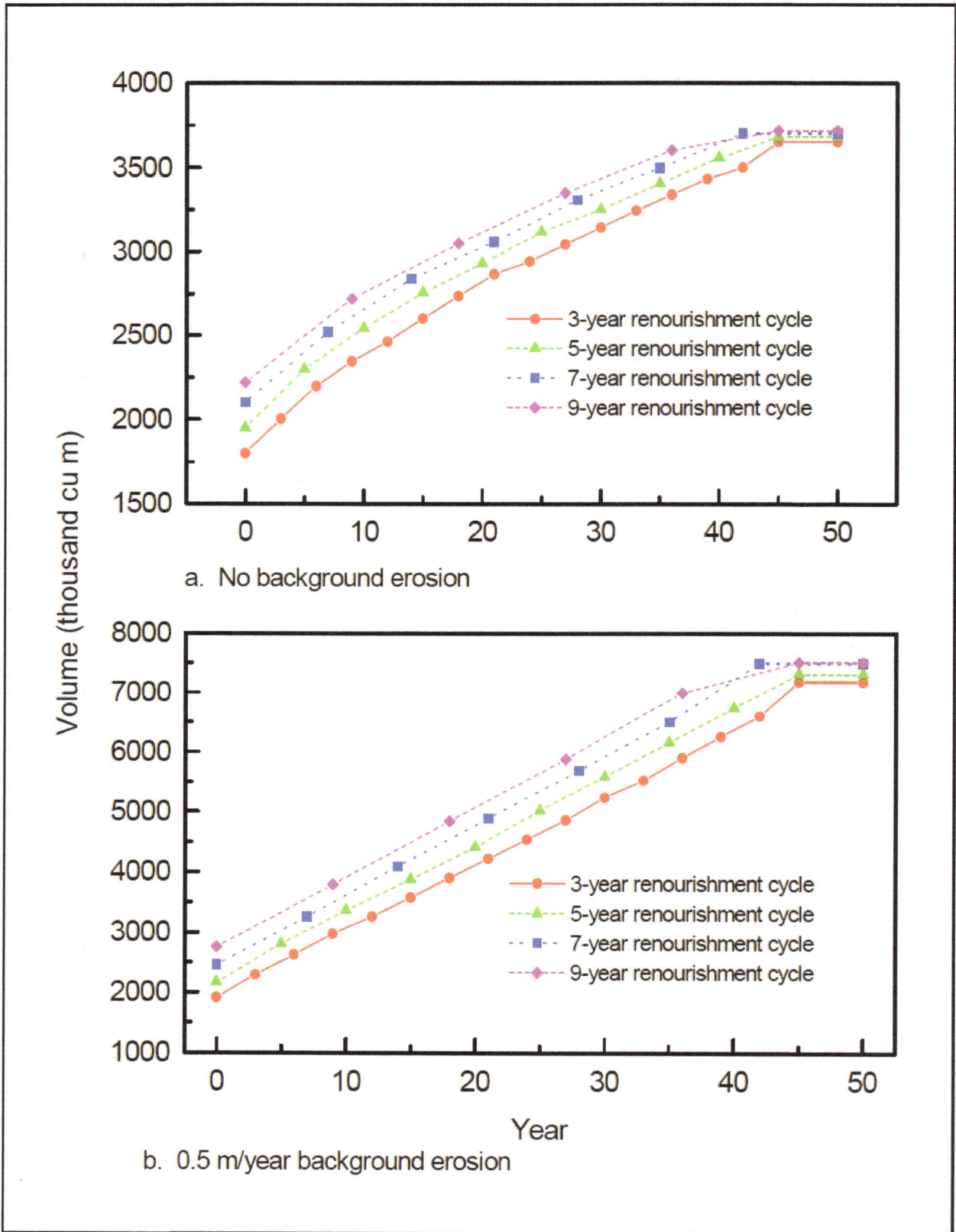

Figure V-4-21. Cumulative beach nourishment volume

volumetric erosion due to a background erosion rate of 0.5 m/year on a 6-km project length with an average berm height of 1.5 m and a depth of closure of 6 m is 1.13 million cu m over a 50-year project life. Adding this volume to the calculated project end loss volume (in the absence of background erosion) of 2.8 million cu m gives a total project volume of 3.93 million cu m which is only about 60 percent of the 6.52 million cu m estimated in the simulations that included a background erosion rate of 0.5 m/year. The explanation for this result is that background erosion outside the project causes the seaward protuberance of the project to become larger each year. At the end of the 50-year project, the design shoreline is 45 m seaward of the adjacent shorelines which are unaffected by the project nourishments; whereas in the absence of background erosion, the protuberance of the design shoreline remains 20 m throughout the 50-year project. The additional nourishment volume is required to provide a natural transition from the eroded adjacent shorelines to the design shoreline. Figure V-4-22 shows the calculated shoreline position at the end of the 50-year nourishment project for each of the project scenarios simulated. It also illustrates the previously discussed requirement for additional nourishment volume in the presence of background erosion. Figure V-4-22 also illustrates the beneficial effects a beach-fill project has on the adjacent shorelines during its lifetime. In the absence of background erosion (Figure V-4-22a) it is seen that the beaches adjacent to the fill project are advanced seaward approximately half the design width for more than a project length on both sides of the fill. With a background erosion rate of 0.5 m/year (Figure V-4-22b) it is seen that the fill project stabilized the shoreline at or seaward of the preproject shoreline position for a distance of approximately one project length on both sides of the fill.

(b) Recommendations for detailed analysis.

- In the previous section, the GENESIS model was applied in a simplified way to estimate volume requirements for an idealized beach-fill project. That application of the GENESIS model was referred to as simplified because the model was not calibrated for the specific project reach nor was a time varying time series of wave information used to represent the environmental forcing. Furthermore, the simplified method of analysis assumes that the project and adjacent beaches can be characterized as long straight beaches with uniform and temporally constant background erosion rates, and that the dominant long-term processes affecting the evolution of the beach-fill project can be captured by examining the alongshore dispersion of the fill by a persistent effective wave condition. However, many if not most projects vary significantly from these assumptions and warrant a more detailed application of the GENESIS model. Gravens, Kraus, and Hanson (1991) and Gravens (1992) provide detailed information concerning the application of the GENESIS model.

- A detailed application involves developing shoreline position and beach profile data sets, analyses of incident wave conditions including detailed nearshore wave transformation, and selection of a time-history of representative wave conditions for use in model calibration, verification, and project forcasting. Typically a detailed wind wave hindcast such as WIS provides the required wave information and serves as the database defining the local and regional characteristics of the offshore wave climate in the vicinity of the project. A long, multiyear record of measured wave data may also suffice.

- A critical aspect of the detailed application is the calibration of the model to site specific project conditions. Calibration involves selection of the domain to be modeled and boundary conditions, evaluation of model accuracy in reproducing historical shoreline change, and net and gross longshore sand transport rates. The calibration and evaluation of the GENESIS model serve to demonstrate the predictive capability of the model at the specific project site. Estimates based on a well-calibrated and verified GENESIS model are considered superior to those based on a simplified application of GENESIS. However, it may be informative to examine the difference between estimates generated

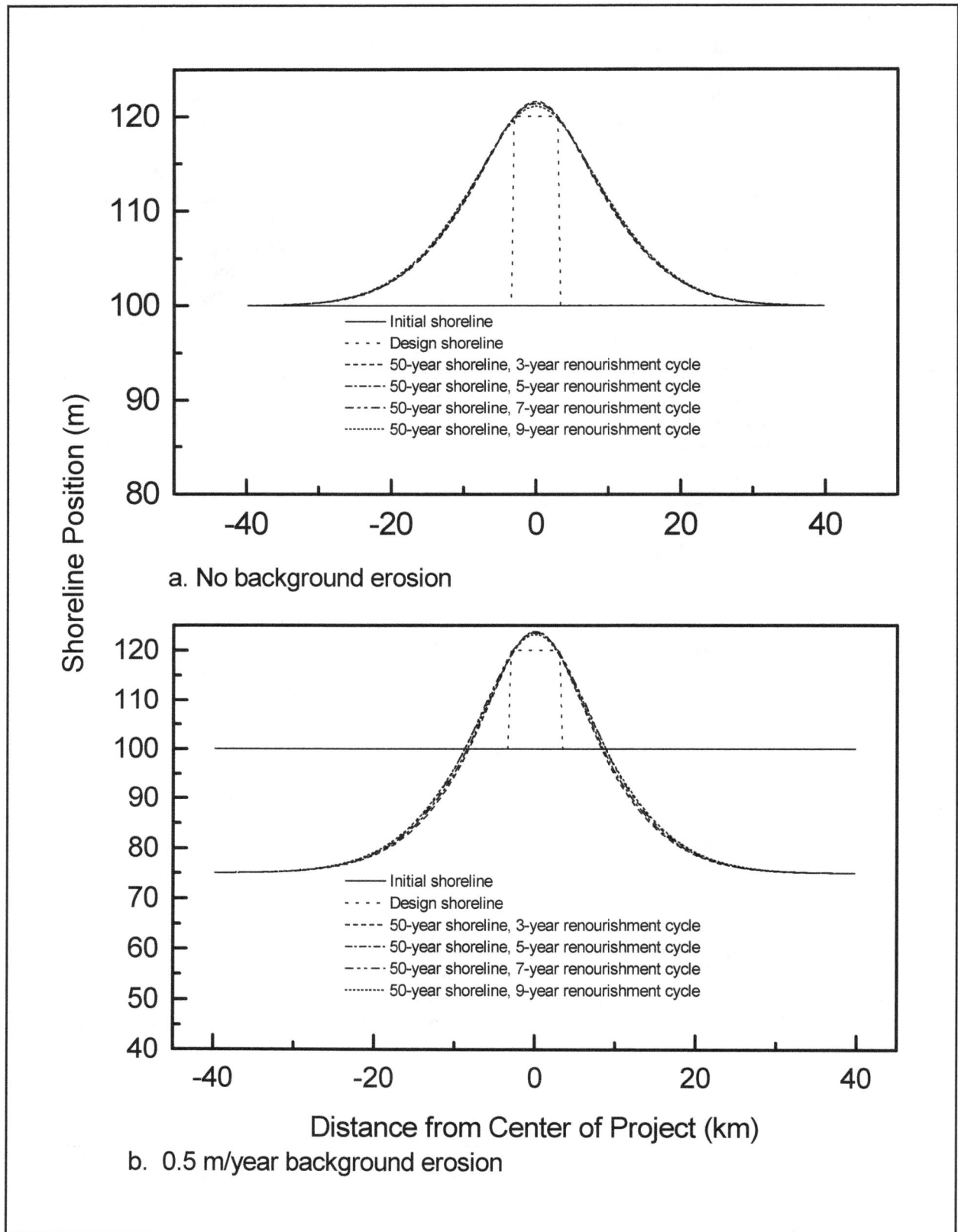

Figure V-4-22. Calculated 50-year shoreline position

through both simplified and detailed applications of GENESIS for a given project to quantify the range of uncertainty between the design methods. Coastal engineering judgement, local coastal experience, and the performance of nearby similarly-situated projects, if any, are important and should be factored into the overall design and specifically the estimates of project volume requirements. The results of a detailed GENESIS model application should always be interpreted in the context of the model calibration and verification. For instance, if the historical rate of erosion was underpredicted in some specific model region of the simulated shoreline in the calibration and verification then it should be expected that the model will continue to underpredict erosion in this area in the simulation of the various design alternatives. Estimates of volume requirements should be adjusted to account for deficiencies in the model calibration and verification.

• One advantage of taking the detailed GENESIS modeling approach in the investigation and prediction of beach-fill project performance is the opportunity to anticipate possible secondary effects that have troubled some past projects. For instance, the excavation of a nearshore borrow area or some preexisting bathymetric feature may significantly alter local breaking wave conditions and result in the development of a localized zone of accelerated erosion along the project shoreline (a hot spot). A detailed application of GENESIS would examine the effects of local bathymetry and project changes in the bathymetry, through the application of nearshore wave transformation models and inclusion of nearshore wave information in the shoreline change calculations. GENESIS model predictions based on this type of data can reveal the potential for the development of shoreline anomalies within the project. Simplified applications of GENESIS or an analysis based on the analytical approach have no chance of revealing possible hot spot development because the processes responsible for their development are ignored or are smoothed over in the analysis.

• If a good calibration of the GENESIS model cannot be achieved because of complex boundary conditions (unstructured inlet, randomly fluctuating shoreline), or insufficient or poor data quality, then a more simplified application of the model is recommended. An alternative to calibrating and verifying the GENESIS model to historical shoreline position data, is to calibrate GENESIS to an accepted existing-condition sediment budget developed through an analysis of the available physical data. In this type of application, a wave time series that reproduces the accepted sediment budget within the model is developed either from hindcast wave information such as WIS, measured wave information, or simply synthesized from observations or general wave and wind information available in a statistical form.

(5) Anticipating hot spots. Hot spots are localized areas within a beach nourishment project that experience a reduction in beach width (corresponding to a loss of sand volume) significantly greater than losses that were predicted and/or are observed throughout the rest of the project. Hot spots are problematic in the sense that the desired level of protection may be compromised locally, whereas the remainder of the project is functioning well. To achieve the desired level of protection, renourishment in the hot spot area may be required on a more frequent basis than the rest of the project. Or, some structural solution may be required to provide the desired level of protection or improve sand retention so that renourishment can be performed on the same schedule as is planned for the rest of the project. Hot spot mitigation measures will probably result in additional unanticipated and undesired costs. The public perception of hot spots that develop shortly after construction can be one of project failure. Therefore, the potential for the occurrence of hot spots should be investigated during the design process. This can be accomplished through, for example, the method used by Smith and Ebersole (1997). The GENESIS model together with detailed nearshore wave conditions can also be used for this purpose (Bodge, Creed, and Raichle 1996).

(a) Hot spots can arise due to a locally strong gradient in longshore sand transport rate, which creates a divergent zone where, in a net sense, more sand is leaving an area than is entering it. Strong longshore transport gradients can develop due to a noticeable change in shoreline/bathymetric contour orientation.

Figure V-4-23 illustrates an example of this type of hot spot at Folly Beach, South Carolina. A beach-fill was constructed along much of the island, and high loss rates were observed locally at the place where the shoreline noticeably changes orientation and creates a convex curve relative to the incident waves (Ebersole, Neilans, and Dowd 1996). Prior to construction of the nourishment project, this area was known as the wash-out area, suggesting an area that previously experienced high erosion rates. Local knowledge or an analysis of historical shoreline change rates may help identify such a zone. However, areas that have historically suffered high sand loss rates may be presently armored with seawalls or revetments, or controlled by some other structures such as groins, thereby pinning the shoreline position and masking the presence of a future hot spot. Any area in which the shoreline has a bulge or convex shape, relative to the incident waves, is a potential candidate for this type of hot spot.

(b) The presence of a submerged offshore shoal, or some other highly irregular bathymetric feature such as an underwater canyon, can also be a cause of erosion hot spots. Morphologic features can alter the propagation of incident waves in a persistent manner, creating strong gradients in net longshore sand transport. Stauble (1994) documented the occurrence of multiple erosional hot spots in a beach nourishment project constructed at Ocean City, Maryland, and hypothesized that the cause of the hot spots was the presence of shore-attached finger shoals that characterize the nearshore bathymetry off Ocean City (see Figure V-4-24). The finger shoals appear as lighter-shaded areas. Stauble also identified corollary cold spots, or areas of unusual sand accumulation, which were formed by the sand that was transported out of the hot spots. Smith and Ebersole (1997) showed that the positions of the hot spots/cold spots were well-correlated to zones of divergent/convergent potential longshore sand transport caused by persistent changes to nearshore wave patterns induced by the shoals. The shoals act to alternately focus and spread wave energy due to the process of wave refraction. Results from the application of a wave transformation model (see Part II-3) were used to compute potential longshore transport rates. Figure V-4-25 shows the variation in transport rates for Ocean City, Maryland. Positive rates are directed to the south, negative to the north. Hot spots correspond to strong uphill gradients (e.g., changes from low southerly transport rates to high southerly transport rates). Cold spots occur at downhill gradients.

(c) Just as natural bathymetric irregularities can lead to the development of a hot spot, an excavated borrow area or a submerged mound constructed out of dredged material can produce the same result. If the manmade feature results in a significant change in bottom relief and is located in water depths where the waves feel the change in depth, persistent longshore sand transport gradients may develop. Hot spots can also form in areas where the incident wave directions are constrained to a narrow window and the area is blocked or sheltered by some land feature or coastal structure, creating a gradient in wave energy and longshore sand transport.

(d) A third type of hot spot can result as a consequence of how a project is designed and built to protect existing structures and infrastructure. Typically beach nourishment projects are built in areas where structures and infrastructure are vulnerable to storm-induced damage. Oftentimes, the project site already has revetments or seawalls that have been built by private landowners. Both protected and unprotected structures may protrude beyond the natural prevailing shoreline position, thereby creating a very irregular, and unnatural, shoreline. It may be difficult to estimate where the prevailing natural shoreline position would be if the structures were not present.

(e) If a project is designed and constructed to "wrap around" protruding structures, in an attempt to create a uniform width of beach in front of each structure, the beach will most likely readjust following construction. The readjustment process will lead to a more stable arrangement of sand in which the design width may not be retained in front of the seaward most structures, whereas the beach width may exceed the design width in other areas. Figure V-4-26 illustrates this process. An offshore depth contour equal to the

Waves from north oblique to both reaches create southerly transport. Breaker angle increases south of shoreline orientation break, increasing transport.

Waves normal to north reach result in no longshore transport. Waves oblique to south reach create southerly longshore transport.

Waves normal to tangent at break in shoreline orientation. Waves oblique to both reaches create longshore transport away from the area.

Waves normal to south reach result in no longshore transport. Waves oblique to north reach create northerly longshore transport.

Waves from south oblique to both reaches create northerly transport. Breaker angle increases north of shoreline orientation break, increasing transport.

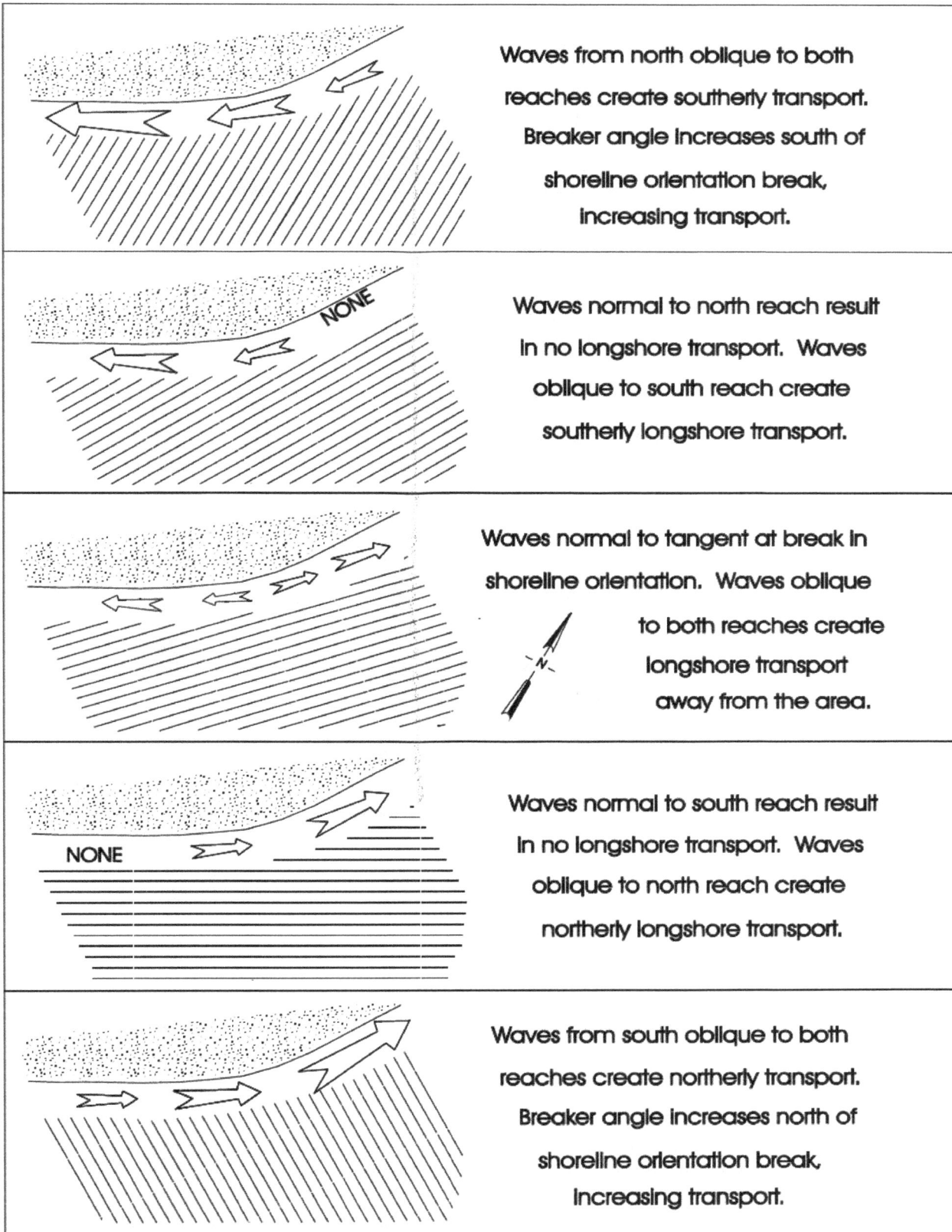

Figure V-4-23. Causes of hot spot formation at a convex shoreline feature

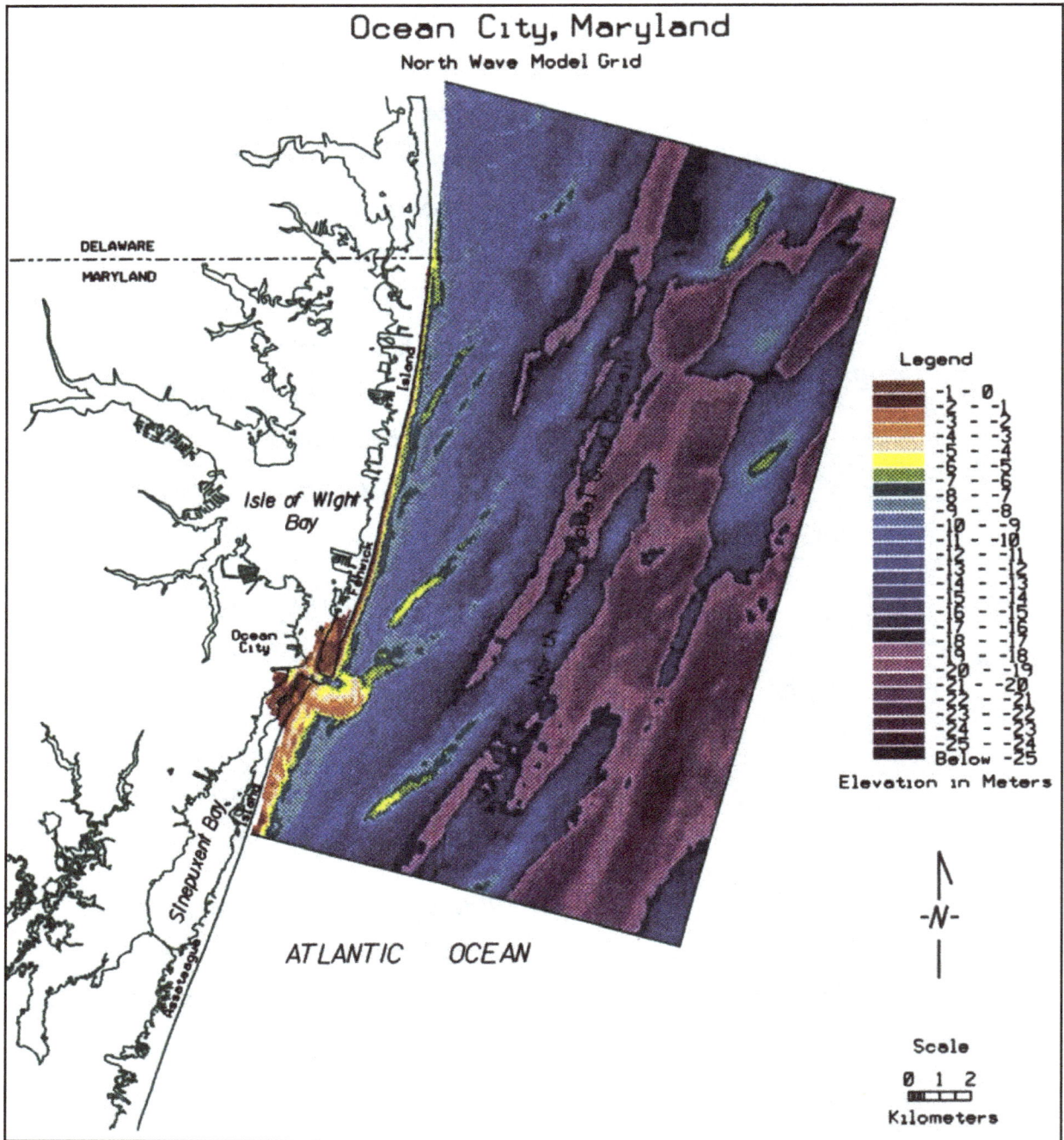

Figure V-4-24. Project shorelines, nearshore bathymetry, wave modeling domain, and position of offshore wave information

breaking wave height associated with frequent storms (2- to 3-m depth contour) may serve as a better indicator of the orientation of the postconstruction, adjusted shoreline.

(f) It may be impossible to maintain the design width in front of some structures without placing a substantially larger volume of sand than is estimated based on "wrapping" the project around the existing structures. It may not be cost-effective to build the entire beach out to the point required to provide the

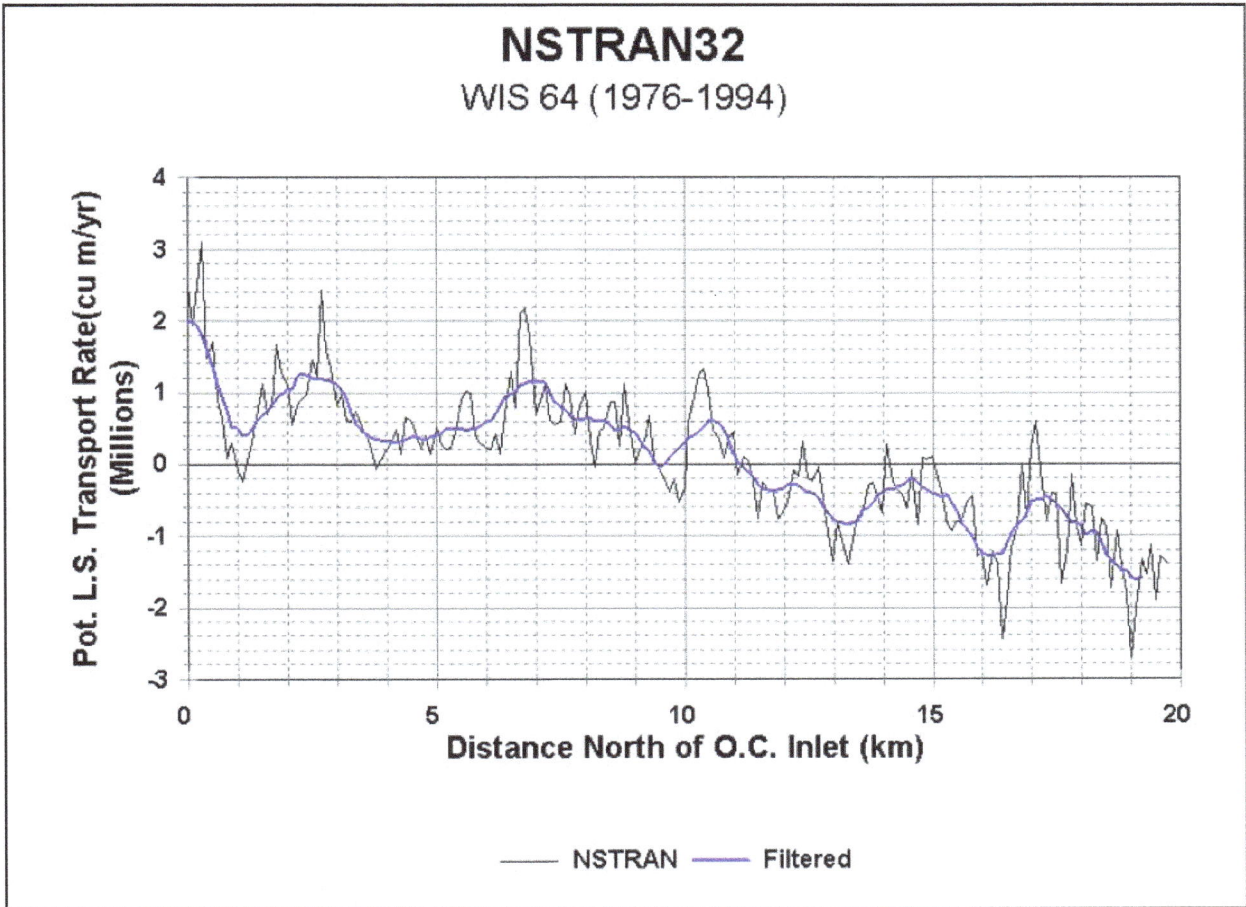

Figure V-4-25. Variations in potential longshore sand transport rates caused by offshore shoals at Ocean City, Maryland

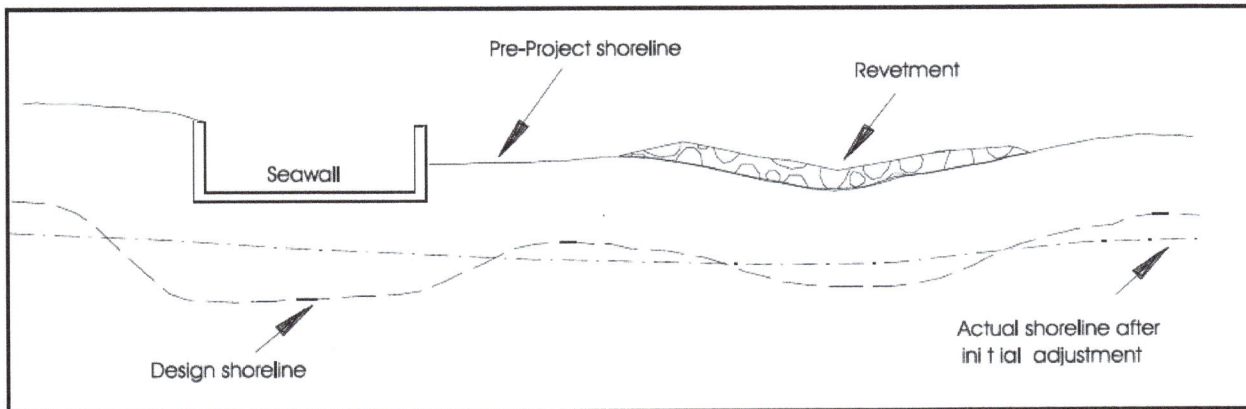

Figure V-4-26. Planform adjustment of a beach-fill project designed to "wrap around" existing coastal structures

desired level of protection to the seaward most structures. If this is the case, other means for achieving the desired level of storm protection in certain local areas, such as more frequent nourishment, use of a feeder beach, a filled groin compartment, or a revetment or seawall, should be considered.

(g) There are probably other less well-understood causes for the development of hot spots. A geologic control such as remnant morphologic feature (relic inlet) with a different sediment composition may cause differential erosion along the project reach. Longshore variation in the sediment characteristics of the borrow material may produce a differential loss rate within sections of the project. However at present, there are no reliable methods for anticipating hot spots caused by these other factors.

   h. Fill transitions.

   (1) Selection of a method for terminating a beach nourishment project depends on several factors. One important consideration is what lies immediately beyond the project limits. For example, is there an open straight beach or an inlet? On open stretches of coast, transition to the adjacent beach can be accomplished using a fill transition section, which was briefly introduced in the previous section, or using hard structures, such as groins, navigation structures, or breakwaters. Structures placed at the boundaries of a project are called terminal structures. If the project is constructed near an inlet, structures may be desirable to minimize movement of fill material into navigation channels. Hard structures will allow an abrupt termination of the beach-fill section. However, these structures can be costly, and can interfere with the natural longshore transport of sediment along the shoreline. If not designed properly, this interference could result in adverse effects along the unrestored beach and subsequent objections by adjacent-beach property owners.

   (2) Another key factor is the purpose for building the project. If the project is built to provide storm protection, what is the reach of shoreline to be protected and what is the desired level of protection within the reach? The desire to maintain a specific design beach width may dictate the design of the transition section. In practice, if a uniform design width is desired for the entire project reach, it will be extremely difficult to maintain that width near the lateral ends of the project, unless a terminal structure is used or the design fill is extended beyond the limits of the design reach. If sand is to be placed outside the design reach, other issues become important. Can the fill provide benefits to the adjacent property, and how will those benefits be factored into the economic justification of design (if at all)? It may be more practical to consider the nourishment project in three sections, one interior section where a desired level of protection will be maintained, and two outer sections of the project where a lesser degree of protection will exist (i.e., the transition sections). In the transition sections, the level of protection will diminish with distance from the main project reach.

   (3) If the project is built primarily for recreational purposes, the design goal may be to maximize retention of fill volume within the limits of the project reach. This design objective may dictate whether or not fill transitions are used at all, and if so, how they are designed.

   (4) Although sometimes approached as an afterthought in project design, beach-fill termination deserves careful consideration. This is particularly true in terms of its relation to the economics of the project and the level of storm protection that is sought/claimed near the limits of the project. This section discusses use of fill transitions at the ends of a nourishment project. The next section, Part V-4-1-i, discusses use of structures in fill stabilization.

   (5) The intent of fill transitions is to minimize the effect of end losses on project integrity. End losses from the main project reach can be reduced by extending the design section past the limits of the reach where protection is desired, or by smoothly tapering the project into the adjacent beaches (i.e., gradually reducing project width from the design width to zero). Tapering the fill ends will decrease the perturbation effects of

the project section. The effect of varying taper length can be estimated with the aid of analytical or numerical shoreline planform response models (Walton 1994; Hanson and Kraus 1989).

(6) The GENESIS model was used in a series of simplified applications to investigate the effect of beach-fill transitions. Figure V-4-27 shows plots of an initial rectangular 6-km-long fill section, and the calculated shoreline at 1-year intervals, for two projects identical in every way except that the project in the top panel has a 400-m-long fill transition and the project in the bottom panel has a 2,000-m-fill transition. The total fill volume is the same in both projects, to allow assessment of how best to utilize a finite volume (or alternatively a fixed cost). The duration for both simulation is 5 years. A single, normally-incident effective wave is considered. The design objective is to maintain a beach width of 20 m throughout the 5-year interval, everywhere within the project reach.

(7) At the end of the 5-year simulation, both projects have beach widths greater than the 20-m design width; and therefore, both projects have satisfied the design objective. It is noted that the average annual volumetric rate of sand loss from the 400-m fill transition project was 148,000 cu m/year; whereas for the 2,000-m fill transition project, the average end losses were 89,000 cu m/year (about a 40 percent reduction in the rate of end losses). However, because approximately 22 percent of the fill volume was placed outside the project reach at the time of construction in the 2,000-m fill transition sections, the percent of total placed fill volume remaining within the 6-km project area is greater for the 400-m fill transition project (58 percent remaining) than for the 2,000-m fill transition project (55 percent remaining). For this case, the example suggests that if there are few or no economic benefits to claim in the transition section, then all the fill volume should be placed within the limits of the project reach (i.e., no use of transition sections). The same conclusion would be reached in this case if maximizing fill volume retention was the primary design goal. However, recall from the previous section that project longevity varies with the square of the project length. Generally, this rule of thumb concerning use of transitions is true for shorter fills. Also, for short fills, it may not be desirable to place sand volume in transition sections which may require a significant percent of the total volume placed within the project.

(8) For longer projects, the design requirement of maintaining the design berm width at the project boundaries throughout the renourishment interval may favor the placement of beach-fill transition sections. Figure V-4-28 provides plots of the initial fill and the calculated shoreline position after 5 years, for 6- and 22-km-long fill projects, with both 400- and 2,000-m fill transition sections. Again, the design berm width is 20 m. In Figure V-4-28a, it is seen that for the 6-km fill project, the 400-m transition project shoreline is seaward of the 2,000-m transition project shoreline after 5 years everywhere within the project domain. However, for the 22-km fill project shown in Figure V-4-28b it is seen that at the project boundaries, the 400-m transition project is landward of the 2,000-m transition project, and more importantly, landward of the design shoreline. This result illustrates the potential value of beach-fill transition sections for long beach-fill projects.

(9) However, even for long projects, the benefit of placing a fill transition section is only realized during the first few years after construction (within the first renourishment cycle, as a rough rule of thumb).

(10) Within only a few years the volume of fill material that has moved from within the project limits onto the adjacent beaches, as a result of alongshore spreading, is likely to be much greater than the volume that was initially placed in the transitions, even with 2,000-m-long transition sections. This conclusion is further illustrated by the results shown in Figure V-4-28, where it is seen that the calculated shoreline position adjacent to the main project after 5 years is for all practical purposes the same for both the project with the 400-m fill transition and the project with the 2,000-m fill transition. The work of Walton (1994) suggests that transition sections have a more lasting significant impact on reducing rates of sand loss from the project only when their length exceeds a value that is 0.25 times the project reach length. In past practice,

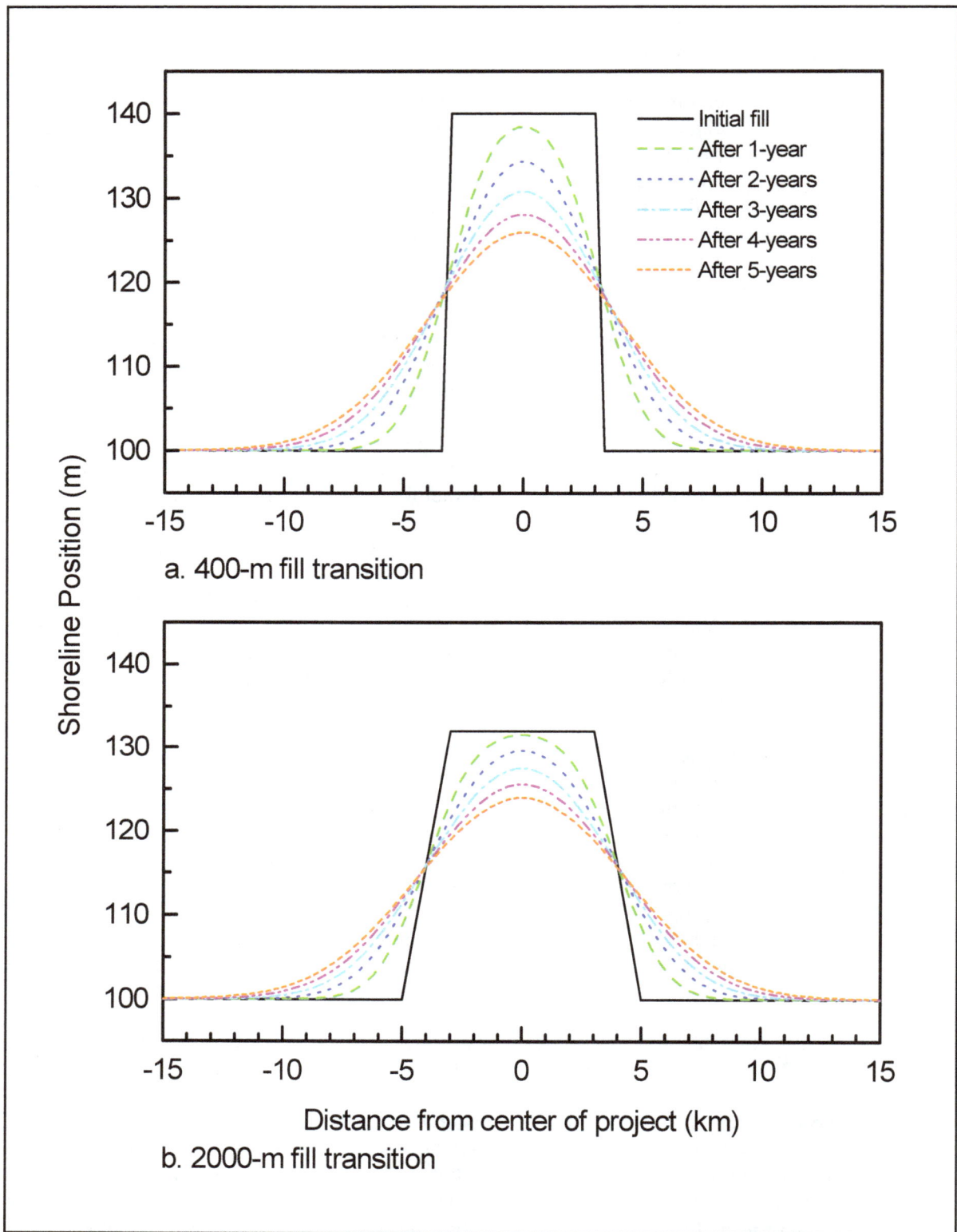

Figure V-4-27. Effect of fill transitions

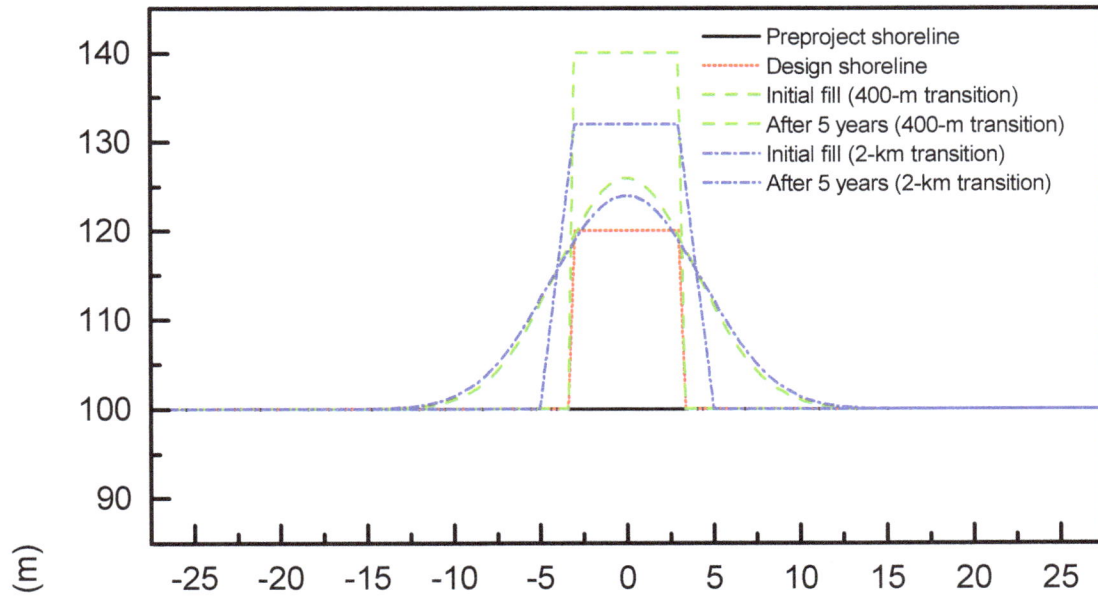

a. 6-km fill project

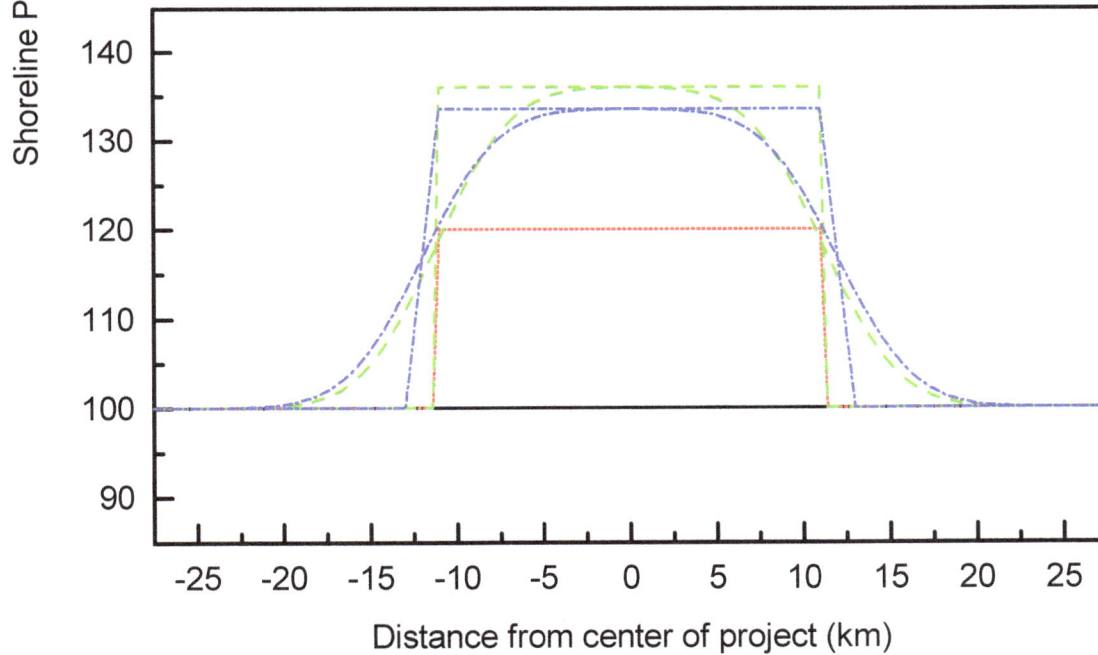

b. 22-km fill project

Figure V-4-28. Effect of fill transitions

Figure V-4-29.   Groin field at Long Beach island, New York

transitions sections have rarely been constructed to this length. In practice, the exact length of the fill transition section does not ultimately matter so long as it is in reasonable proportion to the scale of the project. Typical fill transition lengths for small projects (approximately 1 to 2 km; 100,000 to 200,000 m³) are on the order of 150 to 300 m (500 to 1000 ft) long, while for larger projects they are on the order of 300 to 600 m (1000 to 1,500 ft) long.

(11) Transition sections do reduce the rate of fill loss from the project; but again, this benefit lasts only for the first few years following construction. Some past project designs have inappropriately estimated the beneficial effects of the beach-fill transitions by computing the reduction in the rate of end loss during the first few years of the project and then projecting these benefits throughout the 50-year project life. The analysis presented here suggests that while benefits associated with reduced loss rates are very high during the first few years, they diminish significantly thereafter.

(12) An analysis of beach-fill transitions should first examine the evolution of the project with transition sections, either tapered sections or lateral extensions of the design section, and then examine evolution of the project with the transition volume distributed within the project economic benefit area. The distribution within the project doesn't have to be uniform, as was the case in the examples shown here. If transitions are a desirable feature, they should be optimized by balancing the reduction in the rate of end losses from the project (which will reduce renourishment costs) with the cost of placing fill volume outside of the project's economic benefit area. Then, costs of the transition sections over the project life should be compared to the cost of using a terminal structure or compartmentalizing the beach-fill material with groins or jetties, including an assessment of any impacts that might be caused by a terminal structure. The most cost-effective approach should be selected. Environmental concerns, land ownership constraints, or other factors may also need to be considered in the selection of the optimum fill transition sections.

i. Beach-fill stabilization measures.

(1) Structures. Different types of structures can be used in conjunction with beach-fill projects to retard fill erosion and thereby reduce periodic renourishment costs. As discussed in the previous section, losses are particularly pronounced at the ends of a project where an offset occurs between the fill section and the adjacent unfilled beach. Structures may be needed in these transition zones to keep fill losses at acceptable levels. In some cases, structures may already be in place on the project beach. Depending on type and location of these structures, it may be advantageous to retain them, and perhaps refurbish them. However, some structures may have a negative effect on the beach, or are undesirable from the standpoint of aesthetics or safety. If existing structures are judged to have a probable negative effect, their removal should be considered. Following is a brief discussion of coastal structures in common use in conjunction with beach-fill projects. A more detailed discussion of their characteristics, effects, and design can be found in Part VI.

(a) Groins.

- Groins are low linear structures which are typically built perpendicular to the shoreline, extending from the beach into shallow nearshore water. Their primary purpose is to trap and retain sand moving in the alongshore direction. Groins can be constructed in groups, or fields, consisting of a series of structures spaced at predetermined intervals along a segment of shore to improve retention of beach-fill material, Figure V-4-29, or as an individual structure intended to provide effective termination of a beach-fill project. Weggel and Sorensen (1991) note that the addition and modification of groins within the Atlantic City, New Jersey, nourishment project, constructed in 1986, improved fill performance when compared to that of previous fills for the same location. The groins acted to retain the fill within the project area and prevented fill losses into the adjacent inlet.

•   In the context of beach nourishment projects, the most common use of the groin is as a terminal structure, designed to reduce or eliminate sediment losses out of the project area. The use of a terminal groin is most frequently considered in projects that abut lacks tidal inlets or are adjacent to the beginning or end of a littoral cell. For example, use of a terminal groin at the end of a beach-fill project built next to an unstabilized tidal inlet will not only reduce project end losses but also prevent increased shoaling in the inlet channel caused by deposition of sand lost from the project. Likewise, a terminal groin constructed at the downdrift end of a project built updrift of a submarine canyon can substantially reduce end losses. In both of these cases, the accretionary fillet that develops updrift of the terminal groin can be a renewable source of sediment for maintaining and perhaps renourishing the project. Terminal groins are also appropriate, and should be considered, in projects where the preproject shore lacks existing sandy beach, and in projects where the postproject shoreline will be positioned substantially seaward of the adjacent beaches, particularly in short fills. Federal projects of this type are rare, but private development can often include construction of recreational beaches which depend substantially on terminal structures for their longevity. Figure V-4-30 shows an example of a terminal groin used to terminate a beach nourishment at Oregon Inlet, North Carolina.

Figure V-4-30. Terminal groin, Oregon Inlet, North Carolina

- In littoral cells where there is a dominate net longshore transport direction, a terminal groin centrally located within the cell can exacerbate erosion on the downdrift beach. A terminal groin should not be used within the interior of a littoral cell, unless the downdrift impact of the groin is defined and mitigated. Careful attention must to be given to the design of the groin's profile and length. The terminal groin profile should be a template of the design beach berm and nearshore profile such that the structure blocks littoral transport only to the extent necessary to preserve the design beach cross section, while allowing the ambient net littoral transport to bypass the structure to downdrift beaches. Careful monitoring to assess the structure's impact on the adjacent beaches is recommended. Placing a sacrificial sand fillet downdrift of the terminal groin at the time of initial construction, and perhaps during periodic renourishments, may be necessary to offset any adverse impact to the downdrift beach caused by the groin.

- Groins in general, and groin fields in particular, were commonly used to control erosion and to protect upland development in coastal projects constructed from the 1930s through the 1960s. The experience with groin fields has been checkered with both successes and failures. Groin fields have a proven capacity to provide relief for a local erosion problem by compartmentalization of the shore face and stabilization of the shoreline. However, because the compartments between the groins were rarely filled with beach material at the time of construction, the groin compartments filled over time by capturing, and retaining a portion, if not all, of the longshore sand transport. The result was typically severe erosion downdrift of the groin field. The problem was exacerbated by a chronically increasing deficit of littoral material caused by other upcoast diversions of sand (inlet dredging, breakwaters, hydroelectric dams, etc.). Present coastal engineering knowledge and predictive tools have advanced to the point where competent functional design of groins and groin fields is now possible. However, this design option is often rejected as a matter of policy by many local, state, and Federal agencies with jurisdictional authority.

- Groin fields can provide an effective and economical solution to some coastal erosion problems where shoreline stabilization is the primary project goal. Groins can improve retention of beach fill, but construction of a groin field without concurrent placement of fill is strongly discouraged. As is the case with a terminal groin, groin fields that terminate within a littoral cell, as opposed to at the end of the littoral cell, must provide for sediment bypassing to adjacent shores without interruption. Groin fields should not be terminated in areas with increasing (accelerating) net longshore transport potential. To ensure bypassing, the groin field should be filled during construction and the downdrift groin system should be tapered to provide a smooth transition to the adjacent unprotected shore (see Figure V-4-2). Groin fields have been successfully used together with beach nourishment to stabilize the highly developed southern shorelines of western Long Island at Long Beach Island, New York (see Figure V-4-29). A series of groins were refurbished at Folly Beach, SC, to reduce renourishment requirements for the beach-fill project, which was constructed in 1993 (U.S. Army Engineer District, Charleston, 1987).

- T-head and L-shaped groins are variants of the traditional straight shore-perpendicular groin. These structures include a shore-parallel head section that acts to block and diffract wave energy before it reaches the shoreline. T-head groins can provide a functional improvement over standard groins by reducing the occurrence of rip currents adjacent to the groin and to some extent blocking the offshore movement of sand adjacent to the groin. T-head and L-shape groins are best suited for protecting limited coastal reaches where the mobilizing forces include tidal currents as well as wave-generated currents and where the project objectives are more focused on compartmentalization and stabilization of the coast than on increasing beach width. A series of T-head groins constructed at Ocean Ridge, Florida, downdrift of South Lake Worth Inlet have proven to be effective in local stabilization of the shoreline (U.S. Army Engineer, Charleston, 1987).

(b)   Detached breakwaters.

- Detached breakwaters are linear offshore structures generally oriented more or less parallel to the shoreline to which they have no solid connection. They can be built as a single continuous structure or segmented into a series of short sections with gaps. Detached breakwaters provide protection by reducing the wave energy that reaches the shore through dissipation, reflection, and diffraction. Compared to shore-perpendicular structures, such as groins, which provide protection by impoundment of alongshore-moving sediment, detached breakwaters can be designed to allow continued movement of longshore transport through the project area, thus potentially reducing adverse impacts on downdrift beaches. However, as is the case with groins, if fill material is not placed at the time of breakwater construction, then any sediment that accretes in the lee of the breakwaters is sediment that is denied from downdrift beaches, which can be expected to erode. The use of breakwaters for shore protection and in conjunction with beach nourishment in the United States has been primarily limited to littoral sediment-poor regions characterized by a low-energy wave climate. Most projects have been located on the Great Lakes, Chesapeake Bay, or Gulf of Mexico (Chasten et al. 1993). These projects are typically subjected to short-period, steep waves, which tend to approach the shoreline with limited refraction, and generally break at steep angles to the shore.

- While detached breakwaters have several appealing functional characteristics for shore protection, they also have disadvantages that include the following: limited design guidance, high construction and maintenance costs, limited ability to predict and compensate for structure-related phenomena such as adjacent beach erosion, rip currents, scour at the structures base, structure transmissibility, and effects of settlement on project performance. Like groins, detached breakwaters should only be utilized after careful study of the effects they will have on the project area and adjacent beaches.

- Details concerning the uses of detached breakwaters in beach-fill projects and design considerations can be found in Engineer Manual 1110-2-1617, Pope and Dean (1986); Dally and Pope (1986); Rosati (1990); and Chasten et al. 1993. Case studies of interest are contained in Nakashima et al. (1987); Gorecki (1985), and Gravens and Rosati (1994).

(2)   Evaluation and optimization of stabilization structures.

(a)   As discussed in previous sections, a beach nourishment project requires periodic reconstruction of the beach planform throughout the project life. In this section, use of structures to stabilize a nourishment project and reduce project end losses have been discussed. Structures such as groins and detached breakwaters are comparatively more expensive and permanent than beach-fill material. Project optimization requires an analysis of the cost-effectiveness of incorporating structures to reduce periodic nourishment requirements versus designing the project without the stabilization structures.

(b)   During the plan formulation phase of the project, the performance of alternative plans (with and without fill stabilization structures) can be evaluated using a shoreline/beach change model, such as the GENESIS model, to estimate total volume requirements. A model can also provide information for assessing adjacent beach impacts resulting from the inclusion of fill stabilization structures. The effects of structure length and transmissibility can be estimated using the GENESIS model. The physical characteristics of the structure, therefore, can be optimized to best serve specific project requirements and objectives. Physical models can also be utilized to comparatively evaluate the performance of beach-fills with and without structures such as groins and breakwaters (Bottin and Earickson 1984).

(c)   Techniques for design of a groin system to reduce beach-fill losses and offshore breakwater design are presented in EM 1110-2-1617. Additional guidance for the functional design of offshore breakwaters for shoreline stabilization is presented in Dally and Pope (1986) and Chasten et al. (1993).

(3)   Dune stabilization. Coastal sand dunes are valuable and effective barriers to storm flooding and wave attack; and therefore, they are often an important component of the beach nourishment project. Because of the importance of the dunes to storm protection, and their vulnerability to wind- and wave-induced erosion, it is advisable to stabilize them with sand fences and vegetation. These stabilization measures are relatively inexpensive and serve two beneficial purposes: to enhance the protective nature of the dunes, and to reduce volume losses from the project due to wind blown transport of the fill. Guidelines for estimating the rate of wind blown sediment transport are provided in Part III-4. Beach-fill projects usually involve creating a much wider dry beach, so more sand is available for transport by wind. As a result, if dunes are stabilized with fences and vegetation, they often accrete with time and wind-blown sand losses from the project are minimized.

(a)   Fences.

•   Various types of fences have been used to create, enlarge, and stabilize coastal dunes. To be successful, the barrier must be porous to wind because a solid barrier creates turbulence that may result in scour rather than accretion. Fencing with a porosity (ratio of area of open space to total projected area) of about 50 percent should be used (Savage and Woodhouse 1969). Open and closed areas should be less than 5-cm in width. The standard slat-type wooden snow fence appears to be the most practical and cost-effective, and it has been widely used for dune stabilization and to promote dune growth. Snow fences are usually installed in single or multiple rows aligned parallel to the shoreline or perpendicular to the predominant wind direction and secured to posts about 3 m (10 ft) apart to improve their stability. If accretion fills the fences, additional fencing can be installed at the new level to promote further growth. Field tests of dune building with sand fences under a variety of conditions have been conducted at Cape Cod, Massachusetts, Core Banks, North Carolina, and Padre Island, Texas. The following are guidelines and suggestions based on these tests.

•   Only straight fence alignment is recommended, and the fence should parallel the shoreline. It need not be perpendicular to the prevailing wind direction. Fence will function even if constructed with some angularity to sand-transporting winds. Fence construction with side spurs or a zigzag alignment does not increase the trapping effectiveness enough to be economical (Savage 1962; Knutson 1980). Lateral spurs may be useful for short fence runs of less than 150 m (500 ft), where sand may be lost around the ends (Woodhouse 1978).

•   Placement of the fence at the proper distance shoreward of the berm crest may be critical. The fence must be far enough back from the berm crest to be away from frequent wave attack. Efforts have been most successful when the selected fence line coincided with the natural vegetation or foredune line prevalent in the area.

•   Dunes are usually built with sand fencing in one of two ways: by installing a single fence and following it with additional single fence lifts as each fence fills (Figure V-4-31); or by installing double fence rows with the individual fences spaced about four times the fence height apart and following these with succeeding double-row lifts as each fills (Figure V-4-32). Single rows of fencing are usually the most cost-effective, particularly at the lower wind speeds, but double fences may trap sand faster at the higher wind speeds.

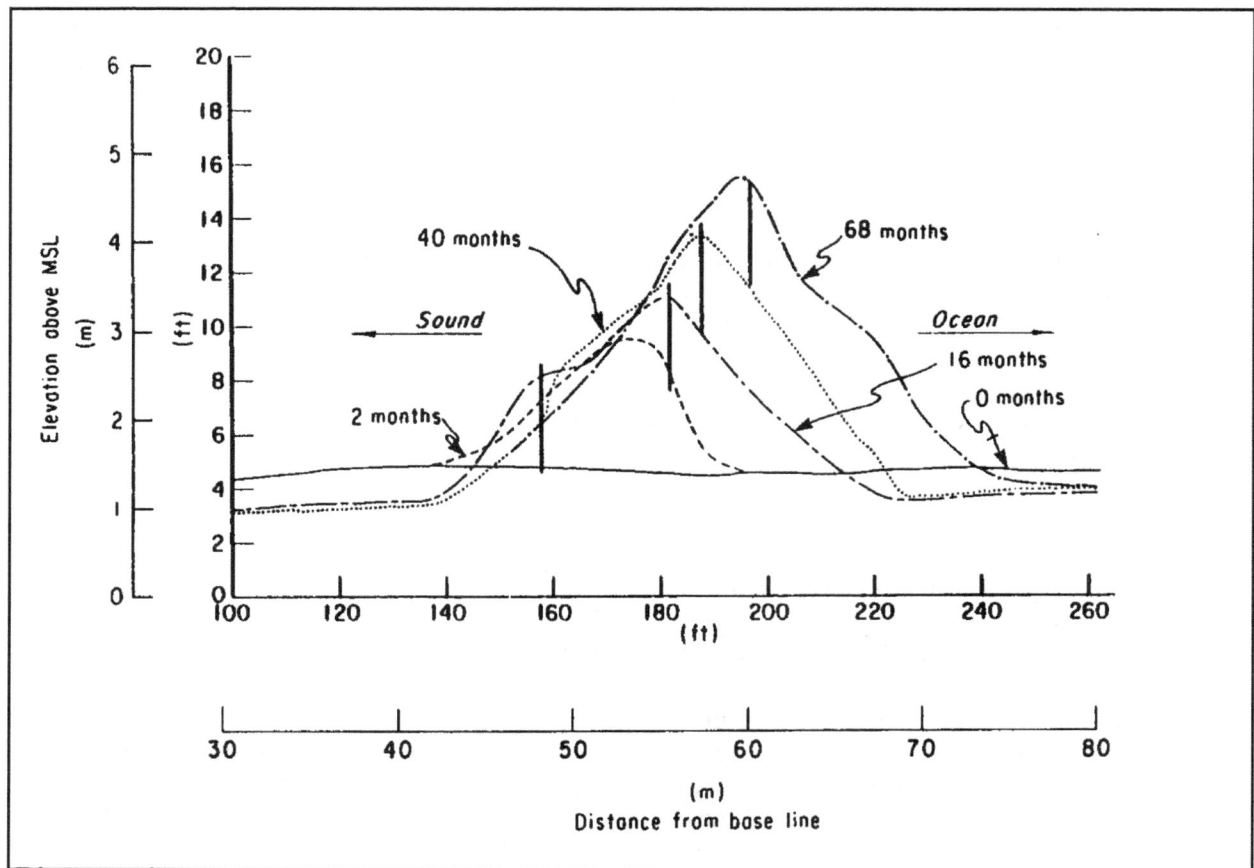

Figure V-4-31. Sand accumulation by a series of four single fence lifts, Outer Banks, North Carolina (Savage and Woodhouse 1969)

- Dune height is increased most effectively by positioning the succeeding lifts near the crest of an existing dune. However, under this system, the effective height of succeeding fences decreases and difficulties may arise in supporting the fence nearest the dune crest as the dune becomes higher and steeper. Dune width is increased by installing succeeding lifts parallel to and about four times the fence height away from the existing fence. The dune may be widened either landward or seaward in this way if the dune is not vegetated.

- Accumulation of sand by fences is not constant and varies widely with the location, the season of the year, and from year to year. Fences may remain empty for months following installation, only to fill within a few days by a single period of high winds. To take full advantage of the available sand, fences must be observed regularly, repaired if necessary, and new fences installed as existing fences fill. With sand moving on the beach, fencing with 50 percent porosity will usually fill to capacity within 1 year (Savage and Woodhouse 1969). The dune will be about as high as the fence. The dune slopes will range from about 1:4 to 1:7, depending on the grain size and wind velocity. The trapping capacity of the initial installation and succeeding lifts of a 1.2-m-high sand fence averages between 5 and 8 cu m/lin m of shoreline (2 to 3 cu yd/lin ft).

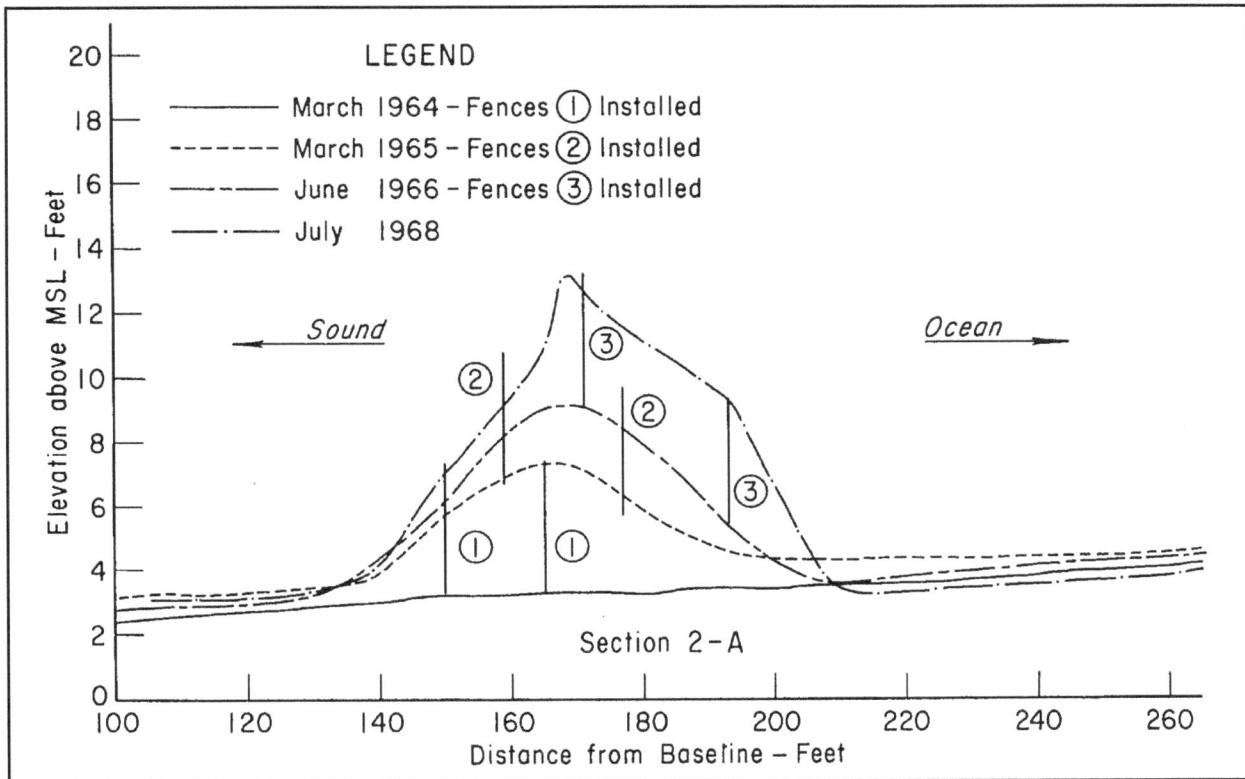

Figure V-4-32. Sand accumulation by a series of three double fence lifts, Outer Banks, North Carolina (Savage and Woodhouse 1969)

- Experience has been that an average of 6 man-hours are required to erect 70 m (230 ft) of wooden, picket-type fence or 60 m (195 ft) of fabric fence when a six-man crew has materials available at the site and uses a mechanical post-hole digger.

- Fence-built dunes must be stabilized with vegetation or the fence will deteriorate and release the sand. While sand fences initially trap sand at a high rate, established vegetation will trap sand at a rate comparable to multiple lifts of sand fence. The construction of dunes with fence alone is only the first step in a two-step operation. Fences have two initial advantages over planting that often warrant their use before or with planting: sand fences can be installed during any season, and the fence is immediately effective as a sand trap once it is installed. There is no waiting for trapping capacity to develop, unlike the vegetative method. Consequently, a sand fence is useful to accumulate sand while vegetation becomes established.

(b) Vegetation. Vegetation is a natural means of shore and dune stabilization that is effective when used under the proper circumstances. Vegetation is relatively economical, and does not detract from, but can enhance environmental quality. On open sea coasts, vegetation is primarily used to enlarge and stabilize dunes. A number of beach grasses and other plants tolerant of a dune environment can be used to create, enlarge, or stabilize dunes. Frequently used species are American beach grass (*Ammophila breviligulata*) in mid- and upper-Atlantic Coast and Great Lakes, European beach grass (*Ammophila arenaria*) on the Pacific Coast, sea oats (*Uniola paniculata*) on the south Atlantic and Gulf Coasts; and the panic grasses (*Panicum amarum*) and (*Panicum amarulum*) on the Atlantic and Gulf Coasts. All of these plants can be propagated by planting suitable stock and are effective in trapping and holding windblown sand. A number of herbaceous and woody plants are also effective in dune areas. The principal considerations in selecting plants for dune building and stabilization are the suitability of the species for growth in the project area, its probable effects on the existing ecology, availability of stock for transplanting, and economics. Detailed information on suitability of plant species for various regions of the United States, methods of propagation and planting, and protection against disease and physical damage can be found in EM 1110-2-1204, "Environmental Engineering for Coastal Protection"; Knutson (1977); Woodhouse (1978); Savage and Woodhouse (1969); and Knutson (1980). Dune stabilization by vegetation has these advantages. It is capable of growing up through the accumulating dune sand and it can repair itself if damaged, provided the damage is not too extensive and the vegetation was well established prior to being damaged. New plantings will require periodic fertilization the first year or two, and watering the first few months, for the young plants to become well established.

(c) Dune cross-over structures. If newly-constructed or existing dunes are to be stabilized with fences and/or vegetation it is important to protect the stabilization efforts by minimizing traffic through or across the dune. Pedestrian and/or vehicular traffic will damage vegetation and fence lines. Dune walk-over structures for pedestrians and controlled vehicular access points to the beach should be provided in the overall design. Controlled access points provide recreational as well as maintenance access to the beach without damaging the stabilized dune area. Barren low spots (such as caused by trampling) within an otherwise healthy vegetated dune can experience wave overwash and further dune deflation during storms (see Part IV-2). These local areas of overwash and deflation are commonly referred to as a "blow-outs." Blow-outs reduce the dune's reservoir of sand and capability to protect upland areas from flooding. Construction guidelines for dune cross-over structures are provided in U.S. Army Engineer Waterways Experiment Station (1981).

j. Construction issues.

(1) Removal, transfer, and placement of borrow material. Removal, transfer, and placement of borrow material is typically the job for the dredging contractor. However, certain aspects of the dredging work should be understood so that practical design and reasonable specifications can be developed prior to bidding out the dredging job. With regard to the removal and transfer phases of the job, the following points are made:

(a) Offshore borrow areas typically have fine material in the top layers. Hence, if the borrow area is shallow, large losses of fines may occur during the placement operation. Typically, thicker borrow areas are more economical for this reason.

(b) The further offshore the borrow area is located, the more expensive the pumping of material becomes. Direct pumping to the beach may not be economically feasible if the borrow area is located too far offshore.

(c) If a hopper dredge or barge is used to transport fill to the site, many of the borrow material fines are washed out with overflow, hence providing a larger grain size distribution of fill material than that based on in situ samples from the borrow site. This fact and the fact that less dirty material (fines) is provided at the subaerial beach may make this type of placement desirable from an environmental and public satisfaction standpoint.

(d) Borrow areas should not be located too close to shore due to the fact that the equilibrium profile of the beach may be downcut. This could lead to offshore transport of material from the nearshore zone to make up for the deficit. If the borrow site is located too close to shore, and in too shallow water, the irregular bathymetric feature created as a result of the excavation can significantly alter the propagation of incoming waves. This could lead to an erosional hot spot. As a rule of thumb, borrow areas should be located in water depths at least twice the estimated depth of closure (see Part III-3-3-b), or a comparable distance offshore.

(e)   For hydraulically-filled projects, there is a minimum practical sectional fill volume. That is, the constructed beach-fill must be of sufficient width for the contractor to erect dikes, move pipeline, and operate equipment for grading the placed fill and building the dune section, etc. Typical sectional fill volumes range from 25 to 63 m^3/m (10 to 25 yd^3/ft) for truck-haul projects and from 75 to greater than 250 m^3/m (30 to greater than 100 yd^3/ft) for hydraulically placed fill projects. Sectional fill volumes between 140 and 215 m^3/m (55 and 85 yd^3/ft) are fairly common; however, Dean and Campbell (1999) recommended a minimum sectional fill volume of 200 m^3/m (80 yd^3/ft) from a performance perspective.

(f)   A good overview of various dredging systems which can be applied for beach nourishment from offshore sources is provided by Richardson (1976). If trailing suction hopper dredges are available (loaded draft ≈ 9 m for large hopper dredges), long borrow areas may be preferable (for long production runs). Cutterhead pipeline dredges may be more economical where borrow areas are more uniform in dimension (i.e., square) and the deposit is relatively thick (1.5 m or greater) so that the dredge and pipeline do not need to be moved far.

(g)   There are four types of material placement for beach-fill purposes. Description and comments on these methods follow:

- Direct placement. In this method, the fill is placed at one time throughout the stretch of shore to be protected. This is the most frequently used method of placement. Usually, fill is pumped as a slurry onto the beach via hydraulic pipeline, then reworked into the desired configuration using earth-moving equipment. Additional pipeline is added in sections to extend the placement zone along the beach. The largest grain sizes in the slurry will settle out closest to the slurry discharge point. Likewise, the finer grain sizes will settle out at greater distances. Hence, locating discharge points in locations of known maximum erosion or hot spot regions may be desirable. Providing a sand dike behind which most of the discharge occurs will reduce loss of fines and provide better water quality in the area.

- Nearshore placement. This method is appealing because large volumes of material can be made available at low costs by hopper dredges or split-haul barges. The principle is that the material, dumped in shallow water is transported towards the beach by wave action. Early tests of this method where the material was dumped in water 6 to 11 m are presented by Hall (1952). In all cases, the results showed that the beaches did not benefit from the dumped sand to any appreciable extent. Hands and Allison (1991) have reviewed a number of offshore dumping projects and found that if disposal depth is less than closure depth, the disposal sediment would be active and move toward the beach. Although nearshore dumping may be more economical, it does not provide the level of protection to upland property that direct placement on the subaerial beach does. It is expected, though, that the nearshore mound will provide some level of wave attenuation. A mound may also change local wave refraction patterns, leading to changes in local longshore sand transport gradients.

- Continuous supply. This method is typically used at a littoral barrier (i.e., navigation channel or inlet) where sand trapped at the updrift side of the barrier is bypassed to the beach on the downdrift side. These operations are more commonly thought of and designed as sand bypassing systems rather than as a method of beach nourishment. The purpose of this approach is to restore the natural flow of littoral transport at the location where such interruption occurs.

- Feeder beaches. Feeder beaches involve the stockpiling of fill at the updrift end of the areas intended to receive the fill as the feeder beach erodes. This method is typically used for smaller projects where sand may be trucked in and/or access to discharge points on the beach is limited. The intent is for the stockpiled material to be distributed by natural littoral processes. Feeder beaches generally work well in areas that serve as a source of material for adjacent beaches. Examples are areas immediately downdrift from inlets or other man-made structures that form a littoral barrier. An erosional hot spot may be another area where a feeder beach is useful as a means for maximizing sand retention.

- Stockpile. In some projects, sand is stockpiled landward of the primary dune (or elsewhere outside of the normally active littoral system) for use in emergency situations. The intent here is to store a quantity of material at favorable costs at a time when the dredges and equipment are already mobilized. Most often this material is used on an as-required basis to repair or restore dunes following storms, or as fill material to close a breach should one be opened by a storm. Having this material readily available at the time of a breach or when needed to restore an eroded dune line can substantially reduce the time and cost associated with completing the emergency repair. For some projects, it may be possible to anticipate that there will be a need for emergency repairs sometime during the life of the project. In these cases, it may be cost-effective to dredge and stockpile excess material for use in emergency situations at the time of initial project construction.

(2) Construction template.

(a) The construction template is the target cross-sectional profile that the contractor is expected to build to ensure that the proper volume is placed. Two construction approaches are generally used. One is to "over-build" the upper part of the beach profile, stacking the fill material to match the specified template. The second method involves placement of the fill volume over more of the active profile. The over-building method is used most often because of the availability and working limitations of equipment used to place and shape the fill, and the desire to place the material in the most economic manner, which usually means minimizing movement of the discharge point. To a degree, the second construction method involves doing what nature will do on its own anyway, redistribute the volume over more of the active profile. The over-building method relies on wave and current action to redistribute the fill to deeper parts of the active profile.

(b) In the overbuilding method, the dune and/or berm of the design profile is built to the desired elevation. However, the constructed berm width will be much greater than the design berm width for two reasons. First, the volume to be placed includes the advance nourishment volume in addition to the volume needed to create the design profile. Second, and most important, the volume is stacked in the nearshore zone only over a portion of the active depth. Figure V-4-33 shows a typical preproject beach profile, the design profile (which includes a dune), a modified design profile that includes the advance nourishment volume, and the construction template. The construction template reflects a different configuration of the sectional fill volume.

(c) The construction template is usually specified as a simple geometry, like that shown in Figure V-4-33. The dune is usually built with a trapezoidal cross section having the desired crest elevation, crest width, and side slopes (see Part V-4-1-f for guidance on selecting these parameters). The berm may include a gentle slope (1:100) that transitions from the toe of the dune to the seaward end of the berm crest on the design profile. Such a gentle slope is found on many natural benches. Additional material seaward of the design berm that arises from the overbuilding technique should be constructed to the design berm elevation. The design berm is either constructed with a uniform elevation equal to the design crest elevation or with a gentle slope from the toe of the dune to the seaward point on the design berm crest. At the seaward terminus of the constructed berm, the fill should be placed at a constant grade, the upper slope, from the berm crest elevation seaward to the elevation that corresponds to mlw. From this point seaward, the fill is placed on a more gentle slope, the lower slope, in an attempt to match the construction template. Earth-moving equipment can be used to shape the above-water portions of the profile to the desired slope. The submerged portions of the profile are more difficult to control.

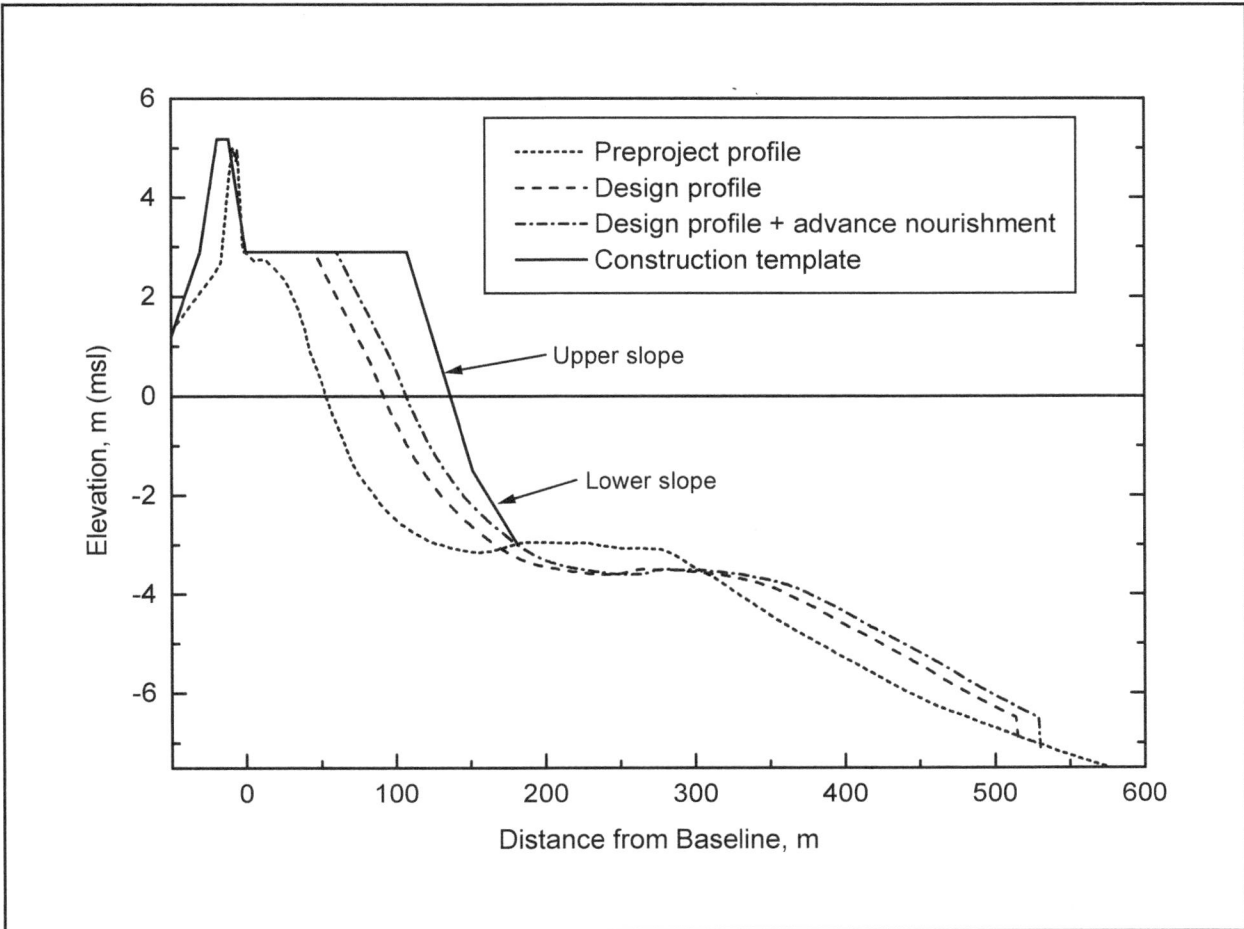

Figure V-4-33. Definition of beach profiles for nourishment projects

(d) To minimize difficulties in constructing the upper and lower slopes of the construction template, they should be defined based on expectations of how the beach will evolve after it is placed. The optimal slopes are primarily a function of the grain size characteristics of the fill and the wave and tide conditions that occur at the site during placement operations. One method for estimating upper and lower slopes involves use of the design profile. Part V-4-1-f-4 discussed methods for estimating the design profile shape. The upper slope of the construction template can be estimated as the average beach slope found on the design profile in the elevation range between the berm crest elevation and the elevation corresponding to mlw. The lower slope of the template can be estimated as an average slope computed from the design profile in the elevation range from mlw to a depth below msl which is approximately equal the typical storm wave height (e.g., 2 to 3 m below msl on the southeastern U.S. coast). The design profile reflects substantial adjustment and equilibration of the placed fill.

• Since the time between fill placement and surveying to assess compliance with the construction template might be short, full equilibration may not be realized. Therefore, the upper and lower slopes of the construction template should be chosen to be slightly steeper than the slopes estimated from the design profile.

• An alternative method is to apply the following rough rules to select both slopes.

Median Grain Size (mm)	Upper Slope	Lower Slope
$D_{50} < 0.2$	1:20 to 1:15	1:35 to 1:20
$0.2 < D_{50} < 0.5$	1:15 to 1:10	1:20 to 1:15
$D_{50} > 0.5$	1:10 to 1:7.5	1:15 to 1:10

(e) The median grain size corresponds to that of the fill material. Coarser material can be constructed to, and maintain steeper slopes; finer material adjusts to gentler slopes. Additional information on the specification of construction slopes, as a function of grain size, based on design experience in the southeastern United States is provided by Creed, Bodge, and Suter (1999). Example V-4-10 illustrates the procedure for developing a construction template.

(f) During placement, the fill should be monitored to determine actual slopes (particularly the lower slope). Strict adherence to the slope of the construction template is not an absolute requirement if the placed material seeks to attain a different adjusted slope. Adjustments can be made to the berm width of the construction template to allow for differences that occur between assumed and actual slopes. This will prevent unnecessary time being expended to mold the beach in a dynamic region of the profile. Analysis of monitoring data should be performed to ensure the prescribed sectional fill volume (volume per unit length of beach) is placed within the active beach zone. Placement of the prescribed sectional fill volume is more important than exactly matching a construction template.

(g) Scarping is one problem that may be encountered in the overbuilding approach. Scarps may develop at the toe of the fill as waves begin the profile adjustment process. Scarps can pose a threat to human safety, and they also present a problem for nesting sea turtles. Scarps can be mechanically smoothed as part of the construction contract or during regular beach maintenance and cleaning. Designing the berm crest elevation to be equal to the natural berm crest elevation, and selecting an upper slope of the construction template similar to the foreshore slope of the design profile, will minimize the likelihood of severe scarping.

(h) The second construction method involves placing more of the fill offshore, i.e., nourishing more of the active profile rather than stacking the material in the inshore zone. In this method, the target construction template is more or less the design profile. However, emphasis is placed on providing the required volume to the active profile, with less emphasis on achieving an actual construction template. Placement at various points across the profile may be more costly depending on the dredge plant used to construct the project. All material should be placed in depths less than the depth of closure. Accurate surveying is needed to monitor the actual volume of material placed, especially in the deeper portions of the active profile.

(3) Postconstruction profile adjustment. As previously above, construction profiles are generally out of equilibrium with the prevailing coastal processes, and they will begin to change shape during and after placement. This is particularly true for fills placed using the overbuilding method, where sand is "stacked" in the nearshore zone. Sand moved to lower elevations on the profile (eventually down to the depth of closure) is required to create and support the volume of sand visible on the dry beach.

EXAMPLE PROBLEM V-4-10

FIND:

Develop and plot the construction template together with the existing and final design profiles for the design condition discussed in Example V-4-2.

GIVEN:

Existing condition profile (Example V-4-2 Figure A).
Final design profile (Example V-4-2 Figure D).
Sectional fill volume is 298 m^3/m.

SOLUTION:

Develop a schematic construction template with berm elevation of 1.5 m (msl), a 1 on 20 slope from the berm crest to -1 m (msl) and a 1 on 35 slope from -1 m (msl) to the intersection with the existing profile. Translate the schematic construction template such that the volume between the existing profile and the construction template equals the volume difference between the existing profile and the final design profile. The results of this procedure are illustrated below. Note that the msl shoreline is advanced 75 m by the construction template whereas the design condition msl shoreline advances the shoreline only 30 m beyond the existing condition.

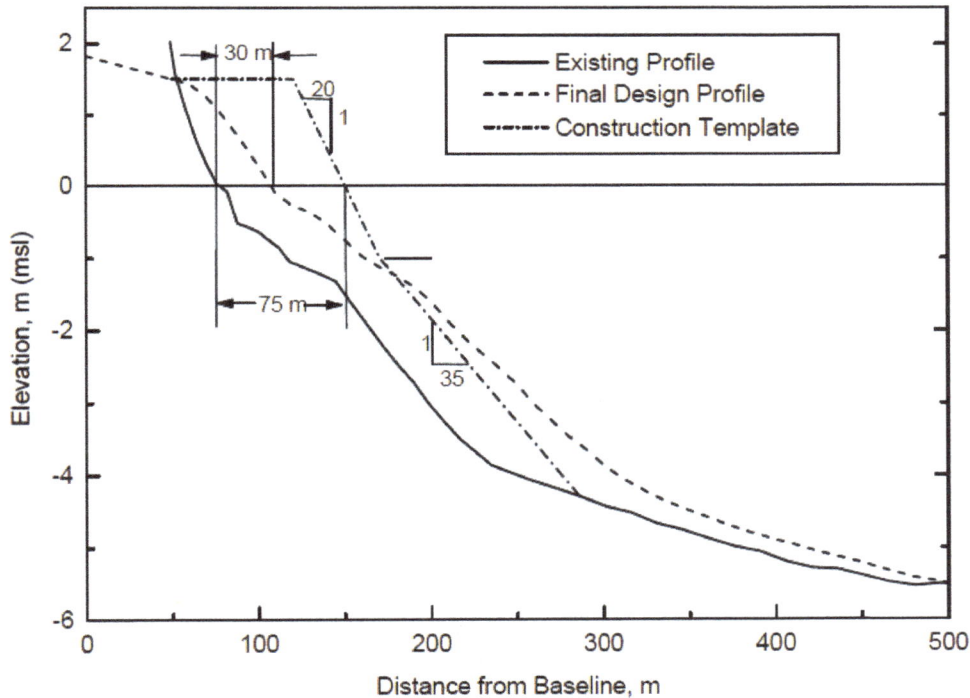

(a)   The high water levels and more energetic wave conditions that accompany storms are very effective in adjusting the fill. Usually, the nourished profile will adjust to a shape that is much closer to the expected equilibrium shape during the first full winter season, at least on the portions of the profile that are shallower than the typical winter storm wave breaking depth. However, unless very severe storms are experienced during that first winter season, the fill material may not adjust to elevations equal to the depth of closure. Through the life of the project, as more severe storms are encountered, the material will be transported to deeper depths.

(b)   For nourishment projects constructed using the overbuilding method, the initial adjustment process and resulting decrease in beach width can be quite dramatic. A rapid decrease in beach width from the constructed width to the design width (including advance renourishment) is often observed, and is to be expected. As a rule of thumb, decreases in beach width of 40 to 75 percent can be expected within a short period of time (months) immediately following construction if the overbuilding method is used. Projects constructed by placing material across much more of the active profile will see less dramatic adjustment in the width of the placed beach.

(c)   The public often perceives this dramatic initial adjustment as project failure, particularly in light of the fact that most of the adjustment occurs either during construction or within a few months after construction. Many projects are built during the winter season (storm season) because of environmental windows imposed on the project, or to avoid the heavy beach-use season. A special effort should be made to educate the public and project sponsors regarding the initial adjustment process, and the differences between constructed profiles and design profiles. If the stacking method of construction is used, it should be explained to the project sponsors and beach users (local community) that the project is less expensive to build utilizing this unnatural construction template, and that the beach berm width will be reduced as natural processes redistribute material to the natural/equilibrium profile. Seasonal, cyclical changes in beach width should also be explained, i.e., loss of beach width during the winter storm season, and subsequent beach recovery during the summer. Seasonal changes in beach width will be masked initially by the much larger changes associated with the initial adjustment process.

k.   Plans and specifications

(1)   Schedules.

(a)   Overall start and completion dates. The contractor is required to commence work under a contract within a specified number of calendar days after the date the contractor receives the notice to proceed. Typically, a period of 10 calendar days is specified. The contractor is directed to prosecute the work diligently and complete all work, ready for use, not later than a specified number of calendar days after receipt of the notice to proceed. The time stated for completion also includes final cleanup of the premises. Time needed to complete the work is directly dependent on the scope and extent of the project, and can vary from as little as 60 to 90 days up to a number of years for large projects. For example, the Atlantic Coast of Maryland Shoreline Protection Project, which included the placement of about 3.5 million cu m of beach-fill and dune construction along about 8 miles of shoreline, required completion within 720 days following the receipt of the notice to proceed (Anders and Hansen 1990). For some projects, start and completion dates may be dictated by environmental considerations such as dredging windows or recreational seasons. Sufficient completion time should be provided to avoid excessively high bids from potential contractors. To enforce the specified completion time for a project, liquidated damages are generally required for each day of delay.

(b)   Start and completion dates for specific subtasks. Interim start and completion dates may be required for specific subtasks depending on the scope of the project. For example, an interim completion date for beach-fill placement between designated stations may be required to enable other project features such as revetment or bulkhead construction to proceed.

- Depending on the scope of the project, the contractor may be required to develop a network analysis system for scheduling the work. The system should consist of diagrams and accompanying analyses that show the order and interdependence of activities and the sequence in which the work is to be accomplished by the contractor. In preparing this system, the scheduling of construction is the responsibility of the contractor. The requirement for the system is included to assure adequate planning and execution of the work and to assist the contracting officer in appraising the reasonableness of the proposed schedule and evaluating progress of the work. An example of one of the numerous acceptable types of network analysis systems is shown in USACE Pamphlet EP 415-1-4 entitled "Network Analysis System Guide."

- A preliminary network defining the contractor's planned operations during the first 60 calendar days after notice to proceed should be submitted soon after the notice to proceed. The contractor's general approach for the balance of the project should be indicated. The complete network analysis system consisting of the detailed network analysis, schedule of anticipated earnings as of the last day of each month, and network diagrams should be submitted within a specified number of calendar days after receipt of notice to proceed. The approved schedule should then be used by the contractor for planning, organizing and directing the work, reporting progress, and requesting payment for work that is completed.

(c) Expenditures. The contractor should submit, at monthly intervals, a report of the actual construction progress. The report should show the activities or portions of activities completed during the reporting period and their total value, which serves as the basis for the contractor's periodic request for payment. Payment made should be based on the total value of activities completed, or partially completed, after verification by the contracting officer. An updated network analysis should be used as a basis for partial payment. The report should state the percentage of work actually completed and scheduled, as of the report date, and the progress along the critical path in terms of days ahead or behind the allowable dates. If the project is behind schedule, progress along other paths with negative slack should also be reported. The contractor should also submit a narrative report which should include but not be limited to a description of the problem areas, current and anticipated factors causing delay, the impact of delay, and an explanation of corrective actions taken or proposed.

(2) Specifications.

(a) Boundaries of the project area. The limit of the geographical area available to the contractor must be shown on the project drawings. Except where indicated, the contractor should confine his work to the area seaward of the construction baseline and between the lateral limits of the project. This area does not generally include access, storage, and staging areas. Access routes and storage and staging areas required to perform the work should be provided by and at the expense of the local sponsor. Sponsors are generally given credit for the cost of obtaining the necessary easements and rights-of-way, which count toward their share of project costs. The contractor should coordinate access to the work area and storage and staging area locations with the contracting officer. Unless otherwise approved by the contracting officer, excess equipment should only be stored in approved storage or staging areas, or in temporary areas, in the immediate vicinity of the fill placement site. Operation of grading and other construction equipment should not be permitted outside the work area limits except for ingress and egress to and from the site at approved locations.

(b) Boundaries of the borrow area. All excavation for beach-fill material should be performed within the borrow area limits shown on the project drawings. Excavation in the borrow areas may be restricted to specified elevations depending on the findings of the geotechnical investigations of the borrow site (see

Part V-4-1-e). For offshore sources, the contractor should be required to set appropriate buoys which should meet U.S. Coast Guard standards to delineate the limits of the borrow areas. The contractor should be required to have electronic positioning equipment capable of achieving Class 1 survey accuracy as specified by EM 1110-2-1003, "Hydrographic Surveying." This accuracy is necessary to locate the dredge when operating in the borrow area. The geographical position of the dredge should be continuously monitored at all times during dredging operations. Locations should be determined with a probable range of error not to exceed 15.24 m (50 ft) to avoid violations of the environmental permits and clearances. Position data should be furnished as a part of the daily report of operations. Prior to initiation of any dredging, the contractor should submit for approval his proposed method of determining the dredge's location. Use of Loran-C positioning systems should be avoided, as output coordinates are not typically repeatable from one machine to the next.

(c) Routes between borrow area and project site. Determination of the route and the method used to transport beach-fill material from the borrow area to the fill site should be at the contractor's option. For offshore borrow sources, the contractor should be required to conduct the work in such manner as to obstruct navigation as little as possible. Upon completion of the work, the contractor should promptly remove his plant, including ranges, buoys, piles, and other marks placed by him under the contract in navigable waters or on shore.

- If a pipeline dredge is utilized in a congested navigation area, the pipeline may have to be submerged except at the dredge or at the location of any booster pumps or pumphouse barges. The contractor should maintain a tight discharge pipeline at all times. The joints of the pipeline should be so constructed as to preclude spillage and leakage. Upon development of a leak, the pipeline should be promptly repaired and the dredge may have to be shut down until a complete repair has been made.

- If a submerged pipeline is placed across a navigable waterway, the contractor should notify the contracting officer in writing. Notification should be received in the District Office prior to the desired closure date. This notification should furnish the following: location and depth (over the top of the pipeline) at which the submerged line should be placed; the desired length of time the navigable water is to be obstructed; the date and hour placement or removal should commence; and the date and hour of anticipated completion.

- It is recommended that a statement concerning submerged pipelines similar to the following be included in the dredging contract:

"Submerged Pipelines. In the event the contractor elects to submerge his pipeline, the top of the submerged pipeline shall be no higher than the required dredging depth for a channel for which the pipeline is placed. The submerged pipeline shall be marked with signs, buoys, and lights as required to the complete satisfaction of the contracting officer."

- Complying with this requirement may require that the contractor excavate a trench in the channel bottom. If it is known during the design phase that a submerged pipeline will cross a navigable waterway, then specifications and provision for the contractor to establish (trench) the crossing should be explicitly provided. The equipment necessary to trench the crossing, and the disposal site for the trenched material will, in all likelihood, be different than that required to construct the beach-fill.

- If the contractor elects to use a hopper dredge or pumpout barge, overflow during loading should be permitted to the extent that designated turbidity and water quality standards are met. The contractor should limit the loading to partial loads, if necessary, to meet turbidity and water quality requirements for the overflow during loading. No overflow or spillage should be permitted during transport to the discharge site.

(d) Placement methods. The contractor should be given the option of starting the beach-fill placement operations at any point and proceeding in any direction along the project beach, unless special conditions exist.

- Acceptance reaches, which are segments of beach measured along the construction baseline between designated stations used to monitor construction of the project, should be defined and identified on the project drawings. For the case of the lateral termini, the acceptance reach is the segment of beach, measured along the construction baseline, between the longitudinal limit of fill and the subsequent designated station.

- Once the contractor begins placement in an acceptance reach, placement in that reach should be completed before proceeding to another acceptance reach. Beach-fill placement operations should proceed in an orderly manner from reach to reach. If more than one dredge and/or pumpout facility is utilized by the contractor, more than one beach-fill operation may be accomplished simultaneously. Placement of beach-fill in multiple locations at any one time should only be allowed if adequate inspection is available.

- Prior to initiation of beach-fill operations, the contractor should submit for approval his proposed plan for beach-fill placement. The plan should include the type of dredge plant to be utilized, the location and type of any booster pump facilities to be utilized, and the sequence of work for beach-fill placement. The contracting officer should reserve the right to reject any scenario which, in his/her opinion, might be detrimental to the stability of the in-place beach-fill, might unduly disrupt access to or use of the beach by the public during placement operations, or for any other credible reason. Excavation of sand from the existing beach for use as beach-fill should not be permitted.

- All materials excavated from the borrow areas should be transported to and deposited on the beach or dune area within the specified lines, grades, and cross sections in a controlled manner so as to maximize sand retention within the beach-fill section and minimize losses to the ocean. This should be accomplished in a manner acceptable to the contracting officer and should include, but not be limited to temporary diking where required, control of the discharge pipe direction and velocity of discharge, and control of the sand and water mixture. Temporary diking included within the dune cross section may be left in place and incorporated into the dune structure.

- For dredged borrow sources, fill placement on the beach can be accomplished by a single or double-pipe system. The double-pipe system consists of a yoke attached to the discharge line, and by use of a double-value arrangement, the discharge slurry is selectively distributed to either one pipe or the other, or to both pipes simultaneously. The beach is built by placing the first discharge pipe at the desired final fill elevation and pumping until the desired elevation is reached. By alternating between the two discharge lines, the beach width is built to the full cross section as they advance. The final placement to the design lines and grades can be accomplished using bulldozers.

- The contractor should be required to maintain and protect the beach-fill in a satisfactory condition at all times until acceptance of the work. Prior to placement of beach-fill, the contractor should remove from the work site all snags, driftwood, and similar foreign debris lying within the limits of the beach-fill section.

- The excavated material should be placed and brought to rest on the beach to the lines, grades, and cross sections indicated on the project drawings, unless otherwise directed by the contracting officer. The beach topography is subject to changes and the elevations on the beach at the time the work is accomplished may vary from the design elevations and the resulting beach-fill quantities may vary from that shown in the unit price schedule. To accommodate this situation, the contracting officer should reserve the right to vary the beach-fill cross sections at any location along the beach.

- The contractor should be responsible for any damage caused by excessive discharge water flowing landward from the beach-fill section. However, the plans and specifications should specify that the contractor will construct sand dikes of a minimum length to control water discharge and promote settlement of the slurry's sediment. Where a pipeline is placed on the beach, sand should be placed around the pipe to form a pedestrian ramp over the pipe at street ends and at midblocks or at locations otherwise directed by the contracting officer. All such ramps should be maintained as long as the pipe is in place.

(e) Final project dimensions. The intent of the contract is to place beach-fill to the lines and grades prescribed in the contract. Tolerances should be provided in the template for the practicality of construction. A tolerance of ±0.2 m (0.5 ft) is usually permitted for the dune and berm design sections. However, in light of the critical nature of beach elevation to shore protection and flood damage reduction, the contractor should be required to provide his best efforts in placing material to the designated lines and grades landward of the influence of waves. Persistent over- or underfilling of the construction template to the plus or minus tolerance should not be permitted. It should be considered that the primary goal of nourishment is to place a specified volume of material per unit length of beach. The required dimensions of the construction template, specifically the width of the construction berm, should not be explicitly specified so that the width of the berm can be adjusted during construction to account for the actual foreshore slope the fill acquires during placement.

- Any material placed above the prescribed cross section, plus the allowable tolerance, should not be included in the pay quantities; however, such material may be left in place at the discretion of the contracting officer. In the event that material placed at any prescribed cross section is below the minus tolerance, the contractor should be required to provide additional sand to the level of the beach-fill template.

- Upon completion of all fill operations in any acceptance reach, the beach should be graded and dressed to eliminate any undrained pockets and abrupt mounds or depressions in the beach-fill surface as necessary to comply with tolerance requirements specified. All temporary dikes not incorporated into the dune cross section should be completely degraded.

- Any material deposited in areas other than those designated or approved by the contracting officer should not be paid for. The contractor should be required to remove such misplaced material and deposit it where directed, at his expense.

(f) Calculating fill volume for payment purposes. Options for calculating fill volumes for payment include: tabulation of the fill delivered by truck load from land borrow sources, comparison of pre- and post-dredging surveys for offshore sources, and measurement of in situ volumes after placement. Of these options, the latter is recommended. With this method, acceptance reaches should be used for the purpose of closely monitoring the cumulative amounts of beach-fill placed. Acceptance reaches should also be used to control the timing for pre- and postplacement surveys. Acceptance for payment by the contracting officer of a reach should not be made until all beach-fill placement is made within the acceptance reach and final surveys have been approved. Separate acceptance for the dune portion of the beach-fill may be made for an acceptance reach upon approval of the contracting officer. In no case, however, should the contractor be paid more than once for sand placed in any space along any acceptance reach should erosion occur before the entire volume of sand is placed. Unless otherwise approved by the contracting officer, acceptance reach stationing should be as shown on the contract drawings.

- Beach fill, satisfactorily placed, may be measured for payment by use of the average end area method. The quantity computations should be verified from survey data submitted by the contractor in accordance with specified procedures. The basis of measurement should be the preplacement cross sections of the beach and dune area taken by the contractor just prior to the placement of fill and a second set of cross sections of the same area taken by the contractor as soon as practicable after completion of beach-fill placement for any acceptance reach. The plans should depict existing conditions and the construction template at intervals along the shoreline not greater than 150 m (500 ft). Measurement for payment is typically surveyed at more frequent intervals, often at about 30-m (100-ft) spacing. Once postplacement surveys have been taken in an acceptance reach, no removal of beach-fill material should be permitted in that reach unless otherwise directed by the contracting officer. Landward of the wave runup limit (in general, landward of the seaward berm crest position), the area of fill material lying above the plus tolerance (an elevation tolerance of less than ± 0.2 m is typical) template should be deducted from the gross area and the net amount used as a basis for payment. Seaward of the wave runup limit, the quantity of fill material used as a basis for payment should be that determined from the area of fill measured between pre- and postfill cross sections, less any material placed seaward of the intersection of the construction template with the existing sand surface (corrected as necessary for the actual slope taken by the fill material, if deemed appropriate by the contracting officer). Compliance with the construction template should be judged reasonably on the submerged portion of the profile. Success of the placement should also consider placement of the required volume of fill per unit length of beach within the active depth (landward of the design depth of closure).

- Payment for beach-fill should be made at the contract price per unit volume. Such payment should constitute full compensation for furnishing all labor and performing all work necessary to excavate, transport, and place beach-fill material, and all other items of work required by the drawings and the specifications for which separate payment is not provided.

- Survey specifications should indicate that the contractor conduct the original and final surveys, and surveys for any interim period, for which progress payments are requested. All these surveys should be conducted under the direction of the contracting officer, unless the contracting officer waives this requirement in a specific instance. The contractor should employ a registered and licensed land surveyor, experienced in land and hydrographic surveying, to perform the work required for quantity surveys. Prior to initiation of any quantitative beach surveys, the contractor should submit to the contracting officer for approval, a description of his method and type of equipment for performing the surveys.

- The contractor should make such surveys and computations as are necessary to determine the quantities of work performed or sand placed. All original field notes, computations, and other records should be furnished to the contracting officer at the work site.

- The contractor should perform his preplacement surveys of an acceptance reach no more than 5 days prior to placement of beach-fill material within that reach. Prior to placement of fill material the contractor should submit to the contracting officer, all field notes, data disc(s), and computations, with sufficient lead time so that control of quantities, and if necessary, adjustment to the berm width can be made.

- Postplacement surveys should be made as soon as practicable after completion of an acceptance reach. The contractor should use the same survey stations that were used in the preplacement surveys. Postplacement surveys for the next reach should not be conducted until the previous reach is accepted by the contracting officer.

- The contractor should prepare and provide to the contracting officer, immediately after completion of an acceptance reach, cross-sectional drawings showing the preplacement conditions, post-placement conditions, and the design beach-fill template for each section surveyed. The survey cross sections should be taken perpendicular to the construction baseline at specified stations, and at the beginning and ending acceptance reach stations. When unusual site or geographical conditions exist, additional stations and elevations should be taken for greater definition. The pre- and postplacement surveys should extend some distance seaward of the intersection of the construction template with the existing sand surface and that distance should be specified. The scale for the plotted cross sections should be on the order of 1 to 60 in the vertical (1 cm = 0.6 m or 1 in. = 5 ft) and 1 to 240 horizontal (1 cm = 2.4 m or 1 in. = 20 ft). All stations and elevation points taken from field books should be clearly indicated on the cross-section plots.

   l.   Project monitoring and data analysis. Beach fills are "soft structures" which respond dynamically to changing waves and water levels, similar to natural beaches. Physical responses of a beach-fill include post construction adjustment of the placed sand, seasonal variation and storm-induced changes of the beach profile, and seasonal and long-term change of the project planform. Functional performance of a beach-fill requires that the design cross section and planform be maintained over the life of the project through scheduled periodic renourishment, and emergency maintenance and/or renourishment after severe storms. The dynamic behavior of beach-fills together with the need to ensure project functionality over the design life requires that a systematic monitoring plan be established for beach-fill projects.

   (1)   Monitoring objectives. Primary objectives of monitoring beach-fill projects are to ensure that project functionality is maintained throughout the design lifetime, and to assess project performance. These objectives involve collection, analysis, and interpretation of data to: evaluate the condition of the project in comparison to design specifications; determine maintenance and renourishment volume requirements; document and assess project performance to determine how well it fulfills the protection requirements for which it was designed; evaluate project impacts on adjacent areas; and address performance problems by identifying causes and developing solutions.

The first two activities listed relate primarily to operational monitoring and provide information required to operate and maintain the project. The latter three activities relate to performance monitoring, and measure the success of the project, identify project problems, and provide information for improving the project design.

   (2)   Physical monitoring components. A physical monitoring plan for a beach-fill project consists of four major components: beach profile surveys, beach sediment sampling, aerial shoreline photography, and wave and water-level measurements. These four components provide information required to document the physical response and condition of a beach-fill project. A recommended schedule of data collection is presented in Table V-4-1. The schedule is divided in two phases. The initial phase is a period of more-intensive monitoring during the first 3 years of the project that focuses on monitoring the performance of the project. During this phase, data are gathered to confirm that the project design is performing as expected and to identify potential design problems such as erosional hot spots, unexpected project impacts, or inadequate design volumes and cross-section dimensions. If problems are identified, the monitoring data provide information for developing solutions and improving project performance based on an understanding of the physical processes. Assuming the project is functioning properly and design/performance deficiencies are addressed, the monitoring plan presented in Table V-4-1 transitions to a second phase after the third year.

Table V-4-1
Physical Data Collection Schedule for Beach-Fill Monitoring

	Phase I: Initial Placement and Project Years 1 to 3[1]								
Monitoring Component	Initial Placement		Year 1 Following Initial Placement				Years 2 to 3		Years 1 to 3 Poststorm Contingency
	Pre-fill	Postfill	Mar	Jun	Sep	Dec	Mar	Sep	
Beach profile surveys	X	X	X	X	X	X	X	X	X
Sediment sampling	X	X			X				
Aerial photography	X				X			X	
Waves and water levels				Continuous					

	Phase II: Nonconstruction Years for Project Years 4 to N (N= Project Life)		
Monitoring Component	Even Number Years Following Each Construction	Odd Number Years Following Each Construction	
	September	September	Years 4 to N Poststorm Contingency
Beach profile surveys	X	X	X
Aerial photography		X	

	Phase II: Beach Renourishment		
Monitoring Component	Renourishment		Year Following Renourishment
	Prefill	Postfill	September
Beach profile surveys	X	X	X
Sediment sampling	X	X	X
Aerial photography	X		X

[1] Following Year 3, decision is made whether to continue Phase I level of monitoring

The second phase of the plan employs an operational level of monitoring to annually assess the condition of the project. This phase of monitoring also provides information for assessing longer term aspects of project performance. Details of the monitoring components and schedules are discussed in the following paragraphs.

(a) Beach profile surveys. Profile surveys provide data which are used to calculate fill volumes and document changes in the beach cross section. Accurate estimates of fill volume are essential during construction to ensure that required design volumes are placed in the construction template, and to verify payment to contractors. Periodic surveying of profile lines over the project lifetime enable calculation of volume of fill remaining in the project bounds and on the subaerial beach and comparison of the present project condition with the required design template.

- Where practical, profile surveys should be conducted using a sea-sled system. Because of the accuracy, simplicity of design, and wide availability of system components, the sea sled is considered to be the best method for profiling beach nourishment projects (Grosskopf and Kraus 1993). An added advantage is that sea-sled surveys can extend from the upper portion of the dry beach out to the depth of closure. In areas where sea-sled surveys are not practical (e.g., in environments with reefs, rock outcrops, submarine canyons, etc.), offshore surveys should be taken by boat with a properly calibrated fathometer and combined with land-based rod and level surveys performed out to wading depth. When using this combined technique, care must be taken to avoid problems in matching the land-based and fathometer surveys.

- Survey locations should be selected to include profiles within the project limits as well as control profiles some distance up and downdrift of the project boundaries to assess benefits or impact on adjacent shorelines due to longshore spreading of the fill. The number and location of profile surveys is site specific and depends on the length of the fill and the degree of longshore uniformity of the beach morphology. On an essentially straight open-coast beach, longshore profile spacing of approximately 300 m should provide adequate resolution for periodic condition profile surveys. Pre- and postconstruction profile surveys are routinely collected at higher resolution, with spacing of 60 m or less, to accurately determine placement volumes for payment purposes. To reduce costs, profile spacing can be evaluated after several monitoring cycles to determine if the number of profile surveys may be reduced while still adequately characterizing the condition of the project.

- Profile locations should be established along a reference baseline from benchmarks that are documented and recoverable in future years. The project baseline should be set at the beginning of a project study, and all surveys should be referenced to this baseline. The surveys should extend across the entire zone of profile change from the upper profile, landward of any dune, out to beyond the depth of closure.

- In practice, exact timing of beach profile surveys depends on the construction sequence and climatic conditions; but in general, surveys should be conducted as indicated in Table V-4-1. Surveys taken immediately prior to and following construction document fill volumes and cross sections. Quarterly surveys taken during the first year following initial construction document the rapid response of the constructed cross section as it forms a more natural shape. Semiannual surveys measured in years 2 and 3 of the project record the ongoing adjustment of the project cross section. After the third year of the project, survey frequency can be reduced to once a year assuming the project is performing satisfactorily. It is recommended that these annual surveys be performed at the end of summer (or the typical period of low-energy wave conditions) to determine project condition prior to winter (or the typical storm season).

- The monitoring plan should include contingency plans to collect surveys immediately after severe storms. Poststorm profile data document storm-induced beach change and poststorm condition of the project.

- Site inspections should be performed concurrently with all beach surveys, and at least once each year. Inspections should document any information relevant to characterizing the condition of the subaerial beach (e.g., level of dune vegetation, evidence of modification or addition of fill material by local property owners, effects of storms such as scarping or overwash, presence of poststorm recovery berm, longshore variability in subaerial beach features). Site inspections should include photographs of the beach, taken looking alongshore, to provide a visual record of the project condition.

(b) Beach sediment samples. Sampling is needed to determine sediment characteristics, such as median grain size and grain-size distribution, which affect the beach profile shape and influence fill volume requirements. Sediment sampling is of particular importance when fill and native material have different characteristics. Sampling of beach sediment provides data that can be used to relate project performance to characteristics of the fill material. This information can then be used to evaluate future borrow-material suitability and determine required fill volumes for renourishment.

- Sediment samples should be collected at selected locations within the project area to account for longshore and cross-shore variability in sediment characteristics. The longshore sample spacing should be approximately 900 m, with sampling locations corresponding to the nearest profile lines.

- Sediment sampling should consist of shallow grab samples at various locations across the beach profile including the dune, berm, midtide level, mlw, and three subaqueous samples spaced uniformly out to the depth of closure (Larson, Morang, and Gorman 1997). This sampling scheme documents cross-shore variability of the median grain size and size distribution which influence profile shape.

- Sediment samples should be collected before and after initial construction and each renourishment, and 1 year following each construction. Sediment sampling should be performed concurrently with beach profile surveying.

(c) Aerial shoreline photography. Aerial photography is an essential monitoring component for documenting long-term performance of beach-fills. Aerial photographs provide a visual record of shoreline position, variations in beach planform, condition of the dune and berm, and subaerial beach width; and do so with a total-project perspective that cannot be obtained by ground photography and beach profile surveys alone. Such information is useful for documenting project planform evolution, evaluating project end effects, and identifying erosional hot spots. Aerial photographs, together with beach profile surveys, provide information on the 3-D characteristics of fill behavior that can be used to better assess condition of the project and renourishment requirements.

- Aerial photography should be taken along a single flight line with 60 percent overlap stereo coverage of the entire project area shoreline, including updrift and downdrift control areas. The scale of the photographs should be sufficient to identify shoreline features. An approximate scale of 1 cm equals 50 m is recommended. All photography should be taken near midday and around low tide to reduce shadows and reflections and to provide the maximum area of exposed intertidal beach.

- Aerial photography should be performed before initial placement and after each construction to document the preproject and postnourishment shoreline. During the first 3 years, aerial photographs should be taken annually to record initial planform spreading and shoreline response to measured wave conditions. After the first 3 years, aerial photographs should be taken once every 2 years between construction and in the year following each construction. It is recommended that aerial photographs be collected in September (prior to the storm season) when the beach is typically in its most-accreted condition. During this time, effects of storms on the observed shoreline are minimized, allowing easier assessment of fill condition from year to year and providing a more consistent measure of long-term project performance.

(d) Wave and water level measurements. Waves and water levels are the principal hydrodynamic forcing parameters controlling beach-fill evolution. Storm waves and water levels erode the upper part of the beach and redistribute sand across the profile. Over the longer term, wave-driven longshore processes reshape the planform and produce shoreline retreat along the project. Establishing a cause-and-effect relationship between actual waves and water levels and measured beach response is essential for understanding project behavior and formulating solutions when problems occur.

- Wave and water level data also provide valuable information for evaluating project design tools and techniques. Application of design tools with monitoring data may enable more accurate assessment of renourishment intervals and quantities and refinement of the project design to improve performance throughout the remainder of the project life.

- Wave and water level data should be collected using a directional wave gage. The gage should provide a continuous record of information from which significant wave height, peak wave period, peak direction, and mean water level can be determined. Water level (depth) measurements from the gage should be compared with area NOS tide gages or other sources to establish vertical datum control. The wave gage should be placed offshore of the center of the project outside the breaking zone. A depth of 10 m is typically sufficient for gage placement. Hemsley, McGehee, and Kucharski (1991) and Morang, Larson, and Gorman (1997) provide further guidelines for collecting wave and water level data.

- Wave data collection should begin prior to project construction and continue for at least 3 years from the time of initial placement. After 3 years, a decision should be made whether to continue data collection based on sufficiency of information obtained in evaluating project performance. At a minimum, data collection should continue long enough to capture at least one significant storm erosion event and to accurately assess trends in project behavior.

(e) Physical monitoring data analysis. A systematic data analysis plan should be developed as part of the monitoring activities. Pre- and postconstruction beach profile surveys should be analyzed to compute constructed fill volumes and verify compliance with construction template specifications. Sand samples should be analyzed to develop composite measures of median grain size and size distribution from which design profile shapes and fill volumes can be estimated for future construction. Seasonal profile surveys collected during Phase I of the monitoring plan should be used together with aerial shoreline photography to document initial fill adjustment and short-term project behavior. If problems arise such as excessive project erosion rates, erosional hot spots, or unanticipated impacts to adjacent shorelines, a coastal processes study should be performed with the monitoring data to identify causes and develop solutions. Poststorm profiles collected after major erosion events should be analyzed together with storm wave and water level data to verify project performance in light of design criteria. Analysis of Phase II monitoring data should focus on assessing project condition and renourishment requirements. The existing cross section should be compared with the design cross section to identify when dune heights or berm widths fall below design specifications. Shoreline data should be analyzed to determined long-term erosion rates and spreading losses for the project which can be used to fine-tune future renourishment activities.

(3) Borrow area monitoring. Monitoring procedures for the borrow area will depend on the type of borrow area being used. Borrow area types include offshore, inlet shoals, sand traps, bay or lagoons, and terrestrial sources. The principal purpose for monitoring borrow sites is to evaluate borrow fill suitability, borrow area bathymetry, continuing changes in the morphology and sediment characteristics, and biology of the area after completion of the borrow operation. Data collection should include bathymetric and sub-bottom surveying, sediment coring and surface sampling, and biological data acquisition. Baseline data should be collected prior to excavation and periodically thereafter. This section primarily addresses borrow sites in water-covered areas. Terrestrial borrow sites generally exhibit little or no change in topography and sediment characteristics after completion of the borrow operation.

(a) Bathymetric and bottom profiling. Once a borrow site is selected, removal of material from the borrow site will affect its morphology. The nature of the modification depends on whether the material was obtained by excavating a thin superficial layer over a large area or deep pits in a comparatively small area. One objective of borrow site monitoring is to determine to what extent marine processes restore the original morphology or create new forms. For this reason bathymetric surveys are needed to monitor the site after the borrow operation.

(b) Borrow area sampling scheme. Borrow area sampling time and collection requirements are presented in Table V-4-2. The borrow area monitoring does not require data collection as often as the project site, however a minimum of 1 year between sampling is recommended.

Table V-4-2
Borrow Area Bathymetry and Sediment Sampling Scheme

Year	Times/year	Number of Samples
pre	1	Cores to characterize borrow material and access fill suitability. Bathymetry and subbottom sampling covering expected borrow sites and control areas.
post	1	Surface sediment grab samples to characterize postdredging borrow area sediment distrbution. Bathymetry of postdredged surface to assess volume removed.
last	1	Cores to characterize infilling sediment grain size distrbution. Bottom surface bathymetry to determine infilling volume.

(c) Changes in processes. Changes in bathymetry due to offshore borrow operations can modify the characteristics of incoming waves. These changes are primarily related to refraction and bottom friction. Dredging fill material from ebb tidal shoals is a likely source of wave modification because these shoals lie close to the shore and their crests are at shallow depths.

- During preproject planning and design, these factors will have been evaluated on the basis of theoretical considerations and indicate wave modification judged to be acceptable. However, it is possible that unforeseen effects may occur. These will usually be indicated by accelerated erosion or accretion of the project beach and/or adjacent shore areas. During postproject monitoring, any unusual erosion or accretion of the project area or adjacent beaches should be investigated with the possibility that it is resulting from modification of offshore borrow sources.

- Another type of process modification can occur where inlets and associated shoals are dredged for borrow material. The strength and distribution of tidal currents in the inlet and shoal areas can be altered by the removal of material. In such cases, provisions should be made for current observations as well as bathymetric and sediment data.

(d) Borrow area data analysis. Analyses should include evaluation of temporal borrow changes, determination of the rate and volume of borrow area infilling, and identification of current patterns in the borrow area channel or basin in cases where inlet shoals are excavated.

(4) Biological monitoring. The excavation and placement of fill material usually impacts the biology of the area that is directly involved. Biological impacts may also be created in adjacent areas from the turbidity created by the excavation process. For this reason, biological surveys of both the beach and borrow area should be performed. Monitoring of the borrow site should include assessment of the infauna, sea grasses, reefs, or other biologically sensitive areas adjacent to the borrow area. The beach project area may also have environmentally sensitive areas such as sea turtle nesting sites, bird nesting areas, beach organisms, nearshore reefs, and sea grasses. Biological sampling should consist of grab samples of the borrow area and quadrate samples of the beach areas to identify the infauna of the borrow and fill locations. Monitoring turbidity in the borrow site and in the surf zone of the fill area may be necessary to assess the impact of

dredging and dumping of fill material on the local biota. A more detailed outline of biological sampling can be found in EM 1110-2-1204, "Environmental Engineering of Coastal Shore Protection."

Data analysis should evaluate fluctuations in the flora and fauna in the beach-fill and adjacent nearshore area, effects of turbidity on fauna at the beach-fill and borrow site, and the effects of the borrow operation on the borrow site organisms. The time and extent of recovery of native organisms should be verified and compared to that of control areas. The absence of native or appearance of new organisms should also be verified and documented.

(5) Structure monitoring. If structures are a component of the project, a periodic inspection and survey report of the structures needs to be performed to assess the condition of the structures. This is especially important after a large storm that might do damage (functional or otherwise) to the structure. If damage is sustained by the structure(s), adequate photography of the structure should be taken to describe the damage and the extent of repairs that will be necessary. As damage photography is best compared to "design built" photography, a baseline set of photography for each structure in the project should be made as soon as possible after the structures are built. Damage photography for any damaged structure should be taken from the same locations as the baseline photography if reasonable and feasible.

At the time of structure inspection, the critical wave conditions (height, period, direction, etc.) under which the structure was damaged (if damaged) or exposed to (if not damaged) should be noted in the report/survey. A detailed survey of the structure may be needed. Last, an assessment of the structure's present and future effectiveness should be made.

m. Operations and maintenance

(1) Purpose. The beach-fill and any structures built for local shore protection, access, and any visitor facilities, must be operated and maintained to obtain the anticipated project benefits. In addition to periodic renourishment, the following types of maintenance work will be needed: mechanical redistribution of sand within the project area, grading, and periodic removal of debris from the project area. Performance and condition monitoring are needed throughout the economic life of the project. As discussed in the previous section, project monitoring and analysis is essential to assuring the project is providing the intended storm protection. In Federal beach nourishment projects, an Operations and Maintenance (O&M) manual specific to the project is prepared upon completion of the initial project construction. The purpose of an O&M manual is to present detailed information to assist the responsible parties in operating and maintaining the project, and to describe the periodic nourishment and monitoring aspects of the project.

(2) Scope of O&M manual. This section will present a possible outline for an O&M manual and briefly describe the contents of each section. A sample outline is provided in Figure V-4-34. The outline should be modified as necessary to meet the specific needs of the project. The manual is divided into four parts. Part I presents general information about the project. Part II provides essential operation and maintenance information necessary to ensure the desired performance of the project. Part III describes the periodic nourishment and monitoring of the project, while Part IV presents information concerning responsibilities of parties involved in the project.

(3) Introduction. This section of the O&M manual provides a concise summary of pertinent information related to the project and generally includes the information discussed in the following paragraphs.

(a) Authority. Cite the authority(ies) which authorized the project construction.

(b) Location. Describe the project location relative to nearby urban centers, water bodies, or other geographic or demographic features. Give the north, east, south, and west project boundaries.

(PROJECT NAME)

(SAMPLE) TABLE OF CONTENTS

PARA.   TITLE   PAGE NO.

SECTION I - INTRODUCTION

SECTION II - OPERATION AND MAINTENANCE

SECTION III - PERIODIC NOURISHMENT

SECTION IV - RESPONSIBILITIES

APPENDICES

Figure V-4-34. Sample Operations and Maintenance manual outline

(c) Brief description. Describe the major features of the project such as dune and berm elevations, widths, and slopes. Give the grain size characteristics of the fill, volume of material placed, type and characteristics of any structures, and lengths of fill, including transitions. Make reference to the availability and location of as-built plans. The anticipated periodic renourishment volume and interval should be briefly discussed.

(d) Protection provided. Discuss the protection provided by the project and if practicable, identify the storm parameters, or combinations thereof, for which the project is expected to limit inland or upland damages to a minor and acceptable level. Alternatively, the expected project response-frequency curves developed from the analysis of the with-project condition could be provided as indicators of expected project responses resulting from the passage of coastal storms.

(e) Local cooperation. Federal beach-fill projects are constructed and maintained by both the Federal government and one or more levels of local government. The partnership established between the various involved parties is detailed in a local cooperation agreement (LCA). The O&M manual should include a copy of the LCA in an appendix and reference to that appendix should be made in this section of the O&M manual. The summary should identify the local sponsor and those represented by the sponsor if more than one entity is involved. The cost-sharing arrangement for periodic nourishment and project monitoring should be stated and the technical document that supports the LCA and cost-sharing agreement should be cited.

(f) Construction history. Review the contracts used in constructing the project indicating the contractor, contract number, award and completion dates, any significant events or circumstances encountered, and the volumes of materials involved.

(4) Operations and maintenance. This part of the O&M manual presents information on general duties and procedures to assist local interests with their responsibilities for operation and maintenance of the beach-fill project (see ER 1110-2-2902, "Prescribed Procedures for the Maintenance and Operation of Shore Protection Works").

(a) Management. The local person or persons that will be responsible for project administration, maintenance, and general operational responsibilities should be identified. The appointment recommendation and approval procedures should be stated.

(b) Duties. Delineate the project management duties related to the project as outlined in ER 1110-2-2902. Some of these duties (in Federal beach-fill projects) include the following: maintain public ownership and use of the beach which formed the basis of Federal participation; prevent unauthorized trespass or encroachment onto the project; ensure alterations are approved by the District Engineer; ensure pedestrian and vehicular traffic are confined to designated access and use areas; and conduct periodic inspections, and operate and maintain the project as specified in this manual.

(c) Periodic inspections. Routine or emergency inspection plans should be identified. The size of the inspection team may vary from the person in charge up to a team of three or four depending on the scale and complexity of the project. Timing and number of routine inspections should be stated along with the features to be inspected, what information to record, and how and when it should be reported. A set of inspection forms should be developed to help ensure needed information is obtained. Inspection procedures to be followed before and after significant storm events should also be included. Notification from the District or some other mechanism should be included to trigger pre- and poststorm action.

(d) Reports. An inspection report is to be completed by the inspection team for each inspection to ensure that no part of the protection project is overlooked. Any item requiring repairs should be noted. Satisfactory items should also be indicated. A completed and signed set of inspection forms, plus any pertinent photographs taken during the inspection or monitoring effort, will accompany and provide the basis for the report content. In the event that repairs have been made, either temporary or permanent, the nature and date of the repair are pertinent and should be included. The address to which the reports are to be submitted should be given along with the timing of the reports. All reports should indicate project deficiencies discovered during the inspection, and the scheduled remedial measures to correct the reported deficiencies.

(e) Improvements or alterations. Drawings or prints of proposed improvements or alterations are to be submitted to the District Engineer sufficiently in advance of initiation of the proposed construction to ensure that the absence of his approval does not delay construction. As-built drawings will be furnished to the District Engineer and maintained with the original plans.

(f) Project features. This part of the O&M manual provides detailed description of the features and their operational and maintenance requirements. The section will typically cover such features as the dune, beach berm, transitions, crossovers and access ways, and, if included as part of the project, groin(s), nearshore breakwater(s), seawalls(s), and bulkhead(s).

(5) Periodic nourishment. This section of the O&M manual provides procedures for monitoring the condition of the beach-fill portion of the storm protection project, analyzing the monitoring data, evaluating when nourishment will be required, and determining the volumes of nourishment needed. The project must be periodically nourished to ensure that the desired protection is provided throughout its life.

(a) Scope. Refer to the design document and LCA to define periodic renourishment, its anticipated volume, and interval of placement. Explain the concept of advanced nourishment. Discuss the parameters and conditions that will trigger a nourishment event. Direct quotes from the design document provide credibility for the need to renourish the project. If "renourishment is triggered when, in effect, the project reaches its design configuration" is quoted from the design document, then those responsible need to understand that the design section is the minimum section required to provide the protection and not the maximum section desired or constructed. It should be emphasized that the profiles discussed are based on the configuration of the project beach that is expected once the beach has reached its equilibrium state. In most cases, this will be quite different from the configuration shown on the plans and specifications or the profile that was actually constructed. The issue of postconstruction profile adjustment should be discussed, as well as what is to be expected in terms of beach width decreases through time. Expected seasonal changes should also be discussed.

(b) Monitoring. There are various components that need to be considered to understand the performance of a beach-fill project and subsequent nourishment requirements. To ensure that the project is providing at least the design level of protection, knowledge of the project conditions via project monitoring is imperative. Consequently, a monitoring program is designed as part of the periodic nourishment of the project. For Federal projects, the monitoring program is typically administered by the Corps Engineering Division. Data collected during project monitoring will be used to assess the condition of the beach-fill and to determine when to initiate a nourishment operation. Part V-4-1-1 provided guidance on the collection of beach profile surveys, sediment samples, aerial photographs, wave data, etc., and additional guidance can be obtained from EM 1110-2-1004, "Coastal Project Monitoring." The application of these monitoring efforts to the project comprise the remaining topic items to be covered in this section of the O&M manual.

(c) Nourishment. Moving material from the foreshore to the higher berm and/or dune area, or from an accreting area within the project limits to an unusually eroded area, is considered maintenance. Artificially adding new material to the beach-fill project is considered renourishment. The need for renourishment is addressed by determining the protection provided by the existing beach-fill project.

(d) Routine monitoring analysis. The O&M manual will require routine inspection and survey of the beach-fill project. Routine analysis will compare existing profile shapes to the design profile shapes. Of specific importance in the comparison of the existing profile with the design profile is the dune elevation and width and the berm elevation and width. If the comparison indicates that the existing condition dune and berm sections are substandard as compared to the design profile, a more detailed and thorough analysis is initiated to determine, for the overall project, the extent of the deficiency and the level of risk associated with delaying nourishment until the next scheduled renourishment. Based on the results of the detailed analysis a decision is made as to whether or not to initiate a project nourishment action or if maintenance (redistribution of sand within the project area) actions can be implemented to alleviate or minimize a localized problem.

(e) Poststorm analysis. This analysis focuses on the protective features of the beach-fill project located on the upper backshore portion of the beach consisting of the dune and/or storm berm. Generally these features, once eroded, are not soon replaced by nature during poststorm beach recovery and therefore must be replaced by maintenance or nourishment of the project beach.

- Inspection and damage assessment will be conducted as soon as possible after the passage of a significant storm. A joint District and local inspection team will assess the project area. Ground photography will be obtained at a minimum and, if warranted, aerial photography and/or video will be obtained to document the poststorm conditions. The inspection will assess the visible part of the project (i.e., dune/berm erosion, damaged fence, destroyed vegetation, etc.)

- If the extent of upper beach erosion is judged to have compromised the integrity of the project, more extensive data collection and analysis will be required. Beach profile surveys should be immediately initiated at the monument locations as described in the monitoring program. Due to the expediency required in reacting to a storm event causing damage to the project and/or upland development, it is recommended that an Indefinite Delivery Type Contract (IDTC) for poststorm surveying be established and maintained.

- Using the water level and wave height data from offshore gages, or other sources, along with other physical data such as storm duration, wind speed and direction, the storm severity should be estimated. Severity should also be assessed in the context of the beach erosion the storm caused, and should include assessment of key project response parameters such as dune crest lowering, landward extent of vertical erosion threshold, volume eroded above National Geodetic Vertical Datum (NGVD), overwash extent and volume, etc. This information should be provided to the local sponsor and used to document the amount of damages that the project prevented. These findings should be reported in an "annual flood damages prevented" report. After collection and analysis of the survey data, a preliminary cost estimate of emergency maintenance and nourishment costs will be made for local use and possible budgeting purposes.

(f) Poststorm maintenance. Using the survey data, volume calculations will be made which will determine the quantity of sand required to restore the dune and/or berm to its design configuration. An assessment will be made as to the vulnerability of specific areas to additional damage during subsequent storms. Appropriate emergency maintenance actions will be identified and performed. Survey data will be used to determine a source of sand within the project boundaries to be used for the repairs. Once the design is completed and a source of material identified, a construction cost estimate will be prepared. Construction will be undertaken by the local sponsor or they may contract with the District to prepare and manage the contract.

(g) Poststorm nourishment. If the above design and survey data indicate a need to obtain material from an outside source and the District and local sponsor determine the vulnerability analysis warrants such action, an out-of-cycle nourishment contracting procedure will be initiated and the proposed contract will be immediately advertised for bids. Design analysis will determine the required sand quantities and placement areas and the associated construction templates. Construction plans and specifications and cost estimates will be prepared, as well as related contract documents. Advertisement and award of the contract should be accomplished as soon as possible to allow as much flexibility as possible in scheduling the construction.

(6) Responsibilities. This section should define the roles and responsibilities of organizations and organizational elements for implementing the provisions of the O&M manual.

V-4-2. References

Required Publications

EM 1110-1-1802
Geophysical Exploration for Engineering and Environmental Investigations

EM 1110-2-1004
Coastal Project Monitoring

EM 1110-2-1617
Coastal Groins and Nearshore Breakwaters

ER 1110-2-2902
Prescribed Procedures for the Maintenance and Operation of Shore Protection Works

Related Publications

EM 1110-2-1003
Hydrographic Surveying

EM 1110-2-1204
Environmental Engineering of Coastal Shore Protection

EP 415-1-4
Network Analysis System Guide

Anders and Hansen 1990
Anders, F. J., and Hansen, M. 1990. "Beach and Borrow Site Sediment Investigation for a Beach Nourishment at Ocean City, Maryland," Technical Report CERC-90-5, Coastal Engineering Research Center, U.S. Army Engineer Waterways Experiment Station, Vicksburg, MS.

Bodge, Creed, and Raichle 1996
Bodge, K. R., Creed, C. G., and Raichle, A. W., 1996. "Improving Input Wave Data for Use with Shoreline Change Models," *Journal of Waterway, Port, Coastal and Ocean Engineering*, ASCE 122 (5), pp 259-263.

Bodge and Rosen 1988
Bodge, K. R., and Rosen, D. S. 1988. "Offshore Sand Sources for Beach Nourishment in Florida: Part I: Atlantic Coast." *Beach Preservation Technology '88 Proceedings.* Gainesville, FL, pp 175-190.

Bottin and Earickson 1984
Bottin, R. R., Jr., and Earickson, J. A. 1984. "Buhne Point, Humboldt Bay, California, Design for the Prevention of Shoreline Erosion; Hydraulic and Numerical Model Investigation," Technical Report CERC-84-5, Coastal Engineering Research Center, U.S. Army Engineer Waterways Experiment Station, Vicksburg, MS.

Creed, Bodge, and Suter (1999)
Creed, C. G., Bodge, K. R., and Suter, C. L., 1999. "Construction Slopes for Beach Nourishment Projects," *Journal of Waterway, Port, Coastal and Ocean Engineering* 126(1), pp 57–62.

Chasten et al. 1993
Chasten, M. A., Rosati, J. D., McCormick, J. W., and Randall, R.A. 1993. "Engineering Design Guidance for Detached Breakwaters as Shoreline Stabilization Structures," Technical Report CERC 93-19, Coastal Engineering Research Center, U.S. Army Engineer Waterways Experiment Station, Vicksburg, MS.

Dally and Pope 1986
Dally, W. R., and Pope, J. 1986. "Detached Breakwaters for Shore Protection," Technical Report CERC-86-1, Coastal Engineering Research Center, U.S. Army Engineer Waterways Experiment Station, Vicksburg, MS.

Dean 1974
Dean, R. G. 1974. "Compatibility of Borrow Material for Beach Fill." *Proceedings, 14th International Conference on Coastal Engineering.* ASCE, 1319-1333.

Dean 1991
Dean, R. G. 1991. "Equilibrium Beach Profiles: Characteristics and Applications," *Journal of Coastal Research* 7(1), pp 53-84.

Dean 2000
Dean, R. G. 2000. "Beach Nourishment Design: Consideration of Sediment Characteristics," UFL/COEL-2000/002, Department of Civil and Coastal Engineering, University of Florida, Gainesville, FL.

Dean and Campbell 1999
Dean, R. G., and Campbell, T. J. 1999. "Recommended Beach Nourishment Guidelines for the State of Florida and Unresolved Related Issues," UFL/COEL-99/022, Department of Civil and Coastal Engineering, University of Florida, Gainesville, FL.

Dean and Yoo 1992
Dean, R. G., and Yoo, C. H. 1992. "Beach-Nourishment Performance Predictions," *Journal of Waterway, Port, Coastal and Ocean Engineering* 118(6), pp 567-585.

Ebersole, Neilans, and Dowd 1996
Ebersole, B. A., Neilans, P. J., and Dowd, M. W. 1996 "Beach-Fill Performance at Folly Beach, South Carolina (1 Year After Construction) and Evaluation of Design Methods," *Shore & Beach* 64(1), pp 11-26.

Field and Duane 1974
Field, M. E., and Duane, D. B. 1974. "Geomorphology and Sediments of the Inner Continental Shelf, Cape Canaveral, Florida," Technical Memorandum No. 42, Coastal Engineering Research Center, U.S. Army Engineer Waterways Experiment Station, Vicksburg, MS.

Gorecki 1985
Gorecki, R. I. 1985. "Evaluation of Presque Isle Offshore Breakwaters for Beach Stabilization." *Proceedings of 42nd Annual Meeting of the Coastal Engineering Research Board.* Chicago, IL, pp 83-126.

Gravens 1992
Gravens, M. B. 1992. "User's Guide to the Shoreline Modeling System (SMS)," Instruction Report CERC-92-1, Coastal Engineering Research Center, U.S. Army Engineer Waterways Experiment Station, Vicksburg, MS.

Gravens, Kraus, and Hanson 1991
Gravens, M. B., Kraus, N. C., and Hanson, H. 1991. "GENESIS: Generalized Model for Simulating Shoreline Change - Report 2 Workbook and System User's Manual," Technical Report CERC-89-19, Coastal Engineering Research Center, U.S. Army Engineer Waterways Experiment Station, Vicksburg, MS.

Gravens and Rosati 1994
Gravens, M. B., and Rosati, J. D. 1994. "Numerical Model Study of Breakwaters at Grand Isle, Louisiana," Miscellaneous Paper CERC-94-16, Coastal Engineering Research Center, U.S. Army Engineer Waterways Experiment Station, Vicksburg, MS.

Grosskopf and Kraus 1993
Grosskopf, W. G., and Kraus, N. C. 1993. "Guidelines for Surveying Beach Nourishment Projects," Coastal Engeineering Technical Note, CETN II-31, U.S. Army Engineer Waterways Experiment Station, Vicksburg, MS.

Hall 1952
Hall, J. V., Jr. 1952. "Artificially Constructed and Nourished Beaches in Coastal Engineering." *Proceedings, 3rd International Conference on Coastal Engineering.* ASCE, pp 119-133.

Hands and Allison 1991
Hands, E. B., and Allison, M. C. 1991. "Mound migration in deeper water and methods of categorizing active and stable berms." *Proceedings, Coastal Sediments '91*, ASCE, pp 1985-1999.

Hanson 1987
Hanson, H. 1987. "GENESIS, A Generalized Shoreline Change Model for Engineering Use," Report No. 1007, Department of Water Resources Engineering, University of Lund, Sweden.

Hanson and Kraus 1989
Hanson, H., and Kraus, N. C. 1989. "GENESIS: Generalized Model for Simulating Shoreline Change," Technical Report CERC-89-19, Coastal Engineering Research Center, U.S. Army Engineer Waterways Experiment Station, Vicksburg, MS.

Hemsley, McGehee, and Kucharski 1991
Hemsley, J. M., McGehee, D. D., and Kucharski, W. M. 1991. "Nearshore Oceanographic Measurements: Hints on How to Make Them," *Journal of Coastal Research* 7(2), pp 301-315.

James 1974
James, J. R. 1974. "Borrow Material Texture and Beach Fill Stability." *Proceedings, 14th International Conference on Coastal Engineering.* ASCE, pp 1334-1344.

James 1975
James, J. R. 1975. "Techniques in Evaluating Suitability of Borrow Material for Beach Nourishment," Technical Memorandum No. 60, Coastal Engineering Research Center, U.S. Army Engineer Waterways Experiment Station, Vicksburg, MS.

Johnson and Nelson 1985
Johnson, R. O., and Nelson, W. G. 1985. "Biological Effects of Dredging in an Offshore Borrow Area," *Biological Sciences* 3, pp 166-188.

Knutson 1977
Knutson, P. L. 1977. "Planting Guidelines for Dune Creation and Stabilization," CETA 77-4, Coastal Engineering Research Center, U.S. Army Engineer Waterways Experiment Station, Vicksburg, MS.

Knutson 1980
Knutson, P. L. 1980. "Experimental Dune Restoration and Stabilization, Nauset Beach, Cape Cod, Massachusetts," TP 80-5, Coastal Engineering Research Center, U.S. Army Engineer Waterways Experiment Station, Vicksburg, MS.

Krumbein 1957
Krumbein, W. C. 1957. "A Method for Specification of Sand for Beach Fills," Technical Memorandum 102, U.S. Army Corps of Engineers, Beach Erosion Board, Washington, DC.

Krumbein and James 1965
Krumbein, W. C., and James, W. R. 1965. "Spacial and Temporal Variations in Geometric and Material Properties of a Natural Beach," Technical Report No. 44, Coastal Engineering Research Center, U.S. Army Engineer Waterways Experiment Station, Vicksburg, MS.

Larson and Kraus 1989
Larson, M., and Kraus, N. C. 1989. "SBEACH: Numerical Model for Simulating Storm-Induced Beach Change, Report 1 - Empirical Foundation and Model Development," Technical Report CERC-89-9, Coastal Engineering Research Center, U.S. Army Engineer Waterways Experiment Station, Vicksburg, MS.

Larson and Kraus 1998
Larson, M., and Kraus, N. C. 1998. "SBEACH: Numerical Model for Simulating Storm-Induced Beach Change, Report 5 - Representation of Nonerodible (Hard) Bottoms," Technical Report CHL-98-9, Coastal Engineering Research Center, U.S. Army Engineer Waterways Experiment Station, Vicksburg, MS.

Larson, Kraus, and Burns 1990
Larson, M., Kraus, N. C., and Byrnes, M. R. 1990. "SBEACH: Numerical Model for Simulating Storm-Induced Beach Change, Report 2 - Numerical Formulation and Model Tests," Technical Report CERC-89-9, Coastal Engineering Research Center, U.S. Army Engineer Waterways Experiment Station, Vicksburg, MS.

Larson, Morang, and Gorman 1997
Larson, R., Morang, A., Gorman, L. 1997. "Monitoring the Coastal Environment; Part II: Sediment Sampling and Geotechnical Methods," *Journal of Coastal Research,* Vol. 13, No. 2, pp. 308-330.

Meisburger 1990
Meisberger, E. P. 1990. "Exploration and sampling techniques for borrow areas," Technical Report CERC-90-18, Coastal Engineering Research Center, U.S. Army Engineer Waterways Experiment Station, Vicksburg, MS.

Meisburger and Duane 1971
Meisberger, E. P., and Duane, D. B. 1971. "Geomorphology and Sediments of the Inner Continental Shelf, Palm Beach to Cape Kennedy, Florida," Technical Memorandum No. 34, Coastal Engineering Research Center, U.S. Army Engineer Waterways Experiment Station, Vicksburg, MS.

Meisburger and Field 1975
Meisberger, E. P., and Field, M. E. 1975. "Geomorphology, Shallow Structure, and Sediments of the Florida Inner Continental Shelf, Cape Canaveral to Georgia," Technical Memorandum No. 54, Coastal Engineering Research Center, U.S. Army Engineer Waterways Experiment Station, Vicksburg, MS.

Morang, Larson, and Gorman 1993
Morang, A., Larson, R., and Gorman, L. 1997. "Monitoring the Coastal Environment; Part I: Waves and Currents," *Journal of Coastal Research,* Vol 13, No. 1, pp. 111-133.

Morang, Mossa, and Larson 1993
Morang, A., Mossa, J., and Larson, R. J. 1993. "Technologies for Assessing the Geologic and Geomorphic History of Coasts," Technical Report CERC-93-5, Coastal Engineering Research Center, U.S. Army Engineer Waterways Experiment Station, Vicksburg, MS.

Nakashima et al. 1987
Nakashima, L. D., Pope J., Mossa, J., and Dean J. L. 1987. "Initial Response of a Segmented Breakwater System, Holly Beach, Louisiana," *Proceedings, Coastal Sediments '87,* ASCE, pp 1399-1414.

National Research Council 1995
National Research Council. 1995. *Beach Nourishment and Protection.* National Academy Press, Washington, DC.

Nelson 1985
Nelson, W. G. 1985. "Guidelines for Beach Restoration Projects," Report SRG-76, Florida Sea Grant College, Gainesville, FL.

Pope and Dean 1986
Pope, J., and Dean, J. L. 1986. "Development of Design Criteria for Segmented Breakwaters." *Proceedings, 20th International Conference on Coastal Engineering.* ASCE, 2,144-2, pp 158.

Prins 1980
Prins, D. A. 1980. "Data Collection Methods for Sand Inventory-Type Surveys," Technical Report CERC-80-4, Coastal Engineering Research Center, U.S. Army Engineer Waterways Experiment Station, Vicksburg, MS.

Richardson 1976
Richardson, T. W. 1976. "Beach Nourishment Techniques: Report 1, Dredging Systems for Beach Nourishment from Offshore Sources," TR H-76-13, U.S. Army Engineer Waterways Experiment Station, Vicksburg, MS.

Richardson 1977
Richardson, T. W. 1977. "Systems for Bypassing Sand at Coastal Inlets," *Proceedings of the Fifth Symposium of the Waterway, Port, Coastal and Ocean Division*, ASCE, November 1977, pp 67-84.

Rosati 1990
Rosati, J. D. 1990. "Functional Design of Breakwaters for Shore Protection: Empirical Methods," Technical Report CERC-90-15, Coastal Engineering Research Center, U.S. Army Engineer Waterways Experiment Station, Vicksburg, MS.

Rosati et al. 1993
Rosati, J. D., Wise, R. A., Kraus, N. C., and Larson, M. 1993. "SBEACH: Numerical Model for Simulating Storm-Induced Beach Change, Report 3 - User's Manual," Instruction Report CERC-93-2, Coastal Engineering Research Center, U.S. Army Engineer Waterways Experiment Station, Vicksburg, MS.

Savage 1962
Savage, R. P. 1962. "Experimental Study of Dune Building with Sand Fences." *Proceedings, 8th International Conference on Coastal Engineering*. ASCE, pp 671-700.

Savage and Woodhouse 1969
Savage, R. P., and Woodhouse, W. W., Jr. 1969. "Creation and Stabilization of Coastal Barrier Dunes," *Proceedings, 11th International Conference on Coastal Engineering,* ASCE, pp 671-700.

Scheffner and Borgman 1999
Scheffner, N. W., Borgman, L. E. 1999. "Users Guide to the Use and Application of the Empirical Simulation Technique," Technical Report CHL-99-00, U.S. Army Engineer Waterways Experiment Station, Vicksburg, MS.

Schwartz and Musialowski 1980
Schwartz, R. K., and Musialowski, F. R. 1980. "Transport of Dredged Sediment Placed in the Nearshore Zone--Currituck Sand-Bypass Study (Phase I)," Technical Paper CERC-80-1, Coastal Engineering Research Center, U.S. Army Engineer Waterways Experiment Station, Vicksburg, MS.

Shore Protection Manual 1984
Shore Protection Manual. 1984. 4th ed., 2 Vol., U.S. Army Engineer Waterways Experiment Station, U.S. Government Printing Office, Washington, DC.

Smith and Ebersole 1997
Smith, S. J., and Ebersole, B. A. 1997. "Numerical Modeling Evaluation of Hot Spots at Ocean City, Maryland." *Proceedings, 10th National Conference on Beach Preservation Technology*. Florida Shore & Beach Preservation Association, pp 230-245.

Sommerfeld et al. 1994
Sommerfeld, B. G., Mason, J. M., Kraus, N. C., Larson, M. 1994. "BFM: Beach Fill Module, Report 1, Beach Morphology Analysis Package (BMAP) - User's Guide," Instruction Report CERC-94-1, U.S. Army Engineer Waterways Experiment Station, Vicksburg, MS.

Stauble and Nelson 1984
Stauble, D. K., and Nelson, W. G. 1984. "Beach Restoration Guidelines: Prescription for Project Success." *The New Threat to Beach Preservation.* L. Tate, ed., Florida Shore and Beach Preservation Society, Tallahassee, FL, pp 137-155.

Stauble 1994
Stauble, D. K. 1994. "Evaluation of Erosion Hot-spots for Beach Fill Performance." *Proceedings, 7th National Conference on Beach Preservation Technology.* Florida Shore and Beach Preservation Association, pp 198-215.

U.S. Army Corps of Engineers 1991
U.S. Army Corps of Engineers. 1991. "National Economic Development Procedures Manual - Coastal Storm Damage and Erosion," Institute of Water Resources Report No. 91-R-6, Institute for Water Resources, Water Resources Support Center, Fort Belvoir, VA.

U.S. Army Engineer District, Charleston 1987
U.S. Army Engineer District, Charleston. 1987. "Final Detailed Project Report, Charleston Harbor South Carolina, Folly Beach, Section 111 of the River and Harbor Act as Amended, Charleston, SC," Charleston, SC.

U.S. Army Engineer Waterways Experiment Station 1981
U.S. Army Engineer Waterways Experiment Station. 1981. "Beach Dune Walkover Structures," Coastal Engineering Technical Note, CETN-III-5, U.S. Army Engineer Waterways Experiment Station, Vicksburg, MS.

Walton 1994
Walton, T. W., Jr. 1994. "Shoreline Solution for Tapered Beach Fill," *Journal of Waterway, Port, Coastal and Ocean Engineering* 120(6), ASCE, pp 651-655.

Weggel and Sorensen 1991
Weggel, J. R., and Sorensen, R. M. 1991. "Performance of the 1986 Atlantic Atlantic City, New Jersey, Beach Nourishment Project," *Shore and Beach* 59(3), pp 29-36.

Wise, Smith, and Larson 1996
Wise, R. A., Smith S. J., and Larson M. 1996. "SBEACH: Numerical Model for Simulating Storm-Induced Beach Change, Report 4 - Cross-shore Transport Under Random Waves and Model Validation with SUPERTANK and Field Data," Technical Report CERC-89-9, Coastal Engineering Research Center, U.S. Army Engineer Waterways Experiment Station, Vicksburg, MS.

Woodhouse 1978
Woodhouse, W. W. 1978. "Dune Building and Stabilization with Vegetation," SR-3, Coastal Engineering Research Center, U.S. Army Engineer Waterways Experiment Station, Vicksburg, MS.

V-4-3. Definition of Symbols

Δy_o	Initial dry beach width (after cross-shore equilibrium) [length]
ε	Shoreline diffusivity parameter (Equation III-2-26) [length^2/time]
$\sigma_{\varphi b}$	Standard deviation of a borrow sediment sample (Equation III-1-3) [phi units]
$\sigma_{\varphi n}$	Standard deviation of a native sediment sample (Equation III-1-3) [phi units]
φ	Sediment grain diameter in phi units ($\varphi = -\log_2 D$, where D is the sediment grain diameter in millimeters) [phi units]
a	Beach-fill half length [length]
a	One-half the length of the rectangular project [length]
A_F	Nourishment material scale parameter (Table III-3-3) [length$^{1/3}$]
A_N	Native sediment scale parameter (Table III-3-3) [length$^{1/3}$]
A_S	Vessel submerged cross-section area (= BT)[length^2]
B	Design berm elevation [length]
B_R	Channel blockage ratio [dimensionless]
C^n_{go}	Deepwater wave group speed of the nth record [length/time]
C^n_*	Wave speed at breaking of the nth record [length/time]
D_{50}	Median grain size [length]
D_C	Depth of closure [length]
E	Historical shoreline recession rate [length]
H_{eff}	Effective wave height [length]
H^m_s	Significant wave height of the nth record in the time series of N wave records [length]
K_s	Shoaling coefficient [dimensionless]
$M_{\varphi b}$	Estimated mean grain size of a borrow sediment sample (Equation (III-1-2) [phi units]
$M_{\varphi n}$	Estimated mean grain size of a native sediment sample (Equation (III-1-2) [phi units]
R_A	Beach nourishment overfill factor [dimensionless]
R_J	Beach nourishment factor [dimensionless]
t	time
T_{eff}	Effective wave period [time]
V	Volume of material per unit length of shoreline required to produce a beach width of W (Equation V-4-6) [length^3/length]
W	Beach width [length]
W_{add}	Added distance of translation (Equation V-4-5) [length]

V-4-4. Acknowledgments

Authors of Chapter V-4, "Beach-Fill Design:"

Mark B. Gravens, Coastal and Hydraulics Laboratory (CHL), U.S. Army Engineer Research and Development Center, Vicksburg, Mississippi.
Bruce A. Ebersole, CHL
Todd L. Walton, Ph.D., CHL
Randall A. Wise, U.S. Army Engineer District, Philadelphia, Philadelphia, Pennsylvania.

Reviewers:

Nicholas C. Kraus, Ph.D., CHL
Kevin R. Bodge, Ph.D., Olsen Associates, Jacksonville, Florida
Donald K. Stauble, Ph.D., CHL

Table of Contents

List of Figures

List of Tables

Chapter V-5
Navigation Projects

V-5-1. Project Assessment and Alternative Selection

a. Introduction

(1) Purpose. The purpose of this chapter is to present information and procedures that help in coastal navigation project planning and design. Both deep-draft ports and small boat harbors are included. Navigation channels, turning basins, anchorage areas, and related structures are discussed. Other areas of port and harbor design, such as docks, facilities, terminals, and other land-side requirements, are not included. These areas are generally nonfederal concerns.

Guidance for navigation projects has traditionally been focused on deep-draft project requirements. Modified guidance based on experience has evolved for shallow-draft projects. This chapter follows a similar philosophy, focusing on deep-draft projects with supplementary material for shallow-draft projects, as appropriate. The chapter provides fairly comprehensive coverage, but it is intended to complement, rather than replace, Engineer Manuals 1110-2-1613, "Hydraulic Design Guidance for Deep-Draft Navigation Projects" (USACE 1998), and 1110-2-1615, "Hydraulic Design of Small Boat Harbors" (USACE 1984).

(2) Contents. This section gives an overview of issues and considerations important in assessing navigation projects and in defining and selecting project alternatives. Since navigation projects are designed to satisfy requirements of a target group of vessels, an understanding of vessel types and behavior is given in Part V-5-2. Determination of design vessel and transit conditions is also discussed. Data needs and sources are reviewed in Part V-5-3, with appropriate reference to other CEM chapters. Part V-5-4 is devoted to a brief discussion of economic analysis, which is crucial to every navigation project. Development of navigation project features is addressed in Parts V-5-5 through V-5-8. Features included are channel depth, width, and alignment, turning basins, anchorage areas, navigation-related structures, and aids to navigation. Post-project activities are discussed in Part V-5-9, including operation, monitoring, and maintenance. Part V-5-10 gives a description of physical and numerical modeling tools and specialized field studies which can assist in planning and designing effective navigation projects. A number of specific project examples are presented to illustrate the capabilities and applications of model and field studies. References are given in Part V-5-11.

(3) Relationship to other chapters and parts. Part V provides general guidance on the Planning and Design Process (Part V-1) and Site Characterization (Part V-2), including data needs and sources and monitoring. Part V also contains chapters with more detailed guidance on particular project types frequently encountered in the U.S. Army Corps of Engineers (USACE). This chapter addresses navigation projects. Part V-6, Sediment Management at Inlets/Harbors, is an important complement to this chapter. It deals with sediment processes in the vicinity of inlets and harbor entrances, engineering methods for managing sediment processes to prevent negative project impacts and/or achieve positive benefits, and project experience and lessons learned. Other chapters of particular relevance are Part II-6, Hydrodynamics of Tidal Inlets; Part II-7, Harbor Hydrodynamics (including a section on vessel interactions); and Part VI, Design of Coastal Project Elements, which provides guidance for the detailed structural design needed for navigation and other coastal projects.

b. Port and harbor facility issues

(1) Motivation. Ports and harbors are vital to the nation. Since ports handle about half of U.S. overseas trade by value and nearly all by weight, waterborne commerce directly affects prosperity and richness of life in the United States. Ports and harbors are also vital for military applications because a large percentage of military goods are transported by ship. Finally, harbors provide launching and berthing facilities for commercial fishing boats and a large number of recreational boaters.

(a) The term *harbor* describes a relatively protected area accessible to vessels. The term *port* indicates a location where ships can transfer cargo. A port may be located in a protected harbor or it may be exposed, such as single-point mooring facilities used for petroleum products.

(b) Ports and harbors must be located so that vessels can penetrate coastal waters and interface with land. Ideally, vessels have a relatively short travel distance between port/harbor areas and open water. Vessels must have sufficient water depth and protection to safely enter and exit the harbor/port area. Thus, a well-maintained, clearly identified channel through any shallow areas is needed.

(c) The requirements for access and protection in harbors and ports often lead to dredged channels and engineered structures, such as jetties and breakwaters. These project features can impact dynamic coastal processes and lead to a range of coastal engineering concerns.

(2) Deep- versus shallow-draft projects. The terms *deep-draft* and *shallow-draft* are often used to distinguish between major commercial port projects and recreational or other small boat harbor projects. USACE definitions for these terms are based on authorized navigation project depth:

- Deep draft   Channel depth greater than 4.6 m (15 ft)

- Shallow draft   Channel depth less than 4.6 m (15 ft)

Defining depth for a deep-draft harbor can vary with context. For example, Federal cost-sharing rules are based on a 6.1-m (20-ft) minimum depth for deep-draft projects. The harbor maintenance tax system is applied to projects with depth greater than 4.3 m (14 ft), while inland fuel taxes apply in shallower-depth projects, excluding entrance channels. Deep-draft U.S. ports serve commercial seagoing ships, Great Lakes freighters, Navy warships, and Army prepositioning ships. Shallow-draft harbors typically serve pleasure craft and fishing boats. The term *small craft* is often used synonymously with *shallow draft*. As part of its mission, USACE has had responsibility for maintaining over 200 deep-draft coastal ports and over 600 shallow-draft harbors. Issues involved in shallow-draft navigation projects have similarities but also significant differences from those in deep-draft projects. For example, shallow-draft boats are typically small and are strongly influenced by wind waves and swell. Thus, wave criteria for safe transit of entrance channels and safe mooring areas are more demanding than for deep-draft vessels. However, large ships typically maneuver with difficulty in confined areas, and channel width is a critical component of deep-draft channels. Deep-draft harbors are more prone to harbor oscillation concerns because resonant periods of moored ship response are typically in the same range as harbor oscillation periods. Flushing is an important issue in many shallow-draft harbors, where numerous users in a confined area can potentially lead to deterioration of water quality. Flushing is usually less critical in deep-draft projects, which require wider entrances and more careful monitoring of vessel discharges. These issues are addressed later in the chapter.

(3) Organizations related to navigation projects. USACE navigation projects involve other organizations, as well. For example, the U.S. Coast Guard is responsible for installing and maintaining the aids to navigation needed to mark Federal channels. Often state and local organizations, such as port authorities, are

part of a navigation project team. Some organizations and acronyms often encountered in navigation projects as team members or information resources are listed in Table V-5-1.

Table V-5-1
Organization Acronyms Related To Navigation Projects

Acronym	Organization
AAPA	American Association of Port Authorities
ABS	American Bureau of Shipping
ASCE	American Society of Civil Engineers
EPA	U.S. Environmental Protection Agency
NAVFAC	U.S. Naval Facilities Engineering Command
NOAA	U.S. National Oceanic and Atmospheric Administration
PIANC	International Navigation Association (formerly, Permanent International Association of Navigation Congresses)
USACE	U.S. Army Corps of Engineers
USCG	U.S. Coast Guard

(4) Trends in port and harbor development. Demand for harbor and port facilities continues to increase while coastal population and other utilization of the coast increases. These competing interests intensify pressures to find mutually agreeable solutions to coastal land and water use. Annual foreign waterborne tonnage (between U.S. and foreign ports) during the years 1987-1996 indicates a clearly increasing trend (Figure V-5-1). Major U.S. ports continue to increase in size and serve larger ships. Dramatic increases in container traffic and a recent trend for container ships to exceed Panama Canal size constraints have helped fuel the need for deeper ports and expanded, modernized terminals. Open terminals and offshore ports have helped accommodate large tankers and bulk carriers, in some markets.

(a) With pressure to serve larger ships, many U.S. ports are faced with costly infrastructure upgrades. Deeper and wider channels, turning basins, and berthing areas are needed. Disposal of large quantities of dredged material, which is often contaminated after many years of harbor operations, can be a major and expensive problem. Dock bulkheads often need to be rebuilt to maintain structural strength in deeper water and with likely higher design loads from ship berthing and apron cargo handling. Landside infrastructure must be capable of efficiently handling cargo from larger ships. This requirement often leads to bigger gantry cranes, single-purpose terminals, larger stockpiling areas, new rail facilities, etc.

(b) Demand for small-craft berthing space is also increasing, mainly to serve recreational boaters. Thus, there is a continuing economic incentive for expansion of existing small-craft harbors and development of new harbors.

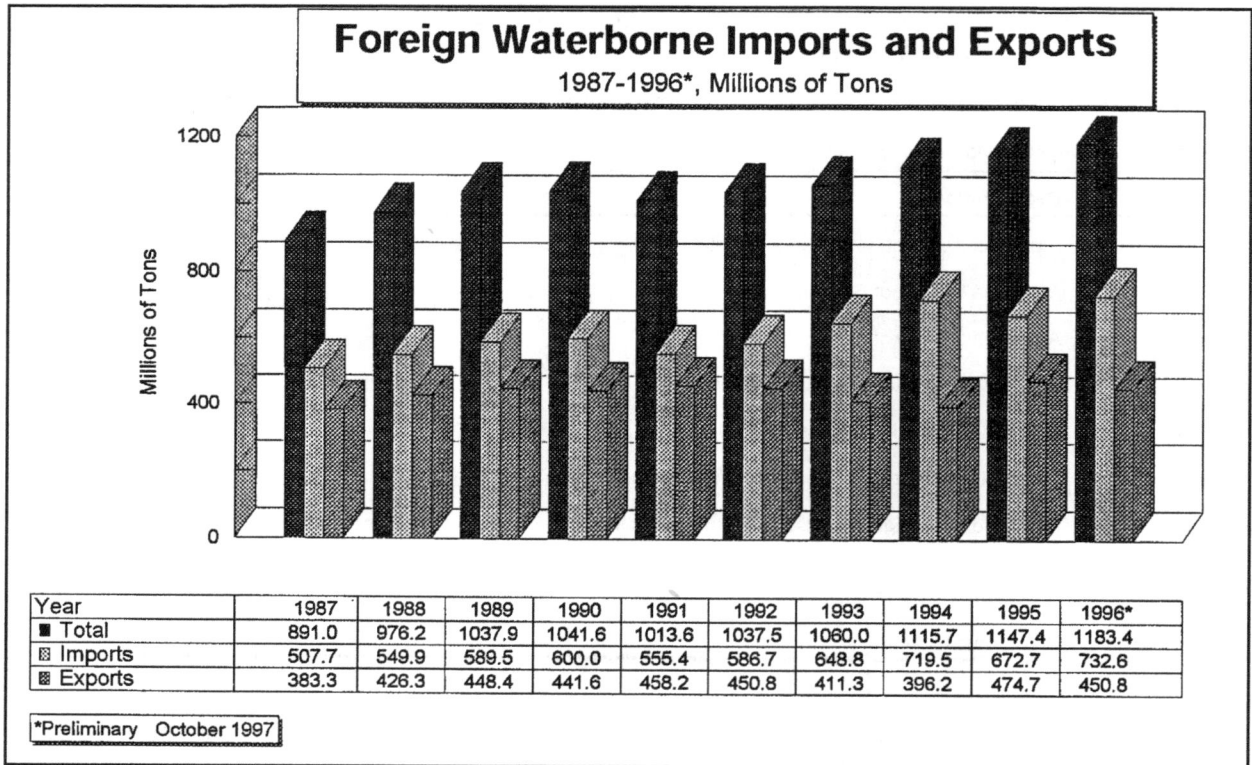

Foreign Waterborne Imports and Exports
1987-1996*, Millions of Tons

Year	1987	1988	1989	1990	1991	1992	1993	1994	1995	1996*
■ Total	891.0	976.2	1037.9	1041.6	1013.6	1037.5	1060.0	1115.7	1147.4	1183.4
▨ Imports	507.7	549.9	589.5	600.0	555.4	586.7	648.8	719.5	672.7	732.6
▨ Exports	383.3	426.3	448.4	441.6	458.2	450.8	411.3	396.2	474.7	450.8

*Preliminary   October 1997

Figure V-5-1.   Foreign waterborne imports and exports (USACE Waterborne Commerce Statistics Center)

c.   *Preliminary planning and design elements.*

(1) Federal coastal navigation projects are focused on channels and maneuvering areas to allow vessels to transit confined nearshore areas and use ports or harbors. Structures needed to accomplish navigation objectives are also included. Preliminary planning and design may include the following considerations, most of which are discussed in subsequent sections of this chapter:

(a) Site characterization.

(b) Design criteria.

(c) Defining vessel requirements.

(d) Entrance channel configuration.

(e) Inner harbor configuration.

(f) Navigation structures.

(g) Harbor and channel sedimentation and maintenance.

(h) Physical and numerical modeling.

(2) Helpful supplementary references for deep-draft projects include McBride, Smallman, and Huntington (1998); PIANC (1997a, 1997b, 1995); Tsinker (1997); Gaythwaite (1990); Turner (1984); Quinn (1972); and U.S. Navy design manuals. For small-craft projects, references include ASCE (1994), Tobiasson and Kollmeyer (1991), State of California (1980), and Dunham and Finn (1974). References with coverage of both deep-draft and small-craft harbors include Herbich (1992) and Bruun (1990). U.S. Army guidance for military ports is given by USACE (1983).

d. Policy considerations. Federal cost-sharing guidelines are a key concern in U.S. navigation projects. Prior to 1986, the Federal Government paid 100 percent of costs for navigation channel deepening and widening. Under present guidelines for *commercial* harbors, the nonfederal share of general navigation feature construction costs is 10 percent for project depth not exceeding 6.1 m (20 ft), 25 percent for project depth greater than 6.1 m (20 ft) but not exceeding 13.7 m (45 ft), and 50 percent for project depth exceeding 13.7 m (45 ft). The nonfederal sponsor must also pay: (1) an additional 10 percent of construction costs that are cost-shared, and (2) for project depths greater than 13.7 m (45 ft), 50 percent of operation and maintenance costs associated with general navigation features. For *recreational* navigation projects or separable recreational elements of commercial navigation projects, the nonfederal share is 50 percent of construction costs and 100 percent of operation and maintenance costs. Partnering between commercial, recreational, and military interests should always be examined. Cost-sharing guidelines are fully described by USACE (1996).

V-5-2. Defining Vessel Requirements

a. Deep-draft ships and shallow-draft vessels.

(1) Vessel dimensions. Navigation projects are designed to accommodate vessels of a desired size. Key vessel dimensions are length, beam (width), and draft. These dimensions are defined in several different ways to characterize the curved, three-dimensional vessel form. Vessel dimensions, especially for commercial ships, are often presented in terms of standard acronyms defined in Table V-5-2. Terms are explained in the following paragraphs.

(a) The shape of a typical commercial ship is depicted in Figure V-5-2. The LOA is an important measure of length for evaluating ship clearances in confined navigation project areas. For example, a turning basin would be sized based on the design ship LOA. The LBP is a more meaningful measure of the effective length for concerns such as ship displacement and cargo capacity.

(b) Definitions of design draft, freeboard, and beam are illustrated in Figure V-5-3. Molded beam is the maximum width to the outer edges of the ship hull, measured at the maximum cross section (usually at the ship waterline at midship). Design draft is the distance from the design waterline to the bottom of the keel. Ship *depth* is a vertical dimension of the hull, as shown in the figure, and it should not be confused with ship *draft*.

(c) Draft may not be uniform along the vessel bottom for both deep- and shallow-draft vessels. For example, draft near the vessel stern (aft) is often greater than near the bow (fore). Two useful indicators of such variations are:

trim - difference in draft fore and aft

list - difference in draft side to side

Table V-5-2
Acronyms Commonly Used to Describe Ship Size and Function

Acronym	Explanation
LOA	Length overall
LBP	Length between perpendiculars (measured at DWL)
LWL	Length along waterline (usually similar to LBP)
DWL	Design waterline (usually represents full load condition)
B	Beam (maximum width of ship cross section)
D	Draft
D_s	Depth of vessel's hull
FB	Freeboard (=D_s - D)
DT	Displacement tonnage (fully loaded)
l.t.	Long ton; = 1016 kg (2240 pounds)
m.t.	Metric ton: = 1000 kg (2205 pounds); $\approx$ 1 l.t.
LWT	Lightship weight (empty)
DWT	Dead weight tonnage (= DT - LWT)
GRT	Gross register ton; 1 register ton = 2.83 cu m (100 cu ft) of internal space (may also be stated in cubic meters)
GT	Gross ton
NRT	Net register ton
NT	Net tons
OBO	Ore/bulk/oil combination carrier
TEU	20-ft equivalent units; standardized 6.1 m x 2.4 m x 2.4 m (20 ft x 8 ft x 8 ft) container units

(d) *Maximum navigational draft* (the extreme projection of the vessel below waterline when fully loaded) is needed for navigation channel depth; *mean draft* is preferred for hydrostatic calculations. *Waterline beam* (width of the vessel at the design or fully loaded condition) is needed for navigation channel width.

(e) Maximum dimensions of the above-water part of a vessel are also critical to ensure adequate clearance. *Maximum beam* is the extreme width of the vessel. For a vessel such as an aircraft carrier, maximum beam is much larger than waterline beam. *Lightly loaded draft* is the minimum vessel draft for stability purposes, from which vertical clearance requirements, such as clearance under bridges, can be determined.

(2) Cargo capacity. Cargo capacity of commercial ships is generally indicated by DWT or, in the special case of container ships, by TEU. Units of measure for weight are usually long tons or metric tons (Table V-5-3). Cargo capacity also provides a convenient indicator of ship size, since ship dimensions for a particular type of ship (e.g. tanker) are usually closely correlated with capacity.

(a) Port duties and shipping costs are often figured in terms of register tons, a measure of *volume*. The GRT indicates total internal volume of the ship; NRT indicates volume available for cargo. The GRT is equal to NRT plus volume of space devoted to fuel, water, machinery, living space, etc. The terms GRT and NRT are currently used for older ships, but the terms GT and NT are favored for newer ships. The LWT

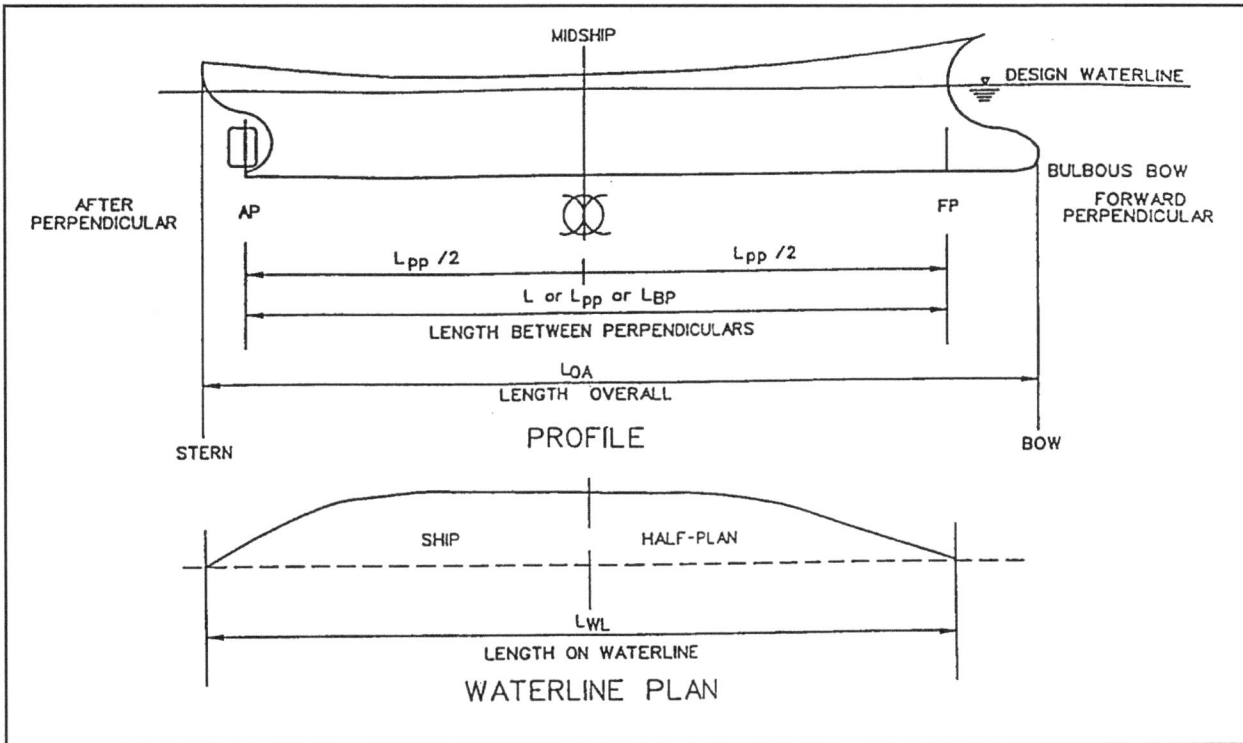

Figure V-5-2. Ship length definitions

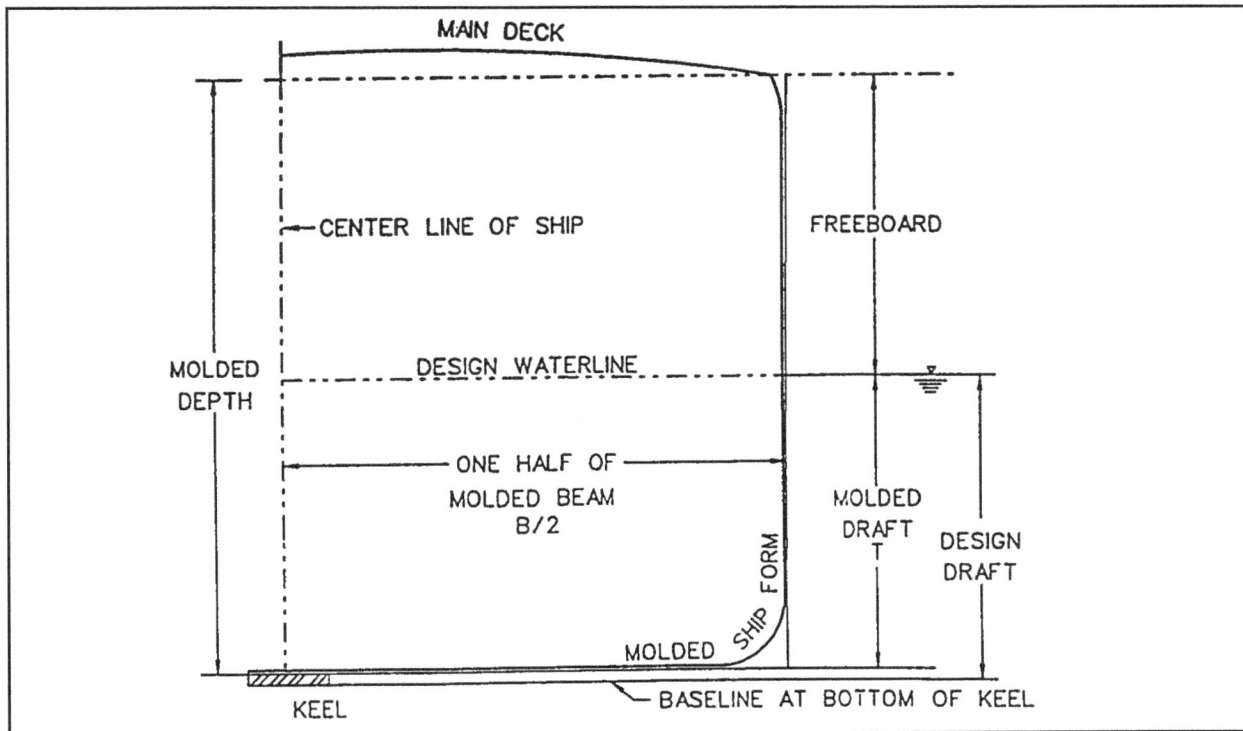

Figure V-5-3. Midship-section molded-form definitions

is the minimum weight a ship can have, such as the weight to be supported in dry dock. In operation, even unloaded ships rarely reach the LWT, as they often take on water, or *ballast*, to increase stability.

(3) Restrictions. Canal and lock sizes can impose distinct restrictions on ship size. The Panama Canal and Suez Canal are the two most critical for oceangoing traffic (Table V-5-3). Ships sized to meet the Panama Canal restrictions are known as *Panamax* vessels. They constitute an important vessel class for navigation projects because many commercial ships fit within the confines of the Panama Canal. Economics associated with some cargos, most notably crude oil, have resulted in ships that cannot pass through the canal. These ships are sometimes referred to as *Post-Panamax* vessels.

Table V-5-3
Canal Restrictions on Ship Size

Canal	Restriction		
	Draft	Beam	Length
Panama	12.0 m (39.5 ft)	32.2 m (105.75 ft)	289.6 m (950.0 ft)
Suez	16.2 m (53.0 ft)	64.0 m (210 ft)	No restriction

(4) Vessel characteristics. Vessels cover a wide range of sizes and shapes. Deep-draft vessels, especially the larger ships that typically dictate navigation project dimensions, may represent a small number of specific ship designs to serve specialized needs and routes. Therefore, deep-draft vessel characteristics are usually presented as a sampling of individual, named vessels. Characteristics of some representative large ships from the world merchant fleet are given in Table V-5-4. Most U.S. ports have controlling depths between 10.7 m (35 ft) and 12 m (40 ft). The deeper ports can accommodate Panamax vessels, but access by larger ships is limited. Common vessel types are briefly reviewed in the following paragraphs. In contrast to deep-draft projects, shallow-draft vessels are usually numerous and their characteristics can be discussed in statistical terms. Also, other factors besides individual vessel characteristics, such as volume of traffic, may be critical to a shallow-draft navigation project.

(a) *Tankers* carry liquid bulk products. Crude oil is by far the most common liquid bulk cargo. Economies of scale have strongly affected tanker design because of the volume and uniformity of product and consistent level of demand. Large tankers are often classified by size (Table V-5-5). The larger vessels far exceed Panamax size, but most can use the Suez Canal in ballast. Loaded tankers less than about 50,000 DWT require a draft of 12 m (40 ft) or less and can enter many U.S. harbors. Supertankers can use partial loading and/or tidal advantage to access U.S. harbors. Navigation projects in the United States generally cannot accomodate the drafts of loaded VLCC and ULCC class tankers. The largest tankers, too big to enter any of the major world ports, ply dedicated trade routes between offshore port facilities.

(b) *Liquid Natural Gas (LNG)* and *Liquid Propane Gas (LPG)* carriers have a highly volatile cargo at very low temperature. They require highly specialized terminals and special safety considerations.

(c) *Dry bulk carriers* carry a wide range of cargoes such as ore, coal, and grain. Size is generally less than 150,000 DWT.

(d) *Combination bulk carriers* are specially configured to carry both liquid and dry bulk cargo. The most common combination is ore/bulk/oil, or *OBO*. Vessel size ranges from 50,000 DWT to 250,000 DWT.

(e) *General cargo ships* carry a wide variety of cargoes packaged in the form of pallets, bales, crates, containers, etc. *Break-bulk* cargo refers to *individually* packaged items that are stowed *individually* in the ship. Size is typically 12,000 to 25,000 DWT.

Table V-5-4
Characteristics of Large Ships

Name	Dead-weight Tonnage	Length		Beam		Draft	
		m	ft	m	ft	m	ft
Tankers							
Pierre Guillaumat	546,265	414.23	1,359.00	62.99	206.67	28.60	93.83
Nisseki Maru	366,812	347.02	1,138.50	54.56	179.00	27.08	88.83
Idemitsu Maru	206,000	341.99	1,122.00	49.81	163.42	17.65	57.92
Universe Apollo	114,300	289.49	949.75	41.28	135.42	14.71	48.25
Waneta	54,335	232.24	761.92	31.70	104.00	12.22	40.08
Olympic Torch	41,683	214.76	704.58	26.92	88.33	12.09	39.67
Ore Carriers							
Kohjusan Maru	165,048	294.97	967.75	47.02	154.25	17.58	57.67
San Juan Exporter	104,653	262.00	859.58	38.05	124.83	15.44	50.67
Shigeo Nagano	80,815	250.02	820.25	36.86	120.92	13.23	43.42
Ore/Oil Carriers							
Svealand	278,000	338.18	1,109.50	54.56	179.00	21.85	71.67
Cedros	146,218	303.51	995.75	43.38	142.33	16.74	54.92
Ulysses	57,829	241.86	793.50	32.39	106.25	12.17	39.92
Bulk Carriers							
Universe Kure	156,649	294.67	966.75	43.33	142.17	17.45	57.25
Sigtina	72,250	250.02	820.25	32.28	105.92	13.36	43.83
Container Ships							
Sally Maersk (6600 TEU)	104,696	347.	1138.	43.	141.	14.5	47.5
Mette Maersk (2933 TEU)	60,639	294.1	964.9	32.3	106.0	13.5	44.3
Korrigan (2960 TEU)	49,690	288.60	946.83	32.23	105.75	13.01	42.67
Kitano Maru (2482 TEU)	35,198	261.01	856.33	32.26	105.83	11.99	39.33
Encounter Bay (1530 TEU)	28,800	227.31	745.75	30.56	100.25	10.69	35.08
Atlantic Crown (TEU)	18,219	212.35	696.67	27.99	91.83	9.24	30.33
Ocean Barges							
SCC 3902	50,800	177.45	582.17	28.96	95.00	12.22	40.08
Exxon Port Everglades	35,000	158.50	520.00	28.96	95.00	9.60	31.50
							(Continued)

Table V-5-4 (Concluded)

Name	Dead-weight Tonnage	Length		Beam		Draft	
		m	ft	m	ft	m	ft
Passenger/Cruise Ships							
Voyager of the Seas	142,000 (DT)	310.50	1,018.70	48.00	157.48	8.84	29.00
Grand Princess	101,999 (DT)	285.06	935.24	35.98	118.04	8.00	26.25
Imagination	70,367 (DT)	260.60	854.99	31.50	103.35	7.85	25.75

Table V-5-5
Large Tanker Classes

Name	Approximate Size	Approximate Draft
Supertanker	50,000-150,000 DWT	11-18 m (35-60 ft)
Very Large Crude Carrier (VLCC)	150,000-300,000 DWT	18-24 m (60-80 ft)
Ultra Large Crude Carrier (ULCC)	Greater than 300,000 DWT	24-30 m (80-100 ft)

(f) *Container ships* are designed to carry cargo packaged in standardized steel container boxes. These ships, increasingly dominant in world trade, travel at high speed, and rely on fast turnaround times at port. Container ship speed and size are correlated. The larger container ships cruise at speeds of 46 km/hr (25 knots). Capacity is expressed in twenty-foot equivalent units (TEU), the number of 20-ft-long containers that can be carried. Ships with 4,000-TEU capacity reach loaded drafts of about 12 m (40 ft). Until fairly recently, container ship sizes were constrained by the Panamax limit. Since economics of shipping and terminal facilities have favored a Post-Panamax size, container ships have rapidly increased in scale. The largest container ships in present operation exceed 8,000 TEU, and have limited access to U.S. ports. Vessels of 15,000 TEU are under consideration. These Post-Panamax ships have necessitated new, longer-reach gantry cranes and other new or updated terminal facilities to handle the longer, wider ships and large volumes of cargo.

(g) Other vessel types include: *LASH* (Lighter Aboard Ship), *SEABEE*, and *BARCAT* vessels, designed to transport barges; *Ro/Ro* (Roll on/Roll off) carriers, essentially large, oceangoing ferries that load and unload wheeled cargo (trailers and/or vehicles) via ramps extending from the vessel; conventional ferries; passenger vessels; barges; etc. The *Integrated Tug/Barge (ITB)* is a special adaptation of barge design in which the barge resembles a vessel hull and a tug can be linked to the barge stern to form, in effect, a single vessel. ITB applications are usually dry and liquid bulk cargo transport.

(h) U.S. military vessels generally have maximum drafts of less than 12 m (40 ft). Nimitz class aircraft carriers have maximum draft of 12.5 m (41 ft). U.S. Navy vessel characteristics are available on the Internet (http://www.nvr.navy.mil) and in the NAVFAC Ships Characteristics Database soon to be on the Internet.

(i) Shallow-draft vessels are typically recreational or small fishing vessels. Recreational boats in the United States can range in length from about 3.6 m to 60 m (12 ft to 200 ft), but they are commonly 9-14 m (30-45 ft) in length with beams of 4.6 m (15 ft) or less. Recreational boats often have features protruding from the bow or stern. Although such features may not be included in the nominal boat length, they should be considered as needed in sizing harbor channel and dock clearances. Shallow-draft vessels may be driven by either engine power or sail. In comparison to powerboats, sailboats have narrow beam and require large maneuvering space when under sail.

(5) Form coefficients. Vessel shape is conveniently represented in terms of simple parameters known as *form coefficients*. The most important of these is the *block coefficient*, defined and illustrated in Figure V-5-4. This coefficient usually represents the fully loaded ship. Values of the block coefficient can normally range from around 0.4 for tapered-form, high-speed ships, such as container ships and passenger ferries, to 0.9 for box-shaped, slow-speed ships, such as tankers and bulk carriers. Small craft, sailboats, and power boats, respectively, represent forms with relatively low and high block coefficient.

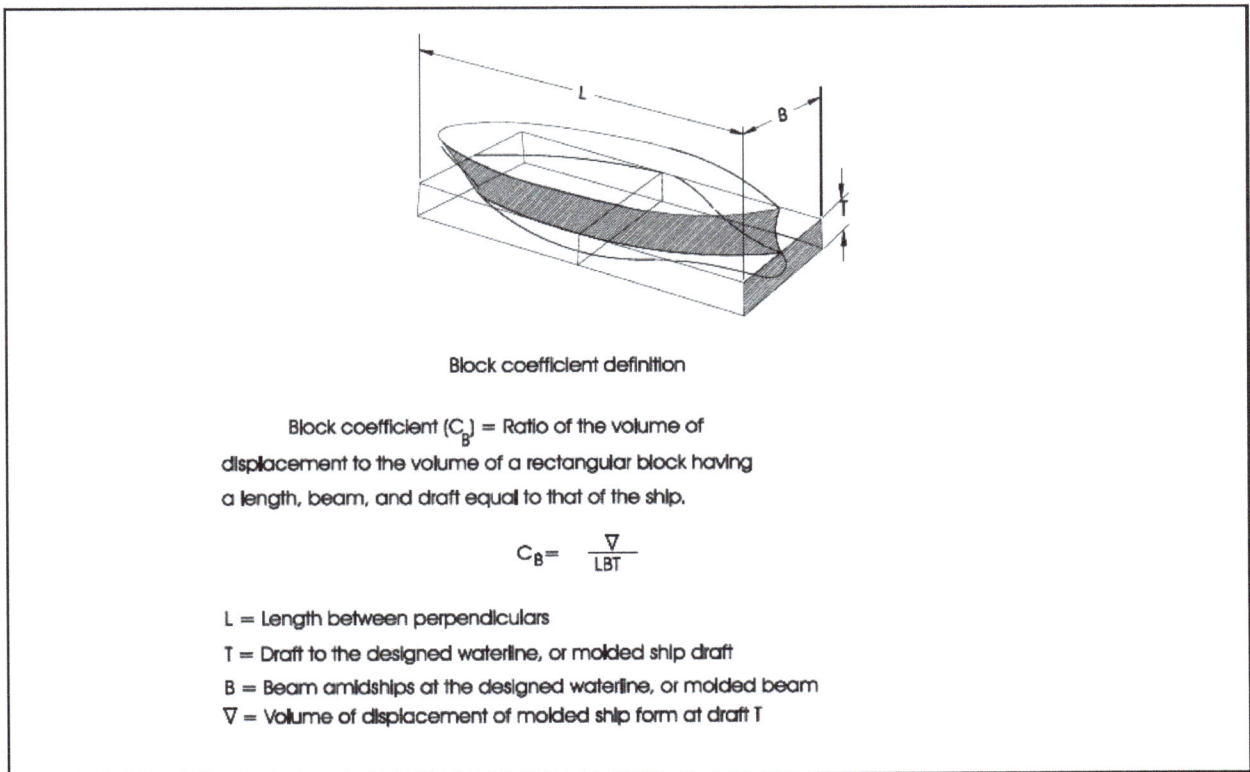

Block coefficient definition

Block coefficient (C_B) = Ratio of the volume of displacement to the volume of a rectangular block having a length, beam, and draft equal to that of the ship.

$$C_B = \frac{\nabla}{LBT}$$

L = Length between perpendiculars

T = Draft to the designed waterline, or molded ship draft

B = Beam amidships at the designed waterline, or molded beam

∇ = Volume of displacement of molded ship form at draft T

Figure V-5-4. Block coefficient definition

(6) Ship speed. Typical transit speeds in deep-draft channels are between 9 and 18 km/hr (5 and 10 knots). Vessel speed in navigation projects often represents a balance between several important considerations, as follows:

(a) Considerations favoring higher vessel speed:

- *Economics.* Vessel productivity increases when transits are faster; loaded vessels may be able to use high tide levels to advantage.

- *Vessel control.* Vessel control in the presence of wind, waves, and/or currents improves when vessel speed is higher.

- *Convenience.* Particularly for small craft, operators and passengers usually prefer quick transits.

(b) Considerations favoring lower vessel speed:

- *Wake effects.* Vessel wakes, directly related to vessel speed, can endanger other vessels and operations and erode banks.

- *Reduction of bank and bottom effects, ship resistance, ship-ship interactions.* Vessels may need to limit speed to avoid creating dangerous, speed-induced pressure differences. The effects are due to constricted clearance between the vessel and other obstacles, such as the bottom, side banks, another vessel in transit, and moored vessels.

- *Safety.* As long as vessels maintain adequate control, lower speeds are generally safer. Typically, vessel speed relative to the water must be at least 4 knots for both deep-draft vessels and small craft.

(7) Maneuverability. Commercial ships are designed primarily for optimum operation in the open ocean. Many of them maneuver poorly in confined areas. A successful navigation project must accommodate the ships using it. Ships are controlled by propellers and rudders at the stern. Some ships are also equipped with bow thrusters or bow and stern thrusters, which aid in control, especially at low speeds. Often, one or more tugs are needed to assist ships in some phases of entering and leaving a port. Control is especially crucial when ships slow to turn, dock, or attach tugs. A navigation project objective is to design ports and approach channels so ships can maintain adequate speed and control and navigate under their own power as much as reasonably possible. Small craft generally respond to engine, sail, and rudder control much more readily than deep-draft vessels. However, as with deep-draft vessels, small craft can encounter conditions in which control is difficult. Factors contributing to loss of control include slow vessel speed, following currents, waves, and cross-wind. Sailboats traveling under sail require extra maneuvering space.

b. Vessel operations. Deep-draft navigation projects are built or improved to enhance safety, efficiency, and productivity of waterborne commerce in U.S. ports and harbors. Shallow-draft projects embody similar concerns and often public recreational access as well. An understanding of vessel operations is critical to successful navigation project design.

(1) Navigation system. Port and harbor operations can be viewed as a system with three main components, as follows:

- Waterway engineering: Navigation channels, environmental factors, dredging and mapping services, shore docking facilities.

- Marine traffic: Operational rules, aids to navigation, pilot and tug service, communications, and vessel traffic services.

- Vessel hydrodynamics: Vessel design, maneuverability and controllability, human factors, navigation equipment.

(a) These components are closely interrelated in a navigation project. Tradeoffs between investment in the components are normal procedure, particularly in deep-draft projects. Thus, for example, channel design is strongly influenced by ship sizes and available accuracy of aids to navigation.

(b) Overall economic optimization of a navigation system can be a complex process. It typically involves crucial tradeoffs between initial investment (e.g. channel dimensions), maintenance, and operational use. For example, a channel that is wide enough for two-way traffic will cost more to dredge and, possibly, to maintain than a one-way channel of the same depth. However, the two-way channel may significantly reduce the amount of time ships must queue while waiting for access to the channel.

(2) Typical operations. Methods of operation must be considered in developing a navigation project. For deep-draft ships, operations depend on interactions between a pilot, captain, crew, and, often, one or more tug captains. On arrival at the entrance to a port, a ship typically is met by a local pilot. The pilot boards the ship near the seaward end of the entrance channel. Boarding is usually accomplished by pulling a small pilot boat next to the ship long enough for the pilot to mount a rope ladder and climb up to the ship deck, a potentially hazardous maneuver during high waves. Local tug services are contacted if needed, and plans finalized for the ship transit. Many tug companies also provide a tug pilot to accompany the local pilot and assist in the tug-aided final phase of transit and docking.

(a) The pilot is stationed on the ship bridge. The pilot effectively takes control of the ship during transit, issuing rudder and engine commands as well as course orders. Transit to a port generally follows a series of straight segments connected by turns. Turn angles greater than about 30 deg require special care because they involve varying currents and changing ship speed and position relative to banks and prevailing wind. Port entrance channels can be especially troublesome due to crosscurrents, waves, shoaling, and wind effects.

(b) A large ship in a confined channel can be difficult to control because ships do not respond quickly to rudder and engine commands. Turning may be sluggish. Bank effects and encounters with passing ships can introduce forces to turn the ship away from the intended travel direction. Such factors, along with human and environmental variability, result in variations in a ship's *swept path* (the envelope of all positions in the channel over which some part of the ship has passed). The swept path is illustrated by a ship simulator study example of ship position at short time increments during transit around a turn, up a channel, and through a turning basin to a dock (see Part V-5-10b for additional discussion of this simulation study) (Figure V-5-5).

(c) The ship must slow down well before approaching the berth or terminal, usually with the assistance of tugs when ship control is lost (at speeds below 6-7 km/hr or 3-4 knots). Often, the ship must pass other port facilities at very slow speed to prevent waves and moored vessel damage. As the ship approaches its berth, tugs typically take full control and push the ship against the dock face while mooring lines are made fast. When the ship departs, operations during a typical outbound run are similar to the inbound run, except in reverse sequence.

(d) Pilots and captains take care to avoid contact between the ship and bottom. However, ship motions and bottom conditions are not entirely predictable, and bottom contact occasionally occurs. Typical consequences are hull abrasion and propeller and rudder damage. Propeller/rudder damage reduces or removes ship control, leaving the ship vulnerable to further damage. It is also costly to repair. Therefore, pilots tend to be very protective of the ship stern when maneuvering in confined channels and turning basins.

(e) Ships may transit in fully loaded, partially loaded, or in ballast condition. The loading condition influences operational concerns. A fully loaded ship has a relatively large fraction of its volume submerged. Hence it is susceptible to currents, shoals, and other bottom influences. A ship in ballast generally has generous bottom clearance, but has a relatively large exposure to wind forces. Some types, including container ships, ferries, and Ro/Ro vessels, have large wind exposure, even in a loaded state.

(f) Sometimes ships approach port with loaded drafts greater than the channel depth, stop in an anchorage area, and offload to smaller vessels (*lighters*). Ships may fully unload to lighters or partially unload to reach an acceptable draft, after which they continue into port.

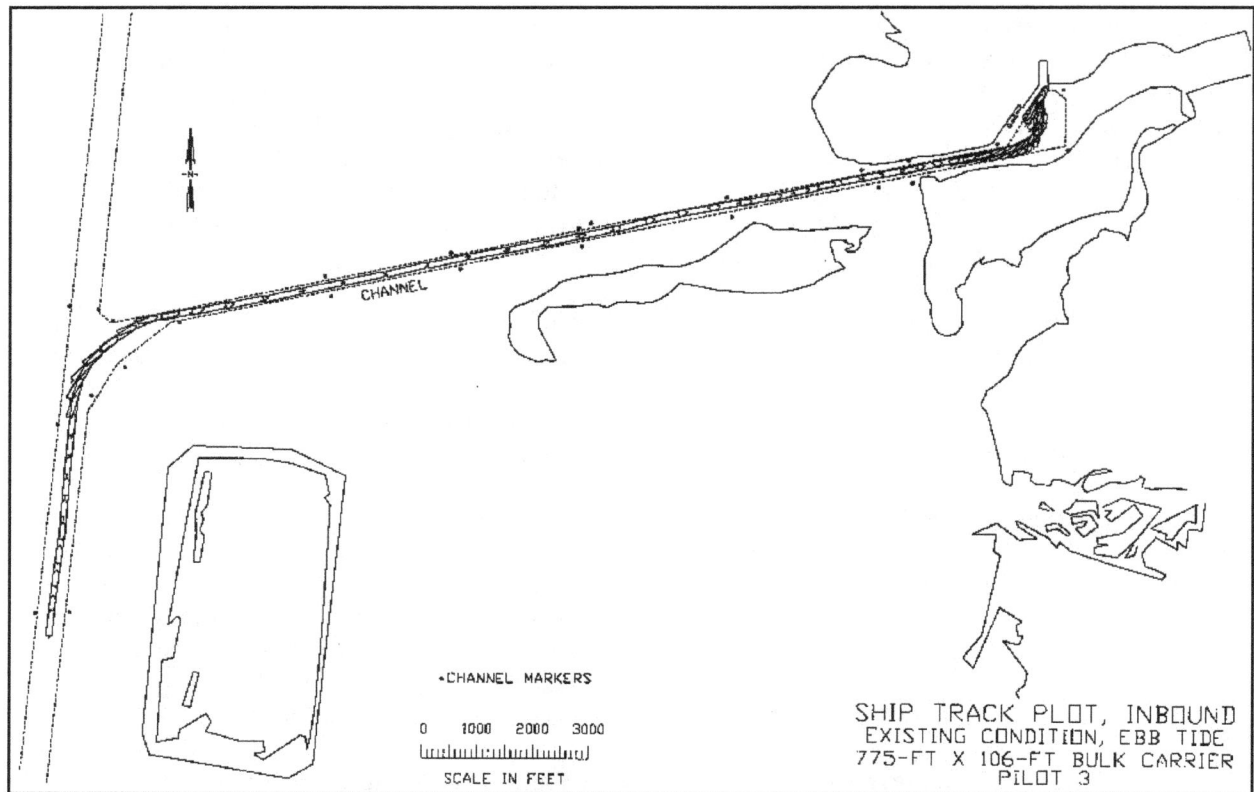

Figure V-5-5. Example ship track, Alafia Channel, Florida

(g) Deep-draft entrance and interior channels are designed for either one-way or two-way traffic. When a ship is entering a port with a one-way channel, any outbound ships must wait until the channel is clear. Two-way channels may accommodate inbound and outbound traffic simultaneously. They may also provide generous horizontal clearance for single ships when passing ships are absent.

(h) For small craft, operational concerns can vary significantly depending on the type of harbor. Small craft travel under their own power, controlled by an operator whose level of expertise and experience can range from novice to seasoned professional. Power boats are typically driven by one or two engines. Sailboats may be equipped with small auxiliary engines for transiting congested harbor areas and for emergency use, but they usually travel under wind power. Depending on wind speed and direction relative to desired travel direction, sailboat operators often must follow a zig-zag course. Small craft operators may take advantage of bays and other protected areas for fishing, sailing, etc., when available, especially if waves along the open coast are rough.

(i) Small craft typically exit the harbor for fishing and/or recreation and return to the same harbor, often on the same day. The number of vessels concurrently using a small-craft harbor during high traffic times can be greater than for deep-draft ports. Level of usage is often affected by holidays, weather, fishing or charter schedules, work schedules, etc. For example, a recreational harbor can be expected to have a higher volume of traffic during fair weather and weekend or holiday times. These same conditions are likely to result in a greater percentage of inexperienced operators and, possibly, a lower level of attention to navigation safety.

(3) Shallow-water, restricted channels. Vessels operating in restricted navigation channels can experience a variety of effects (Figure V-5-6). Large, loaded ships can block much of the channel cross section

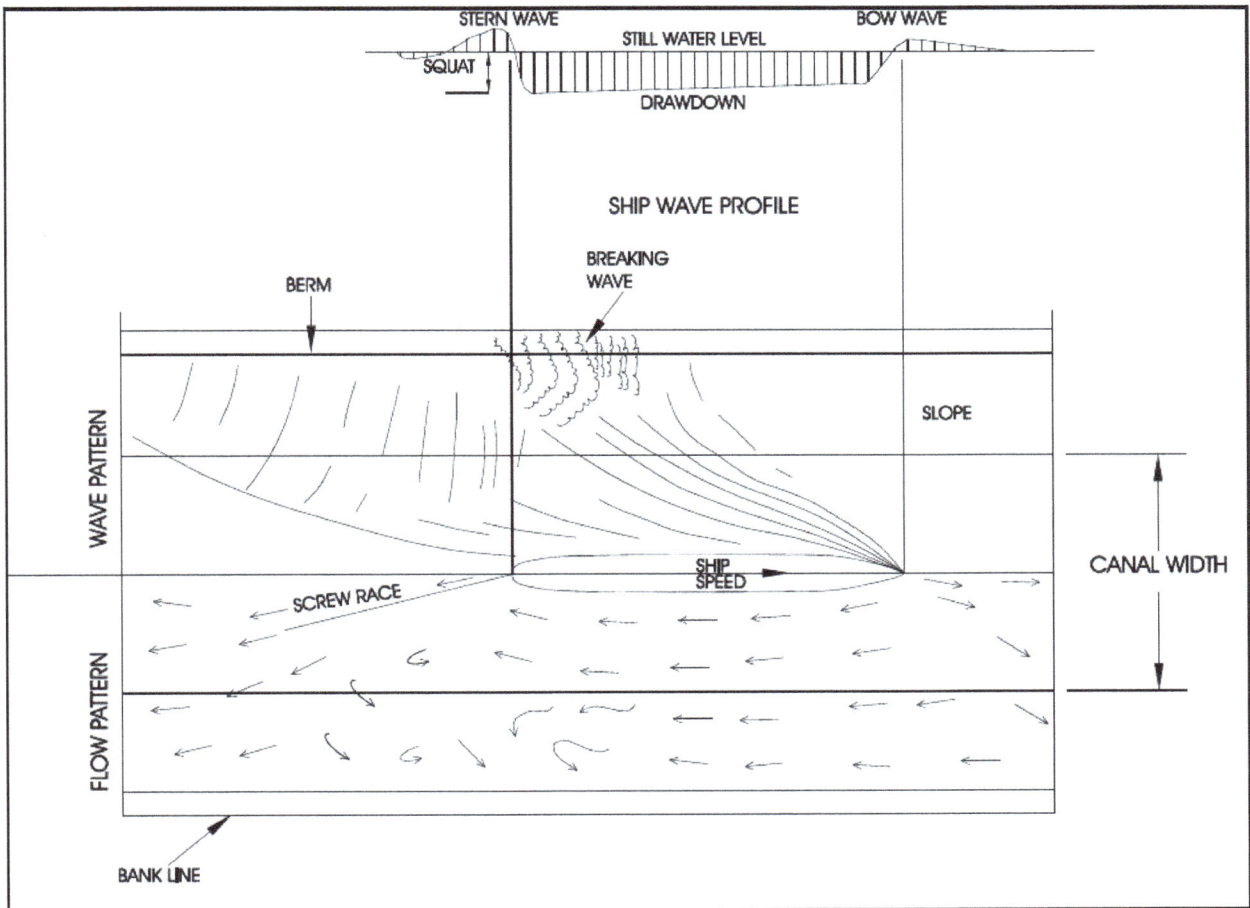

Figure V-5-6. Ship wave and flow pattern in a canal

and encounter very significant hydrodynamic resistance. Much of the water in the channel cross section must flow around the passing ship through highly confined space under the hull and between the hull and channel side slopes. Also, the ship experiences resistance due to the waves it creates as it moves forward.

(a) A moving vessel in a shallow waterway drops in the water relative to its at-rest level. The drop, referred to as *squat*, is due to a reduction in pressure exerted by water flowing around the moving vessel. Squat includes an overall lowering of the vessel *(sinkage)* and a motion-induced trim. Squat further reduces clearance between the vessel hull and the bottom.

(b) The maximum speed a vessel can attain in shallow water is significantly reduced from the typical deepwater speed. An important parameter governing shallow-water effects on a moving vessel (both deep- and shallow-draft vessels) is the depth Froude number

$$F_h = \frac{V}{\sqrt{gh}} \qquad (V-5-1)$$

where

F_h = channel depth Froude number

V = vessel speed

h = depth of channel or shallow-water area

g = acceleration due to gravity; = 9.80 m/sec^2 (32.2 ft/sec^2)

(c) Consistent units must be used for V, h, and g in the equation. Vessel resistance becomes very high as F_h approaches unity. In practice, a normally self-propelled merchant ship would never operate at F_h greater than about 0.6.

(d) The effect of a restricted, shallow *channel* configuration is to further increase wave effects, squat, and vessel resistance. Relative channel restriction is characterized by the *channel blockage ratio*

$$B_R = \frac{A_C}{A_S}$$

(V-5-2)

where

B_R = channel blockage ratio

A_C = channel cross-section area

A_S = vessel submerged cross-section area, = BT

B = vessel beam at midship

T = vessel draft

(e) The channel blockage ratio is illustrated in Figure V-5-7 for the extreme case of a canal with vertical sides. In this case, $A_C = Wh$, where W is channel width. The limiting ship speed for self-propelled vessels in a canal, represented as F_h, is shown as a function of B_R in Figure V-5-7. Typical B_R values range from 2 for very restricted cases to 20 or more for open channels, giving F_h values between 0.2 and 0.7. This effect significantly limits vessel speed in restricted channels. The limit is known as the *Schijf limiting speed*. For example, with $B_R = 3$ at a 12-m (40-ft) water depth, the maximum ship speed is 11.9 km/hr (6.4 knots). In practice, a ship's engine would not have enough power to drive the ship at the Schijf limiting speed.

(f) Bank effects in a channel can make ship control more difficult by creating suction and/or higher pressures along vessel hulls when vessels are off the channel center line. Forces may differentially affect vessel bow and stern and act to turn the vessel toward a potentially hazardous crosswise position relative to the channel. Bank effects become stronger as *channel overbank depths* (water depth of the natural bottom adjacent to the channel) decrease. Pilots sometimes take advantage of bank effects to assist in turning. Similar differential pressure effects arise when ships pass in a channel and when a ship passes a moored ship adjacent to the channel.

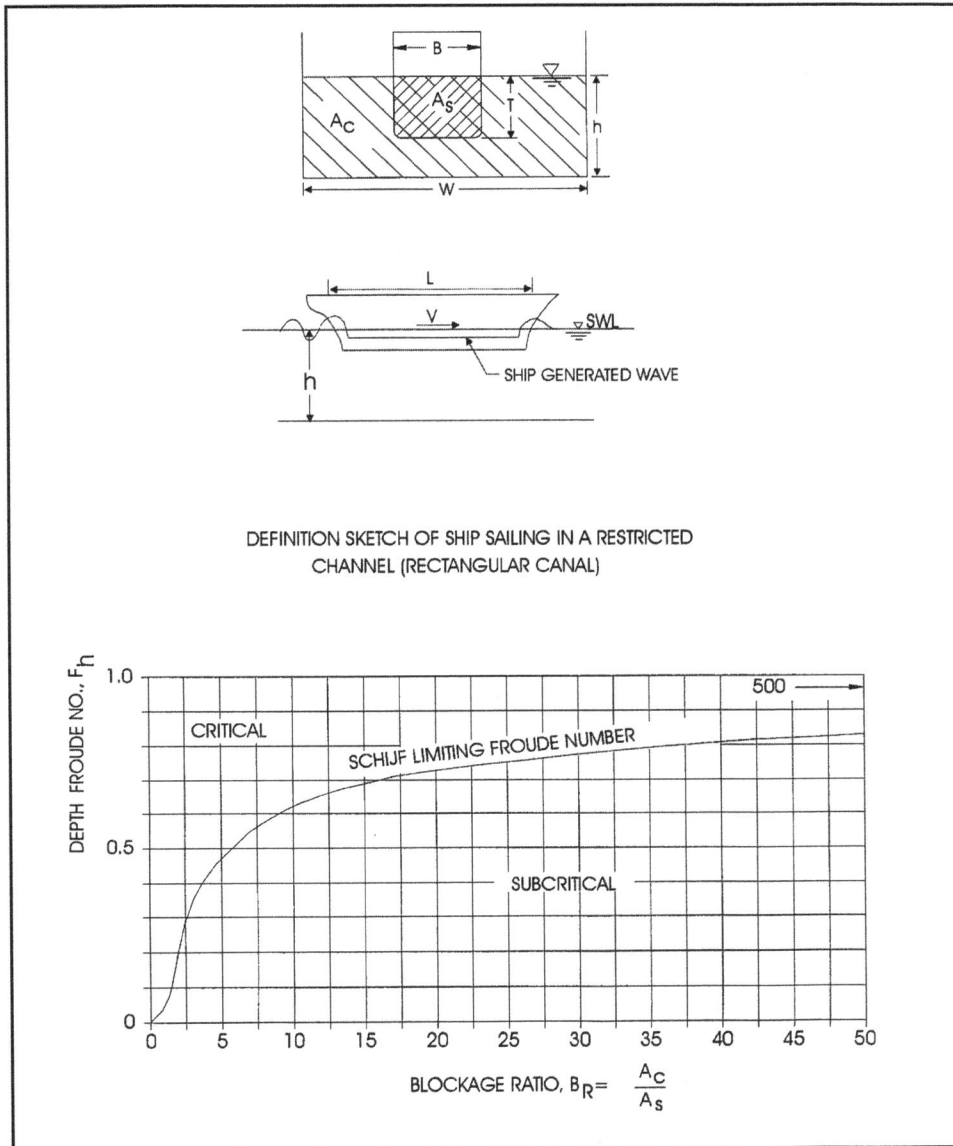

Figure V-5-7. Ship-limiting speed in a canal

(4) Ice navigation. Winter conditions along northern sea coasts, estuaries, and large lakes can cause ice to be an occasional, if not chronic, concern for safe and efficient navigation (PIANC 1984, USACE 1990a). About 42 percent of the earth experiences temperatures below freezing during the coldest month of any year (Figure V-5-8). The presence of ice is accompanied by longer nights and increased fog and precipitation (Figure V-5-9). Shipboard mechanical equipment, instruments, and communications apparatus are less efficient and more prone to failure in cold temperatures. Aids to navigation become less effective, and maneuvering in ice is much more difficult. Navigation projects in northern areas should be designed with consideration of these difficult conditions.

(a) Sea ice nomenclature and map notation symbols are defined by the World Meteorological Organization (WMO 1970). Ice thickness and structure are key concerns. *Multi-year ice*, sea ice more than 1 year in age, can be over 3 m thick, but it is found only in the Arctic Ocean and its marginal seas and near Antarctica.

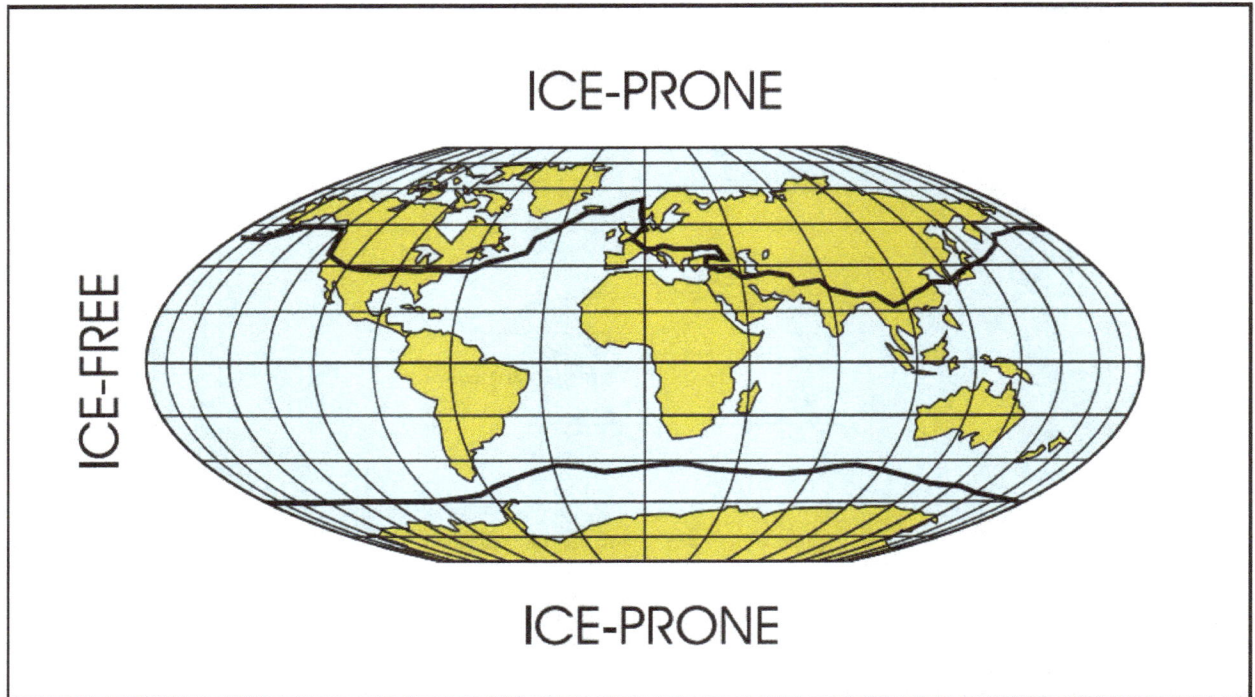

Figure V-5-8. Cold regions of the world

Figure V-5-9. Fog over frozen waterway (photograph, courtesy of Orson P. Smith, University of Alaska, Anchorage, Alaska)

First-year ice is generally classified as *young ice* (0-10 cm), *thin* (10-30 cm), *medium* (30-70 cm), or *thick* (70-120 cm).

(b) Ships that regularly navigate icy waters must have exceptional structural strength and propulsion power for safety of the crew, equipment, cargo, and environment. Special hull, propeller, and rudder designs reduce resistance and help clear lanes. Icebreakers may be needed to escort ice-strengthened cargo vessels or to periodically clear shipping routes. Shallow- and deep-draft ice-strengthened ships are in service around the world, classified for ice navigation by the ABS or several other agencies in Canada, Russia, and Europe.

(c) Factors to be considered for ship operation in ice rather than temperate conditions are:

- Ship maneuverability is retarded.

- Ice forces can divert ships from their intended course.

- Darkness is more common.

- Low visibility is more common (fog and precipitation).

- Winds can be very strong.

- Visual aids to navigation are less effective.

- Shipboard instruments are more prone to malfunction.

- Assistance or rescue by tugs is more difficult.

- Crews are more strained and fatigued in the face of these challenges.

c. Design considerations. Deep-draft navigation projects are typically formulated to provide safe and efficient passage for a selected ship under specified transit conditions. The design ship and transit conditions may be selected to represent the "maximum credible adverse situation," the worst combination of conditions under which the project would be expected to maintain normal operations. A project that successfully accommodates this situation can be expected to perform well with a full range of smaller ships and less difficult transit conditions. If future needs require it, the project may also accommodate ships larger than the design ship under milder transit conditions than the design scenario. Shallow-draft navigation projects are designed to safely accommodate the variety of small craft anticipated during the project design life, typically around 50 years.

(1) Design vessel. For deep-draft projects, the design ship or ships are selected on the basis of economic studies of the types and sizes of the ship fleet expected to use the proposed channel over the project life. The design vessel or vessels are chosen as the maximum or near maximum size ships in the forecast fleet based on the characteristics (length, beam, draft) of the ships being most representative of the potential economic advantage to be found in the forecast ship fleet.

(a) For small craft projects, the design vessel or vessels are selected from comprehensive studies of the various types and sizes of vessels expected to use the project during its design life. Often, different design vessels are used for various project features. For example, sailboats, with relatively deep draft, may determine channel depth design; and fishing boats, with relatively wide beam, may dictate channel width design.

(2) Design transit and mooring area conditions. Operational conditions selected for design can strongly affect a navigation project. A deep-draft project should be designed to allow the design ship to pass safely under design transit conditions. Normally, extreme events are not considered in specifying design transit conditions. Ship operators can usually suspend operations during these rare events without undue hardship. Some important exceptions for which extreme events may need to be considered include ships under construction or in repair facilities, inactive vessels, and USCG vessels.

(a) Operational factors to be specified for design transit include:

- Wind, wave, and current conditions.

- Visibility (day, night, fog, and haze).

- Water level, including possible use of tidal advantage for additional water depth.

- Traffic conditions (one- or two-way, pushtows, cross traffic).

- Speed restrictions.

- Tug assistance.

- Underkeel clearance.

- Ice.

(b) The use of tidal advantage in specifying design transit conditions allows for reduced channel depth, since the design ship would be constrained to transit during a high tide level. Channel length and vessel speed determine vessel transit time. The channel must provide the necessary water depth during at least the transit time period. If tidal advantage is included in design, the optimum design depth may be based on an analysis of water level probabilities, costs (vessel delays, dredging and disposal, etc.), and benefits.

(c) Normal operational conditions are strongly influenced by individual, local pilot, and pilot association rules and practices. For example, pilots usually guide a transit only when conditions allow adequate tug assistance. There may be operational wind, wave, or current limitations on the ability to safely moor a ship at a terminal or berth. Turning operations and maneuvering into a side finger slip may be limited by tide level and current conditions, including river outflow. Energetic wave conditions at the seaward end of the entrance channel may prohibit pilots from safely transferring between the pilot boat and ship. Such operational limitations may well be controlling factors in determining whether or not a safe transit is possible, and navigation project design should be consistent with these limitations.

(d) Design transit conditions for small-craft projects include vessel maneuverability (particularly if parts of the project will accommodate vessels under sail), traffic congestion, wind, waves, water levels, and currents. Design wave criteria are usually expressed as significant wave height, in probabilistic terms, for the entrance and access channels and mooring areas. Typical criteria are:

- Mooring areas: Significant wave height will not exceed 0.3 m (1 ft) more than 10 percent of the time.

- Access channels: Significant wave height will not exceed 0.6 m (2 ft) more than 10 percent of the time.

(e) Final design criteria for small-craft projects should be determined on the basis of economic optimization of the complete project (Part V-5-4).

V-5-3. Data Needs and Sources

a. Introduction. Planning and design of a navigation project are based on a wide range of information about the project area. The necessary data and sources are briefly reviewed in this section. More detailed information about data types and sources is given in Part V-2, "Site Characterization," and Part II-8, "Hydrodynamic Analysis and Design Conditions."

b. Currents. Currents in navigation channels and other project features can strongly affect vessels. Currents may also be important in navigation structure design. Currents of concern are usually tidal circulations or river flows.

c. Water levels. Water levels are essential for determining design depth in channels and other navigation areas. Water level variations are usually due to tides, river discharge, or lake levels. NOAA predicts tides and tidal currents at primary reference stations by the method of harmonic analysis. Phase and amplitude corrections to reference station predictions are available for many other secondary stations (NOAA *Tide Tables* and *Tidal Current Tables*, annual). Interpolation between secondary stations is often practical, if no major constrictions or confluences are present. Commercial software is available that can predict tides at hourly intervals at secondary stations. A similar analysis can be performed by applying the methods of Harris (1981).

d. Wind. Wind forces can strongly influence both vessels under way and moored vessels. Often an airport wind station located in the general area of a project can be used as a source of representative wind data.

e. Waves. Waves can have a major impact on navigation in exposed channels, particularly vertical excursions of the vessel and channel depth requirements. Deep-draft vessels respond to wave periods typically found in exposed ocean and Gulf of Mexico waters. Small craft respond to a wide range of wave periods, and waves can be especially troublesome if they break in the channel. Waves are also important for navigation structure design and for predicting channel shoaling.

f. Water quality. Water quality and potential changes in water quality may become issues for projects creating more enclosed harbor areas, where circulation and flushing may be reduced as a result of the project.

g. Sediments. Sediment characteristics in project and adjacent areas are needed. Bottom materials in any areas to be dredged are especially important. Sediment quality can dictate disposal options and costs ranging from beneficial uses such as beach fill to confined or capped disposal of contaminated material.

h. Bathymetry and sediment transport processes. Bathymetry and sediment transport processes are needed to determine baseline conditions, optimum project location, initial dredging quantities, channel shoaling rate and maintenance needs, and potential project impact on adjacent shorelines. Methods for predicting channel sedimentation are reviewed by Irish (1997).

i. Ecological processes. Navigation projects can have an impact on ecological processes. Baseline conditions may need to be established.

j. Local coordination. Local people familiar with the project area and/or using an existing project regularly can provide valuable insight about present conditions and impact of any modifications.

(1) Pilot interviews. Deep-draft navigation project planners/designers should develop strong coordination with local pilot groups. Pilot interviews can be used to determine users' opinions about existing

channel navigation safety, suitability of design transit conditions, and feasibility and safety of proposed channel design alternatives.

(2) U.S. Coast Guard. The local USCG office should also be contacted early in the project development to solicit their views on channel dimensions and alignment for safe navigation. They can also provide guidance on placement of aids to navigation.

(3) Accident records. Marine accident records are available from the USCG annual compilation of casualty statistics in an automated system called Coast Guard Automated Main Casualty Data Base (CASMAIN). Accident data on existing navigation channel projects proposed for enlargement or improvement should be studied. USCG and National Transportation Safety Board special investigation reports are available for some accidents, which can provide insight on navigation problems.

V-5-4. Economic Analysis

a. A number of alternatives can usually be defined to meet navigation design requirements. Alternatives should generally include a range of channel depths, since depth is one of the major cost-determining parameters. For example, a proposed navigation channel may suffice for the design ship and design transit, but lead to undesirable ship delays and queueing because of heavy ship traffic and limited channel capacity. The adaptability of each alternative for meeting future navigation requirements should also be considered. The final design is usually selected from among the alternatives to maximize economic benefits. USACE navigation project evaluation procedures are described by USACE (1990b).

b. A complete design approach includes consideration and optimization of the full navigation system described in Part V-5-2.b. The comprehensive design process, beyond the scope of this chapter, is discussed by PIANC (1997a).

c. For economic optimization, the cost of each alternative design is estimated. Costs associated with all the elements of developing and maintaining the project should be included. Normally, several alternative channel alignments and widths, as well as depths, are represented. Navigation structures may also be part of some or all alternatives. Costs include initial construction (dredging, dredged material disposal, aids to navigation, breakwaters, jetties, etc.), replacement cost, and operation and maintenance.

d. Benefits for deep-draft projects are determined by transportation savings, considering ship trip time (including loading/unloading time, which may include lightering), cargo capacity, and delays due to project limitations. Benefits are evaluated by determining the transportation costs per ton of commodity for each increment of channel depth. Transportation costs are based on ship annual operating cost for each type of ship, including fixed cost and annual operating expenses. Data on ship operating costs are periodically compiled by the USACE Water Resources Support Center. Benefits may result from:

(1) Use of larger ships.

(2) More efficient use of large ships.

(3) More efficient use of present ships.

(4) Reductions in transit or delay times.

(5) Lower cargo handling costs.

(6) Lower tug assistance costs.

(7) Reduced insurance, interest, and storage costs.

(8) Use of water rather than land transport mode.

(9) Reduction of accident rate and cost of damage.

e. The USACE evaluation procedure to estimate navigation benefits includes nine individual steps (Figure V-5-10). Accurate projection of commodity movements over the proposed alternative project design (steps 3 and 7) is key to evaluation. Details of the procedure are given by USACE (1990b).

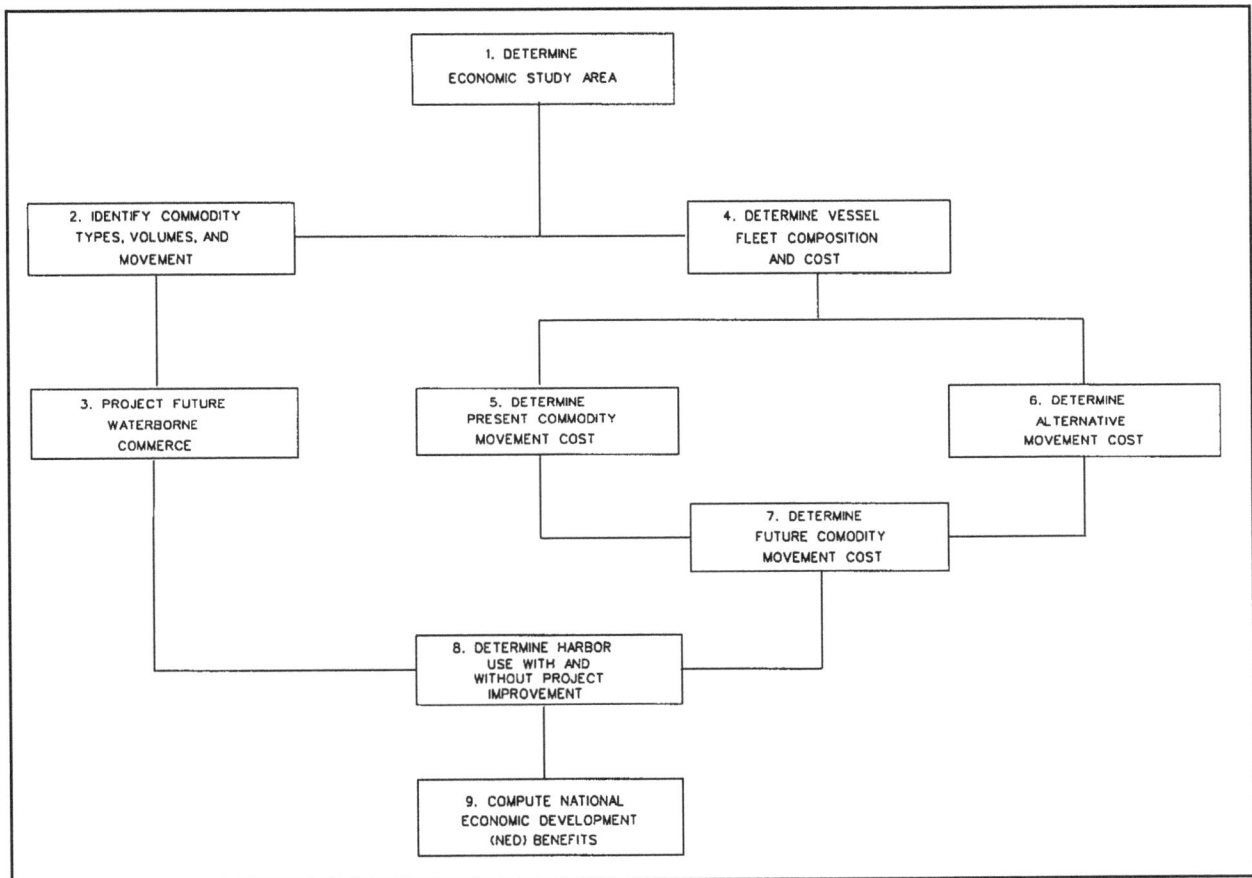

Figure V-5-10. Deep-draft navigation benefit evaluation procedure

f. Small-craft harbor projects follow a similar economic optimization procedure. Generally, several entrance channel and basin configurations are identified as alternatives. The alternatives should provide varying protection and accommodations so an optimal alternative can be selected. The cost for each alternative is estimated, including initial cost, maintenance cost, and social and environmental aspects. Benefits are also estimated. The alternative that maximizes net benefits (difference between benefits and cost) is usually the preferred project plan. After the alternative with optimal level of protection and size has been determined, then the most economical way of providing that protection and size should be developed.

V-5-5. Channel Depth

a. Introduction. Channel depth is a key factor in the cost and usability of a navigation project. It should be adequate to accommodate present and expected traffic. Typically it is chosen on the basis of economic optimization. The design channel depth need not be constant throughout the project. It can, and often does, vary in segments as needed to allow the design vessel or vessels to make safe and efficient transits in a cost-effective project.

(1) Vessels navigating in shallow water encounter a variety of channel cross sections over the length of a navigation project. Since channel cross section can significantly affect natural processes and vessel behavior, it is useful to define characteristic types (Figure V-5-11). A *canal* has an enclosed cross section with exposed land adjacent to both sides of the channel. A *trench* is a deepened passage with submerged overbanks on either side. A *fairway* is a passage with no lateral constraints.

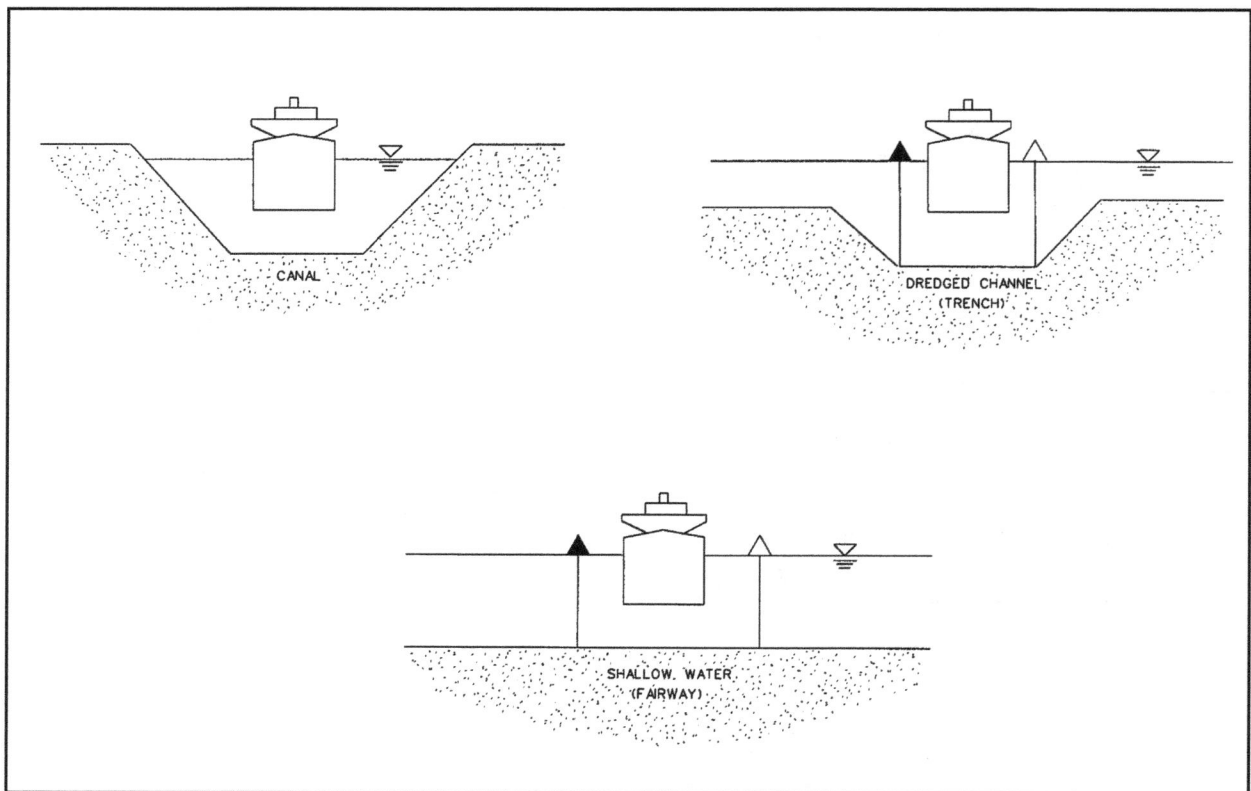

Figure V-5-11. Definition of channel types

(2) Channel depth for both deep- and shallow-draft navigation projects may be determined by figuring a depth increment for each of the important factors affecting vessel underkeel clearance requirements and adding those to the design vessel draft (Figure V-5-12). This depth, required for safe vessel passage, provides a basis for Congressional authorization of Federal channel depth, referred to as the *authorized channel depth*. The dredged channel depth, or *contract depth*, generally exceeds the authorized depth to accommodate potential sedimentation and maintain navigability. In some projects, consideration must also be given to the *permitted depth*, the extreme dredging depth permitted by regulators. The same factors generally apply for both deep- and shallow-draft projects, but their relative importance and estimation procedures differ somewhat, as discussed in the following paragraphs.

b. Effect of fresh water. The nominal draft of seagoing ships usually represents the seawater environment. When ships enter channels and ports in brackish or fresh water, ship draft increases due to the lower water density. The draft increase between ocean and fresh water is 2.6 percent. In USACE practice, a maximum depth allowance of 0.3 m (1 ft) may be included for this effect (Table V-5-6) (see Figure V-5-12). This maximum allowance corresponds to the draft increase between ocean and fresh water for a vessel with an 11.6-m (38-ft) draft. The freshwater effect is generally not considered in small craft navigation projects.

Table V-5-6
Depth Allowance for Freshwater Effect

Port Location	Allowance	
	m (ft)	Percent
Brackish water	0.15 (0.5)	1.3
Fresh water	0.3 (1.0)	2.6

c. Vessel motion from waves. Vessel vertical motion in response to waves must be considered in design of channel depth at exposed locations (see Figure V-5-12). Entrance channel design depth is typically greater than interior harbor channel depth because of the need to accommodate wave-induced vertical vessel motions. Wave effects tend to increase as wave height increases and decrease with longer vessel length. Maximum vessel response occurs with wavelengths approximately equal to vessel length. Most deep-draft ships are relatively unaffected by very short-period waves but respond when periods are longer than around 6-8 sec.

(1) Vessel motions that affect channel depth are roll, pitch, and heave (Figure V-5-13). Roll is most important when waves are perpendicular to the vessel travel direction (*beam seas*). Pitch and heave are most important when the vessel and wave travel directions are colinear (*head sea* or *following sea*). These motions can have a large impact on deep-draft channel depth requirements. For example, a pitch angle of 1 deg increases the extreme excursion of a 300-m- (1000-ft-) long ship by 2.7 m (9 ft). A 5-deg roll of a ship with a 46-m (150-ft) beam can increase extreme excursion by 2.1 m (7 ft).

(2) Vessel response to waves depends on the combined effects of:

(a) Wave height, period, wavelength, and propagation speed.

(b) Wave direction relative to vessel.

(c) Vessel length and beam.

(d) Vessel speed.

(e) Natural periods of vessel roll, pitch, and heave.

(f) Vessel draft and underkeel clearance.

(g) Channel depth and overbank depth.

(h) Wind and currents (speed and direction relative to vessel).

(i) Pilot strategy.

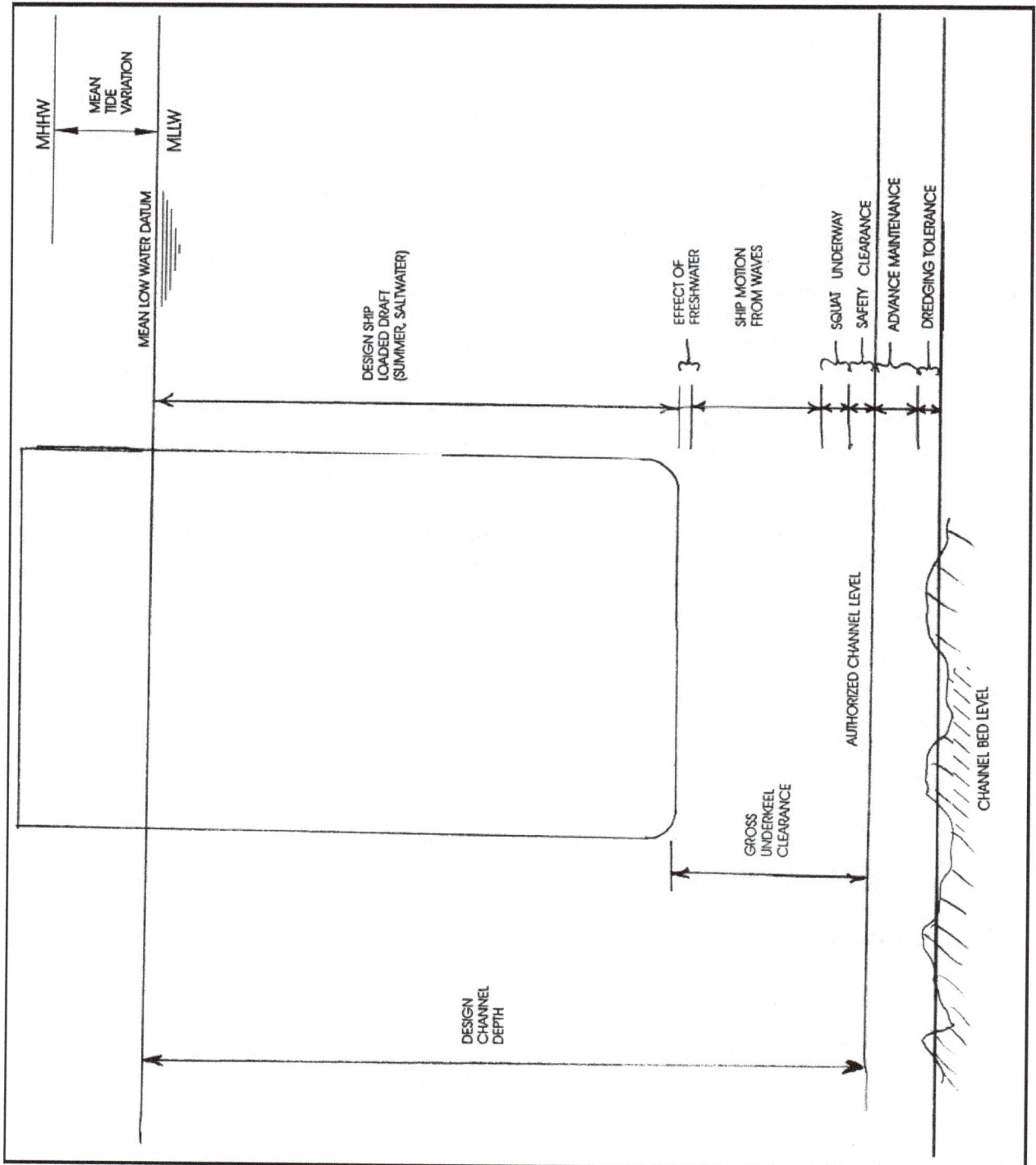

Figure V-5-12. Channel depth allowances

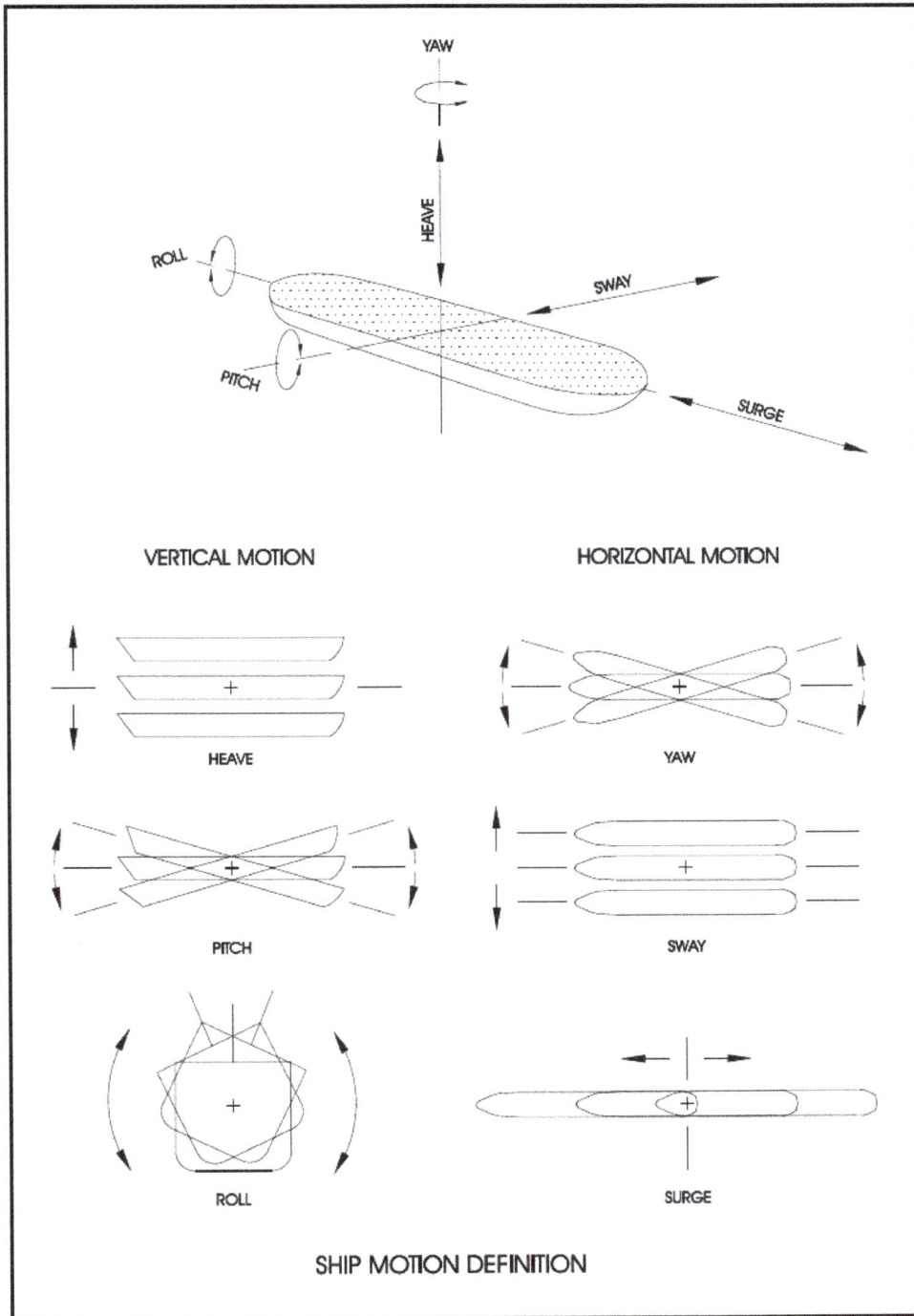

YAW

HEAVE

ROLL

PITCH

SWAY

SURGE

VERTICAL MOTION

HORIZONTAL MOTION

HEAVE

YAW

PITCH

SWAY

ROLL

SURGE

SHIP MOTION DEFINITION

Figure V-5-13. Ship motion definitions

(3) Many of these factors, most notably wave height, vary as a vessel transits a channel. The net effect of these factors on a vessel is difficult to estimate analytically. The design objective is to achieve a channel depth that allows the design vessel to transit the project with a very low probability for any damaging contact with the channel bottom.

(4) For deep-draft projects, the depth allowance needed to accommodate wave-induced ship motions is a major concern for which no easy, accurate solutions are available. Because of the magnitude of depth allowance for waves in many exposed entrance channels and because waves are highly variable, even within a specified sea state, it is prudent for final design to review existing data for similar ships and, in many cases, to conduct studies to develop realistic estimates. Options include (Part V-5-10):

(a) Analytic studies, using strip theory or other theoretical calculation methods as developed by naval architects.

(b) Interactive, real-time ship simulator studies.

(c) Physical model studies, using radio-controlled, free-running scaled ship models with wave response measurements.

(d) Direct, onboard ship measurements while transiting through the entrance channel.

(5) An example of physical model studies to aid in probabilistic navigation channel design is presented in Section V-5-10.b.(1).

(6) Direct field measurements of ship motion are a valuable addition to channel design studies, but extreme conditions controlling design are not easily captured. Field measurements are dependent on available ships and environmental conditions during the limited duration of the measurement program. Figure V-5-14 provides an example of results from a large field measurement program in a high-wave-energy entrance channel. Data were collected over a 2-year period at the mouth of the Columbia River, at the Oregon/Washington border (Wang et al. 1980). The *average* ratio of ship bow/stern response amplitude to wave amplitude on each transit varied between about 0.5 and 2.0 over 29 instrumented voyages.

(7) Simple general guidelines for minimum depth clearance requirements in channels influenced by waves are given by PIANC (1997) as

$$\frac{Water\ depth}{Ship\ draft} \geq 1.3 \quad \text{when } H \leq 1\text{m (3.3 ft)}$$

$$\frac{Water\ depth}{Ship\ draft} \geq 1.5 \quad \text{when } H > 1\text{m (3.3 ft) and wave periods and directions are unfavorable}$$

where H = wave height

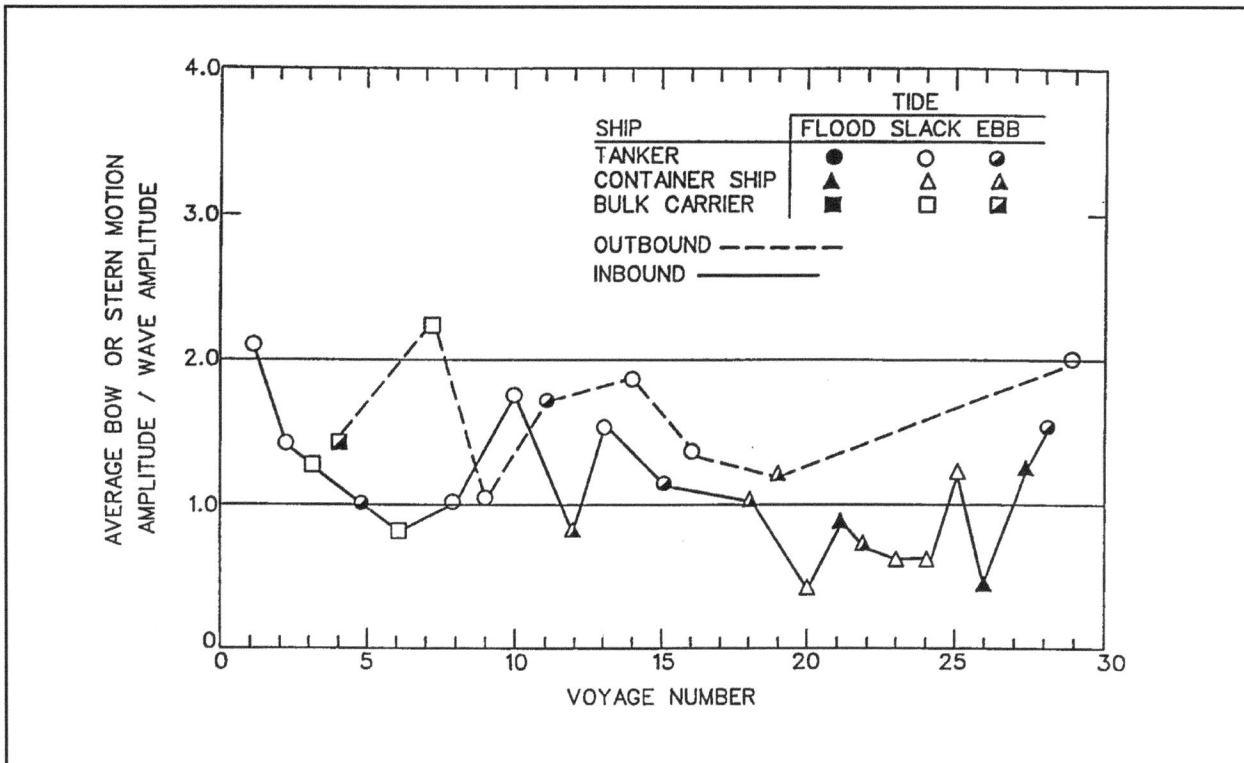

Figure V-5-14. Ship motion response, mouth of Columbia River

(8) For small recreational craft, a depth allowance for waves is also important but difficult to estimate accurately. As with deep-draft vessels, model and measurement studies may be conducted; but such project-specific studies are usually impractical for small-craft channel design. Small craft length and beam are often small relative to wavelengths important for design. It is realistic to consider the maximum vertical drop experienced by a small boat in a wave to occur when the boat is fully contained in the wave trough. The magnitude of this drop below the Swl is then the wave amplitude $H/2$. In practical design, typically one-half the design significant wave height is used.

d. Vessel squat. A depth allowance for squat experienced by vessels under way is included in channel depth design (see Figure V-5-12). The amount of squat experienced by a vessel depends strongly on speed and relative blockage of the channel cross section by the vessel. A small, fast-moving boat in a small channel may experience as much or more squat than a large, slower-moving ship. Squat increases when ships meet and pass in a channel because the total blockage is increased. Squat is an important consideration in both deep- and shallow-draft navigation projects. As with vertical vessel motion due to waves, squat is difficult to estimate accurately and is a subject of present research.

(1) Simplified methods for estimating squat are available. The method presented here is based on equations for squat in fairway and canal channel configurations, with a simple interpolation between these limiting cases to accommodate trench configurations (USACE 1998). A definition sketch of parameters is given in Figure V-5-15.

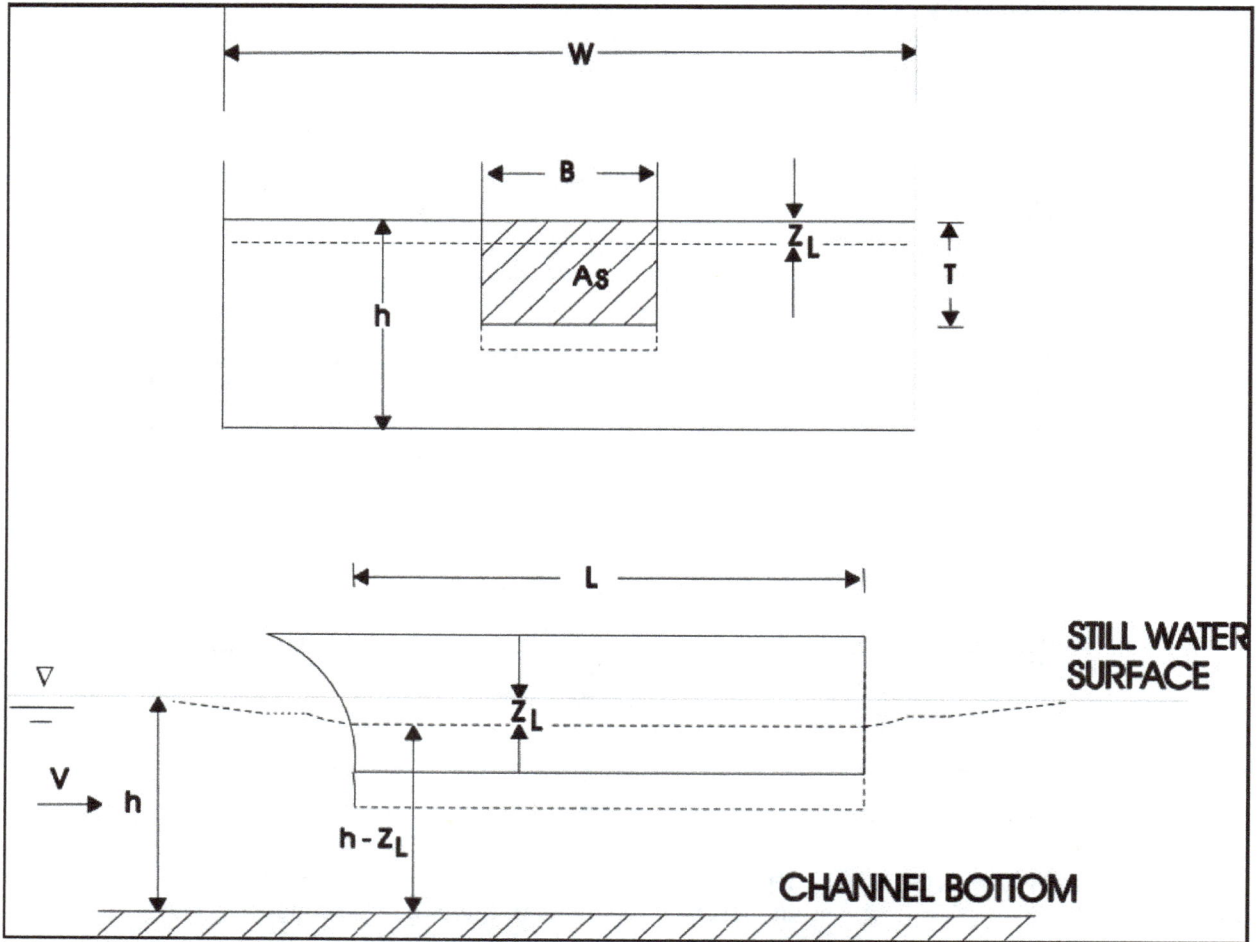

Figure V-5-15. Squat analysis definition

(2) Squat in a fairway can be estimated as (Norrbin 1986)

$$Z = 0.2125 \, C_B \, \frac{B}{L} \, \frac{T}{h} \, V^2 \qquad V \text{ in knots, } Z \text{ in ft}$$

$$\text{(fairway)} \tag{V-5-3}$$

$$Z = 0.01888 \, C_B \, \frac{B}{L} \, \frac{T}{h} \, V^2 \qquad V \text{ in km/hr, } Z \text{ in m}$$

where

Z = maximum ship squat

C_B = block coefficient (Figure V-5-4)

B = ship beam at midship (Figure V-5-15)

T = ship draft (Figure V-5-15)

L = ship length (Figure V-5-15)

h = depth of shallow-water area (Figure V-5-15)

V = ship speed

This equation is applicable when Froude numbers are less than 0.4.

(3) Squat in a rectangular canal can be related to the Schijf limiting Froude number, which corresponds to the Schijf limiting speed (Part V-5-2-b-(3)). The limiting Froude number is given by (Huval 1980)

$$F_L = \frac{V_L}{\sqrt{gh}}$$

$$\approx \left\{ 8 \cos^3 \left[\frac{\pi}{3} + \frac{1}{3} \cos^{-1} \left(1 - \frac{1}{B_R} \right) \right] \right\}^{\frac{1}{2}} \quad \text{(rectangular canal)}$$

(V-5-4)

where

F_L = Schijf limiting Froude number

V_L = Schijf limiting ship speed in squat analysis

g = acceleration due to gravity; = 9.80 m/sec^2 (32.2 ft/sec^2)

h = depth of canal (Figure V-5-15)

B_R = channel blockage ratio (Equation V-5-2 and Figure V-5-7)

(4) Maximum ship squat at the Schijf limiting Froude number is given by

$$Z_L = h \left[\frac{F_L^2}{2} \left(F_L^{1/3} - 1 \right) \right] \quad \text{(rectangular canal)}$$

(V-5-5)

where

Z_L = maximum squat at Schijf limiting Froude number

An approximate analysis for nonrectangular canal cross sections can be made by replacing the channel depth by a cross-section mean depth.

(5) Trench channels can assume a variety of shapes, including asymmetric overbank depths and different lane depths in two-way channels (Figure V-5-16). The trench configuration is intermediate between the fully

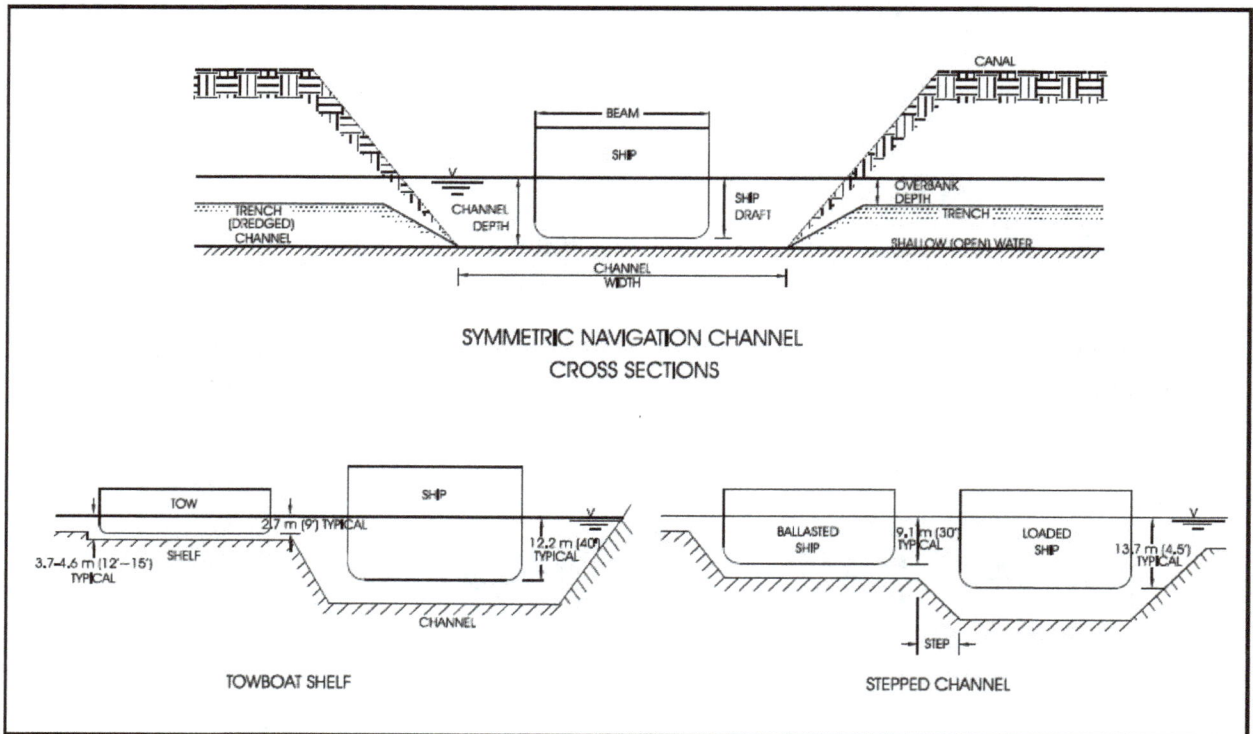

Figure V-5-16. Example channel cross sections

open fairway and fully restricted canal configurations. A first approximation to squat may be made by interpolating between those two extremes, based on the ratio of average overbank depth-to-channel depth (Figure V-5-17)

$$ Z_T = \left(\frac{h_1 + h_2}{2h} \right) Z + \left(1 - \frac{h_1 + h_2}{2h} \right) Z_L \quad \textbf{(trench)} \tag{V-5-6} $$

where

Z_T = maximum squat in a trench channel

h_1, h_2 = overbank depths

Computational results from a computerized version of this model illustrate squat variation with ship speed and channel type (Figure V-5-18) (Huval 1993). The figure illustrates how a narrow, confined channel can significantly increase squat and, hence, required channel depth.

(6) Squat for small recreational craft is generally less critical than for large displacement, deep-draft ships. However, it should be included in channel design. The usual procedure is to use a fixed depth allowance for squat in entrance channels, where boat speed is relatively high, and a smaller fixed allowance for interior areas, where boat speed is generally low (Table V-5-7).

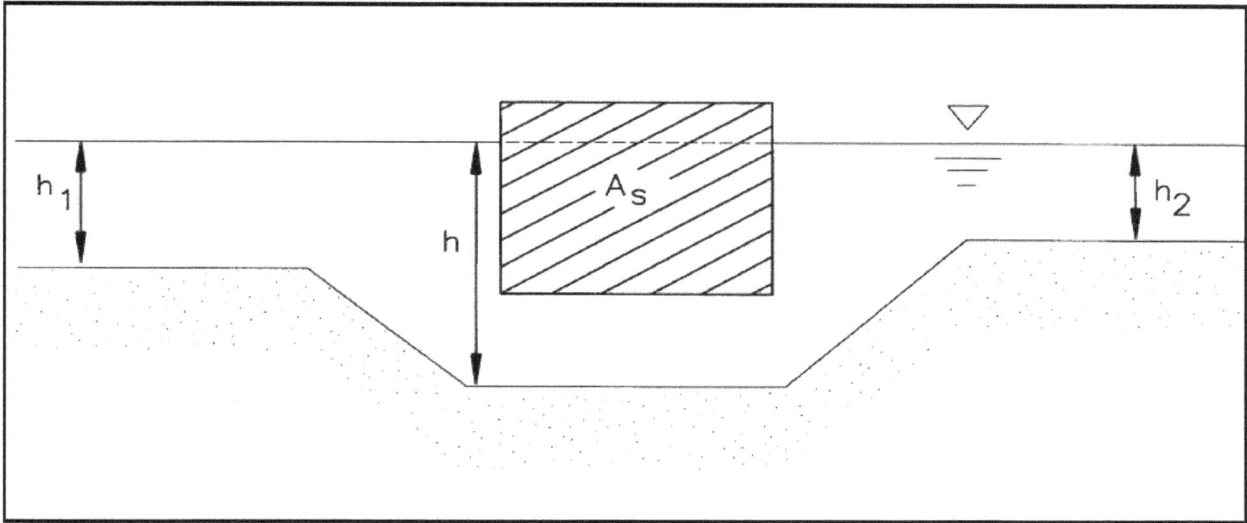

Figure V-5-17. Trench channel definitions

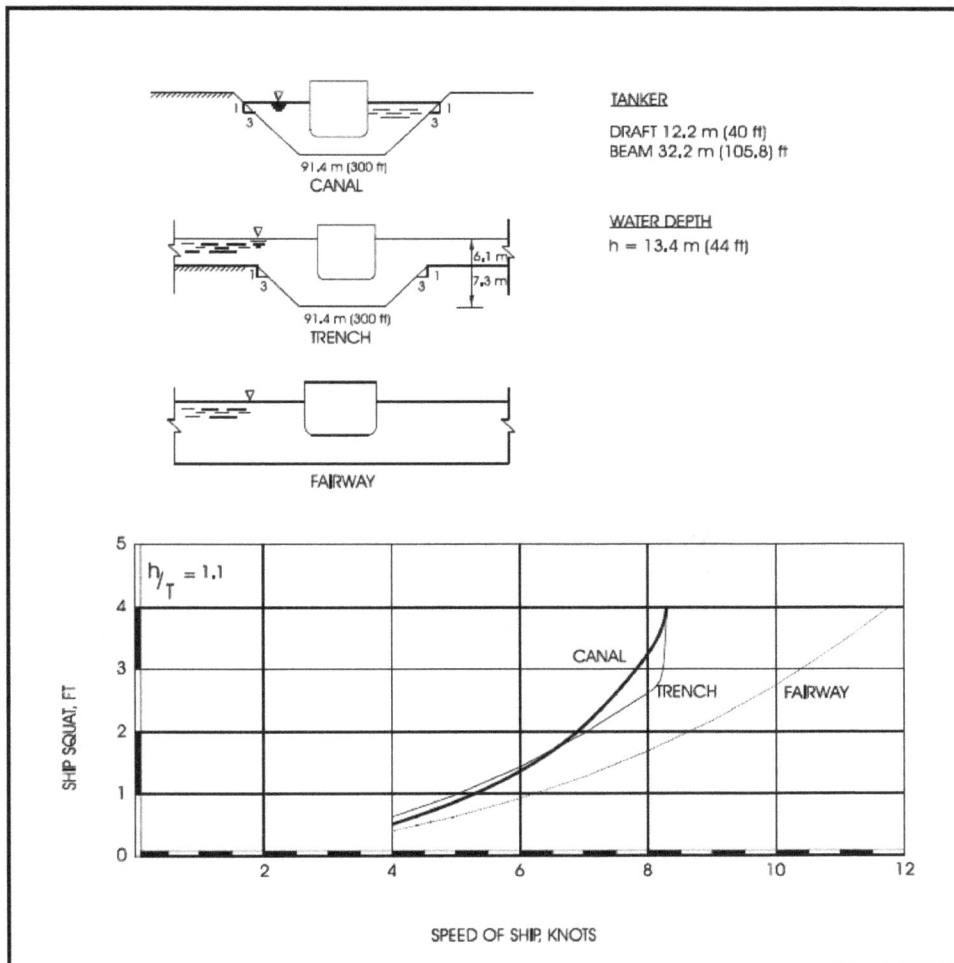

Figure V-5-18. Example squat calculations

Table V-5-7
Squat Allowance for Small Recreational Craft

Location	Allowance
Entrance channels	0.3 m (1.0 ft)
Interior channels, moorage areas, turning basins	0.15 m (0.5 ft)

e. Trim. Deep-draft ships often operate with trim for a variety of reasons. For example, ships may be loaded or ballasted to lower the stern a small amount deeper than the bow, which can improve maneuverability. Ships under way tend to change from the static trim. Small craft in motion can experience significant trim, with the bow rising high and stern dropping low in the water. Because vessel trim conditions are mainly determined by operational decisions, a channel depth allowance for trim is not included.

f. Shallow-water effects. Even when deep-draft ships have sufficient channel depth to avoid hitting bottom, they may experience adverse safety and efficiency effects due to small underkeel clearance. Steering and turning become significantly more difficult, more power is required to maintain speed, and potential for bottom scour and bank failure increases considerably as propeller speed is augmented. Ship cooling systems may ingest benthic organisms or sediment if intakes are too near the bottom. Although these effects can be significant, no depth allowance is included in general channel design to lessen their impact. In some cases, it may be prudent to quantify the location and size of cooling system intakes, evaluate the impacts of small clearances, and perhaps impose a minimum clearance as an added constraint on project design. For example, the U.S. Navy requires a 1.5-m (5-ft) clearance beneath aircraft carrier intakes.

g. Safety clearance. To protect vessel hull, propellers, and rudders from bottom irregularities and debris, a channel depth allowance for safety is included (Table V-5-8). A larger clearance is needed when the channel bottom is hard, such as rock, consolidated sand, or clay (see Figure V-5-12).

Table V-5-8
Safety Clearance

Bottom Type	Minimum Safety Clearance
Soft	0.6 m (2 ft)
Hard	0.9 m (3 ft)

h. Advance maintenance. An additional increment beyond the channel design depth is added to maintain reliable navigable depth between dredge events. This depth increment is to provide for accumulation and storage of sediment. The depth allowance for advance maintenance should be determined by considering several different increments and choosing that which minimizes total channel maintenance cost. Dredge mobilization costs and safety concerns must be balanced against the tendency for a deeper channel to shoal more rapidly. A sediment trap near the entrance may be an economic alternative to reduce advance maintenance requirements in the channel. Depth increments of 0.6-0.9 m (2-3 ft) are normal advance maintenance allowances (see Figure V-5-12).

i. Dredging tolerance. Another depth increment is added beyond the design channel depth to compensate for the inherent mechanical inaccuracies of dredges working in the hostile environment of adverse currents, fluctuating water surface, and non-homogeneous bottom material. A dredging tolerance of 0.3-0.9 m (1-3 ft) is typical (see Figure V-5-12).

j.   Tidal shoals and ship transit concerns example. Coastal entrances, bays, and river estuaries typically have meandering natural channels and intermittent shoals. When deeper ship drafts are anticipated, the impact of tidal shoals on schedules and/or margin of safety must be evaluated. Increasing design channel depth over the shoals may eliminate tidal constraints altogether or just improve access to a more economical level. The question, "How deep is deep enough?" must be answered by weighing transportation cost savings against the cost of channel excavation and maintenance. The following example illustrates considerations involved in answering this question and developing an optimum design channel depth.

(1)   The example port is located in an estuary, 35 km from the ocean entrance (Figure V-5-19). At a distance of 30 km from port, ships must pass over a shoal with controlling depth of 8 m below mllw (*Shoal 1*). At a distance of 10 km from port, ships encounter another shoal with controlling depth of 7 m below mllw (*Shoal 2*). The port itself has 13 m depth mllw, which is about the average depth available between the two shoals. The diurnal tidal range between mllw and mhhw in this example is amplified from 3 m at the ocean entrance to 4 m at the port, which might correspond to tides in a gradually narrowing bay. Conversely, friction in narrow river estuaries often results in reduced tidal ranges upstream. Tidal datums (e.g., mllw and mhhw) also change as the tidal wave is transformed by the waterway.

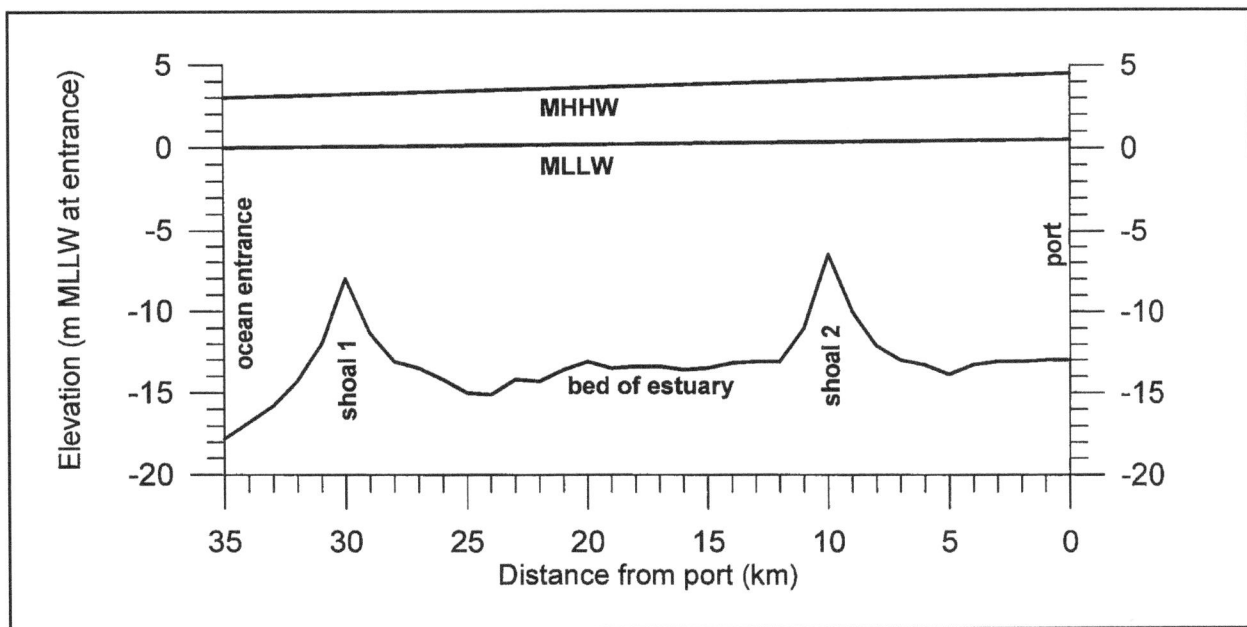

Figure V-5-19.   Hypothetical route to a port at the head of an estuary (cross section along channel center line)

(2)   The distribution of tidal water levels is estimated by using NOAA primary and secondary tide station data (Part V-5-3). Interpolation between secondary stations is often practical, as needed, if no major constrictions or confluences are present. Figure V-5-20 illustrates the predicted distribution of hourly depths for 1 year at the shallower shoal, Shoal 2. A similar analysis can be performed by applying the methods of Harris (1981).

(3)   River discharges and wind-induced water level changes affect water depth over the shoals. Waterway confluences or constrictions and hydrodynamic effects of the shoal itself affect water levels in the vicinity. Seasonal or storm-related changes in elevation of the shoal crest can also affect the depth available to ships. These complications may call for a program of site-specific water level measurements. Numerical modeling

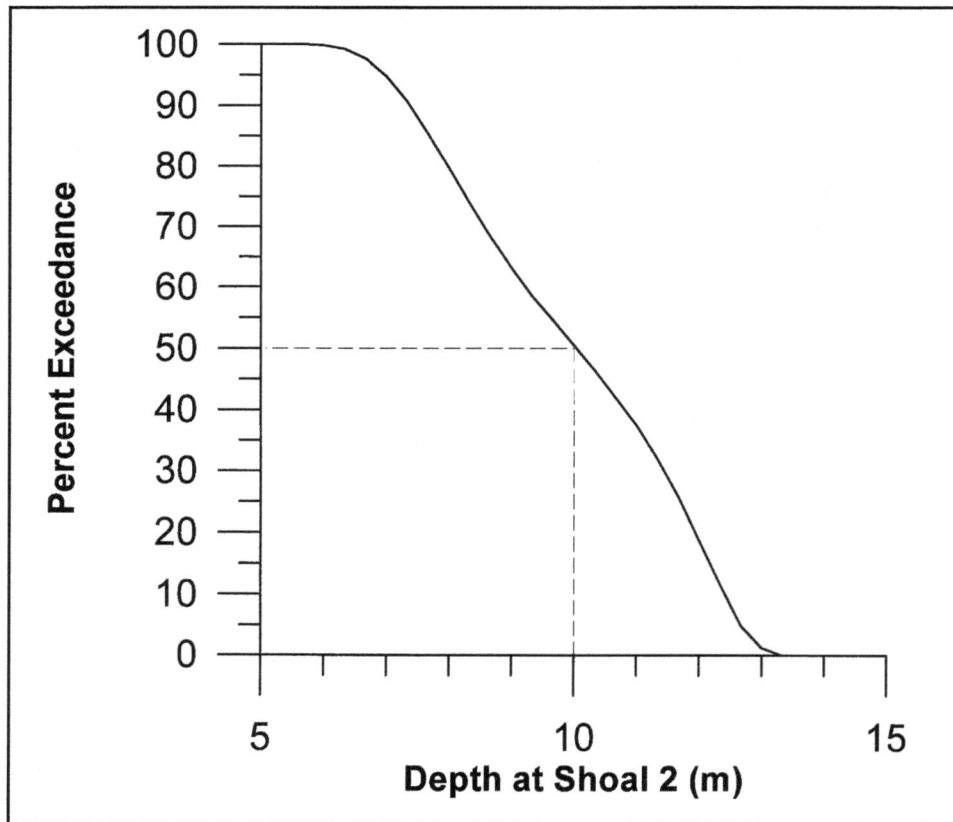

Figure V-5-20. Hypothetical distribution of depths for Shoal 2

of historical conditions (hindcasting) may also be effective for precisely describing water level changes. Figure V-5-21 illustrates water level variations on the main branch of the lower Fraser River near Vancouver, British Columbia, Canada, including tidal and river discharge variations (Ferguson 1991). The transect begins at the river mouth at Sand Heads and continues 35 km up the estuary to New Westminster. Shaded areas show the impact of river outflow on water level for two tide levels (high and low tide). River outflow has little impact on water level at the mouth, but significant impact up into the estuary, especially at low tide. River outflow can add up to 2 m to low tide water depth along this channel. It also affects current. With high river outflow, water level at New Westminster becomes nearly constant over the tidal cycle. Similar processes may need to be considered at the example port.

(4) Figure V-5-20 indicates that ships approaching the example port with draft and keel clearance requirements for a minimum 10-m depth must wait until the upper half of the tidal cycle to cross Shoal 2. Since ships generally arrive on schedules independent of the tide, roughly half the ships of this draft will be delayed. Time spent waiting for the tide, or more likely spent at slow speed offshore, will vary from one arrival to another. Time will also be lost on departure, since ships loaded and ready to depart will have to wait for favorable tide. Deeper ships will lose more time. The shoals of the example are about an hour's sailing apart, so the dangerous possibility exists of crossing one shoal on ebb tide only to find insufficient depth at the second.

(5) The variability of arrival and departure times, ship draft and speed, and water depths is difficult to resolve by multi-variate probability analysis based only on recorded port data. Port arrival and departure times and drafts are usually recorded, but details of shoal-related delays are not. Historical and future delays

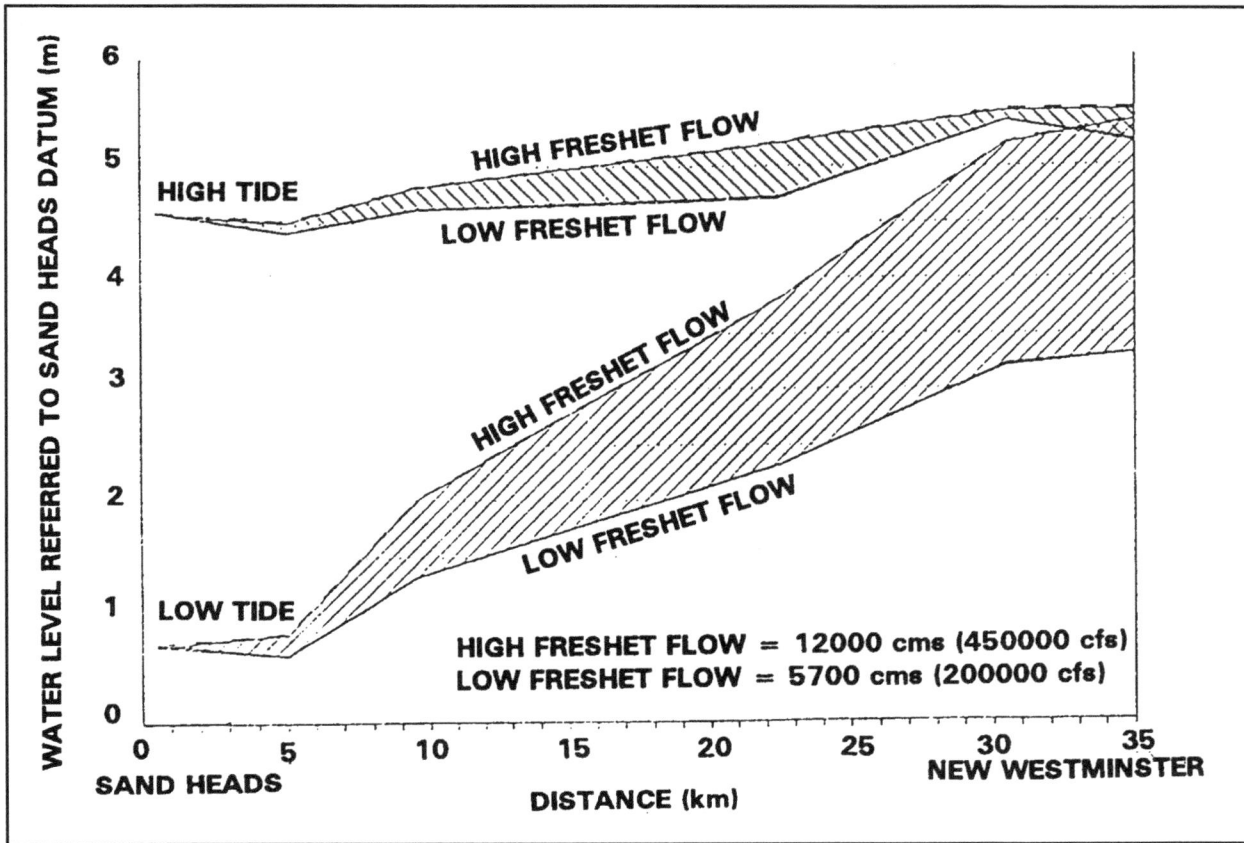

Figure V-5-21. Range of water levels in Fraser River, British Columbia (after Ferguson 1991)

can often be estimated more accurately with numerical simulations. For example, a risk-based system for evaluating channel depth requirements (and for evaluating operational transit safety when coupled with real-time field environmental measurements) is discussed by Silver (1992) and Silver and Dalzell (1997).

(6) Figure V-5-22 illustrates the conceptual approach of a computer program for time-and-motion simulations which applies predicted tides and tidal currents, historical or projected ship cargoes, drafts, and arrival times at the ocean entrance, and cargo transfer rates at port. The same model can simulate ship transit times with various dredged channel geometries. Waiting times are computed as the difference between transit time simulated across the shoals and transit time without shoal restrictions.

(7) Randomly occurring combinations of variables affecting transit times, such as strong winds, river discharge- and wind-induced depth changes, high waves, low visibility, and ice conditions can be added using a Monte Carlo approach. This method requires enough repeated simulations of the same input variables with random values of stochastic variables to encounter the full range of combined extremes. Statistics of transit time, waiting time, and associated costs can be computed from these data. A more complete discussion of vessel traffic flow simulation models is given by PIANC (1997a).

(8) Ship costs are generally proportional to operating time, either under way or at berth. Reductions in time navigating the port approach, at the dock, and departing translate into cost savings. Other cost factors include impacts of ship arrivals on longshoremen, mechanical equipment, and cargo staging at the port. Transportation costs without channel improvement must be estimated as a baseline against which to measure cost-effective optimization of channel excavation. The savings realized by a range of excavation depths and

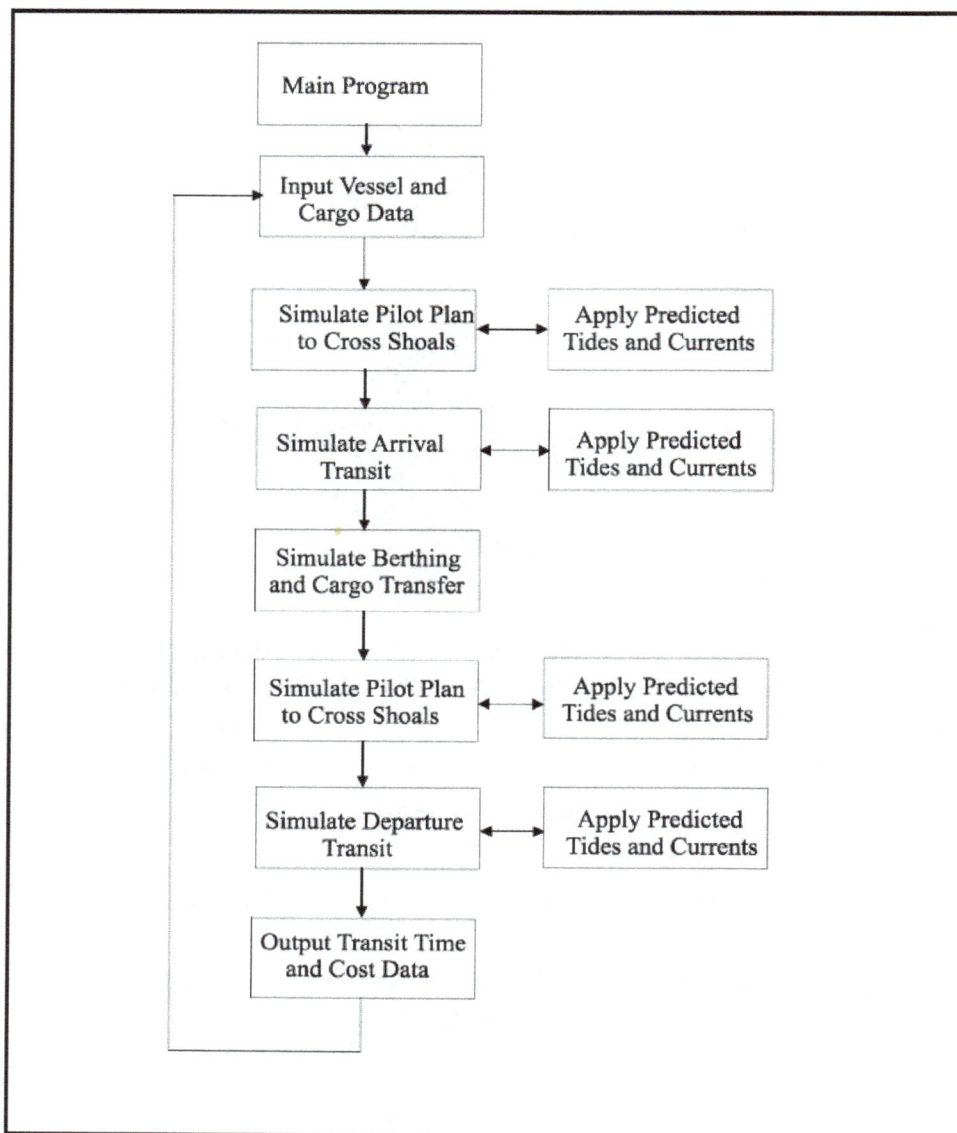

Figure V-5-22. Components of a ship transit simulation program

channel configurations should be compared with corresponding dredging and disposal costs. The ideal optimum will achieve the maximum net savings, but environmental quality effects, financial capabilities, and user preferences can affect the final project design.

V-5-6. Channel Alignment and Width

a. Alignment. Navigation channels are normally aligned as much as possible with natural channels in the pre-project bottom contours. This approach has several important advantages: initial and maintenance dredging are usually minimized; and currents typically take this path in line with the channel, a preferred condition for navigation. The effect of predominant winds and waves, as well as currents, on navigation should be considered. A channel oriented in line with these forces typically serves navigation best.

(1) A straight channel is preferred over a channel with bends. If turning is required, straight reaches with turns between channel segments are preferred over curved alignments. This type of alignment allows the channel to be clearly marked with aids to navigation. Straight segments in a deep-draft channel should be at least five times the length of the design ship. Few turns and small turning angles are best for navigation. Typically, the number of turns introduced in the channel alignment must be balanced against the turning angles to achieve an optimum alignment for navigation purposes. Both deep- and shallow-draft channels should be aligned so that vessels can maintain speed and controllability through areas where they are exposed to potentially damaging winds, waves, and currents. This consideration generally precludes sharp turns in exposed areas.

(2) The entrance channel to San Juan Harbor, Commonwealth of Puerto Rico, helps illustrate the difficulty of turning large ships in design transit conditions (Figure V-5-23). The exposed entrance, or "bar," channel makes a 57-deg turn into Anegado Channel just inside the harbor entrance. A ship simulator was used to model navigation channels in the harbor (Webb 1993). Local pilots from San Juan Harbor conducted simulation runs with two design ships, a tanker and a container ship. Ship tracks from inbound runs with the tanker are shown, where the ship hull outline is plotted at short intervals along each run. The variability of ship track and differences in pilot strategy for making the turn are evident in the figure. Ship track plots such as this clearly show the ship's swept path relative to channel boundaries. The envelope of multiple ship tracks gives valuable information about the navigability of the channel being simulated.

(3) Channel alignment option studies should consist of selecting several alternate routes when viable alternatives are available. Construction and maintenance costs are developed for each alternative. A comparison of annual project costs and benefits then determines the optimum channel alignment.

   b. *Inner channels (protected waters and harbor areas).*

(1) Width. Harbor access channels leading from the bar or entrance channel to the port or harbor area are referred to as *interior channels*. For straight deep-draft channels, the required channel width is based on the following factors, listed in decreasing order of importance:

   (a) Traffic pattern (one-way or two-way).

   (b) Design ship beam and length.

   (c) Channel cross-section shape.

   (d) Current speed and direction.

   (e) Quality and accuracy of aids to navigation.

   (f) Variability of channel and currents.

   • Design channel width is defined as the width measured at the bottom of the side slopes on each side of the channel at the design depth. For one-way deep-draft channels, channel width has traditionally been figured as the sum of a maneuvering lane width and bank clearance increments on either side (Figure V-5-24). For two-way channels, an additional maneuvering lane and a ship clearance lane dividing the two lanes of traffic are added. The required width for each increment was given as a factor applied to the design ship beam (Table V-5-9). Factors vary with ship controllability and judgment. Special judgment is required when vessels are exposed to yawing forces.

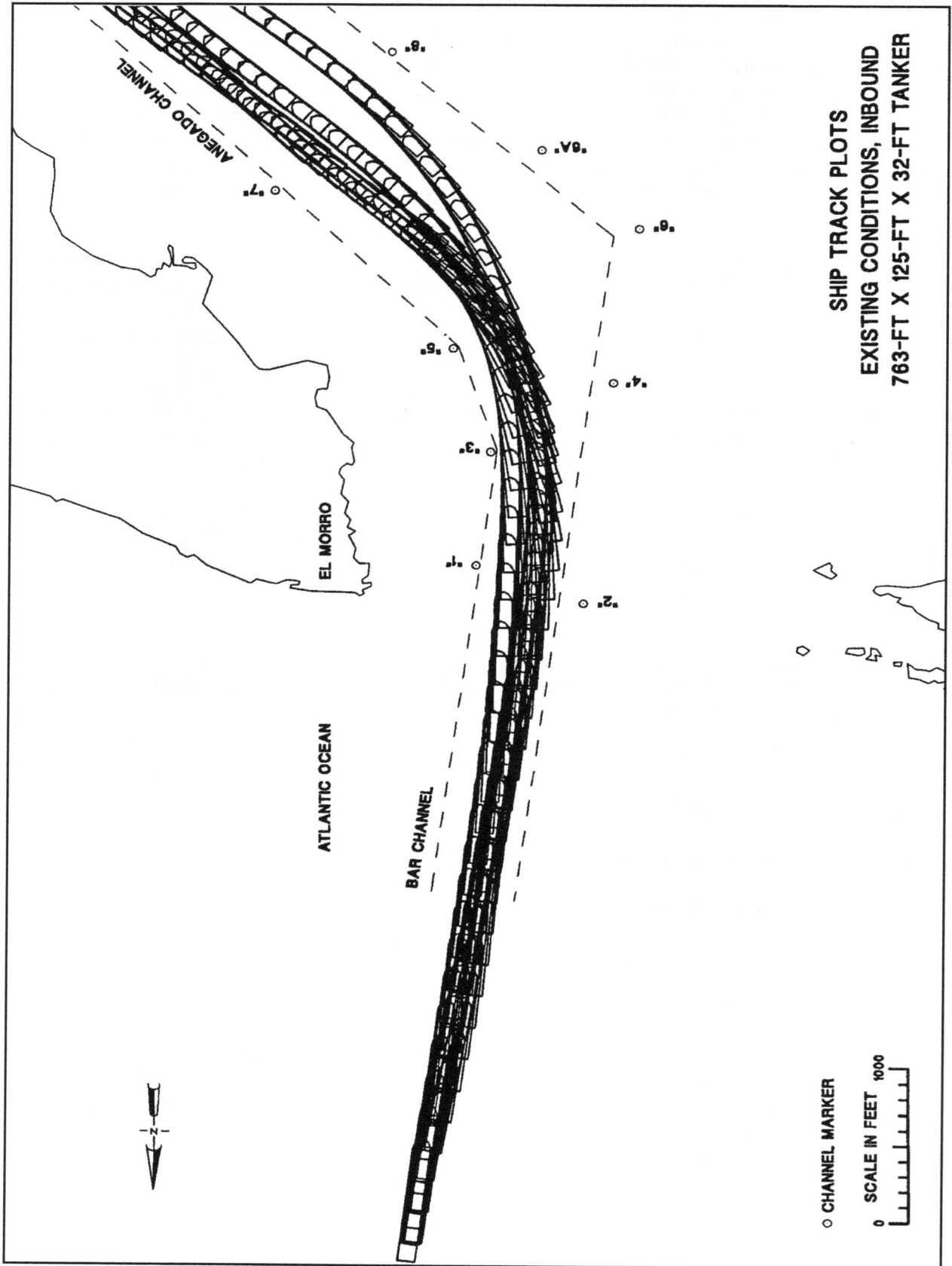

Figure V-5-23.   Ship simulation of deep-draft entrance channel, San Juan Harbor, Commonwealth of Puerto Rico

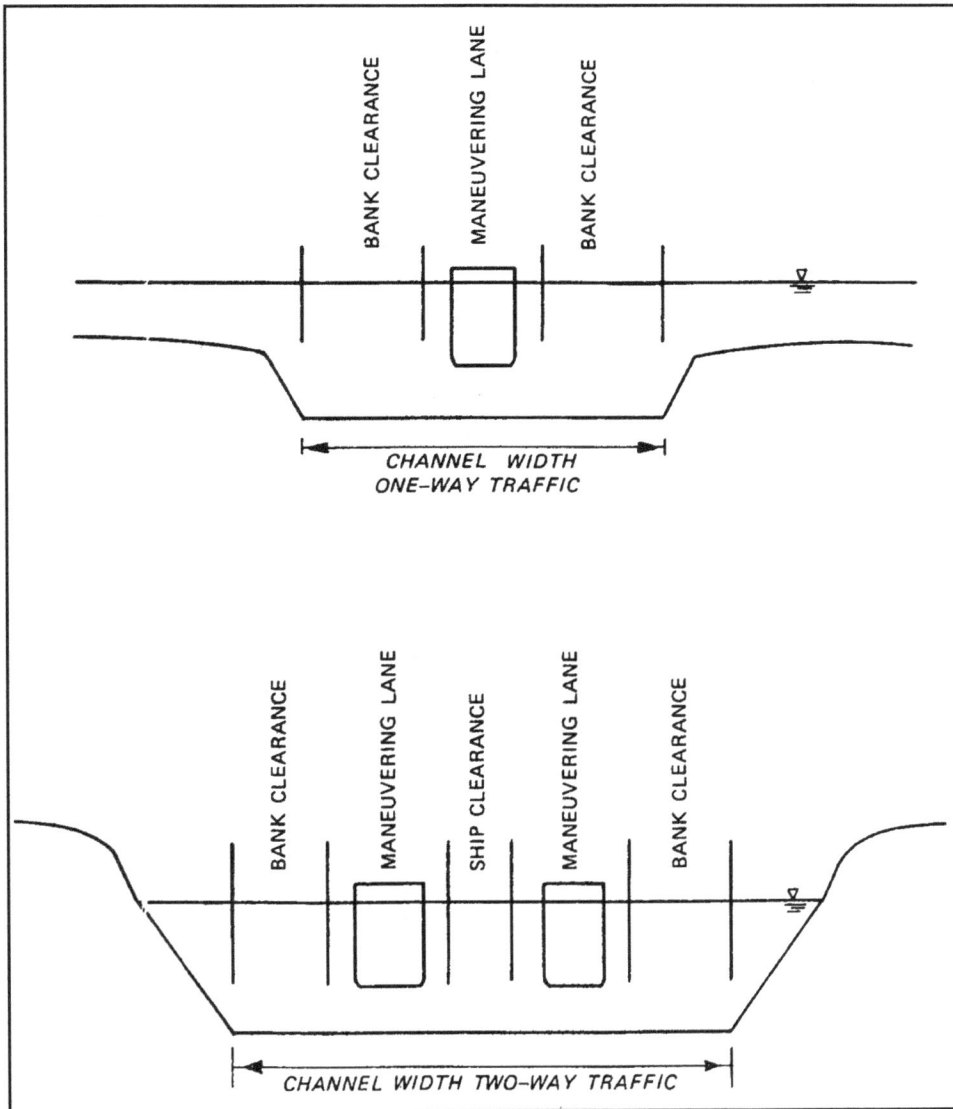

Figure V-5-24. Traditional interior channel width elements

Table V-5-9
Traditional Criteria for Deep-Draft Channel Width Design[1]

Location	Vessel Controllability			Channels with Yawing Forces
	Very Good	Good	Poor	
Maneuvering lane, straight channel	1.60	1.80	2.00	Judgment[2]
Bend, 26-deg turn	3.25	3.70	4.15	Judgment[2]
Bend, 40-deg turn	3.85	4.40	4.90	Judgment[2]
Ship clearance	0.80	0.80	0.80	1.00 but not less than 30 m (100 ft)
Bank clearance	0.60	0.60+	0.60+	1.50

[1] Criteria expressed as multipliers of the design ship beam; i.e., W = (factor from table) $\times B$
[2] Judgment is based on local conditions at each project.

- Professional pilots control ships in a way that makes the traditional channel width divisions illogical. For example, they routinely move the ship off center line to use bank effects as a cue in determining ship position. Also, they sometimes use bank effects to assist in turning. Guidance for deep-draft channel width is best expressed as a total channel width based on the design ship beam (e.g., Figure V-5-25). The quality of aids to navigation, type of channel cross section, and current strength impact the required width. Experience with ship simulator studies has indicated that traditional channel width design criteria are overly conservative. Interim guidelines have been developed based on simulator studies (USACE 1998) (Tables V-5-10 and V-5-11). If current speeds are greater than 2.9 m/sec (3.0 knots), design channel width should be developed with the assistance of a ship simulator study (Part V-5-10).

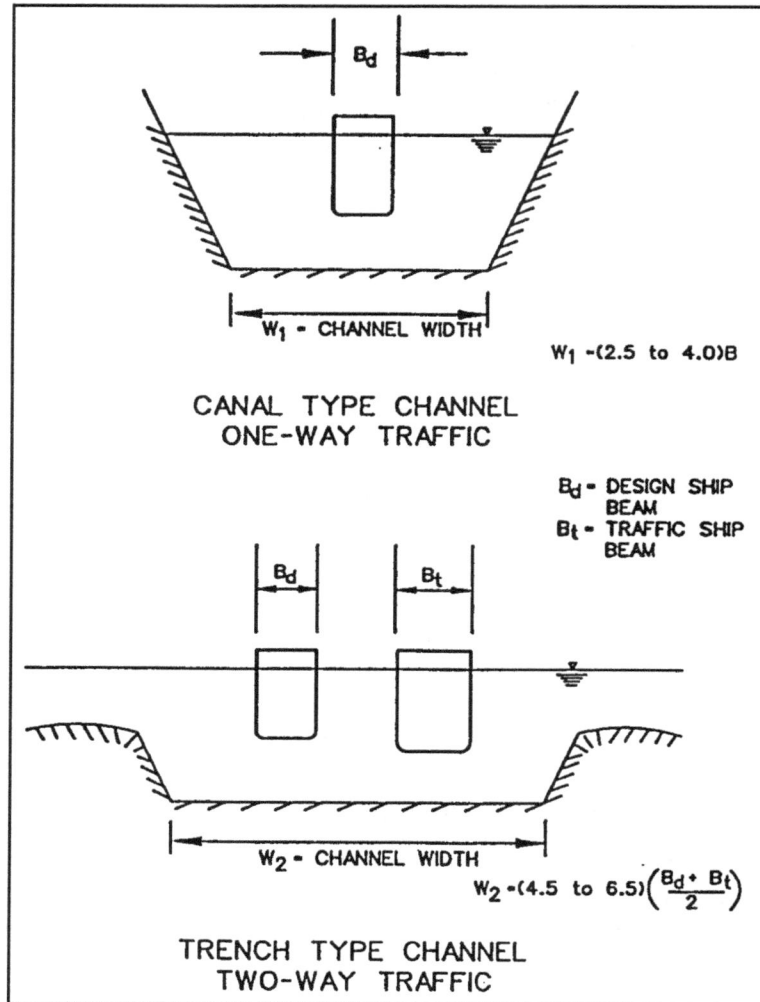

Figure V-5-25. Interior channel design width

- Navigation is more difficult when channel cross section (overbank depths, channel depth and width) varies significantly. Bank effects and currents become less predictable and extra care is needed for vessel control. Table V-5-10 gives channel width factors for a somewhat challenging navigation scenario, with variable cross section and average aids to navigation, and for an ideal scenario with constant cross section and excellent aids to navigation. These two scenarios bracket most channel design projects likely to be encountered in the United States.

Table V-5-10
One-Way Ship Traffic Channel Width Design Criteria[1]

Channel Cross Section	Maximum Current		
	0.0 to 0.3 m/sec (0.0 to 0.5 knots)	0.3 to 0.8 m/sec (0.5 to 1.5 knots)	0.8 to 1.5 m/sec (1.5 to 3.0 knots)
Constant Cross Section, Best Aids to Navigation			
Shallow	3.0	4.0	5.0
Canal	2.5	3.0	3.5
Trench	2.75	3.25	4.0
Variable Cross Section, Average Aids to Navigation			
Shallow	3.5	4.5	5.5
Canal	3.0	3.5	4.0
Trench	3.5	4.0	5.0

[1] Criteria expressed as multipliers of the design ship beam; i.e., W = (factor from table) × B

Table V-5-11
Two-Way Ship Traffic Channel Width Design Criteria[1]

Channel Cross Section	Maximum Current		
	0.0 to 0.3 m/sec (0.0 to 0.5 knots)	0.3 to 0.8 m/sec (0.5 to 1.5 knots)	0.8 to 1.5 m/sec (1.5 to 3.0 knots)
Constant Cross Section, Best Aids to Navigation			
Shallow	5.0	6.0	8.0
Canal	4.0	4.5	5.5
Trench	4.5	5.5	6.5

[1] Criteria expressed as multipliers of the design ship beam; i.e., W = (factor from table) × B

- Shallow-draft interior channels should be designed to safely handle the expected volume of two-way traffic. Traditional guidance for channel width is the same as for deep-draft channels (Table V-5-9). However, an approach that takes account of the traffic congestion may be preferable. Small-craft channel design guidance is being reevaluated in present research studies. Present recommended simple guidance is a minimum width, generally based on average vessel beam, and an additional width increment based on the number of boats using the project (ASCE 1994, Dunham and Finn 1974)

$$W = W_{\min} + 0.03\ N_B \quad \text{in meters (interior channels)}$$

$$W = W_{\min} + 0.10\ N_B \quad \text{in feet}$$

(V-5-7)

where

W = design small-craft channel width

W_{min} = minimum width; = 5 B or 15 m (50 ft), whichever is greater

B = average beam

N_B = number of boats using the project

Thus, an interior channel serving 500 small boats with average beam of 5 m would have a minimum width of 5*5 + 0.03*500 = 40 m.

(2) Berthing areas. Although entrance and interior harbor channels and turning basins are usually part of federal navigation projects, berthing areas are typically nonfederal concerns except in military harbors. Normally, a berthing area must have sufficient depth to accommodate the design vessel to be using the berths, which may be smaller than the design vessel for sizing harbor channels, under all or nearly all expected water levels. Typically the design water level for berthing areas is extreme low water. Berthing areas must have sufficient space for safe maneuvering of the appropriate design vessel, often with tug assistance in deep-draft facilities. Guidance for sizing berthing areas to provide adequate access may be found in other references (e.g., Tsinker (1997), Gaythwaite (1990), and U.S. Navy (1981) for deep-draft ports; ASCE (1994), Tobiasson and Kollmeyer (1991), Dunham and Finn (1974), and State of California (1980) for small-craft harbors).

(3) Special considerations due to ice. Increased navigation difficulties in areas with ice may introduce special considerations for channel depth and width. Policies of pilot associations, shipping companies, or vessel insurance underwriters may call for additional keel and bank clearances beyond those allowed in temperate ice-free conditions. These groups and other marine interests, such as the USCG, NOAA, and military operators, should be solicited in the planning process for their views on these matters.

(a) The magnitude and direction of ice forces on ships are random in nature. These forces are combinations of impacts and frictional resistance. Impacts of wind- or current-driven ice forces can be oblique to the course of the ship, causing a sudden diversion. Ship response to such a diversion is slowed by additional oblique impacts and frictional resistance.

(b) Design width for a channel that will be navigated with ice should be increased by as much as 50 to 100 percent over the conventional width. Design depth may also need to be increased. The conventional ice-free design depth including either a standard wave allowance or an additional 0.5-m depth increment, whichever is greater, should be used. Wave allowance can be neglected in ice navigation because ice usually suppresses waves.

c. *Entrance channels*. Entrance channels are generally wider than interior channels because of many factors complicating navigation at entrances, particularly waves and currents. Intensified waves and currents at entrances make navigation difficult, but they also result in potentially treacherous sediment movement and dynamic shoaling patterns. Often vessels navigate this difficult environment in close proximity to rock or concrete breakwater or jetty structures. Loss of control can result in disastrous collision with these solid structures.

(a) Because of the complexity of processes and vessel responses in entrance channels, experience at the project site or related sites can be an especially helpful guide to determining an acceptable channel width. The traditional simple design approach is to use the guidance in Table V-5-9, choosing factors for a level of vessel controllability appropriate to the difficulty of the particular entrance channel. Deep-draft entrance channel width design can benefit significantly from comprehensive physical model studies, navigation simulation studies, and field measurements of ship motions (Part V-5-10).

(b) For small-craft harbors, entrance channel width should be a minimum of 23 m (75 ft) (ASCE 1994). Guidance in Table V-5-9 has traditionally been applied. These factors typically exceed those for interior channels, leading to a widened entrance channel design. Small-craft harbor entrance channel design is the subject of a present research study.

(c) USACE practice in southern California, where sailboat usage is accommodated in design, is a minimum width of 90 m (300 ft). Consideration is also given to the number of boats using the project. USACE practice is to increase channel width over the 90-m (300-ft) minimum if the number of boats exceeds 1,000.

$$W = 90 + 0.03 \, (N_B - 1000) \qquad \text{in meters}$$
$$W = 300 + 0.1 \, (N_B - 1000) \qquad \text{in feet}$$

(V-5-8)

Thus, an entrance channel serving 2,000 small boats would have a minimum width of 90 + 0.03*(2,000-1,000) = 120 m (300 + 0.1*(2,000-1,000) = 400 ft), based on practice in southern California.

d. Turns and bends. Channels with turns and bends are more difficult to navigate than straight channels. Vessel control is reduced and width of the vessel swept path (the envelope around ship hull positions in the horizontal plane) is naturally increased when turning. Therefore, additional width is required in turns and bends for large vessels. Key parameters in a turn are defined in Figure V-5-26. The recommended configuration for widening a turn in a deep-draft channel depends on the deflection angle (Figure V-5-27 and Table V-5-12). Bank conditions, not included in the simple guidance presented here, are important in turn design since pilots often use them to assist in turning.

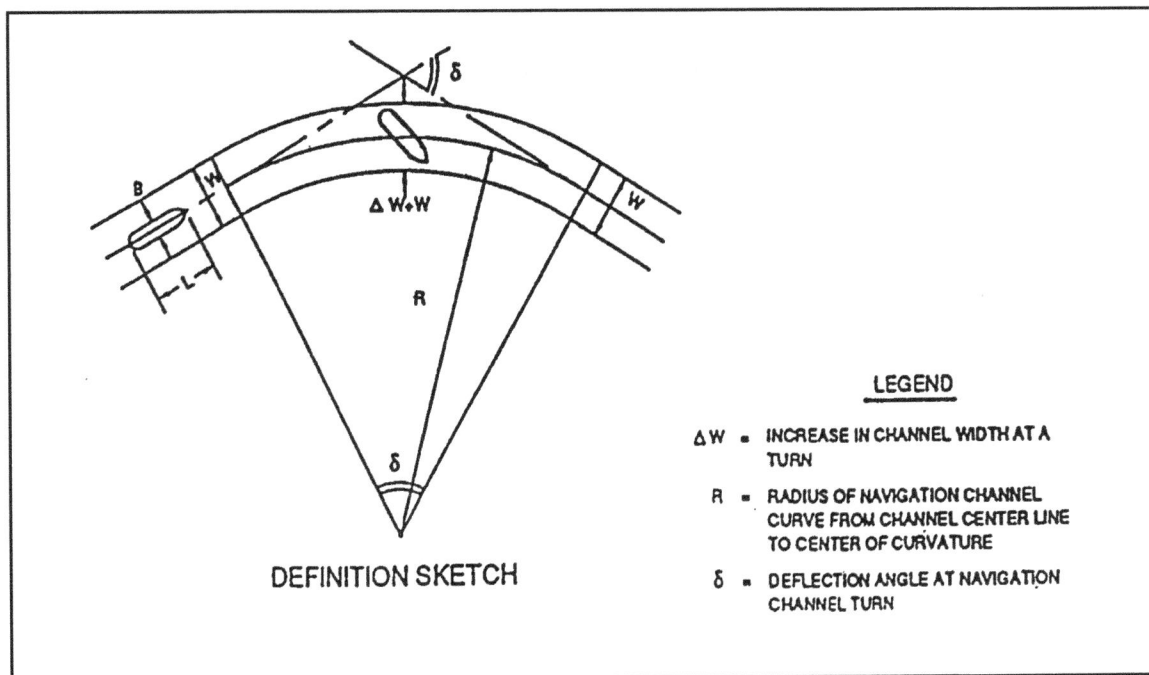

Figure V-5-26. Definition of parameters in channel turn

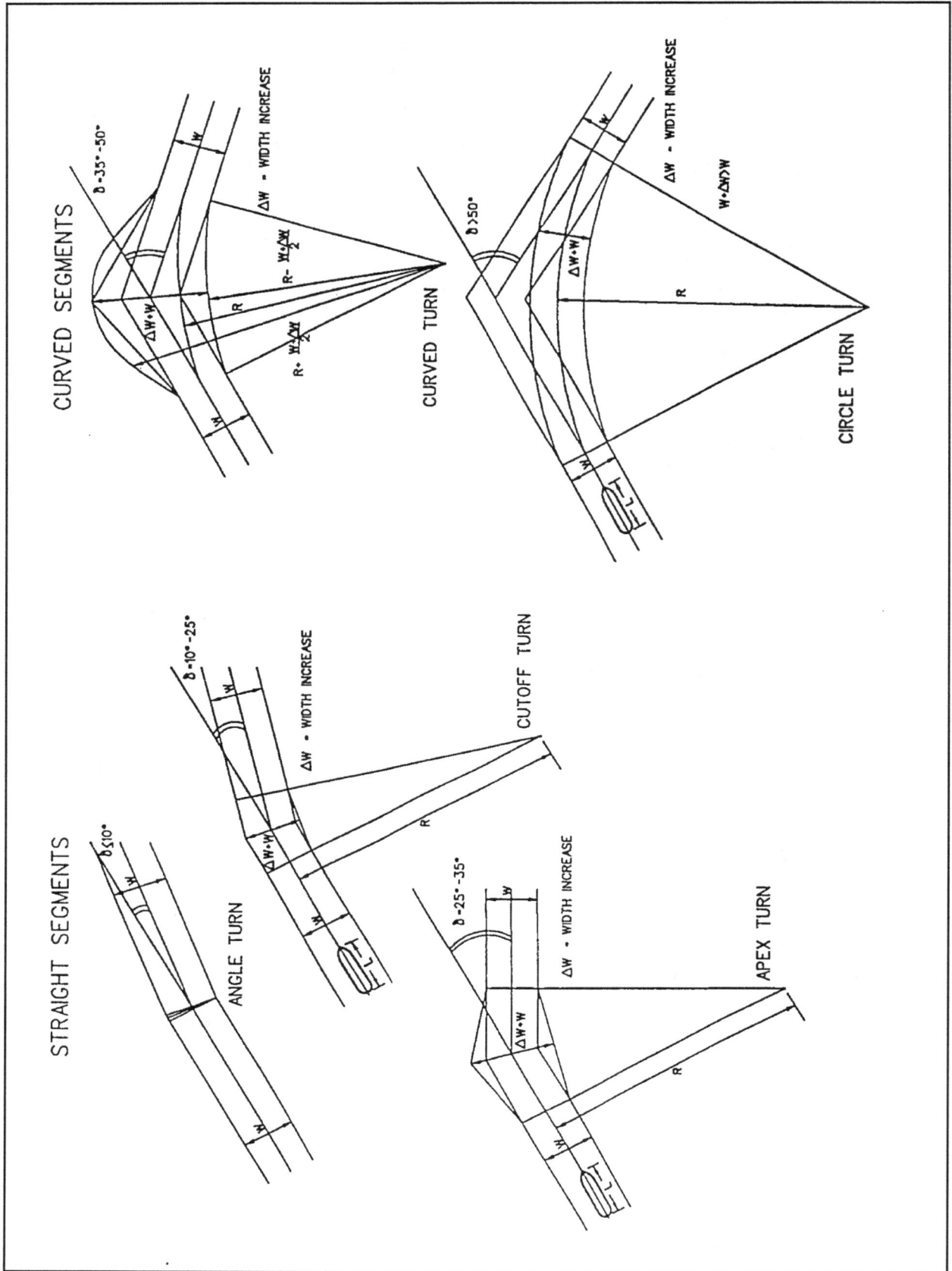

Figure V-5-27. Channel turn configurations

Table V-5-12
Recommended Deep-Draft Channel Turn Configurations

Turn Angle, deg	R/L[1]	Turn Width Increase Factor[2]	Turn Type
0-10	0	0	Angle
10-25	3-5	2.0-1.0	Cutoff
25-35	5-7	1.0-0.7	Apex
35-50	7-10	0.7-0.5	Curved
>50	>10	0.5	Circle

[1] R = curve radius; L = design ship length (see Figure V-5-26).
[2] Expressed as a multiplier of the design ship beam; i.e., ΔW = (factor from table) × B

(1) Apex or cutoff configurations are commonly used because they are simple, easily defined shapes that serve most channel turn requirements. They are easiest to control for dredging, easiest to mark with aids to navigation (usually with two range marker pairs and buoys), and easiest to monitor for maintenance. Some dimensions that are useful for computer drafting are included in Figure V-5-28. Turns should be placed, when possible, in locations easily visible for range markers. A drawback to the apex and cutoff methods is that they can produce difficult current patterns for navigation. In high current areas and/or canals, turns with parallel circular arcs gradually transitioning from straight channel segments into the turn may be warranted.

(2) The impact of channel turns on currents and shoaling and vessel response to wind and waves can be significant. The navigability of a turn design and its shoaling tendencies may merit study with numerical models and a navigation simulator.

(3) Small craft are sufficiently maneuverable that extra channel width in turns is generally not as critical as for deep-draft ships. For large turn angles in the presence of difficult conditions, extra maneuvering space should be provided. A typical situation would be a sharp turn in an exposed entrance channel.

V-5-7. Other Project Features

a. Turning basins. Turning basins are generally provided to allow vessels to reverse direction without having to go backward for long distances. Turning basins provide the extra width needed to comfortably turn. Turning basins are usually located at the upstream end of interior access channels. In long channels accommodating many dock facilities, an extra turning basin may be placed at the upstream end of each group of docks. In normal operations, larger ships turn with pilot and tug assistance.

(1) The minimum turning basin size should allow a turning circle with diameter of 1.2L, where L is the design ship length (Figure V-5-29). Turning difficulty increases significantly when currents are present. If current speed exceeds 0.5 m/sec (0.5 knot), turning basin diameter should be increased as indicated in the figure. If current speed exceeds 0.8 m/sec (1.5 knot), a circular shape no longer suffices. The turning circle should be elongated in the current flow direction, as shown in the figure. Dimensions of the high-current turning basin configuration should be determined with a ship simulator (Part V-5-10). If turning operations will include ships with high sail areas and design wind speeds of greater than 12.9 m/sec (25 knots), a ship simulator design study is also needed.

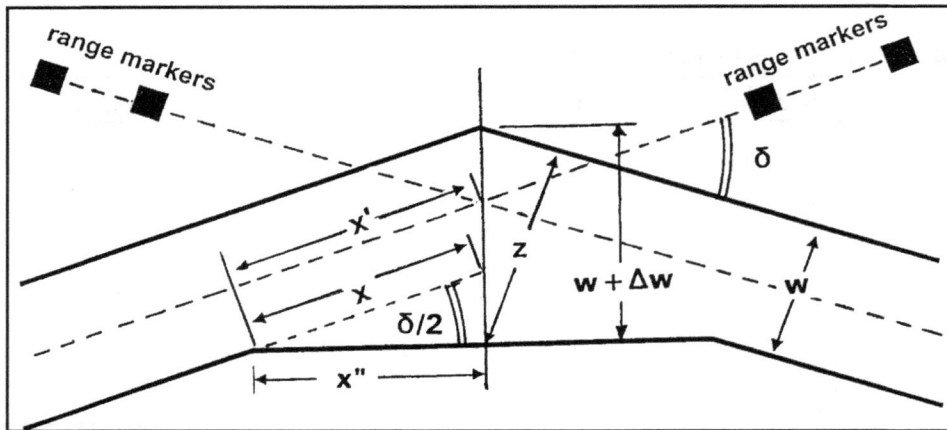

Figure V-5-28. Layout of apex-style turn

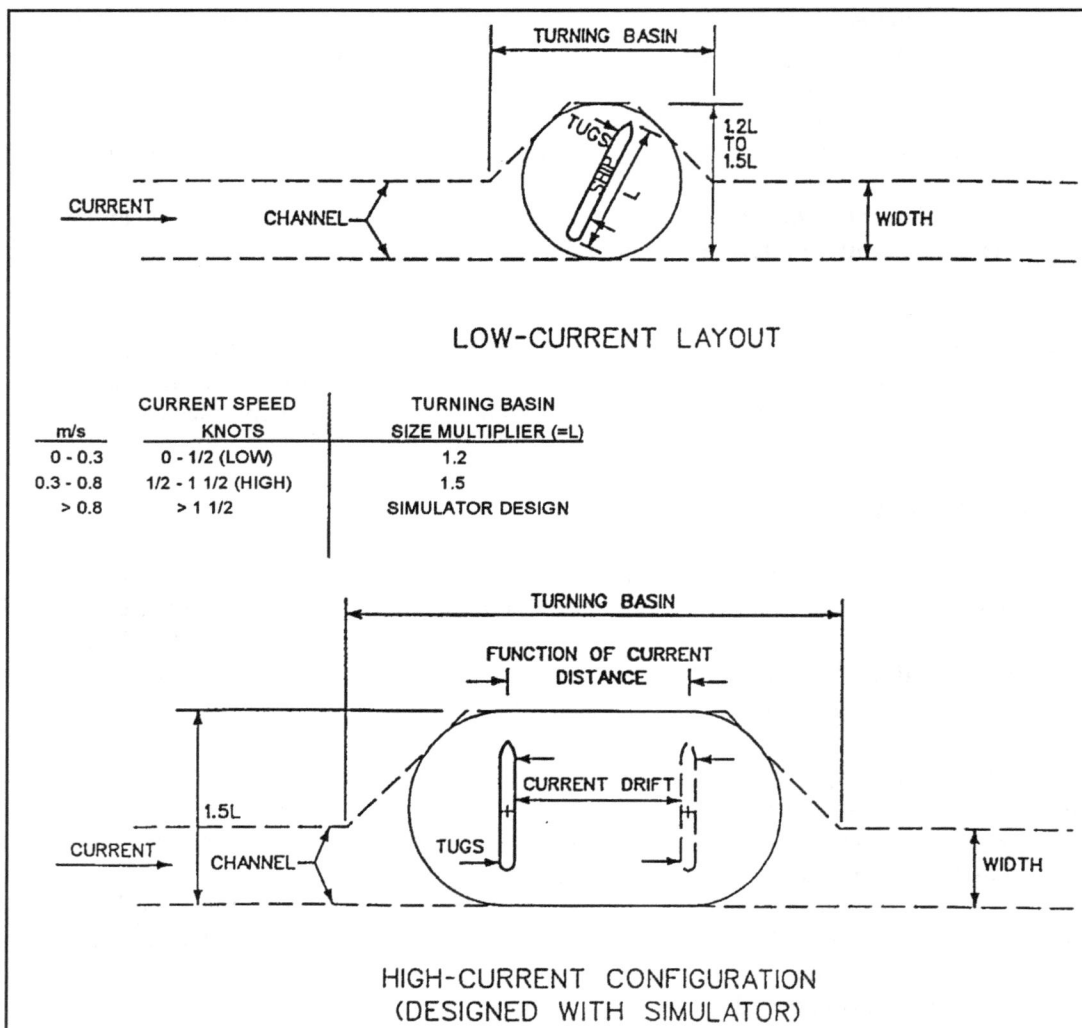

m/s	CURRENT SPEED KNOTS	TURNING BASIN SIZE MULTIPLIER (=L)
0 - 0.3	0 - 1/2 (LOW)	1.2
0.3 - 0.8	1/2 - 1 1/2 (HIGH)	1.5
> 0.8	> 1 1/2	SIMULATOR DESIGN

Figure V-5-29. Turning basin alternative designs

Navigation Projects

(2) Turning space for small boats should be sufficient to allow turning without backing or assistance. The required space depends on the types of boat using the project, but the design channel width usually accommodates turning.

(3) Turning basin design depth normally matches the adjacent channel depth. Turning basins tend to trap sediment because they are wider and typically have lower currents than the navigation channel. Also, being located at the upstream end of the navigation channel, the turning basin is often the first dredged area to intercept river-borne sediment. The potential shoaling problem may be reduced by careful choice of turning basin location and shape to reduce sediment availability and promote flushing.

b. Moorage/anchorage areas. Anchorage areas are provided in some deep-draft ports, typically near the entrance, to accommodate ships awaiting berthing space, undergoing repairs, receiving supplies and crews, awaiting inspection, and lightering off cargo. Anchorage areas may also serve as a refuge for ships during severe storms, since most facilities are designed with the assumption that no ships will be in port during a hurricane or typhoon. Ships may be anchored at the bow and allowed to swing freely or be moored against fixed dolphins. Design guidance is given in Figure V-5-30. Guidance for free-swinging anchorage is approximate, based on a 15-m (50-ft) depth and design ship length of 213 to 305 m (700 to 1,000 ft). The fixed mooring alternative requires a much smaller area for each ship than the free-swinging alternative. If available space is limited or dredging is required, fixed mooring may be the preferred design. Design guidance is available from the U.S. Navy (1998).

c. Basin flushing and water quality. Water quality within a harbor is often a concern with local, state, and Federal agencies. Water quality in a harbor depends on three key factors: quality outside the harbor, substances introduced into the water inside the harbor, and exchange of water between the inside and outside. Although outside water quality is typically beyond the scope of a navigation project, harbor design and operation can have a major impact on the other key factors. Contaminants introduced inside a harbor may include sewage discharge, shower/dishwashing water, bottom paint leaching, fuel and oil spillage, and deck and hull washing. Persistent contaminants may affect bottom sediments as well as water quality, which can increase future dredging costs.

(1) Water movement within a harbor is often restricted by the perimeter design. The extent to which the basin is enclosed, placement of the opening(s), and basin shape all affect water movement in the harbor. Tide, wind, and river flows can help promote circulation in a harbor and exchange between harbor and outside water. For example, a large tidal prism combined with a small low-tide volume gives excellent flushing in a harbor. Flushing processes in harbors are discussed in Part II-7.

(2) Many harbors are located in areas where natural water movement is relatively weak, a desirable condition for navigation but a potential problem for basin flushing. For example, when outside water has a high content of treated sewage, a typical concern, then water should not remain trapped anywhere in the harbor for many days. Methods to promote flushing and prevent stagnation include gaps between perimeter structures and shore, openings within the perimeter structure itself (e.g., segments, baffles, culverts), overlapping wave protective structures to act as scoops to passing water flow, and mechanical agitation. A harbor entrance centrally located along the perimeter is usually preferred over an entrance at one end, which may lead to stagnation at the opposite end.

(3) The flushing characteristics and water quality in a harbor can be predicted with physical and numerical models, as discussed in Part II-7. Both water exchange rates and oxygen or other chemical constituent reactions and replenishment rates should typically be addressed.

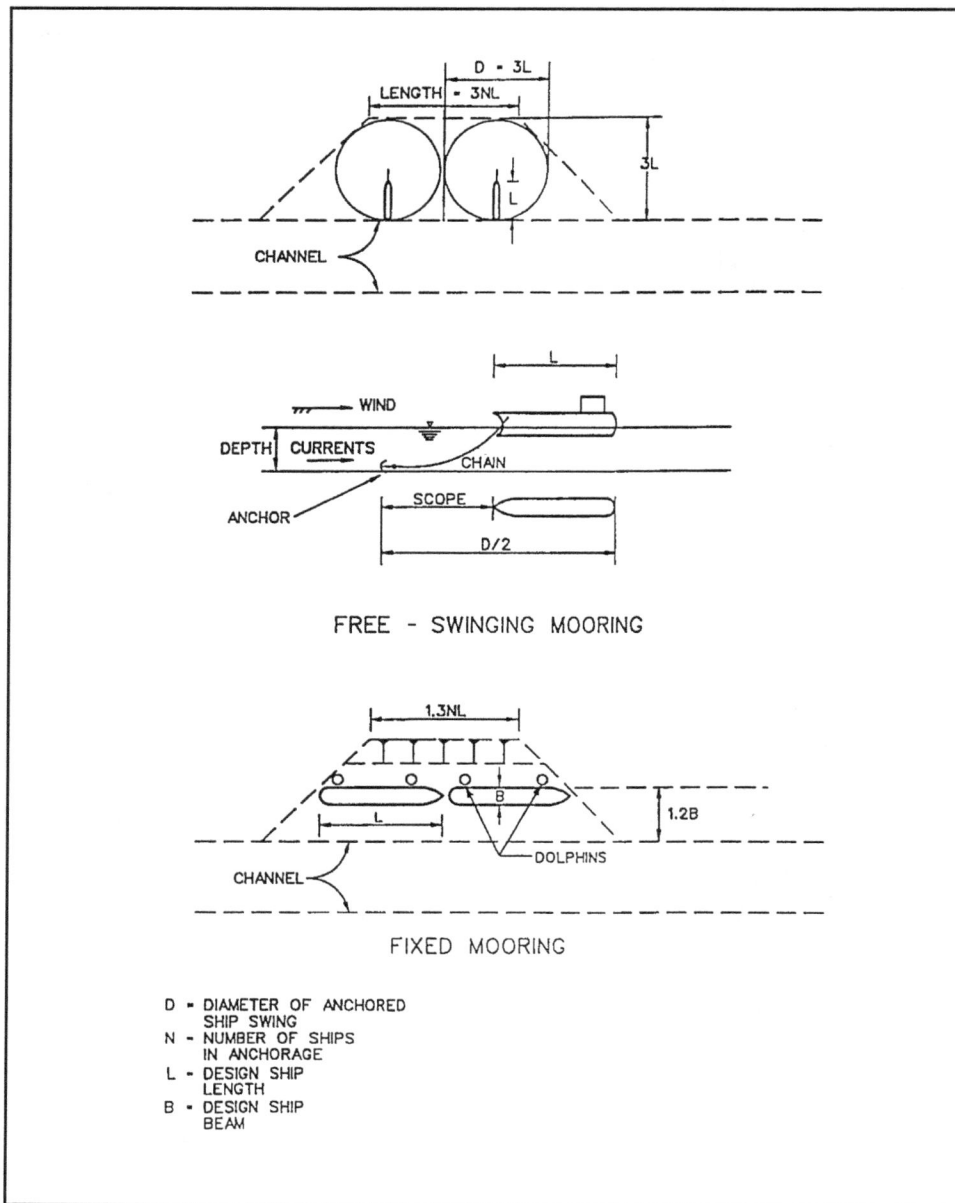

Figure V-5-30. Alternative anchorage designs

d. Navigation structures. Often a navigation project requires one or more engineered structures to accomplish its objectives. Structures can serve a variety of purposes. However, their presence also establishes a major hazard for vessels. Hence, a navigation structure must be designed with regard to several functional concerns. Basic types of structures and functions involved in navigation projects are briefly discussed in this section. Sediment processes and management at inlets and harbors are discussed in Part V-6. Detailed guidance on structure design is given in Part VI.

(1) Breakwaters. Breakwaters are used to protect a harbor, anchorage, basin, or area of shoreline from waves. Breakwaters reflect or dissipate wave energy and thus prevent or reduce wave action in the protected area. Breakwaters must be designed to effectively serve competing requirements for wave blockage and safe vessel passage from fully exposed waters through a constricted entrance into tranquil harbor waters.

Navigation Projects

(1) For navigation projects, breakwaters are frequently shore-connected and constructed to provide calm waters in a harbor (Figure V-5-31). Shore-connected breakwaters allow access from land for construction, operation, and maintenance, but may have an adverse impact on water quality or sediment movement along the coast.

Figure V-5-31. Harbor with shore-connected breakwater, Waianae Small Boat Harbor, Oahu, Hawaii (June 1980)

(2) If the harbor to be protected is on the open coast and predominant wave crests approach parallel to the coast, a detached offshore breakwater may be a good option. Water quality is preserved with this type structure, but access for construction and maintenance is more difficult than for a shore-connected structure. Offshore structures are sometimes used to provide protection to existing harbor entrances (Figure V-5-32). Accumulation of sediment in the lee of an offshore breakwater must be considered in design (Part V-6).

(3) Many breakwater systems utilize a combination of breakwater types to protect anchorage or mooring areas, such as a shore-connected and a detached structures (Figure V-5-33). Some structures have been constructed with arrowhead entrance configurations, but experience has shown them to be of questionable benefit, and some have been modified to improve performance. The cellular arrowhead breakwater configuration in Figure V-5-33 was modified by adding an overlapping rubble-mound extension to the seaward end of the shore-connected breakwater. In some cases, the primary breakwaters do not provide sufficient protection, and interior breakwaters are needed (Figure V-5-34).

(4) Most breakwaters built on open coasts of the United States are of rubble-mound construction. It is important for harbor tranquility to design such structures to be high enough to prevent excessive wave overtopping and sufficiently impermeable to deter wave transmission through the structure. However, some wave overtopping and/or transmission may be beneficial for harbor flushing. Other structural types include

Figure V-5-32. View of offshore detached breakwater used to protect harbor entrance, Marina del Rey, California (August 1966)

Figure V-5-33. Harbor with both shore-connected and detached breakwaters, Barcelona Harbor, New York (April 1986)

Figure V-5-34.   Inner breakwater provides protection to small-craft mooring area, Port Washington Harbor, Wisconsin (September 1983)

concrete caisson, timber crib, sheet pile, cellular steel sheet pile, composite (rubble-mound with concrete cap for stability, etc.), and floating.

(5) The optimum layout of breakwaters for harbor protection is difficult to determine because of the complex conditions typically involved. Waves refract on approaching the entrance, often in the presence of shoals as well as a navigation channel. Wave energy propagates through the entrance, diffracts, and reflects from inner harbor structures. Currents are usually present due to tides, wind, waves, and river flows.

(6) Because they can accurately reproduce many of the complex, interacting hydrodynamic effects on a harbor, physical models provide the most reliable method for optimizing breakwater layout (Part V-5-10). Physical modeling and subsequent monitoring of prototype performance has led to general guidance relative to harbor layout (Bottin 1992). Lessons learned relative to navigation entrance channel and mooring area protection in small boat harbors are:

(a) *Align entrances toward a perpendicular to the incoming wave crests.* It is very dangerous for small craft to travel parallel to high incoming wave crests (*beam seas*). Harbors should be designed to minimize the chance of this condition.

(b) *Block wave energy from the harbor.* It is preferable to prevent wave energy from entering a harbor than to try to dissipate excessive wave energy once inside. Energy entering a harbor can be minimized by using overlapping breakwaters at the entrance, reducing entrance width, minimizing breakwater overtopping, and using impermeable breakwater cores. Breakwaters seaward of the entrance may also be incorporated into a design. If the harbor is at a river mouth, care must be taken to prevent upstream flooding due to flow restriction and/or ice jamming in the harbor.

(c) *Absorb wave energy inside the harbor if necessary.* When physical limitations or costs prevent blocking sufficient wave energy out of the harbor, some energy can be absorbed inside the harbor with judiciously placed rubble slopes and/or spending beaches. Concrete absorber units (e.g. *igloos*) can also be helpful. For long-period wave energy, absorbers installed along harbor slips are essentially ineffective.

(d) *Anticipate Cross Currents on Reefs.* Waves breaking across reefs generally result in very strong currents alongshore. These currents may be hazardous to small craft entering and navigating channels cut through the reef and into harbors. Breakwaters may be used to deflect these currents offshore away from the entrance (Figure V-5-35). Currents also tend to enter the harbor through the entrance. These currents can be used to advantage by laying out interior channels to promote circulation and flushing, as was done with the aid of physical modeling for Agana Small Boat Harbor, Territory of Guam.

Figure V-5-35. Breakwaters protecting small boat harbor from waves and wave-generated crosscurrents on reef, Agana, territory of Guam (May 1978)

(e) *Locate Harbor Facilities and/or Boat Ramps Away from the Entrance Opening.* Wave energy propagating through and diffracting around entrance structures can affect nearby facilities. Facilities can be given further protection by including interior breakwater structures or revetted moles. Although these structures can be very effective, they are expensive and can limit future expansion.

(f) *Avoid vertical walls in high-energy areas of the harbor.* Vertical wall breakwaters and harbor structures are highly reflective. Waves reflecting off entrance structures can result in very confused and hazardous navigation conditions. Reflections inside harbors can cause hazardous anchorage and mooring conditions. Reflected waves from vertical structures have also been found to induce erosion.

(g) *Orient entrances away from the direction of predominant longshore transport.* It is very desirable to promote natural sand bypassing of the harbor entrance. Entrances facing the predominant longshore transport direction are likely to serve as a sediment trap. A typical successful design consists of an outer curved breakwater that overlaps a short shore-connected structure (e.g. Figure V-5-31). The shorter downcoast structure helps to prevent sediment moving along the shoreline opposite to the dominant direction from coming into the entrance.

(h) *Consider Using Segmented Structures.* Segmented breakwaters are effective in providing wave protection while still allowing tidal circulation through the breakwater openings. They can be effective substitutes for floating or baffled breakwaters. Segmented rubble absorbers inside a harbor, as opposed to a continuous absorber, also have proven to be effective in terms of both performance and cost.

(2) Jetties. A jetty is a shore-connected structure, generally built perpendicular to shore, extending into a body of water to direct and confine a stream or tidal flow to a selected channel and to prevent or reduce shoaling of that channel. Jetties at the entrance to a bay or a river also serve to protect the entrance channel from storm waves and crosscurrents. When located at inlets through barrier beaches, jetties help to stabilize the inlet.

(a) Jetties are usually built in pairs, one on either side of an entrance (Figure V-5-36). Sometimes, jetties are used in combination with a breakwater (Figure V-5-32). A single jetty may also be used, located on the updrift side of the entrance. A disadvantage of the single jetty is that the navigation channel is unconfined and will likely migrate. A typical single jetty problem is the case of an impermeable jetty where the navigation channel migrates to a position immediately beside the jetty, with consequent threats of passing vessels colliding with the structure and undermining of the structure itself. Some jetties have been designed with a low-crested weir section near shore to pass sediment into a deposition basin (Part V-6). The semi-protected waters of the deposition basin are then periodically dredged and sediment is bypassed to the downdrift beach.

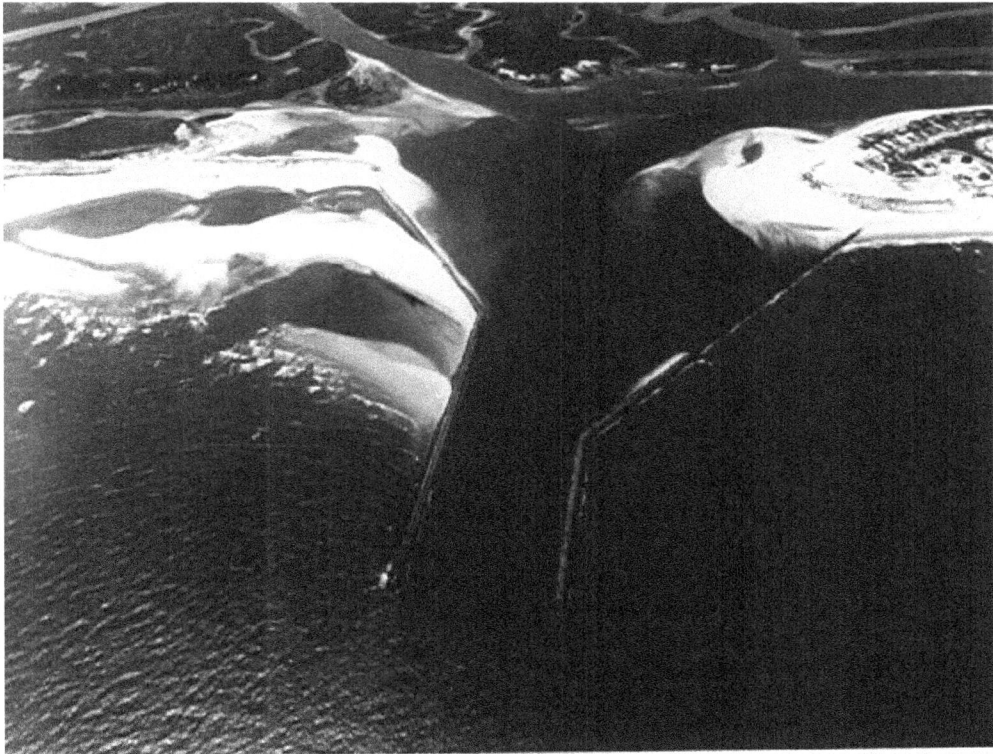

Figure V-5-36. Dual jetty configuration at a tidal inlet, Murrells Inlet, South Carolina (March 1982)

(b) Though jetties have a different function than breakwaters, jetty structural design is similar to breakwaters. Most jetties built on open U.S. coasts are rubble-mound structures. Materials used for jetty construction include stone, concrete, steel, and timber. Unlike breakwaters, jetties are usually designed to allow some wave overtopping. Also, jetty cores may be lower and more permeable than breakwater cores, provided the jetty sufficiently blocks passage of sediment into the navigation channel.

(c) Jetties should be designed to use available construction materials efficiently and effectively to accomplish functional objectives without adversely impacting other physical processes or the environment. Jetties at river or creek mouths, not considering tidal effects, should be spaced close enough together to allow normal river currents to maintain required navigation depths, yet far enough apart to prevent backwater effects and flooding upstream during high river flows. In cold regions, jetties should be oriented to avoid contributing to ice jamming in the entrance, which could cause upstream flooding. In some cases, overlapping jetties have proven to contribute to natural sand bypassing, and they reduce wave energy entering the mouth and lower reaches of the stream. If large quantities of sediment are moving in the area, sand bypassing schemes should be included in the design (Part V-6).

(d) Jetty layout in tidal areas is much more difficult. Interaction of wave-induced currents and tidal currents, sometimes with freshwater discharges, through an inlet connecting the ocean with an embayment, is very complex. A fairly uniform distribution of flow across the entrance is one objective. Dual jetties of equal length usually serve best. The jetties should be parallel if practical; otherwise training dikes or spurs should be added to divert or concentrate flow through the desired channel alignment. If jetty spacing is too wide, shoaling and channel meandering are likely to occur. If jetty spacing is too narrow, structure toes may be undercut and hazardous navigation conditions may occur.

(e) Parallel jetties tend to confine flood and ebb flow, raising flow velocities and providing adequate sediment flushing into the flood and ebb deltas. Arrowhead jetties frequently allow channel shoaling and meandering because ebb flow is not confined enough to produce nondepositional velocities in the widest area between jetties. Barnegat Inlet, New Jersey, is an example where both jetty configurations have been used (Sager and Hollyfield 1974; Seabergh, Cialone, and Stauble 1996) (Figure V-5-37). Arrowhead jetties were built in 1939, and a new south jetty, nearly parallel to the north jetty, was built in 1991. As another general alternative for jetty layout, curved jetties may be designed to produce nondepositional velocities, but flow concentrations on the outside of the curve may cause jetty undermining and a difficult channel alignment for navigation (McCartney, Hermann, and Simmons 1991).

(f) Jetties should be long enough to prevent littoral transport around the jetty ends and into the navigation channel. Jetty orientation for navigation purposes should ensure that the channel is approximately aligned with the approach direction of the more severe waves. Typically, a jetty alignment perpendicular to shore serves this purpose. The ideal jetty alignment for navigation is often a poor alignment for sheltering interior areas from waves. However, waves lose a significant amount of energy in traveling between parallel jetties (energy loss increases with jetty length) or passing through an entrance gap between breakwaters or jetties (Part II-7). The height of waves traveling between parallel jetties may be estimated by treating the jetty entrance as a breakwater gap (Melo and Guza 1991). The inter-jetty propagation distance corresponds to the normal interior distance from the gap and wave height can be estimated from height diffraction contours as given in Part II-7.

(g) Spacing between dual jetties should be determined with consideration of tidal processes, wave protection requirements, river flood discharge requirements, and safe navigation requirements. Jetty spacing for tidal concerns depends on the tidal prism volume or the actual tidal flow exchange through the inlet. Relationships between minimum cross sectional area required at inlet throats as well as detailed design of jettied entrances are presented in Part II-6. Wave protection requirements of interior shorelines and facilities

(a) Arrowhead jetties (November-December 1965)

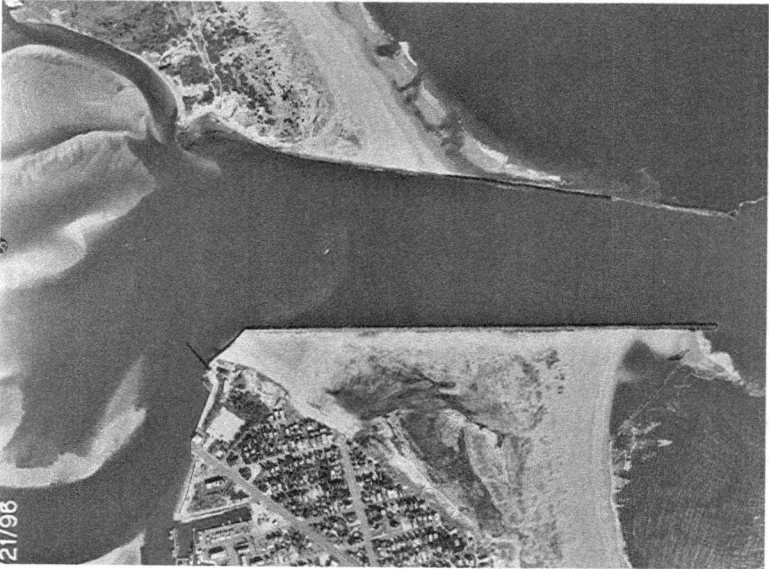

(b) Parallel jetties (June 1996)

Figure V-5-37. Jetty types, Barnegat Inlet, New Jersey

are specific to each project. A rule of thumb for safe deep-draft navigation is that an entrance width equal to the design ship *length* is satisfactory for two-way traffic. An overbank area between the channel and jetty is needed to protect the structure toes.

(3) Training dikes. Training dikes serve to direct current flow in a desired path. A common application is the use of training dikes interior to a jettied inlet to confine currents to the navigation channel and help prevent channel shoaling and erosion of nearby banks and shores. Dikes are usually constructed of stone, timber pile clusters, or piling with stone fill.

(4) Wave absorbers. Wave energy that penetrates into small boat harbors through entrances or by transmission or overtopping of breakwaters may create serious problems for navigation in interior channels, moored boats, and bank erosion. Excessive wave energy in deep-draft ports can cause similar problems. Wave absorbers are sometimes constructed inside harbors to dissipate this short-period wave energy. Absorbers are generally placed in harbor areas with high concentrations of wave energy.

(a) Beaches are the most effective wave absorbers. Beach slopes dissipate most wind wave and swell energy and reflect very little energy. However, beaches also occupy a relatively large area and may not be practical. More commonly, rubble slopes serve as wave absorbers in navigation projects. For best efficiency, a stone wave absorber should have a rock layer thickness equal to three times the representative diameter of individual rocks used and porosity should be about 30 percent (LeMéhauté 1965).

(b) Sometimes rubble slope absorbers are integrated into a dock to provide wave dissipation without interfering with harbor use. Typical applications are an armored slope under a pile-supported dock face or an absorbing quay wall (PIANC 1997b) (Figure V-5-38). Another approach is a perforated wall along the dock face with a vertical solid-wall fill some distance behind the dock face (Figure V-5-39). The optimum porosity for the perforated wall should be around 30 percent and the optimum set-back of the solid wall should be about one tenth of the wave length to be absorbed (LeMéhauté 1965). Molded concrete wave absorber units may also be used along a vertical dock face. Such units have proven effective in model tests for a Great Lakes site (Bottin 1976) (Figure V-5-40), but have not been used in U.S. harbors.

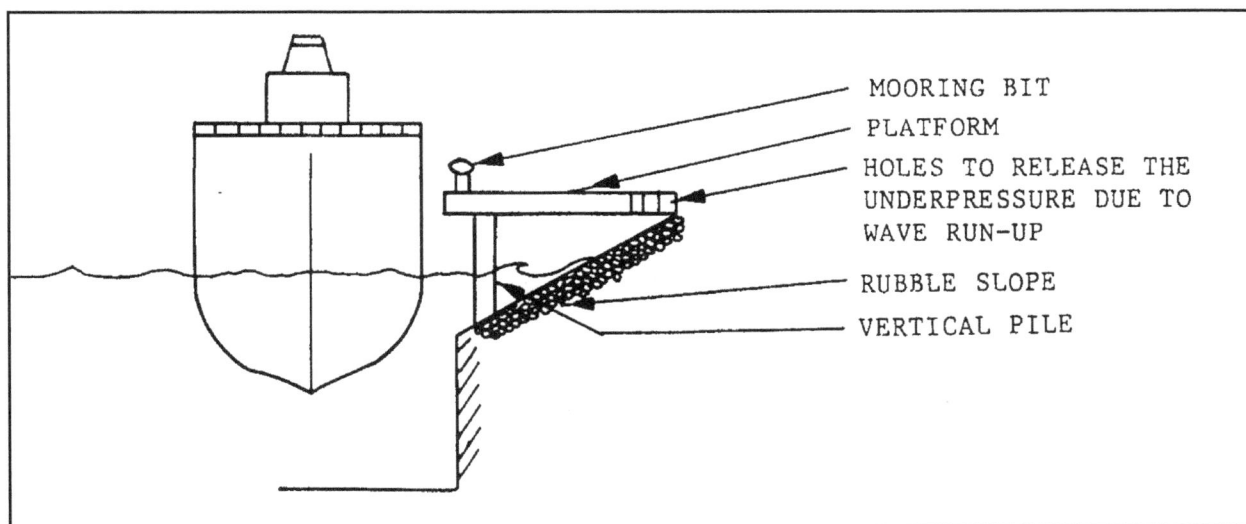

Figure V-5-38. Rubble wave absorber at a dock

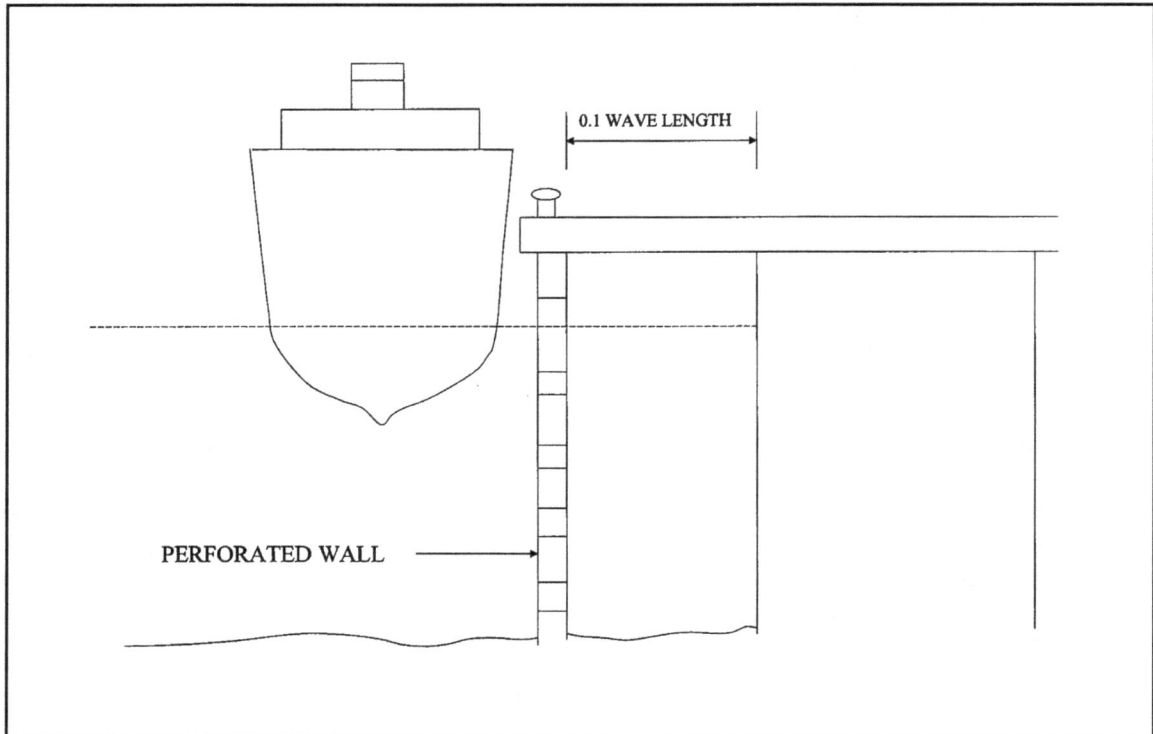

Figure V-5-39.   Perforated wall wave absorber at a dock

Figure V-5-40.   Stacked Igloo model wave absorber units

(5) Revetments, seawalls, and bulkheads. Though not actually navigation structures, revetments, seawalls, and bulkheads are included in, or adjacent to, most harbors. They are briefly reviewed here and discussed in detail in Parts V-3, VI-2, and VI-7. In general, vertical structures are classified as either seawalls or bulkheads, according to their function, while protective materials laid on slopes are called revetments. A seawall is a massive structure that is designed primarily to reduce wave energy and provide wave protection along coastal property. Bulkheads are retaining walls whose primary purpose is to hold or prevent backfill from sliding while providing protection against light-to-moderate wave action.

(a) Revetments, seawalls, and bulkheads are structures placed parallel, or nearly parallel, to the shoreline to separate a land area from a water area. The purpose of the structure dictates which type is used. In harbors, these structures may improve or worsen wave conditions and land access to beaches, depending on the location and design.

(b) Vertical structures (seawalls and bulkheads) are useful as quay walls or docking/mooring areas. Because of potential wave reflection problems, bulkheads to provide vessels with direct access to shore should only be placed in areas with little exposure to wave energy. Wave reflection from vertical structures may create hazardous conditions for small craft, erosion of adjacent shorelines, and beach profile changes.

(c) Revetments are typically less reflective than vertical structures. Rubble revetments may be effective wave absorbers but may hinder access to a beach. Smooth revetments built with concrete blocks generally present little difficulty to pedestrians, but are more reflective than rubble revetments.

(d) On seawalls and bulkheads, convex-curved face and smooth slopes are least effective in reducing wave runup and overtopping. Concave-curved face structures are most effective for reducing wave overtopping when onshore winds are light. Where the structure crest is to be used as a road, promenade, or other such purpose, concave-curved may be the best shape for protecting the crest and reducing spray. If onshore winds occur with high waves, a rubble slope should be considered to reduce runup and overtopping. A stepped-face wall provides the easiest access to beach areas from protected areas, and reduces the scouring of wave backwash. Some seawalls and bulkheads may create access problems and may require the building of stairs.

V-5-8. Aids to Navigation

(1) Aids to navigation are the markers and signals vessels require to safely use a navigation project. The navigation safety of a project is directly related to the clarity and visibility of aids to navigation. Channel design must be planned so that the layout, dimensions, and alignment facilitate clear marking. A reduced width may be possible in a well-marked channel as compared to a poorly marked channel, so a tradeoff between channel widening cost and aids to navigation cost should be considered in design.

(2) The U.S. Coast Guard is responsible for the design, establishment, and maintenance of all aids to navigation in Federal interstate waters (U.S. Coast Guard 1981, 1988a, 1988b). Figure V-5-41 gives two examples of typical devices used to mark U.S. navigation channels. They are uniquely identified by color, shape, and number or letter. They may also include lights, sound, radar reflectors, and electronic signals. Beacons are fixed structures, generally on pilings in shallow water up to about the 5-m (15-ft) depth. Buoys are floating, anchored to the bottom with a chain connected to a concrete block. They mark channel boundaries, hazards, and channel curves or turns, especially in areas where water depth makes beacons impractical. Height above the water, and hence visibility, is more limited for buoys than for beacons. Another limitation of buoys is that their location relative to the channel is imprecise. It can vary over a small distance because buoys are free to move about the anchor point in response to environmental forces.

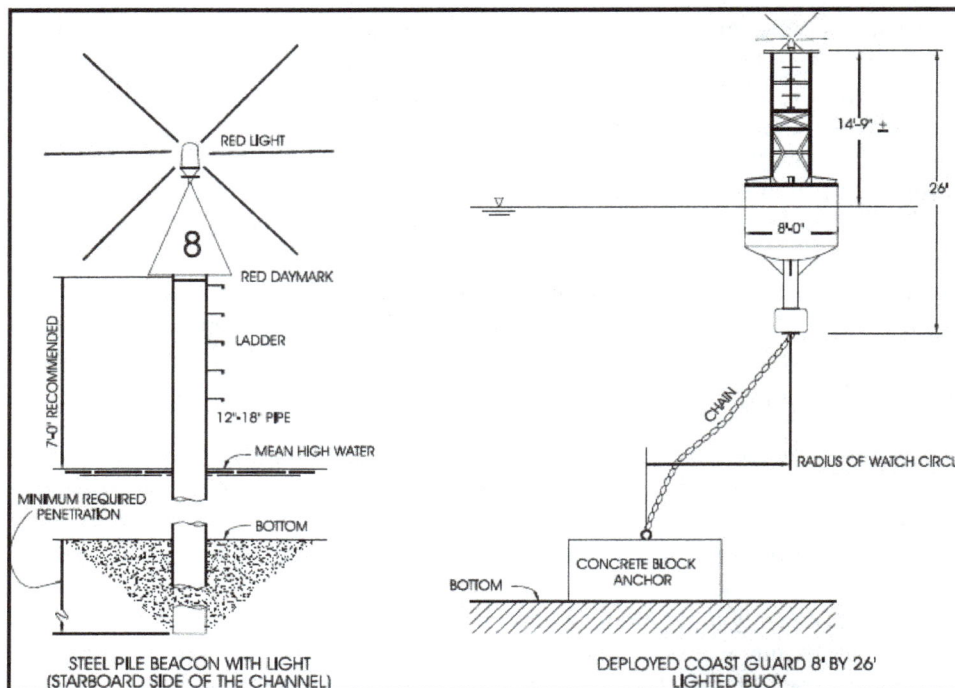

Figure V-5-41. Examples of aids to navigation

(3) Occasionally, buoy/anchor systems are completely moved out of position by strong environmental forces or by vessel impacts. Buoys are also susceptible to sinkage or drifting if mooring connections are lost.

(4) The channel marking system used in U.S. Federal waters is nearly standardized. Conventions for color, shape, numbers/letters, and light characteristics are well-established. Left and right channel sides are relative to an inbound vessel coming from open water into a harbor. Basically, the left side of the channel is signified by green (paint color and/or lights), squares, and odd numbers (Figure V-5-42). The right side is signified by red, triangles, and even numbers. Numbering begins at the seaward end of the channel.

(5) Ranges are pairs of fixed structures usually aligned with the channel center line at one or both ends of straight reaches. They are usually on shore or in very shallow water. The rear marker is always higher than the front marker. They are typically marked with rectangular signs, designated by letters, high-intensity lights, and red and white vertical stripes, as indicated in Figure V-5-42. By observing the placement of front and rear markers relative to each other, mariners can determine vessel position relative to the channel center line (Figure V-5-43).

(6) Additional important aids to navigation along the seacoast include major lights and sea buoys. One or more major lights are located near each harbor entrance. The high-intensity, well-maintained lights are located on fixed structures or towers at heights of up to 60 m (200 ft), sufficient to be visible over a long distance. Electronic aids to navigation are often collocated with major lights. Sea buoys are large, easily visible buoys marking the ocean end of most deep-draft harbor entrance channels. A typical sea buoy is 12 m (40 ft) in diameter and 9 m (30 ft) or more in height, with high-intensity light, electronic aids, and a sound signal. Sea buoys are usually located in deep water on the channel center line extended 2 to 4 km (1 to 2 miles) seaward beyond the channel's seaward end. Often the sea buoy marks an area where inbound ships await local pilot assistance.

RANGE MARKERS

WHITE LIGHT

WHITE | RED | WHITE

A

LEFT OR PORT SIDE
ODD NUMBERED AIDS
GREEN COLOR

3

UNLIT CAN BUOY

GREEN LIGHT

9

LIGHTED BUOY

1

GREEN DAYMARKS

CHANNEL

INBOUND FROM
SEA OR LAKE

RIGHT OR STARBOARD SIDE
EVEN NUMBERED AIDS
RED COLOR

6

UNLIT NUN BUOY

RED LIGHT

10

LIGHTED BUOY

2

RED DAYMARKS

LATERAL DESIGNATION SYSTEM OF
AIDS TO NAVIGATION

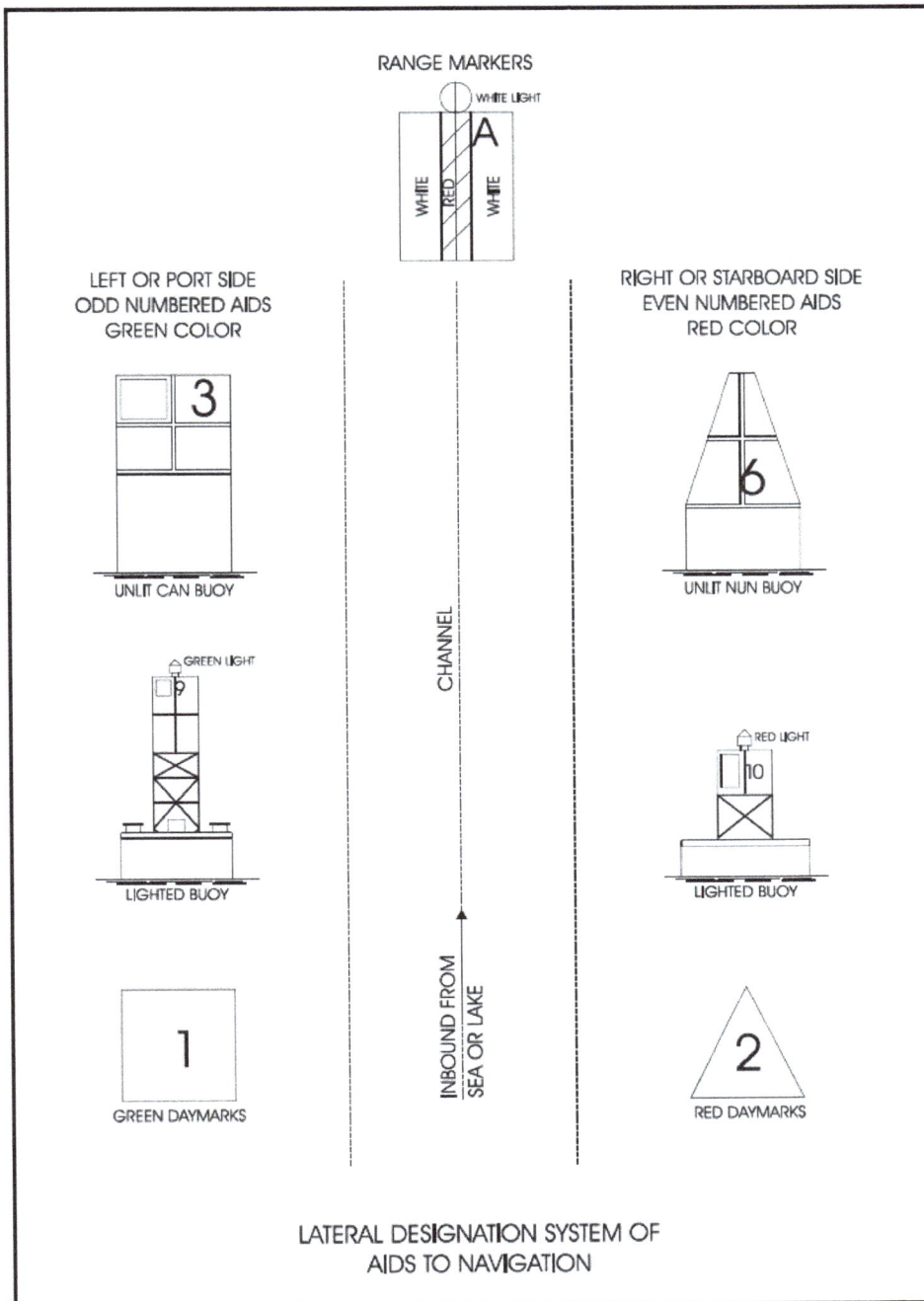

Figure V-5-42. Designation system of aids to navigation

(7) Aids to navigation are normally placed along straight channel reaches so that at least two on either side are always visible. This consideration leads to a practical maximum spacing of about 2.3 km (1.25 n.m.). Range markers must be visible along the entire reach. Practical limitations on range marker height, visibility through fog, and earth curvature effect on line of sight dictate that straight channel reaches should be no longer than about 8 to 10 km (5 to 6 miles). It is good practice to have redundant aids to navigation, such as both range markers and side channel markers, to ensure that failure of a marker will not create a navigation crisis.

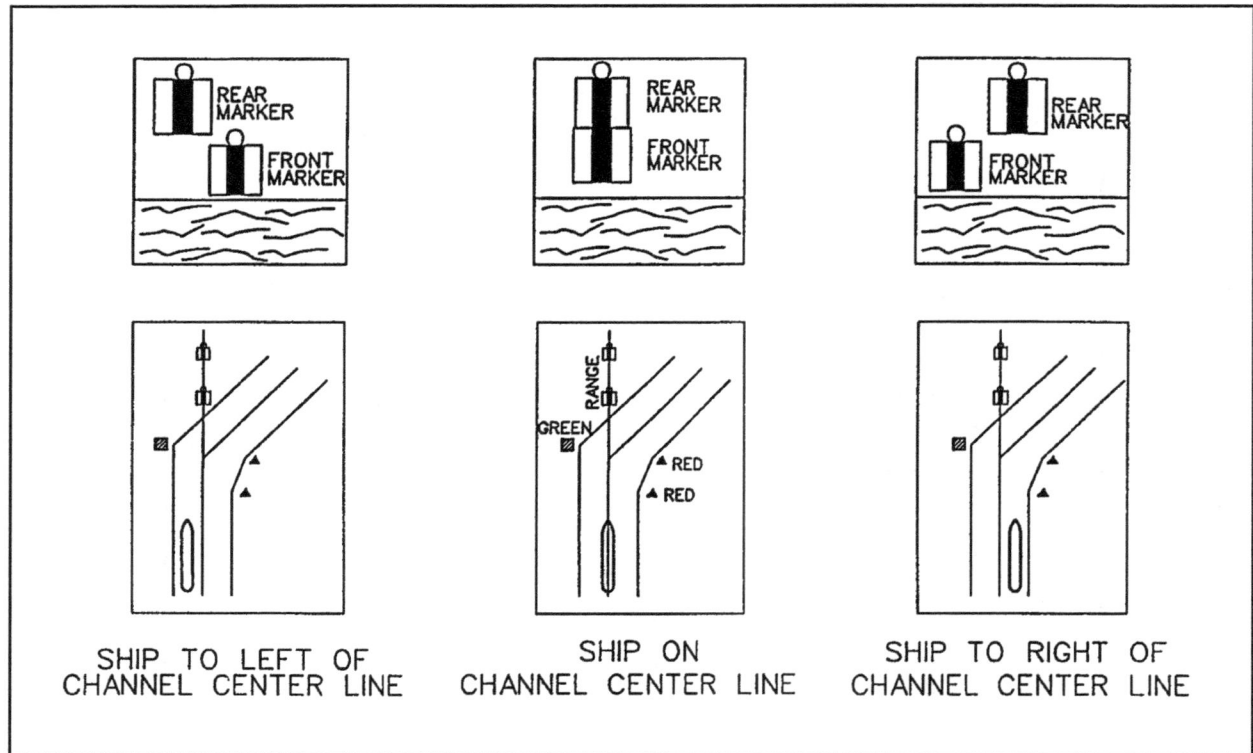

Figure V-5-43. Use of ranges for channel position

(8) An example deep-draft navigation channel at Brunswick Harbor, Georgia, is shown in Figure V-5-44 (Huval and Lynch 1998). The entrance channel begins near a sea buoy in exposed Atlantic Ocean waters, passes between St. Simons and Jekyll Islands, makes a severe turn to pass behind Jekyll Island, and continues with additional turns up the Brunswick River to port facilities at Brunswick. The channel also passes under the Sidney Lanier Bridge. Navigation buoys and range markers are shown.

V-5-9. Operation, Monitoring, and Maintenance

a. After a navigation project has been designed and constructed, operation and maintenance are required to sustain safe and efficient use of the project. Operation and maintenance requirements and costs can be substantial. They are typically estimated with care and optimized against initial construction costs in planning and designing a navigation project. Anticipated maintenance costs are based on predictions of physical changes after the project is constructed.

b. A completed navigation project must be monitored to ensure safe operation and to plan for maintenance activities as needed (see Part V-2-17). Monitoring typically includes hydrographic surveys, beach profile surveys, tide and wave data collection, and navigation structure condition surveys. Surveys are typically done on a planned schedule, such as annually, and before and after periods of maintenance and repair. Surveys should be analyzed comparatively to determine rates of erosion, shoaling, and structure deterioration.

c. Often, periodic dredging to maintain project depths is the major maintenance need. Maintenance dredging intervals depend on factors such as shoaling rate, dredge availability, and dredge mobilization costs.

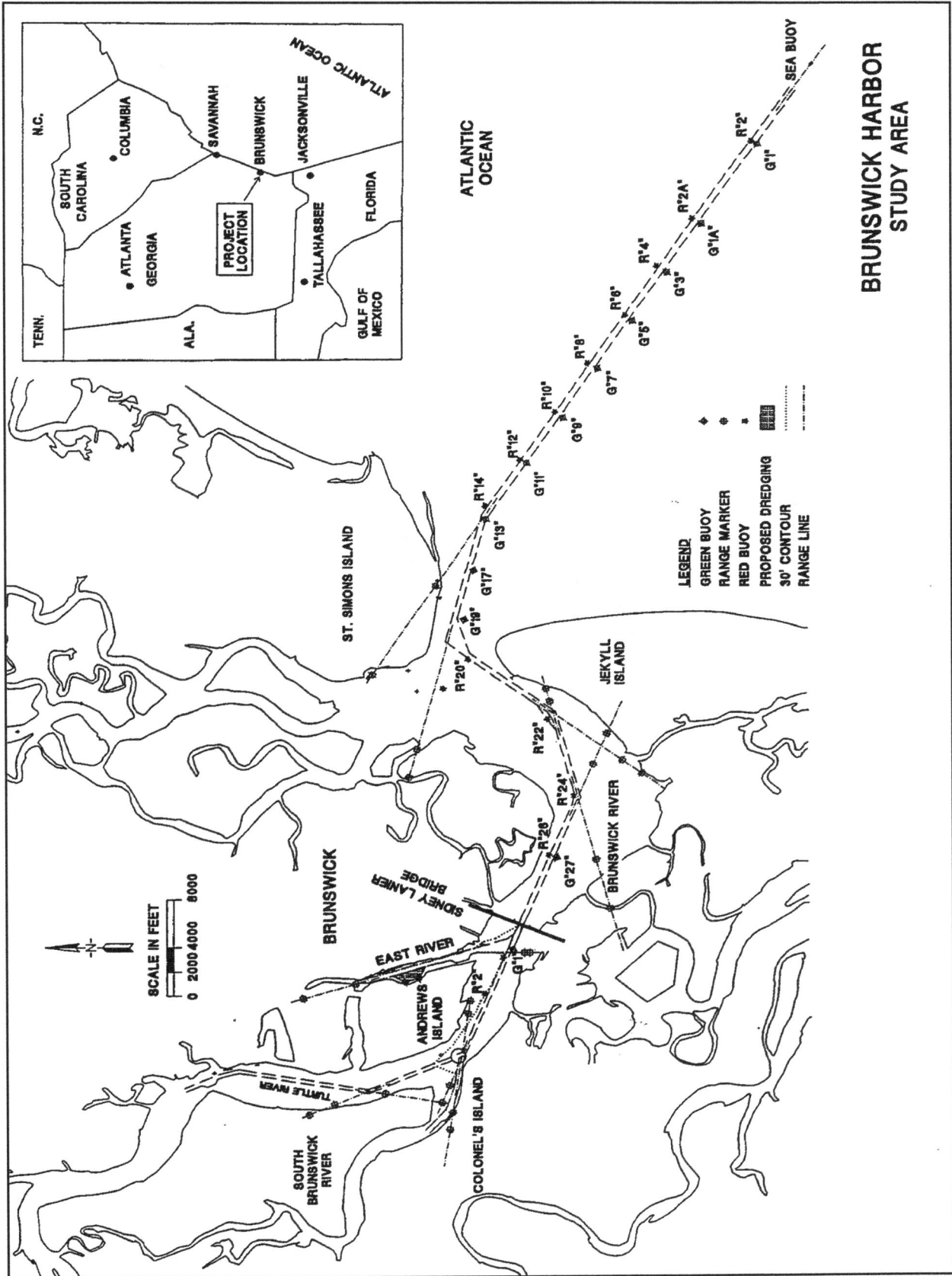

Figure V-5-44. Deep-draft navigation channel, Brunswick Harbor, Georgia

d. Typical maintenance intervals are on the order of 1-3 years at some projects, but maintenance needs are often strongly influenced by storm events. However, environmental forces impacting a navigation project are highly variable. The number and intensity of storms affecting a project each year can only be predicted in terms of probabilities. A single severe storm can cause major shoaling and structure damage. Consequently, monitoring and maintenance activities may occasionally need to respond quickly to maintain project integrity.

e. Project performance should be assessed, based on monitoring data. Actual project performance should be evaluated relative to project expectations during original design. In particular, actual maintenance costs should be compared to those originally predicted. Coordination with local interests, including boaters, pilots, port authorities, etc., should also be part of project performance assessment. Monitoring, maintenance, and performance assessment should continue at an appropriate level for the life of the project.

V-5-10. Model and Specialized Field Studies

a. Harbor modeling. Physical and numerical harbor models are important design tools that can help the designer: Locate the project to ensure maximum wave protection; locate and design breakwaters and/or jetties to provide adequate protection and maintain entrance navigation channels; and locate, orient, and dimension navigation openings to provide vessels safe and easy passage into and out of a harbor without sacrificing wave protection; position spending beaches and other forms of wave absorbers inside the project area. Physical and numerical modeling tools are helpful in developing and optimizing harbor designs. When compounded with problems caused by nearby or adjacent rivers, and/or shoaling problems resulting from littoral transport, and/or harbor oscillation problems relative to long-period wave energy, the designer encounters difficulty in obtaining adequate answers strictly by analytical means. One or both of these tools should be applied when a study has large economic consequences. Even small harbor studies generally benefit from model studies.

(1) Physical modeling as a design tool. Hydraulic scale models are commonly used to plan harbors and to design and lay out breakwaters, jetties, groins, absorbers, etc., to obtain optimum harbor protection and verify suitable project performance. A detailed description of physical modeling related to coastal ports and harbors is given by Hudson et al. (1979). Physical hydraulic model studies may be used to study the following:

- The most economical breakwater and/or jetty configurations that will provide adequate wave protection and navigation channel control for vessels using the harbor.

- Wave heights in the harbor.

- Undesirable wave and current conditions in the harbor entrance.

- Proposals to provide for harbor circulation and/or flushing.

- Qualitative information on the effects of structures on the littoral processes.

- Flood and ice flow conditions.

- Shoaling conditions at harbor entrances.

- River flow and sediment movement in rivers that may enter in or adjacent to the harbor.

- Long-period oscillations (Part II-7).

- Tidal currents or seiche-generated currents in the harbor (Part II-7).

- Inlet entrances.

- Remedial plans for alleviation of undesirable conditions as found necessary.

- Possible design modifications to significantly reduce construction costs and still provide adequate harbor protection.

(a) To ensure accurate reproduction of short-period wave and current patterns (i.e., simultaneous reproduction of both wave refraction and wave diffraction), undistorted models (i.e., vertical and horizontal scales are the same) are necessary for harbor studies. Physical hydraulic models are designed and operated in accordance with Froude's model law (Stevens et al. 1942). Scale relations commonly used for undistorted physical models are shown in Table V-5-13. A scale of 1:100 is used for illustrative purposes.

Table V-5-13
Typical Physical Model Scales for Harbors

Characteristic	Dimension[1]	Scale Relations
Length	L	$L_r = 1:100$
Area	L^2	$A_r = L_r^2 = 1:10,000$
Volume	L^3	$\forall_r = L_r^3 = 1:1,000,000$
Time	T	$T_r = L_r^{1/2} = 1:10$
Velocity	L/T	$V_r = L_r^{1/2} = 1:10$
Roughness (Manning's coefficient, n)	$L^{1/6}$	$n_r = L_r^{1/6} = 1:2,154$
Discharge	L^3/T	$Q_r = L_r^{5/2} = 1:100,000$
Force (fresh water)	F	$F_r = L_r^3 \gamma_r = 1:1,000,000$
Force (salt water)	F	$F_r = L_r^3 \gamma_r = 1:1,025,641$

[1] Dimensions are in terms of length (L), time (T), and force (F).

(b) Selection of a suitable model scale is an important step in model design. It involves a trade-off between scale effects and construction costs. For short-period wave studies, the model area generally includes enough offshore area and bathymetry to allow waves to refract properly on approaching the harbor and enough upcoast/downcoast area to allow the littoral current to form. Model waves must be large enough to be free from excessive friction and surface tension and to be measured with reasonable accuracy. Typical scales are between 1:75 and 1:150. For long-period wave studies (periods longer than about 25 sec), larger prototype areas are generally needed to include relevant interactions with bay or coastal shelf bathymetry. Also, long-period waves tend to reflect from model basin boundaries and the model harbor should be far from these boundaries. Long-period wave studies usually require distorted-scale models, with unequal horizontal and vertical scales. For example, a USACE model of Los Angeles/Long Beach Harbor, in use since the 1970's has a vertical scale of 1:100 and a horizontal scale of 1:400.

(c) Small-scale models must be constructed very accurately to reproduce conditions in the prototype. The model should reproduce underwater contours to model wave transformation. Shoreline details and irregularities also are important to simulate diffraction, runup, and reflection. The model bed should be as smooth as possible to minimize viscous scale effects. However, models involving estuary tidal flows and/or river flows typically require the addition of bottom roughness in those areas to correctly simulate flow conditions. When a model involves breakwater or jetty structures, model armor and underlayer stone sizes are adjusted, based on previous research and experience, to reproduce prototype transmission and reflection characteristics. Structure stability is generally not reproduced in harbor models.

(d) The reproduction of river discharges and steady-state tidal flows often is required in wave action model studies. These flows generally are reproduced using circulation systems (i.e., for a river discharge, water is normally withdrawn from the perimeter of the model pit area and discharged in a stilling basin that empties into the upper reaches of the river and flows downstream in the model).

(e) Reproducing the movement of sediment in small-scale coastal model investigations is very difficult (Hudson et al. 1979). Ideally, quantitative, movable-bed models best determine the effectiveness of various project plans with regard to the erosion and accretion of sediment. This type of investigation, however, is difficult and expensive to conduct and entails extensive computations and prototype data. In view of these complexities and due to time and funding constraints, most models are molded in cement mortar (fixed-bed) and a tracer material is selected to qualitatively determine the degree of movement and deposition of sediments in the study area. In past investigations, tracer was chosen in accordance with the scaling relations of Noda (1972), which indicates a relation or model law among the four basic scale ratios: horizontal scale, vertical scale, sediment size ratio (d_{50} model tracer material divided by d_{50} prototype sediment), and relative specific weight ratio. These relations were determined experimentally using a wide range of wave conditions and bottom materials, and they are valid mainly for the breaker zone. This procedure was initiated in the mid-1970's, and has been successful in reproducing aspects of prototype sediment movement as evidenced by the performance of completed projects that have been studied (Bottin 1992). Currently, research is being performed to better understand aspects of sediment movement and improve methods to model it, and scaling relations have been developed for mid-scale, two-dimensional model tests (Hughes and Fowler 1990).

(f) After model construction, representative test conditions must be selected. Wave height and period characteristics, direction of wave approach, and frequency of occurrence are typically needed. Refraction analyses are normally required to transform deepwater waves to shallow water at the location of the model wave generator. From this point, model bathymetry will transform the waves to the harbor area. Still-water levels (swl's) are also important test conditions. Normally more wave energy reaches a harbor with the higher swl's. Lower swl's may result in more seaward movement of longshore sediment (i.e. around a jetty head), since the breaker zone would be moved farther offshore. Dominant movement of wave-induced currents and sediment transport patterns are required for verification of the model. River discharge and/or tidal flow information is also required, if applicable to the study.

(g) Data collected in a physical model include time series measurements at selected points of water surface elevation and, when needed, current. Spatial current patterns and velocities can be estimated by timing the progress of weighted floats over known distances on the model floor. Photographs and videotapes of experiments in progress provide valuable visual documentation of wave transformation patterns, wave breaking, sediment movement, etc.

(2) *Example: Dana Point Harbor, California.* Dana Point Harbor, located on the southern California coast about 64 km (40 miles) southeast of the Los Angeles/Long Beach Harbors, is an example of a small craft harbor designed with the aid of physical modeling (Figure V-5-45). The harbor occupies a small cove in the lee of Dana Point. The harbor consists of a 1,676-m- (5,500-ft-) long west breakwater, a 686-m- (2,250-ft-) long east breakwater, and inner harbor berthing areas partially enclosed by the shoreline and mole sections. The harbor encloses an area of about 0.85 sq km (210 acres) and provides berthing facilities for about 2,150 small boats.

Figure V-5-45. Dana Point Harbor, California (May 1969)

(a) The area is exposed to storm waves from directions ranging from southwest counterclockwise to south-southeast and to ocean swells from the south. Damaging wave energy may reach the berthing areas by passing through the outer navigation entrance and by overtopping and/or passing through the rubble-mound breakwaters.

(b) As part of the original design effort, a 1:100-scale hydraulic model investigation was conducted to determine the optimum breakwater plan and location and size of the navigation opening that would provide adequate protection for mooring areas during storms (Wilson 1966). Waves with periods ranging from 9 to 18 sec and heights ranging from 2.4 to 5.5 m (8 to 18 ft) were generated from eight deepwater directions using an swl of +2.0 m (+6.7 ft) mllw. A wave height acceptance criterion of 0.46 m (1.5 ft) was established in the harbor berthing areas by the sponsor and waves in the fairway were not to exceed 1.2 to 1.5 m (4 to 5 ft).

(c) Experiments were conducted for existing conditions and 13 plans. Results for existing conditions indicated rough and turbulent conditions in the area even for low-magnitude storm waves. The proposed improvement plan involved construction of outer breakwaters and inner-harbor development consisting of east and west basin berthing areas partially enclosed by the shoreline on the north and a mole section on the south, southeast, and southwest (Figure V-5-46).

Figure V-5-46. Physical model view of Dana Point Harbor, California, under storm wave attack

(d) Experimental results indicated that wave conditions in the berthing areas were acceptable; however, wave heights in the fairway were about 2.0 m (6.5 ft) for severe storm wave conditions. It was noted that these conditions were due to a standing wave system caused by reflected waves from the mole slopes. Experimental results revealed that modifying the mole slope flanking the fairway, to include a berm, would reduce wave action considerably in the fairway.

(e) The harbor was constructed in accordance with recommendations from the physical model investigation. Post-construction monitoring has shown the harbor is performing as predicted. Mooring areas have experienced no wave problems. The outer west breakwater is overtopped by storm waves, which propagate across the interior channel to the outer revetted mole slope. Vessels in mooring areas behind the moles remain protected. The harbor has successfully endured intense storms, when other southern California small- craft harbor facilities have been severely damaged.

(3) Numerical modeling as a design tool. Numerical modeling requires computerized solution of equations that approximate harbor response to imposed natural forces. Numerical models and computer technology have evolved to the point where useful modeling of actual harbors can often be conveniently done on microcomputers. Numerical models are helpful in harbor studies, even for relatively simple harbor shapes. A combination of physical and numerical modeling is usually preferred for investigating the full range of conditions in a harbor. Numerical modeling related to harbors is best discussed in terms of the natural phenomenon to be modeled, such as waves, circulation, and shore response. Each model's equations and input/output forms are developed for application to particular phenomena. Model systems are now available that provide convenient access to a variety of modeling options under a single user-friendly interface (e.g. SMS 1994).

(a) Wind waves, swell, and harbor oscillations. Numerical wave models can be effectively applied to wave periods ranging from wind waves to long-period harbor oscillations. Numerical models have been useful for: Very long-period wave studies; initial evaluations of harbor conditions; comparative studies of harbor alternatives; and revisiting harbors documented previously with field and/or physical model data.

- For example, numerical models have been used effectively to select locations for field wave gauges (to achieve adequate exposure and avoid oscillation nodes) and to identify from many alternatives a few promising harbor modification plans for fine-tuning in physical model tests. Lillycrop et al. (1993) suggested that numerical modeling is preferable to physical modeling for oscillation periods longer than 400 sec. Both modeling tools can be used effectively for shorter period oscillations.

- Numerical wave modeling concerns are discussed in the following paragraphs, followed by an illustrative example. Additional information on numerical modeling of waves in harbors is available from a number of sources (e.g., Panchang, Xu, and Demirbilek (1998)).

- An initial step in modeling a harbor is to define the area to be covered and required horizontal resolution in the model grid. The coverage area should include the harbor and an area seaward encompassing bathymetry important for waves approaching the harbor. The seaward boundary should be a minimum of several wavelengths away from the harbor entrance, based on the longest wave periods to be modeled. Horizontal resolution is determined by the wavelength of the shortest wave period to be modeled. Typical resolution requirements are between $L/6$ and $L/15$ as the maximum grid element width. If grid elements are uniform in size over the entire grid, L should be based on the shallowest depths of interest. Depending on the particular study and grid-building software available, it may be preferable to build a grid with element sizes varying according to water depth, but still satisfying the wavelength-based maximum size criterion. Computer demands (memory, processing time, storage) in running a harbor model are directly linked to the total number of grid elements. Most harbor studies require a trade-off between coverage area and grid resolution to achieve a workable grid.

- Numerical models applied to harbors are usually based on a form of either the mild slope equation (MSE) or Boussinesq equations. Development of the equations is given by Dingemans (1997), Mei (1983), and others. MSE models are typically steady-state. The MSE model calculates an amplification factor (ratio of local wave height to incident wave height) and phase (relative to the incident wave) for every node in the grid. The MSE does not incorporate spectral processes. Typically, MSE models are run with a representative set of wave height, period, and direction combinations, based on knowledge of incident wave climate. If the MSE is linear, a single wave height for each period/direction combination will suffice. For wind wave and swell applications, regular wave results from the MSE model may be linearly combined, with appropriate weightings, to simulate harbor response to directional wave spectra.

- Boussinesq models are nonlinear and time-dependent. They are forced with an incident wave time series on the seaward boundary and produce a time series of wave response at each node in the grid. The time series may represent regular or irregular wave conditions. Boussinesq models are capable of more accurate representation of harbor wave response than MSE models, at the price of considerably greater computational demands. They are warranted in some practical studies and, with continuing intensive research and development, are likely to become a more workable option in the near future.

- Results from numerical harbor models are in the form of information at selected points or over the entire grid. Point information from Boussinesq models is comparable to time series from field or physical model wave gauges and may be analyzed in similar ways. Spatial information from Boussinesq models can provide animated displays of waves approaching, entering, and interacting with the harbor. Snapshots of waveforms over the harbor at selected times can easily be extracted for still displays. Spatial information from MSE models is in the form of snapshots of amplification factor and phase over the harbor area. Animated displays can be created by expanding amplification factor and phase information into sinusoidal wave time series, if desired.

(b) *Example: Kikiaola Harbor, Kauai, Hawaii.* Kikiaola Harbor is a small, shallow-draft harbor, located along the western part of the Kauai's south shore (Figure V-5-47). The original harbor consisted of west and east breakwaters. The harbor experienced excessive waves, resulting in the addition of inner and outer stub extensions to the east breakwater and a short inner breakwater. A wharf and boat ramp are located along the north boundary of the harbor, east of the inner breakwater.

- Prevailing northeast tradewinds result in a strong predominance of winds from the northeast, east, and southeast at the harbor. Typical wind speeds are 5 to 10 m/sec (10 to 20 mph). Winter storms can generate strong winds from the south. The harbor is exposed to waves approaching from a sector between the 134- and 278-deg azimuths, though the small island of Niihau creates some sheltering in the western exposure. Southern swell, generated by storms in the southern Pacific and Indian Oceans, is a significant part of the wave climate. Also, waves generated by storms in the North Pacific can wrap around the western side of Kauai and affect Kikiaola Harbor. Hurricanes can attack the harbor, which is important for structure design; but they are rare and do not impact operational concerns.

- Use of the existing harbor is limited by two primary factors. First, the harbor is quite shallow. Sediment movement along the local coast, predominantly from east to west, has resulted in shoaling of the entrance and inner harbor. Second, the existing entrance experiences breaking wave conditions that are hazardous to navigation. These two factors are interrelated. Breaking waves are more likely in the existing, shoaled entrance than they would be in a deeper, maintained entrance channel.

- Two plans for modifying the breakwater structures and navigation channels were defined, as follows (Figure V-5-48):

- *Plan 1.* Remove outer stub of east breakwater; remove and reconstruct inner stub of east breakwater a small distance further east; raise crest elevation of exposed portions of east breakwater by 1 m (3-4 ft) and flatten seaward slope to 1:2; widen outer 67 m (220 ft) of west breakwater; dredge 221-m- (725-ft-) long entrance channel with width varying from 32 to 62 m (105 to 205 ft) and maneuvering area to facilitate a 90-deg turn into access channel; dredge 98-m- (320-ft-) long access channel varying in width from 21 to 32 m (70 to 105 ft).

- *Plan 6.* Remove outer and inner stubs of east breakwater; raise crest elevation of exposed portions of east breakwater by 1 m (3-4 ft) and flatten seaward slope to 1:2; extend east breakwater further west to a distance of 33 m (100 ft) past the existing west breakwater location; shorten west breakwater to allow space for access channel; dredge entrance and access channels comparable to those in Plan 1.

- A numerical model study was initiated to investigate wave conditions in the proposed plans relative to the existing harbor and to USACE criteria for channels and berthing areas (Thompson et al. 1998b). An MSE-based numerical model was used to analyze the harbor area (Chen and Houston 1987). Because water depths are shallow (on the order of 1 m in the existing harbor) and the shortest wave period to be modeled was 6 sec, a dense grid was required. To maintain a workable grid size for the model being used, the offshore extent of the grid was significantly limited and could not reach deep water.

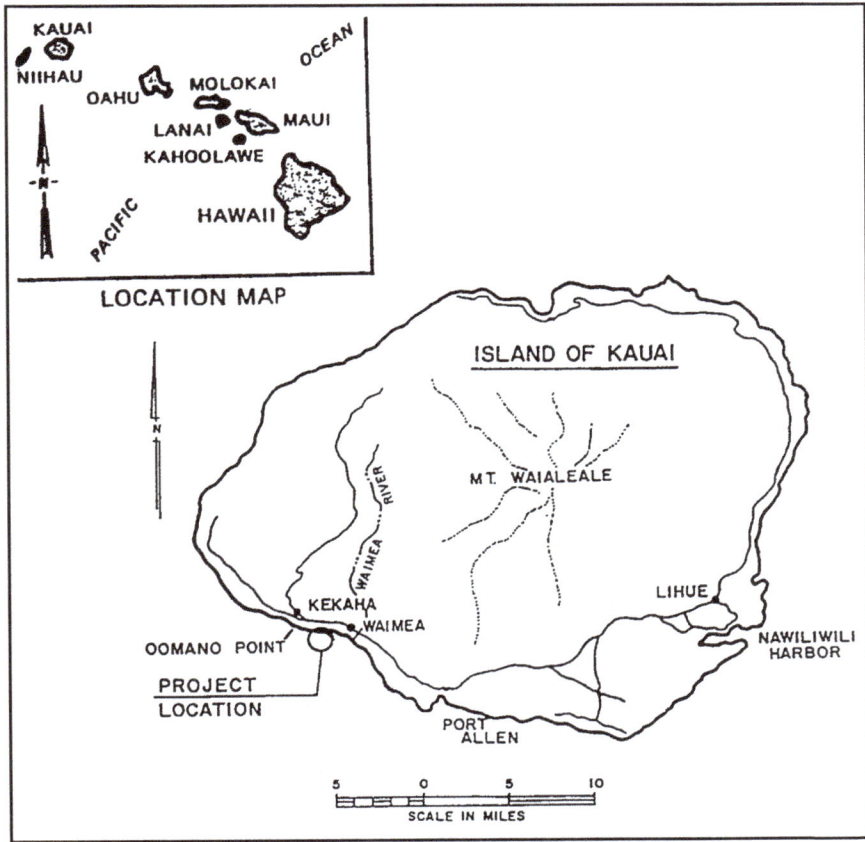

(a) Location map

(b) Photograph (1998)

Figure V-5-47.   Kikiaola Harbor, Kauai, Hawaii

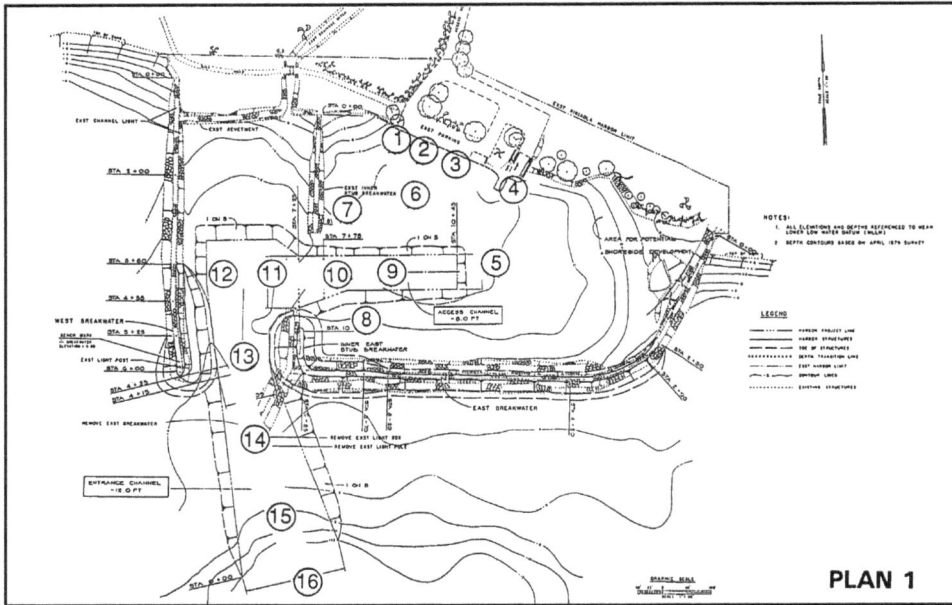

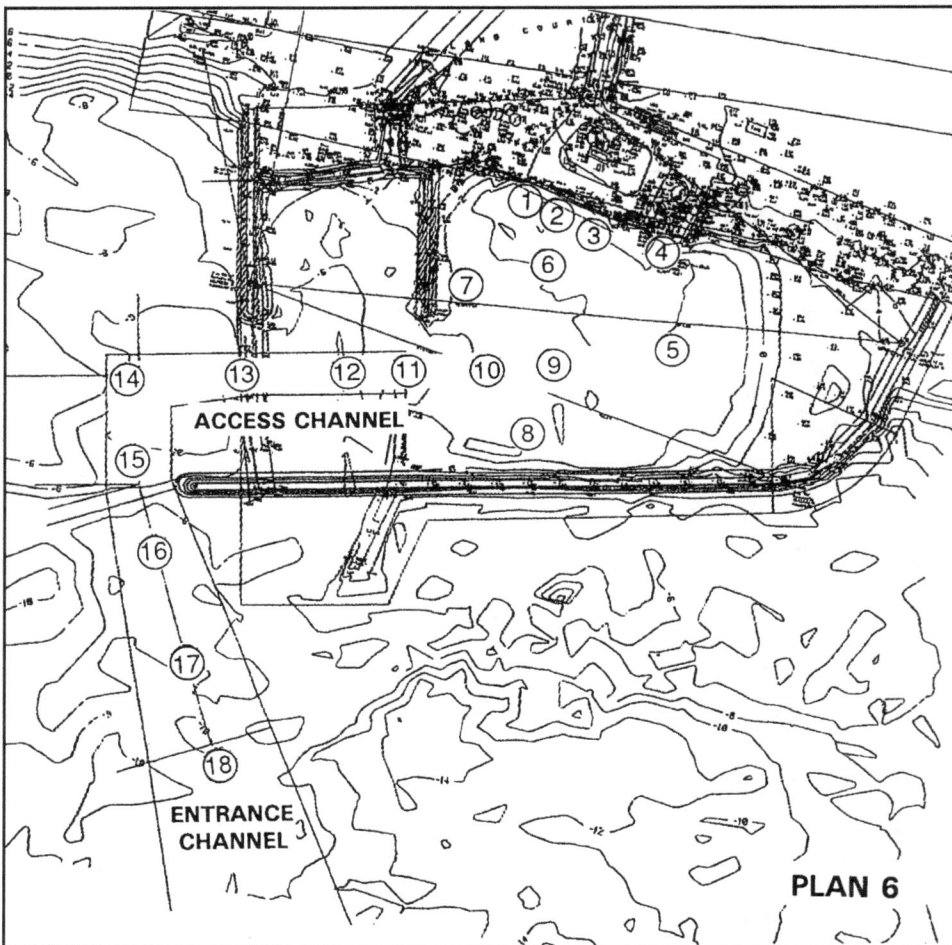

Figure V-5-48.   Kikiaola Harbor alternative plans and model stations

- Wave climate at the seaward boundary of the harbor model grid was developed from updated WIS hindcasts in the Pacific OceanAugust 7, 2000. An additional modeling step accounted for sheltering by the islands of Kauai and Niihau as waves approach the harbor. Deepwater wave information was estimated at a point 1.6 km (1 mile) offshore from Kikiaola Harbor. Then, a shallow water transformation model was used to provide wave estimates at the seaward harbor model boundary, at about the 4-m (13-ft) water depth.

- The harbor model grid for the existing harbor consisted of 24,227 elements and 12,461 nodes. Model parameters, including boundary reflection coefficients and bottom friction, were set appropriate to the harbor configuration and model requirements. The tide range at Kikiaola Harbor is about 0.3 m (1 ft). Since harbor response is unlikely to vary much with water level over this small range, a water level of +0.3 m (+1 ft) mllw was used in all runs, representing a high tide condition.

- Wind wave and swell cases were periods ranging from 6 to 22 sec, in 1-sec increments, and approach directions of 164-, 184-, and 204-deg azimuths, representing the range of incident wave directions and entrance exposures. A linear form of the model was used (bottom friction set to zero), so a nominal 0.3-m (1-ft) wave height with each period/direction combination was sufficient. Thus, a total of (17 periods)x(3 directions) = 51 cases was run in the harbor model. Spectral results for each T_p and θ_p needed to represent the incident wave climate were simulated by linearly combining the 51 cases with appropriate weightings based on a JONSWAP spectrum with cos^{2s} directional spreading.

- Snapshots of amplification factor and phase for one incident wave condition illustrate harbor response (Figure V-5-49). A nonspectral condition is shown so that phases can be presented. The amplification factor increases over shoal areas just west of the entrance channel and then steadily decreases as the waves progress through the entrance and into the inner harbor. Plans 1 and 6 provide more shelter to the inner harbor than does the existing plan. Phase lines in the figure show the alignment of wave crests. They give a visual representation of diffraction and shoaling effects on wave direction and length as the 12-sec waves interact with harbor structures and bathymetry.

- Standard operational criteria used by USACE for wind waves and swell in small-craft harbors are:

- H_s in berthing areas will not exceed 0.3 m (1 ft) more than 10 percent of the time.

- H_s in access channels and turning basins will not exceed 0.6 m (2 ft) more than 10 percent of the time.

- To compare with USACE wave criteria, between 15 and 18 stations were selected in each plan, to include the wharf area, berthing area, access channel, and entrance channel (Figure V-5-48). Stations in the existing plan are identical to those shown for Plan 1 except the entrance channel stations are shifted appropriately. Spectral amplification factors were computed and applied to each incident H_s to give a wave climate at each station. The value of H_s exceeded 10 percent of the time was computed at each station. Results for Stations 1-9 are compared to the USACE berthing area criterion and the remaining stations to the USACE access channel criterion (Figure V-5-50). All plans, including the existing, satisfied the berthing criterion at all stations. The inner channel satisfies the channel criterion in all plans. The existing entrance channel does not meet the criterion; and the seaward portions of the Plan 1 and Plan 6 entrance channels slightly exceed the criterion. In conjunction with the increased width of the outer part of the plan entrance channels, the small exceedance of the USACE channel criterion is unlikely to interfere with safe navigation.

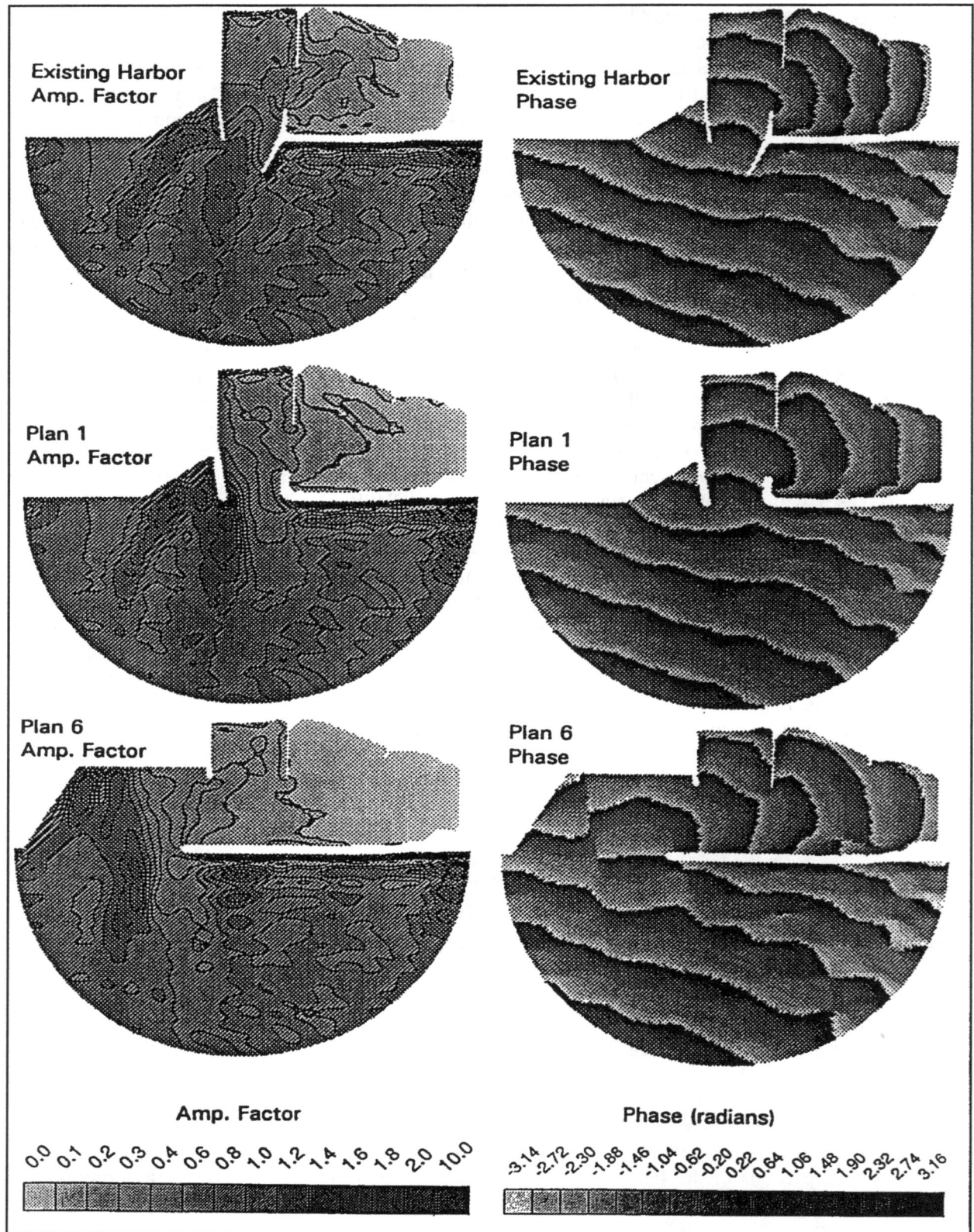

Amp. Factor

0.0 0.1 0.2 0.3 0.4 0.6 0.8 1.0 1.2 1.4 1.6 1.8 2.0 10.0

Phase (radians)

-3.14 -2.72 -2.30 -1.88 -1.46 -1.04 -0.62 -0.20 0.22 0.64 1.06 1.48 1.90 2.32 2.74 3.16

Figure V-5-49. Amplification factor and phase contours, 12-sec wave period, 200-deg azimuth approach direction, Kikiaola Harbor

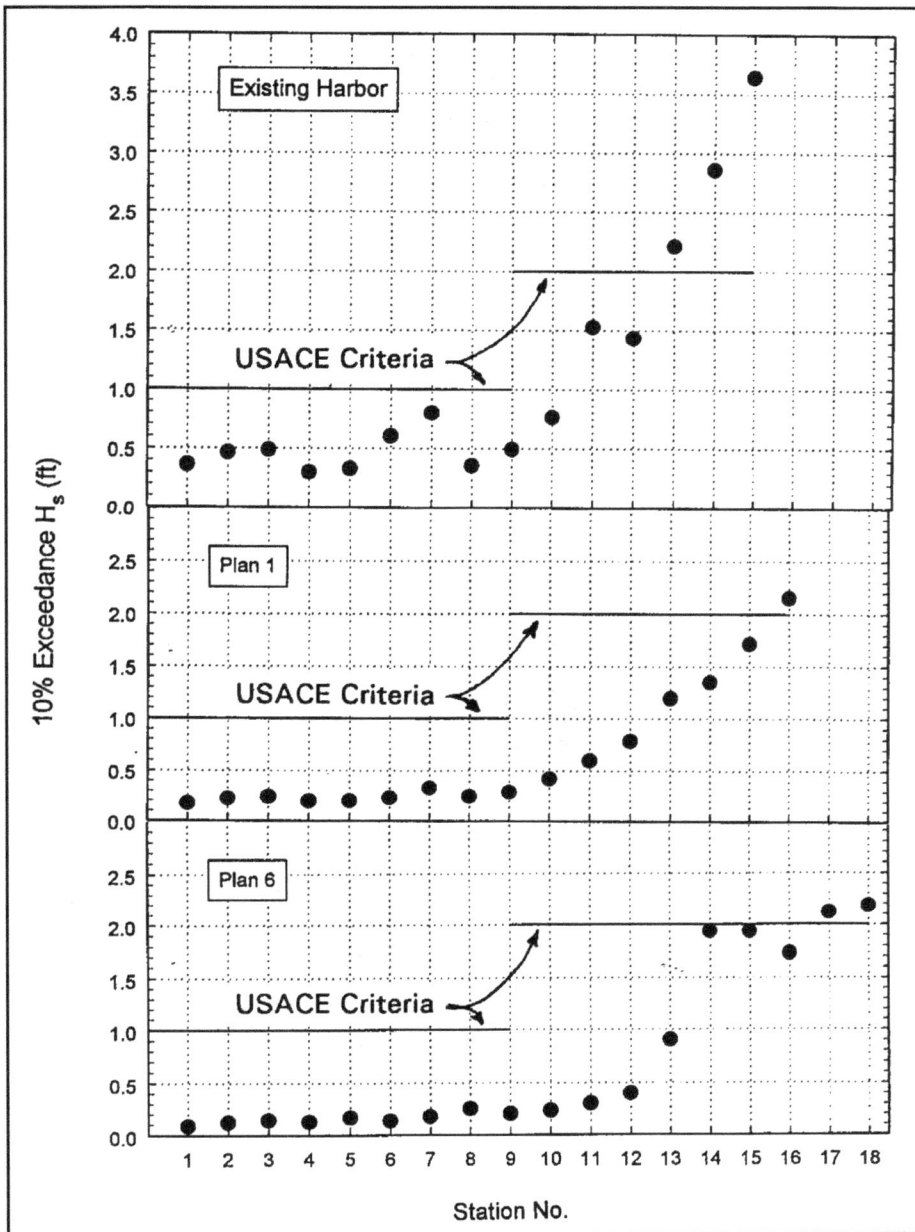

Figure V-5-50. Comparison of H_s exceeded 10 percent of the time, Kikiaola Harbor

- Harbor oscillation characteristics of the existing and plan harbors were also investigated to ensure the plans would not have operational problems due to oscillations. Model parameters were changed to give constant, nonzero bottom friction and full reflection from harbor boundaries. A total of 451 long-wave periods were run, ranging from 25 to 500 sec. The frequency increment between periods was 0.0001 Hz up to a period of 80 sec and 0.00006 Hz for longer periods. Fine resolution in frequency is needed to ensure that resonant peaks are captured. One long wave height is used, representing a moderately energetic long wave case, based on measurements at another Hawaiian harbor. One long wave direction, directly approaching the harbor entrance, is used, since past studies have indicated that harbor response is relatively insensitive to incident long wave direction.

- A snapshot of amplification factors in the existing harbor for a long-period resonance at a period of 150.6 sec represents a simple oscillation between the outer harbor and the east part of the inner harbor (Figure V-5-51). A node (indicated by very low amplification factor) is located a little east of the inner harbor entrance. Considering the most active areas of operational concern, amplification factors at the boat ramp (sta 4) and in the outer harbor (sta 12) were 2 and 3, respectively. A simple basis for judging the operational importance of harbor oscillation amplification factors is given in Table V-5-14 (Thompson, Boc, and Nunes 1998). Thus, the 150.6-sec resonance is not expected to be a problem in operational areas for boats. All amplification factors in all plans across the full range of long-wave periods were significantly less than 5, indicating that harbor oscillations are not a problem in the plan harbors. Additional information on numerical modeling of harbor oscillations is presented in Part II-7-5-f.

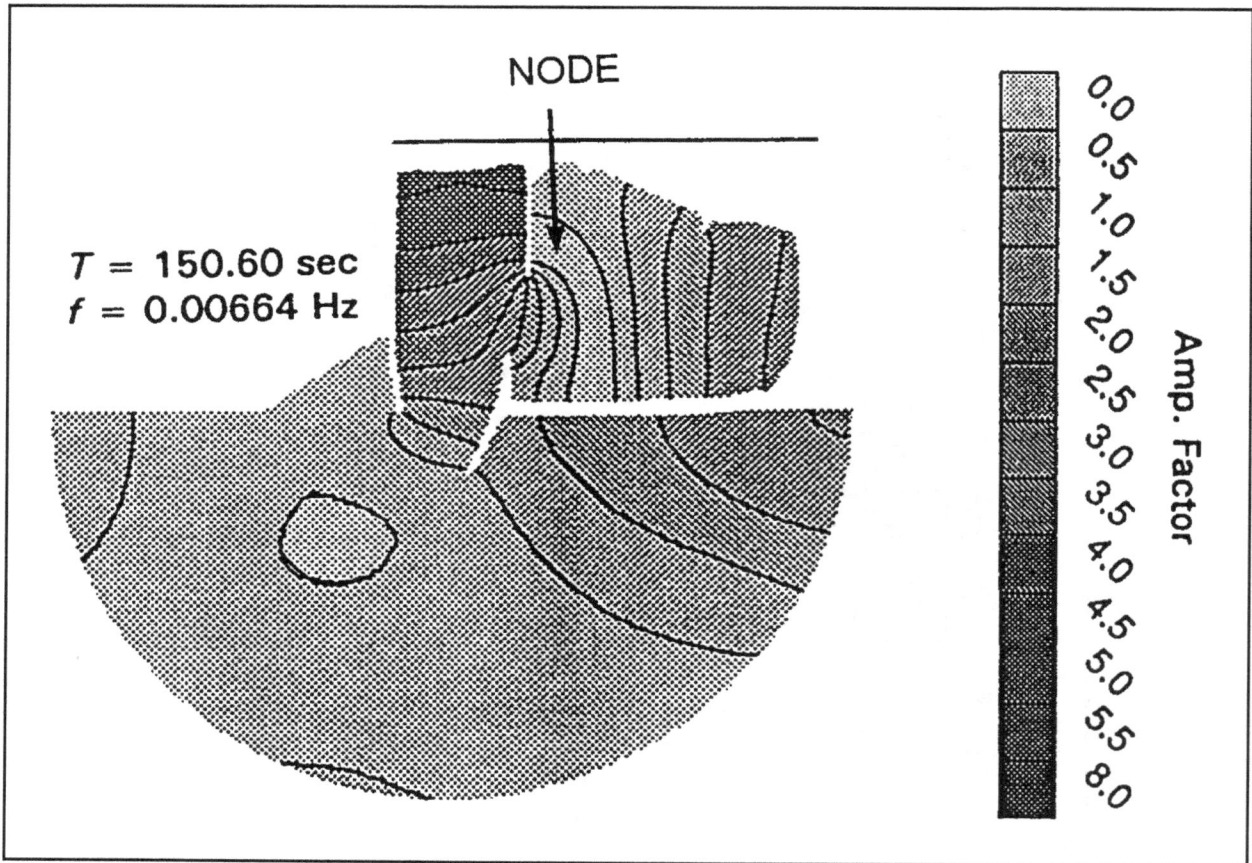

Figure V-5-51. Resonant long wave amplification factor contours, existing Kikiaola Harbor

Table V-5-14
Simple Criteria for Assessing Operational Impact of Harbor Oscillations

Amplification Factor	Operational Impact
> 5	Some problems
> 10	Major problems

(c) Flushing and circulation. Numerical models are effective for evaluating flushing and circulation in harbors and entrances due to forces such as tides and wind. This application is discussed in Part II-7. Numerical models also provide detailed currents along navigation channels needed in ship simulations.

(d) *Example: Maalaea Harbor, Maui, Hawaii.* Maalaea Harbor is a south-facing small-craft harbor located on the southwest coast of the Island of Maui (Figure V-5-52). The harbor facility consists of a 27-m- (90-ft-) wide, 3.7-m- (12-ft-) deep entrance channel and a 0.05-sq-km (11.3-acre) dredged basin. The harbor is protected by a 30-m- (100-ft-) long, 27-m- (90-ft-) wide breakwater on the south side and a 265-m- (870-ft-) long breakwater on the east side. A 91-m- (300-ft-) long paved wharf is located at the shore opposite the entrance. The west and central parts of the harbor are small-craft berthing areas.

Figure V-5-52.   Maalaea Harbor, Maui, Harbor (from Air Survey Hawaii, March 1984)

- In response to needs for increased berthing space and better protection during severe wave conditions, the U.S. Army Engineer Division, Pacific Ocean, conducted studies to develop and evaluate harbor modification plans. To evaluate potential impacts on water quality in the harbor, a numerical model circulation and flushing study was conducted for the existing harbor and two proposed plans (Wang and Cialone 1995).

- Circulation in Maalaea Harbor is forced by tide and persistent, often strong winds from the north and northeast. Prototype data were collected over a 9-day period, including currents at two locations and tide. Concurrent wind measurements were available from a nearby airport.

- The numerical model used a curvilinear boundary-fitted coordinate system to generate a computational grid with two vertical layers. It provides current vectors over the harbor and a larger area outside the harbor (Figure V-5-53). The average horizontal grid cell size is 15 m (50 ft). The model is time-dependent, driven with time series of surface elevation along the seaward boundary and wind over the grid surface. Prototype data were used to calibrate the model. Parameters adjusted during calibration include friction, drag, and mixing coefficients.

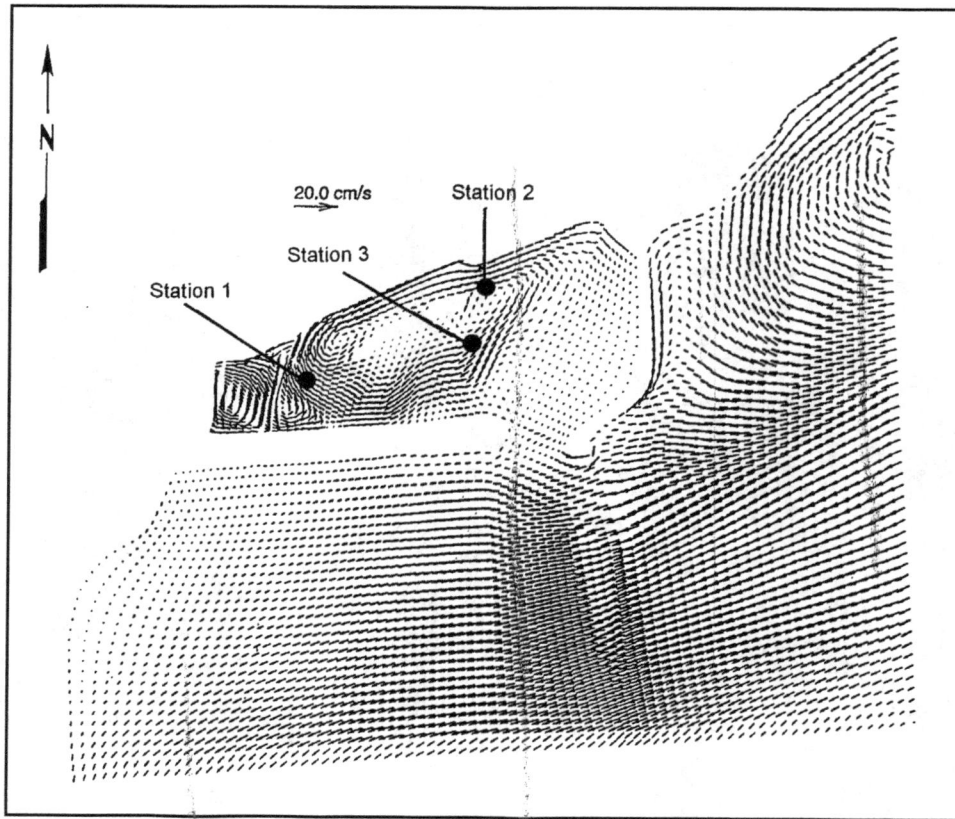

Figure V-5-53. Station locations and surface layer circulation snapshot at Day 3, Maalaea Harbor, existing plan

- Flushing time was defined as the time required for a conservative tracer to decrease to 36.8 percent ($1/e$, $e = 2.71828$) of its initial concentration. This time was evaluated in each plan by beginning a simulation with constant concentration of 100 ppt (parts per thousand) in the harbor and 0 ppt outside the harbor, running the simulation for multiple days, and extracting a concentration time series at three interior harbor stations (Figure V-5-53). Flushing time in the existing harbor was longest at sta 1, in the west part of the harbor and most distant from the entrance (Figure V-5-54). The 2.9-day flushing time was considered acceptable, based on flushing times of 2-4 days as acceptable for design, 4-10 days as marginal, and greater than 10 days as unacceptable (Clark 1983).

(e) Shore response. Possible changes in adjacent shoreline configuration and nearshore bathymetry in response to navigation structures and channel dredging are often a significant concern in navigation projects. Shoreline evolution with and without the project in place can be predicted and analyzed with numerical modeling tools. Guidance on sediment processes at inlets and harbors and numerical modeling options is given in Parts III-2 and V-6.

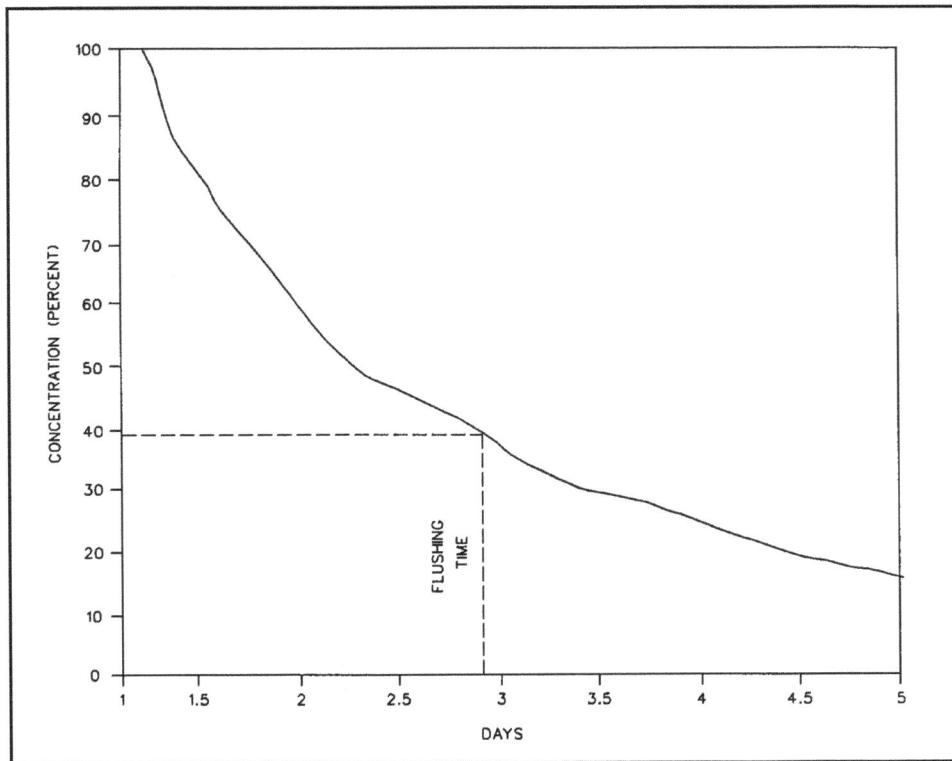

Figure V-5-54.   Time series of conservative tracer concentration at sta 1, Maalaea Harbor, existing plan

b.   Navigation modeling.

(1) Physical models. Physical models have been used for a variety of navigation studies, with vessel controls such as autopilot, human pilot steering, and free-running vessels with remote control. Physical models are particularly useful for evaluating the behavior of a vessel in the presence of intense, interacting forces, typically involving an entrance channel with ocean waves and possibly harbor or alongshore cross-currents. Waves and human control decisions are statistical processes. Free-running vessel motions in response to many samplings of those processes provide valuable design information about channel depth, width, layout, etc. Vessel position is tracked with high precision relative to channel boundaries in the model. The use of physical models for designing navigation projects is illustrated in the following example.

(a) *Example: Barbers Point Harbor, Oahu, Hawaii.* Barbers Point Harbor is a deep-draft commercial harbor located near the southwest corner of the Island of Oahu, Hawaii (Figures V-5-55 and V-5-56). The harbor was constructed along a previously uninterrupted coastline in 1982. The harbor complex includes a barge basin and small craft marina in addition to the deep-draft basin. The design ship was a general cargo vessel 219 m (720 ft) long with a beam of 29 m (95 ft) and a loaded draft of 10.4 m (34 ft). The entrance channel, designed for one-way traffic, is a constant 137 m (450 ft) in width and 12.8 m (42 ft) mllw in depth over its full length. Just past the coastline, channel depth transitions to 11.6 m (38 ft) mllw, the design depth of the inner channel and deep-draft harbor. The deep-draft basin is approximately 671 m x 610 m (2,200 ft x 2,000 ft) in size, covering an area of 0.37 sq km (92 acres).

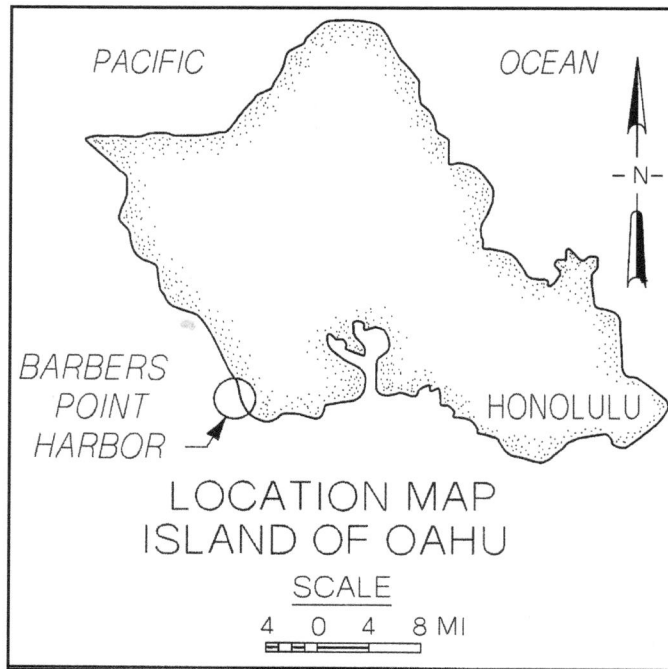

Figure V-5-55.   Barbers Point Harbor, Oahu, Hawaii, location map

Figure V-5-56.   Barbers Point Harbor, Oahu, Hawaii (August 1994)

• Changing economic conditions have created a need for the harbor to serve larger ships. In response to this need, the State of Hawaii and the U.S. Army Engineer Division, Pacific Ocean, sponsored physical and numerical model studies to assist in designing harbor modifications (Briggs et al. 1994, Harkins and Dorrell 1998). The primary study task was to evaluate the navigability of proposed channel and harbor configurations for a larger design ship unaided by tugs.

• A physical model of the harbor complex and adjacent coastal areas was constructed (Figure V-5-57). The model scale, 1:75 undistorted, was selected for proper reproduction of important harbor features, storm waves and longshore currents, and the design ship. Model bathymetry extended to the 30-m (100-ft) mllw bottom contour and a distance of about 1,067 m (3,500 ft) along the coast on either side of the entrance channel. Total area covered by the model was over 1,000 sq m (3,500 sq ft). A directional spectral wave maker was placed seaward of the modeled bathymetry. Longshore currents, which affect navigation in the existing harbor, were created in the model with a system of PVC pipe extending along each lateral boundary (with diffuser ports) and meeting at a pump station located behind the model, landward of the coastline. Pump controls allowed generation of longshore currents in either direction. Diffuser ports were open or plugged as needed to achieve desired current patterns.

Figure V-5-57. Physical model of Barbers Point Harbor

• Two design ships were identified, based on anticipated use of the harbor for container and bulk coal traffic. Existing ships were selected as representative of future harbor traffic, the *President Lincoln*, a C9 container ship with capacity of 2,900 TEU operated by American President Lines, and the *Bunga Saga Empat*, a bulk carrier (Figure V-5-58). The design bulk carrier was a modified version of the *Bunga Saga Empat*, with length increased by 30 m (100 ft). Design ship dimensions are summarized in Table V-5-15.

Figure V-5-58. Bulk carrier *Bunga Saga Empat*

Table V-5-15
Design Ship Dimensions for Barbers Point Harbor Studies

	Prototype Ship Dimensions	
	Container Ship	Bulk Carrier
Length Overall	262 m (860 ft)	259 m (850 ft)
Beam	32 m (106 ft)	32 m (106 ft)
Fully Loaded Draft	11.9 m (39 ft)	13.7 m (45 ft)

- Model ships were constructed to match the harbor model scale, 1:75 (Figures V-5-59 and V-5-60). Model ships were self-powered by onboard batteries. Forward and reverse speeds, rudder angle, and, for the container ship, bow thruster direction and speed were remote-controlled.

- A set of design transit conditions was selected for simulation. Prototype measurements of waves and currents near the harbor entrance were available. Wave data were collected over a period of approximately 4 years; currents were collected over a 65-day period. For harbor plan evaluation, the following conditions were used. The highest measured H_s values and a representative range of T_p and θ_p were selected, a total of eight wave conditions. The range of H_s and T_p values was 2.1 to 3.0 m (7.0 to 10.0 ft) and 6 to 18 sec, respectively. Longshore currents were selected to represent average, normal, and extreme conditions from both directions. Extreme currents were 0.41 m/sec (0.80 knot) from the north and 0.33 m/sec (0.65 knot) from the south. Based on data from a nearby airport, severe wind speeds of 10.3, 12.9, and 20.6 m/sec (20, 25, and 40 knots) were selected.

Figure V-5-59. Model bulk carrier

Figure V-5-60. Model container ship

- Six harbor plans were studied with varying combinations of the waves, current, and wind, as selected for design transit conditions. Wind forces were simulated with a ship-mounted fan. Model ships were guided by remote control between deep water and the protected harbor. Both inbound and outbound runs were made. Two experienced local pilots assisted in verifying the model setup and conducting some of the runs. Inbound runs were significantly more difficult than outbound runs. The ship must slow in approaching the entrance and it becomes more difficult to control. Typical inbound ship speeds are 13.0 km/hr (7 knots) at the seaward end of the entrance channel, 7.4-9.3 km/hr (4-5 knots) in the vicinity of the coastline, and 3.7-5.6 km/hr (2-3 knots) in the harbor. After a recommended harbor plan was identified, a number of channel/harbor depth variations were studied to optimize design depths. A total of nearly 2,000 runs were made, of which the majority were inbound.

- Navigability was evaluated by several methods during the course of the model studies. Ship operators recorded their observations after each run, with particular attention to any difficulties during the run. An overhead video camera recorded each run. A commercial motion analysis system was used to collect and analyze model ship motions. The system uses digital cameras and strobes to track reflecting balls. Six balls were mounted on the model ship (e.g., Figure V-5-59) and four were placed at fixed locations around the channel. After processing, the system provides a time series of clearance between ship hull and bottom.

- Physical model data on ship horizontal and vertical clearance in the channel were evaluated in a probabilistic assessment of channel design. The design then includes a consideration of the natural variability of wave, current, wind, ship track, and ship response, which is crucial in realistically assessing the probability of a momentary grounding event during ship transit. Thus risk of design ship contact with channel sides or bottom can be incorporated into the design process. The expected time interval between C9 container ship grounding events as a function of number of transits per year illustrates risk information available for design (Figure V-5-61). Since several different methods for estimating probabilities were applied to the physical model tests, the average from all methods is shown, bracketed by best and worst expected performance based on variability in the methods. Additional details are given by Briggs, Bratteland, and Borgman (2000).

- The recommended plan differs from the existing harbor in the following ways:

  - Entrance channel is deepened and flares out at the seaward end to allow ships more maneuvering space during initial approach.

  - Transition from entrance channel depth to harbor depth is moved from the coastline to the inner harbor basin opening. This change moves the transition to a lower wave energy environment and gives pilots more space to correct when vessel shear occurs at the depth discontinuity.

  - Harbor is deepened and expanded in size by excavation in the east part of the harbor.

(b) Additional information, based on numerical model studies of Barbers Point Harbor oscillation characteristics, is available in Part II-7.

(2) Ship simulations. Increasingly, deep-draft channels are being designed using ship simulators. For example, the USAERDC ship simulator is schematized in Figure V-5-62. Ship simulations typically have pilots operate the steering wheel and ship controls and navigate a realistic course in real time. Pilots give verbal commands for tug assistance as needed, and an assistant operates tug controls. Pilots are drawn from professional pilot associations serving the project area. Their experience and intuition aids in evaluating existing projects as well as refining and studying new alternatives for safe and optimum channels and/or

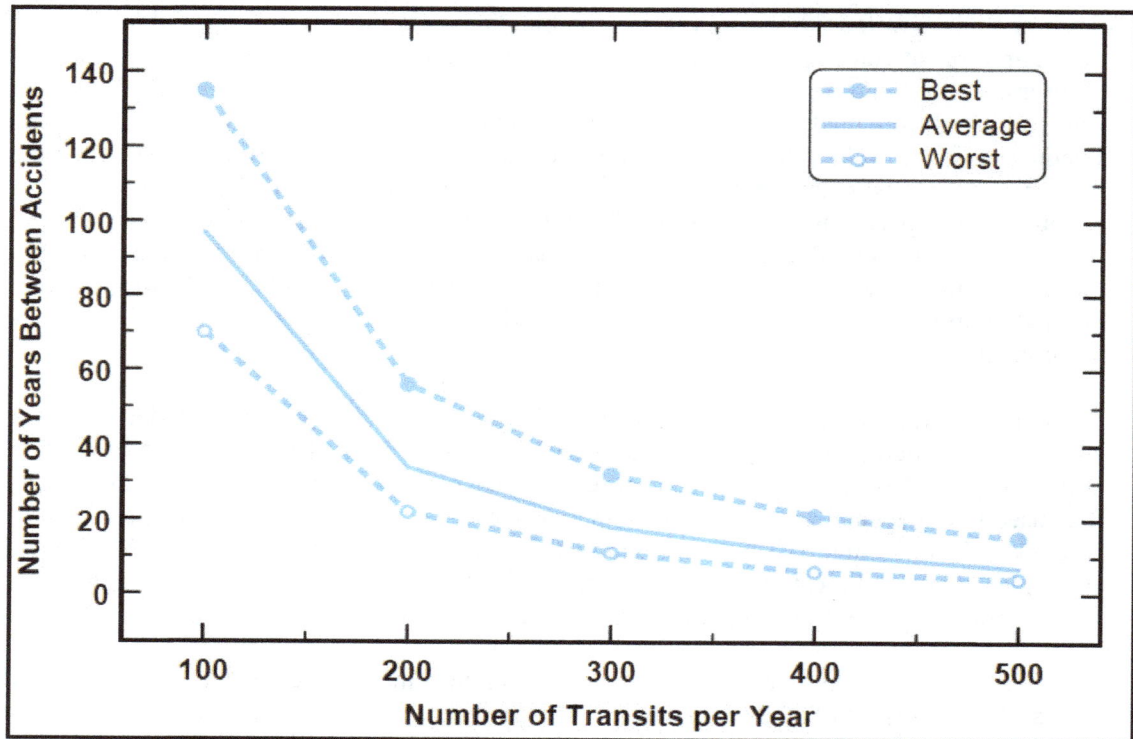

Figure V-5-61. Probability assessment for C9 container ship navigating recommended entrance channel, Barbers Point Harbor

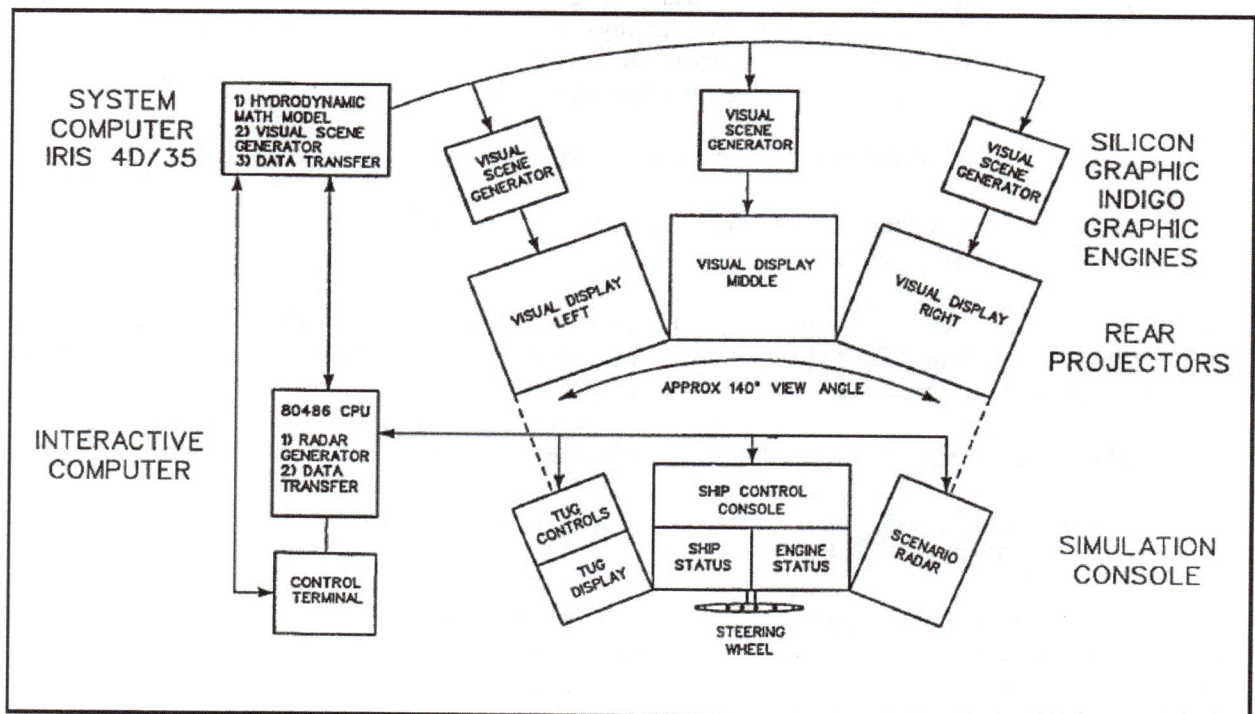

Figure V-5-62. USAERDC ship simulator system

turning basins. At some levels of project design, simulators may be used advantageously for fast-time runs with either autopilot or human control instead of a more comprehensive and costly real-time pilot evaluation program.

- Simulators are special numerical models involving representations of a ship, navigation channel, currents, wind, visual scene (including view over the ship, aids to navigation, bridges, docks, and other visual features needed for piloting cues and adequate realism), radar image, tugs and thrusters, ship bridge controls, and typical bridge instruments. Simulated forces and effects are depicted in Figure V-5-63. The ship model(s) experiences these forces and effects in ways similar to the prototype ship(s).

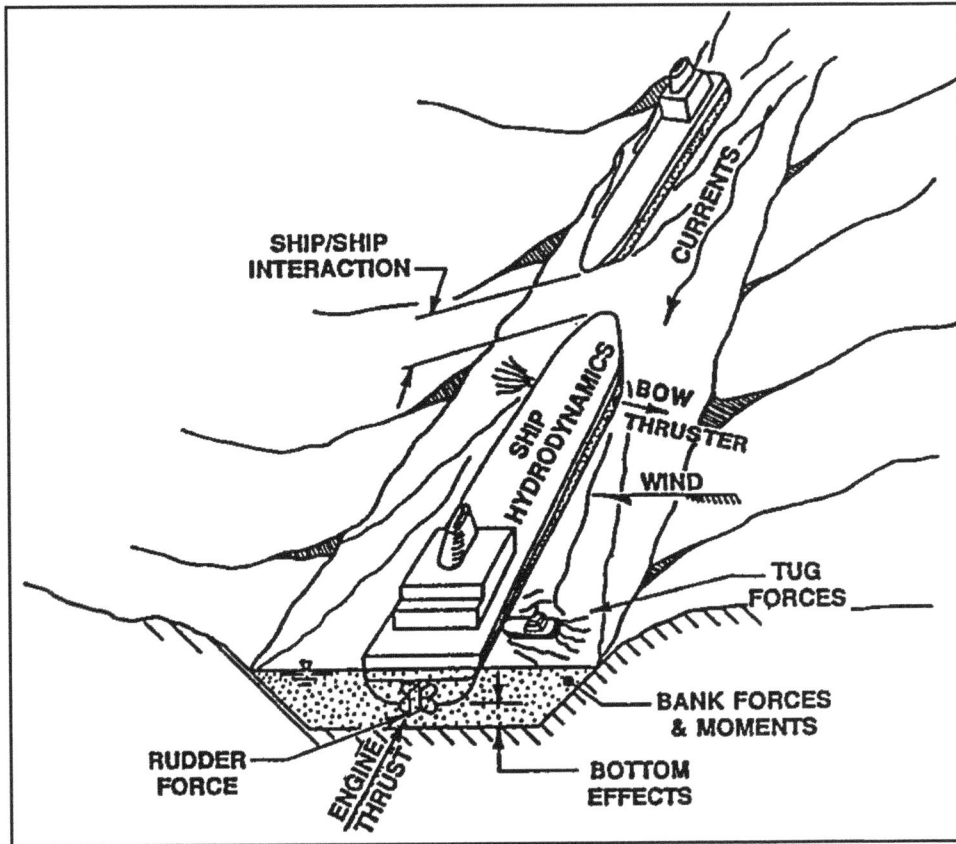

Figure V-5-63. Ship simulator forces and effects

- The key steps in a real-time simulation are shown in Figure V-5-64. Output information saved at selected short time intervals during a simulation includes ship position, engine and rudder settings, ship movement information (speed, heading, rate of turn, drift angle), and minimum clearance relative to channel boundaries. If tugs are used, information on tug forces imposed on the ship may also be saved.

- Two example ship simulator studies are discussed in the following paragraphs. More information on ship simulators is available from USACE (1998), Webb (1994), and PIANC (1997a).

(a) *Example: Alafia River Harbor, Florida.* The Alafia River Harbor is located along the eastern shore of Hillsborough Bay, about 13 km (8 miles) southeast of Tampa, Florida (Figure V-5-65). The

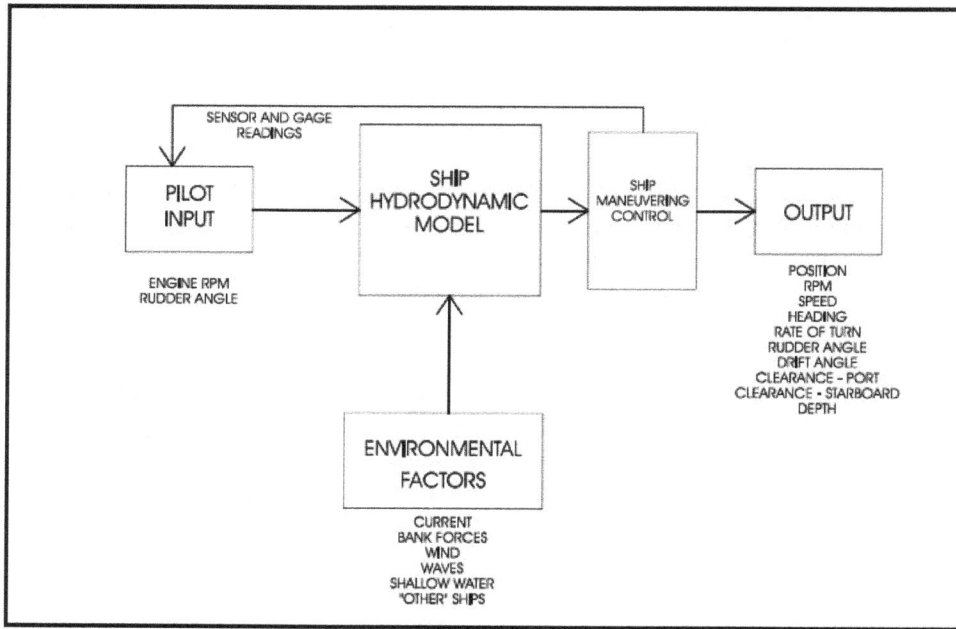

Figure V-5-64. Real-time simulation

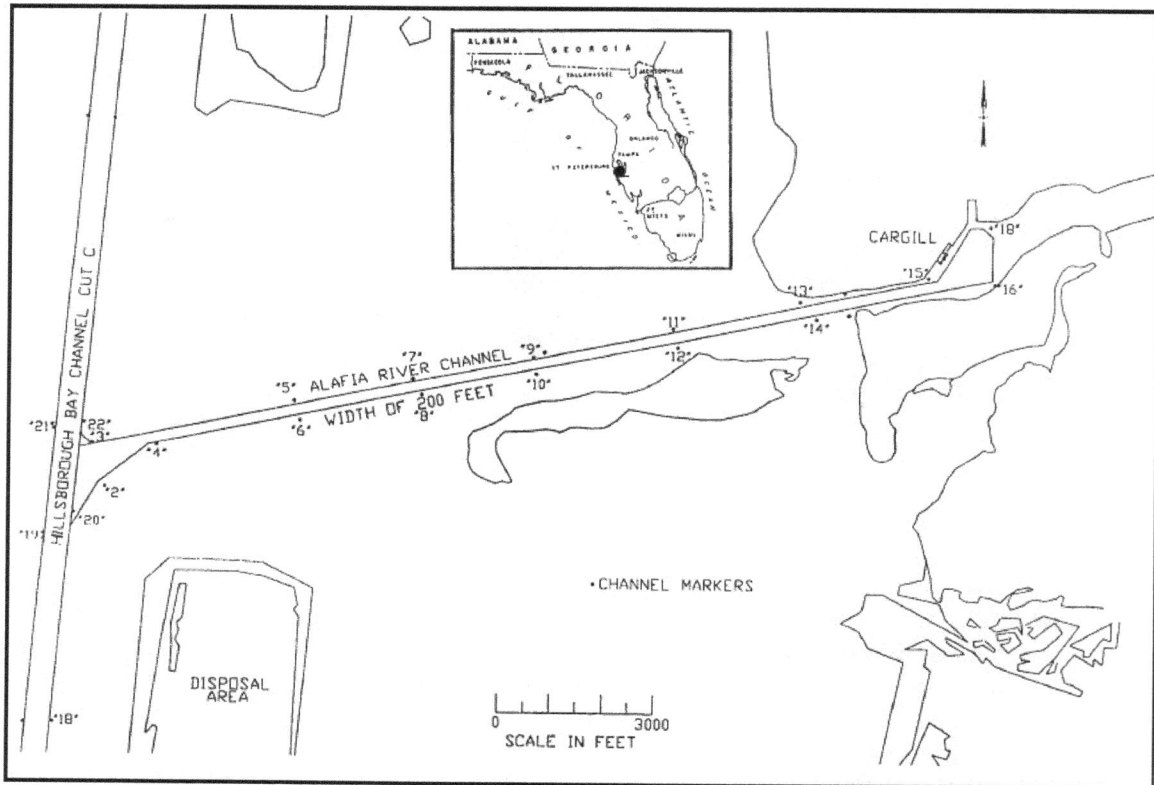

Figure V-5-65. Existing Alafia River Channel and turning basin

existing federally maintained project consists of a turning basin adjacent to the dock facilities and a channel connecting the turning basin to Hillsborough Bay Channel Cut C, the primary north-south shipping channel in Hillsborough Bay. Total length of the federal project is 5.8 km (3.6 miles). Channel depth is 9.1 m (30 ft) mllw. Channel width is 61 m (200 ft). The turning basin is 213 m (700 ft) wide and 366 m (1,200 ft) long.

- Alafia River Harbor is used mainly to ship phosphate rock and bulk phosphate products. Ships typically enter the harbor in ballast and load bulk materials until the ship draft reaches the limit allowed in Alafia River Channel or until the ship is fully loaded. Ships turn in the turning basin at the start of the outbound run, in a loaded condition.

- The U.S. Army Engineer District, Jacksonville, funded ERDC to conduct a ship navigation simulation study to investigate performance of two proposed plans for upgrading the Alafia River Channel and turning basin to accommodate larger ships. A notable part of the study is the detailed visual scene developed to provide pilots with realistic visual cues. The cues are a crucial part of slowing the ship on approach to the turning basin, approaching the dock, and turning the ship for the outbound run. Figure V-5-66 shows two pilots operating a bulk carrier. One pilot is guiding the ship, the other is operating tug controls on command. The ship has just entered the turning basin and turned toward the dock. The view direction (which is easily selected by the pilot) is to starboard, with the ship bow visible at the right side of the scene. The Alafia Channel heading out to Hillsborough Bay is visible at the left side, including a channel marker in the foreground. The dock and dock-side loading facilities are just to the left of the ship bow. Further left are numerous small trees and a line of rail cars. This scene adjusts continuously as ship position or pilot view direction change. Additional details of the study are given by Thompson et al. (1998a).

Figure V-5-66. Visual scene of Alfia River Harbor ship simulation study, inbound bulk carrier approaching dock

(b) *Example: San Juan Harbor, Commonwealth of Puerto Rico*. San Juan Harbor is located on the north coast of Puerto Rico, with open exposure to the Atlantic Ocean (Figure V-5-67). It is the largest port in Puerto Rico and a major container port. Noncontainerized cargo, such as petroleum products, lumber, grain, automobiles, and steel, is also imported to the island by sea. Rum, Puerto Rico's principal export, is shipped in containers. Cruise vessels frequently call on San Juan Harbor.

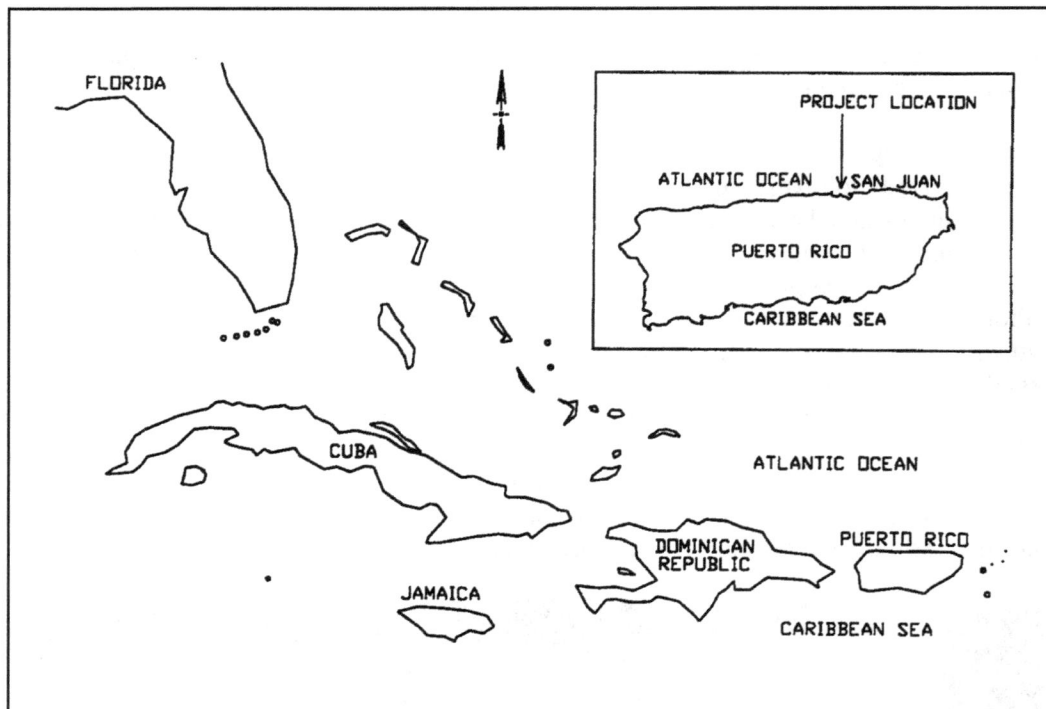

Figure V-5-67. San Juan Harbor, Commonwealth of Puerto Rico, location map

- Federally maintained channels include an entrance channel (Bar Channel), a main interior approach channel to the harbor complex (Anegado Channel), and three interior channels forming a triangular path accessing the principal dock areas (Army Terminal, Puerto Nuevo, and Graving Dock Channels) (Figure V-5-68). Design depth of the outer Bar Channel is 13.7 m (45 ft). The deepest approach to the harbor is the S-shaped path along Bar, Anegado, and Army Terminal Channels, with a controlling depth of 11.0 m (36 ft). Puerto Nuevo and Graving Dock Channels have design depths of 9.8 and 9.1 m (32 and 30 ft), respectively. Bar and Anegado Channel widths are 152 and 305-366 m (500 and 1,000-1,200 ft), respectively. The other three channels have design widths of between 91 and 122 m (300 and 400 ft).

- Wind and waves strongly affect the harbor entrance. Winds are usually steady and are described as being between 8 and 10 m/sec (15 and 20 knots) predominantly from the east and northeast. Waves typically approach the entrance from the north, northeast, and east, with significant heights up to 6-7 m (20-22 ft) during severe events.

- Pilots typically board inbound ships when they are 4.8 km (3 miles) from the harbor entrance. The entrance channel can be difficult to navigate in the presence of wind and wave conditions. Ships must maintain speed in the entrance channel for control, yet they must slow to make the relatively sharp turn into Anegado Channel. All documented accidents in recent years are groundings that have occurred on the south side of this turn. The turn is difficult for outbound ships, too, because of the

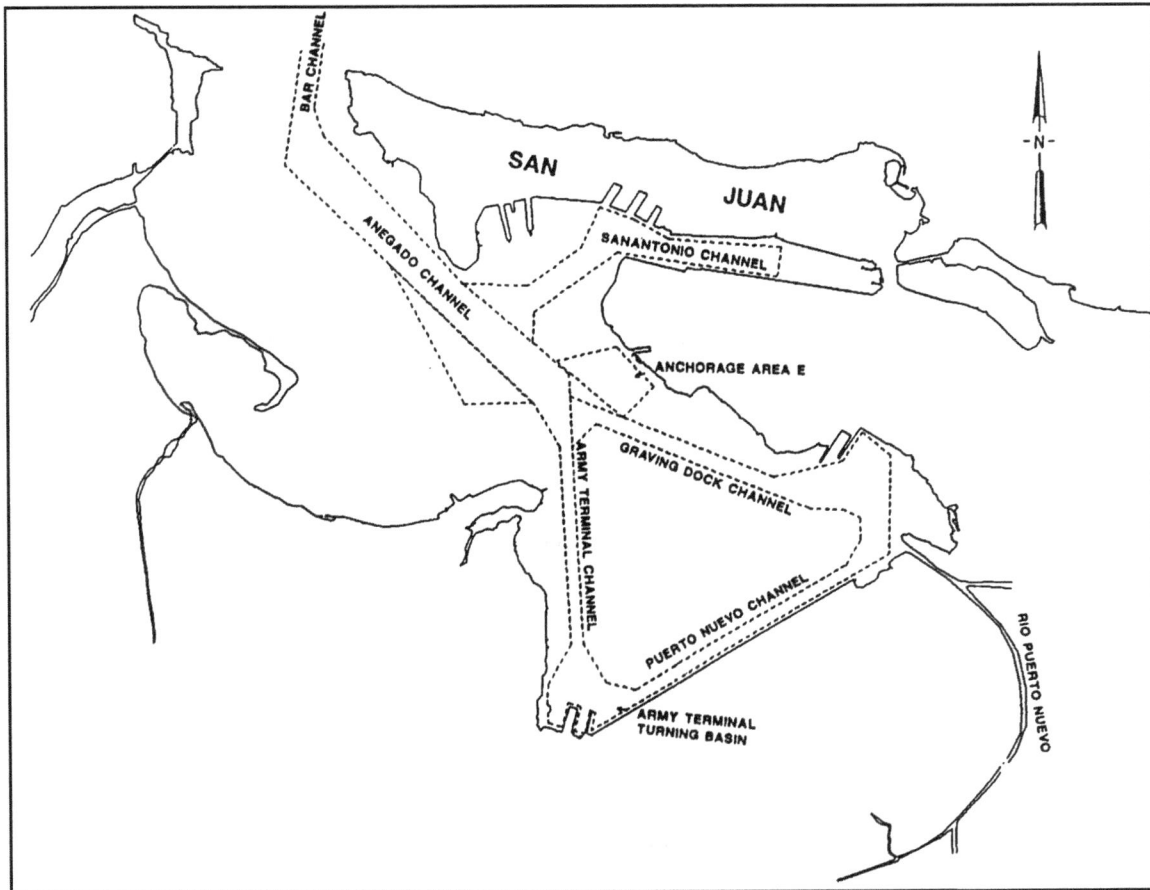

Figure V-5-68. San Juan Harbor channels

relatively narrow entrance channel. Sharp turns associated with the relatively narrow interior channels can also be difficult to navigate.

- The U.S. Army Engineer District, Jacksonville, funded a real-time ship simulator study of existing and two proposed alternative plans to address navigation concerns in San Juan Harbor channels and to allow access to deeper draft ships (Webb 1993). Controlling depth in the proposed plans is 11.9 m (39 ft) in Anegado, Army Terminal, and Puerto Nuevo Channels and 11.0 m (36 ft) in Graving Dock Channel. The purposes of the simulator study were to determine effects of the proposed improvements on navigation, to optimize channel width and alignment for safe and efficient navigation, and to determine necessary depths in Bar and Anegado Channel sections affected by waves.

- Design transit conditions were developed. A wind from the northeast was used with a speed of 10.3 m/sec (20 knots) in the outer entrance. Wind speed was decreased to between 0 and 7.7 m/sec (15 knots) in interior areas sheltered by bluffs and/or tall buildings. Wave information from a 20-year hindcast was used to define incident wave conditions for moderate and heavy seas. A numerical model transformed the selected incident wave conditions to the harbor and through the entrance, giving wave estimates along the channel. For simulation, incident H_s was 4.6 m (15 ft), coming from the northeast. This H_s is about the practical upper limit for ships to enter the harbor. The H_s progressively decreased along Bar Channel to 1.2 m (4 ft) and then to 0 after the turn into Anegado Channel. Tidal currents in the channels were determined with a numerical model of the harbor embayment. Currents are very small. Since flood tide tends to reduce control of ships entering the harbor, flood tide currents were used with all simulations. A wave-driven cross-current of 0.3 m/sec was added in the more exposed section of the Bar Channel, based on pilot comments.

- Two design ships were used, a tanker 232.6 m (763 ft) in length (LBP) with a 38.1-m (125-ft) beam and a container ship 246.9 m in length (LBP) with a 32.3-m (106-ft) beam. The inbound tanker draft (loaded) was 9.8 m (32 ft) for the existing channels and 11.0 m (36 ft) for proposed channels. The outbound tanker draft (in ballast) was 7.9 m (26 ft). For both inbound and outbound runs, the container ship draft was the same as the inbound loaded tanker draft.

- The simulation was validated with the assistance of two pilots from the San Juan Harbor Pilots Association. Simulations were conducted in three 1-week periods. A total of six licensed San Juan Harbor pilots conducted the simulations (two per week), giving a representative range of experience and piloting strategies. Pilots completed a written questionnaire immediately after each run, including a rating scale of key project features. Some desirable modifications to the proposed plans emerged after the first week of simulations. The plans were adjusted to improve navigation in localized areas with difficult clearance and/or to reduce dredging in areas not needed for navigation.

- Design depths for the wave-influenced Bar and outer Anegado Channels were developed in a separate study component. A range of wave conditions and ship speeds were considered. The sum of vertical ship motion and squat was used to define the required underkeel clearance to be added to the 11.0-m (36-ft) ship draft. Because wave height decreases along the Bar Channel, a stepped design depth was recommended, with depth of 16.8 m (55 ft) at the seaward end of Bar Channel progressively decreasing in three steps to 13.7 m (45 ft) through the turn into Anegado Channel.

- Results from the simulations were summarized to evaluate proposed plans. For example, average pilot ratings for inbound container ship runs indicate the plans will significantly improve harbor access, especially in Puerto Nuevo Channel and at the turn separating it from Army Terminal Channel (Figure V-5-69). The wider entrance channel in the plans gives a significant improvement. Ship track plots from all runs show how the increased width and gentler turn would be used in navigation (Figure V-5-70). An unused area on the outside of the proposed turn is defined by the envelope of ship tracks. Simulations indicated that a ship could not enter this area and still turn safely. Therefore, one study recommendation was that the unused area be deleted from the plan, reducing dredging requirements. Complete results, conclusions, and recommendations are given by Webb (1993).

c. Specialized field studies.

(1) Harbors. As field measurement and data collection techniques have advanced, field studies have become powerful and reliable for documenting the behavior of existing harbors. However, the cost for comprehensive field studies is significant, and such studies are generally practical only for large projects with high economic impact. Typically, field data are used for calibration and validation in physical and numerical model studies. Field data helpful for harbor studies include incident directional waves, water levels, waves and currents at several interior locations, and winds. The data provide valuable information about harbor response to wind waves and swell, entrance channel wave conditions, harbor oscillations, and circulation and flushing. A representative, but extensive, field data collection program in the massive Los Angeles and Long Beach, California, harbor complex is described by Seabergh, Vemulakonda, and Rosati (1992). The program was aimed at enhancing physical and numerical models used in harbor planning.

(a) *Example: Kahului Harbor, Maui, Hawaii.* Kahului Harbor, located on the north shore of the Island of Maui, is the island's only deep-draft harbor and the busiest port in Hawaii outside of the Island of Oahu

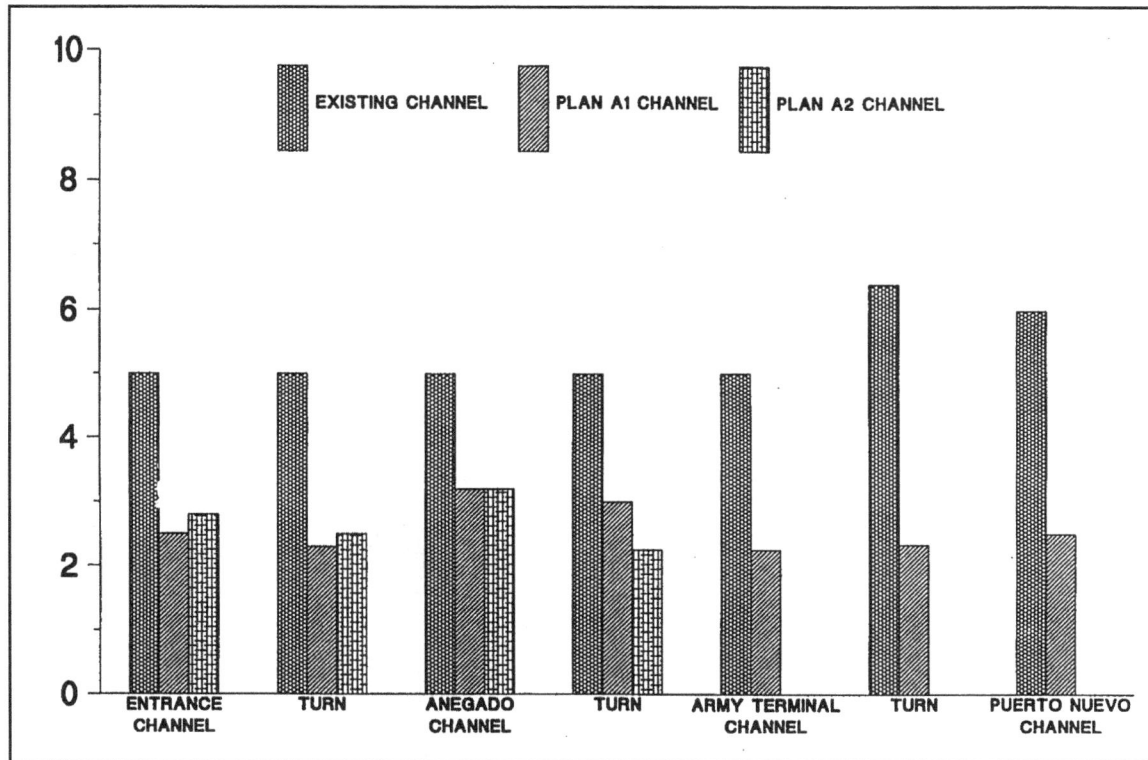

Figure V-5-69. Pilot degree of difficulty ratings, inbound container ship, San Juan Harbor

(Figure V-5-71). Commercial piers are presently on the east side of the harbor. In conjunction with long-term planning for expanded harbor usage, a field wave data collection program was established in the harbor. The program included an offshore directional array to measure incident waves and four pressure gauges in the harbor interior (Figure V-5-72). Interior gauge locations were determined with the assistance of a preliminary numerical model study of harbor wave response (Okihiro et al. 1994).

(b) Data from more than 1 year were collected and proved to be very helpful in subsequent harbor wave response, modeling, and planning studies (Thompson et al. 1996, Okihiro and Guza 1996).

(2) Ship tracking. The optimum depth and width of proposed navigation channel improvements may be determined with increased accuracy by measuring actual ship motions. The measurement program should encompass a significant number of transits of the route during adverse conditions. Availability of differential Global Positioning System (DGPS) apparatus for recording accurate ship fixes (±3 m) at a rapid rate (0.2 Hz or faster) makes this an affordable component of feasibility studies. Commercial software is available for data recording and display in formats applicable to channel design. These systems use standard DGPS receivers compatible with the U.S. Coast Guard network of DGPS radio beacons, as illustrated in Figure V-5-73. A time series of fixes is recorded with concurrent gyrocompass headings and other data, such as engine rpm, rudder angle, and relative wind speed and direction.

(a) Commercial gyrocompasses aboard seagoing cargo vessels usually provide heading accuracy of ±0.3 deg or better. Concurrent time series of position and heading define the swept path of the vessel. Comparison of ship tracks with tidal currents, winds, waves, water levels, visibility conditions, and other environmental conditions present at the time of recording, provide channel designers with realistic parameters for width computations.

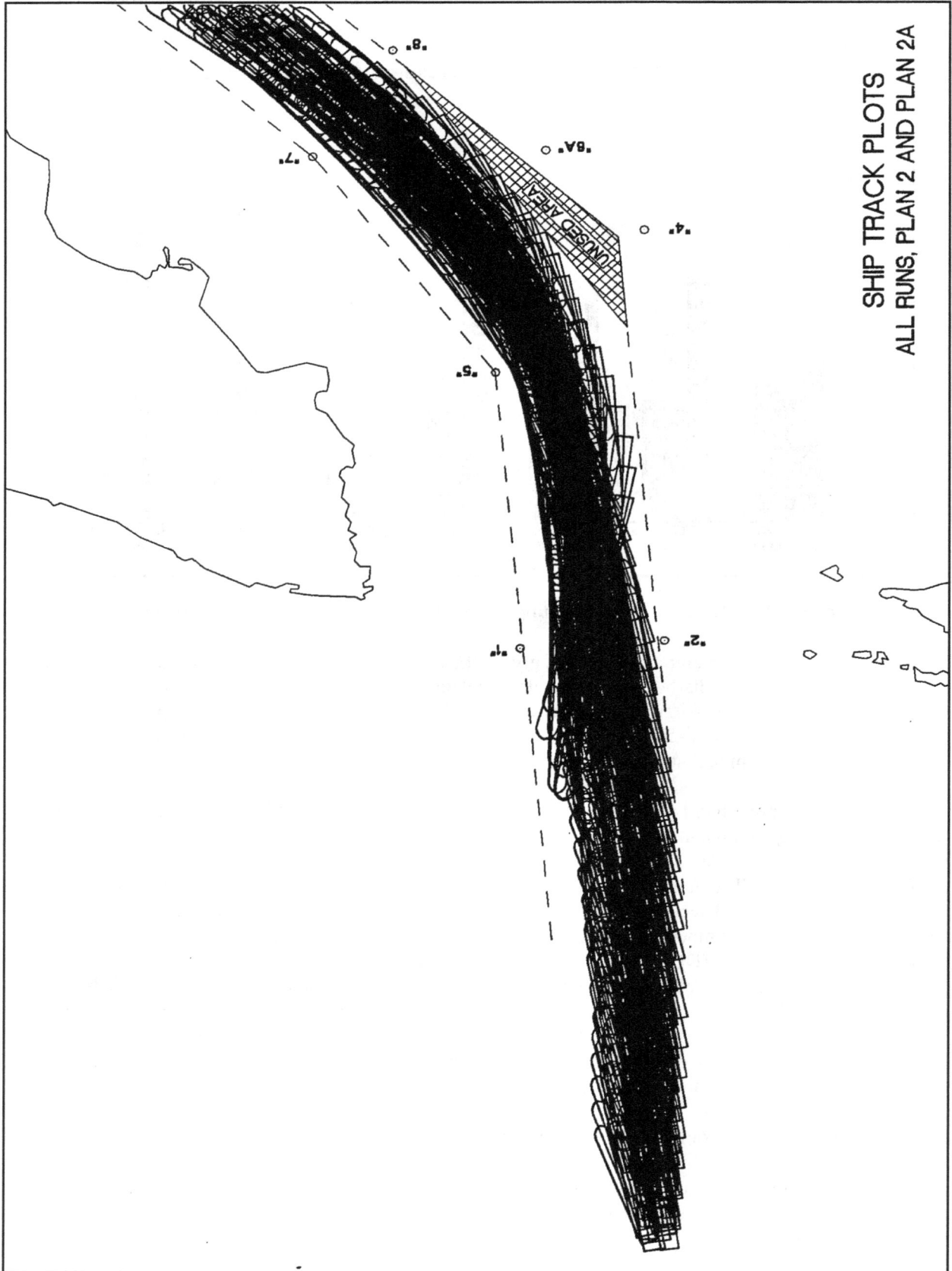

Figure V-5-70.   Ship tracks, proposed Bar Channel and turn into Anegado Channel, San Juan Harbor

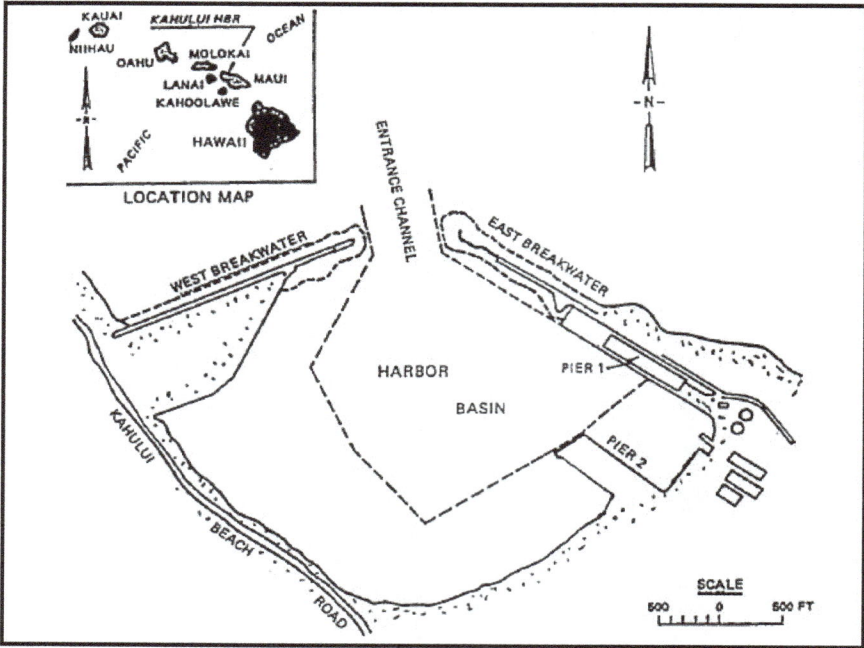

Figure V-5-71. Kahului Harbor, Maui, Hawaii, location map

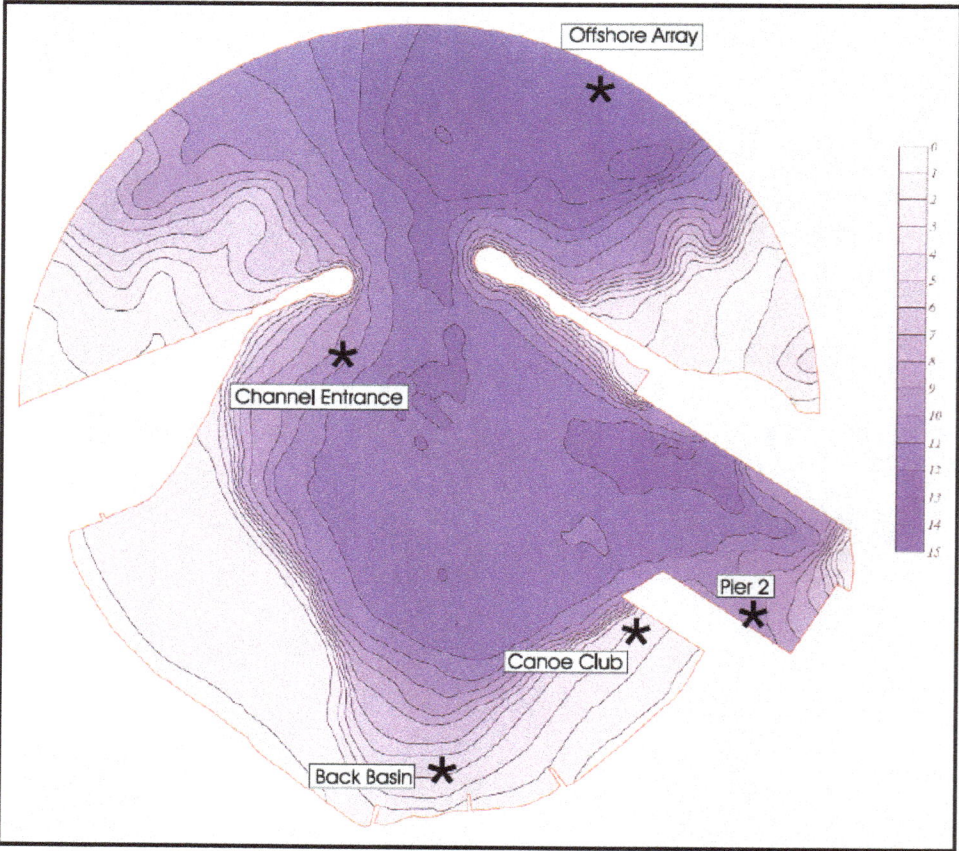

Figure V-5-72. Field gauge locations and bathymetry, Kahului Harbor

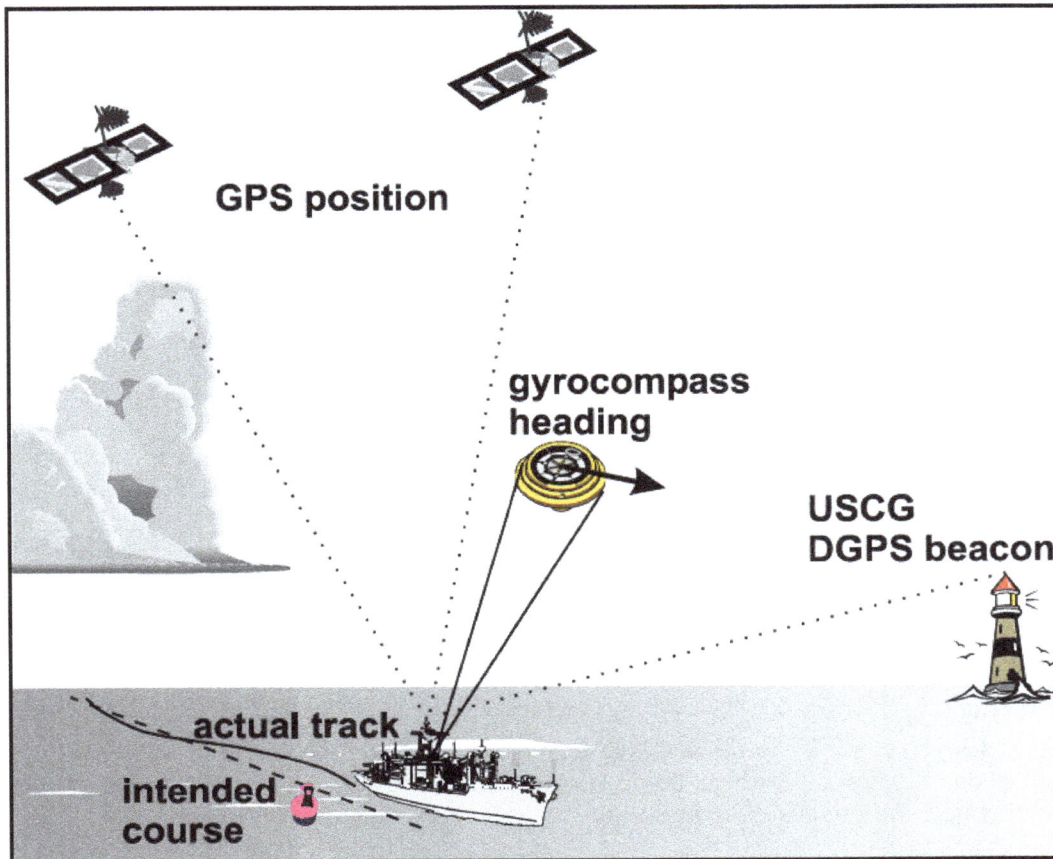

Figure V-5-73. Components of ship track measurements

(b) A dual-frequency DGPS system that measured horizontal and vertical location of ship bow and stern with 1 cm accuracy is described by Webb and Wooley (1998). The data, along with concurrent measurements of current and water level, provided direct calculation of ship squat. The data are useful for evaluating existing navigation conditions, designing modifications, and/or validating ship simulation models to study proposed conditions.

V-5-11. References

EM 1110-2-1615
EM 1110-2-1615, "Hydraulic Design of Small Boat Harbors"

EM 1110-8-1
EM 1110-8-1, "Winter Navigation on Inland Waterways"

EM 1110-2-1613
EM 1110-2-1613, "Hydraulic Design Guidance for Deep-Draft Navigation Projects"

ER 1105-2-100
ER 1105-2-100, "Guidance for Conducting Civil Works Planning Studies"

EP 1165-2-1
EP 1165-2-1, "Digest of Water Resources Policies and Authorities"

American Society of Civil Engineers 1994
American Society of Civil Engineers (ASCE). 1994. *Planning and Design Guidelines for Small Craft Harbors*, ASCE Manual and Reports on Engineering Practice No. 50, New York, NY.

Bottin 1976
Bottin, R. R., Jr. 1976. "Igloo Wave Absorber Tests for Port Washington Harbor, Wisconsin; Hydraulic Model Investigation," Miscellaneous Paper H-76-22, U.S. Army Engineer Waterways Experiment Station, Vicksburg, MS.

Bottin 1992
Bottin, R. R., Jr. 1992. "Physical Modeling of Small-Boat Harbors: Design Experience, Lessons Learned, and Modeling Guidelines," Technical Report CERC-92-12, U.S. Army Engineer Waterways Experiment Station, Vicksburg, MS.

Briggs, Bratteland, and Borgman 2000
Briggs, M. J., Bratteland, E., and Borgman, L. E. 2000. "Probability Assessment of Deep Draft Navigation Channel Design," submitted to *Coastal Engineering Journal*, Elsevier Science, The Netherlands.

Briggs et al. 1994
Briggs, M. J., Lillycrop, L. S., Harkins, G. S., Thompson, E. F., and Green, D. R. 1994. "Physical and Numerical Model Studies of Barbers Point Harbor, Oahu, Hawaii," Technical Report CERC-94-14, U.S. Army Engineer Waterways Experiment Station, Vicksburg, MS.

Bruun 1990
Bruun, P. 1990. *Port Engineering*, Fourth Edition (2 Vols.), Gulf Publishing, Houston, TX.

Chen and Houston 1987
Chen, H. S., and Houston, J. R. 1987. "Calculation of Water Oscillation in Coastal Harbors: HARBS and HARBD User's Manual," Instruction Report CERC-87-2, U.S. Army Engineer Waterways Experiment Station, Vicksburg, MS.

Clark 1983
Clark, J. R. 1983. *Coastal Ecosystem Management: A Technical Manual for the Conservation of Coastal Zone Resources*, John Wiley & Sons, New York, NY.

Dingemans 1997
Dingemans, M. W. 1997. *Water Wave Propagation over Uneven Bottoms*, 2 vols., World Scientific, Singapore.

Dunham and Finn 1974
Dunham, J. W., and Finn, A. A. 1974. "Small-Craft Harbors: Design, Construction, and Operation," Special Report No. 2, U.S. Army Engineer Research and Development Center, Coastal and Hydraulics Laboratory, Vicksburg, MS.

Ferguson 1991
Ferguson, A. 1991. "Navigation, Dredging and Environment in the Fraser River Estuary," Navigation and Dredging Workgroup Report, Fraser River Estuary Management Program, British Columbia, Canada.

Gaythwaite 1990
Gaythwaite, J. W. 1990. *Design of Marine Facilities for the Berthing, Mooring, and Repair of Vessels*, Van Nostrand Reinhold, New York, NY.

Harkins and Dorrell 2000
Harkins, G. S., and Dorrell, C. C. 2000. "Barbers Point Harbor Physical Model Navigation Study," Technical Report ERDC/CHL TR-00-2, U.S. Army Engineer Waterways Experiment Station, Vicksburg, MS.

Harris 1981
Harris, D. L. 1981. "Tides and Tidal Datums in the United States," Special Report No. 7, U.S. Army Engineer Waterways Experiment Station, Vicksburg, MS.

Herbich 1992
Herbich, J. B. 1992. *Handbook of Coastal and Ocean Engineering, Vol. 3, Harbors, Navigational Channels, Estuaries, Environmental Effects*, Gulf Publishing, Houston, TX.

Hudson et al. 1979
Hudson, R. Y., Herrmann, F. A., Jr., Sager, R. A., Whalin, R. W., Keulegan, G. H., Chatham, C. E., Jr., and Hales, L. Z. 1979. "Coastal Hydraulic Models," Special Report SR-5, U.S. Army Engineer Research and Development Center, Coastal and Hydraulics Laboratory, Vicksburg, MS.

Hughes and Fowler 1990
Hughes, S. A., and Fowler, J. E. 1990. "Midscale Physical Model Verification for Scour at Coastal Structures," Technical Report CERC-90-8, U.S. Army Engineer Waterways Experiment Station, Vicksburg, MS.

Huval 1980
Huval, C. J. 1980. "Lock Approach Canal Surge and Tow Squat at Lock and Dam 17, Arkansas River Project; Mathematical Model Investigation," Technical Report HL-80-17, U.S. Army Engineer Waterways Experiment Station, Vicksburg, MS.

Huval 1993
Huval, C. J., ed. 1993. "Planning, Design, and Maintenance of Deep Draft Navigation Channels," PROSPECT Training Course Notes, U.S. Army Engineer Waterways Experiment Station, Vicksburg, MS.

Huval and Lynch 1998
Huval, C. J., and Lynch, G. C. 1998. "Ship Navigation Simulation Study, Brunswick Harbor, Georgia," Technical Report CHL-98-18, U.S. Army Engineer Waterways Experiment Station, Vicksburg, MS.

Irish 1997
Irish, J. L. 1997. "Sensitivity of Channel Sedimentation Prediction to Wave-field Characterization," Bulletin No. 95, PIANC, Brussels, Belgium, pp 5-19.

LeMéhauté 1965
LeMéhauté, B. 1965. "Wave Absorbers in Harbors," Contract Report No. 2-122, U.S. Army Engineer Waterways Experiment Station, Vicksburg, MS, prepared for National Engineering Science Company, Pasadena, CA, under Contract No. DA-22-079-CIVENG-64-81.

Lillycrop et al. 1993
Lillycrop, L. S., Briggs, M. J., Harkins, G. S., Boc, S. J., and Okihiro, M. S. 1993. "Barbers Point Harbor, Oahu, Hawaii, Monitoring Study," Technical Report CERC-93-18, U.S. Army Engineer Waterways Experiment Station, Vicksburg, MS.

McBride, Smallman, and Huntington 1998
McBride, M. W., Smallman, J. V., and Huntington, S. W. 1998. "Guidelines for Design of Approach Channels," *Proceedings, PORTS '98*, ASCE, pp 1315-1324.

McCartney, Hermann, and Simmons 1991
McCartney, B. L., Hermann, F. A., Jr., and Simmons, H. B. 1991. "Estuary Waterway Projects - Lessons Learned," *Journal of Waterway, Port, Coastal, and Ocean Engineering* 117(4), American Society of Civil Engineers, pp 409-421.

Mei 1983
Mei, C. C. 1983. *The Applied Dynamics of Ocean Surface Waves*, John Wiley & Sons, New York, NY.

Melo and Guza 1991
Melo, E., and Guza, R. T. 1991. "Wave Propagation in Jettied Entrance Channels. II: Observations," *Journal of Waterway, Port, Coastal, and Ocean Engineering* 117(5), pp 493-510.

NOAA (Annual)
National Oceanic and Atmospheric Administration (NOAA). (Annual). *Tide Tables* and *Tidal Current Tables*, Rockville, MD.

Noda 1972
Noda, E. K. 1972. "Equilibrium Beach Profile Scale-Model Relationship," *Journal of the Waterways, Harbors, and Coastal Engineering Division* 98(WW4), ASCE, pp 511-528.

Norrbin 1986
Norrbin, N. H. 1986. "Fairway Design with Respect to Ship Dynamics and Operational Requirements," SSPA Research Report No. 102, SSPA Maritime Consulting, Gothenburg, Sweden.

Okihiro and Guza 1996
Okihiro, M. S., and Guza, R. T. 1996. "Observations of Seiche Forcing and Amplification in Three Small Harbors," *Journal of Waterway, Port, Coastal, and Ocean Engineering* 122(5), ASCE, 232-238.

Okihiro et al. 1994
Okihiro, M. S., Guza, R. T., O'Reilly, W. C., and McGehee, D. D. 1994. "Selecting Wave Gauge Sites for Monitoring Harbor Oscillations: A Case Study for Kahului Harbor, Hawaii," Miscellaneous Paper CERC-94-10, U.S. Army Engineer Waterways Experiment Station, Vicksburg, MS.

Panchang, Xu, and Demirbilek 1998
Panchang, V. G., Xu, B., and Demirbilek, Z. 1998. "Wave Models for Coastal Applications," *Handbook of Coastal Engineering*, J. B. Herbich, ed., Vol. 4 (to be published).

PIANC 1984
PIANC. 1984. "Ice Navigation," Supplement to Bulletin No. 46, Permanent International Association of Navigation Congresses, Brussels.

PIANC 1995
PIANC. 1995. "Approach Channels, Preliminary Guidelines," First Report of the Joint PIANC-IAPH Working Group II-30, Permanent International Association of Navigation Congresses, Brussels, Belgium.

PIANC 1997a
PIANC. 1997a. "Approach Channels, a Guide for Design," Supplement to Bulletin No. 95, Permanent International Association of Navigation Congresses, Brussels, Belgium.

PIANC 1997b
PIANC. 1997b. "Guidelines for the Design of Armoured Slopes Under Open Piled Quay Walls," Supplement to Bulletin No. 96, Permanent International Association of Navigation Congresses, Brussels, Belgium.

Quinn 1972
Quinn, A. D. 1972. *Design & Construction of Ports & Marine Structures*, McGraw-Hill.

Sager and Hollyfield 1974
Sager, R. A., and Hollyfield, N. W. 1974. "Navigation Channel Improvements, Barnegat Inlet, New Jersey," Technical Report H-74-1, U.S. Army Engineer Waterways Experiment Station, Vicksburg, MS.

Seabergh, Cialone, and Stauble 1996
Seabergh, W. C., Cialone, M. A., and Stauble, D. K. 1996. "Impacts of Inlet Structures on Channel Location," *Proceedings, 25th International Conference on Coastal Engineering*, ASCE, pp 4531-4544.

Seabergh, Vemulakonda, and Rosati 1992
Seabergh, W. C., Vemulakonda, S. R., and Rosati, J., III. 1992. "Los Angeles-Long Beach Harbors Model Enhancement Program," *Proceedings, PORTS '92*, ASCE, pp 884-897.

Silver 1992
Silver, A. L. 1992. "Environmental Monitoring and Operator Guidance System (EMOGS) for Shallow Water Ports," *Proceedings, PORTS '92*, ASCE, pp 535-547.

Silver and Dalzell 1997
Silver, A. L., and Dalzell, J. F. 1997. "Risk Based Decisions for Entrance Channel Operation and Design," *Proceedings, 7th International Offshore and Polar Engineering Conference*, International Society of Offshore and Polar Engineers, pp 815-822.

SMS 1994
SMS. 1994. "Surface-Water Modeling System," developed by Brigham Young University and USACE, updated periodically.

State of California 1980
State of California. 1980. "Layout and Design Guidelines for Small Craft Berthing Facilities," Department of Boating and Waterways.

Stevens et al. 1942
Stevens, J. C., Bardsley, C. E., Lane, E. W., Straub, L. G., Wright, C. A., Tiffany, J. A., and Warnock, J. E. 1942. *Hydraulic Models*, Manuals of Engineering Practice, No. 25, American Society of Civil Engineers, New York, NY.

Thompson, Boc, and Nunes 1998
Thompson, E. F., Boc, S. J., Jr., and Nunes, F. S. 1998. "Evaluating Operational Impact of Waves Along Proposed Harbor Piers," *Proceedings, PORTS '98*, ASCE, pp 860-869.

Thompson et al. 1996
Thompson, E. F., Hadley, L. L., Brandon, W. A., McGehee, D. D., and Hubertz, J. M. 1996. "Wave Response of Kahului Harbor, Maui, Hawaii," Technical Report CERC-96-11, U.S. Army Engineer Waterways Experiment Station, Vicksburg, MS.

Thompson et al. 1998a
Thompson, E. F., Fong, M. T., Van Norman, P. S., and Brown, B. 1998a. "Ship Navigation Simulation Study, Alafia River, Tampa Bay, Florida," Technical Report CHL-98-13, U.S. Army Engineer Waterways Experiment Station, Vicksburg, MS.

Thompson et al. 1998b
Thompson, E. F., Lin, L., Hadley, L. L., and Hubertz, J. M. 1998b. "Wave Response of Kikiaola Harbor, Kauai, Hawaii," Miscellaneous Paper CHL-98-5, U.S. Army Engineer Waterways Experiment Station, Vicksburg, MS.

Tobiasson and Kollmeyer 1991
Tobiasson, B. O., and Kollmeyer, R. C. 1991. *Marinas and Small-Craft Harbors*, Van Nostrand Reinhold, New York, NY.

Tsinker 1997
Tsinker, G. P. 1997. *Handbook of Port and Harbor Engineering*, Chapman & Hall, New York, NY.

Turner 1984
Turner, H. O., Jr. 1984. "Dimensions for Safe and Efficient Deep-Draft Navigation Channels," Technical Report HL-84-10, U.S. Army Engineer Waterways Experiment Station, Vicksburg, MS.

USACE 1983
U.S. Army Corps of Engineers. 1983. "Engineering and Design of Military Ports," Technical Manual TM 5-850-1, Headquarters, Department of the Army, Washington, DC.

U.S. Coast Guard 1981
U.S. Coast Guard. 1981. "Aids to Navigation Manual - Administration," COMDTINST M16500.7, U.S. Coast Guard, Washington, DC.

U.S. Coast Guard 1988a
U.S. Coast Guard. 1988a. "Aids to Navigation Manual - Seamanship," CG-222.2, U.S. Coast Guard, Washington, DC.

U.S. Coast Guard 1988b
U.S. Coast Guard. 1988b. "Aids to Navigation Manual - Technical," COMDTINST M16500.3, U.S. Coast Guard, Washington, DC.

U.S. Navy 1981
U.S. Navy. 1981. "Harbors," Design Manual 26.1, Naval Facilities Engineering Command, Alexandria, VA.

U.S. Navy 1998
U.S. Navy. 1998. "Handbook for Mooring Design," MIL-HDBK 1026/4, Naval Facilities Engineering Command, Alexandria, VA.

Wang and Cialone 1995
Wang, H. V., and Cialone, A. 1995. "Numerical Hydrodynamic Modeling and Flushing Study at Maalaea Harbor, Maui, Hawaii," Miscellaneous Paper CERC-95-8, U.S. Army Engineer Waterways Experiment Station, Vicksburg, MS.

Wang et al. 1980
Wang, S., Butcher, C., Kimble, M., and Cox, G. D. 1980. "Columbia River Entrance Channel Deep-Draft Vessel Motion Study," Tetra Tech Report No. TC-3925, U.S. Army Engineer District, Portland, OR.

Webb 1993
Webb, D. W. 1993. "Ship Navigation Simulation Study, San Juan Harbor, San Juan, Puerto Rico," Technical Report HL-93-17, U.S. Army Engineer Waterways Experiment Station, Vicksburg, MS.

Webb 1994
Webb, D. W. 1994. "Navigation Channel Design Using Real-Time Marine Simulator," *Proceedings, Dredging '94*, ASCE, pp 1081-1090.

Webb and Wooley 1998
Webb, D. W., and Wooley, R. T. 1998. "Using DGPS Ship Transit Data for Navigation Channel Design," *Proceedings, PORTS '98*, ASCE, pp 1325-1332.

Wilson 1966
Wilson, H. B. 1966. "Design for Optimum Wave Conditions, Dana Point Harbor, Dana Point, California; Hydraulic Model Investigation," Technical Report 2-724, U.S. Army Engineer Waterways Experiment Station, Vicksburg, MS.

WMO 1970
World Meteorological Organization (WMO). 1970. "WMO Sea-ice Nomenclature," WMO/OMM/BMO-No.259.TP.145, as amended, Geneva, Switzerland.

V-5-12. Definitions of Symbols

α_b	Breaking wave angle [deg]
A_C	Channel cross-section area [length^2]
B	Vessel beam at midships [length]
C_B	Block coefficient (Figure V-5-4) [dimensionless]
F_h	Channel depth Froude number [dimensionless]
F_L	Schijf limiting Froude number (Equation V-5-7) [dimensionless]
g	Gravitational acceleration [length/time^2]
h	Depth of channel [length]
h_1, h_2	Overbank depths [length]
L	Ship length [length]
N_B	Number of boats using the project
T	Vessel draft [length]
V	Vessel speed [length/time]
V_L	Schijf limiting ship speed in squat analysis [length/time]
W	Width of channel [length]
Z	Maximum ship squat [length]
Z_L	Maximum ship squat at Schijf limiting Froud number (Equation V-5-8) [length]
Z_T	Maximum squat in a trench channel (Equation V-5-9) [length]

V-5-13. Acknowledgments

Authors of Chapter V-5, "Navigation Projects:"

Edward F. Thompson, Ph.D., Coastal and Hydraulics Laboratory (CHL), U.S. Army Engineer Research and Development Center, Vicksburg, Mississippi.
Carl J. Huval, CHL (retired)
Ray R. Bottin, CHL
Orson P. Smith, Ph.D., U.S. Army Engineer District, Alaska, Anchorage, Alaska, (retired)

Reviewers:

David Curfman, Naval Facilities Engineering Command, Norfolk, Virginia
Lee E. Harris, Ph.D., Department of Marine and Environmental Systems, Florida Institute of Technology, Melbourne, Florida
Mona King, U.S. Army Engineer District, Seattle, Seattle, Washington
John H. Lockhart, Headquarters, U.S. Army Corps of Engineers, Washington, DC, (retired)
J. Richard Weggel, Ph.D., Department of Civil and Architectural Engineering, Drexel University, Philadelphia, Pennsylvania

Table of Contents

List of Figures

List of Tables

Chapter V-6
Sediment Management at Inlets

V-6-1. Introduction

   a. Overview. The U.S. Army Corps of Engineers (USACE) has a mission to provide military and civil works engineering to the nation. Congress authorized the navigation part of this mission in 1824, when USACE was directed to remove sandbars and snags from major navigable rivers. Today, USACE's dredging program involves the planning, design, construction, operation, and maintenance of riverine, estuarine, and coastal passages to meet navigation needs. The most dynamic of these waterways are inlets, entrances, and harbors on oceans, bays, and lakefronts, at which wind waves and tides or seiching of water may combine to transport sediment. The foremost navigation goal is to provide safe passage in these channels for vessels of a given design draft at the least cost. In fiscal year (FY) 1999, USACE dredged approximately 360 million cu m from Federal channels. Maintenance dredging accounted for an average of 85 percent of this volume while new work and emergency dredging made up the remainder. The most efficient means of managing sediment that enters these navigable waterways is the subject of this chapter.

   b. Definitions.

   (1) For convenience, the term "inlets" will be used herein to encompass all these navigation passages. Inlet generally refers to a short, narrow waterway that connects a lagoon or bay with a larger parent body of water (i.e., ocean or lake), and experiences the inflow and outflow of water due to tidal or seiching. Inlets may be stabilized with shore-perpendicular structures at the mouth, called jetties. Waves and currents at an inlet combine to form ebb- and flood-tidal shoals (Figures V-6-1a, b, and c).

   (2) An entrance is the opening to a navigation channel (Figure V-6-1d). A harbor refers to any protected body of water that affords a place of safety for vessels (Figure V-6-1e). Well-situated harbors do not experience significant currents due to tides or seiching and therefore do not have notable ebb- or flood-tidal shoals.

   (3) The term "sediment" includes all types of minerals or organic materials that have been deposited by water, wind, or ice. Many navigation channels shoal because of the inflow of sand and finer-grained sediments. If this material impedes navigation, it may be dredged to provide safe passage through the inlet. Littoral pertains to a lake, sea, or ocean. As applied herein, the term "littoral material" refers to sediments (sand, silt, clay, gravel, etc.) that are transported by waves and currents along or near the coast. Note that wind-blown, or eolian, sediment may also contribute to shoaling in navigable waterways. Beach-quality sediment generally refers to sand that is of a sufficient size, color, and composition to be placed on the beach or nearshore. Nonbeach-quality materials can be used in other ways to benefit the coastal zone, e.g., creating wetlands and environmental habitat.

   c. Background.

   (1) The need to create navigable passage is a direct consequence of the dynamic nature of inlets. As an illustration, imagine a new inlet that is breached from the ocean side of a barrier island during a storm. Waves, an elevated ocean water level, and tidal currents scour sediments from the barrier to form the new inlet. The material is carried landward and seaward to form incipient flood- and ebb-tidal shoals. As tidal cycles pass, a channel is incised into the former barrier bed, thereby establishing the inlet as a coastal feature. Sand and other sediments moving alongshore are entrained by inlet ebb and flood currents, and

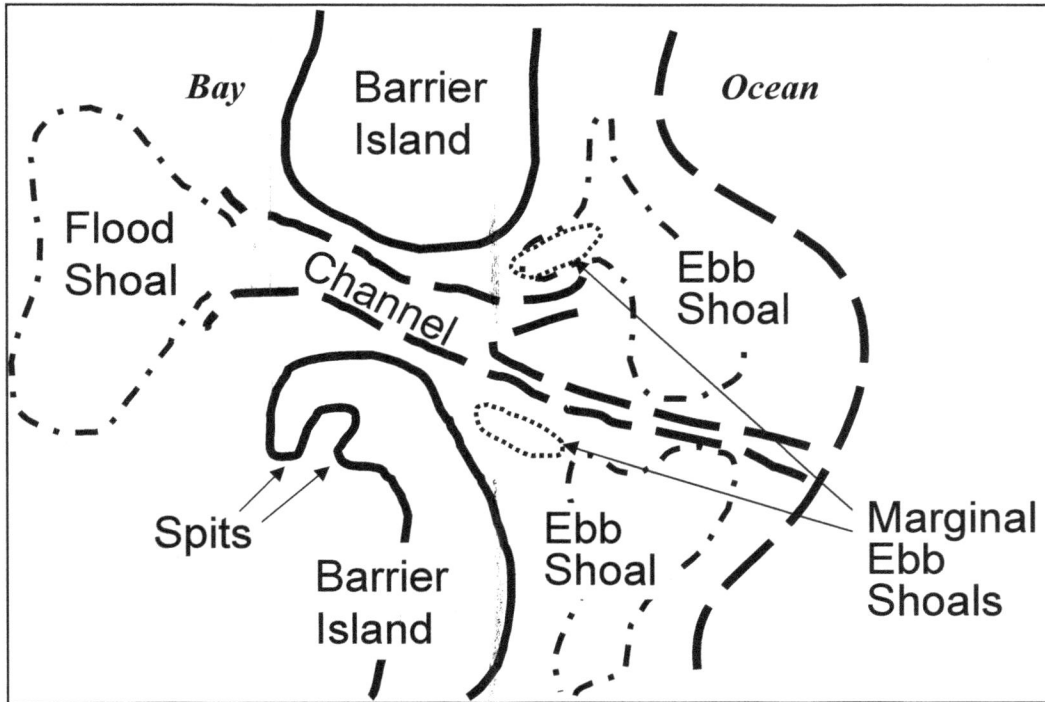

Figure V-6-1a. Conceptual illustration of unstabilized inlet

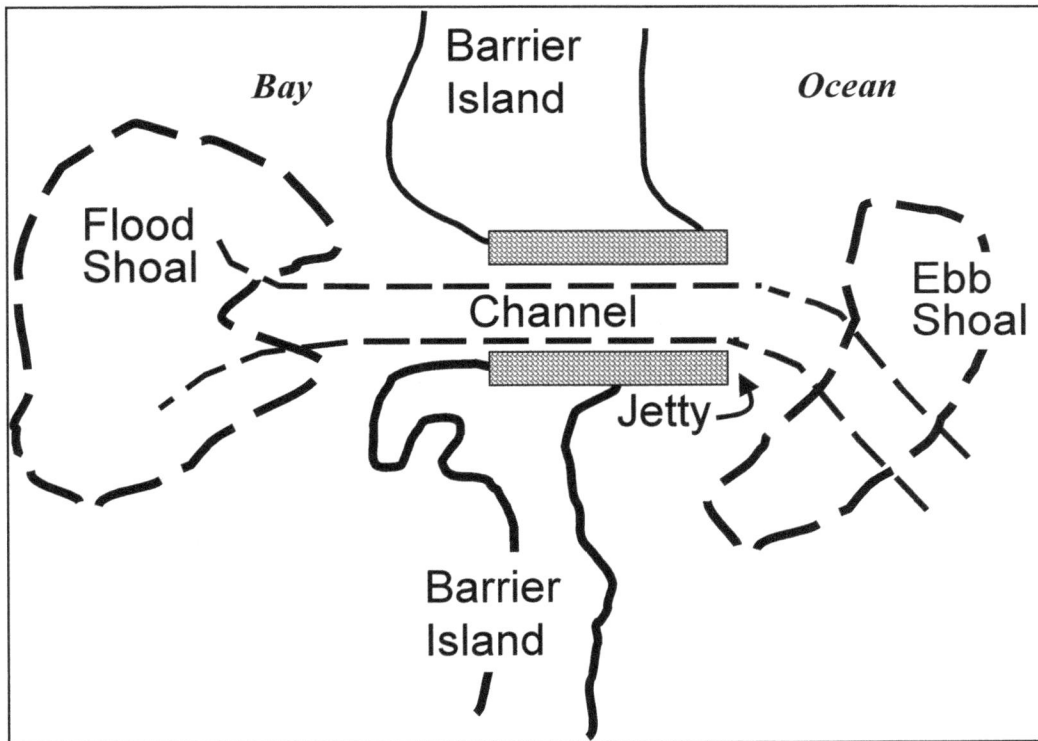

Figure V-6-1b. Conceptual illustration of stabilized inlet

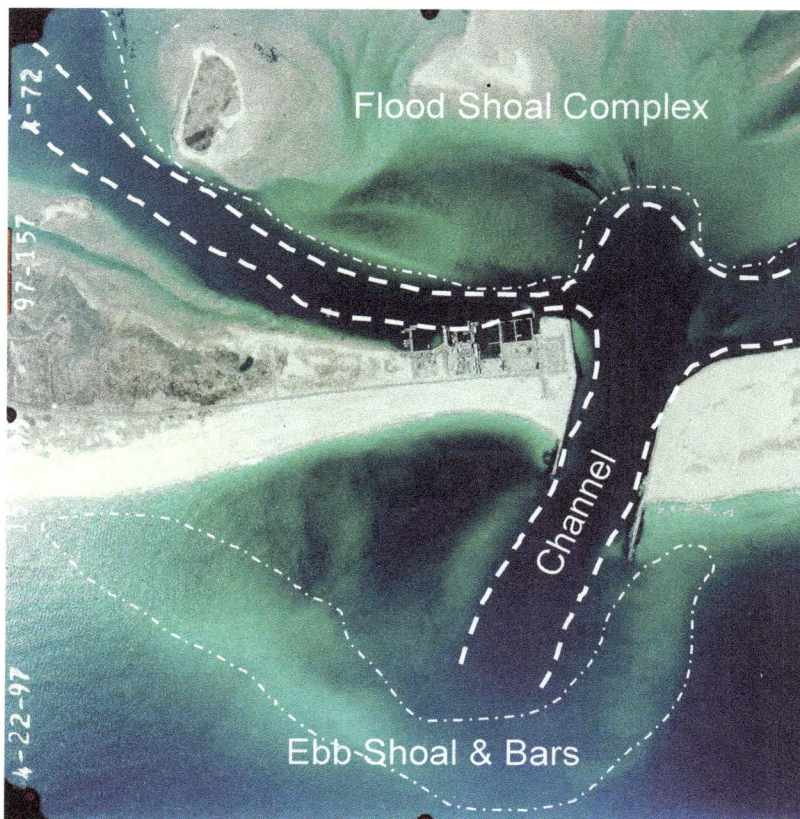

Figure V-6-1c. Example of stabilized inlet: Shinnecock Inlet, Long Island, New York (22 April 1997)

Figure V-6-1d. Example of entrance: Port Canaveral, Forida, (date unknown)

Figure V-6-1e. Example of harbor: Oceanside, California, (12 February 1999, photograph by Eagle Aerial)

may be deposited in the inlet channels. The sediment may remain in the channels or may be jetted to the ebb- and flood-tidal shoals. As the inlet matures, the adjacent beaches may advance and retreat due to the changing waves, currents, and bathymetry in the vicinity of the opening. Volumetrically, the effect of this new inlet is a net loss to the adjacent beach system. Navigability of this new inlet is typically treacherous. Channels may meander and their depths may vary rapidly. Waves may break due to shoaling and due to current-induced wave steepening during ebb flows. Depending on the coastal processes and geologic constrictions, the inlet may migrate alongshore.

(2) The construction of jetties and dredging of channels are navigation controls that seek to promote safe passage through the inlet while minimizing construction and maintenance costs. Jetties stabilize the inlet's location and confine the tidal currents within the channel. In addition, the constricted currents help to clean the channel of shoaled sediment. Dredging improves channel navigability by increasing the depth as required to meet the needs of the design vessel, and straightening channel meanders that make passage through the inlet dangerous. However, the effective removal and disposal of sediment that is removed from the channel and the ebb- and flood-tidal shoals present an ongoing challenge.

(3) The goals of inlet sediment management are multifaceted:

(a) Through design of inlet structures, channels, and other controls (e.g., bypassing plants), to minimize the volume of sediment that accumulates within the navigable waterway.

(b) To remove shoaled sediment and morphologic features that inhibit safety at least cost.

(c) To optimize channel depth such that dredging frequency is reduced while the stability of inlet structures and adjacent beaches are not compromised.

(d) To keep the beach-quality dredged material within the nearshore beach system, effectively bypassing sand around the inlet.

(e) To place beach-quality material within the nearshore system and on adjacent beaches such that it does not cycle back into the inlet.

(f) To assist with other missions of USACE as appropriate, such as using dredged material to enhance environmental habitat, provide flood and storm damage protection, and establish or restore wetlands.

d. Summary. The purpose of this chapter is to discuss engineering works directed towards effective inlet sediment management and review analytical methods that can be applied in the design and evaluation of these projects.   Experiences at several sites are reviewed to illustrate approaches to evaluating sediment management problems, the solutions that were implemented, and their performance.

V-6-2. Regional Sediment Management and Coastal Inlet Processes

a. Overview. This section discusses the effects that sediment management activities at inlets have both in the vicinity of the inlet and regionally (on the scale of tens of kilometers along the coast). First, regional sediment management is defined. The relationship between local, inlet-specific activities, and the response of other features within the region is discussed. Next, sediment transport processes at natural and engineered inlets are reviewed, with reference to material contained in Parts II-6 and IV-3.

b. Regional sediment management.

(1) Overview. Regional sediment management refers to the use of littoral, estuarine, and riverine sediment resources in an environmentally beneficial and economical manner. Regional sediment management strives to maintain or enhance the natural exchange of sediment within the boundaries of the physical system. A region may include a variety of geologic features, uplands, beaches, inlets, rivers, estuaries, and bays, and is defined by the sediment transport paths within this physical system (Figure V-6-2). Implementation of regional sediment management recognizes that the physical system and embedded ecosystems are modified and may respond to natural forcing and engineering activities beyond the formal dimensions and time frames of individual projects.

(a) USACE has a central role in the implementation of regional sediment management. The mission areas of USACE include navigation, environmental restoration, storm-damage reduction, and flood reduction. In particular, the mission area ensuring the navigability of our nation's waterways involves removing, transporting, and placing sediment. Some of this material can be used to achieve goals in other mission areas. As opposed to individual homeowners, county and state governments, and other Federal agencies, USACE's role spans both political and geographic boundaries. Although the term "regional sediment management" is new, recognition of the regional nature of coastal processes and the regional

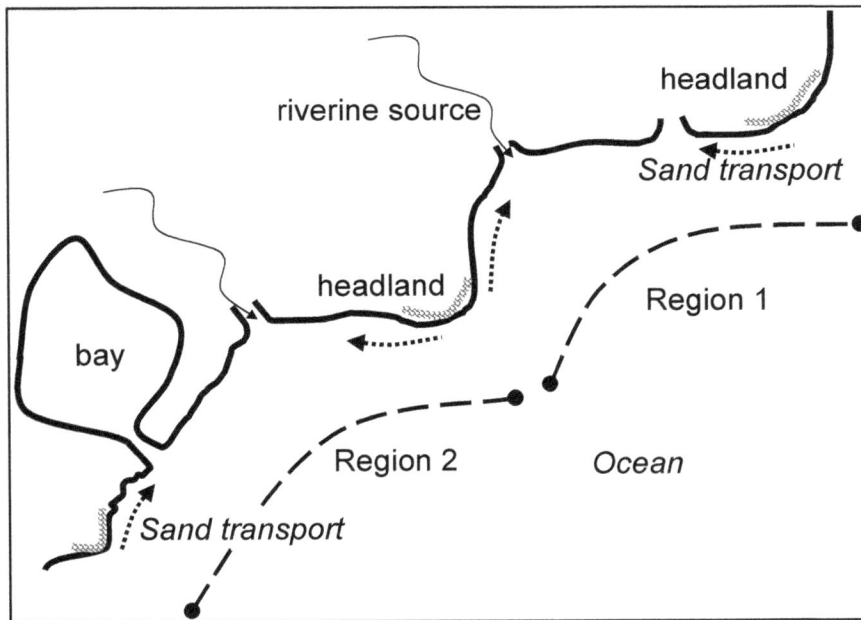

Figure V-6-2. Conceptual illustration of regions

influence of engineering works is not. The interrelationship between coastal navigation projects and contiguous beaches became a Federal interest at least as early as the 1930s (Brooke 1934). The first sand bypassing systems at navigation projects, designed to reinstate net longshore sand transport to downdrift beaches, were put into operation in the mid-1930s at Santa Barbara, California (mobile plant) (Penfield 1960) and South Lake Worth Inlet, Florida (fixed plant) (Caldwell 1951).

(b) Today, USACE is pursuing regional sediment management by collaborating with local and state governments to manage sediments over regions encompassing multiple projects. Schmidt and Schwichtenberg (2000) describe the Jacksonville District's application of regional sediment management principles for proposed improvements to Port Everglades, a deep-draft port on the east coast of Florida. These improvements include widening and deepening existing channels and turning basins, and creating a new turning basin to accommodate larger container ships. Federal shore protection projects along the adjacent three counties will require almost 59 million cu m of beach-quality sand. The District is working with county and state agencies to satisfy the navigational goals at Port Everglades while using the dredged sediments to nourish the adjacent beaches and restore environmental habitat. It is envisioned that beach-quality dredged material can be identified prior to dredging and placed directly on the beach. For mixed sediments, a temporary deposition site on an adjacent beach has been considered. After the dredged slurry has been deposited on the beach, hydrocyclones could separate fine sediments from beach quality sand. The beach-quality sand would be trucked to adjacent beaches, and the silt could be placed and vegetated to build natural berms and dunes.

(2) Critical role of inlets. Inlets are significant elements in controlling sediment movement and its distribution within local and regional domains. They may serve as a source if they funnel riverine, estuarine, and bay sediments to the open coastline. They may also act as a sink by diverting littoral material into the inlet and its bay, harbor, and estuaries. In a manner analogous to water flow through a tidal inlet (see Part II- 6), littoral material may be transported into the inlet on the flood (rising) tide and out of the inlet on the ebb (falling) tide. Flood or ebb dominance of water flow through an inlet, however, does not necessarily determine the net sediment transport direction through the inlet (Stauble et al. 1988; see also Part IV-3-4, or along the adjacent beaches. An inlet in a state of "dynamic equilibrium" indicates

that there is a net long-term (years to decades and longer) balance between water flow through the inlet and sediment transport to and from the inlet. The net, long-term effect is a hydrodynamic field that is sufficient to maintain flow through the inlet and maintain an approximate balance of sediment volume on the beaches to either side of the inlet. In the short term (days to years), cyclical events such as seasonal changes in waves and currents or storms may temporarily disturb the balance. Therefore, although an inlet may be in a state of dynamic equilibrium, it may require dredging to maintain navigability. Inlets that are not in a state of dynamic equilibrium include those that: migrate alongshore, are closing because of shoaling of the channels (either from an upland or ocean source of sediment), and are widening or scouring due to an increasing tidal prism or unstable geologic substrate. For an inlet to remain open, sediment entering the inlet must be diverted from the inlet's principal channels, or the channels must move. If sediment is not bypassed across the inlet, it must be transported by waves or currents away from the inlet throat and into interior shoals, exterior shoals, or beach impoundments; or, it must be removed by dredging. The following section describes hydrodynamic and sediment transport processes at inlets.

(3) Coastal inlet processes.

(a) Inlet features. The following is an overview of principal inlet flow fields and sedimentary features. Parts II-6-1 and IV-3-4 present more detailed descriptions of inlet hydrodynamics and sediment transport/morphology.

- Typical elements of nonstabilized and stabilized tidal inlets are illustrated in Figures V-6-1a and b, respectively. An inlet connects a larger, open-water body (the ocean, sea, lake, etc.) with a generally smaller, protected water body (a bay, lagoon, estuary, or river) through a land body such as a barrier spit. The inlet technically includes the entire seabed between the banks of the adjacent barrier shores, measured from the seaward edge of the ebb shoal to the landward edge of the flood shoal (discussed in the following paragraph). The inlet throat is the region with the smallest cross-sectional flow area and highest velocities. The channel (geologically known as a gorge or thalweg) defines the deepest parts of the inlet. There may be several channels in a natural inlet, which may migrate, vanish, or alternate in dominance. The navigable channel is the one that is deepest at a given time or is maintained for navigation by dredging.

- A tidal inlet (also known as a lagoonal inlet) connects a tidal water body with the bay, and features significant flow directed both into and out of the inlet throat as a function of water level fluctuations in the ocean and bay. (Riverine mouths or fluvial inlets are dominated by river discharge and minimally affected by flood tide flow). During flood tide, currents are funneled toward the inlet mouth from the ocean (Figures V-6-3a and V-6-4a). In the absence of high, impermeable jetties (Figure V-6-3a), the flood flow is typically most dominant along the ocean shorelines of the adjacent barrier islands, sometimes creating and/or following marginal flood channels. The flood flow converges and accelerates in the throat. Upon its discharge, the flood flow diverges and decelerates beyond the inlet's bayside mouth, depositing sediment in a flood shoal. The flow reverses during ebb current, although rarely symmetrically (Figures V-6-3b and V-6-4b). The ebb flow sweeps sediment from the flood shoal and inlet channels and carries it through the inlet throat. The ebb flow discharges from the inlet's ocean mouth similarly to a free jet, depositing its sediment load in an ebb shoal. The ebb shoal may protrude significantly into the deeper ocean, forming the ebb shoal plateau or terminal lobe.

- The coast in the direction from which the majority of offshore wave energy arrives is termed the updrift shoreline (top barrier island in Figures V-6-3c and V-6-4c). The opposite coast is the

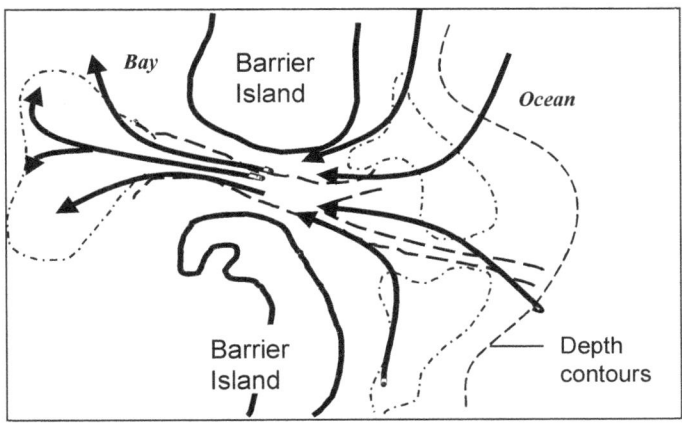

(a) Flood flow

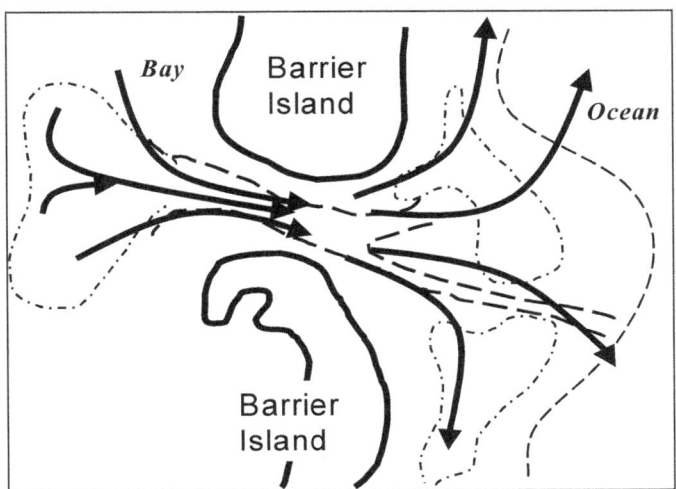

(b) Ebb flow

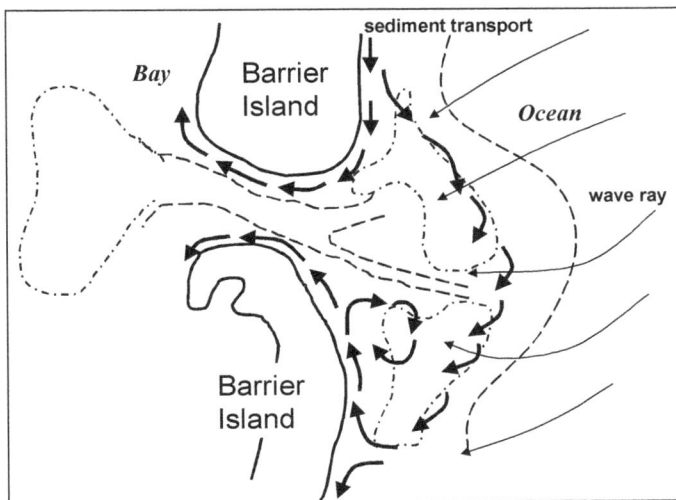

(c) Waves

Figure V-6-3. Idealized sediment transport pathways for unstabilized inlet

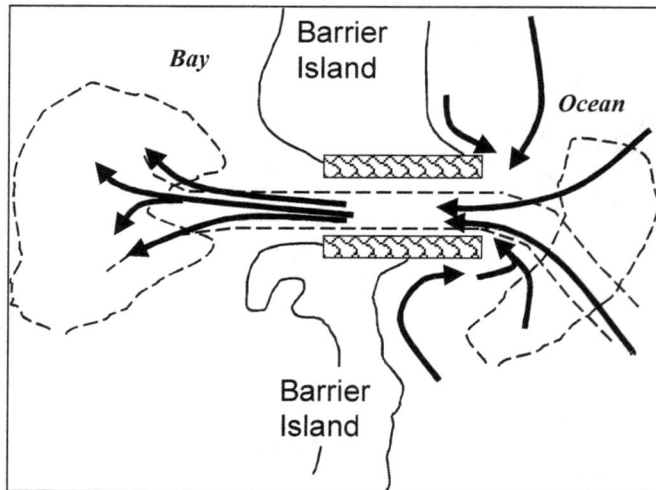

(a) Flood flow

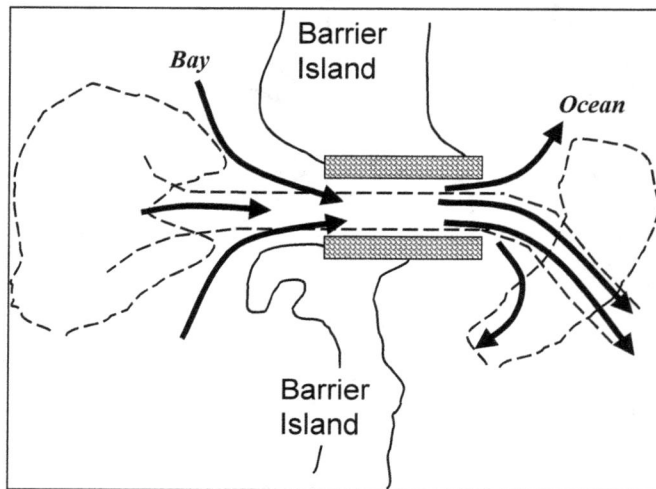

(b) Ebb flow

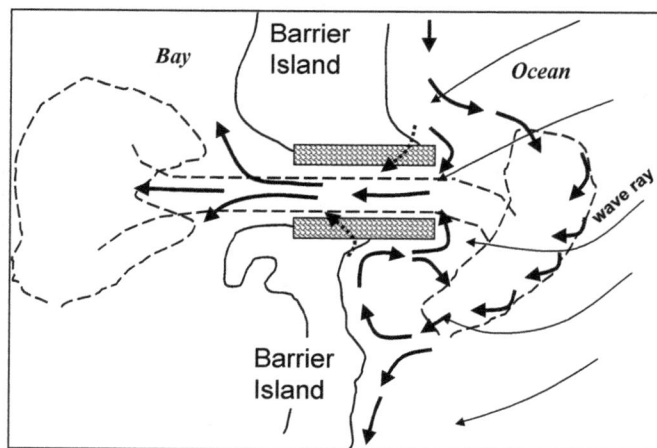

(c) Waves

Figure V-6-4. Idealized sediment transport pathways for stabilized inlet

downdrift shoreline. Along the updrift shoreline, obliquely incident waves transport sediment alongshore and directed toward the inlet. Across the inlet mouth, the waves breaking upon the ebb shoal transport sediment both toward (into) the inlet and toward the downdrift shoreline. The ebb shoal may, in turn, be skewed toward or elongated beyond the downdrift shoreline. Downdrift of the inlet (a distance of anywhere between one-half and 10 km), wave refraction across the ebb plateau decreases the obliquity of the wave's angle and the associated magnitude of the downdrift-directed transport. Closest to the inlet, transport along the shoreline (leeward of the ebb shoal) may even be reversed, so that transport is directed toward the inlet from the downdrift shoreline. This transport is augmented by diffraction currents (resulting from decreased wave energy and water level setup) directed toward the inlet along the immediate downdrift shoreline. The sediment transported toward the inlet from the downdrift shore may be carried into the inlet and/or diverted toward the ebb shoal by gyres or eddies (circular currents) associated with the inlet's tidal currents. If the jetties are not sand-tight, sediment may be transported over or through these structures (dotted lines in Figure V-6-4c).

- As illustrated in Figure V-6-5a, inlets where longshore currents dominate are more likely to form an ebb shoal with a shallow, arcuate form across the mouth. If tidal currents dominate littoral transport processes (Figure V-6-5b), marginal ebb shoals (also known as linear bars) are likely, aligned with and bordering the inlet channels. Mixtures of both types of shoals are not uncommon. These characteristics are typically observed at unstabilized inlets.

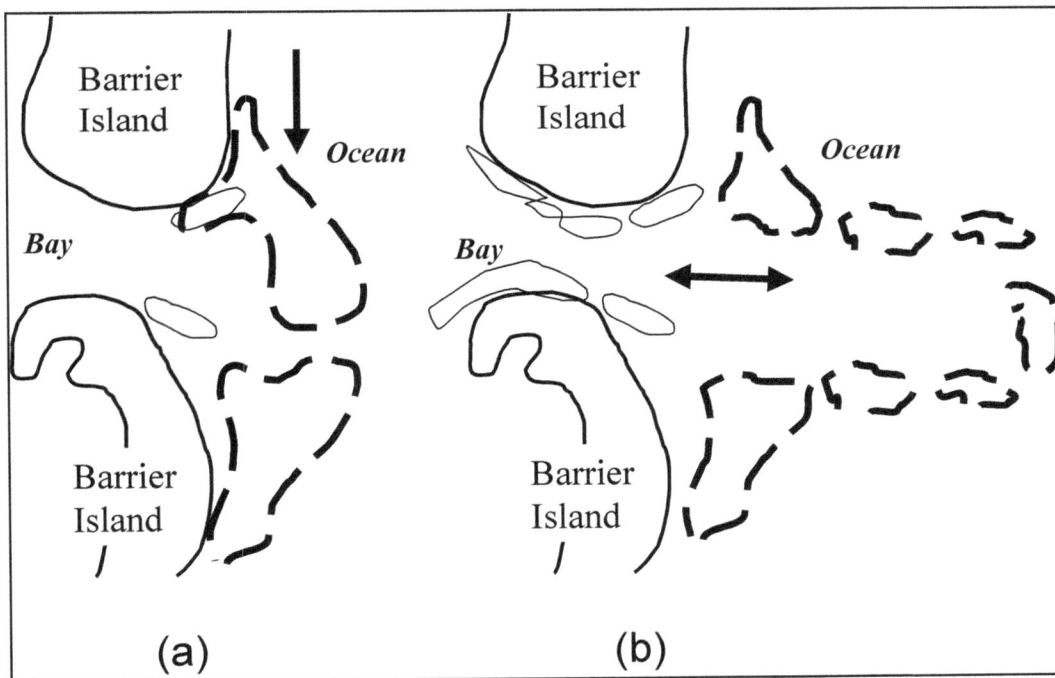

Figure V-6-5. Unstabilized tidal inlets for which (a) longshore current effects dominate, and (b) tidal current effects dominate (adapted from Oertel 1988)

- Other sedimentary features include interior spits and channel shoals (see Figure V-6-1a). Spits are emergent landforms that result from the deposition of littoral drift from the adjacent shores. Channel shoals are submerged deposits that encroach upon or cover parts of the inlet seafloor. Spits and channel shoals are usually attached to the barrier island at, or just landward of, the point at which the transport spills from the shoreline into the inlet.

Sediment Management at Inlets

- Processes other than just tidal-induced currents and waves can heavily influence inlet dynamics and morphology. These include wind, salinity differences between the ocean and bay waters, wave-current interactions, river discharge, underlying geology, and inlet orientation. These concepts are outlined in Parts II-6 and IV-3.

(b) Bypassing processes. Sediment may be trapped at an inlet or it may also be bypassed by a variety of natural forces. Bruun and Gerritsen (1959) described three principal mechanisms of bypassing for unstabilized inlets: Wave-induced transport along the ebb tidal shoal (bypass bar). Transport into and out of the inlet by tidal currents. Migration and welding of sandbars.

- Figure V-6-6 illustrates wave-induced transport along the ebb tidal shoal (termed, in this case, a bypassing bar). Waves transport sand (littoral drift) from the updrift shoreline toward the inlet. Some of the sand is transported (ramped) directly to the ebb shoal (or bar). This sand is temporarily or permanently stored in the shoal, or is transported downdrift by waves. Once outside the principal influence of the inlet's ebb flow, waves transport the sand both landward and further downdrift, where it can rejoin the downdrift shoreline and littoral stream.

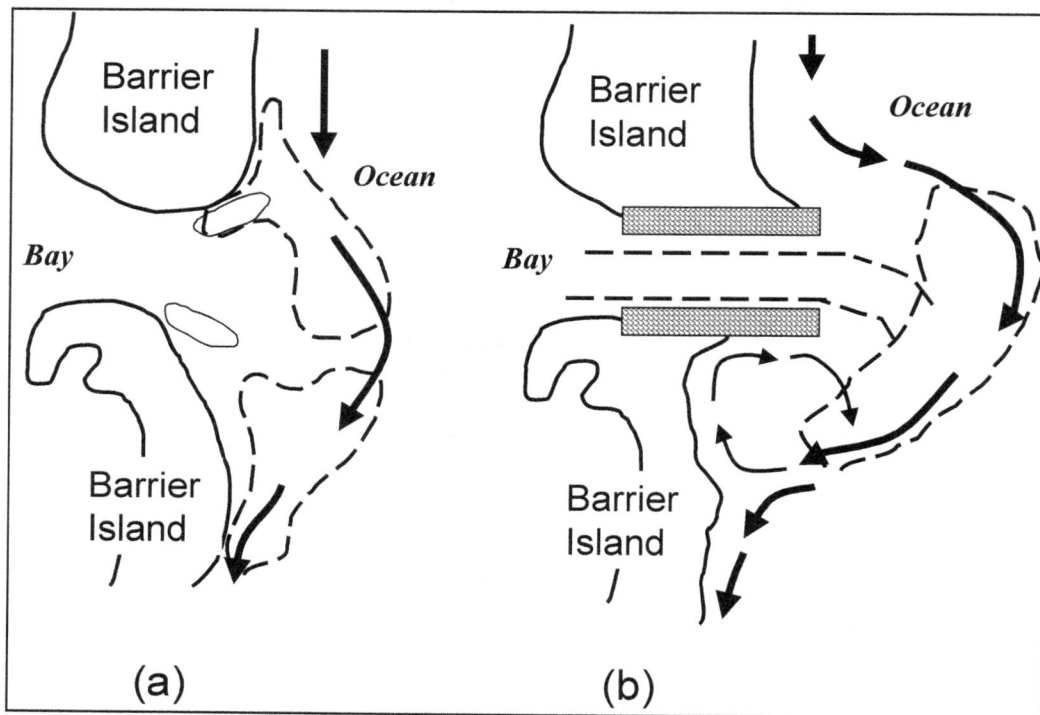

Figure V-6-6. Idealized wave-induced bypassing along the ebb tidal shoal ("bypass bar") for (a) unstabilized and (b) stabilized inlets

- Figure V-6-7 illustrates idealized wave and current bypassing. A portion of the littoral drift from the updrift shore may be transported toward the inlet mouth. There, it may be temporarily or permanently stored as part of the updrift beach, creating a bulbous shoreline shape or a shallow plateau at the shoreline's toe, or it may be impounded at the updrift jetty. Or, waves and marginal flood currents may transport the sediment directly into the inlet. Flood currents sweep sand bayward from the inlet's ebb shoals, channels, and interior shoals, and deposit the sand on the flood shoal and along shoals and spits which line the inlet's interior shoreline. Ebb currents, in turn, sweep sand seaward from the flood shoal (and the channels and interior shoals) and deposit it in the ebb shoal. There, the sand may be again swept bayward by flood currents and returned by ebb currents, or may be transported along the ebb shoal by waves, where it can eventually rejoin the downdrift shoreline as previously described.

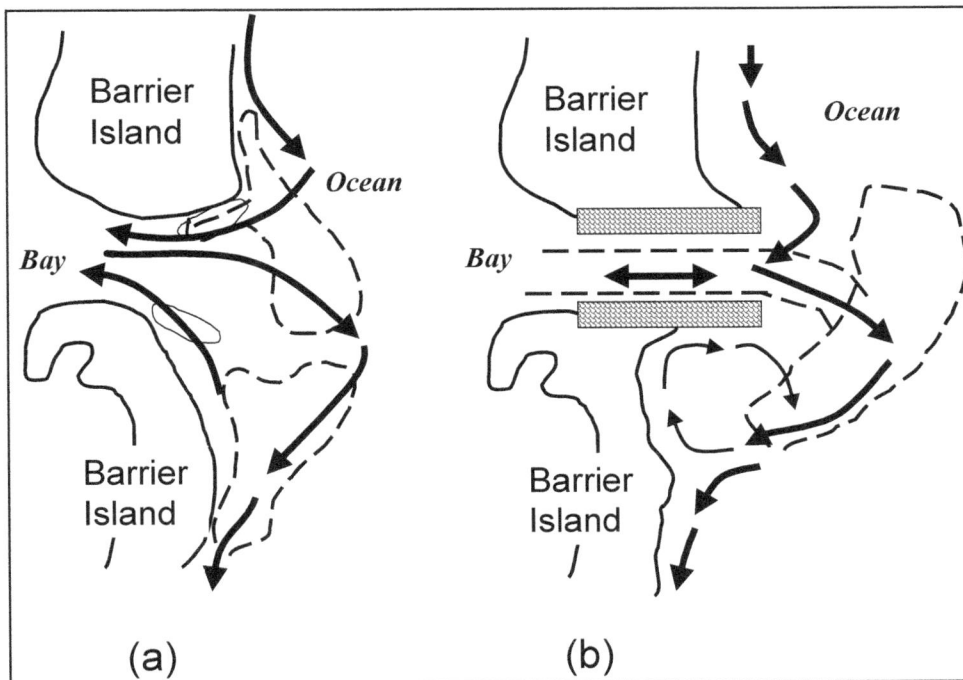

Figure V-6-7. Idealized wave and tidal current bypassing for (a) unstabilized and (b) stabilized inlets

- Sand can also be bypassed directly across the ebb shoal in a zig-zag pattern by the combined action of waves and tidal currents. During transport reversals (when the waves approach from the opposite direction), the bypassing pattern can reverse; however, the methods and efficiency of the bypassing are rarely identical, particularly if the inlet and ebb shoal morphology is not symmetrical about the inlet center line. (Lack of symmetry is frequently encountered when one direction of transport dominates). These bypassing mechanisms are generally considered to be a more-or-less steady process, modulated in intensity by periods of increased littoral transport (storms, high waves, etc.).

- In Figure V-6-8, increasing numbers indicate the time sequence of bypassing through migration and welding of bars. This process may occur to some degree at stabilized inlets, but it is more common for unstabilized inlets. Waves again transport sand from the updrift shore toward the inlet. Large shoals then form at the edge of the shoreline and the inlet, and/or a spit may develop

Sediment Management at Inlets

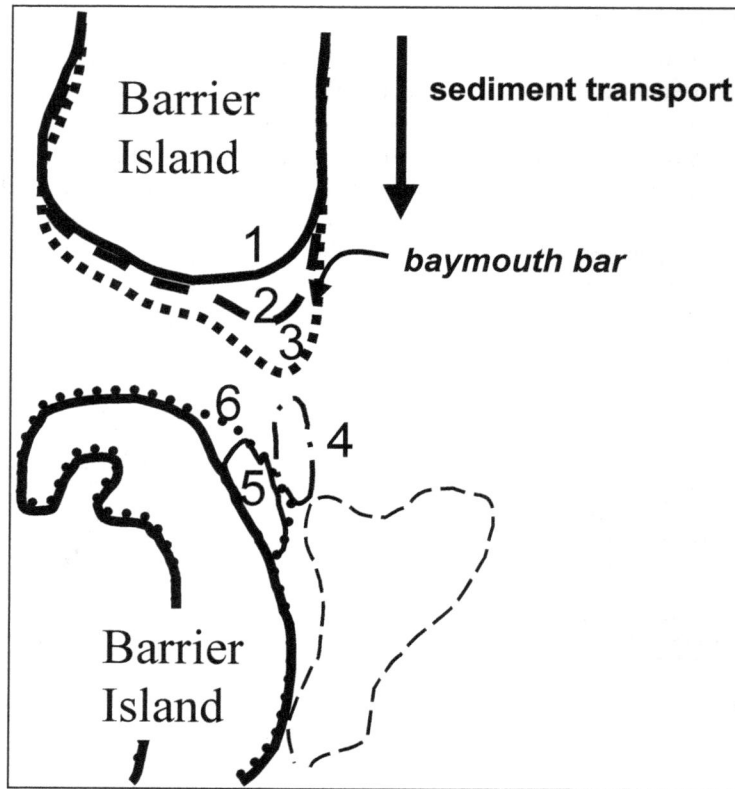

**Figure V-6-8. Idealized bypassing through migration and welding
of bars for an unstabilized inlet (increasing numbers indicate time sequence)**

across the inlet throat (a baymouth bar). Growth of the shoals and spit force the inlet channel to, migrate downdrift. This, in turn leads to channel encroachment upon the downdrift shoreline and conesquent beach erosion. Eventually, the channel's hydraulic efficiency becomes sufficiently poor such that the inlet will close, or the channel will breach the shoals or spit (usually during a storm or flood) near the channel's original location. The shoals or spit, now severed from their direct updrift sediment source, become subject to downdrift-directed transport by waves. These sand bodies, thus, migrate downdrift and shoreward where they eventually weld onto the downdrift shoreline, nourishing the beach which was previously eroded by the inlet channel's earlier migration.

- In this way, sand can be naturally bypassed in a cyclical manner, by which the updrift and downdrift shorelines retreat and advance in periodic cycles. The channel shoals and spits are the means of the bypassing mechanism. Their creation, severance, and downdrift attachment can be gradual (quasisteady) or episodic. Time scales of the cycle may range anywhere from 3 to 50 years (FitzGerald 1984, 1988; Fenster and Dolan 1996; Gaudiano and Kana 2001).

- In a variation of this pattern for unstabilized inlets (not shown in Figures V-6-6 through V-6-8), waves transport sand from both adjacent shorelines toward the inlet mouth. Strong ebb and flood currents transport the sand into and out of the mouth. The sand is deposited in the ebb tidal plateau as well as large marginal ebb shoals that line the inlet channel. The emerging marginal shoals are ultimately subject to the landward-directed transport of wave action, and eventually migrate toward shore and weld to the beach. The result is a pattern of erosion and deposition along barrier islands marked by the formation of large intertidal bars along the inlet channel, and apparently little (or at least, uncertain) net sediment bypassing. FitzGerald (1984) and Gaudiano and Kana (2001) described this process as controlling South Carolina's barrier island/inlet shorelines, noting cycles of ebb-tide delta growth and decay (15 to 20 percent changes in volume) which last from 4 to 8 years. Further elaboration of these processes is presented by FitzGerald (1988) and in Part IV-3 and Part IV-4.

- Inlets, particularly unstabilized ones, can bypass sand through a combination of the bypassing mechanisms. As part of the bypassing process, unstabilized inlets may migrate, usually in the net downdrift direction, but they can also migrate updrift (see Aubrey and Speer (1984) and Douglass (1991), among others).

- Inlets' natural bypassing processes also serve to illustrate the dynamic nature of the shorelines adjacent to an inlet. Roughly, the scale of the bypassing features (and therefore of the directly affected shoreline lengths) can range from 1/2 km updrift and 1-2 km downdrift for a small inlet (for example, 200 m wide), to several kilometers updrift and tens of kilometers downdrift for a large inlet (for example, >1,000 m wide) (The alongshore extent of inlets' influence is discussed in Part V-6) Beach erosion is often manifest if upland development is sited within an inlet's zone of migration or cyclical accretion/erosion associated with natural bypassing and ebb delta fluctuation. This inopportune timing of oceanfront development refers to the platting of land or establishment of construction setback lines, along the oceanfront or the inlet's interior shorefront, at times when the shoreline is at an anomalously advanced location associated with the inlet's cycle.

c. Inlet operation and maintenance activities. To maintain navigable waterways, USACE as well as state and local agencies construct jetties and dredged channels. These activities may change inlet processes (see also Part IV-4).

(1) Dredging of shoals or channels within an inlet, whether with or without structural (jetty) improvements, can increase the inlet's potential to act as a sediment sink. Most simply, the response is analogous to the natural tendency for a depression in the seabed to fill with sediment from the surrounding seabed and the active littoral system. If sand dredged from the inlet during the initial work or its subsequent maintenance is not placed upon the adjacent beaches or otherwise within the active littoral system, then the net long-term effect is expected to be erosion of the adjacent beach on one or both sides of the inlet.

(a) Inlet dredging may also increase the inlet's flow velocities or its tidal prism (the volume of water that flows into or out of the inlet in one-half tide cycle). (See Part II-6.) This may result in enlargement of the inlet shoals, and seaward displacement of the ebb shoal. Enlargement of shoals will remove sediment from the open coast to the inlet channels and on to the flood and ebb shoals. Seaward displacement of the ebb shoal is expected to reduce bypassing capacity of the inlet.

(b) An example of this effect is the 1892 creation of St. Lucie Inlet, Florida, by local residents, who, desiring a convenient channel between Indian River and the Atlantic Ocean, dredged through the sandy barrier island (Sargent 1988). Beach sand was rapidly diverted from the adjacent shores to form large ebb and flood tidal shoals. Within 37 years after the initial cut, the updrift shoreline (Hutchinson Island to the north) retreated up to 540 m from its preinlet location, and the downdrift beach (Jupiter Island) retreated over 650 m (USAED, Jacksonville, 1973). The construction of a 1,010 m-long north jetty in 1929 partially restored the north shoreline's location over the next 30 years by impounding littoral drift, but not before millions of cubic meters of sand had been (mostly irretrievably) diverted from the littoral system (Olsen Associates, Inc. 1995). A 490-m jetty was constructed at the south side in 1980, but does not nearly extend to that shoreline's preinlet location.

(2) Dredging a channel through the ebb tidal platform can also change bypassing mechanisms via the ebb shoal. Initially, portions of the shoal adjacent to the channel are subject to shoreward-directed transport by waves and can simultaneously be deflated by wave- and current-induced erosion. These portions of the shoal may migrate landward and weld to the up- or downdrift shore, resulting in significant accretion of the beach. The welded shoal and its seaward plateau, severed from its bypassing-related sand source, eventually erode and cause the beach to recede from its artificially advanced position. An example of this process may include Bald Head Island, North Carolina, located directly east of the Cape Fear River entrance. The initial stabilization and deepening of the Cape Fear River entrance channel beginning in the 1870s, by dredging, is thought to have severed the inlet's large outer bypassing plateau. The eastern lobe of the ebb shoal (totaling about 4.6 million cu m) migrated shoreward and resulted in a 370- to 600-m advance of the Bald Head Island shoreline directly east of the inlet between 1890 and 1920. The shoreline position then remained relatively stable for about 50 years. At the same time, however, the submerged toe of the severed shoal, which had welded to the shoreline, steadily eroded. Resort and residential development along the accreted, quasi-stable shoreline began in the early 1970s. But by 1974, the shoreline began to rapidly erode toward its earlier location, receding at an average rate of 20 to 30 m/year (Cleary and Hosier 1988; Olsen Associates, Inc. 1989; Jarrett 1990). Similar examples of deflation and landward migration of severed ebb tidal deltas are described for inlet improvement projects at Murrell's Inlet, South Carolina, and Little River Inlet on the North Carolina/ South Carolina border (Douglass 1987; Hansen and Knowles 1988; Chasten 1992).

(3) Structural stabilization of an inlet channel by jetties can have numerous consequences to the sediment-sharing system and adjacent shorelines. Sand-tight jetties may perform much like groins and impound littoral material. Sand may accrete the beach on the updrift shore and, during reversals in longshore transport direction, to a lesser degree on the downdrift beach. If the ebb shoal connects to the beach, construction of the jetty may extend through the bypassing platform, thereby interrupting this bypassing mechanism. Or, jetties may block the formation of bypassing spits and shoals that would otherwise cyclically grow from the updrift shoreline and migrate to the downdrift shoreline.

(a) Jetties can also transport littoral material seaward via offshore-directed (rip) currents that can develop along their updrift face. In some cases, the diverted sand may deposit in the ebb shoal, and it may or may not ever return to the downdrift beach through natural mechanisms.

(b) Jetties displace an ebb shoal seaward as a function of their length and the degree to which they increase the inlet's tidal prism and velocity. For example, the construction of the 3.2-km-long jetties at St. Mary's entrance in the late 1800s is estimated to have caused the net deposition of 92 million cu m of sand across an ebb tidal platform. This platform has been displaced about 4,000 m seaward of the original shoal (Olsen 1977; USAEWES 1995).

(c) Weir jetties, or structures that are low or porous, may increase the degree to which sand is diverted from the adjacent shores. Weir jetties constrict flood currents toward the shoreline, increasing the flow velocity and sediment transport directed into the inlet (Figure V-6-9). The subsequent ebb flow is minimally reduced by the weir's presence and transports the sediment seaward beyond the limit of the weir and jetty structure. (This process can be minimized if the sand falls into a sediment trap and is regularly bypassed by mechanical means to the adjacent shores.) Similarly, low or porous jetties allow flood flow to transport sediment into the inlet from the adjacent seabed, but then act to transport the sediment further seaward during ebb flow by their jetting action upon the ebb current. Dean (1988) has described the latter phenomenon at St. Mary's entrance.

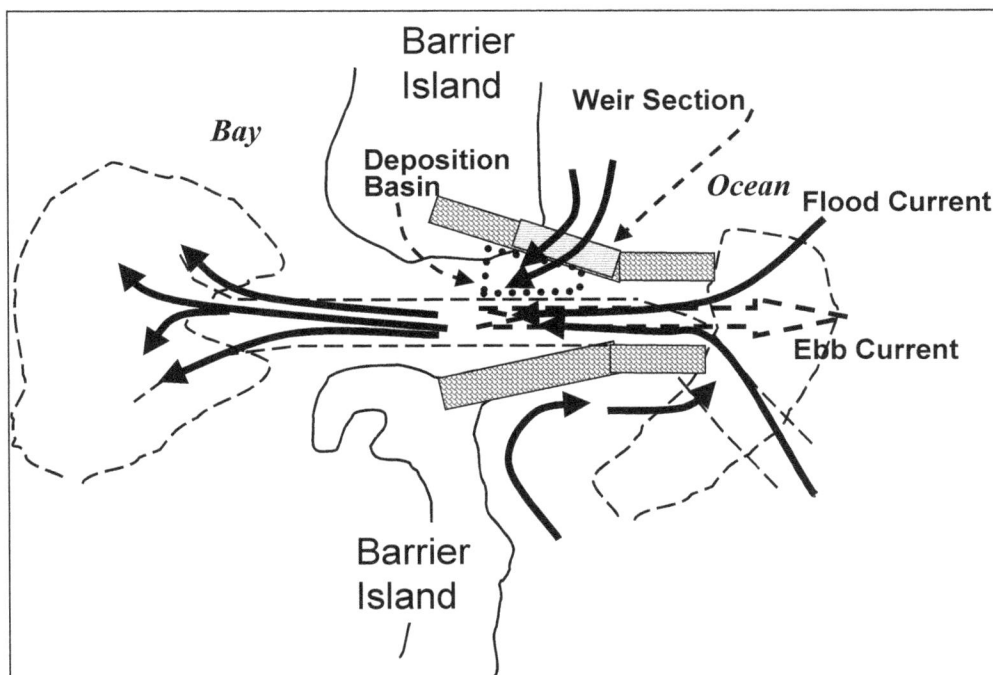

Figure V-6-9. Flood and ebb tidal flow in a weir jetty system

(d) Jetties also influence the littoral transport pattern along adjacent shorelines by developing shadow zones in which the refracted wave field is modified (see Figure V-6-4c). On the downdrift side of a jetty (or in the downdrift lee of an ebb shoal modified by the jetty's presence), sand transport is generally directed into the shadow (i.e., toward the jetty) from the adjacent shoreline by:

- Refraction and diffraction of waves about the structure's seaward end.

- Diffraction currents associated with a decrease in wave energy within the shadow zone.

The processes can cause a leeward impoundment fillet against the jetty.

d. *Inlet modifications of longshore transport.*

(1) As discussed in Part III-2, coastal engineering analyses often involve two types of longshore transport rates. (Note that the discussion herein defines rates as positive quantities; the presentation in Part III-2-2(a) applies an alternate definition.) The net longshore transport rate is defined as the difference between the right-directed and left-directed longshore transport over a specified time interval for a seaward-facing observer:

$$Q_{net} = Q_R - Q_L \qquad (V\text{-}6\text{-}1)$$

where the leftward-directed transport rate Q_L and rightward-directed rate Q_R are taken as positive.

(2) The gross longshore transport rate is defined as the sum of the right-directed and left-directed transport rates over a specified time interval for a seaward-facing observer:

$$Q_{gross} = Q_R + Q_L \qquad\qquad\qquad (V\text{-}6\text{-}2)$$

Inlets can trap longshore sediment transport from both directions, resulting in a greater amount of transport interruption than simply the net transport rate. That is, inlets are a potential sink to the gross longshore sediment transport in the region. Additionally, tidal currents, ebb-shoal bathymetry, and shoreline morphology can each increase the gross transport directed toward the inlet, increasing the inlet's diversion of littoral drift from the adjacent system. Examples are described in the following paragraphs.

(3) Figure V-6-10(a) depicts a fully-bypassing, equilibrated inlet in which 100 units and 40 units of longshore transport, respectively, are directed toward the inlet from the updrift and downdrift coastlines. The net and gross transport quantities are thus 60 units and 140 units, respectively. As the inlet is assumed to be fully bypassing, there is no net change to the adjacent shorelines.

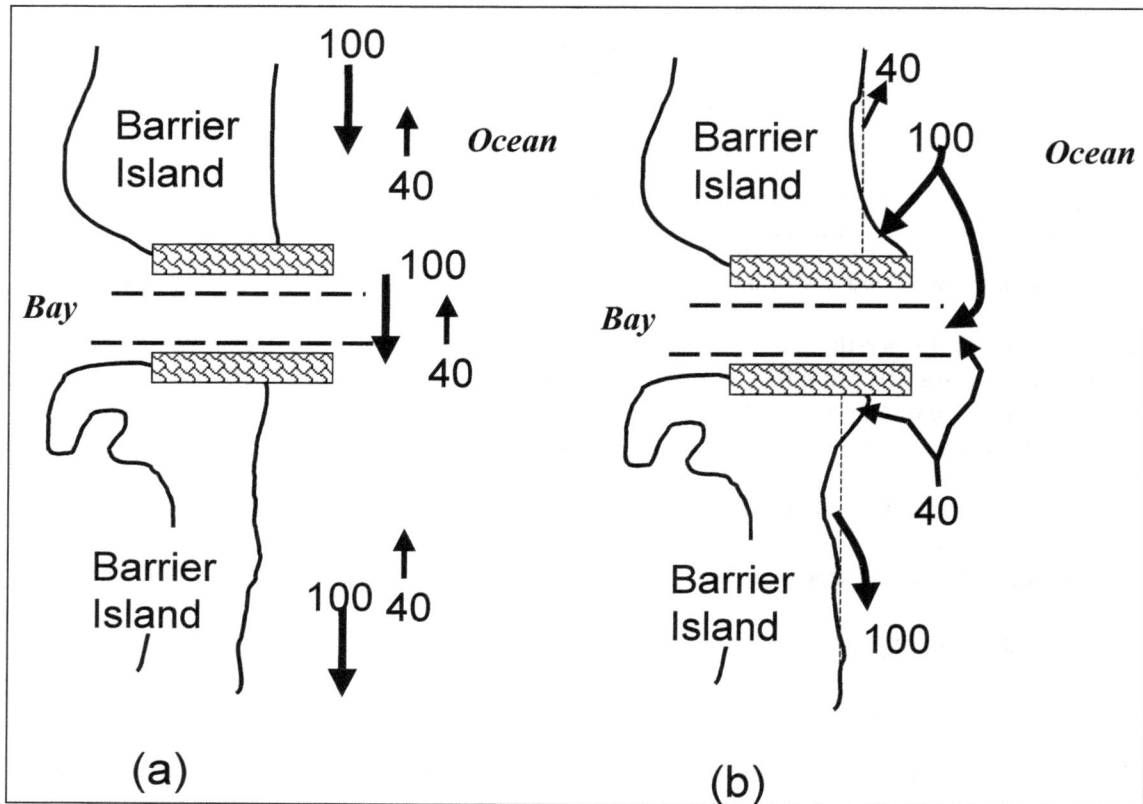

Figure V-6-10. Example of inlet-adjacent shore erosion associated with (a) a fully-bypassing inlet, and (b) an inlet that is a complete littoral sink

(4) Figure V-6-10(b) depicts an inlet which is a complete littoral sink under the same transport conditions as previously described. An inlet may be a complete littoral sink immediately after formation; as it matures, bypassing via the shoal system increases. In Figure V-6-10(b), the incident, updrift transport (100 units) is assumed to be diverted to an updrift impoundment fillet, the inlet channel, and/or shoals. The downdrift shoreline, beyond the jetty's wave shadow and deprived of the 100 units of incident transport, must yield 100 units of beach material in response to the longshore transport potential. During transport reversals, the incident transport (40 units) is likewise assumed to be diverted to a minor impoundment fillet on the other side of the inlet, the

inlet channel, and shoals. If the updrift impoundment fillet is wholly located in the jetty shadow of the reversal waves, or otherwise removed from the littoral stream by refraction effects, then the updrift shoreline, beyond the shadow, must yield 40 units of beach material in response to the longshore transport potential. If impoundment fillets form, the inlet will not shoal and the net effect is the same as the net longshore transport rate: 60 units of accretion (updrift) and erosion (downdrift). However, if no impoundment fillets form, or if the impoundments are already at capacity, the inlet channel and/or shoals may capture the gross transport (140 units), with 40 units of erosion (updrift) and 100 units of erosion (downdrift).

(5) In Figure V-6-10(b), the potential for net updrift erosion associated with the inlet is counterintuitive. Its ultimate manifestation depends upon the degree to which the updrift impoundment fillet is capable of yielding sand to the adjacent shore during transport reversals. In the limit, when there is no impoundment at the updrift jetty and when all of the updrift transport directed toward the inlet is lost therein (and there is no bypassing from the downdrift shoreline), the potential for updrift inlet-related erosion is totally realized.

(6) Wave refraction across the ebb shoal plateau may locally increase the rate of longshore transport directed toward the inlet mouth. Inlet-directed transport may likewise be induced from diffraction currents that result from gradients in breaking wave energy within the wave-shadow of the ebb shoal plateau. Additionally, a shoreline that is curved toward the inlet's mouth, such as that which results from local beach erosion, further exaggerates the local rate of inlet-directed longshore transport. The latter is a result of the increased angle with which waves break along the shoreline. Each of these phenomena can potentially increase the rate at which sand is transported toward the inlet and mouth.

(7) Figure V-6-11 illustrates potential alteration of the pattern of longshore transport by an ebb shoal. Waves 1.0 m in height and 8 sec in period were refracted from deep water across the bathymetry illustrated in the bottom graphic. The incident wave conditions were: From the left (+23 deg), 40 percent annually; from shore-normal (0 deg), 35 percent annually; from the right (-23 deg), 25 percent annually.

(8) The bathymetry is idealized, but is representative of an inlet for which littoral drift dominates tidal flow and for which the net longshore transport direction is from left to right. The upper graphic depicts the potential longshore transport rate computed from each of the three refracted wave cases. The middle graphic depicts the total right-directed, left-directed, and net transport rates that result from the weighted superposition of the individual transport curves. The component transport rates (upper graphic) are normalized by the maximum transport value among the three cases. In the middle graphic, the transport values are normalized by the maximum value of the net transport rate. Only the effect of obliquely incident waves is considered, and transport induced by alongshore gradients in wave height (diffraction currents) and tidal currents are omitted. The methods employed to compute these transport rates are described in Part III-2-2, and, later in this chapter.

(9) In this example, waves from each of the three offshore directions result in transport directed toward the inlet mouth within the alongshore reach of the ebb shoal's refractive influence. Within this region (indicated by asterisks in the middle graphic), no transport is directed away from the inlet. Additionally, the transport magnitude increases significantly beyond that of the ambient transport (i.e., the transport associated with obliquely incident waves outside of the shoal's refractive influence). The result is an increase in transport toward the inlet mouth – with

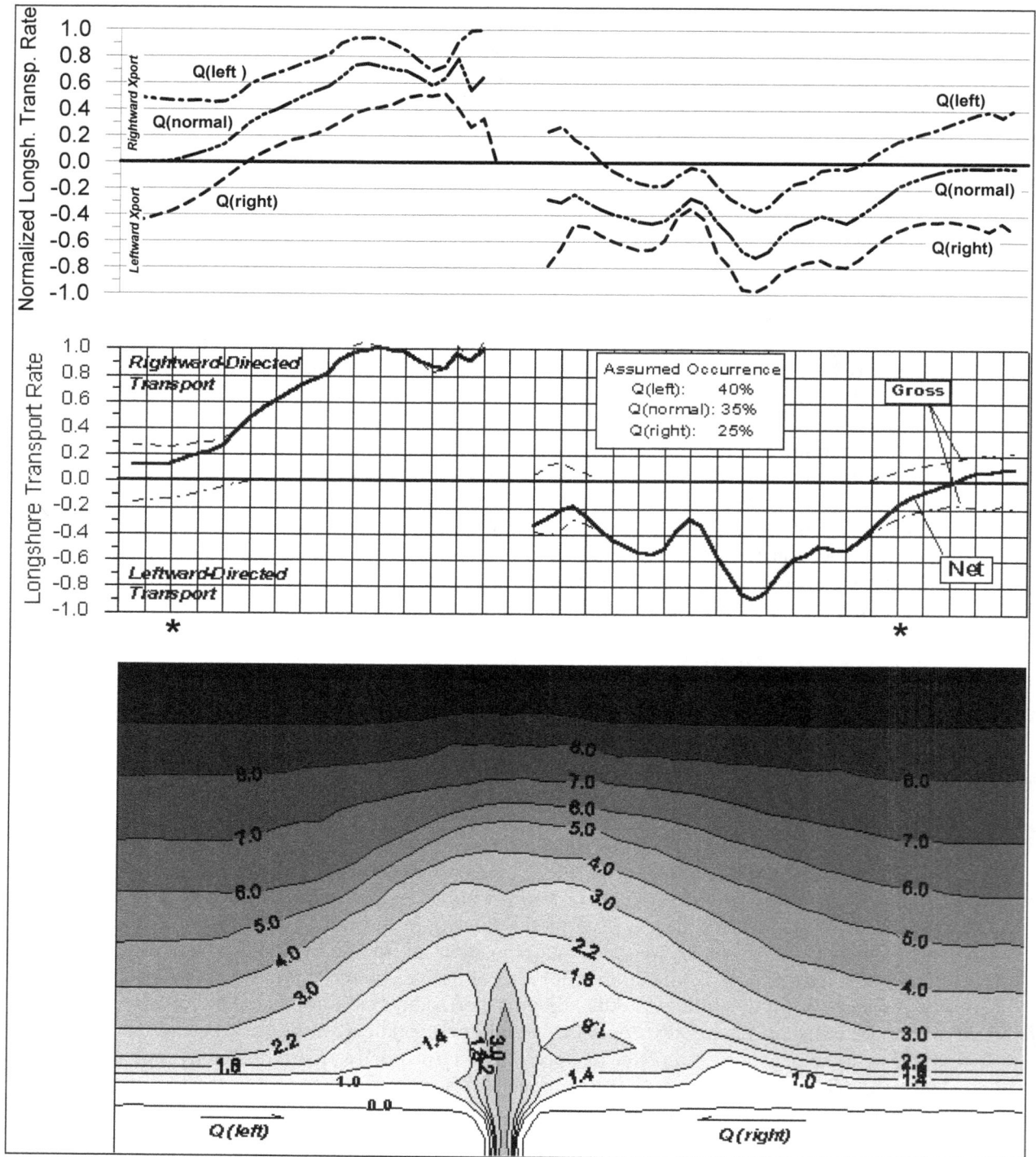

Figure V-6-11. Illustration of longshore sediment transport potential computed for three wave conditions incident to inlet's ebb shoal plateau (depths in meters)

EXAMPLE PROBLEM V-6-1

GIVEN:

The right- and left-directed volume transport rates (volume per unit of time) entering and leaving sections of shoreline upcoast and downcoast of an inlet induced by obliquely incident waves and currents are shown in the following tabulation for each section and for the inlet. Assume the inlet is a total littoral barrier (i.e., no bypassing; material can move into the inlet, but material cannot move out of the inlet).

A	B	C	D	Inlet	E	F	G	H	I
100 100 →	120 →	125 →	125 →	0 →	0 →	0 →	30 →	80 →	100 →
30 ←	30 ←	10 ←	0 ←	0 ←	60 ←	40 ←	50 ←	40 ←	30 30 ←

FIND:

The potential volume rate of change at shoreline sections A through I associated with these transport rates, and the total net volume rate of change on the left (updrift) and right (downdrift) sides of the inlet. Material transported <u>into</u> a section is positive (accretion = +) and material transported <u>out</u> of a section is negative (erosion = -).

SOLUTION:

The volume rate of change at each section A through I is equal to the algebraic sum of all volume transport rates entering and leaving each section. For example, at section C, the volume rate of change is the difference between what goes into section C (+120 + 0 = +120) and what comes out of section C (-125 + (-10) = -135). The algebraic sum of these two numbers (+120 + (-135) = -15) is the volume change per unit of time at section C.

Table V-6-1
Updrift Volume Change

Updrift	Vol. Change
A	0
B	-40
C	-15
D	0
Subtotal	-55
Inlet	+125
Total	+70

Table V-6-2
Downdrift Volume Change

Downdrift	Vol. Change
E	-20
F	+10
G	-40
H	-60
I	-20
Subtotal	-130
Inlet	+60
Total	-70

The total volume rate of transport on the left (updrift) side of the inlet (sections A through D) is equal to the volume rate of material transport entering and leaving through the left face of section A (+100 + (-30) = + 70) plus the volume rate of material transport entering and leaving through the right face of section D (-125 + 0 = -125). The algebraic sum of these two values (+70 + (-125) = -55) is the subtotal of all the changes occurring on the updrift side of the inlet (stations A through D). Thus, the left side of the inlet is eroding at a rate of -55 per unit of time.

(Continued)

Sediment Management at Inlets

EXAMPLE PROBEM V-6-1 (Concluded)

The total volume rate of transport on the right (downdrift) side of the inlet (sections E through I) is equal to the volume rate of material transport entering and leaving through the left face of section E (+0 + (-60) = -60) plus the volume rate of material transport entering and leaving through the right face of section I (-100 + (+30) = -70). The algebraic sum of these two values (-60 + (-70) = -130) is the subtotal of all the changes occurring on the downdrift side of the inlet (sections E through I). Thus, the right side of the inlet is also eroding, but at a rate of -130 per unit of time.

The total volumetric rate of change across the shoreline from sections A through I equals that material entering and leaving the left face of section A (+100 + (-30) = +70) plus that material entering and leaving the right face of section I (-100 + (+30) = -70. Thus, the total volumetric rate of change across sections A through I equals zero since the net amount of material entering the left face of section A is equal to the net amount of material leaving the right face of section I. However, the left (updrift) side of the inlet is eroding at a rate of -55 per unit of time, and the right (downdrift) side of the inlet is eroding at a rate of -130 per unit of time. Since the total erosion rate across sections A through I equals (-55 + (-130) = -185 per unit of time), and the net material rate entering the left face of section A is equal to the net material rate leaving the right face of section I, the material being eroded from the sections to the left and right of the inlet must be accumulating in the inlet. And, indeed, this is the amount of material rate passing through the right face of section D (+125) into the inlet (no material leaves the inlet), plus the amount of material rate passing through the left face of section E (+60) into the inlet (again, no material leaves the inlet), for a total of +185 per unit of time accumulating in the inlet.

---

no mechanism for reversal. The larger transport gradient erodes sand from the beach and transports it toward the inlet. An inlet's total potential effect is therefore equal to the sum of the ambient gross transport plus additional transport locally directed toward the inlet that is induced by the perturbation caused by the inlet on the local geometry.

e. Alongshore extent of inlet influence.

(1) The length of coastline directly influenced by an inlet is manifest, in large part, by alongshore changes in the nearshore bathymetry (and associated wave field), and alongshore changes in the beach profile shape and magnitudes of historical shoreline fluctuation. The ultimate degree to which an inlet modifies adjacent shores is a function of the volume of sand contained in its other features (Fitzgerald 1988; Bodge 1994b; Fenster and Dolan 1996); that is: impounded by jetties and other structures; captured in the flood and ebb tidal shoals; dredged; placed on adjacent beaches (e.g., dredged material or other source of fill).

(2) Absolute distances of inlet influence remain a subject of study, but may extend further than were previously thought. Shoreline response immediately adjacent to an inlet is typically obvious and erroneously led earlier investigators to conclude that the extent of inlet improvements' influence was limited to short reaches downdrift; usually some multiple (about 3 to 10) of the jetty length away from the inlet. Many, if not most, estimates in the literature of the length of inlet influence were truncated by the limited extent of coast that the investigators examined (Bruun 1995).

(3) Based upon analysis of shoreline change rates, Fenster and Dolan (1996) report the spatial influence for natural inlets along the wave-dominated North Carolina coast as up to 13.0 km, and up to 6.1 km along the mixed-energy, tide-dominated Virginia barrier islands. Dean and Work (1993) report downdrift inlet influence for inlets along Florida's central and southern Atlantic coastlines as at least 2 km

to more than 6 km; however, their analysis was limited to 6 km or less. The erosive effect of the entrance to Lagos, Nigeria, may extend at least 40 to 50 km downdrift (Bruun 1995). Migniot and Granbaulan (1985) report inlet-influenced erosion reaching 50 to 80 km downdrift of river entrances on the southwest coast of France. Bruun (1995) describes inlet-attributable erosion occurring 30 to 35 km downdrift of the Hirtshals navigation works on the North Sea coast of Denmark, 20 to 25 years after the construction of entrance breakwaters and dredging increases. Inlet-adjacent erosion may be particularly severe when the beaches are restricted by geological features, such as pocket beaches on the United States Pacific coast, but this hypothesis needs further research. Terich and Komar (1974) and Komar (1976) describe this effect at the entrance to Tillamook Bay, Oregon.

(4) Outside of the direct influence area (i.e., within the shadow of bypassing bars, jetties, etc.), an inlet's influence appears to decrease exponentially away from the inlet. One example is Port Canaveral Entrance (Bodge 1994a). This effect is also predicted by analytic shoreline-change models, including, for example, the one-line model of Pelnard-Considére (1956) described in Part III-2-2. According to Pelnard-Considére's solution for shoreline change adjacent to a sediment-trapping structure, the shoreline recession y at a distance x downdrift of the structure, at the time t_f at which the structure becomes filled to capacity and begins to bypass sand, is:

$$\frac{y}{Y} = \left[\exp\left(-u^2\right) - \sqrt{\pi} u \; erfc\left(u\right) \right] \tag{V-6-3}$$

where

$$u = \frac{x}{y} \frac{1}{\sqrt{\pi}} \tan \alpha_b \tag{V-6-4}$$

and

$$t_f = \pi Y^2 / 4g \tan^2 (\alpha_b) \tag{V-6-5}$$

(5) The parameter Y is the structure length, $erfc(\ )$ the error function complement, and α_b is the angle of the breaking wave crests relative to the shoreline (assumed to be quasi-steady). Both y and Y are measured relative to the no-structure shoreline. The longshore diffusivity, ε,

$$\varepsilon = \frac{K H_b^{25} \sqrt{g/\kappa}}{8(s-1)(1-n)(h_* + B)} \tag{V-6-6}$$

is a measure of the tendency of wave action to spread out a shoreline perturbation, where K is an empirical coefficient from the longshore transport Equation III-2-5 (of order 1), H_b is the breaking wave height, g is the acceleration of gravity, κ is the ratio of local breaking wave height to local depth (about 0.8), s is the ratio of sediment to water specific gravity (about 2.65), and n is sediment porosity (about 0.4). The value h_* is the depth of active sediment transport (depth of closure) and B is the beach berm height, where B and h_* are positive values above and below the water level, respectively.

(6) Figure V-6-12 presents solutions to Equation V-6-3 for various wave breaking angles. For typical angles on the order of 2 to 8 deg, shoreline recession on the order of 5 percent of the structure's length is predicted between 15 and 50 jetty lengths downdrift. In the case of a 200-m-long jetty, this corresponds to 10 m of recession at locations between 3 and 12 km downdrift of the jetty.

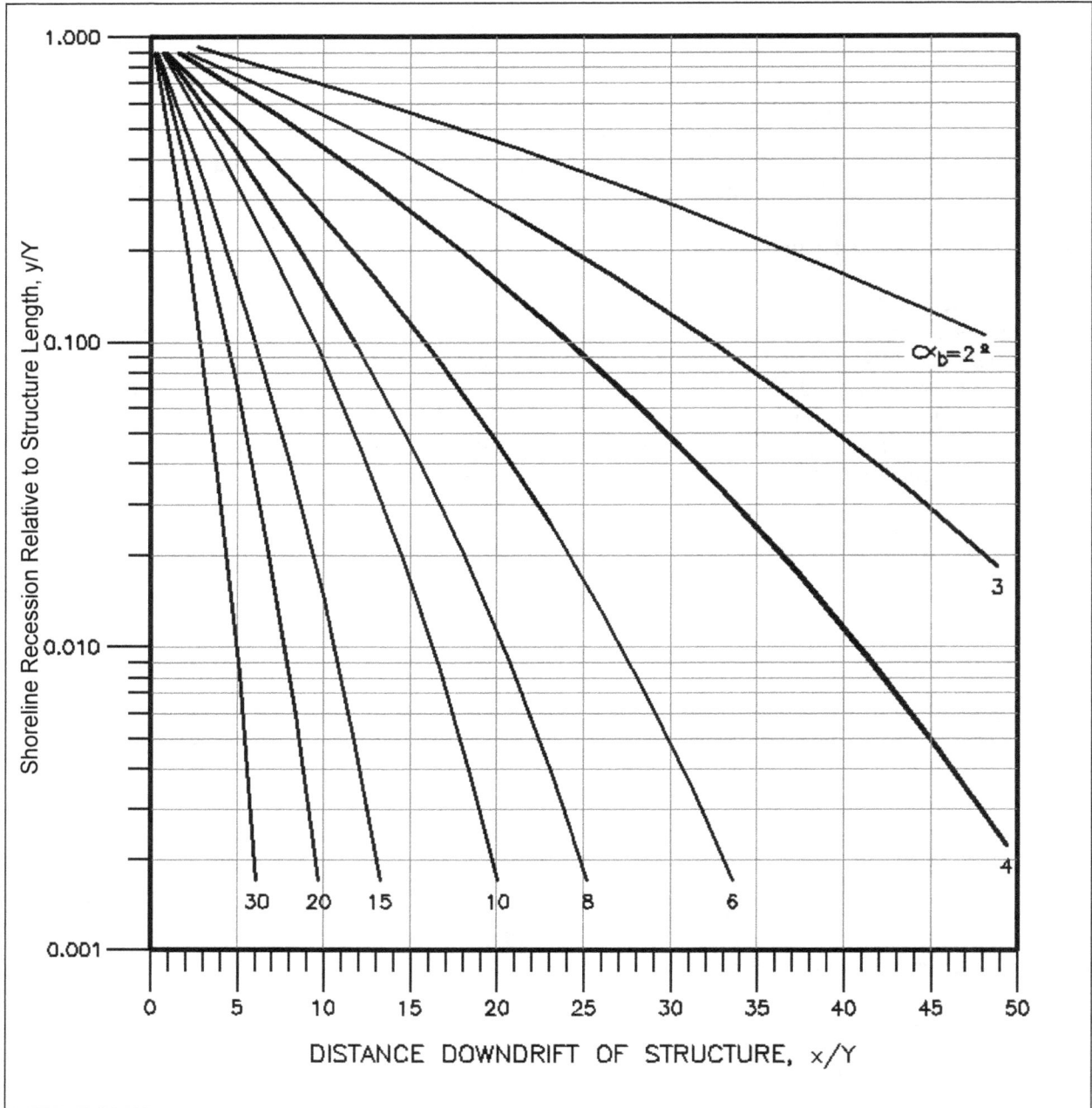

Figure V-6-12. Solutions of Equation V-6-3 (from Bodge 1994b)

(7) Dean and Work (1993) present a solution for Pelnard-Considère's equation for the case where sand is removed from the system at a rate Q_I at the inlet location $x = 0$:

$$y(x,t) = \frac{2p\,Q_I\sqrt{t}}{(h_* + B)\sqrt{\pi\,\varepsilon}}\,y' \tag{V-6-7}$$

where p describes the degree to which the inlet's sand removal alters each of the inlet's two adjacent shorelines; i.e.,

$$p = \frac{(\text{sand lost to inlet from shoreline of interest})}{(\text{sand lost to inlet from both shorelines})} \tag{V-6-8}$$

$$y' = \left[\sqrt{\pi}\, v\, erfc\,(v) - \exp\left(-\left(v^2\right)\right) \right] \tag{V-6-9}$$

$$v = \frac{|x|}{\sqrt{4gt}} \quad ; \quad |x| = v\sqrt{4gt} \tag{V-6-10}$$

(8) In Equation V-6-8, a value of $p = 0.5$ implies that the inlet's sink effect (Q_I) is evenly distributed between the two adjacent shorelines; whereas a value of $p = 1.0$ implies that all of the inlet's sink effect occurs from the shoreline of interest. As another example, $p = 0.3$ implies that 30 percent of the inlet's sink effect occurs from the shoreline of interest (whereas the other 70 percent occurs as sand losses from the other shoreline).

(9) Figure V-6-13 shows nondimensional solutions of Equation V-6-9. Additional examples and applications of Pelnard-Considère's shoreline change solution are presented in Part III-2-2. As described in that chapter, solutions for $y(x,t)$ can also be linearly added to yield the combined response of shoreline perturbations.

(10) Although Equations V-6-3 and V-6-7 provide useful theoretical insight to the potential response of a shoreline near an inlet, they are generally of limited practical value in assessing real world response. The equations do not readily account for alongshore variabilities in breaking wave angle nor sand impoundment adjacent to inlet jetties. Likewise, cross-shore transport and seasonal reversals in longshore transport direction are not taken into account. Nonetheless, these equations can be examined for order-of-magnitude effects which might be expected for a given inlet situation.

(11) From the Pelnard-Considère equations, Dean (1993) likewise presents solutions of shoreline change adjacent to an inlet for the case where sand is directed toward a dredged inlet by transport induced by ambient oblique waves and by the inlet's perturbative effect to the local shoreline. Dean observes that in the absence of effective jetties, the rate of sand that enters the inlet decreases with the reciprocal square root of time (i.e., $1/T_d^{1/2}$); however, the cumulative amount of sand entering the inlet over time from the adjacent shorelines is infinite. Dean also notes that the equations predict that the location of maximum shore erosion migrates with time away from the inlet. That is, measurements taken at one fixed point along the shoreline would first demonstrate no inlet erosion effect, then a rapid rate of erosion, then a decreasing rate as the shoreline asymptotically approaches an equilibrium, eroded position. Alternately stated, the peak erosion zone would migrate like a wave away from the inlet, and the magnitude of this peak rate would decrease with time.

(12) Bruun (1995) describes numerous examples where erosion downdrift of an inlet appears as an obvious near field effect and as a less obvious, but faster-migrating, far field effect. Bruun concluded that the downdrift shoreline development at a littoral drift barrier may in general, but not always, be described by a short (near field) effect as well as a long distance (far field) effect, both of which move downdrift at various rates. The short distance influence occurs first. After it develops to a certain extent, the long distance effect may appear, gradually moving downdrift faster than the short distance but fading out with distance. The short distance peak of the erosion occurs close to the littoral drift barrier, and moves downdrift at a rate of about 0.3 to 0.5 km/year compared to a rate of 1 to 1.5 km/year for the long distance erosional front. Bruun suggests that the short-distance effect is a coastal geomorphological feature, whereas the long-distance effect is a materials-deficit feature.

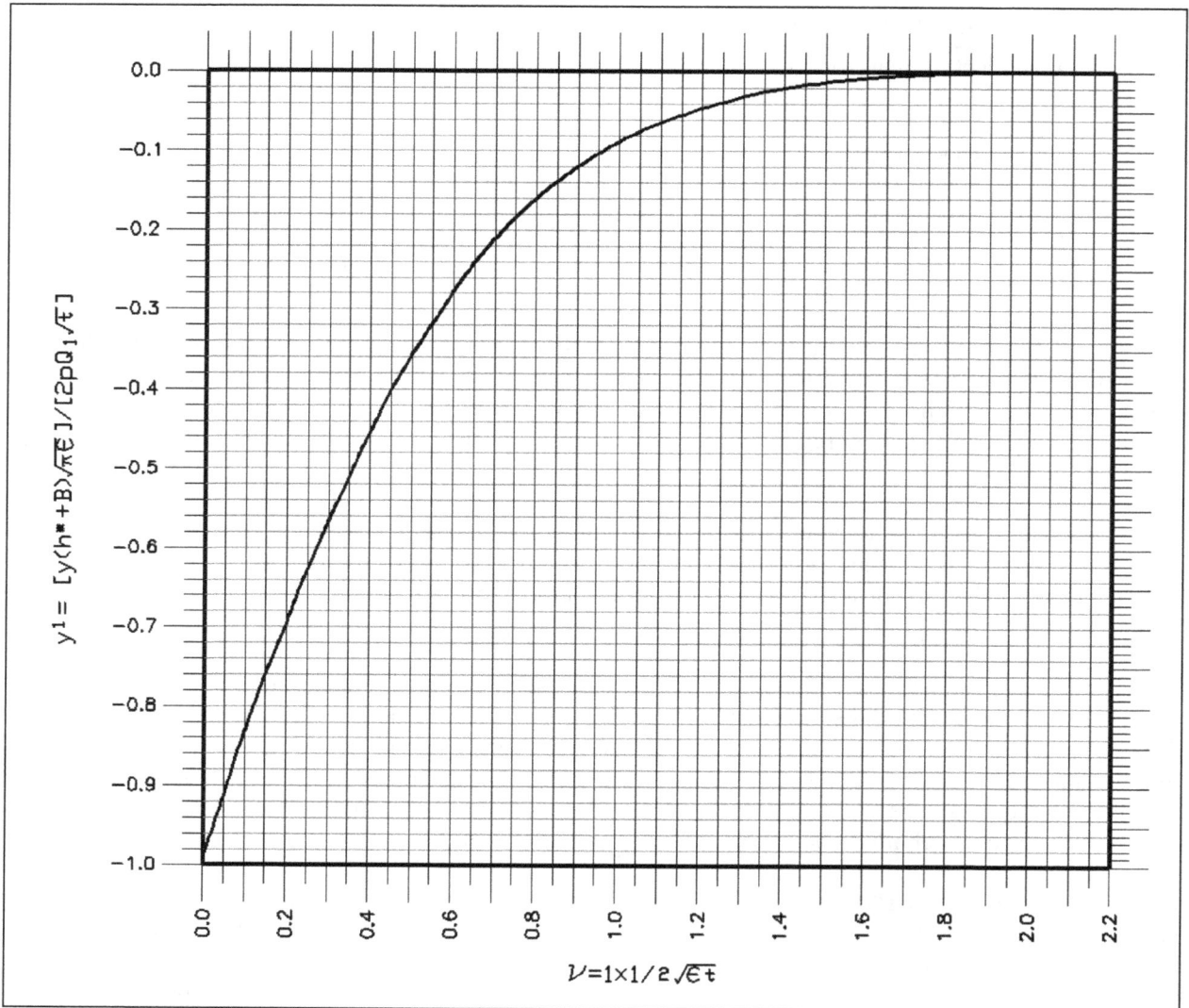

Figure V-6-13. Solution of Equation V-6-9

f. Estimating alongshore extent of inlet influence. Five methods to estimate the alongshore extent of inlet influence are as follows: Examination of historical shoreline changes; even-odd analysis; alongshore variations in beach morphology; wave refraction analysis; examination of inlet's net sink effect.

(1) Of these, Methods 1 and 2 (and to a lesser extent, 5) rely principally upon shoreline change data. Methods 3 and 4 (and, principally 5) rely upon data that are mostly distinct from shoreline change or location data. Methods 2, 3, and 4 typically yield only estimates of direct inlet effect; whereas Methods 1 and 5 potentially yield estimates of both direct and indirect (near field and far field) inlet effects. Method 5 may be the most powerful approach because it first assesses the inlet's littoral impact within the inlet, and then attempts to identify the adjacent shoreline length along which the inlet's volumetric impact is manifest. In practice, some combination of all five methods is typically necessary to assess the volumetric and lineal extent of an inlet's effect upon the adjacent shores.

EXAMPLE PROBLEM V-6-2

GIVEN: An inlet's north jetty is 240 m long (Y). It is estimated that 90 percent (p) of the inlet's net shoaling rate of 55,000 m^3/year (Q_l) is attributable to sand losses from the nonjettied downdrift shoreline. The inlet was created at time t = 0 year, at which time the shoreline was initially straight and in equuilibrium. Assume a beach of typical quartz sand of specific gravity S = 2.65, and with berm elevation B = +2 m and a depth of closure h_* = 6 m, both relative to the same datum. The average breaking wave height H_b = 0.8 m. Assume the ratio of local breaking wave height to local depth to be κ = 0.8, and the constant from the longshore transport Equation III-2-5 is K = 0.5. Assume also that the breaking wave crests typically approach with an angle of breaking α_b = 4 deg relative to the shoreline, directed to the south.

FIND: The theoretical shoreline response downdrift of this inlet at the time the updrift jetty initially begins to bypass sand. This response consists of two parts: the impoundment effect, and the sand loss effect. These two components are computed separately and added to determine the total shore recession.

SOLUTION: The maximum recession downdrift of the inlet which is stabilized only by one jetty on the updrift side of the inlet will exist at the time t_f when bypassing begins to occur around the jetty. To determine t_f, the diffusity, ε, must first be obtained from Equation V-6-6:

$$\varepsilon = \frac{KH_b^{2.5}(g/\kappa)^{0.5}}{(8)(S-1)(1-n)(h_* + B)}$$

where the gravitational constant g = 9.81 m/sec^{-2} and the sediment porosity n = 0.4 (a typically representataive value).

$$\varepsilon = \frac{(0.5)(0.8m)^{2.5}(9.81m\sec^{-2}/0.8)^{0.5}}{(8)(2.65-1)(1-0.4)(6m+2m)}$$

$$\varepsilon = 0.0158m^2/\sec = 498,600m^2/year$$

The time at which bypassing is expected to begin can now be determined by Equation V-6-5:

$$t_f = \pi Y^2 / 4\varepsilon(\tan\alpha_b)^2$$

$$t_f = \frac{(3.1416)(240m)^2}{(4)(0.0158m^2/\sec)(\tan 4.0)^2}$$

$$t_f = 5.85 \times 10^8 \sec = 18.5 \; years$$

First, compute the erosion due to the impoundment effect. Erosion downcoast of the inlet will be a maximum at the inlet and occur at the time   that sand begins to pass around the jetty of length Y = 240 m. This erosion will become asymptotic to the original downcoast shoreline. For convenience, arbitrary multiples of the jetty length in a downcoast x-direction are considered (i.e., x/Y = 1, 2.5, 5, 10, ..., 40 times the jetty length, Y). These computations are tabulated in column 1 of Table V-6-3.

(Continued)

EXAMPLE PROBLEM V-6-2

Table V-6-3
Calculations for Impoundment and Sand Loss Effects at Inlet

	Impoundment Effect				Sand Loss Effect		
1	2	3	4	5	6	7	8
x/Y (select)	y/Y (Figure V-6-12 for α_b=4°)	x (m) (= col. 1 x 240 m)	y (m) (= col. 2 x 240 m)	v = $|x|/(4\varepsilon t)^{1/2}$	y'(from Fig. V-6-13)	y (m) (Equation V-6-7)	y_{tot} (m) (Cols. #4 + #7)
1	0.95	240	-228	0.040	-0.93	-40	-268
2.5	0.85	600	-204	0.099	-0.83	-35	-239
5	0.70	1200	-168	0.197	-0.70	-30	-198
10	0.45	2400	-108	0.395	-0.45	-19	-127
15	0.29	3600	-70	0.592	-0.28	-12	-82
20	0.18	4800	-43	0.789	-0.16	-7	-50
25	0.092	6000	-22	0.987	-0.09	-4	-26
30	0.05	7200	-12	1.184	-0.06	-3	-15
35	0.024	8400	-6	1.381	-0.04	-2	-8
40	0.011	9600	-3	1.579	-0.02	-1	-4

For the arbitrary values of x/Y of column 1, the corresponding values of y/Y for α_b = 4 deg are determined from Figure V-6-8. Here y is the impoundment effect component of the total recession. The appropriate y/Y values for the arbitrary x/Y values with α_b = 4 deg are shown in column 2. Column 3 shows the x-distances downcoast for the arbitrary x/Y values of column 1. Next, the Impoundment Effect, y, is obtained by multiplying the y/Y values of column 2 by Y = 240 m, and these results are shown in column 4.

It is now necessary to compute the erosion recession distance associated with the sand loss effect from the downdrift shoreline into the inlet over the 18.5-year period during which the updrift impoundment fillet formed. This erosion distance is a function of two additional parameters (v and y') which must first be determined.

The parameter v is defined by Equation V-6-8 as:

$$v = \frac{|x|}{(4\varepsilon t_f)^{0.5}}$$

$$v = \frac{|x|m}{\left[(4)(0.0158 m^2/\sec)(5.85 \times 10^8 \sec)\right]^{0.5}}$$

$$v = \frac{|x|}{6080}$$

(Continued)

EXAMPLE PROBLEM V-6-2 (Concluded)

For each of the values of x in column 3, and for the time t_f = 18.5 years (5.85 x 10^8 sec), the corresponding dimensionless value of υ is computed, and displayed in column 5.

The parameter y' can be determined from Equation V-6-7 or from Figure V-6-9, for each of the values of υ from column 5. These values of y' are shown in column 6.

For each of the y' values shown in column 6, the corresponding value of y due to the sand loss effect can be determined from Equation V-6-7:

$$y(x,t) = \frac{2p\,Q_I(t)^{0.5}\,y'}{(h_* + B)(\pi\varepsilon)^{0.5}}$$

At t = t_f = 18.5 years, and for each y' value which is also a function of distance x,

$$y = \frac{(2)(0.9)(55,000m^3 / year)(18.5\,years)^{0.5}\,y'}{(6m + 2m)\left[(3.1416)(498,600m^2 / year)\right]^{0.5}}$$

$$y = 42.5y'(m)$$

Thus, for each y' value computed at the alongshore distances x of column 3, the corrresponding recession due to the sand loss effect has been determined, and is shown in column 7.

The total recession at an alongshore location from the inlet, x, is the sum of the recession due to the impoundment effect (column 4) plus the recession due to the sand loss effect (column 7). These total recession values are shown in column 8. For example, at location x = 1,200 m downdrift of the inlet (here x/Y = 5, since Y = 240 m), the shoreline recession is predicted to be -168 m + (-30 m) = -198 m.

---

(2) The practical utility of the methods that rely upon shoreline change data can be significantly degraded by the limitations of historical shoreline data. Shoreline change data contain temporal and/or survey noise; or, may be biased by artificial manipulation; or, may not adequately characterize the beach profile's behavior as a whole. Three precautions are thus warranted when using adjacent shoreline change data to ensure the best quality conclusions:

(a) *Identify the season and/or storm events that characterize the survey data.* Shoreline locations developed from surveys, charts, or photographs that contain the record of a particular wave event (such as a storm or a period of reversed transport) may not represent the net long-term behavior of the local coastline. If a time series of shoreline positions is available, this problem is often obvious as an anomalous advance of the shoreline followed immediately by a more or less equivalent retreat (or vice versa). Remediation of this problem often requires deleting of the suspect data set.

(b) *Identify the effect of seawalls, beach nourishment, and other artificial manipulations upon the data.* Seawalls or other structures bias data by maintaining the shoreline at more or less constant location. This is often evidenced by an anomalously low (or zero) rate of shoreline change amidst an otherwise erosional coastline. The placement of beach fill by nourishment or bypassing obviously results in a large initial advance of the shoreline. Less obvious, however, is the fact that shoreline change rates measured within 1 or 2 years after such activity will be anomalously high because of profile equilibration. Failure to recognize the effects of seawalls and beach-fill placement will result in gross overestimation of shoreline stability. Also, failure to recognize the accelerated rate of profile adjustment after beach-fill placement will result in gross overestimation of erosion. Remediation of this problem generally requires deleting (or adjusting) those data that are artificially biased. In the case of beach fill, the estimated effect of the fill's presence can be subtracted from the data; or better, if profile data are available, changes in total beach volume should be considered instead of shoreline location (see the following paragraph).

(c) *Determine the degree to which the elevation selected as the shoreline characterizes the profile behavior as a whole.* The translation of a discrete elevation at which the shoreline is identified may not accurately represent the overall profile's behavior. For example, lower elevations (such as the mean low water line) typically exhibit seaward advance if a profile erodes because of profile flattening. The flatter slopes of lower elevations are highly sensitive to profile change when computing the horizontal location at a given elevation. The mean high water line (mhwl) is typically selected to represent shoreline changes. Particularly along coasts that feature high bluffs or dunes, however, mhwl changes may not reflect beach erosion manifest by dune retreat. For example, 36 percent of the State of Florida's historical beach profiles examined along two of that state's east coast counties exhibited retreat of the dune face and concurrent advancement of the mhwl. For another 36 percent of the profiles, both the dune face and the mhwl exhibited retreat, but in two-thirds of those cases, the rate of dune face recession was over twice as great as that of the mhwl (Savage 1990).

(3) If local shoreline change is not representative of the profile change as a whole, net changes in beach volume should be considered instead of shoreline location. Here, volume changes above the depth of closure (or other sufficiently deep elevation) should be considered; i.e., the volume change per unit alongshore length of shoreline (cu m/m or cu yd/ft, etc.). Alternately, where shoreline change must be specified, as in the case of one-line analytic or numerical models, the local volume change can be converted to an equivalent, artificial shoreline change. This is computed by dividing the volume change (volume/alongshore length) by a specified profile depth (height), where the latter is usually taken as the vertical distance between the berm elevation, B, and the depth of closure, h_*. Precautions regarding the transformation of shoreline changes to volume changes are discussed in the following paragraphs (see Methods 3 and 5).

(a) *Method 1: Examination of historical shoreline changes.* In this method, temporal fluctuations in shoreline location adjacent to an inlet are quantified by comparing (overlaying) historical charts, aerial photographs, profile surveys, etc. Specifically, the cross-shore location of a specific beach elevation (mean high water, etc.) is identified at constant locations along the coast for each time at which data are available. The rates of shoreline change at each location, and between consecutive data sets, are then computed. (Alternately, the change in beach volume above a certain elevation can be quantified at each location for each data set.) Using these methods, Fenster and Dolan (1996) describe three criteria to identify the spatial extent of inlet processes on adjacent shorelines:

- The cessation of abrupt changes in the rates of change alongshore, and/or the reduction in variability of these rates alongshore. Rate-of-change values are deemed no longer abrupt if:

  - The difference in rate values between adjacent transects over an X-meter-long reach does not increase or decrease by more than x meters per year (where X and x may be on the order of about 100- to 500 m and 0.2- to 0.3 m, respectively, depending upon the general coastline of interest).

  - The standard deviation of a subset of along-the-shore values (neglecting the transects nearest the inlet) is minimized.

- The slope of a regression line drawn through a subset of along-the-shore values (neglecting transects nearest the inlet) most closely equals zero.

- Changes in the sign of the rate value from erosion to accretion (or vice versa).

- A change from less erosional to more erosional (or vice versa); or, from less accretionary to more accretionary (or vice versa).

- A change in slope of the cumulative shoreline change or volume change computer along the shoreline.

- For natural inlets located on the wave-dominated North Carolina coast and the mixed-energy, tide-dominated Virginia barrier coast, Fenster and Dolan (1996) found that the first criterion revealed the greatest lineal extent of inlet-related shoreline impact (≤6.1 km). The second criterion generally yielded the next greatest degree of impact (≤5.2 km), and third criterion yielded the most conservative estimate (≤4.3 km). Paraphrasing their results, Fenster and Dolan concluded that there are zones in which inlet-related processes dominate shoreline trends (estimated from the third criterion), and where inlet-related processes influence shoreline trends (estimated from the first criterion). The second criterion yields estimates between the two zones.

- The fourth criterion is a potentially useful synthesis of the first three, particularly where shoreline change data are noisy. In this approach, the shoreline change (or more meaningfully, the local volume change (volume per unit alongshore beach width)), is integrated along the shoreline, starting at the inlet ($x = 0$). The process of integration smoothes fluctuations between adjacent profiles. This allows improved visualization of large-scale trends and easier discrimination of the data points that are dominating the data set. Integrating away from the inlet (and for a positive-valued shoreline axis), positive slopes in the cumulative curve represent shoreline accretion while negative slopes represent erosion.

(b) *Method 2: Even-odd analysis.* Dean and Work (1993) describe the application of the so-called "even-odd" analysis of profile change data (Berek and Dean 1982) for shoreline changes adjacent to inlets in Florida. The total shoreline (or volume) change y at an alongshore distance x from the inlet is considered to be composed of an even (symmetric) component, $y_E(x)$, and an odd (antisymmetric) component, $y_O(x)$:

$$y(x) = y_E(x) + y_O(x)$$
(V-6-11)

where $y_E(x) = y_E(-x)$, and $y_O(x) = -y_O(-x)$. The even and odd components are extracted from the total (measured) shoreline (or volume) change signal by

$$y_E(x) = [\, y(x) + y(-x)\,] / 2$$
(V-6-12a)

$$y_O(x) = [\, y(x) - y(-x)\,] / 2$$
(V-6-12b)

- The even component is a change that is symmetric about the inlet center line ($x = 0$). Physically, it is interpreted as an ambient, or background change that is common to both shorelines adjacent to the inlet (e.g., storm erosion, relative sea level change), equal placement of fill on both sides of the inlet, or equal transport from the two shorelines into the inlet.

EXAMPLE PROBLEM V-6-3

GIVEN:

Historical charts from 1927 and 1956, sampled at 250-m intervals (x), were used to develop mean high water shoreline changes (y) both north and south of an inlet. These data are shown in columns 1 and 2, respectively, in the following table for directions both north and south of the inlet. Positive values reflect shoreline seaward advance (accretion = +), and negative values reflect shoreline retreat (erosion = -). North of the inlet, the profile's active vertical height $h_* + B = 8.0$ m. Within 1,500 m south of the inlet, $h_* + B = 6.5$ m. Further south, $h_* + B = 7.5$ m. These data are shown in column 3. Here h_* is the depth of profile closure, and B is the berm thickness. The sum of $h_* + B$ is the thickness of sand movement.

Table V-6-4
Volume Change Calculations North and South of Inlet

North of Inlet						South of Inlet					
Given Data			Find (compute)			Given Data			Find (compute)		
1	2	3	4	5	6	1	2	3	4	5	6
Station x (m)	y (m)	h_*+B (m)	ΔV cu m/m	$x\,\Delta V$ 1000's of cu m	Σ $x\Delta V$	Station x (m)	y (m)	h_*+B (m)	ΔV cu m/m	$x\,\Delta V$ 1000's of cu m	Σ $x\Delta V$
2+50	+75.3		602	150.6	150.6	2+50	-8		-52	-13.0	-13.0
5+00	+56.2		450	112.5	263.1	5+00	-86		-559	-139.8	-152.8
7+50	+55.0		440	110.0	373.1	7+50	-99	6.5	-643.5	-160.9	-313.7
10+00	+28.5		228	57.0	430.1	10+00	-66		-429	-107.3	-421.0
12+50	+17.5		140	35.0	465.1	12+50	-42		-273	-68.3	-489.3
15+00	+21.2	8.0	170	42.4	507.5	15+00	-9		-58.5	-14.6	-503.9
17+50	+25.6		205	51.2	558.7	17+50	+3.4		25.5	6.4	-497.5
20+00	+5.4		43.2	10.8	569.5	20+00	-10.		-75	-18.8	-516.3
22+50	+7.4		59.2	14.8	584.3	22+50	-12	7.5	-90	-22.5	-538.8
25+00	+2.1		16.8	4.2	588.5	25+00	-27		-202.5	-50.6	-589.4
27+50	-4.4		-35	-8.8	655	27+50	-27		-202.5	-50.6	-640.0
30+00	+2.0		16	4	659	30+00	-16		-120	-30.0	-670.0
32+50	+3.2		25.6	-6.4	653	32+50	-24		-180	-45.0	-715.0
35+00	-7.5		-60	-15	638	35+00	-18		-135	-33.8	-748.8
37+50	-4.5		-36	-9	629	37+50	-12		-90	-22.5	-771.3

(Continued)

EXAMPLE PROBLEM V-6-3 (Continued)

FIND:

The updrift and downdrift limits of inlet influence upon the adjacent shoreline based upon these historical shoreline change data. Identify the total net volume gained or lost from each adjacent shoreline.

SOLUTION:

The updrift and downdrift inlet effects on the adjacent shorelines are apparent from the data of columns 1 and 2. Here are shown distances alongshore from the inlet, and the corresponding mean high water shoreline changes.

The volume of material V either gained or lost per beach cross section taken at 250-m intervals can be estimated by approximating the cross-sectional area A at each interval and multiplying by the interval distance alongshore x. The cross-sectional area A at a beach section can be approximated as the product of the mean horizontal high water change (y, from column 2) and the section profile's active vertical height ($h_* + B$, from column 3), such that $A = y(h_* + B)$, shown in the Figure V-6-14. When multiplied by $x = 1$ m alongshore, the resulting volume ΔV is the amount of material being gained or lost per meter of shoreline in the vicinity of that particular beach cross section, and is shown in column 4. When the cross-sectional A is multiplied by the interval distance alongshore (x, from column 1), the resulting volume V is the amount of material being gained or lost between the adjacent cross sections, and is shown in column 5. The summation of the volume of material on both the north and south sides of the inlet are shown in column 6, and these volumes are in Figure V-6-14.

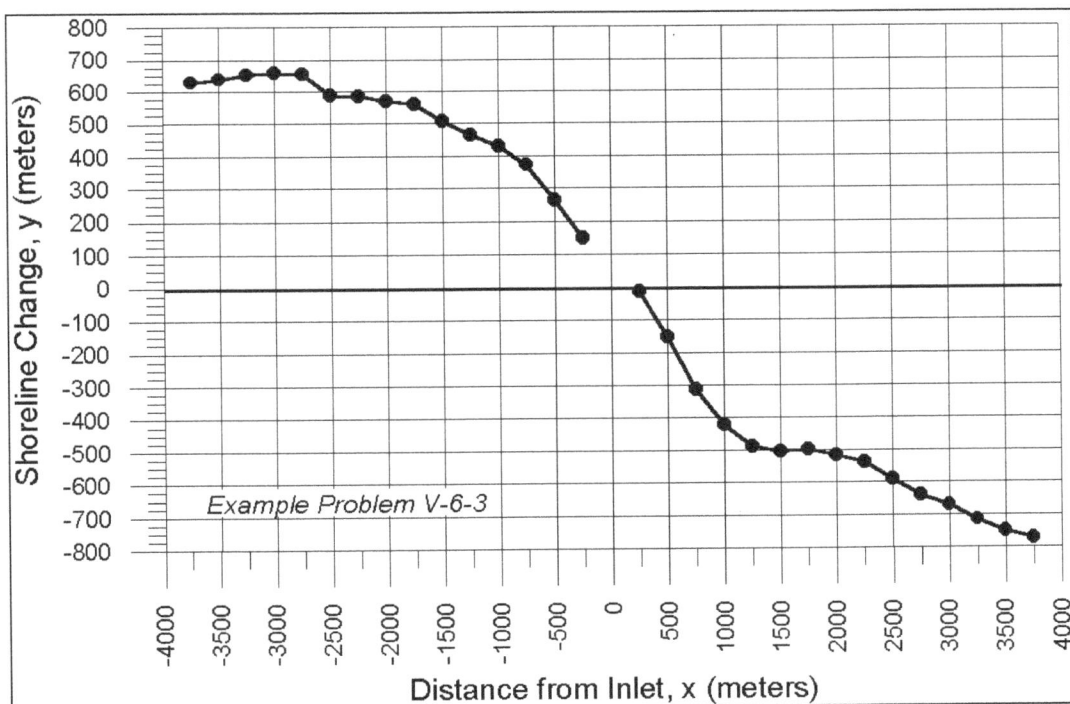

Figure V-6-14. Shoreline change as a function of distance from inlet (Example Problem V-6-3)

EXAMPLE PROBLEM V-6-3 (Concluded)

From this graph, it can be seen that the limit of direct inlet influence is about 2,500 to 2,700 m north of the inlet, and totals about 660,000 cu m of impoundment. Over the period of record, this equates to 660,000 cu m/29 years = 22,800 cu m/year, on average. There is a primary (near field) inlet effect within about 1,500 to 1,700 m south of the inlet, totaling about 500,000 cu m of erosion. There is an apparent far field inlet effect beginning about 2,000 m south of the inlet and extending beyond the 3,750-m limit of data. Within the limit of data, the net volume change downdrift of the inlet is 774,000 cu m of erosion, or about 26,700 cu m/year, on average. The actual downdrift deficit may be greater, as the inlet's influence apparently extends beyond the data limit (indicated by the continuing downward slope of the cumulative-volume curve at the right end of the graph). The fact that the downdrift erosion is greater than the updrift accretion (-774,000 cu m vs. +660,000 cu m) suggests that sand is being lost from the downdrift side by both interruption of the net southerly-directed transport and losses of northerly-directed transport from the south shoreline into the inlet.

- The odd component is antisymmetric about the inlet center line. It can be interpreted as the inlet's interruption effect upon the net littoral drift (e.g., impoundment along the updrift shoreline and erosion along the downdrift shoreline), placement of fill on one side of the inlet, accretion of the shoreline due to an attachment of the ebb tidal shoal, etc. Again, volume change (instead of shoreline change) is as aptly used in these equations. Likewise, the difference in shoreline or volume changes before and after a given event in time can be considered by this approach. In this case, the given event may be the creation of, or some modification of, the inlet.

- By its nature, the even-odd analysis assumes that shoreline changes (and transport processes) are symmetric/antisymmetric across the inlet center line. Accordingly, the degree to which results from the even-odd analysis accurately reflect actual inlet processes depends upon the degree to which the inlet-adjacent processes are, indeed, symmetric. Imbalances of volume change across the inlet are inherently spread across the inlet (i.e., split between the two shorelines) by the even-odd analysis. In Example Problem V-6-4, the impoundment effect of both sides of the inlet is equal, while in Example Problem V-6-5, erosion of the downdrift shoreline greatly exceeds the accretion (impoundment) of the updrift shoreline. The result is a significant net deficit of sand in the system.

- In Example Problem V-6-5, the total, net value of the inlet's effect is correct for the shoreline lengths considered (at least for the assumed background erosion rate); however, the distribution of the inlet's computed effect may not be correct. That is, the computed distribution assumes that the net inlet-induced loss from the littoral system (-995,000 cu m) is evenly divided between the updrift and downdrift shores. In reality, the net inlet-induced loss from the system may be much greater for one shore than for the other.

- The fact that neither the odd- nor even-components tend to zero at the limits of the measurements reveals the net volume deficit in the system. The non-zero values at the analysis' limits also illustrate that the inlet's sediment influence extends beyond the ±4,000-m distance considered in the analysis.

- The odd-component is often (incorrectly) assumed to solely represent the inlet's effect. Note, however, that even if the odd-component vanished to zero at the limits of the analysis, the fact that the even-component does not tend to zero indicates that the inlet effect may extend beyond the limits of the analysis. (This is easily demonstrated in this example by setting the measured shoreline change at +4,000 m equal to the shoreline change value at –4,000 m.)

(c) *Method 3: Alongshore variations in beach morphology.* Changes in the beach profile shape and sediment texture may provide evidence of the extent of the inlet's direct effect. For example, in order that adjacent beach profiles close at a common depth across a littoral barrier, accretional shorelines updrift of the barrier must be steeper than erosional shorelines downdrift. The limit of inlet's direct effect may be marked by the point at which variations in the profile shape (steepness) decrease, or when the profile reaches a shape or steepness similar to the remainder of the shore far from the inlet. In practice, profile steepness can be simply expressed as the horizontal length between two fixed elevations on the profile (usually measured from the berm to a depth between the beachface and closure), relative to the fixed elevation difference.

- A useful way to compare profile similarity is to compute, for each profile station, the relationship between the change in profile volume and the change in shoreline locations (for some fixed elevation):

$$G_p = \frac{\Delta V}{\Delta y} = \frac{\text{volume change per unit shoreline length}}{\text{change of shoreline location}} \tag{V-6-13}$$

- The S.I. units of G_p are cu m/m of shoreline change per meter along the shore; cu m/m/m = m. If the profile shape remains identical as the shoreline advances or retreats, G_p is equal to the height of the active profile; i.e., the vertical height between the top of the berm and the depth of profile closure, $(B + h_*)$. As previously noted, however, profiles steepen as sand is impounded updrift of a barrier (such as a jetty), and flatten as the shoreline recedes due to sediment deficits or the inlet's sink effect. The value of G_p is typically greater for impoundment-type shorelines, and lesser for receding-type shorelines.

- The overall practical value of this method is limited because it identifies only the inlet's direct influence. It therefore yields a conservative (short) estimate of inlet effect. Moreover, natural changes in beach morphology, wave climate, or sediment type may obscure those variations that are attributable to the inlet.

(d) *Method 4: Wave refraction analysis.* An inlet's perturbative effect upon nearshore bathymetry strongly influences the transformation (refraction, shoaling, diffraction, etc.) of waves as they approach and break along the shoreline. This, in turn, influences the littoral drift patterns. (See, for example, Figure V-6-11.) Wave refraction analysis is a useful means by which to investigate this effect.

- Chapter II-3-2 describes methods to compute wave transformation. Of central importance in the current application is the computation, along the shore, of the wave height and angle at the point of incipient breaking (i.e., immediately at the breakpoint), and the associated longshore sediment transport potential. Use of a grid-based wave refraction model (such as RCPWAVE) is assumed. In order to capture the effect of the inlet's ebb shoals, it is important that the wave transformation be computed all the way to the shoreline or final wave breaking point. That is, computation of breaking wave conditions at the shoreline from the wave conditions at some nearshore reference depth that is along the seaward edge of the ebb shoal plateau will greatly underestimate the inlet's perturbative effect on the wave field (Bodge, Creed, and Raichle 1996).

EXAMPLE PROBLEM V-6-4

GIVEN:

Shoreline surveys spaced 500 m apart both north (+x) and south (-x) of a small inlet were used to identify changes in the location of the mean high water shoreline between 1976 and 1980. These data are shown in columns 1 and 2, respectively, of the table below. Background erosion (i.e., outside the inlet's influence) is thought to be about 0.33 m (1 ft) over this period.

Table V-6-5
Data and Calculations for Even-Odd Analysis

1	2	3	4	5	6	7	8
x (m)	y measured	y_E even	y_O odd	y background	y erosion	y impoundment	y net inlet effect
-4000	-4	-3.5	-0.5	-0.33	-3.17	-0.5	-3.67
-3500	-5	-3.5	-1.5	-0.33	-3.17	-1.5	-4.67
-3000	-5	-3.5	-1.5	-0.33	-3.17	-1.5	-4.67
-2500	-1	-6	5	-0.33	-5.67	5	-0.67
-2000	5	-2	7	-0.33	-1.67	7	5.33
-1500	22	-2	24	-0.33	-1.67	24	22.33
-1000	35	2	33	-0.33	2.33	33	35.33
-500	50	10	40	-0.33	10.33	40	50.33
-1	45	11.5	33.5	-0.33	11.83	33.5	45.33
1	-22	11.5	-33.5	-0.33	11.83	-33.5	-21.67
500	-30	10	-40	-0.33	10.33	-40	-29.67
1000	-31	2	-33	-0.33	2.33	-33	-30.67
1500	-26	-2	-24	-0.33	-1.67	-24	-25.67
2000	**-9**	**-2**	**-7**	**-0.33**	**-1.67**	**-7**	**-8.67**
2500	-11	-6	-5	-0.33	-5.67	-5	-10.67
3000	-2	-3.5	1.5	-0.33	-3.17	1.5	-1.67
3500	-2	-3.5	1.5	-0.33	-3.17	1.5	-1.67
4000	-3	-3.5	0.5	-0.33	-3.17	0.5	-2.67
Net vol. change (1000's of cu m)							
Vol (-x)	358.5	-8.5	367	-8.5	0	367	367
Vol (+x)	-375	-8.5	-367	-8.5	0	-367	-367
Vol (tot.)	-17	-17	0	-17	0	0	0

(Continued)

EXAMPLE PROBLEM V-6-4 (Continued)

FIND:
The even- and odd-components of the shoreline change signals adjacent to this inlet, and the net effect of the inlet along the shoreline beyond ambient (background) changes.

SOLUTION:
In this example, the even-odd analysis is applied under the assumption that shoreline changes (and transport processes) are symmetric across the inlet center line (i.e., it assumes the shoreline on one side of the inlet is a reverse mirror image of the other side). Therefore, shoreline change data should be obtained at the same corresponding distances on both sides of the inlet to apply the even-odd analysis (i.e., data should be compiled at distance +x on one side of the inlet and -x on the other side of the inlet).

The even and odd components of the shoreline change can be determined by Equations V-6-12a and V-6-12b, respectively, and from the data of columns 1 and 2.

$$y_E(x) = \frac{y(x) + y(-x)}{2}$$

$$y_O(x) = \frac{y(x) - y(-x)}{2}$$

For example, at x = 2,000 m:

$$y_E(2,000) = \frac{(-9) + (5)}{2} = -2$$

$$y_O(2,000) = \frac{(-9) - (5)}{2} = -7$$

At x = -2,000 m:

$$y_E(-2,000) = \frac{(5) + (-9)}{2} = -2$$

$$y_O(-2,000) = \frac{(5) - (-9)}{2} = 7$$

The even and odd components for all the x-stations are shown in columns 3 and 4, plotted in Figure V-6-15.

(Continued)

EXAMPLE PROBLEM V-6-4 (Continued)

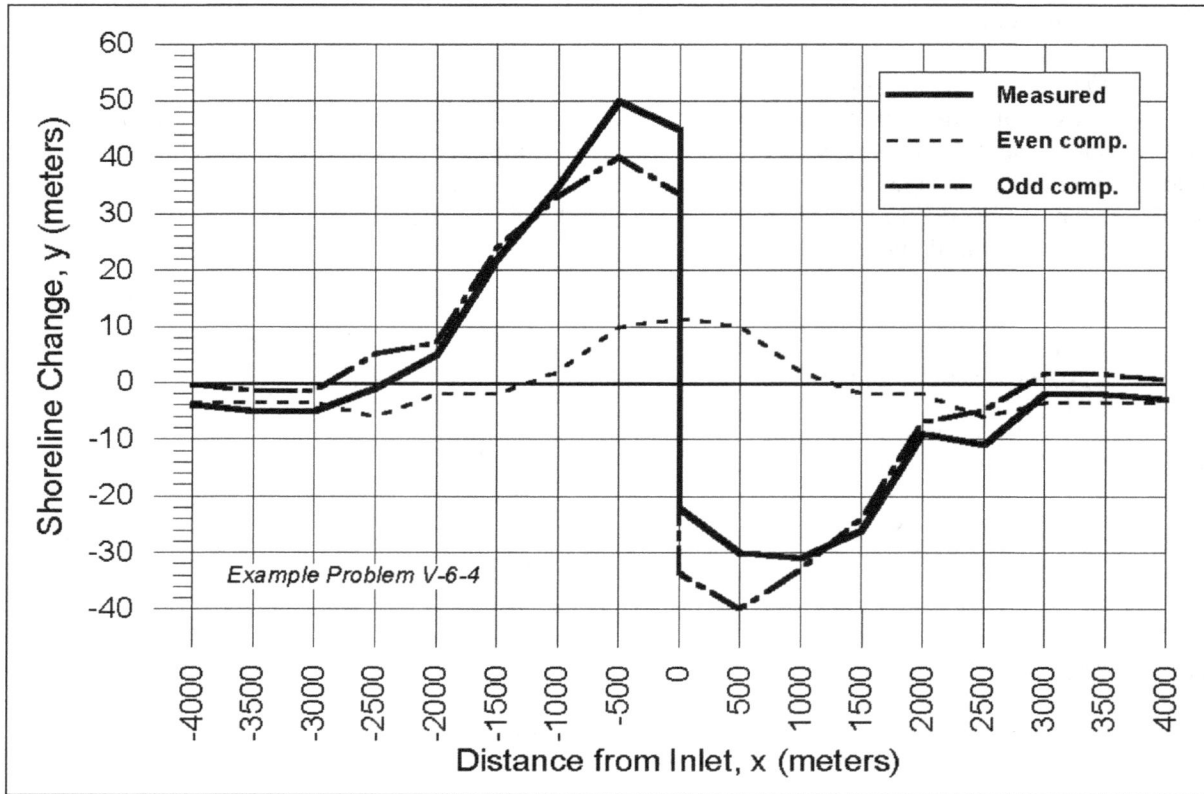

Figure V-6-15. Even-odd analysis for Example Problem V-6-4

From this graph, the positive portion of the even components suggests that there is transport directed towards the inlet within −1,250 to +1,250 m. The odd component reveals the presence of an impoundmnet fillet north of the inlet and a corresponding sediment deficit south of the inlet.

(Continued)

EXAMPLE PROBLEM V-6-4 (Concluded)

The estimated effect of sand transport direct toward the inlet along both shorelines (column 6) is computed as the even component (column 3) minus the estimated background signal of -0.33 m (column 5). For example, at x = +2,000 m, this effect is -2.0 - (-0.33) = - 1.67 m. Values for all the stations are listed in column 6. The estimated impoundment effect of the inlet is simply equal to the odd component (column 4), and is listed again in column 7.

The total net inlet effect relative to the estimated background signal (column 8) is computed as the measured shoreline change (column 2) minus the background signal of -0.33 m (column 5). For example, at x = +2,000 m, the total net inlet-induced effect is -9 - (-0.33) = -8.67 m. Values for all the stations are listed in column 8.

The volume changes associated with the shoreline change data are computed as the shoreline change value multiplied by the active profile depth (6.0 m) multiplied by the alongshore length of beach represented by each station. For example, the measured volume change estimated for the -x stations is:

$$[(-4 - 5 - 5 - 1 + 5 + 22 + 35 + 50)(500) + (45)(250)](6) = 358,500 \text{ cu m}$$

The net volume changes are listed at the bottom of each column for the updrift (-x) and downdrift (+x) shorelines, and for the sum of both.

In this example, the total net measured change across both shorelines (-17,000 cu m) is attributed to the area's background erosion rate (column 5). No net erosional effect is computed from transport toward the inlet (column 6), although there is a redistribution of 55,700 cu m of sand from the zone further than 1,000 m away from the inlet to the area within 1,000 m of the inlet. (This volume is computed as the sum of the changes in column 6 within +/-1,000 m of the inlet.) The net volume of the inlet's impoundment effect is computed to be 367,000 cu m (column 7). Neglecting the re-distribution of sand caused by inlet-induced transport (column 6), and neglecting any net accumulation of sand within the inlet's shoals which is not manifest as shoreline erosion within the +/-4-km study area, the inlet's net downdrift deficit beyond background erosion is therefore 367,000 cu m. This effect appears to extend about 3,000 m south of the inlet.

EXAMPLE PROBLEM V-6-5

GIVEN:

Shoreline surveys spaced 500 m apart both north (+x) and south (-x) of a small inlet document changes in the location of the mean high water shoreline between July 1956 and December 1968 (12.5 years). These data are shown in columns 1 and 2 of Table V-6-6. Background retreat (outside of the inlet's influence) is assumed to be 0.8 m/year over this time period, for a total retreat of 10 m.

Table V-6-6
Data and Calculations for Even-Odd Analysis and Net Volume Change

1	2	3	4	5	6	7	8
x (m)	y measured	y_E even	y_O odd	y background	y erosion	y impoundment	y net inlet effect
-4000	-11	-19	8	-10	-9	8	-1
-3500	-10	-18.5	8.5	-10	-8.5	8.5	0
-3000	-8	-19.5	11.5	-10	-9.5	11.5	2
-2500	-1	-30.5	29.5	-10	-20.5	29.5	9
-2000	-2	-38.5	36.5	-10	-28.5	36.5	8
-1500	22	-36.5	58.5	-10	-26.5	58.5	32
-1000	35	-45	80	-10	-35	80	45
-500	50	-30	80	-10	-20	80	60
-1	45	-26.5	71.5	-10	-16.5	71.5	55
1	-98	-26.5	-71.5	-10	-16.5	-71.5	-88
500	-110	-30	-80	-10	-20	-80	-100
1000	-125	-45	-80	-10	-35	-80	-115
1500	-95	-36.5	-58.5	-10	-26.5	-58.5	-85
2000	-75	-38.5	-36.5	-10	-28.5	-36.5	-65
2500	-60	-30.5	-29.5	-10	-20.5	-29.5	-50
3000	-31	-19.5	-11.5	-10	-9.5	-11.5	-21
3500	-27	-18.5	-8.5	-10	-8.5	-8.5	-17
4000	-27	-19	-8	-10	-9	-8	-17
Net vol. change (1000's of cu m)							
Vol (-x)	293	-752	1045	-255	-497	1045	548
Vol (+x)	-1797	-752	1045	-255	-497	-1045	-1542
Vol (total)	-1505	-1505	0	-510	-995	0	-995

(Continued)

EXAMPLE PROBLEM V-6-5 (Continued)

FIND:
    The even- and odd-components of the shoreline change signals adjacent to the inlet, and the net effect of the inlet beyond ambient (background) shoreline change.

SOLUTION:
    As stated earlier, the degree to which results from the even-odd analysis accurately reflect actual inlet processes depends upon the degree to which inlet and adjacent processes are, indeed, symmetric. In this example, retreat of the downdrift shoreline greatly exceeds the advance due to impoundment of the updrift shoreline. The result is a significant deficit of sand in the system.

The even and odd components of the shoreline change are again determined by Equations V-6-12a and V-6-12b, respectively, and from the data of columns 1 and 2.

$$y_E(x) = \frac{y(x) + y(-x)}{2}$$

$$y_O(x) = \frac{y(x) - y(-x)}{2}$$

For example, at x = 2,000 m:

$$y_E(2,000) = \frac{(-75) + (-2)}{2} = -38.5$$

$$y_O(2,000) = \frac{(-75) - (-2)}{2} = -36.5$$

At x = -2,000 m:

$$y_E(-2,000) = \frac{(-2) + (-75)}{2} = -38.5$$

$$y_O(-2,000) = \frac{(-2) - (-75)}{2} = 36.5$$

(Continued)

EXAMPLE PROBLEM V-6-5 (Concluded)

The even and odd components for all the x-stations are shown in columns 3 and 4, and plotted in Figure V-6-15.

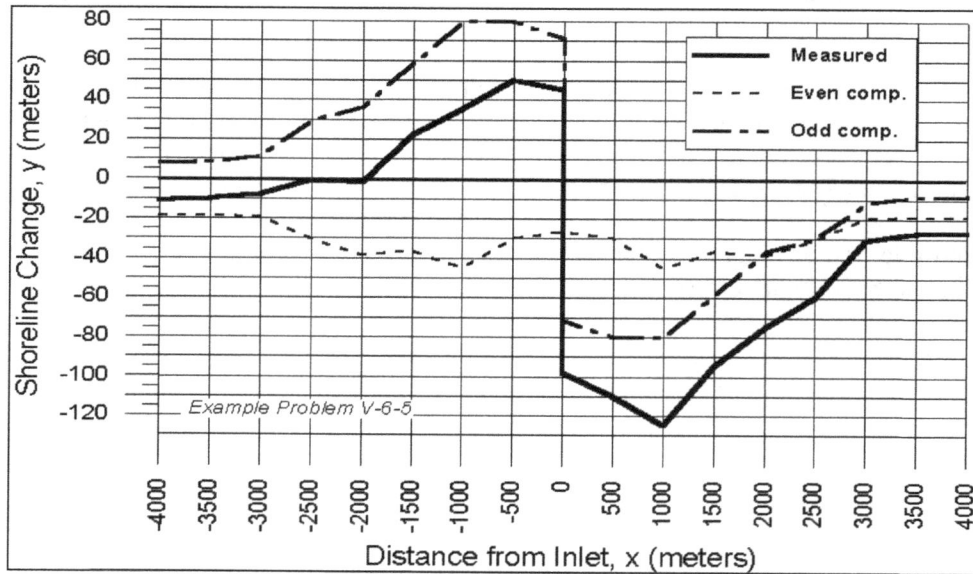

Figure V-6-16. Even-odd analysis for Example Problem V-6-5

The background erosion rate is 0.8 m/year x 12.5 years = 10 m. This value (column 5) is subtracted from the even component (column 3) since this component represents the common (symmetric) signal across both shorelines. The result is listed in column 6.

The even component, adjusted for background erosion effects, suggests that there is transport directed toward the inlet and lost therein. The net effect of this loss is computed as -497,000 cu m from each shoreline, for a total of -995,000 cu m (column 6). The odd component reflects the inlet's gross impoundment effect and the associated downdrift deficit associated with the impoundment's interruption of net littoral drift. This value is computed as -1,045,000 cu m (column 7).

Of the total net volume change estimated from the measured shoreline change data, -1,505,000 cu m (column 2), the -10 m background change accounts for -510,000 cu m (column 5). The balance is the inlet's net effect, -995,000 cu m (column 8). The even-odd analysis distributes this net deficit even across the inlet (i.e., -497,000 cu m for each shoreline, column 6). When added to the impoundment effect, the computed net result is 548,000 cu m of net accretion along the updrift shoreline and -1,542,000 cu m of net erosion along the downdrift shoreline (column 8).

- For each alongshore column of the refraction grid, and for each case of interest, the shoreward-most occurrence of a nonbroken wave is identified; i.e., where the wave height H is less than κ times the water depth, h. The breaking index κ is typically equal to about 0.8, or as otherwise specified in the refraction model. Assuming shallow water wave conditions, the breaking wave height H_b and α_b can be estimated from:

$$H_b \sim \frac{A_1}{1 - \frac{1}{5} A_1 A_2} \qquad (V\text{-}6\text{-}14)$$

$$\alpha_b = \sin^{-1}\left[\frac{\sin \hat{\alpha}}{C} \sqrt{g H_b / \kappa} \right] \qquad (V\text{-}6\text{-}15)$$

where

$$A_1 = (\kappa / g)^{1/5} H^{4/5} (C_g \cos \hat{\alpha})^{2/5} \qquad (V\text{-}6\text{-}16a)$$

$$A_2 = g (\sin \hat{\alpha})^2 / C^2 \qquad (V\text{-}6\text{-}16b)$$

- The values C and C_g are the wave celerity and group celerity, respectively, at the reference (non-broken wave) location. The value $\hat{\alpha}$ is the wave angle at the reference location measured relative to the local shoreline orientation:

$$\hat{\alpha} = \alpha - \beta \qquad (V\text{-}6\text{-}17)$$

where α is the wave angle relative to the grid at the reference location (i.e., the value output by the wave refraction analysis) and β is the shoreline angle relative to the grid at the alongshore column of interest.

- From Part III-2-2, the potential longshore transport rate is computed for each column, and for each wave case, from

$$Q = K' H_b^{5/2} \sin(2\alpha_b) \qquad (V\text{-}6\text{-}18)$$

where Q has units of volume per time; and where

$$K' = \frac{K \sqrt{g}}{16 (s-1)(1-n)} \qquad (V\text{-}6\text{-}19)$$

and K is the dimensionless coefficient from the CERC Formula (Equation III-2-5).

EXAMPLE PROBLEM V-6-6

GIVEN:

Local water depths, wave heights, and angles were computed by a numerical wave transformation model for five grid elements of one column of a numerical computational grid. The wave period T is 9.0-sec, and the shoreline angle (β) at this column is +2.0 deg relative to the grid. The numerical model's breaking index is $\kappa = H_b/h_b = 0.8$, where H_b is the wave height at breaking and h_b is the water depth where the 9.0 sec wave breaks. Assume the sand specific gravity $S = 2.65$, the in-place sediment porosity n = 0.4, and the imprecise dimensionless coefficient from the CERC Formula (Equation III-2-18) is taken to be $K = 0.5$.

Table V-6-7
Numerical Model Results for Longshore Transport Potential Calculations

	GIVEN: Numerical Model Results			Compute
Grid Row #	Depth h (m)	Wave Ht H (m)	Angle α (deg)	$\kappa = H/h$
1	0	0	0	--
2	0.3	0.24	3.5	0.8
3	0.7	0.56	5.4	0.8
4	1.7	0.57	5.6	0.34
5	2.9	0.52	5.8	0.18

FIND:

The incipient breaking wave height and angle, and the longshore transport potential from nearshore wave conditions computed across this numerical wave transformation grid.

(Continued)

EXAMPLE PROBLEM V-6-6 (Continued)

SOLUTION:

The incipient breaking wave height and angle occur in the grid row closest to the shoreline where the breaking index $\kappa = H/h$ is less than 0.8. Here H is the wave height and h is the water depth at a particular cell of the grid. Hence, it is necessary to compute κ for each row of this column of cells, starting at the shoreline and working seaward. At grid row 1, the water depth is 0.0, and H/h is undefined. At grid row 2, H/h = 0.24/0.3 = 0.8 (a broken wave). At grid row 3, H/h = 0.56/0.7 = 0.8 (also a broken wave). At grid row 4, H/h = 0.57/1.7 = 0.34 (a non-broken wave). Thus, the shoreward-most occurrence of a non-breaking wave is at grid row 4 where the water depth is 1.7 m. (Note: The wave height and angle in the above table for grid row 4 are not breaking wave characteristics. Breaking, however, does occur somewhere within grid row 4. Grid row 4 becomes the reference location from which breaking conditions, H_b and α_b, will be computed.) Here this 9.0 sec wave has a non-broken wave height H = 0.57 m and a non-broken wave angle $\alpha = 5.6$ deg relative to the grid. A definition sketch of the layout of the grid relative to the shoreline and approaching wave crests is presented in Figure V-6-17.

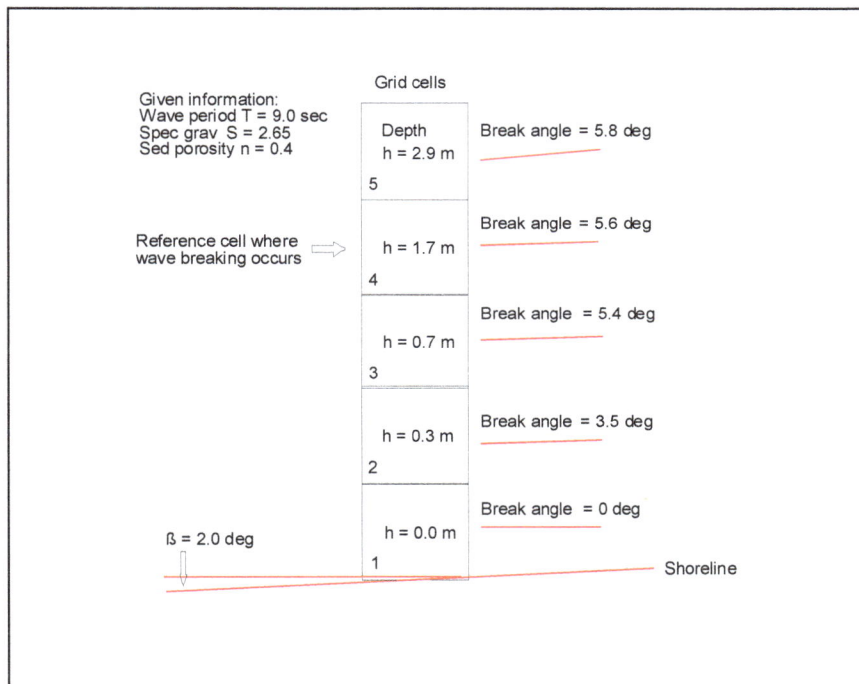

Figure V-6-17. Sketch of numerical model results for Example Problem V-6-6

(Continued)

Sediment Management at Inlets

EXAMPLE PROBLEM V-6-6 (Continued)

The value $\hat{\alpha}$ is the wave angle at the reference location measured relative to the local shoreline orientation. From Equation V-6-17:

$$\hat{\alpha} = \alpha - \beta$$

where α is the wave angle relative to the grid at the reference location (i.e., the output value from the numerical wave transformation model), and β is the shoreline angle relative to the grid at the grid column of interest.

Assuming shallow-water wave conditions, the breaking wave height (H_b) and breaking wave angle α_b can be estimated from Equations V-6-14 and V-6-15, respectively:

$$H_b = \frac{A_1}{1 - \dfrac{A_1 A_2}{5}}$$

$$\alpha_b = \sin^{-1}\left[\frac{\sin\hat{\alpha}}{C}(gH_b / \kappa)^{0.5} \right]$$

A_1 and A_2 are defined by Equations V-6-16a and V-6-16b, respectively.

where

$$A_1 = (\kappa / g)^{0.2} H^{0.8} (C_g \cos\hat{\alpha})^{0.4}$$

and

$$A_2 = \frac{g(\sin\hat{\alpha})^2}{C^2}$$

The acceleration due to gravity is $g = 9.81$ m/sec^2. The values C and C_g are the wave celerity and group celerity, respectively, at the reference (nonbreaking) location where the wave period $T = 9.0$ sec and the water depth $h = 1.7$ m. From Equation II-1-7:

$$C = L/T$$

where L is the local wavelength. Hence, it becomes necessary to determine the local wavelength L at this location. From Equation II-1-10:

$$L = \frac{gT^2}{2\pi}\tanh\left(\frac{2\pi h}{L}\right)$$

(Continued)

EXAMPLE PROBLEM V-6-6 (Continued)

The solution of this expression for the local wavelength L involves some difficulty because L occurs on both sides of the equation. There are, however, a number of ways to solve this equation. It can be solved by iteration, or from tabulated values of L and C by Goda (1985), or from tabulated values of h/L_0 and h/L by Wiegel (1954), or by the procedure of Fenton and McKee (1990), or by use of Figure II-1-5 (Part II). From either of these methods, with $T = 9$ sec and h = 1.7 m, it can be determined that $L = 36.27$ m. Applying the CEM method (for example), $L_0 = gT^2/2\pi = 9.81(9)^2/2\pi = 126.46$ m. Thus, $h/L_0 = 1.7/126.46 = 0.01344$. From Figure II-1-5 at $h/L_0 = 0.01344$, it can be determined that $\sinh(2\pi d/L) = 0.2988$, from which $2\pi d/L = 0.2945$, and from which L = 36.27. Because $C = L/T$, $C = 36.27$ m/9 sec = 4.03 m/sec.

By Equation II-1-49:

$$C_g = \frac{1}{2}\frac{L}{T}\left[1 + \frac{\dfrac{4\pi h}{L}}{\sinh\dfrac{4\pi h}{L}}\right]$$

$$C_g = \frac{1}{2}\frac{36.27m}{9\sec}\left[1 + \frac{\dfrac{(4)(\pi)(1.7m)}{36.27m}}{\sinh\dfrac{(4)(\pi)(1.7m)}{36.27m}}\right]$$

$$C_g = 3.91\,m/\sec$$

Angle $\alpha = 5.6$ deg (the wave angle relative to the grid), and was determined by the numerical model output at grid row 4. Angle $\beta = 2.0$ deg (the shoreline angle relative to the gird) was given. Hence:

$$\hat{\alpha} = \alpha - \beta = 5.6\,\text{deg} - 2.0\,\text{deg} = 3.6\,\text{deg}$$

$$A_1 = (\kappa/g)^{0.2} H^{0.8} (C_g \cos\hat{\alpha})^{0.4}$$

Here κ is the model's breaking wave index, $\kappa = H_b/h_b = 0.8$.

$$A_1 = (0.8/9.81m\sec^{-2})^{0.2}(0.57m)^{0.8}(3.91m\sec^{-1}\cos 3.6\,\text{deg})^{0.4}$$

$$A_1 = 0.677m$$

$$A_2 = g(\sin\hat{\alpha})^2/C^2$$

(Continued)

EXAMPLE PROBLEM V-6-6 (Concluded)

$$A_2 = 9.81 m \sec^{-2} (\sin 3.6 \deg)^2 / (4.03 m \sec^{-1})^2$$

$$A_2 = 0.00238 \ m^{-1}$$

$$H_b = \frac{A_1}{1 - \frac{A_1 A_2}{5}} = \frac{0.0667 m}{1 - \frac{(0.667m)(0.00238m^{-1})}{5}}$$

$$H_b = 0.067 \ m$$

$$\alpha_b = \sin^{-1} \left[\frac{\sin \hat{\alpha}}{C} (gH_b / \kappa)^{0.5} \right]$$

$$\alpha_b = \sin^{-1} \left(\sin 3.6 \deg / 4.03 m \sec^{-1} \right) \left[(9.81 m \sec^{-2})(0.67m) / 0.8 \right]^{0.5}$$

$$\alpha_b = 2.56 \ deg$$

The potential longshore sediment transport rate can now be determined by Eq. V-6-16:

$$Q = K' H_b^{2.5} \sin(2\alpha_b)$$

where

$$K' = \frac{Kg^{0.5}}{16(S-1)(1-n)}$$

$$K' = \frac{(0.5)(9.81 m \sec^{-2})^{0.5}}{(16)(2.65 - 1)(1 - 0.4)}$$

$$K' = 0.099 m^{0.5} \sec^{-1}$$

from which

$$Q = (0.099 m^{0.5} \sec^{-1})(0.67m)^{2.5} [\sin(2)(2.56 \deg)]$$

$$Q = 0.0033 m^3 \sec^{-1}$$

EXAMPLE PROBLEM V-6-7

GIVEN:

The potential longshore transport rate (Q, in units of volume per time) for each of five alongshore columns of a refraction grid, and for three different wave conditions each during the time period of interest, are presented as lines 1-3 in the Table V-6-8. The percentages of occurrence (P) of the three wave conditions are 22, 25, and 16 percent. There were no waves (calm) during 37 percent of the time period of interest. These percentages of occurrence are shown as lines 4-6 in the table.

Table V-6-8
Potential Longshore Transport Rate Information for Fire Grid Columns

		Grid Column				
Line	Parameter	1	2	3	4	5
1	Q_1 (cu m/hr)	2.9	3.3	5.2	6.4	6.2
2	Q_2 (cu m/hr)	-0.7	-0.4	1.6	1.8	2.1
3	Q_3 (cu m/hr)	-6.8	-6.5	-1.5	0.8	0.2
4	P_1 (%)	0.22	0.22	0.22	0.22	0.22
5	P_2 (%)	0.25	0.25	0.25	0.25	0.25
6	P_3 (%)	0.16	0.16	0.16	0.16	0.16
7	P_1Q_1 (cu m/hr)	0.64	0.73	1.14	1.41	1.36
8	P_2Q_2 (cu m/hr)	-0.18	-0.10	0.40	0.45	0.53
9	P_3Q_3 (cu m/hr)	-1.09	-1.04	-0.24	0.13	0.03
10	Q_R (cu m/hr)	0.64	0.73	1.54	1.99	1.92
11	Q_L (cu m/hr)	-1.27	-1.14	-0.2	0	0
12	$Q_{(NET)}$ (cu m/hr)	-0.63	-0.41	1.30	1.99	1.92
13	Q_{RN} (cu m/hr)	0.32	0.37	0.77	1.00	0.96
14	Q_{LN} (cu m/hr)	-0.64	-0.57	-0.12	0	0
15	$Q_{(NET)N}$ (cu m/hr)	-0.32	-0.21	0.65	1.00	0.96

FIND:

The total rates of right-directed (+), left-directed (-), and net transport potential for the five grid columns along this shoreline. Normalize the results by the net transport maxima.

SOLUTION:

The potential longshore transport rate (Q) can be expressed as some function of the breaking wave height (H_b) and the breaking wave angle (α_b), as given by Equation V-6-18,

$$Q = K' H_b^{2.5} \sin(2\alpha_b)$$

with units of volume per time (e.g., cu m/hr).

(Continued)

EXAMPLE PROBLEM V-6-7 (Concluded)

A numerical simulation wave refraction model was used to determine the breaking wave height (H_b) and breaking wave angle (α_b) for each of the five columns of the refraction grid. Equations V-6-14 through V-6-19 were then used to determine the potential longshore transport rate Q (cu m/hr) for these conditions, and these are GIVEN values shown as lines 1-3. The expression for the potential longshore transport rate Q (cu m/hr) (Equation V-6-18) inherently assumes there is an unlimited supply of sand available for movement. The total amount of sand moved during the time interval of interest is, therefore, the product of the rate of movement Q (cu m/hr) multiplied by the time interval (which may be more or less than 1 hr). The three wave conditions under consideration here existed for $P_1 = 22$ percent, $P_2 = 25$ percent, and $P_3 = 16$ percent of an hour, respectively. These percentages of occurrence are presented as lines 4-6.

The rate of actual material movement per hour by a particular wave climate is determined as the product of the potential longshore transport Q (cu m/hr) and the percentage of occurrence of that particular wave. Hence, the rates of actual material movement are shown as P_1Q_1, P_2Q_2, and P_3Q_3 on lines 7-9, respectively. These are the weighted values of actual (potential) transport rate, weighted based on the percent of time (hour) that a particular wave climate actually existed.

The weighted sum of the actual (potential) transport rates for each column is the sum of all components of transport moving to the right (+), and to the left (-). These values are shown as lines 10 and 11, respectively. For example, for column 1, the weighted sum of the three transports produced by the three different wave climates are 0.64, -0.18, and -1.09 cu m/hr (Lines 7-9 of column 1). The transport to the right Q_R (+) is the only positive value of the three (0.64 cu m/hr), shown as Line 7. The transport to the left Q_L (-) is the sum of the two negative transports (-0.18 and -1.09 cu m/hr), shown as lines 8 and 9. The weighted sum of the actual (potential) transport rates for the five columns of the numerical grid are shown as lines 10 and 11, for right transport Q_R and left transport Q_L, respectively. The weighted net transport $Q_{(NET)}$ is the algebraic sum of the right transport Q_R and the left transport Q_L, and is shown as line 12.

The weighted sums of the actual (potential) longshore transports are then normalized by the maximum value of $Q_{(NET)}$ for all columns as shown in line 12. This maximum value occurs at column 4 (1.99 cu m/hr). All values of lines 10-12 are normalized (divided) by this value, and these right normalized values Q_{RN}, left normalized values Q_{LN}, and net normalized values $Q_{(NET)N}$ are shown as lines 13-15, respectively. These values indicate the relative actual (potential) longshore transport quantities per column of the numerical grid with respect to the maximum value of these parameters occurring within the region of interest.

- The total right-directed and left-directed potential transport rates at a given alongshore column are, respectively, the weighted sum of the positive-valued and negative-valued transport estimates for that column:

$$Q_R = \sum p_i Q_i \quad for \quad Q_i \geq 0 \tag{V-6-20}$$

$$Q_L = \sum p_i Q_i \quad for \quad Q_i \leq 0 \tag{V-6-21}$$

where p_i is the fractional occurrence of each wave case i. Likewise, the net transport potential is the weighted sum of all transport values at a given column:

$$Q_{NET} = \Sigma \left(p_i Q_i \right) = Q_R + Q_L \tag{V-6-22}$$

- The net, right- and left-directed potential transport rates for each grid column along the shore can then be examined or plotted to assess the alongshore extent and nature of the inlet's influence upon the refracted wave field and the associated littoral transport patterns (see, as an example, Figure V-6-11). The results are also useful in elucidating the inlet's sand transport pathways and for subsequent development of the sediment budget.

- This approach is limited in that it does not address transport induced by tidal currents or gradients in wave height. Likewise, it addresses only the inlet's direct effect upon the shoreline and does not reveal far field erosion caused by the inlet's sink effect upon the littoral transport.

(e) *Method 5: Examination of inlet's net sink effect.* Whereas methods 1, 2, and 3 discern an inlet's effect through examination of shoreline change, this method primarily relies upon changes measured at and within the inlet. This is potentially advantageous because shoreline position data are often ambiguous and may reflect engineering activities such as dredged material placement, while, in some cases, the inlet's impoundment of sand adjacent to and within the inlet is more readily discerned (usually with some level of quantifiable uncertainty). Additionally, the assessment of the inlet's impacts is not a priori biased (limited) by the length of shoreline selected for examination.

- An inlet's net sink effect is defined as the quantity of material that the inlet has taken from the littoral system. In most cases, natural and stabilized inlets remove sand from the littoral system through accretion of adjacent shores, shoaling in channels, and accretion of ebb- and flood-tidal shoals. However, inlets with riverine input may be the source of littoral material for the coast (e.g., Columbia River, Washington/Oregon).

- The net sink effect, or volumetric impact, is first computed by adding: The volume (or rate) of impoundment adjacent to the inlet entrance; the volume (or rate) of net sand accumulation within the inlet's channels and shoals; the volume (or rate) of sand removed from the littoral system by dredging and offshore (or out-of-system) disposal. (If dredged material is placed on the adjacent beaches, it remains within the adjacent littoral system and thus is not added to the total).

- Then, the following is subtracted from this total: The volume (or rate) of riverine (or other upland) sedimentary input; the volume (or rate) of barrier removed due to creation (through dredging or breach) of the inlet (if this event is within the time period of consideration).

- The resulting value is the volume (or rate) of sand which has been removed from the adjacent shores' littoral systems over the period of examination. Inlet-adjacent volume changes are then examined to discern the minimum distance away from the inlet along which this volumetric impact is manifest.

EXAMPLE PROBLEM V-6-8

GIVEN:

Measured shoreline changes adjacent to an inlet are presented in columns 1 and 2 of Table V-6-9 for the period 1975-1989. Surveys from 1980 and 1988 suggest that the inlet's ebb and interior shoals outside the limits of dredging accreted by 152,000 cu m ±48,000 cu m. The correlation G_p between profile volume change per unit of shoreline length (alongshore) and cross-shore beach width change is estimated to be:

    a. Updrift (east) of the inlet: 7.5 cu m/m of shoreline length per meter of cross-shore beach width change (7.5 cu m/m/m).

    b. Downdrift (west) of the inlet: 7.2 cu m/m of shoreline length per meter of cross-shore beach width change) (7.2 cu m/m/m).

Profile erosion between 1.5 and 3.5 km downdrift of the inlet is restricted by the presence of seawalls. An average of about 192,000 cu m is dredged from the inlet's entrance channel every 3 years, all of which is disposed in deep water offshore. In 1987, 360,000 cu m of beach fill were placed between 0 and 4.5 km downdrift (west) of the inlet from an offshore sand source. Neglect riverine input of sand. Assume that erosion induced by the inlet along the updrift shoreline is negligible. The total rate at which the inlet removes sand from the littoral system is the sum of:

    a. The updrift impoundment rate.

    b. The inlet's net shoaling rate (after dredging).

    c. The rate at which maintenance-dredging sand is disposed outside the littoral system.

FIND:

The lineal and volumetric extent for which the inlet's net sink effect can be attributed to measured shoreline erosion.

SOLUTION:

The natural inlet and shoreline processes are assumed to be similar over the entire period of interest (1975-1989). The change in shoreline volume per unit of shoreline length (alongshore) ΔV (m³/m), per unit of change of cross-shore beach width change Δy (m), is denoted by Equation V-6-11:

$$G_p = \frac{\Delta V \text{ (volume change per unit of shoreline length alongshore)}}{\Delta y \text{ (cross - shore beach width change)}}$$

This relationship G_p was previously estimated to be 7.5 m³/m of shoreline length (alongshore) per meter of cross-shore beach width for the coastal segment immediately updrift of the inlet, and 7.2 m³/m of shoreline length (alongshore) per meter of cross-shore beach width for the coastal segment immediately downdrift of the inlet. The average annual cross-shore shoreline change data of column 2 with units of m/year can be converted to average annual local volume change data with units of m³/m/year by using Equation V-6-11.

(Continued)

EXAMPLE PROBLEM V-6-8 (Continued)

Table V-6-9
Measured Shoreline Changes and Calculations for Net Sink Effect

GIVEN			COMPUTE		
1	2	3	4	5	6
Distance from inlet, x (km)	Avg. ann'l shoreline change, dy/dt 1975-89 (m/year)	Avg. annual local vol. change, dV/dt (cu m/m/year)	Reach Length (m)	Avg. annual reach vol. change (cu m/year)	Cum. vol. change - average annual (cu m/year)
7	-0.1	-0.75	1,000	-750	46,825
6	0.2	1.5	1,000	1,500	47,575
5	-0.2	-1.5	1,000	-1,500	46,075
4	0.5	3.8	1,000	3,800	47,575
3	1.2	9	1,000	9,000	43,775
2	1.5	11.3	1,000	11,300	34,775
1	2.0	15	750	11,250	23,475
0.5	2.3	17.3	500	8,650	12,225
0 (east)	1.9	14.3	250	3,575	3,575
0 (west)	-6.2	-44.6	250	-11,150	-11,150
-0.5	-5.3	-38.2	500	-19,100	-30,250
-1	-2.4	-17.3	500	-8,650	-38,900
-1.5	0	0	500	0	-38,900
-2	0	0	500	0	-38,900
-2.5	1.7	12.2	500	6,100	-32,800
-3	3.9	28.1	750	21,075	-11,725
-4	2.1	15.1	1,000	15,100	3,375
-5	0.3	2.2	1,000	2,200	5,575
-6	-1	-7.2	1,000	-7,200	-1,625
-7	-1.7	-12.2	1,000	-12,200	-13,825
-8	-1.9	-13.7	1,000	-13,700	-27,525
-9	-1.3	-9.4	1,000	-9,400	-36,925
-10	-0.6	-4.3	1,000	-4,300	-41,225
-11	-0.3	-2.2	1,000	-2,200	-43,425
-12	-0.4	-2.9	1,000	-2,900	-46,325
-13	-0.5	-3.6	1,000	-3,600	-49,925

(Continued)

EXAMPLE PROBLEM V-6-8 (Continued)

Since
$$G_p = \frac{\Delta V}{\Delta y} \; m^3 / m / m$$

then
$$\left(G_p \; m^3 / m / m \right) \left(\frac{\Delta y}{\Delta t} \; m / year \right) \; = \; m^3 / m / year$$

For example, at x = +7 km,

$$(7.5 \; m^3/m/m)(-0.1 \; m/year) = -0.75 \; m^3/m/year$$

These equivalent average annual local volume change data are shown in column 3.

The effective shoreline reaches for each section updrift and downdrift of the inlet are shown in column 4. Next, the total rate of change along each reach of shoreline was determined by the simple trapezoidal rule (i.e., by multiplying the local volume change rate (column 3) by the local effective shoreline reach (column 4)). For example, the volume change at x = +7 km is

$$(-0.75 \; m^3/m/yr)(1,000 \; m) = -750 \; m^3/year$$

These results are shown in column 5. Then the cumulative volume changes along the updrift (east) and downdrift (west) shorelines were determined, starting at the inlet for each direction. These results are shown in column 6. From examination of column 6, the updrift impoundment appears to extend about 4 km east of the inlet, and totals about 47,000 m³/year (updrift impoundment rate).

The inlet's net shoaling rate between 1980 and 1988 was transformed to an annual average rate as (152,000 m³ ±48,000 m³)/8 year = 19,000 ±6,000 m³/year (inlet net shoaling rate).

The rate of dredging and out-of-system disposal is annualized as 192,000 m³/3 year = 64,000 m³/year (maintenance dredging).

Again, the total rate at which the inlet removes sand from the littoral system is the sum of: the updrift impoundment rate; the inlet's net shoaling rate (after dredging); and the rate at which maintenance-dredging sand is disposed outside the littoral system.

Hence, the total rate at which the inlet removes sand from the littoral system is:

47,000 cu m/year + (19,000 ± 6,000) cu m/year + 64,000 cu m/year = 124,000 to 130,000 cu m/year

Hence, in the absence of the beach nourishment project, the downdrift beach is expected to exhibit a loss of between 124,000 and 136,000 cu m/year. Instead, primarily due to the 360,000-cu m-fill, the downdrift beach exhibited a net measured loss of only 49,925 cu m/year along the 13-km reach for which data are given.

To more carefully assess the inlet's actual impact, the beach fill must be included in the analysis. This fill, which was placed on the shore between 0 and 4.5 km downdrift of the inlet, was also removed (eroded) from the beach; therefore, it must be included in the total amount of material removed from the beach. This analysis is performed in the following table. Only the downdrift (west) shoreline is listed in the table since it is only this shoreline that is affected by the beach fill.

(Continued)

EXAMPLE PROBLEM V-6-8 (Continued)

The average annual cumulative volume change downdrift of the inlet (column 6) has been converted to the equivalent total cumulative volume change over the 14-year period between the 1975 and 1989 survey period by multiplying column 6 by 14 years. These data are shown in column 7.

The placed beach fill was also removed (eroded) from the beach. We assume that the 360,000 cu m of beach fill was uniformly distributed over the 4.5-km placement area (i.e., 80 cu m/m of shoreline. Column 8 shows the corresponding fill volume placed within each reach (between each station). From these results, the cumulative alongshore fill volume was computed, and is shown in column 9 as the cumulative placed fill volume along the beach downdrift of the inlet.

The cumulative alongshore volume change over the 14-year survey period, with the beach fill considered, was computed by adding the cumulative placed beach fill volume (column 9) to the cumulative measured beach fill volume change (column 7). These sums are shown in column 10 as the cumulative total volume change. The equivalent average annual cumulative total volume change is computed by dividing these values (column 10) by the survey interval (14 years). These results are shown in column 11. The cumulative updrift and downdrift volume changes, with and without consideration of the fill, are shown in Figure V-6-18.

Table V-6-10
Calculations for Cumulative Volume Change West of Inlet

From previous table			Downdrift volume change with beach fill considered (included)				
1	4	6	7	8	9	10	11
Distance from inlet, x (km)	Reach length (m)	Cum. vol. change - average annual (m³/year)	Cum. vol. Change 1975-1989 (m³)	Fill vol. placed in reach (m³) [assumed]	Cum. Placed fill vol. (m³)	Cum. total vol. change with fill considered; 1975-1989 (m³)	Cum. total vol. change with fill considered; avg. annual (m³/year)
0 (west)	-6.2	-11,150	-156,100	20,000	20,000	-176,100	-12,580
-0.5	-5.3	-30,250	-423,500	40,000	60,000	-483,500	-34,540
-1	-2.4	-38,900	-544,600	40,000	100,000	-644,600	-46,040
-1.5	0	-38,900	-544,600	40,000	140,000	-684,600	-48,900
-2	0	-38,900	-544,600	40,000	180,000	-724,600	-51,760
-2.5	1.7	-32,800	-459,200	40,000	220,000	-679,200	-48,510
-3	3.9	-11,725	-164,150	60,000	280,000	-444,150	-31,725
-4	2.1	3,375	47,250	80,000	360,000	-312,750	-22,340
-5	0.3	5,575	78,050	0	360,000	-281,950	-20,140
-6	-1.0	-1,625	-22,750	0	360,000	-382,750	-27,340
-7	-1.7	-13,825	-193,550	0	360,000	-553,550	-39,540
-8	-1.9	-27,525	-385,350	0	360,000	-745,350	-53,240
-9	-1.3	-36,925	-516,950	0	360,000	-876,950	-62,640
-10	-0.6	-41,225	-577,150	0	360,000	-937,150	-66,940
-11	-0.3	-43,425	-607,950	0	360,000	-967,950	-69,140
-12	-0.4	-46,325	-648,550	0	360,000	-1,008,550	-72,040
-13	-0.5	-49,925	-698,950	0	360,000	-1,058,950	-75,640

(Continued)

EXAMPLE PROBLEM V-6-8 (Concluded)

With the effect of the fill included (column 11), the downdrift volume change within the 13-km limit of data equals -75,640 cu m/year. This value is only about 56 to 61 percent of the total net impact (124,000 to 136,000 cu m/year) estimated from the inlet's diversion of sand from the littoral system (i.e., via updrift impoundment, net shoal growth, and out-of-system maintenance dredging disposal). Neglecting any inlet-induced erosion updrift of the inlet, the remaining 39 to 44 percent of the inlet's impact (48,360 to 60,360 cu m/year) must therefore occur further downdrift than 13 km.

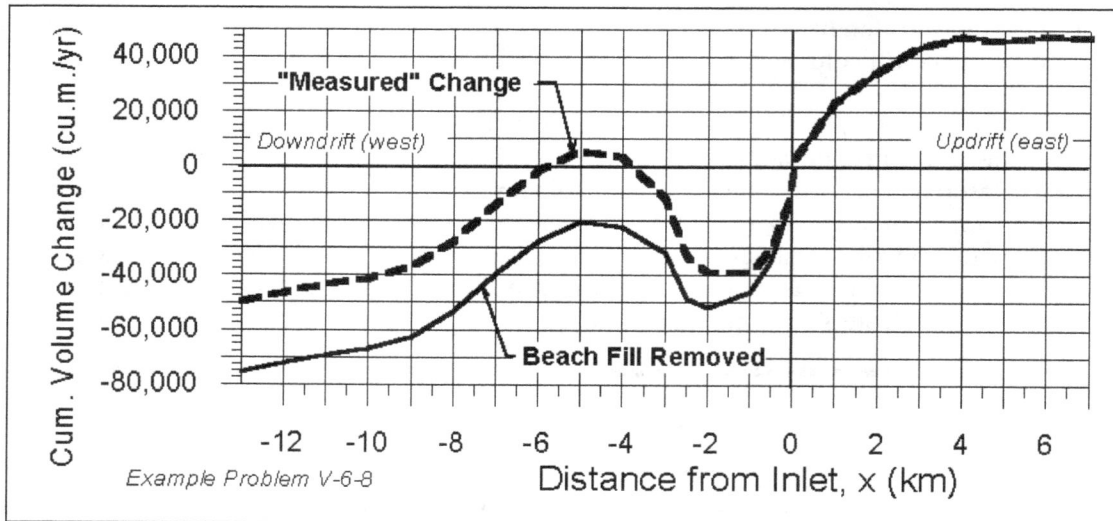

Figure V-6-18. Cumulative volume change with and without beach fill (Example Problem V-6-8)

In summary, the inlet's volumetric impact to the adjacent shorelines is estimated to range from 124,000 to 136,000 cu m/year. About 56 to 61 percent of this quantity (about 75,640 cu m/year) is manifest as erosion within 13 km downdrift (west) of the inlet. The total littoral impact, therefore, extends beyond 13 km from this inlet.

- In applying this method, it must be realized that both natural and stabilized inlets have a net volumetric impact. The tendency is to assume that the volumetric impact of a stabilized inlet is entirely due to its stabilization and engineering activities. However, the same inlet in its natural state would also have an impact. The equivalent value for a natural inlet can be estimated by examining volumetric changes at a natural inlet with similar longshore transport rates, tidal range, and wave climate. Then, the volumetric change due to stabilization and engineering activities of the study site can be estimated by subtracting the two volumes.

- In practice, the volume of sand impounded at the inlet's entrance (usually by jetties) can be estimated from profile surveys or shoreline-change data using the along-the-shore cumulative approach described above in method 1, criterion No. 4. (See Example Problem V-6-3.) The net volume of sand retained within the inlet shoals is estimated through comparison of bathymetric survey data; or at worse, through comparison of aerial photography with an assumed rate of vertical shoal thickness or accumulation. The volume of sand removed by dredging is estimated from records. Care should be taken to: Avoid double-counting of shoaling volumes and dredging volumes; exclude dredging quantities for new work associated with first-time construction or harbor deepening; exclude nonlittoral fractions (i.e., silt, clay, rock) from the dredging quantities.

- Account for dredging quantities placed upon the adjacent shorelines from inlet/harbor maintenance and new work, or offshore sources.

- In this fairly typical example, the downdrift shoreline changes, by themselves, do not reveal a clear picture of the inlet's signature. Methods 1 and 2 would therefore be of little value in this example, and methods 3 and 4 would not likely reveal the complete extent of the inlet's influence.

g. Inlet interactions with adjacent beaches.

(1) General.

Sediment may accumulate outside and inside an inlet. Sand storage exterior to the inlet is observed in the form of impoundment fillets along the shorelines adjacent to the inlet entrance. Also, a mature inlet with a well-developed ebb tidal shoal may bypass sand via the ebb shoal, resulting in an attachment zone, or bulge, on the downdrift beach. Interior sand storage occurs in the form of shoals, spits, and shoaling in channels. Mechanisms for this interior storage include (a) leakage of sand over, through, and around the inlet's terminal structures (jetties), and (b) wave- and current-driven transport to the inlet shoals. The effect of inlet-related bathymetry upon the wave field and sediment transport patterns can be observed through examination of wave refraction (discussed in the previous section), and by bathymetric and shoreline influences upon tidal currents.

Understanding the interaction of the inlet with its adjacent beaches is based upon:

(a) The degree to which the inlet's jetties impound, leak, or bypass sand.

(b) The direction, distribution and relative strength of the inlet's tidal currents.

(c) The effect of the inlet's ebb shoal plateau and adjacent bathymetry upon wave transformation (refraction) and littoral transport potential.

(d) The pattern in which shoals form within the inlet.

(e) The location and frequency of maintenance dredging of littoral sediments, and the placement history (location and frequency) and quality (e.g., sediment type and size) of this material.

Methods by which these factors may be examined are outlined in the following paragraphs.

- Site observations and data.

- There is no substitute for site visits and hours spent observing, walking, and wading along the inlet and its adjacent beaches. Observations should be preferably made during both strong ebb flow and flood flow, and at least once during energetic wave conditions from each of the principal directions of offshore incidence. Site visits should be initiated before commencing detailed studies to better identify those processes and data sources that bear examination. During or after the office analysis, additional site visits are necessary to assess the applicability of the analysis' results and to identify processes that require further study.

- Important sources of historical data include:

- Aerial photography.
- Bathymetric surveys and charts.

- Beach profiles and historical shoreline maps.
- Dredging records.
- Discussions with dredging contractors experienced with the inlet's shoaling patterns.
- History of man's improvements and alterations.
- History of storm events.
- Existing reports and anecdotal histories of the inlet and adjacent shores.

- If adequate data are not already available, researchers should consider the following new data:

- Controlled, rectified aerial photography.

- Shoreline location survey.

- Bank-to-bank bathymetric survey of the inlet throat, entire flood and ebb shoals, marginal shoals, and nearshore seabed.

- Beach profiles for characterizing the shape of the existing beach as a function of distance from the inlet, and direct comparison to previously collected beach profile data.

- Physical surveys of existing inlet structures (topography, elevations, permeability, etc.).

- Tidal current data focusing upon the horizontal distribution of principal flows and gyres on both the ebb flow and flood flow (dye injections into the water with concomitant aerial photography can be useful in this regard).

- Dye studies to discern jetty leakage (if not otherwise obvious).

- Directional wave data for a meaningful time period (includes hindcast data).

- Jet-probing, vibracore, core-boring, surface grab samples, periodic samples from the dredge plant, and other geotechnical means by which to discern those portions of the inlet's shoals and channels which contain littoral sediments derived from the adjacent shorelines.

- Environmental studies necessary to determine possible effects of disturbing portions of the inlet system for sand management (not discussed herein).

To the greatest extent practical, bathymetric surveys and beach profile surveys should be planned so as to facilitate comparison with existing, historical data. EM 1110-2-1810, "Coastal Geology," lists sources of geographic data in the United States. Other summaries of inlet-related data sources include Barwis (1975, 1976), Weishar and Fields (1985), and Chu, Lund, and Camfield (1987). The local sponsor (in the case of a Federal navigation project) or local interest(s) responsible for managing an inlet, and their private-sector contractors often hold the greatest repository of data specific to a given inlet, most of which is otherwise unavailable.

(2) Detection.

Approaches by which to discern how an inlet interacts with the adjacent beaches are described in the following paragraphs.

(a) Impoundment fillets and attachments. The degree to which sand is stored at an inlet's jetties can be detected by the shoreline signature (anomalous accretion) or the profile shape (anomalous steepness). Identification of these signals is described in Part V-6-2, methods 1, 2, and 3. Impoundment can occur on either or both sides of a stabilized inlet. Sand may also be stored in the attachment zone, the region in which the inlet bypasses sand via the ebb-tidal shoal. This can be observed through a bulge in the shoreline and an extension of the ebb shoal towards the downdrift beach.

(b) Jetty leakage. Three elements demonstrate a jetty's permeability: its elevation relative to the adjacent beach and seabed; its length relative to the adjacent beach planform; and its porosity.

- Elevation. The elevation of most old jetties is typically less than a meter above high water. Jetty crests below the elevation of the berm on the adjacent beach (usually more than 1 to 2 m above high water) allow wave runup to transport sand over the structure and into the inlet. This is typically shown by a high-water shoreline, berm shoreline, or dune line that curves landward immediately adjacent to the inlet (see Figure V-6-19). This local curvature is the result of a sluice that forms between the beach and the jetty. Because the jetty is lower than the beach, a downhill gradient exists from the beach toward the jetty. This promotes return flow of wave runup from the beach toward the jetty, which, in turn, accelerates sand transport from the beach into the inlet beyond that which would be normally transported by tidal currents and oblique incident waves.

- Length. In the limit, beach profiles for which the closure depth (the limiting depth of sediment transport) falls seaward of the jetty's end will experience transport from the beach, around the jetty, and into the inlet. Field observation suggests that this effect becomes particularly significant when the beach profile adjacent to the jetty has accreted to the point where the average-annual, low-tide breaking depth (i.e., $h = H_{avg}/\kappa$, low-water datum) is at or seaward of the end of the jetty. In such a condition, it is likely that most, if not all, of the littoral drift that reaches the jetty passes around it. In this instance, the jetty is considered to be saturated.

- At the same time, even in the saturated-jetty case, some fraction of the incident littoral drift may still be impounded further updrift of the jetty by the existing impoundment fillet. In this instance, the impoundment fillet continues to grow in a planform shape that may be similar to a backwater curve in classical hydraulics. This signature is sometimes evident in the shoreline's planform, and is predicted by analytic models such as Pelnard-Considère (1956) (see Part III- 2).

(c) Porosity. Jetty porosity is defined by sand transport through the structure. For bulkhead-type jetties, porosity results from cracks, corrosion, nonsealed joints, deterioration of backing (filter) cloth, or other sources of seepage. For rock or other armor-and-core structures, porosity results from insufficient chinking of stable, small rock within the voids between the large rocks or armor units. While this effect may be present within the structure's core, it is most commonly encountered near the top (waterline) of the structure. Here, there is generally minimal or no core, and small stones may have been washed out of the structure. Jetty porosity should be suspected for all rock structures that are low in elevation and/or narrow in profile (i.e., less than three or four armor units' width at the high waterline).

- Jetty porosity is also demonstrated by the transmission of waves or current, however small, through the structure. Where transmission is uncertain, dye can be injected on the upwave or upcurrent side of the structure and observed. Porosity may also be evidenced by the occurrence of shoals immediately adjacent to the downwave / downcurrent side of the structure. Similarity between the shoal sediment and that of the adjacent beach is additional evidence of porosity.

- In addition to the three factors previously listed, the appearance and growth rate of shoals and spits within the inlet is additional evidence of jetty permeability. This is discussed in item (e) in the following paragraphs.

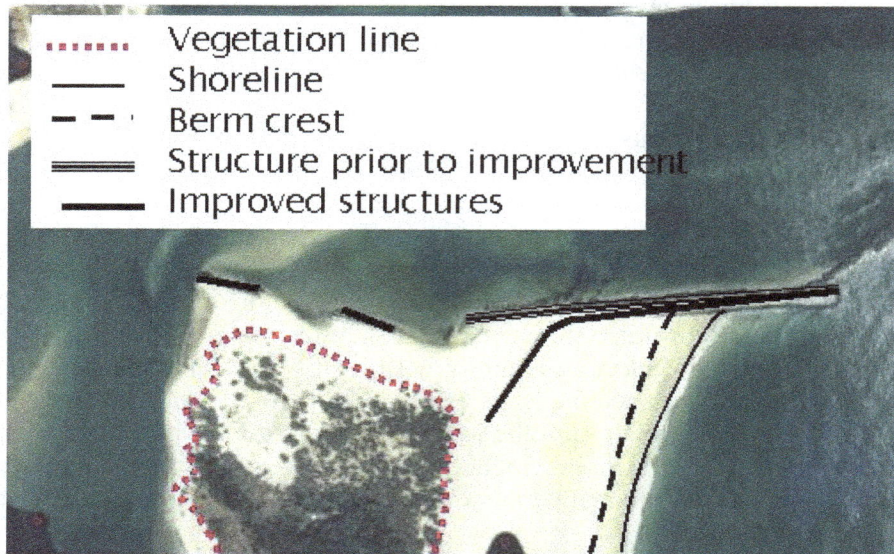

Figure V-6-19. Aerial photograph of Ocean City Inlet, MD, showing high-water shoreline, berm shoreline, and duneline

(d) *Jetty rips.* Offshore-directed rip currents can form along the upwave side of shore-perpendicular structures (such as jetties). These rip currents divert sand from the beach offshore, where the sand may or may not be transported to or along the inlet's ebb shoal. Rips are visually evident by a seaward-directed flow of water and the steepening of waves that encounter the rip. A modest sluice within 5 to 50 m adjacent to the structure may be visible at low tide. Lifeguards and surfers familiar with the area can be useful sources of data regarding local current and wave patterns.

(e) *Inlet bathymetry and shoaling patterns.* The location, rate of growth, and direction of movement of inlet shoals and spits, as well as the inlet's overall bathymetry, are excellent indicators of an inlet's sediment transport pathways. Use of bathymetry and shoaling patterns as a diagnostic tool requires, at a minimum, a bank-to-bank bathymetric survey of the inlet throat, flood shoal(s), and entire ebb shoal plateau. The latter should ideally extend outward and alongshore from the inlet entrance to capture the apparent point at which the ebb shoal attaches to the adjacent shorelines (if at all). Diagnostic ability is greatly improved when multiple, comparative surveys collected over time are available. Whenever possible, navigation channel condition surveys should be expanded to include bank-to-bank coverage of the inlet and as much of the inlet's shoals as possible. Pre-dredging, condition surveys are ideal because the shoals have had maximum opportunity to form prior to their removal by dredging. Reliance upon condition surveys that only include the navigation channel risks missing critical elements of the system's sediment transport pathways.

- Charts and surveys of the inlet should be contoured at frequent depth intervals in order to illuminate shoaling patterns. Even using traditional, coarse-resolution nautical charts (for which only limited contours are routinely plotted), this technique can offer an enlightening view of the inlet's transport patterns. In the United States, the original National Ocean Service "T-sheets," upon which the printed nautical charts are based, provide additional data. The location and shape of shoals can reveal the potential source of the shoals' sand and the apparent direction of transport. The location of deep areas (outside of dredged areas) indicates apparent, preferred orientation of flow; and, in turn, areas where current-induced erosional stress may be exercised upon the adjacent shore. Figure V-6-20 illustrates an example of an off-the-shelf nautical chart of an inlet, before and after the addition of higher-resolution depth contours.

- A particularly effective approach is to digitize, contour, and color-enhance the bathymetric data to more clearly reveal shoaling patterns. Pertinent upland features (jetties, shoreline, channel markers, etc.) should be added to the graphic to allow better visual correlation between the shoals, channels, and the inlet improvements. A prototype example is shown in Figure V-6-21, where lighter shading corresponds to shoals. Shading-enhanced depth contours readily show the leakage of sand through and around both jetties, the subsequent transport of the sand toward the flood shoal, and transport from the flood shoal toward the inlet entrance.

- The location, growth, and chronic reappearance of shoals can highlight the paths by which sand enters an inlet. Shoals located near a jetty's seaward end indicate a saturated jetty or jetty rip currents. Shoals near a jetty's landward end suggest a structure that is porous or too low. Shoals and spits that form by sand transport through and over the jetty's landward section are most generally displaced well landward of the adjacent shoreline. These shoals are subject to landward-directed transported by incident waves that sweep along the inlet's interior shorelines. An emergent, sandy shoreline along the inlet's interior, connecting the jetty to the interior shoals, is further evidence of transport through or over the landward section of the jetty, or around the end of a short, saturated jetty. The possibility should also be considered that interior shoals may have been created (or enhanced) by riverine sand that enters the inlet and settles in quiescent water.

- Figure V-6-22 illustrates an example in which the chronic reappearance of four distinct shoals along the inlet channel clearly reflects jetty leakage. The seaward two shoals correspond to known transport around the ends of the jetties. The landward two shoals correspond to transport through and over the jetties along the adjacent beaches' shorelines. Diving inspection, core boring, and dredging confirmed that the sediment from all four shoals is distinct from the silts and clays of the channel, but closely matches the adjacent beach sand.

- Comparisons of ebb and flood shoal surveys are necessary to estimate the net growth rate of these features. The degree to which these shoals contain nonlittoral material (e.g., silts and clays from rivers) should be considered so that the contributions from the riverine and coastal sources can be properly attributed. Identifying material type is typically based on examining samples from core-boring, dredged material, box-cores, or recovery methods. If only surface samples are available, care should be taken to consider whether silts and clays overlay sandy shoals that originated from the adjacent beaches.

- In some locations, sand from shoals (particularly interior shoals) has historically been dredged for upland use. This practice, infrequently recorded or quantified – may have removed significant quantities of sand from many inlet systems (sand which originated from adjacent beaches). Evidence of interior shoal dredging may be offered by the increase of adjacent landfills (readily noted in historical aerial photography) and anecdotal information.

(f) Dredging records. Dredging records may be available from the inlet's managing agency or local sponsor, USACE district offices, and dredging contractors. The date(s), locations of dredging and disposal, purpose (maintenance or new work), and quantity are of central importance. The quantity of dredged littoral (sand) material vs. nonlittoral (silt-clay) material must be identified to differentiate coastal versus upland or riverine origin. Shoaling records (condition surveys) combined with geotechnical data (USAEWES 1995) and/or discussions with dredging contractors and contract managers can be employed for this purpose. Additionally, the location of, and cross-sectional shape of, the shoaling pattern are strong indicators of the source. Shoals that form near shallow, sandy portions of the ambient seabed (such as the ebb shoal) are more likely to be of littoral origin. Shoals that form against one side of the channel, as opposed to those that form uniformly across the channel bottom, are also more likely to be of littoral (beach) origin.

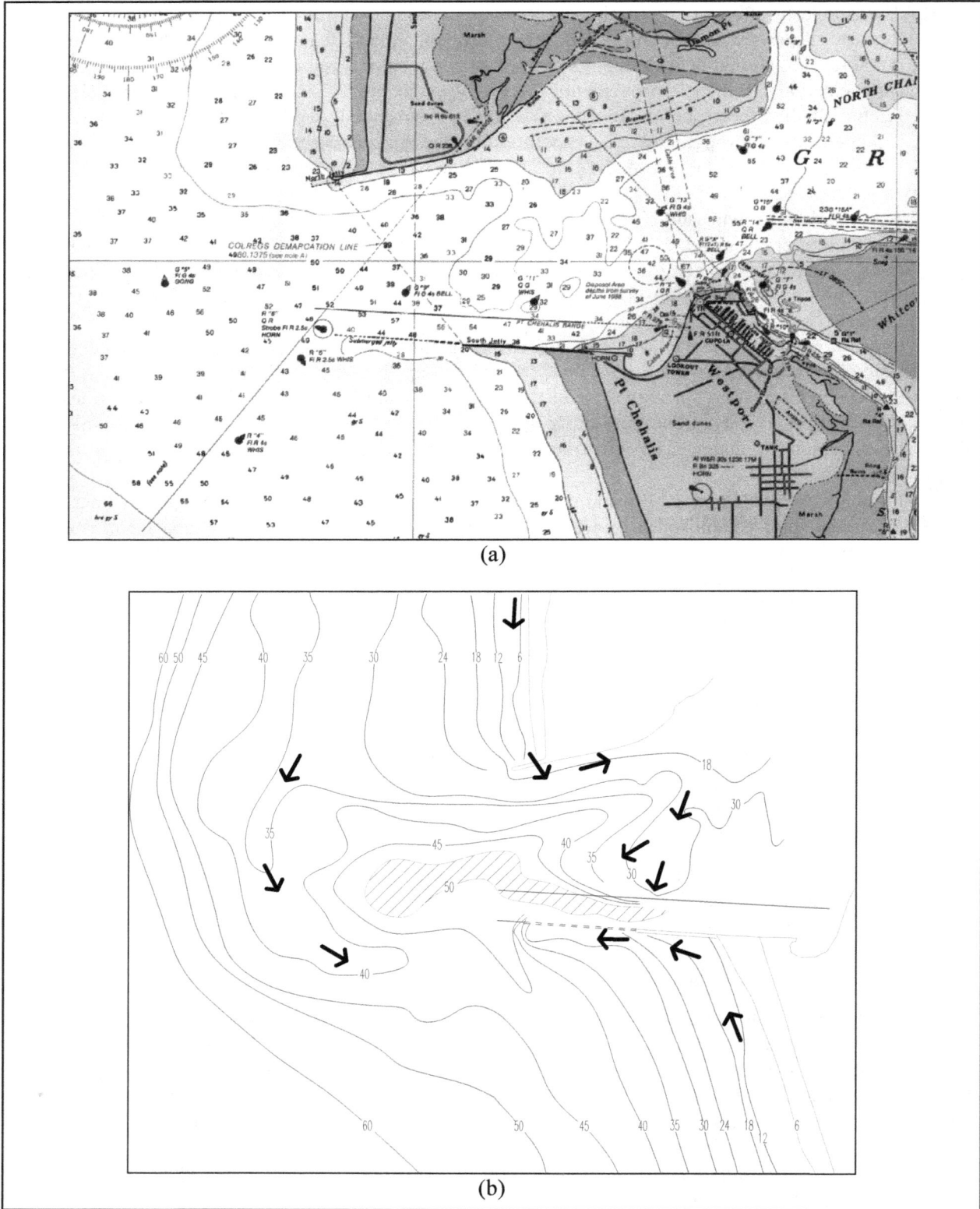

(a)

(b)

Figure V-6-20. Nautical chart with and without addition of higher-resolution contours. Arrows indicate sediment transport paths suggested by bathymetry

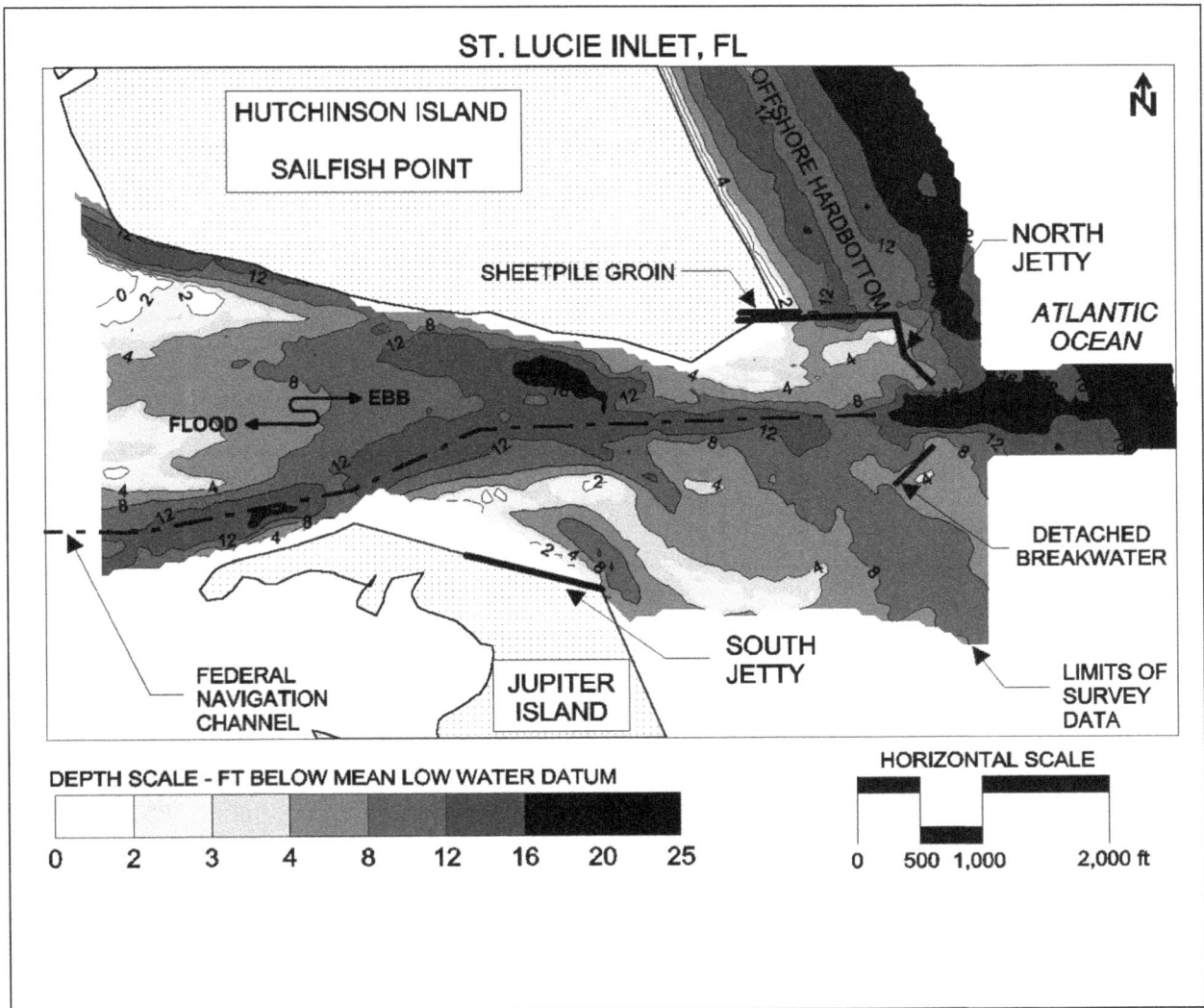

Figure V-6-21. Bathymetric chart of St. Lucie Inlet, Florida (c. 1995), digitized and gray-scale enhanced to better define the presence of shoals and channels

- Figure V-6-23 illustrates a classic example of such shoaling signatures. The bar cut of the Brunswick Harbor Federal navigation channel is maintained at a depth of about 10 m (mlw) through an ambient seabed that was historically only 1 to 5 m deep. Material dredged from along the north bank and channel center line in the vicinity of Section A is typically beach-quality sand originating from the area's net southerly littoral drift. Material from the vicinity of Section B is typically silt and clay that settles into the deep channel from the surrounding seabed (slumping), riverine transport (terrigenous source), and oceanic deposition (pelagic settling).

- We should note that not all of the littorally-supplied material at an inlet is necessarily sand. Some fraction of the dredged material may consist of silt, clay, gravel or organics of littoral origin. If this material is deposited offshore, it still represents removal of littoral material from the system, and must be included in computations of the inlet's littoral impact.

Sediment Management at Inlets

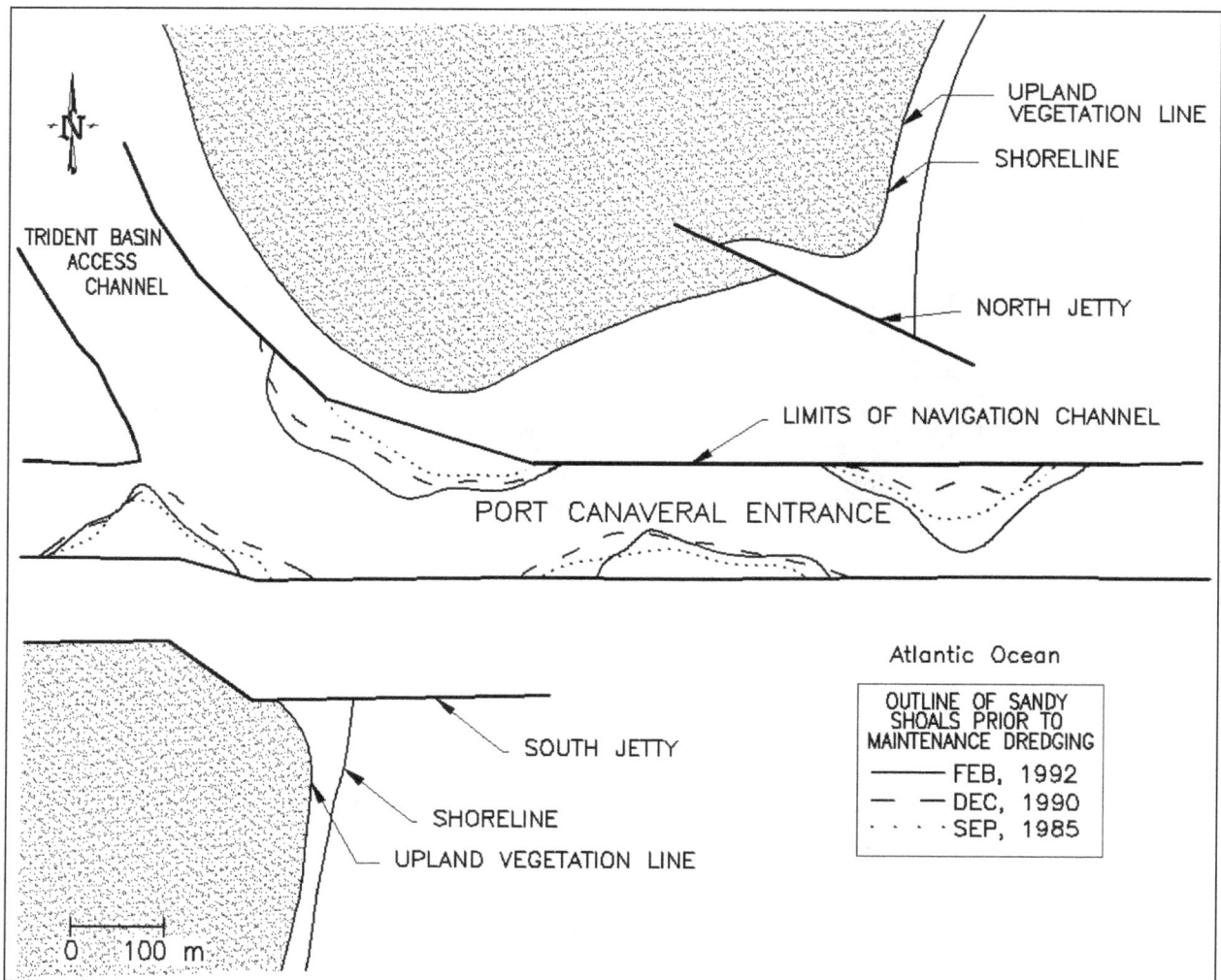

Figure V-6-22. Historical locations of sandy shoals within Port Canaveral Entrance, Florida prior to implementation of jetty/inlet improvements (Adapted from Bodge 1994a)

- Dredging for new work (increased channel or harbor dimensions) is typically not included in inlet impact analysis because the material that is dredged is considered to have been a static part of the littoral system. In reality, this exclusion underestimates inlet impacts because some portion of the requisite dredging usually involves removal of beach sand that sloughs into the new channel during initial construction. However, the portion of advance maintenance dredging which features littoral material is included in the inlet impact analysis.

- The degree to which increased channel dimensions or other inlet modifications alter dredging requirements can also reveal the degree to which the inlet diverts sand from the littoral system. This, again, requires discrimination of that part of the dredging quantity that was of littoral (beach sand) origin and undertaken for maintenance. Plotting the average-annual or cumulative quantity of maintenance-dredged littoral material as a function of time, and noting those times at which inlet modifications were undertaken, is a potentially useful technique to examine this effect (see Figure V-6-24).

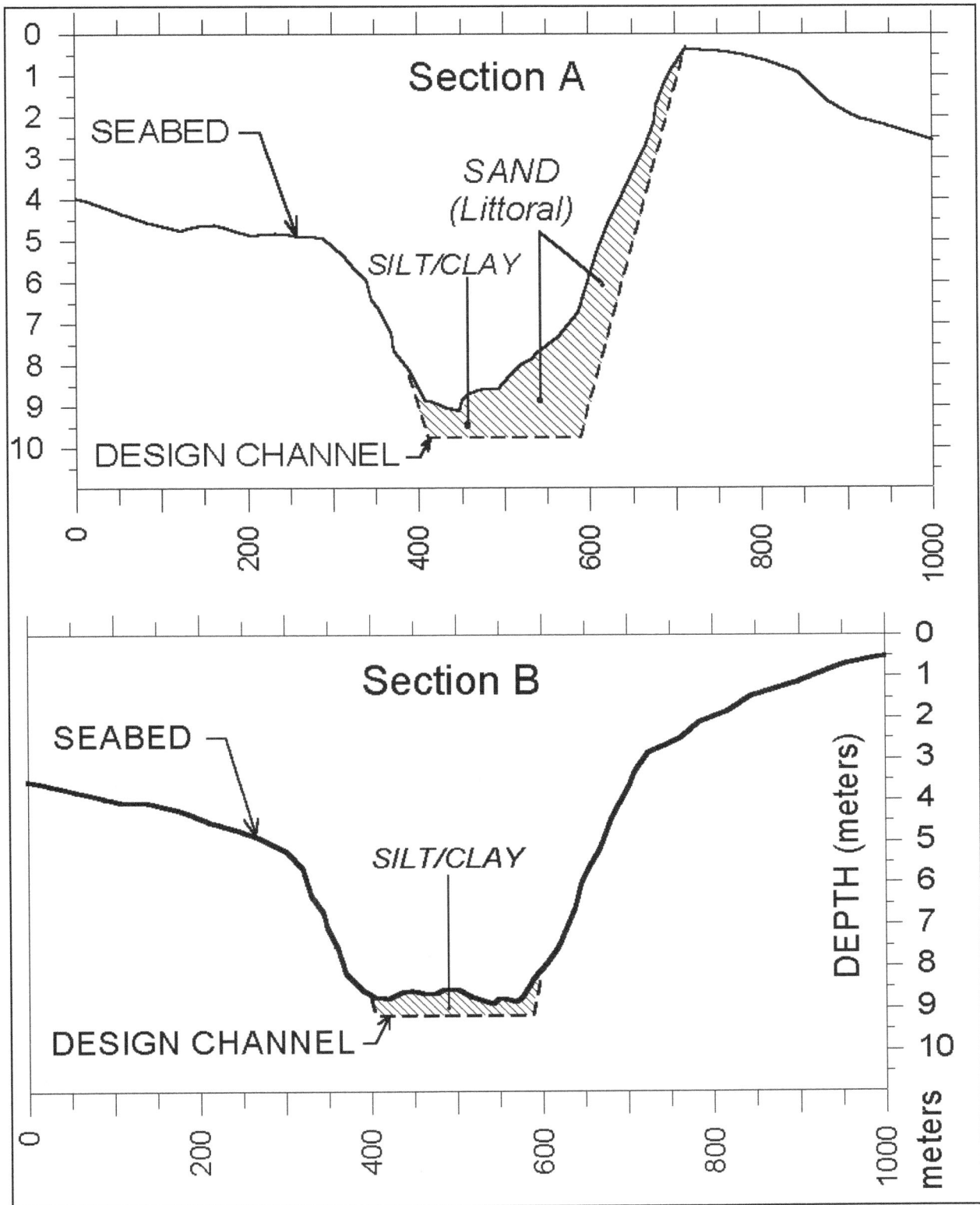

Figure V-6-23. Bathymetric contours and typical channel sections at Brunswick Harbor Federal Navigation Project, Glynn County, Georgia

Sediment Management at Inlets

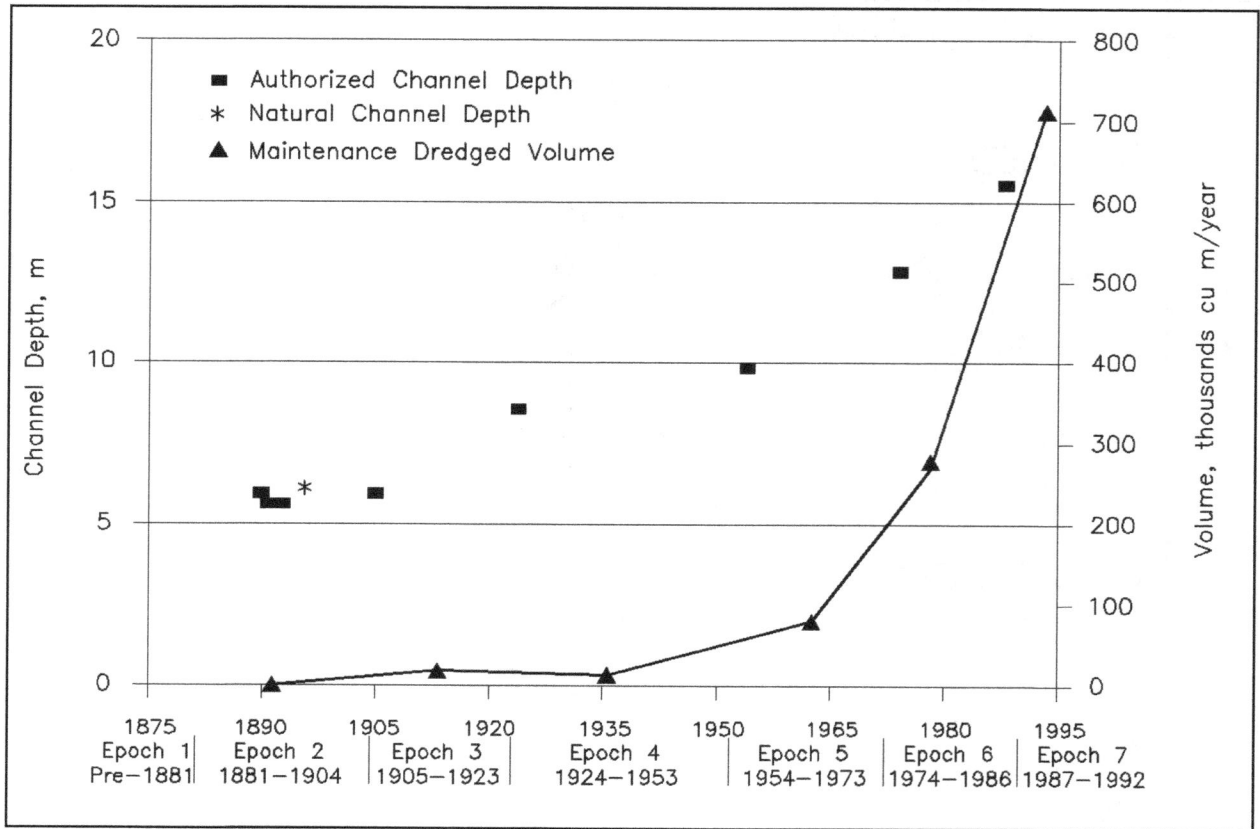

Figure V-6-24. Average annual volume of maintenance dredging at St. Mary's Entrance Channel and authorized channel depths; 1870 to 1992 (from USAEWES 1995)

h. Conclusion. This part of Chapter V-6 has discussed inlet processes and engineering activities at inlets. Inlets and associated engineering activities not only interact with local beaches and morphology, but also have the potential to alter sediment transport patterns and rates over the regional scale, especially over periods of many years. Methods discussed herein allow the coastal researcher to evaluate the impacts of these processes on the inlet morphology and adjacent beaches. These evaluations, together with inlet and adjacent beach sediment budgets (discussed in Chapter V-6-3), will improve the general understanding of the specific inlet's operation, and allow the engineer to optimize the timing and scope of engineering activities at the inlet and along the adjacent beaches.

V-6-3. Inlet and Adjacent Beach Sediment Budgets

a. Overview. This chapter discusses sediment budgets at inlets and their adjacent beaches. First, a history of sediment budgets is provided, followed by the theory of sediment budgets and required data sets to develop an inlet and adjacent beach sediment budget. Examples of previous sediment budgets are presented, and a series of example problems are given. The chapter concludes with a description of recent sediment budget methodologies.

b. Introduction.

(1) A sediment budget is an accounting or tabulation of the inflows and outflows of sediment together with the change in sediment volume within specified boundaries and for the time interval covered by the data. It is similar to a home budget that balances income (source of money), expenses (sinks of money), savings (positive amount), and debts (negative amount). Formulation of a realistic inlet and adjacent beach sediment budget is an essential prerequisite to the development of a successful inlet sand management strategy. Because an inlet's effect upon the adjacent shorelines is a function of the gross transport rates (not just the net rate), it is important to include both right- and left-directed transport in the sediment budget, as well as local, inlet-induced transport. In those cases where transport reversals occur at some time during the period of consideration (i.e., the majority of practical cases), inclusion of only the net transport rate will result in a potentially serious misinterpretation of the inlet's sediment transport pathways and associated volumetric rates of change.

(2) Development of the sediment budget requires that the rates and directions of sand transport adjacent to, within, and through the inlet be "deconvolved" from measured rates of volumetric change within the inlet system; specifically, the updrift and downdrift shorelines, the inlet shoals, and the channels. Allowances for the effects of dredging, beach fill, and other engineering of the inlet features must also be made. Development of the sediment budget is complicated by uncertainties in alongshore sand transport rates, typically incomplete or "noisy" volume-change data, and the fact that the number of transport rates for which values are unknown is greater than the number of equations which define the transport paths. The latter point means that the equations for an inlet sediment budget are almost always underconstrained. Accordingly, a unique solution is rarely possible. The goal, instead, is to develop a suite of possible solutions that is based upon reasonable assumptions of the inlet's processes. In this way, estimates of the practical ranges (or limiting values) of the rates at which sand is transported into or across the inlet from each shoreline can be made for various assumptions of the incident, alongshore transport rate or other physical parameters.

c. Theory.

(1) Since the 1950s, sediment budgets have been created to define the magnitudes and direction of sediment transport within a defined region and to understand, for example, inlet channel sedimentation and patterns of erosion and accretion along the coast. A balanced sediment budget yields an integrated picture of sediment (typically sand) motion, associated beach change, dredging and infilling of navigation channels at inlets, and other engineering activities within the reach covered by the analysis. Typically, the more reliable data form the foundation for the sediment budget, and lesser-known or more uncertain parameters are calculated to balance the budget by applying the principle of conservation of mass of sand (converted to volume or volumetric rate). A balanced sediment budget is a valuable tool for investigating observed coastal change and for forecasting the overall future state of the coast and consequences of management alternatives. Examination of an unbalanced sediment budget provides basic and useful information about the coastal system. An unbalanced budget may indicate a deficiency in the data set forming the budget (Dolan et al. 1987), reveal a misunderstanding in certain physical processes and assumptions (Inman 1991), or give bounds on the uncertainty range for the data sets. Whether or not formally developed, the sediment budget concept is fundamental to coastal engineering and science, usually providing the backdrop by which processes and projects are evaluated and alternatives considered. For additional background discussion, see Komar (1996, 1998).

(2) A sediment budget is a tallying of sediment gains and losses, or sources and sinks, within a specified control volume (or cell), or series of connecting cells, over a given time. There are numerous ways of formulating a sediment budget (e.g., Jarrett 1991; Bodge 1999). The difference between the sediment sources and the sinks in each cell, hence, for the entire sediment budget, must equal the rate of

change in sediment volume occurring within that region, accounting for pertinent engineering activities. The sediment budget equation can be expressed as,

$$\sum Q_{source} - \sum Q_{sink} - \Delta V + P - R = \text{Residual} \tag{V-6-23}$$

in which all terms are expressed consistently as a volume or as a volumetric change rate, Q_{source} and Q_{sink} are the sources and sinks to the control volume (expressed as positive values), respectively, ΔV is the net measured change in volume within the cell, P and R are the amounts of material placed in and removed from the cell, respectively, and Residual represents the degree to which the cell is balanced. For a balanced cell, the residual is zero. Note that the notation for Q_{source} and Q_{sink} due to Longshore Sand Transport (LST) may differ, depending on the application. In Equation V-6-23, Q_{source} and Q_{sink} are expressed as positive values (this differs from that specified in Longshore Sand Transport discussion of Part III-2-2).

(3) Figure V-6-25 schematically illustrates the parameters appearing in Equation V-6-23. For a reach of coast consisting of many contiguous cells, the budget for each cell must balance in achieving a balanced budget for the entire reach.

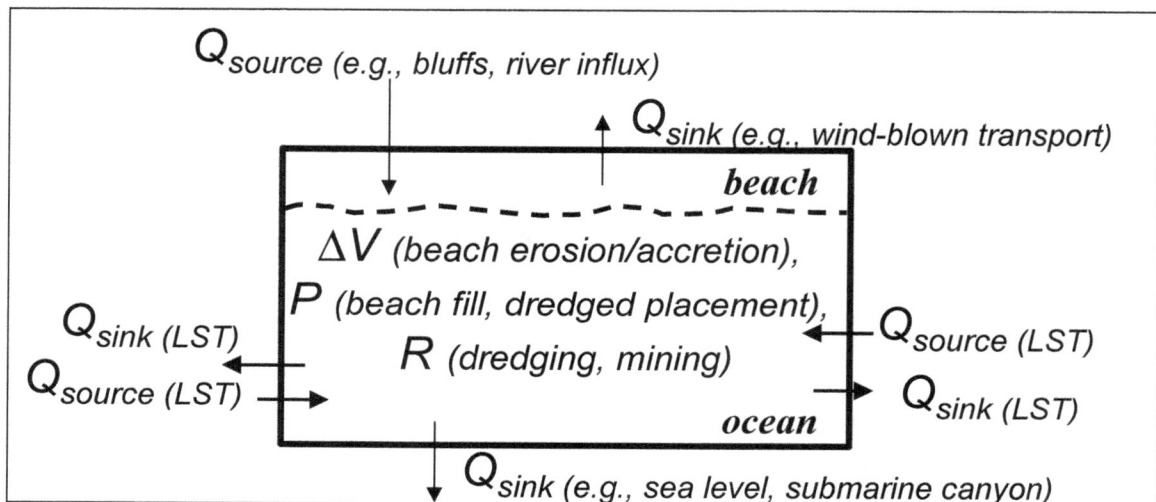

Figure V-6-25. Sediment budget parameters as may enter Equation V-6-23

(4) As noted in Figure V-6-25, sources to the sediment budget include longshore sediment transport, erosion of bluffs, transport of sediment to the coast by rivers, erosion of the beach, beach fill and dredged material placement, and a decrease in relative sea level. Examples of sediment budget sinks are longshore sediment transport, accretion of the beach, dredging and mining of the beach or nearshore, relative sea level rise, and losses to a submarine canyon. If inlets are located within the domain of a coastal sediment budget, they present significant challenges because inlets and the adjacent beaches are connected. Inlets increase the complexity of sediment budgets for several reasons. First, sediment-transport magnitudes and pathways are difficult to define at inlets even in a relative sense. Flood and ebb currents, combined waves and currents, wave refraction and diffraction over complex bathymetry, and engineering activities complicate transport rate directions and may increase or decrease their magnitudes. Because the pathways of sediment movement in the vicinity of an inlet can be circuitous, equations describing the sediment budget of regions directly adjacent to the inlet are not unique (that is, different formulations are possible).

(5) An inlet channel has the potential to capture the left- and right-directed components of the gross longshore transport of sediment, and the inlet system may bypass left- and right-directed longshore transport. Thus, knowledge of the net and gross transport rates, as well as the potential behavior of the inlet with respect to the transport pathways, may be required to correctly represent transport conditions within the vicinity of inlets, as emphasized by Bodge (1993).

d. Project applications.

(1) Sediment budgets can enter at any of four stages in project development:

(a) Existing condition. A sediment budget for the existing condition is the most common type considered. This budget forms the basis for evaluating the impacts of future engineering activities and the natural evolution of the inlet or coast.

(b) Historical (pre-engineering activity) condition. This budget is typically constructed for comparison with the existing-condition budget. A common application using the two budgets is to estimate shore erosion on adjacent beaches attributable to the impacts of inlet-related engineering activities that may warrant mitigation.

(c) Forecast future condition. Adapting and extrapolating the existing-condition sediment budget can assess the potential response to future projects or modifications.

(d) Intermediate condition. Sediment budgets representing other periods create a model of inlet or coastal evolution through time, which may lend insight to interpreting present or future evolution. As examples, intermediate-condition sediment budgets may document evolution of the inlet from initial formation to a quasi-equilibrium state, or may provide a picture of long-term natural bypassing through a cycle of channel migration and welding of a portion of the ebb-tidal shoal to the adjacent beach.

e. History of and procedure for sediment budget formulation.

(1) Consider a regional approach.

(a) For accurate representation of a local project area, especially in the vicinity of inlets, a sediment budget is formulated with its lateral boundaries located well beyond the apparent (expected) local project boundaries. A regional sediment budget might include multiple barrier islands, several inlets or several headlands and pocket beaches to fully capture the past and potential future changes in the sediment transport rate magnitude and direction.

(b) Inlets that have been stabilized by jetties for periods ranging from decades to centuries have the potential to influence the transport of sediment on the adjacent beaches for many kilometers, a distance that may extend well beyond what is considered the direct area of the inlet. Thus, a regional sediment budget that incorporates adjacent barrier islands, bay regions, underlying geology, estuarine and riverine impacts, and perhaps several inlets, may be required to assess the impacts of past and future projects. Engineering activities at navigable inlets and other data required for an inlet sediment budget have a high degree of uncertainty. Examples of these data include dredging quantity, location, and littoral quality; adjacent beach-fill volumes, initial cross-shore and longshore adjustment, and littoral quality; limited ebb-and (more commonly) flood-tidal shoal bathymetric coverage; and assessing the degree to which structures block, reduce, and modify the sediment transport pathways and magnitudes (Kraus and Rosati 1998a, 1999).

EXAMPLE PROBLEM V-6-9

Given:

Referring to figure V-6-26, balance the sediment budget equation given that $Q_{1\ source}$ = 250, $Q_{1\ sink}$ = 100, $Q_{2\ source}$ = 100, $Q_{2\ sink}$ = 200, $Q_{4\ source}$ = $Q_{4\ sink}$ = 0, ΔV = -70, P = 130, and R = 250 (all units are thousands of cu m/year).

Find:

Find the value of $Q_{3\ sink}$ that balances the sediment budget (see Equation V-6-23 and Figure V-6-26).

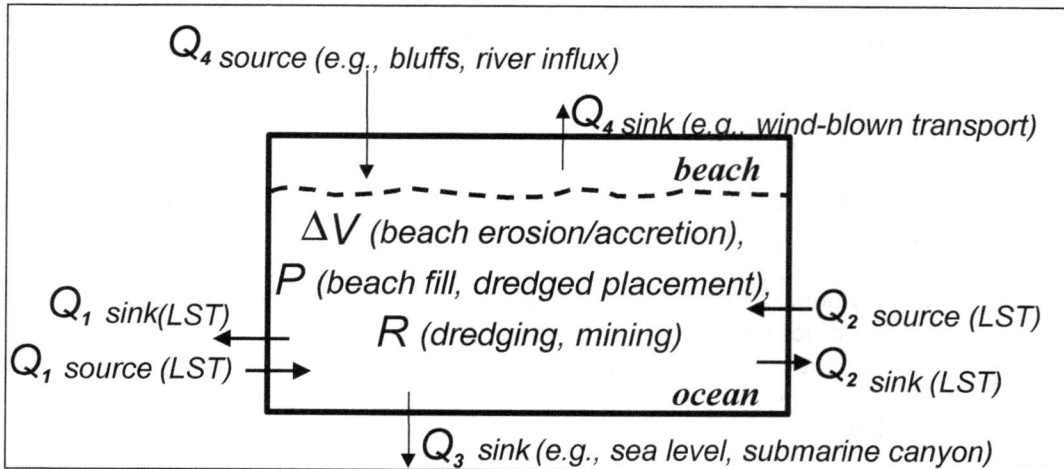

Figure V-6-26. Sediment budget for Example V-6-9

Substituting into Equation V-6-23 gives

$$\sum Q_{source} - \sum Q_{sink} - \Delta V + P - R = Residual$$
$$(250 + 100) - (100 + 200 + Q_{3\,sink}) - (-70) + (130) - 220 = 0$$
$$Q_{3\,sink} = 30$$

The value of $Q_{3\ sink}$ indicates that 30,000 cu m/year are transported out of the littoral cell, in the offshore direction.

(c) In one of the earliest works that may be considered a regional sediment budget, Caldwell (1966) summarizes a study performed in the 1950s by the U.S. Army Engineer District, New York, for the north New Jersey coast (USACE, 1957, 1958). The budget, formulated by examining changes in shoreline position, served as a field laboratory of shore processes with the objective of examining alternatives to mitigate for erosion. This celebrated study deduced a regional divergent nodal point in net longshore transport direction at Mantoloking, located just north of Dover Township. Net longshore transport to the north increased with distance north from Mantoloking because of the sheltering by Long

Island, New York. The budget considered net and gross longshore sand transport for this 190-km reach including 10 inlets over time intervals of 50 to 115 years.

(d) Another example of a regional sediment budget is that of Jarrett (1977, 1991) for the North Carolina shoreline, including three barrier islands and two inlets. Mann (1993) discussed use of a near field versus regional sediment budget. The near field sediment budget represents local (project area) sediment sources, sinks, and pathways. The regional sediment budget combines the near field budget with the sediment-transport processes occurring on the adjacent shorelines. For development of inlet sand management strategies (and estimating the inlet's littoral impacts), Mann recommends consideration of a regional sediment budget so that interactions of the inlet (and any proposed modifications) on the adjacent shorelines can be assessed. Although it may be difficult to define and balance all sources, sinks, and sediment transport pathways within a regional context, this comprehensive approach may allow the practitioner to recognize a source or sink of sediment hundreds of kilometers away that has a potential significance to the project area (Komar 1998).

(2) Develop a conceptual budget. Kana and Stevens (1992) introduced a conceptual sediment budget, which they recommend developing in the planning stage prior to making detailed calculations of individual sources and sinks. The conceptual sediment budget is a qualitative model giving a regional perspective of the inlet interaction with beach processes, containing the effects of offshore bathymetry (particularly shoals and, therefore, wave-driven sources and sinks), and incorporating natural morphologic indicators of net (and gross) sand transport. The conceptual model may be put together in part by adopting sediment budgets developed for other sites in similar settings, and incorporates all sediment sinks, sources, and pathways. The conceptual model should be developed initially, perhaps based upon a reconnaissance study at the site as part of the initial data set. Once the conceptual sediment budget has been completed, data are assimilated to validate the model rather than to develop the model.

(3) Ensure compatibility of temporal and spatial scales. In a discussion of the planning process for coastal projects, Kraus (1989) advocated the concept that the temporal and spatial scales of data used to develop and drive a model (whether a numerical, analytical, physical, or conceptual model) must be commensurate with these scales of the model itself. For example, a sediment budget developed based on pre- and post-storm data representing a day-to-month-length temporal scale within the immediate vicinity of the inlet should not be extrapolated to forecast to temporal scales of years and decades for a region extending over several barrier islands. Similarly, a sediment budget developed based on a 50-year period cannot adequately bracket the seasonal fluctuation observed locally at the project site. Sediment budgets are commonly required to represent periods of engineering and geomorphic significance, from 3-5 years (dredging cycle at inlets) to 30-50 years (project life span; time scale for cyclic ebb shoal welding). Data sets reflecting the longer durations are required to develop the sediment budget for spatial scales reflecting a regional approach. However, seasonal and year-to-year variability should be considered and can contribute to the uncertainty in a sediment budget, or form the basis for a sensitivity analysis.

(4) Delineate littoral cells. A littoral cell (or control volume) defines the boundaries for each sediment budget calculation, and denotes the existence of a complete self-contained sediment budget within its boundaries (Dolan et al. 1987). Bowen and Inman (1966) introduced the concept of littoral cells (Inman and Frautschy 1966) within a sediment budget. The southern California coast lends itself to this concept, with evident sources (river influx, sea cliff erosion), sinks (submarine canyons), and coastal geology (rocky headlands) defining semicontained littoral cells and subcells (Komar 1996, 1998). A littoral cell can also be defined to represent a region bounded by assumed or better-known transport conditions, or natural and engineered features such as the average location of a nodal region (zone in which $Q_{net} \sim 0$) in net longshore transport direction or a long jetty.

(5) Consider net and gross transport rates. For cells of the regional budget that may capture a portion of the left- or right-directed transport, both of these components must enter in the formulation. Examples include submarine canyons and inlet channels that capture both left- and right-directed transport; inlet weirs which may trap a portion of the left- or right-transport rate; and initial beach response at a long groin or headland feature, which may indicate accretion associated with left- and right-directed transport. Caldwell (1966) considered the gross transport rate as a potential indicator of shoaling for inlets in the vicinity of Cape May, New Jersey. Jarrett (1977, 1991) balanced potential longshore energy flux calculations with measured beach and tidal inlet change to solve for net and gross rates of longshore sand transport. Bodge (1993) focussed on the inlet and its adjacent beaches, and emphasized the importance of considering the gross components of longshore sediment transport, especially for inlets that act as sediment sinks. The gross transport rate can also provide an upper limit for the net, left-, and right-directed rates.

(6) Assign values and uncertainties.

(a) Known, estimated, or easily-obtained values and their associated uncertainties are assigned to source, sinks, and engineering activities within the sediment budget. This step should represent a low level of effort to quickly assess the integrity of the macrobudget (discussed in the following paragraphs) and to flush out any problems before detailed analysis begins.

(b) Every measurement has limitations in accuracy and contains a certain error. For coastal and inlet processes, typically direct measurement of many quantities cannot be made, such as the long-term longshore sand transport rate or the amount of material bypassing a jetty. Values of such quantities are obtained with predictive formulas or through estimates based on experience and judgement, which integrate over the system. Therefore, measured or estimated values entering a sediment budget can be considered as consisting of two terms, expressed schematically as

$$\textit{Reported Value} = \textit{Best Estimate} \pm \textit{Uncertainty} \tag{V-6-24}$$

Uncertainty, in turn, consists of error and true uncertainty. A general source of error is limitation in the measurement process or instrument. True uncertainty is the error contributed by unknowns that may not be directly related to the measurement process. Significant contributors to true uncertainty enter through natural variability and unknowns in the measurement process. These include temporal variability (daily, seasonal, and annual beach change); spatial variability (longshore and across shore); definitions (e.g., shoreline orientation, direction of random seas); and inability to quantify a process, such as the volume of material pumped to a beach, or the sediment pathways at an inlet. Other unknowns can enter, such as grain size and porosity of the sediment (especially true in placement of dredged material).

(c) As an example, suppose the variable X is a sum or difference of several independent parameters as $X = x + y - z + ...$, then the two common estimates for the uncertainty are:

$$\delta X_{max} = \delta x + \delta y + \delta z + ... \tag{V-6-25a}$$

and

$$\delta X_{best} = \sqrt{(\delta x)^2 + (\delta y)^2 + (\delta z)^2 + ...} \tag{V-6-25b}$$

In both expressions, the errors add, whether or not the variables enter the quantity being reported as a sum or difference. Equation V-6-25a represents an extreme bound for the plausible error. Equation V-6-25b is the root-mean-square (rms) error, and the validity of this expression rests on the assumptions that the individual uncertainties are independent and random. The rms error accounts for the uncertainty in uncertainty by giving a value that is not an extreme, such as δX_{max}.

(d) Suppose the quantity entering the budget is expressed as a product or quotient of independent variables as $X = xyz$ or as xy/z. In either case, the uncertainty in X is:

$$\left(\frac{\delta X}{X}\right)_{max} = \frac{\delta x}{x} + \frac{\delta y}{y} + \frac{\delta z}{z} \tag{V-6-26a}$$

and

$$\left(\frac{\delta X}{X}\right)_{best} = \sqrt{\left(\frac{\delta x}{x}\right)^2 + \left(\frac{\delta y}{y}\right)^2 + \left(\frac{\delta z}{z}\right)^2} \tag{V-6-26b}$$

and it is seen that the errors are additive whether a variable enters as a product or quotient. These equations state that the relative uncertainty of a product or quotient is equal to the sum of relative uncertainties of each term forming the product or quotient. Kraus and Rosati (1998a, 1999) further discuss uncertainty in coastal-sediment budgets.

(7) Formulate a macrobudget. A macro-budget is a quantitative balance of sediment inflows, outflows, volume changes, and engineering activities within the regional conceptual budget. Essentially, the macrobudget solves the budget with one large cell (perhaps by temporarily combining many interior cells) that encompass the entire longshore and cross-shore extents of interest. Balancing the macro-budget reduces the possibility of inadvertently including potential inconsistencies in a detailed or full budget (Kraus and Rosati 1999).

(8) Refine estimated values and uncertainties. Once the macrobudget has been balanced, detailed analysis for all inflows, outflows, volume changes, and engineering activities pertaining to each individual cell may commence. Originally estimated values used in formulating the macrobudget are used to provide reasonable ranges for each quantity.

(9) Use residuals to balance individual cells. As presented in Equation V-6-23, both a balanced sediment budget cell and the macrobudget have the sum of all sources, sinks, and engineering activities equal to zero. Inman (1991) discussed recording an unbalanced sediment budget cell (a cell with a non-zero residual in Equation V-6-23) as a region requiring more definition and investigation of the unknown processes. Knowledge of the residual may also be useful to bracket the uncertainty range for the data sets (Kraus and Rosati 1999).

(10) Conduct sensitivity testing. Once a sediment budget has been formulated, it can be copied and modified to evaluate the impact of any assumptions on the final sediment budget. Different data sets for the same project site can be applied to evaluate seasonal variations in beach change and transport rate direction and magnitude. A balanced budget representing a historical time period can be copied and altered to represent a potential future with-project condition.

f. Data required. The types of data sets that are available for refining sediment budget quantities are discussed in the following paragraphs. This section supplements the data required for inlet analysis described in Section V-6-2(f).

(1) Aerial photography. Interpretation of aerial photographs offers the best means of obtaining broad qualitative understanding of the site. As examples, photographs of sites with relatively clear water can identify the planform shape of the flood-tidal shoal to estimate its volume if more quantitative data are unavailable. The pattern of wave breaking over the ebb-tidal shoal indicates the planform shape of this feature, which might lend qualitative understanding of its interaction with adjacent beaches and of sediment pathways. Overwash fans on adjacent barrier islands indicate pathways for loss of sediment to the coastal littoral system, from which quantification of volumes might proceed. Shoals adjacent to jetties might indicate sediment-transport over and through the structure as a potential sediment transport pathway. In a more quantitative analysis, controlled and rectified aerial photographs are commonly interpreted to identify the berm or high-water line (HWL) shoreline position (Kraus and Rosati 1998b).

(2) Beach-profile surveys. Volume change, ΔV, in the beach can be obtained accurately through repetitive surveys of the beach profile. The volume change for a given profile is typically assumed to represent the region of beach of length Δx between adjacent profile lines. Both the elevations B of the berm and of the profile closure depth D_C can be estimated from beach-profile surveys if the profile data are sufficiently accurate and well controlled. The active berm crest is a discernible morphologic feature on the profile representing the upward limit reached by the water under normal tide and water-level conditions. The profile may have two berm crests if the beach has recently accreted, and the elevation of the seaward-most feature should be noted. The depth of closure is located where no significant depth changes occur over times of engineering significance (typically, 10 to 50 years) (see Hallermeier 1978; Birkemeier 1985; Wise 1998). Kraus, Larson, and Wise (1999) discuss the depth of closure in detail and extend its definition to cover varied conditions as encountered in engineering practice. Profile surveys performed near structures may indicate their condition. For example, a jetty that allows sediment transport over and through it might be indicated by a berm-crest elevation adjacent to the structure that is comparatively lower than the berm crest further away from the structure. Similarly, surveys close to structures reveal whether the profile deviates from the average shape far from the structures, improving estimates of sand volume. The investigator should be cautious in interpreting beach-profile data near the inlet because of migrating shoal features that may affect the profile shape.

(3) Shoreline-position data. Shoreline-position data may be obtained from analysis of topographic and HWL surveys, aerial photographs, beach-profile surveys, and bathymetric data (Anders and Byrnes 1991; Byrnes and Hiland 1995). In a qualitative manner, beach morphology indicated by shoreline position may imply sediment-transport pathways or controls. As examples, a salient or bulge-type feature in the shoreline downdrift of an inlet may represent the location for ebb shoal bypassing to the adjacent beach. Rocky headlands and outcroppings indicate geologic controls on sediment-transport pathways.

(a) As quantified in a sediment budget, the change in shoreline position Δy averaged over a given longshore distance Δx can be converted to a volume change by assuming that the shoreline translates parallel to itself over an active depth D_A, given by

$$D_A = B + D_c \tag{V-6-27}$$

in which B is the elevation of the seaward-most active berm relative to a datum, and D_C is the depth of closure measured from the same datum. The volume change[1] ΔV over a time interval Δt is given by

$$\Delta V = \frac{\Delta y \Delta x D_A}{\Delta t} \qquad (V\text{-}6\text{-}28)$$

(b) If available, the impoundment rate at a shore-perpendicular structure such as a groin or jetty that is sand tight gives an estimate of the longshore sediment-transport rate.

(4) Bathymetry. Historical and recent bathymetric data sets are a valuable resource for determining the rate of volume change in the inlet channel and on the ebb and flood-tidal shoals. If coverage is sufficient, differences in bathymetric surfaces give the subaqueous volume change on the adjacent beaches and channel and ebb- and flood-tidal shoals. It is noted that, in the past, typical bathymetric coverage has been limited to the inlet channel. However, the benefits of increasing the survey area to include the ebb- and flood-tidal shoals far outweigh the additional costs, particularly in view of reductions in the cost of bathymetric surveys (e.g., SHOALS bathymetric survey system, Parson and Lillycrop 1998). Bathymetric data can also indicate sediment-transport pathways. As examples, a finger shoal extending from the tip of a jetty likely indicates a dominant sediment transport pathway, and the morphologic form of an ebb-tidal shoal that connects to the adjacent beaches may indicate inlet bypassing. Aerial photography of flood-tidal shoals at different, but known tidal elevations can be referenced to create a contour map of the shoals, and thereby to estimate a shoal volume.

(5) Engineering history.

(a) Engineering activities of significance to a sediment budget fall into two categories – those that are of a descriptive nature and must be quantified within the sediment budget; and those that are a priori quantified. Rehabilitation of a jetty is an example of a descriptive activity that requires quantification. The morphology of the inlet and adjacent beach before and after structure rehabilitation, as well as the type of rehabilitation (e.g., raising the jetty crest elevation, inserting a sand-tight core, adding armor stone), and other pertinent data sets indicate the effectiveness of the structure. Consideration should be given as to the degree of sediment transport through, over, and around the structure before and after rehabilitation. Another example of descriptive data is the grain size of dredged material placed on the adjacent beaches. From this information, the engineer must estimate percentage of material that would remain in the active littoral zone.

(b) Engineering activities that are a priori quantified (although sometimes only partially) include the following: volumes, locations, and times of dredged and placed material; volume of material mined from ebb- and flood-tidal shoals, the locations and times of mining; configuration of the placement; volume of fill on adjacent beaches and its placement location and time period of placement; and records of mechanical bypassing (volume, placement location, and time periods). These quantities will enter the sediment budget calculations by adjusting measured volume changes to account for either the removal or placement of material through engineering activities. The adjustment of an initial beach fill can be used to infer rates of longshore and cross-shore sediment transport.

---

[1] Comparison of a shoreline position derived from aerial photography with a shoreline position derived from beach-profile surveys should account for possible differences in the vertical datums to which each is referenced. For example, it is likely that an aerially-derived shoreline position represents a berm crest or HWL position, whereas a beach-profile shoreline may represent a zero elevation relative to a standard datum (e.g., National Geodetic Vertical Datum, or Mean Sea Level). See Kraus and Rosati 1998b or CEM, Part II, Chapter 5.

(6) *Coastal processes.* Data on the acting coastal processes are a resource for understanding and quantifying inlet- and sediment-transport pathways and quantitites. Examples are discussed here.

(a) Net, left-, and right-directed potential longshore sand-transport rates can be calculated from wave gauge, Wave Information Study (WIS), and Littoral Environment Observation (LEO, see Schneider (1980)) wave height, period, and direction data. Gravens (1989) discusses the methodology for calculating net potential longshore sand-transport rates from WIS data. The components of the net transport, directed to the left or right as noted by a shore-based observer, can be calculated by using the left- or right-directed waves, respectively, with the methodology as outlined in Gravens (1989). Often the magnitudes of the calculated net, left-, and right-directed potential longshore sand transport rates do not agree with accepted values for the site. However, the relative magnitude between the left- and right-directed transport can be applied in a sediment budget with an accepted value for net longshore sediment transport to adjust the magnitudes. Wave height, period, and direction data allow construction of wave rays or orthogonals as indicators of pathways of sediment transport.

(b) Inlet flow speed and direction data as indicated by current meters or drogue movement give the relative magnitude of sand-transport rates and pathways. For example, measurements of the current from Ocean City Inlet, Maryland, indicated that the flow to the northern part of the bay was considerably greater than that to the southern part (Dean, Perlin, and Dally 1978). This information can be adapted to proportion the relative magnitude of the bay-directed sand transport to different parts of the bay.

(c) The rate of relative sea-level rise may represent a contributing factor to the observed beach change. The long-term beach loss Δy_{sl} because of an increase S in relative sea level is (Bruun 1962; 1988; Komar 1998).

$$\Delta y_{sl} = \frac{L_c}{B + D_c} S \qquad\qquad (V\text{-}6\text{-}29)$$

for which L_c is the cross-shore distance from datum to the long-term depth of closure D_c .

(d) Other types of coastal process data useful for formulation of a sediment budget include:

- River-flow speed, fluvial sediment grain size and sediment availability as a possible sediment source to the coastal environment.

- Wind speed and direction, sediment grain size and availability as a potential aeolian sediment source to or a sink from the coastal environment.

- Sediment characteristics (e.g., median size, size distribution, mineral content) as natural tracers for sediment movement.

g. Sediment budget methods and tools. This section discusses sediment budget methods and tools available for inlets and regional studies. The discussion is organized with the simpler methods presented first, followed by the more complex.

EXAMPLE PROBLEM V-6-10

GIVEN:

Develop a conceptual sediment budget for the period 1938 to 1979 for the regional littoral system in the vicinity of Shinnecock Inlet, Long Island, New York. This region extends east of Shinnecock Inlet to Montauk Point and west of Shinnecock Inlet to Moriches Inlet.

BACKGROUND INFORMATION:

Shinnecock Inlet, located on the south shore of Long Island, New York, was formed during a hurricane in September 1938 (Figure V-6-27). The west jetty was initially constructed by New York State in 1947 and was extended from 1953 to 1955, and the east jetty was constructed from 1952 to 1953. Shinnecock Inlet's littoral system is bounded to the east by Montauk Point, a location at which net longshore sediment transport is negligible because of its shoreline orientation and fetch distance from the mid-Atlantic Coast. West of Montauk Point, 10- to 21-m-high bluffs extend for 8 km and are a source of sediment roughly estimated as 35,000 cu m/year based on analysis of profile data. The U.S. Army Engineer District, New York, formulated a sediment budget for the inlet (USAED, New York, 1987). Estimates are available for the net longshore sand-transport rate 1 km east of the inlet (230,000 cu m/year), the ebb-shoal volume change (77,000 cu m/year), the flood-shoal volume change (15,000 cu m/year), and the net longshore sand-transport 1.8 km west of the inlet (189,000 cu m/year).

Nersesian and Bocamazo (1992) developed a preliminary sediment budget in which the net transport east of Shinnecock was 281,000 cu m/year, the ebb and flood shoal captured 77,000 and 15,000 cu m/year, respectively; and transport west of the inlet was 189,000 cu m/year. Kana (1995) estimated net transport rates 3 km east and 2 km west of Shinnecock Inlet as 219,000 and 104,000 cu m/year, respectively. A seaward bulge located approximately 2 km downdrift of Shinnecock Inlet is apparent in the 1979 shoreline position, indicating a possible region of sediment exchange between the ebb-tidal shoal and the downdrift beach. West of Shinnecock Inlet, the Westhampton barrier island extends for 25 km to Moriches Inlet. Moriches Inlet was formed in March 1931 and migrated 1,200 m to the west before it closed naturally in May 1951. Jetties were constructed in 1952 to 1953 at the position of the former inlet, and through dredging and a minor storm, the inlet reopened. Taney (1961a,b) estimated net transport rates at Moriches Inlet as 229,000 cu m/year.

(Continued)

---

(1) SBAS2000.

(a) Overview. The Sediment Budget Analysis System (SBAS2000) is a PC-based program that utilizes a graphic-based interface to apply Equation V-6-23. The SBAS was developed to allow the engineer and scientist to formulate sediment budgets in areas with complex sediment pathways, such as at inlet entrances, and over a wide regional extent that might encompass several inlets with beaches and infrastructure in between. SBAS can zoom out to reveal a large regional extent, and combine (collapse) sediment budget cells for a macroscale interpretation of sand transport along the coastlines and through rivers. For project-level sediment budgets, SBAS can zoom in to the site-specific area and

EXAMPLE PROBLEM V-6-10 (Concluded)

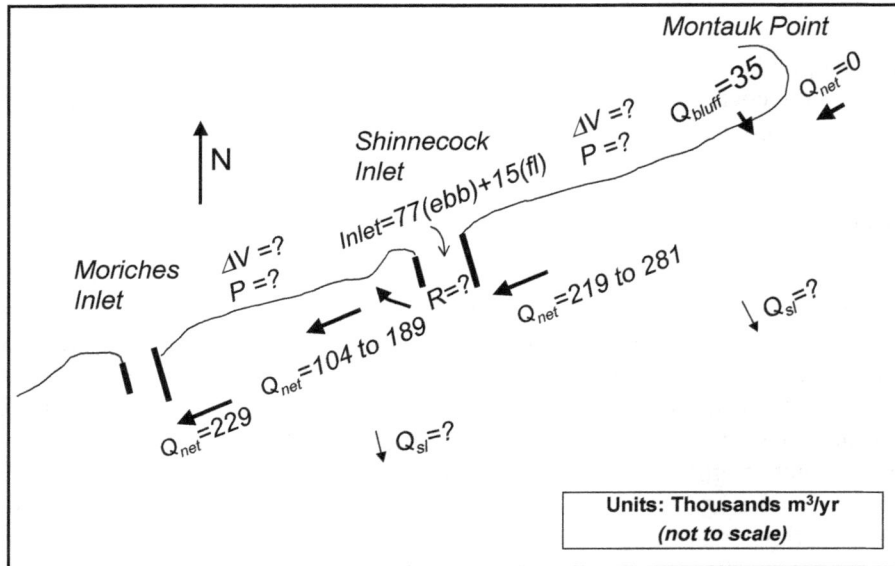

Figure V-6-27. Conceptual sediment budget for Shinnecock Inlet, New York

CONCEPTUAL SEDIMENT BUDGET:

Figure V-6-27 shows the conceptual sediment budget developed from the information presented. Applying this information with Equation V-6-23 indicates that the beaches between Shinnecock Inlet and Montauk Point and between Moriches and Shinnecock Inlets most likely have eroded during the subject period, unless a significant quantity of beach fill was placed. Some volumes are not quantified (e.g., beach losses because of relative sea level rise Q_{sl} ; beach-fill placement rate P; dredging (removal) rate R), but are represented for completeness.

reinstate (explode) sediment budget cells. Through SBAS, numerous sediment budgets can be established, copied, and modified for different project or forcing conditions with internal checks provided by the system. Although SBAS was developed to support regional coastal sediment management, its algorithmic structure and commercial-grade PC interface are independent of application. For example, SBAS can be applied to inland navigation projects or as a general ledger to track funds or products in multiple interactive accounts. Data entry can be accomplished visually or through a spreadsheet, and results can be displayed graphically in a number of ways or as lists, depending on the background and needs of the viewer (management review; engineering detail; overviews to sponsors). Because of the intuitive interface, SBAS can be operated with just a few minutes of training. The system also includes ways to estimate uncertainty, to consider "what if" questions easily, to cut-and-paste graphics, and to obtain hard copy reports.

EXAMPLE PROBLEM V-6-11

GIVEN:

Refine the conceptual budget for Shinnecock Inlet and the beaches ± 3.2 km east and west of the inlet. For this example, uncertainty in the sediment budget will be omitted (refer to Equation V-6-25 and V-6-26 for estimating uncertainty within a sediment budget).

BACKGROUND INFORMATION:

The refined conceptual budget is shown in Figure V-6-28, and details of its formulation are presented here. At 3.2 km east of the inlet, wave refraction modeling indicated that the ratio of Q_R to Q_L was approximately 1.9. The same ratio west of the inlet, also estimated from wave refraction modeling, was 1.8. These ratios indicate a westerly directed net transport that is slightly greater at the eastern boundary as compared with the western boundary. Based on profile-survey data, the berm-crest level was 3.5 m relative to National Geodetic Vertical Datum (NGVD), and the depth of closure was 7.0 m NGVD. The average shoreline change rate $\Delta y \Delta/t$ for Adjacent Beach 1 (from inlet to 3.2 km east, hereafter noted as A1) was 1.40 m/year, and the same quantity for Adjacent Beach 2 (from inlet to 3.2 km west, noted as A2) was −1.43 m/year. Beach-fill placements for A1 and A2 were 13,000 and 25,000 cu m/year, respectively. The rate of relative sea-level rise was 0.003 m/year, and the distance from datum to the depth of closure L_c was approximately 760 m for A1 and A2. The inlet channel and shoals had a net volume change of 111,000 cu m/year, with dredging averaging 2,400 cu m/year (Moffatt and Nichol Engineers and URS Consultants 1999).

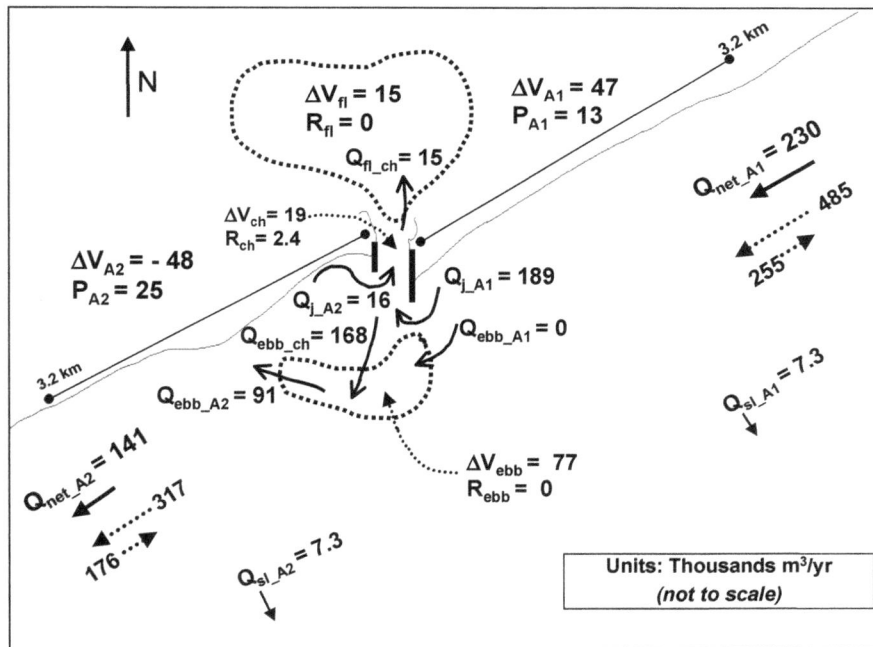

Figure V-6-28. Refined sediment budget for Shinnecock Inlet, New York

(Continued)

EXAMPLE PROBLEM V-6-11 (Continued)

CALCULATIONS:

Applying Equation V-6-27 gives an active depth for A1 and A2,

$$D_A = B + D_C = 3.5 + 7.0 = 10.5 \text{ m}$$

The rate of volume change for A1 and A2 can be calculated with Equation V-6-28,

$$\Delta V_{A1} = \frac{\Delta y \Delta x D_A}{\Delta t} = (1.40 \text{ m/year}) \ (3,200 \text{ m}) \ (10.5 \text{ m}) = 47,000 \text{ cu m/year}$$

$$\Delta V_{A2} = \frac{\Delta y \Delta x D_A}{\Delta t} = (-1.43 \text{ m/year}) \ (3,200 \text{ m}) \ (10.5 \text{ m}) = -48,000 \text{ cu m/year}$$

Losses due to relative sea-level rise can be estimated by Equation V-6-29,

$$\Delta y_{sl\_A1} = \Delta y_{sl\_A2} = \frac{L_c}{B + D_c} \ S = \frac{760}{3 + 7.5} \ (0.003) = 0.22 \text{ m/year}$$
$$\text{or,}$$
$$Q_{sl\_A1} = Q_{sl\_A2} = (\Delta y_{sl\_A1} \text{ or } \Delta y_{sl\_A2})(\Delta x)(D_A) = (0.22 \text{ m/year}) \ (3,200 \text{ m}) \ (10.5 \text{ m}) \sim 7,300 \text{ cu m/year}$$

The total change in volume for the inlet channel and shoals was given as 111,000 cu m/year. To fully develop the inlet sediment budget, this quantity will be proportioned between the ebb shoal, inlet channel, and flood shoal following the conceptual budget as guidance. Table V-6-11 lists the rate of measured volume change ΔV, beach fill placed P, dredging (removal) R, and losses because of relative sea-level rise Q_{sl}, for A1, each region of the inlet, and A2.

Table V-6-11
Rates of Volume Change for Shinnecock Inlet Sediment Budget, 1938 to 1979 (1,000s cu m/year)

Control Volume	ΔV	P	R	Q_{sl}
A1 (Adjacent Beach 1)	47	13	0	7.3
Inlet: Ebb Shoal	77	0	0	0
Inlet: Channel	19	0	2.4	0
Inlet: Flood Shoal	15	0	0	0
A2 (Adjacent Beach 2)	-48	25	0	7.3

(Continued)

EXAMPLE PROBLEM V-6-11 (Continued)

REFINING CONCEPTUAL SEDIMENT BUDGET:

To formulate the inlet sediment budget, one can assume a rate of net transport at the updrift boundary, Q_{net_A1} = 230,000 cu m/year. This value is within the range identified in the conceptual sediment budget. In a more expanded analysis than presented here, a range of values for Q_{net_A1} can be applied in the sediment budget to examine fully the sensitivity of the inlet sediment-transport magnitudes and pathways to this parameter. The ratio of Q_R and Q_L was given as 1.9, and entering this value into Equation V-6-23 gives,

$$Q_{net\_A1} = Q_{R\_A1} - Q_{L\_A1} = 1.9 \; Q_{L\_A1} - Q_{L\_A1} = 0.9 \; Q_{L\_A1}$$

$$230 = 0.9 \; Q_{L\_A1}$$

$$Q_{L\_A1} = 255 \text{ and } Q_{R\_A1} = 485$$

Considering the entire reach as the control volume forms a macrobudget. Applying Equation V-6-23 gives,

$$\sum Q_{source} - \sum Q_{sink} - \sum \Delta V + \sum P - \sum R = \text{Residual} = 0$$

$$\left(Q_{net\_A1} \right) - \left(Q_{sl\_A1} + Q_{sl\_A2} + Q_{net\_A2} \right) - \left(\Delta V_{A1} + \Delta V_{A2} + \Delta V_f + \Delta V_{ch} + \Delta V_{ebb} \right) + \left(P_{A1} + P_{A2} \right) - \left(R_{fl} + R_{ch} + R_{e\,bb} \right) = 0$$

$$(230) - (7.3 + 7.3 + Q_{net\_A2}) - (47 + -48 + 15 + 19 + 77) + (13 + 25) - (0 + 2.4 + 0) = 0$$

$$Q_{net\_A2} = 141$$

Applying Equation V-6-23 at the western boundary gives,

$$Q_{net\_A2} = Q_{R\_A2} - Q_{L\_A2}$$

$$141 = 1.8 \; Q_{L\_A2} - Q_{L\_A2}$$

$$Q_{L\_A2} = 176 \quad and \quad Q_{R\_A2} = 317$$

Now, the control volume A1 can be considered. There are two unknowns, the rate of sediment transport around the east jetty, Q_{j_A1}, and sediment transport from A1 to the ebb-tidal shoal, Q_{ebb_A1}. Inspection of bathymetric charts and aerial photography shows no evidence of morphologic pathways (e.g., shoal features) from A1 to the ebb-tidal shoal. Thus, one can assume that $Q_{ebb\_A1} \sim 0$ and solve for Q_{j_A1} in Equation V-6-23,

$$\sum Q_{source} - \sum Q_{sink} - \sum \Delta V + \sum P - \sum R = \text{Residual} = 0$$

$$\left(Q_{net\_A1} \right) - \left(Q_{sl\_A1} + Q_{j\_A1} + Q_{ebb\_A1} \right) - \left(\Delta V_{A1} \right) + \left(P_{A1} \right) = 0$$

$$(230) - (7.3 + Q_{j\_A1} + 0) - (47) + (13) = 0$$

$$Q_{j\_A1} = 189$$

(Continued)

EXAMPLE PROBLEM V-6-11 (Continued)

gives,

$$\sum Q_{source} - \sum Q_{sink} - \sum \Delta V + \sum P - \sum R = Residual = 0$$

$$(Q_{net\_A1}) - (Q_{sl\_A1} + Q_{sl\_A2} + Q_{net\_A2}) - (\Delta V_{A1} + \Delta V_{A2} + \Delta V_{fl} + \Delta V_{ch} + \Delta V_{ebb}) + (P_{A1} + P_{A2}) - (R_{fl} + R_{ch} + R_{e\,bb}) = 0$$

$$(230) - (7.3 + 7.3 + Q_{net\_A2}) - (47 + -48 + 15 + 19 + 77) + (13 + 25) - (0 + 2.4 + 0) = 0$$

$$Q_{net\_A2} = 141$$

Next a control volume for A2 is formulated, excluding the ebb-tidal shoal. There are also two unknowns for this control volume, the rate of sediment transport around the west jetty, Q_{j_A2}, and the rate of sediment transport bypassed from the ebb-tidal shoal to A2, Q_{ebb_A2}. A more detailed analysis of the shoreline position and beach-fill placement records for A2 indicates that $\Delta V - P = -16$ for the region east of the bulge in the 1979 shoreline position, and $\Delta V - P = -58$ for the region west of the bulge. As a first estimate, one can set $Q_{j\_A2} = 16$, implying that all sediment lost from the region east of the bulge moved around the west jetty. This assumption also implies that this morphologic feature represents a long-term nodal zone for net longshore sand transport. Using Equation V-6-23 to solve for Q_{ebb_A2} gives,

$$\sum Q_{source} - \sum Q_{sink} - \sum \Delta V + \sum P - \sum R = Residual = 0$$

$$(Q_{ebb\_A2}) - (Q_{net\_A2} + Q_{j\_A2} + Q_{sl\_A2}) - (\Delta V_{A2}) + (P_{A2}) = 0$$

$$(Q_{ebb\_A2}) - (141 + 16 + 7.3) - (-48) + (25) = 0$$

$$Q_{ebb\_A2} = 91$$

in units of thousands of cubic meters.

The control volume for the ebb-tidal shoal now has one unknown, the rate of sediment transport from the channel to the ebb-tidal shoal, Q_{ebb_ch}. Applying Equation V-6-23 gives,

$$\sum Q_{source} - \sum Q_{sink} - \sum \Delta V + \sum P - \sum R = Residual = 0$$
$$(Q_{ebb\_A1} + Q_{ebb\_ch}) - (Q_{ebb\_A2}) - (\Delta V_{ebb}) = 0$$
$$(0 + Q_{ebb\_ch}) - (91) - (77) = 0$$
$$Q_{ebb\_ch} = 168$$

(Continued)

EXAMPLE PROBLEM V-6-11 (Concluded)

The final unknown is the rate of sediment transport from the channel to the flood-tidal shoal, Q_{fl_ch}. Equation V-6-23 applied to the inlet channel control volume gives,

$$\sum Q_{source} - \sum Q_{sink} - \sum \Delta V + \sum P - \sum R = Residual = 0$$

$$\left(Q_{j\_A1} + Q_{j\_A2}\right) - \left(Q_{ebb\_ch} + Q_{fl\_ch}\right) - \left(\Delta V_{ch}\right) - \left(R_{ch}\right) = 0$$

$$\left(189 + 16\right) - \left(168 + Q_{fl\_ch}\right) - \left(19\right) - \left(2.4\right) = 0$$

$$Q_{fl\_ch} = 15.6 \sim 15$$

The calculated value of $Q_{fl\_ch} = 15.6$ approximately agrees with the assumed change in volume for the flood-tidal shoal, $\Delta V_{fl} = 15$, indicating that there are no other significant sediment sources contributing to the growth of the flood-tidal shoal.

DISCUSSION OF EXAMPLES.
    These example problems illustrate one approach that can be taken for formulating a sediment budget. The following assumptions entered:

- The net longshore sand transport rate at the updrift boundary of the control volume was assumed to be 230,000 cu m/year.

- The rate of sediment transport from A1 to the ebb-tidal shoal was assumed to be negligible.

- The rate of sediment transport from A2 around the west jetty was assumed to be 16,000 cu m/year.

- Volume change rates for the ebb-tidal shoal, inlet channel, and flood-tidal shoal were assumed to be 77,000, 19,000, and 15,000 cu m/year, respectively.

- Uncertainties in quantities forming the sediment budget were omitted for this example problem.

    (b) Operation. SBAS has been designed to organize the user's work space and facilitate visualization of sediment budget alternatives. Within the right-hand side of the screen, called the Topology Window, SBAS formulates a sediment budget by allowing the user to create a series of cells and arrows representing sources and sinks that characterize the budget. The user selects items from the upper tool bar to generate elements of the sediment budget. The lower tool bar allows the user to import georeferenced images and data into a desired coordinate system and viewed with the sediment budget in the Topology Window. The user can zoom in to show a detailed area, and zoom out to view the regional sediment budget. The left-hand side of the screen organizes alternatives within a particular project. Alternatives may represent various time periods, different boundary conditions for the same time period, or modifications to assumptions within the budget reflecting a sensitivity analysis. Once a sediment budget alternative has been defined, and the user has created sediment budget cells with sources and sinks, values can then be assigned to the various components of the sediment-budget topology. Figure V-6-29a shows a screen capture with a georeferenced image of the regional project area, and Figure V-6-29b shows a zoom-in image of a local sediment budget. As shown in Figure V-6-29b, the SBAS indicates by color-coding whether a cell is balanced or not.

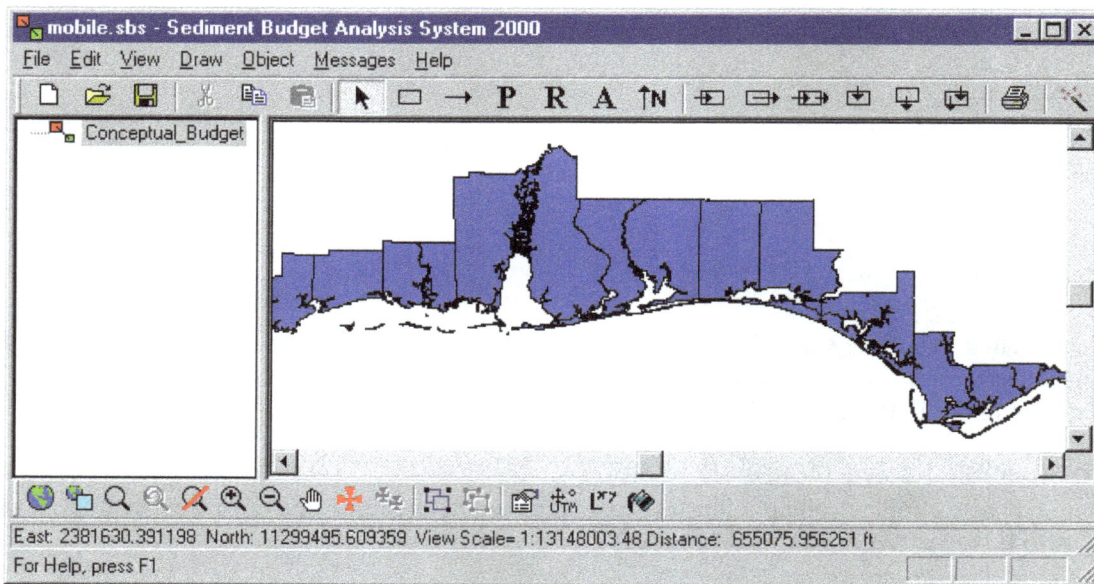

Figure V-6-29a. SBAS Alternative Window (left) and Topology Window (right)

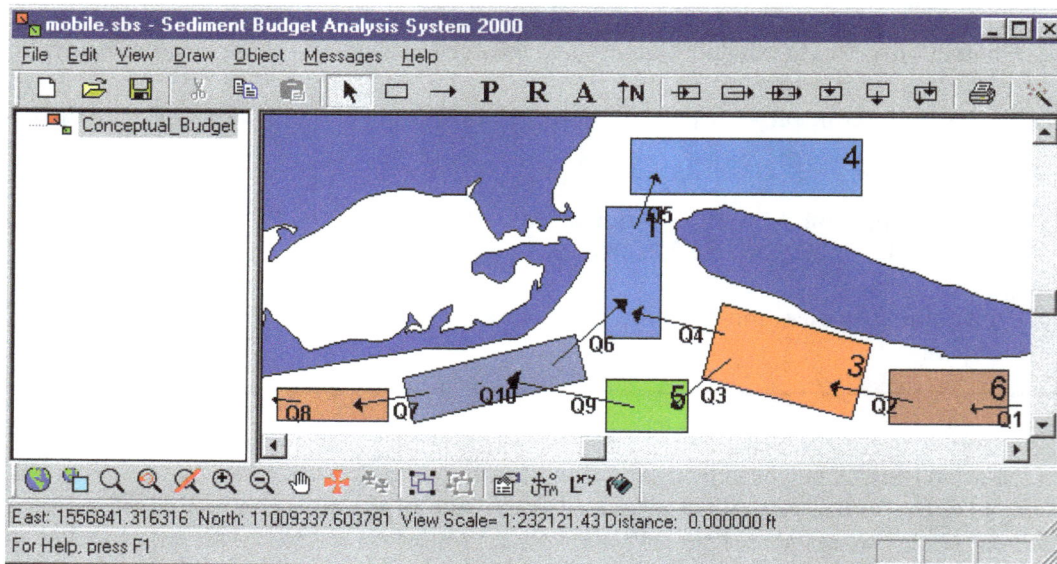

Figure V-6-29b. Zoom-in showing sediment budget cells. Color coding indicates varying degrees of imbalance

- The lower tool bar allows the import of georeferenced images and data into a desired coordinate system (Figure V-6-30) and viewed with the sediment budget in the Topology Window. The user can zoom out to view the regional sediment budget (see Figure V-6-31, which shows the Gulf of Mexico shore of Alabama and Florida panhandle), and zoom in to show a detailed area (Figure V-6-32). By dragging the cursor over any combination of cells, sediment budget cells can be combined (collapsed), as shown in Figure V-6-32. Collapsed cells are useful for regional views of a sediment budget. For local, project-level applications, selecting the collapsed cell and choosing an icon from the bottom tool bar will reinstate (explode) the collapsed cell (Figure V-6-33).

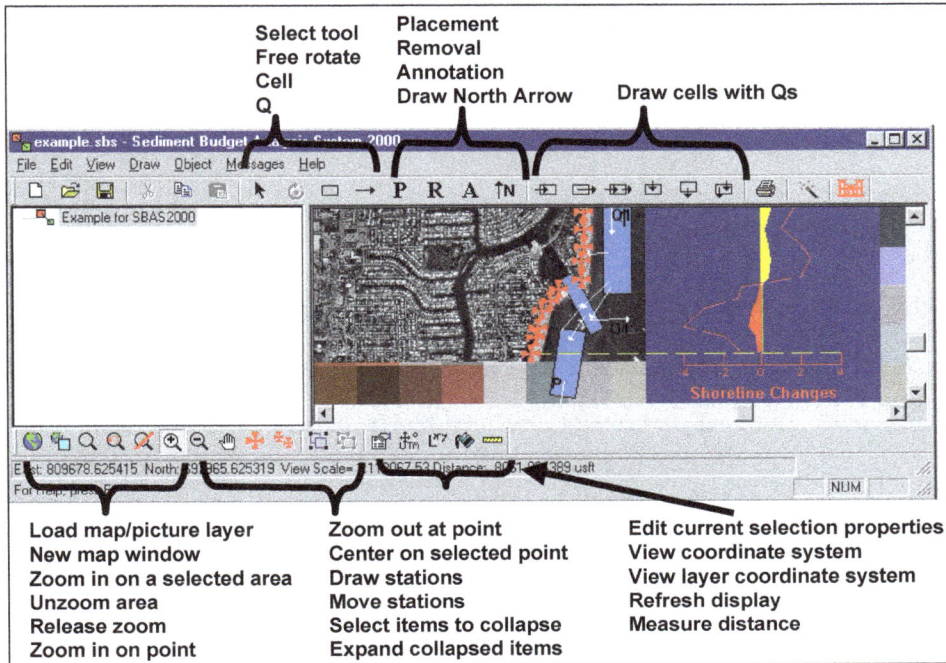

Figure V-6-30. SBAS tool bars and button functions

SBAS can also display data on shoreline change and the rate of volume change associated with stations located along the coast, bay, or river (Figure V-6-34). Location of stations and associated data may be entered directly into SBAS's spreadsheet (Figure V-6-35), or data may be imported using a user-specified format.

- Georeferenced images can be imported into SBAS by means of the *Load map/picture layer* on the bottom tool bar. This menu (Figure V-6-36) allows the user to indicate the coordinate system (group, system, datum, and linear unit) in which the sediment budget project will be defined. Next, the maps to be imported are selected, and their coordinate system is selected. SBAS converts the map's coordinate system to the project coordinate system. Multiple images (maps, aerial photographs, contour maps, etc.) can be layered in the sediment budget. SBAS allows for 55 different map coordinate systems.

- A demonstration that guides the user through the operation and features of SBAS Version 1.02 available for viewing on the PC, as described in a later section.

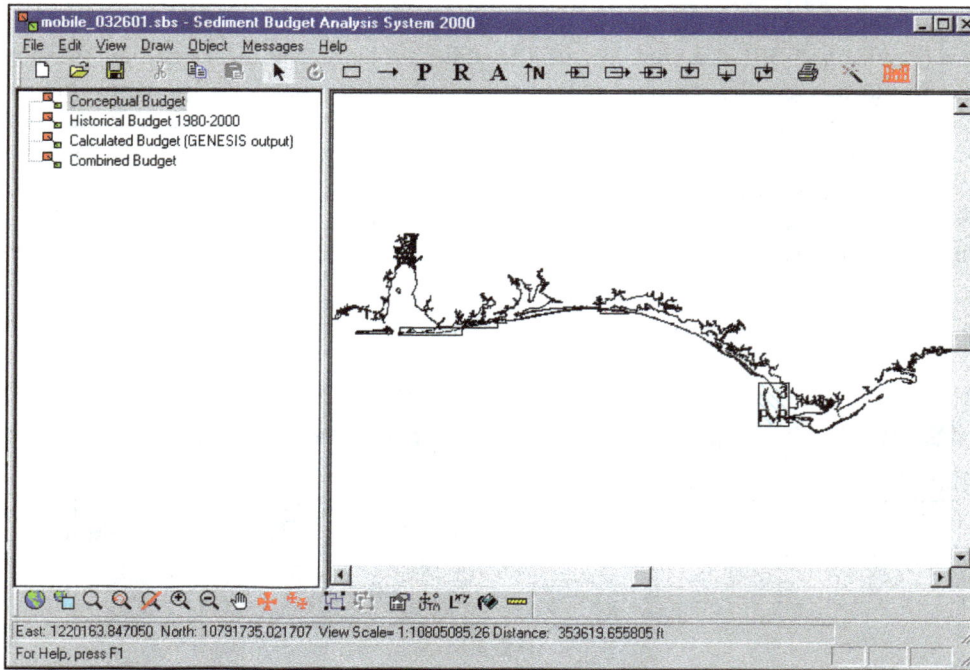

Figure V-6-31. Regional view of a sediment budget (zoom-out), with four alternatives

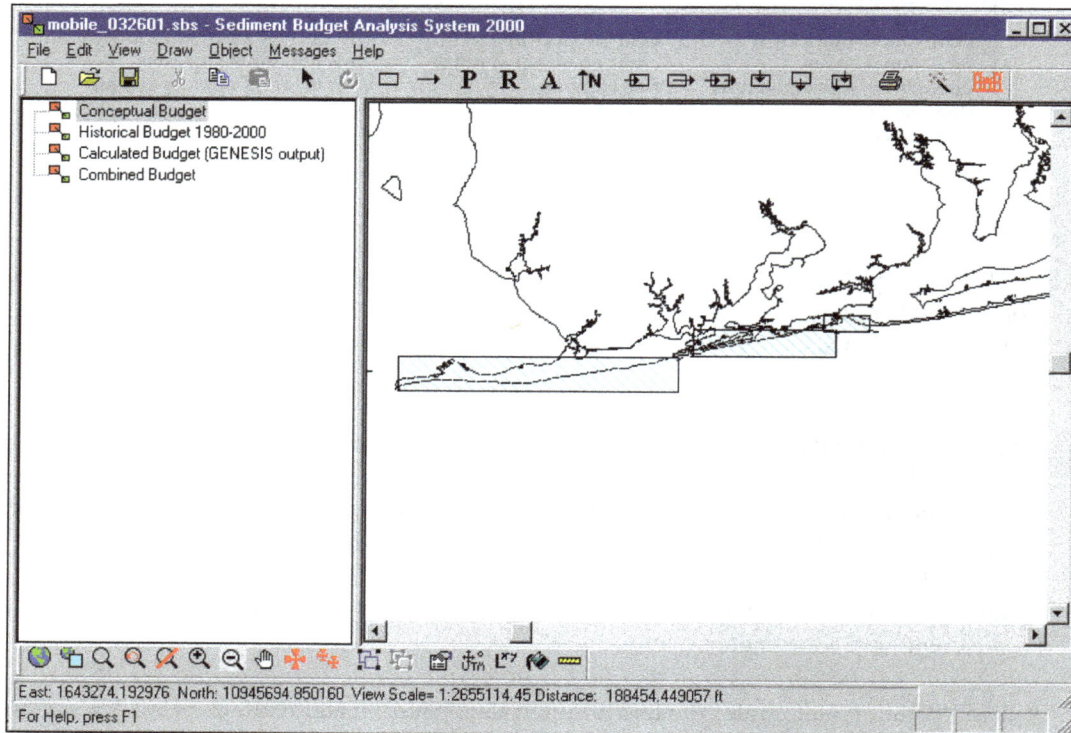

Figure V-6-32. Local view of a sediment budget (zoom-in) showing combined (collapsed)
sediment budget cells

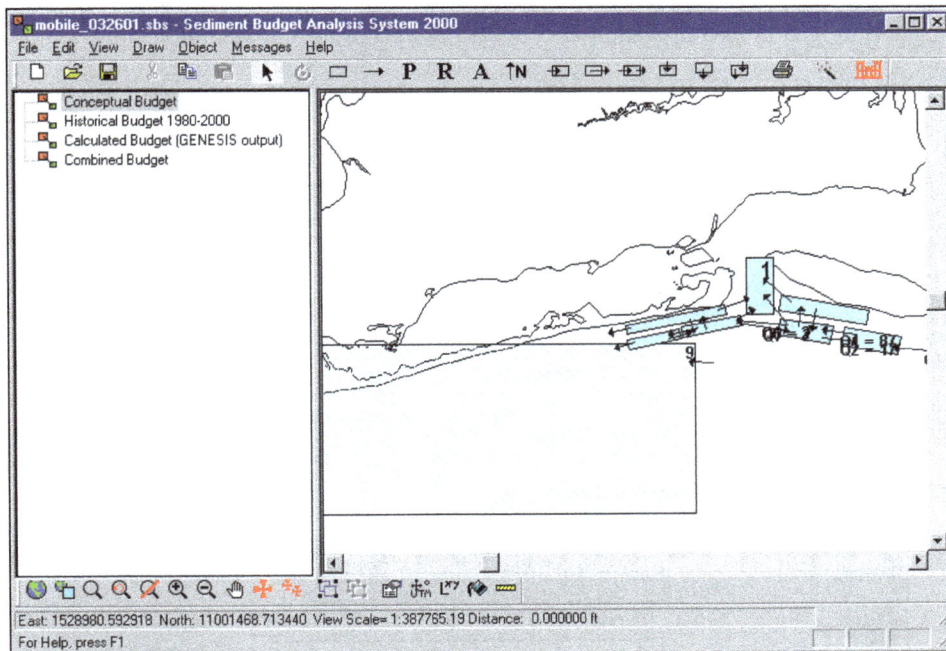

Figure V-6-33. Sediment budget with reinstated (exploded) cells

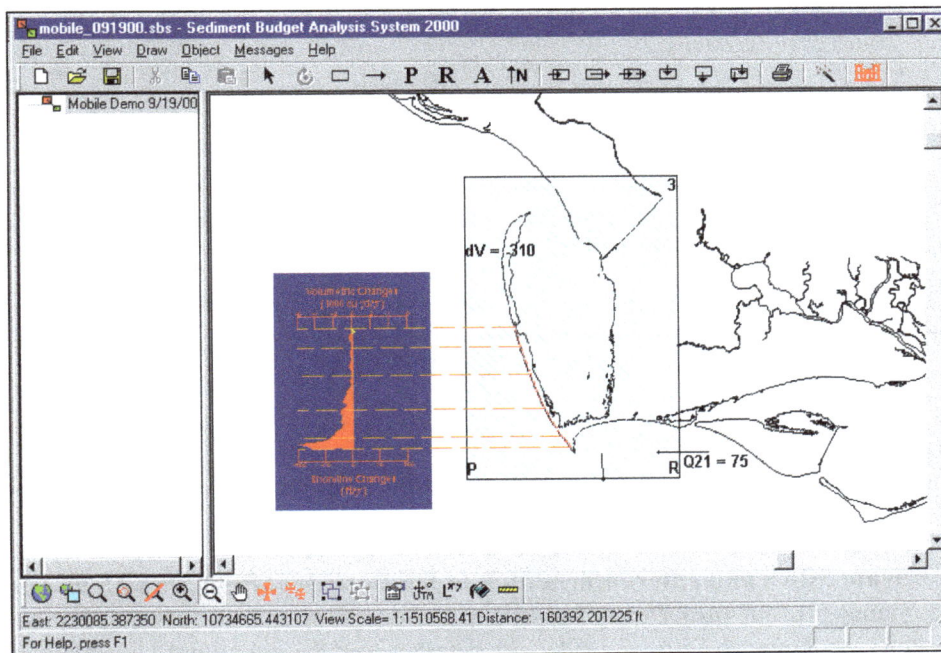

Figure V-6-34. Rate graph feature of SBAS

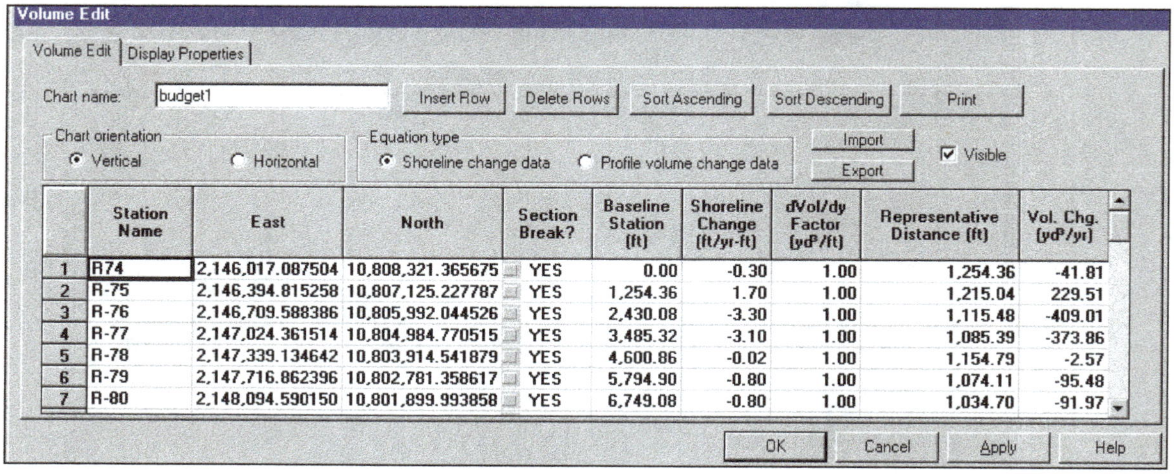

Figure V-6-35. *Volume Edit* spreadsheet in SBAS

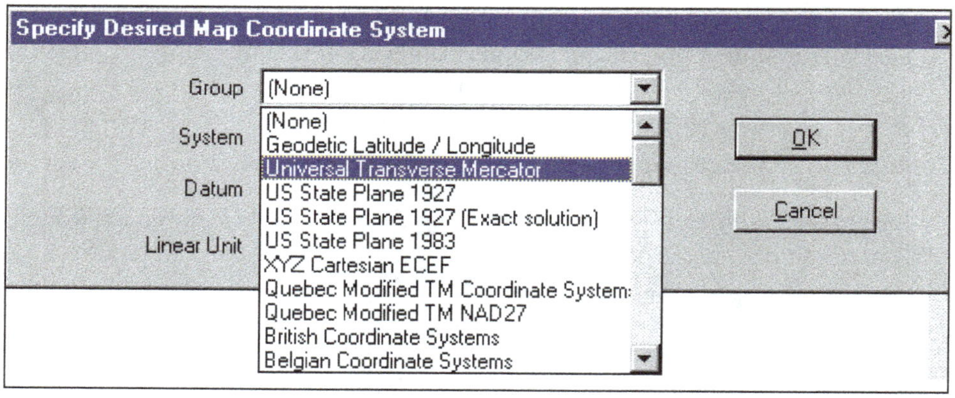

Figure V-6-36. **Entering project coordinate system with the *Load map/picture layer* feature on the bottom tool bar**

EXAMPLE PROBLEM V-6-12

EXAMPLE:
   The SBAS installation includes an example file with georeferenced images, a sediment budget, and rate graph. The following example problem guides the use of these files.

Step 1: Activate SBAS and enter information to the *Document Properties Menu*. Information such as the manager of the project, keywords, units, and type of uncertainty calculation are entered into this menu (see Kraus and Rosati 1998a, for details about the uncertainty calculation). Note: the user should choose the same project coordinate system and consistent units for the transport rate or volume change (e.g., if the project coordinate system will be in U.S. Customary Units, rates should be expressed as cu yd/year and volume as cu yd).

(Continued)

EXAMPLE PROBLEM V-6-12

Step 2: Import georeferenced images. To import the map provided in the installation, activate the Topology Window (right side of the screen) by clicking in this area. Choose the globe on the bottom tool bar. The user first defines the desired project coordinate system as follows:

> Group: US State Plane 1927
> System: Florida East 901
> Datum: NAD 27
> Units: USFEET
> Select **OK**

Then load the map provided in the installation with the *Connect Map Layer* menu. Select the following files:

> Conus12.tif (image) and Conus12.tab (reference file)

On the next menu, indicate that these files are in

> Group: Geodetic Latitude/Longitude
> System: Latitude/Longitude
> Datum: WGS84

(If the image is not visible, select release zoom on the bottom tool bar.)

Step 3: Import a photograph. Choose the globe on the bottom tool bar and import another map layer. Select the following files:

> Br21-38.tif (image) and Br21-38.tfw (reference file) (photograph of Hillsboro Inlet, Florida and associated world file)

On the next menu, indicate that these files are in the following coordinate system (same as the project coordinate system):

> Group: US State Plane 1927
> System: Florida East 901
> Datum: NAD27
> Units: USFEET

The coordinate system for the layers that have been loaded can be viewed by choosing the **Edit Current Selection Properties** button on the bottom tool bar.

Step 4: Create a sediment budget for Hillsboro Inlet. Accessing the top toolbar, select the cell button and drag and pull sediment budget cells. Cells may be rotated to better represent different areas of the budget (the channel, for instance) by choosing the rotate button on the top tool bar and rotating a corner of a selected cell. Placement and removal in cells may be indicated by choosing the **P** and **R** buttons on the top tool bar, respectively, then clicking on the cell that has the engineering activity. Sediment sources and sinks to and from each cell may be indicated by selecting the **arrow** button on the top tool bar, and dragging arrows into and out of each cell as appropriate. Double clicking on a cell will bring up the *Cell Properties* menu, and values for each sediment budget element can be entered.

(Continued)

Sediment Management at Inlets

EXAMPLE PROBLEM V-6-12 (Continued)

Cells can be collapsed by selecting the **Select Items to be Collapsed** button on the bottom tool bar, and dragging the cursor over the items to be combined. The cells may be reinstated by selecting the collapsed cell, then choosing the **Expand Collapsed Items** on the bottom tool bar. Note: colors of cell balance, gain, and loss, arrows, and other sediment budget elements may be changed by selecting the **Object-Colors** button at the top tool bar.

Hillsboro Inlet Sediment Budget. The sediment budget can be formulated with three cells: updrift, channel, and downdrift. The following text summarizes the sediment budget (Personal Communication, Thomas D. Smith, April 2001, U.S. Army Engineer District, Jacksonville), which has been conceptualized as shown in Figure V-6-37.

Updrift Cell. Net longshore sand transport entering the north boundary of the updrift cell is approximately 120,000 cu yd/year. Sinks from this cell include accretion of the updrift beach at a rate of 6,000 cu yd/year, 54,000 cu yd/year transported over the north jetty weir section to the channel cell, and 60,000 cu yd/year moving around the jetty into the channel cell.

Channel Cell. The source of sediment to this cell is 30,000 cu yd/year moving into the channel from the downdrift cell. Sinks from this cell include 4,000 cu yd/year transported to deep water on ebb flow, 110,000 cu yd/year dredged from the channel and placed within the downdrift cell, and 30,000 cu yd/year which is naturally bypassed from the channel to the downdrift cell.

Downdrift Cell. Dredged material placement into this cell is 110,000 cu yd/year, which is also the value of net longshore transport exiting this cell.

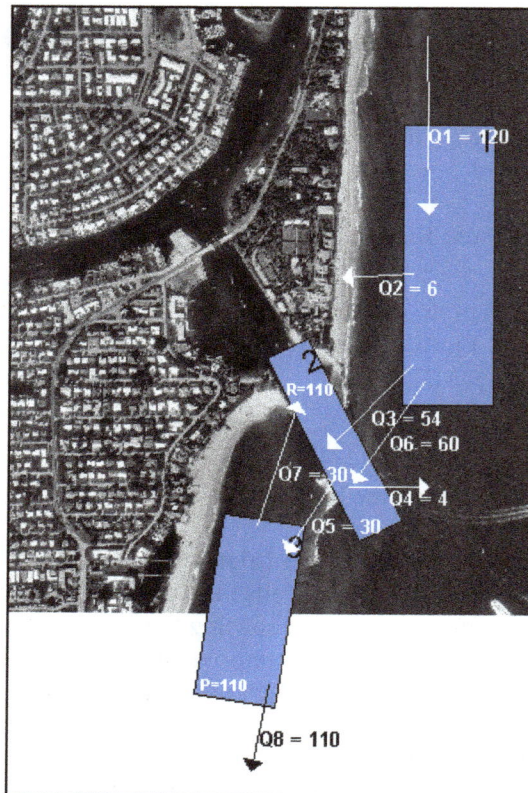

Figure V-6-37. Sediment budget for Hillsboro Inlet, Florida

(Continued)

EXAMPLE PROBLEM V-6-12 (Concluded)

Step 5: Enter shoreline change rate data for Hillsboro Inlet. Select the **Draw Stations** button on the bottom tool bar, and visually place benchmarks for shoreline data on the photograph. (If you have a pre-existing file with data, these station locations may also be imported to the SBAS Rate Graph spreadsheet). Once all stations have been entered, choose the **View-Rate Graph** button on the top tool bar, select the rate graph (named *Untitled* at this point), and choose **Edit**. Shoreline change (or volume change) data may be entered for each benchmark position. SBAS automatically plots these data in the rate graph. Selecting and dragging will move the graph, and deselecting the graph after clicking the **View-Rate Graph** button on the top tool bar can turn it off. An example is shown in Figure V-6-38.

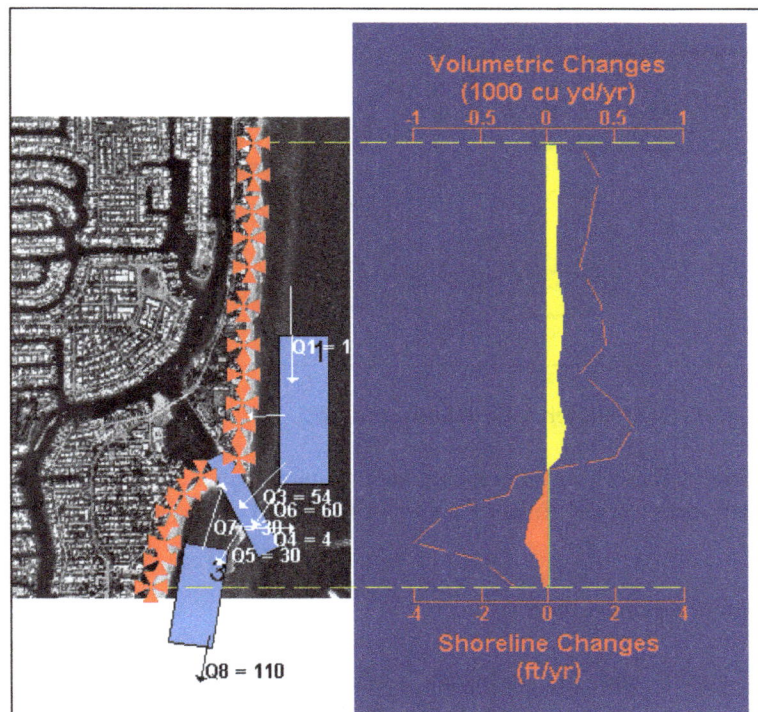

Figure V-6-38. Hypothetical shoreline change data for beaches adjacent to Hillsboro Inlet

- SBAS2000 may be obtained by contacting the Coastal and Hydraulics Laboratory, U.S. Army Engineer Research and Development Center, Vicksburg, MS.

(2) Reservoir Model (Kraus 2000)

(a) Kraus developed a time-dependent model of inlet morphology that is mathematically analogous to a reservoir system with known (or estimated) equilibrium volumes for each reservoir and defined transfer rates between reservoirs. The reservoirs represent morphologic features, including the ebb-tidal shoal, bypassing bar, attachment bar, and flood tidal shoal (Figure V-6-39). Once calibrated, the model predicts a time-dependent sediment budget based on the equilibrium volume of each morphologic feature, and the longshore sand transport characteristics in the region. The following discussion is summarized from Kraus (2000).

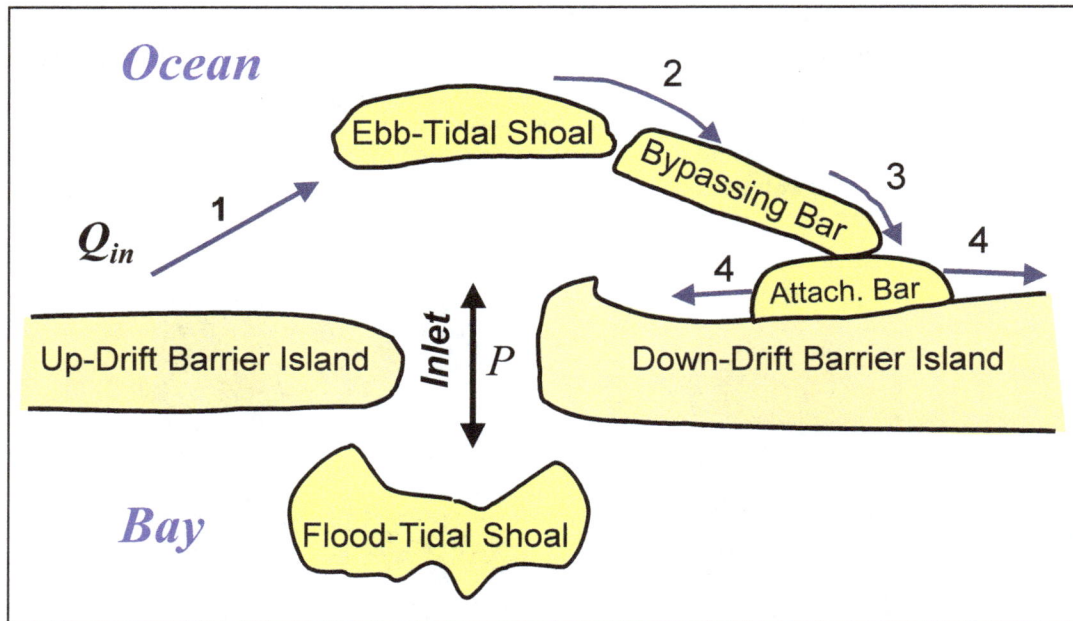

Figure V-6-39. Definition sketch for inlet morphology

(b) The time-dependent Reservoir Model operates on the temporal and spatial scales associated with the entire (aggregated) morphological form of an ebb-tidal shoal. Five assumptions are invoked to arrive at the model:

- Mass (sand volume) is conserved.

- Morphological forms and the sediment pathways among them can be identified, and the morphologic forms evolve while preserving identity.

- Stable equilibrium of the individual aggregate morphologic form(s) exist.

- Changes in mesomorphological and macromorphological forms are reasonably smooth.

- Material composing the ebb-tidal shoal is predominately transported to and from it through longshore transport.

(a) The ebb-shoal complex is defined as consisting of the ebb shoal proper, one or two ebb-shoal bypassing bars (depending on the balance between left- and right-directed longshore transport), and one or two attachment bars. These features are shown schematically in Figure V-6-39 and the pattern of wave breaking on the crescentic ebb-shoal complex at Ocean City Inlet, Maryland, is shown in Figure V-6-40.

Figure V-6-40. Pattern of wave breaking on ebb shoal and bar, Ocean City Inlet, Maryland, November 1991

The model distinguishes between ebb-tidal shoal proper (hereafter, ebb shoal), typically located in the confine of the ebb-tidal jet, and the ebb-shoal bypassing bar (hereafter, bypassing bar) that grows toward shore from the ebb shoal, principally by the transport of sediment alongshore by wave action. The bar may shelter the leeward beach from incident waves so that a salient might form – similar to the functioning of a detached breakwater (Pope and Dean 1986), initiating creation of the attachment bar.

(b) Previous authors have combined the ebb shoal and the bar(s) protruding from it into one feature referred to as the ebb shoal. Here, the shoal and bypassing bars are distinguished because of the different balance of processes. When a new inlet is formed, the shoal first becomes apparent within the confine of the inlet ebb jet, and bypassing bars have not yet emerged. Bypassing bars are formed by sediment transported off the ebb shoal through the action of breaking waves and the wave-induced longshore current (tidal and wind-induced currents can also play a role). A bar cannot form without an available sediment source, similar to the growth of a spit (as modeled mathematically by Kraus 1999). In this sense, bypassing bars are analogous to the spit platform concept of Meistrell (1972) in which a subaqueous sediment platform develops from the sediment source prior to the visually observed subaerial spit. Bypassing bars grow in the direction of predominant transport as do spits. At inlets with nearly equal left- and right-directed longshore transport or with a small tidal prism, two bars can emerge from the ebb shoal creating a nearly concentric halo about the inlet entrance. As the bypassing bar merges with the shore, an attachment bar is created, thereby transporting sand to the beach. At this point in evolution of the ebb-shoal complex, substantial bypassing of sand can occur from the updrift side of the inlet to the downdrift side.

(c) In this conceptualization, if an inlet is created along a coast, the littoral drift is intercepted to deposit sand first in the channel and ebb shoal. This material joins that volume initially jetted offshore when the barrier island or landmass was breached. Over time, a bar emerges from the shoal and grows in the predominant direction of drift. After many years, as controlled by the morphologic or aggregate scale of the particular inlet, an attachment bar may form on the downdrift shore. At this stage, significant sand bypassing of the inlet can occur, re-establishing in great part the transport downdrift that existed prior to formation of the inlet. The following model can describe the evolution of an ebb-shoal complex from initial cutting of the inlet, as well as changes in morphologic features and sand bypassing resulting from engineering actions such as mining or from time-dependent changes in wave climate (for example, seasonal shifts).

(f) Reservoir aggregate model. The conceptual model of the ebb-shoal complex described in the preceding section is represented mathematically by analogy to a reservoir system, as shown in Figure V-6-41. It is assumed that sand is brought to the ebb shoal at a rate q_{in}, and the volume v_e in the ebb shoal at any time increases while possibly leaking or bypassing some amount of sand to create a downdrift bypassing bar. The input rate q_{in} typically is the sum of left- and right-directed longshore sand transport. For the analytic model presented, a predominant (unidirectional) rate is taken, but this constraint is not necessary and is relaxed in a numerical example in the following paragraphs.

(g) The volume V_E of sand in the shoal can increase until it reaches an equilibrium volume V_{Ee} (the subscript e denoting equilibrium) according to the hydrodynamic conditions such as given by Walton and Adams (1976). As equilibrium is approached, most sand brought to the ebb shoal is bypassed in the direction of predominant transport. Similarly, the bypassing bar volume V_B grows as it is supplied with sediment by the littoral drift and the ebb shoal, with some of its material leaking to (bypassing to) the attachment bar. As the bypassing bar approaches equilibrium volume V_{Be}, most sand supplied to it is passed to the attachment bar V_A. The attachment bar transfers sand to the adjacent beaches. When it reaches its equilibrium volume V_{Ae}, all sand supplied to it by the bar is bypassed to the downdrift beach. The model thus requires values of the input and output rates of transport from each feature, and their respective equilibrium volumes.

(h) Analytical model. Simplified conditions are considered here to obtain a closed-form solution that reveals the parameters controlling the aggregated morphologic ebb-shoal complex. In the absence of data and for convenience in arriving at an analytical solution, a linear form of bypassing is assumed. The amount of material bypassed from any of the morphological forms is assumed to vary in direct proportion to the volume of the form (amount of material in a given reservoir) at the particular time. Therefore, the rate of sand leaving or bypassing the ebb shoal, $(Q_E)_{out}$, is specified as:

$$(Q_E)_{out} = \frac{V_E}{V_{Ee}} Q_{in} \qquad (V-6-30)$$

in which Q_{in} is taken to be constant (average annual rate), although this is not necessary.

• The continuity equation governing change in V_E can be expressed as:

$$\frac{dV_E}{dt} = Q_{in} - (Q_E)_{out} \qquad (V-6-31)$$

where t = time. For the present situation with Equation V-6-30, it becomes:

$$\frac{dV_E}{dt} = Q_{in}\left(1 - \frac{V_E}{V_{Ee}}\right) \qquad (V-6-32)$$

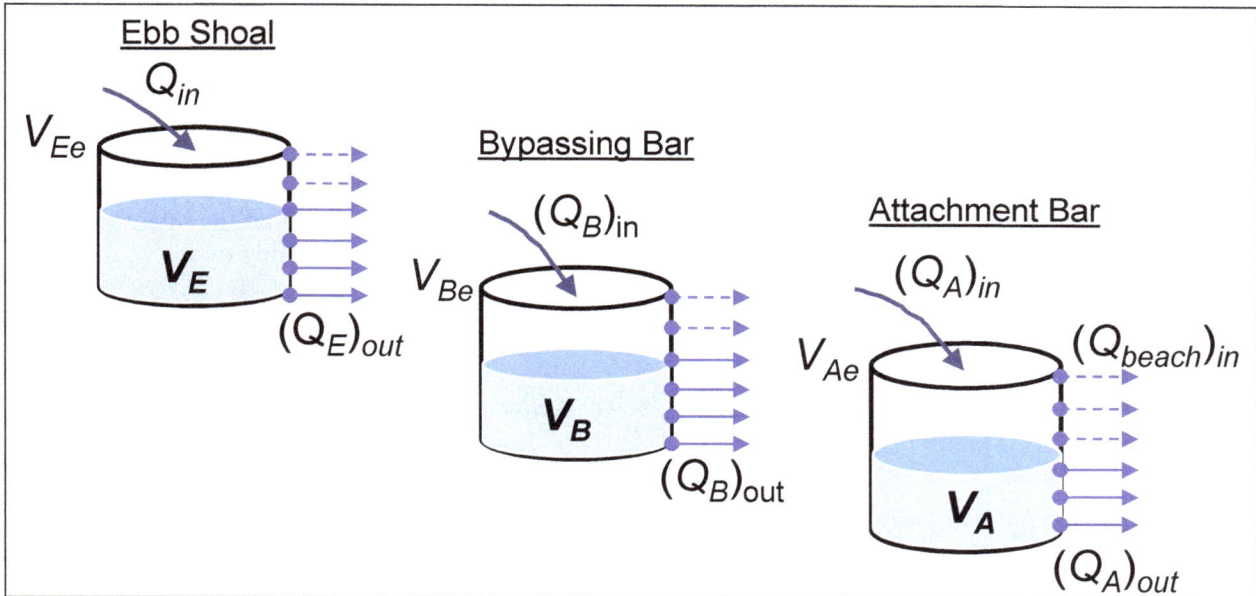

Figure V-6-41. Conceptual design for reservoir model

- With the initial condition $V_E(0) = 0$, the solution of Equation V-6-32 is:

$$V_E = V_{Ee}\left(1 - e^{-\alpha t}\right) \qquad \text{(V-6-33)}$$

in which

$$\alpha = \frac{Q_{in}}{V_{Ee}} \qquad \text{(V-6-34)}$$

- The parameter α defines a characteristic time scale for the ebb shoal. For example, if $Q_{in} = 1 \times 10^5$ cu m/year and $V_{Ee} = 2 \times 10^6$ cu m, which are representative values for a small inlet on a moderate-wave coast, then $1/\alpha = 20$ years. The shoal is predicted to reach 50 and 95 percent of its equilibrium volume after 14 and 60 years, respectively, under the constant imposed transport rate. These time frames are on the order of those associated with development of inlet ebb shoals.

- The characteristic time scale given by α has a physical interpretation by analogy to the well-known model of bar bypassing introduced by Bruun and Gerritsen (1960) and reviewed by Bruun, Mehta, and Johnsson (1978). Bruun and Gerritsen (1959; 1960); introduced the ratio r as

$$r = \frac{P}{M_{tot}} \qquad \text{(V-6-35)}$$

in which P = tidal prism, and M_{tot} = average annual littoral sediment brought to the inlet. Inlets with a value of $r > 150$ (approximate) tend to have stable, deep channels and are poor bar bypassers from updrift to downdrift, whereas inlets with $r < 50$ (approximate) tend toward closure and are good bar bypassers.

Because the equilibrium volume of the ebb-tidal shoal is approximately linearly proportional to the tidal prism, $V_{Ee} \propto P$ (Walton and Adams 1976), α is proportional to $1/r$. The reservoir aggregate model therefore contains at its center a concept widely accepted by engineers and geomorphologists. Established here through the continuity equation, the reservoir model gives theoretical justification for the Bruun and Gerritsen ratio by the appearance of α.

- The volume of sediment $(V_E)_{out}$ that has bypassed the shoal from inception of the inlet to time t is the difference between the amount that arrived at the shoal and that remaining on the shoal:

$$(V_E)_{out} = Q_{in}t - V_E \tag{V-6-36}$$

- The rate of sand arriving at the bypassing bar $(Q_B)_{in}$ equals the rate of that leaving the shoal $(Q_E)_{out} = d(V_E)_{out}/dt$, or

$$(Q_B)_{in} = (Q_E)_{out} = Q_{in} - \frac{dV_E}{dt}$$
$$= \frac{V_E}{V_{Ee}}Q_{in} \tag{V-6-37}$$

which recovers (1) by volume balance. The right side of Equation V-6-37 can be expressed as αV_E, again showing the central role of the parameter α.

- Continuing in this fashion, the reservoir aggregate model yields the following equations for the volume of the bypassing bar,

$$V_B = V_{Be}\left(1 - e^{-\beta t'}\right), \qquad \beta = \frac{Q_{in}}{V_{Be}}, \qquad t' = t - \frac{V_E}{Q_{in}} \tag{V-6-38}$$

and for the volume of the attachment bar,

$$V_A = V_{AE}\left(1 - e^{-\gamma t''}\right), \qquad \gamma = \frac{Q_{in}}{V_{AE}}, \qquad t'' = t' - \frac{V_B}{Q_{in}} \tag{V-6-39}$$

- The quantities β and γ are analogous to α in representing time scales for the bypassing bar and attachment bar, respectively.

- The quantities t' and t'' in Equation V-6-38 and V-6-39 can be interpreted as lag times that delay development of the bar and attachment, respectively. To see this explicitly, Taylor expansions for small relative time give $t' \approx \alpha t^2/2$ and $t'' \approx \alpha^2\beta\, t^4/8$, as compared to growth of the ebb shoal given by αt.

- The interpretation is that after creation of an inlet, a certain time is required for the bypassing bar to receive a significant amount of sand from the shoal and a longer time for the attachment bar or beach to receive sand. Similarly, modification of, say, the ebb shoal as through sand mining will not be observed immediately in the bypassing at the beach because of the time lags in the system.

- A unique crossover time t_c occurs when the volume of material leaving the shoal equals the volume retained, $(V_E)_{out} = V_E$. After the crossover time, the shoal bypasses more sediment than it retains, characterizing the time evolution of the ebb shoal and its bypassing functioning. The crossover time is determined from Equation V-6-36 to be:

$$t_c = \frac{1.59}{\alpha} \qquad \text{(V-6-40)}$$

- Finally, by analogy to (8), the following equations are obtained for the bypassing rate of the bar $(Q_B)_{out}$, which is equal to the input of the attachment $(Q_A)_{in}$, and the bypassing rate of the attachment $(Q_A)_{out}$, which is also the bypassing rate or input to the beach, $(Q_{beach})_{in}$:

$$\left(Q_B\right)_{out} = \frac{V_E}{V_{Ee}} \frac{V_B}{V_{Be}} Q_{in} = \left(Q_A\right)_{in} \qquad \text{(V-6-41)}$$

$$\left(Q_A\right)_{out} = \frac{V_E}{V_{Ee}} \frac{V_B}{V_{Be}} \frac{V_A}{V_{Ae}} Q_{in} = \left(Q_{beach}\right)_{in} \qquad \text{(V-6-42)}$$

- The quantity $(Q_A)_{out}$ describes the time dependence of the amount of sand reaching the downdrift beach and is, therefore, a central quantity in beach nourishment and shore-protection design.

(i) Validation of model for Ocean City, Maryland. Calculations are compared with observations of the growth in the ebb shoal at Ocean City Inlet, Maryland. Ocean City Inlet was opened by a hurricane in August 1933. Stabilization of the inlet began 1 month later by placement of jetties (Dean and Perlin 1977), with the south and north jetties constructed during 1934 and 1935. Rosati and Ebersole (1996) estimated that between 4.3×10^5 and 9.7×10^5 cu m of sediment were released during the island breach. This material would be apportioned to the flood shoal, ebb shoal, and adjacent beaches. Assateague Island, located to the south and downdrift, began to erode in a catastrophic manner because of interruption of sediment formerly transported from the beaches of Ocean City. Erosion of Assateague Island and growth of the ebb shoal have been well documented (e.g., Dean and Perlin 1977; Leatherman 1984; Underwood and Hiland 1995; Rosati and Ebersole 1996; Stauble 1997).

- The location and shape of the ebb shoal and bar at Ocean City can be inferred from the locations of wave breaking, as shown in Figure V-6-40. Numerous bathymetry surveys (Underwood and Hiland 1995; Stauble 1997) confirm this inference. The bypassing bar is skewed to the south and has continued to move to the south (Underwood and Hiland 1995). Several independent authors have noted that much of the longshore sand transport moving to the south along Ocean City is diverted to the ebb shoal. Dean and Perlin (1977) concluded that the north jetty area was fully impounded, and the U.S. Army Engineer District, Baltimore (Personal Communication 1999, G. Bass, Senior Engineer, U.S. Army Engineer District, Baltimore, noted that recent growth of the northern edge of the ebb shoal may be composed of beach fill material placed on Ocean City beaches.

- Dean and Perlin (1977) estimated the long-term net (southward) longshore sand transport rate as between approximately 1.15×10^5 and 1.50×10^5 cu m/year, based on impoundment at the north jetty. Underwood and Hiland (1995) estimated the equilibrium volume of the ebb-shoal complex as between 5.8×10^6 and 7.2×10^6 m^3 based on the tidal prism of 2.3×10^7 cu m given by Dean and Perlin (1977) and calculation methods given in Walton and Adams (1976). With M_{tot} estimated by the upper value of net transport, one finds $r \cong 150$, consistent with lack of a bar across the entrance channel and good navigation. In fact, the entrance channel is rarely dredged.

- Bathymetry survey data were available to this study for the dates of 1929/1933 to define the preinlet condition, 1937, 1961/1962, 1977/1978, 1990, and a composite of various surveys conducted in 1995. Underwood and Hiland (1995) developed data sets prior to 1995, and Stauble (1997) assembled the 1995 data set from various sources for the Baltimore District. As part of the present work, the raw data sets were reviewed and vertical datums made consistent.

- The seaward boundary of lines unambiguously defining depositional features (ebb shoal, bypassing bar, and attachment bar) were found to be located at the 6-to 7-m National Geodetic Vertical Datum (NGVD) depth contour. Contours lying deeper than 7 m exhibited randomness and loss of identity of the particular feature. Here, to avoid the necessity of employing color to denote relatively complex contours, the landward portions of the bypassing bar and attachment bar polygons are defined by the zone of deposition as given by comparisons of bathymetry change. The lateral and landward boundaries of the ebb shoal polygons are defined by deposition from 1937 to 1962, and by a combination of deposition and depth contours for later time periods. This procedure accounts for the observation that the ebb shoal was fully developed by 1962, whereas changes were observed for the bypassing bar and attachment bar. Based on inspection of several depth contours and the differences in bathymetric surfaces, the ebb shoal was defined as a polygon within the area occupied by the 1962 shoal. As shown in Figure V-6-42a, the 1937 survey revealed a small ebb shoal, evidently located within the confines of the ebb-tidal jet. By 1962, a bypassing bar had formed that emanated southward from the ebb shoal.

- The bypassing bar was defined by a polygon located to the south of the ebb shoal, and the attachment bar was defined by the position of the high-water shoreline in the later data sets. By this means, volume change could be calculated for the distinct morphological features as differences between successive surveys, and the evolution of these features is plotted in Figure V-6-42b. Limited data sets were available for the attachment bar. Figure 4b also shows accretion of the shoreline near the south jetty, promoted by sand tightening of the jetty in 1985. In both Figure V-6-42a and V-6-42b, no notable growth to the north of a bypassing bar is evident.

- For evaluation of the analytic model, input values were specified based on coastal-processes information from the aforementioned studies, with no optimization of parameters made. The four required parameters were specified as $Q_{in} = 1.50 \times 10^5$ cu m/year corresponding to the upper limit of expected net longshore transport to the south, $V_{Ee} = 3 \times 10^6$ cu m, $V_{Be} = 7 \times 10^6$ cu m, and $V_{Ae} = 5 \times 10^5$ cu m.

- The value of V_{Ae} was arbitrarily assigned as an order-of-magnitude estimate. Calculations were made for the 100-year interval 1933-2032.

- The measured and calculated volumes of the ebb shoal are plotted in Figure V-6-43, with the dashed lines calculated for values of α for $Q_{in} \pm 50,000$ cu m/year to demonstrate sensitivity of the solution to α and to estimate the range of predictions that may be reasonably possible. The trend in the data is well reproduced by Equation V-6-33. Although the Q_{in}-value chosen is at the upper range for the net transport rate, this value must also account for sediment sources other than the net drift to the south. In particular, prior to 1985 when the south jetty was tightened, some sand moving north could pass through it and into the navigation channel (Dean and Perlin 1977), where a portion would be jetted offshore.

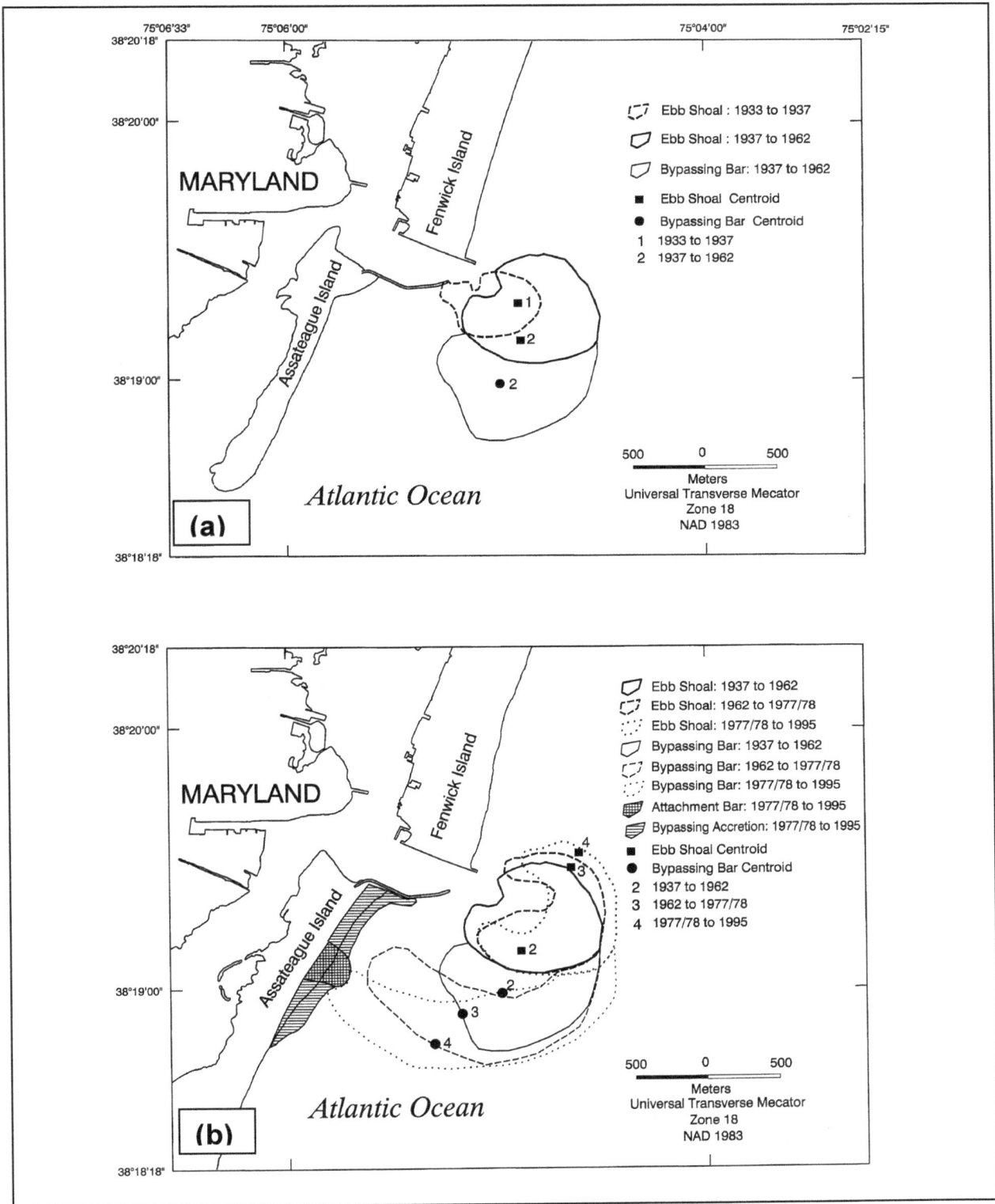

Figure V-6-42. Ebb-shoal planform determined from interpretation of 7-m contour, Ocean City Inlet

Sediment Management at Inlets

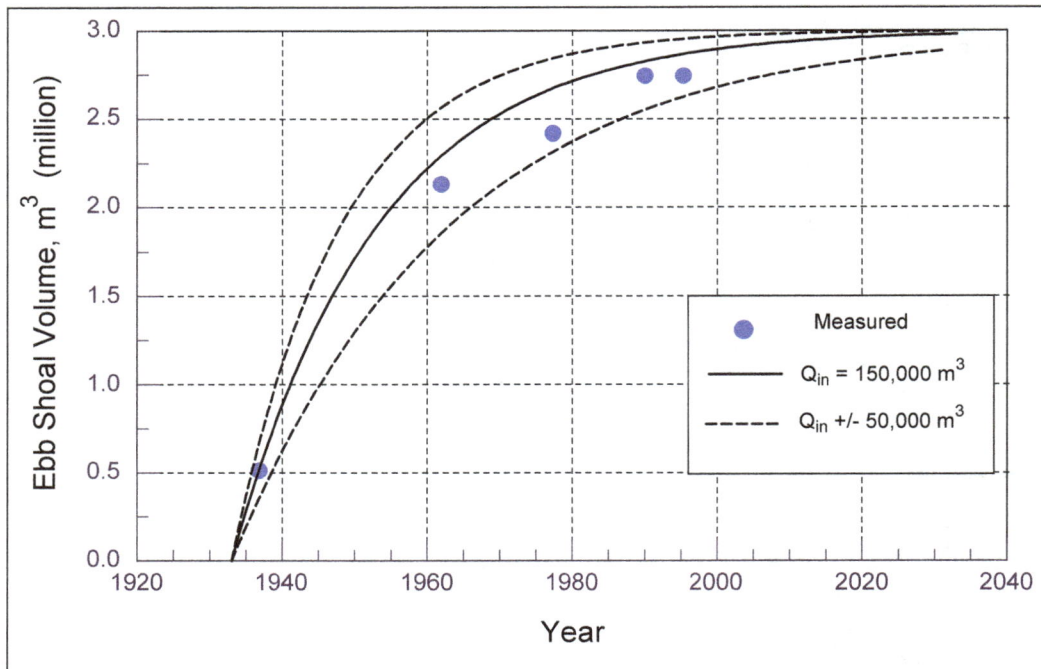

Figure V-6-43. Volume of ebb shoal, Ocean City Inlet

- Calculated and measured volumes of the ebb shoal, bypassing bar, and attachment bar are plotted in Figure V-6-44. The calculations exhibit lags in development of the bypassing bar and attachment bar. Based on examination of aerial photographs, Underwood and Hiland (1995) concluded that the attachment had occurred by 1980, when a distinct bulge in the shoreline was seen. The data and model indicate that, although the ebb shoal has achieved equilibrium, the bypassing bar is continuing to grow, so that natural bar bypassing from north to south has not achieved its full potential for sand storage. Bypassing rates calculated with the model, normalized by Q_{in}, are plotted in Figure V-6-45 and indicate the approach to full potential.

- The calculations show a substantial lag in sand reaching the bypassing and attachment bars. The bypassing rate at the attachment equals the rate of sand reaching the downdrift beach. Its magnitude as shown in Figure V-6-45 should be interpreted with caution, because the equilibrium volume of the attachment is not presently known. Under the given model input parameters, it appears that in 1999 approximately 60 percent of the net transport is reaching northern Assateague Island. With the stated values, as a simple estimate one can define an effective α for Ocean City through the sum of the ebb shoal and bypassing bar equilibrium volumes. Then the crossover time at which the shoal and bar are predicted to bypass more volume than they retain is $t_c = 77$ years or in the year 2000, in accord with the 60 percent estimate of bypassing.

(j) Discussion. This section explores sensitivity of the reservoir model to selection of the input and initial conditions, and its extension by numerical solution.

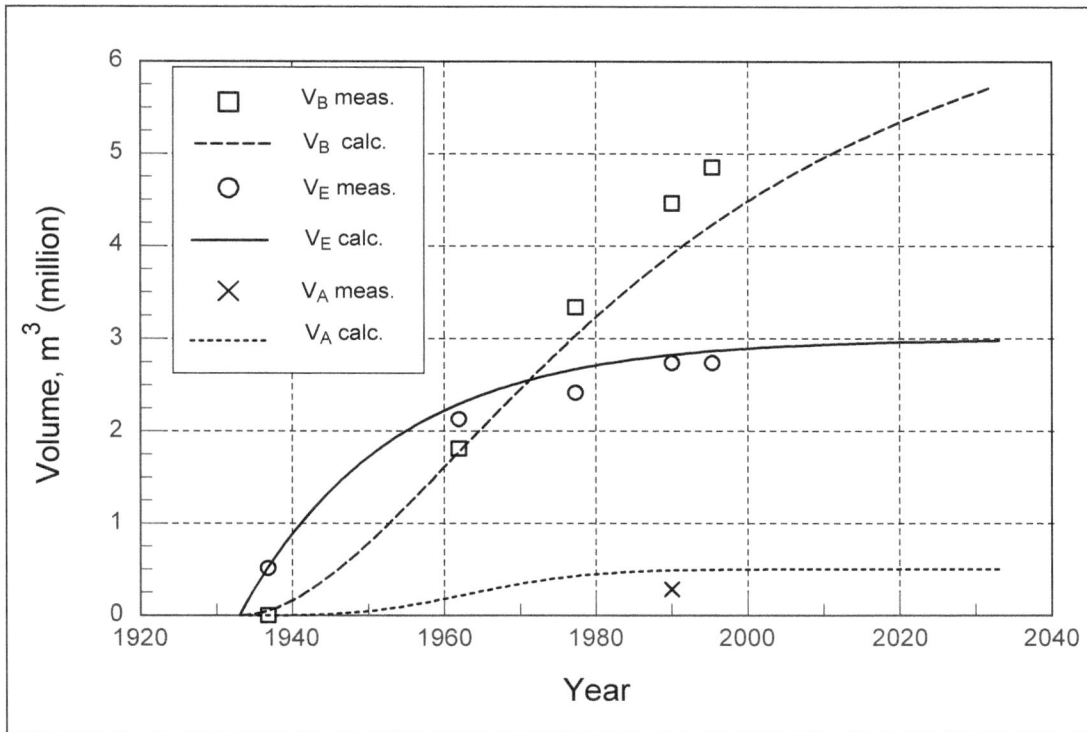

Figure V-6-44. Volumes of ebb shoal, bypassing bar, and attachment bar, Ocean City Inlet

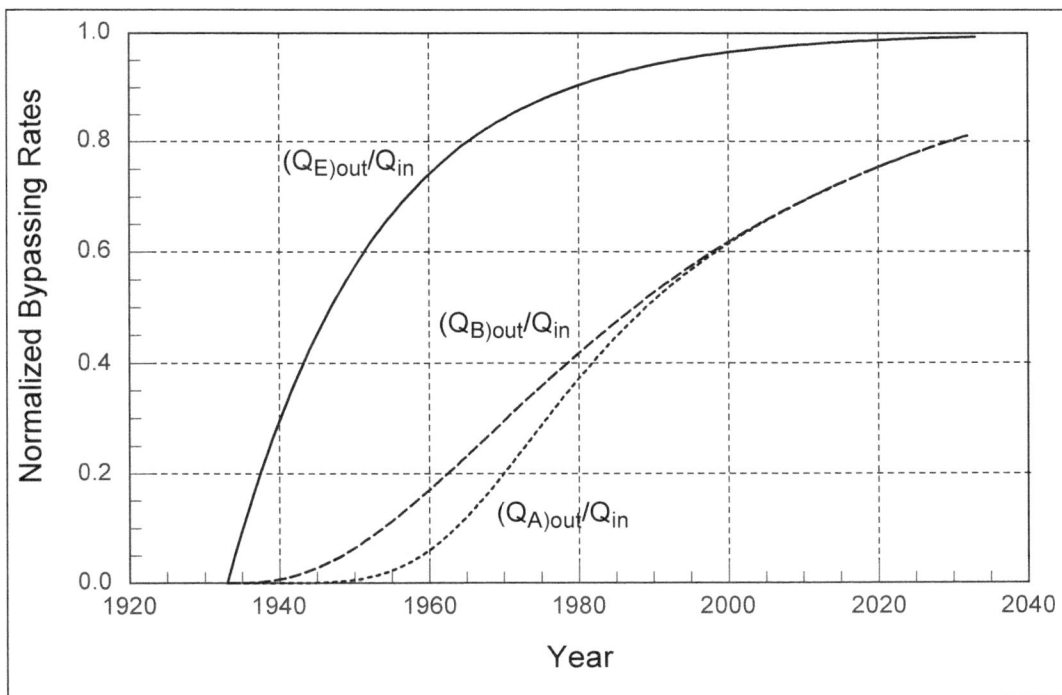

Figure V-6-45. Calculated bypassing rates, Ocean City Inlet

- Sensitivity to Q_{out} specification.

- The manner in which the equilibrium volume of an ebb shoal is approached depends upon Q_{out}. Choices other than the linear form of Equation V-6-30 might be made, leading to consideration of the sensitivity of the solution on Q_{out}. As a possible alternative for a quadratic dependence as $Q_{out} = (V_E/V_{Ee})^2 Q_{in}$ can be specified. Then one finds:

$$V_E = V_{Ee}\tanh(\alpha t) \qquad\qquad\qquad (V\text{-}6\text{-}43)$$

- Equations V-6-33 and V-6-43 are compared in dimensionless form in Figure V-6-46. The quadratic dependence version of Q_{out} produces a more rapid approach to equilibrium. However, the general forms of the solutions are similar, indicating that a substantial change in the manner in which the shoal bypasses sand does not cause a notable change in approach of the shoal to equilibrium. There appear to be no data available to distinguish among such solutions, but specification of Q_{out} is available for improving simulations of all the inlet morphologic forms once adequate observations are made.

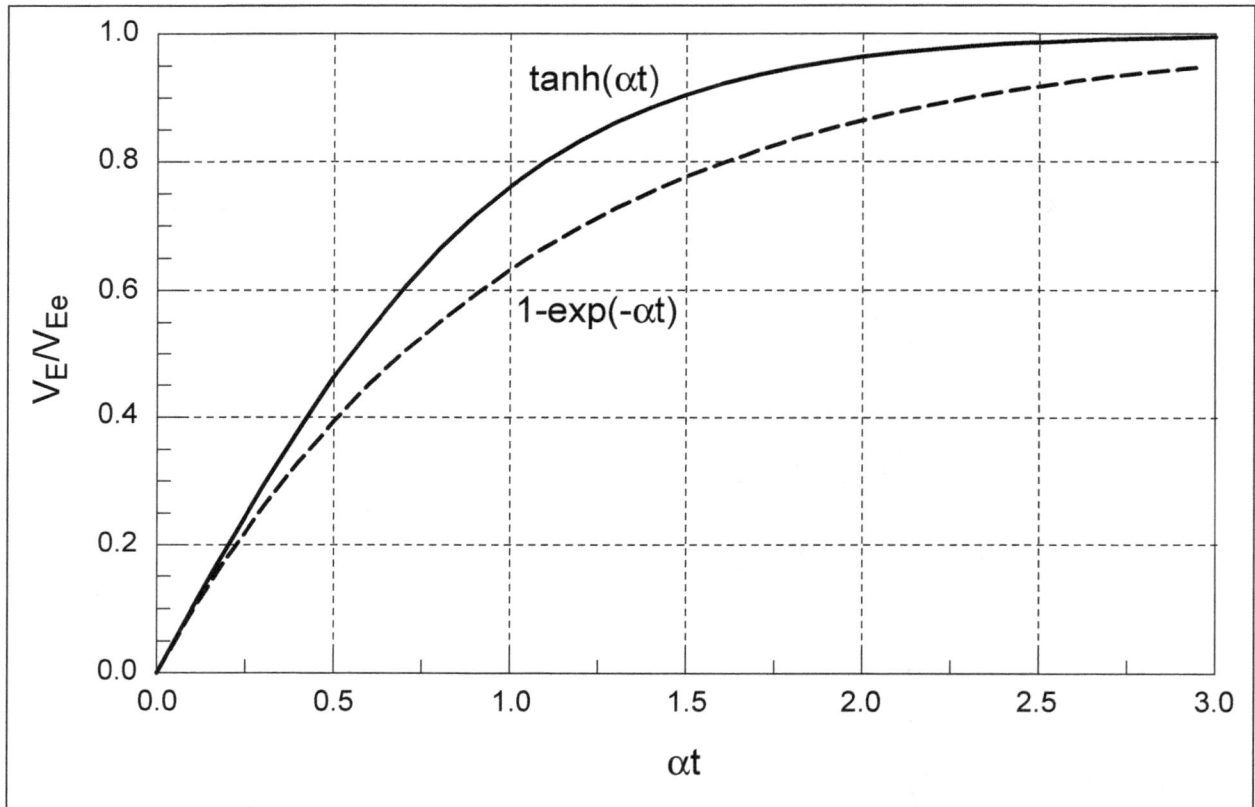

Figure V-6-46. Solutions based on linear and quadratic forms for Q_{out}

- Initial breach volume.

- At the initial breach of an inlet, tidal and littoral currents will distribute the released material to adjacent beaches and form a flood shoal and an ebb shoal. Distribution of material among these three areas will depend on the strength and asymmetry of the tidal current and on the incident wave height and direction, among other factors. The time over which the initial distribution occurs is expected to be much shorter than the time scale $1/\alpha$ and can be approximated by an initial ebb-shoal volume V_{E0} at $t = 0$. With this initial condition, Equations V-6-30 and V-6-31 yield:

$$V_E = V_{Ee}\left(1 - e^{-\alpha t}\right) + V_{E0}e^{-\alpha t} \qquad\qquad\text{(V-6-44)}$$

- For a probable overestimate such as $V_{E0} \approx 0.1V_{Ee}$, inclusion of an initial ebb-shoal volume from the initial breach does not significantly alter the trend of evolution of the shoal, especially considering survey accuracy and temporal and spatial variability in the morphologic system that would obscure minor changes.

- Numerical solution.

- The governing equations, such as Equation V-6-31 and related initial condition, can be solved numerically to represent an arbitrary initial condition and time-varying forcing by Q_{in}. For example, assuming Q_{in} is time dependent, a second-order accurate, unconditionally stable solution of Equation V-6-31 is:

$$V_E' = \frac{\Delta t}{2\left(1 + \frac{\Delta t}{2V_{Ee}}Q_{in}'\right)}\left[Q_{in}' + Q_{in} + \left(1 - \frac{\Delta t}{2V_{Ee}}Q_{in}\right)V_E\right] \qquad\qquad\text{(V-6-45)}$$

where quantities denoted with a prime indicate values at the next time-step, and Δt = time-step. To validate the solution method, Equation V-6-45 was implemented for Ocean City Inlet. With $\Delta t = 0.1$ year, the analytical and numerical solutions plotted on top of one another. In exploration of the solution scheme, reasonable accuracy was maintained with $\Delta t = 5$ years for the constant input transport rate.

- As an example, engineering application of the numerical model, recovery of the ebb-tidal shoal and alteration of bypassing rates at Ocean City Inlet are calculated in response to hypothetical mining of the bypassing bar. Limited quantitative work has been done to estimate the consequences of ebb-shoal mining (e.g., Mehta, Dombrowski, and Devine 1996; Cialone and Stauble 1998). Walther and Douglas (1993) reviewed the literature and applied an analytical model to three inlets in Florida to estimate recovery time of the shoal and bypassing rates. No time lag was included in their model, however, which yields different responses depending upon depth and location of the mining, factors not included in the present aggregate model.

- For the present example, 750,000 cu m were removed from the bypassing bar in the year 2000, which will be about 25 percent of the material comprising the bar at that time. Figure V-6-47 shows plots of the evolution of bar volume and bypassing rates from the bar and from the attachment bar to the beach. The volume was normalized by V_{Be}, and the bypassing rates by Q_{in}. Mining at the year 2000 stage of development is predicted to effectively translate the bar growth and bypassing rates approximately 20 years back in time, with the bypassing rate to the beach moving from about 0.6 to 0.45 of the potential maximum value of Q_{in}. This example with simplified conditions is not adequate for design, but it does indicate possible applicability of the model in comparison of alternative mining plans. More rigor could be introduced through inclusion of a time-dependent Q_{in} in the present model and estimation of the consequences of ranges of variability in the governing parameters.

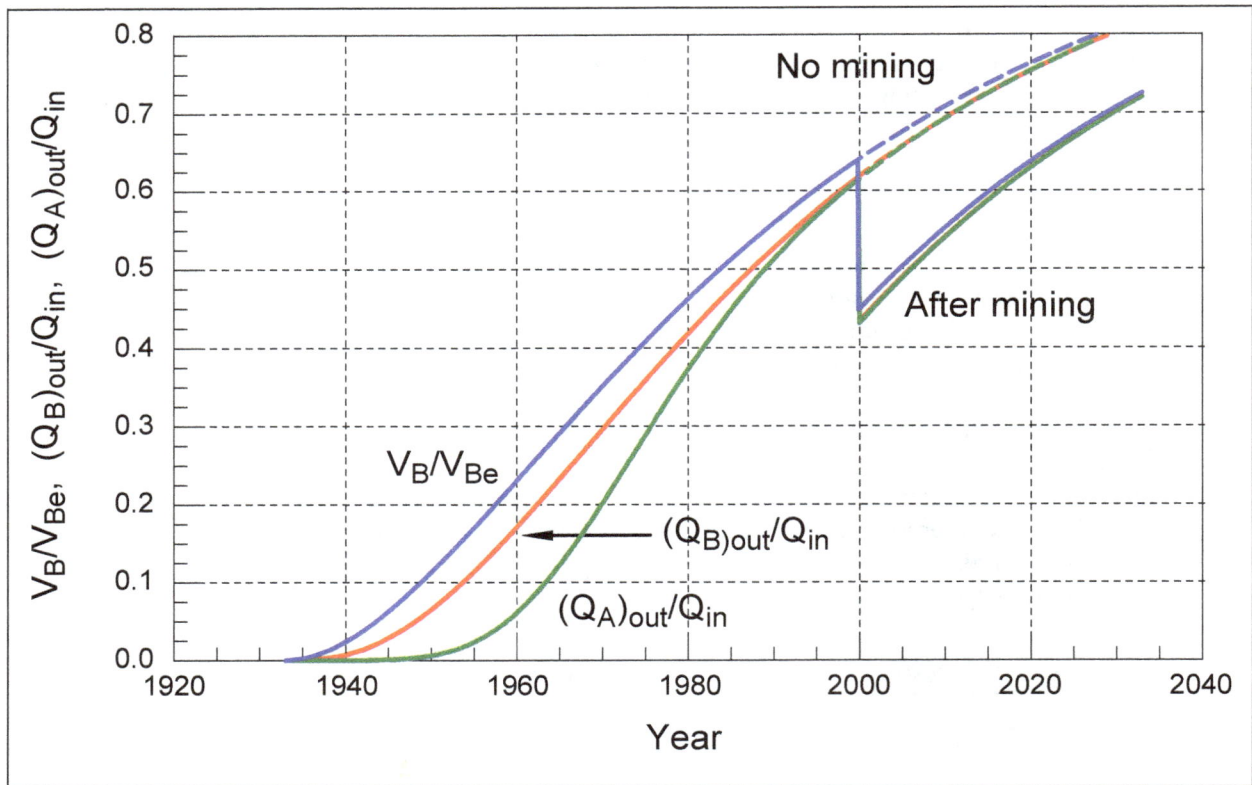

Figure V-6-47. Bypassing bar volume and bypassing rates with/without mining

- Finally, an example involving idealized bidirectional longshore transport is presented. To interpret results readily, the equilibrium volumes of the ebb shoal, bypassing bar, and the attachment bar were set at $V_e = 1 \times 10^6$ cu m, and the magnitude Q the input longshore transport rate was one-tenth of this amount, whether from directed to the right or to the left. Starting from an initial condition of no inlet features, the transport was directed to the right for 25 years and then to the left for 25 years. Time evolution of the normalized volumes and bypassing rates are shown in Figure V-6-48a and V-6-48b, respectively, where subscripts "R" and "L" denote quantities associated with right- and left-directed transport.

- Under the stated condition of equal magnitude but opposite transport, the volume of the ebb shoal grows without discontinuity, because the shoal accepts sand from either direction. With transport directed to the right, the volume of the bypassing bar V_{BR} and of the attachment bar V_{AR} grow while experiencing the characteristic time lag. These features would emerge on the right side of the ebb shoal, for a viewer standing on the shore and facing the water. When the transport rate shifts after 25 years, the volumes V_{BL} and V_{AL} of features on the left side of the ebb shoal begin to grow, but the bypassing bar on the left experiences no time lag because the ebb shoal has a sand supply to contribute immediately. The attachment bar on the left experiences a shorter time lag as compared to its counterpart on the right because transported sand is delayed only by formation of its (left side) bypassing bar and not by the ebb shoal. The bypassing rates show behavior similar to the volumes. In particular, the left-directed bypassing rate on the ebb shoal starts at a large value because of the existence of the shoal created by the right-directed transport. Also, the bypassing rate from the left bypassing bar begins immediately after the switch in transport direction, and the attachment shoal on the left experiences a much shorter lag in receiving sand to bypass than did the attachment shoal on the right.

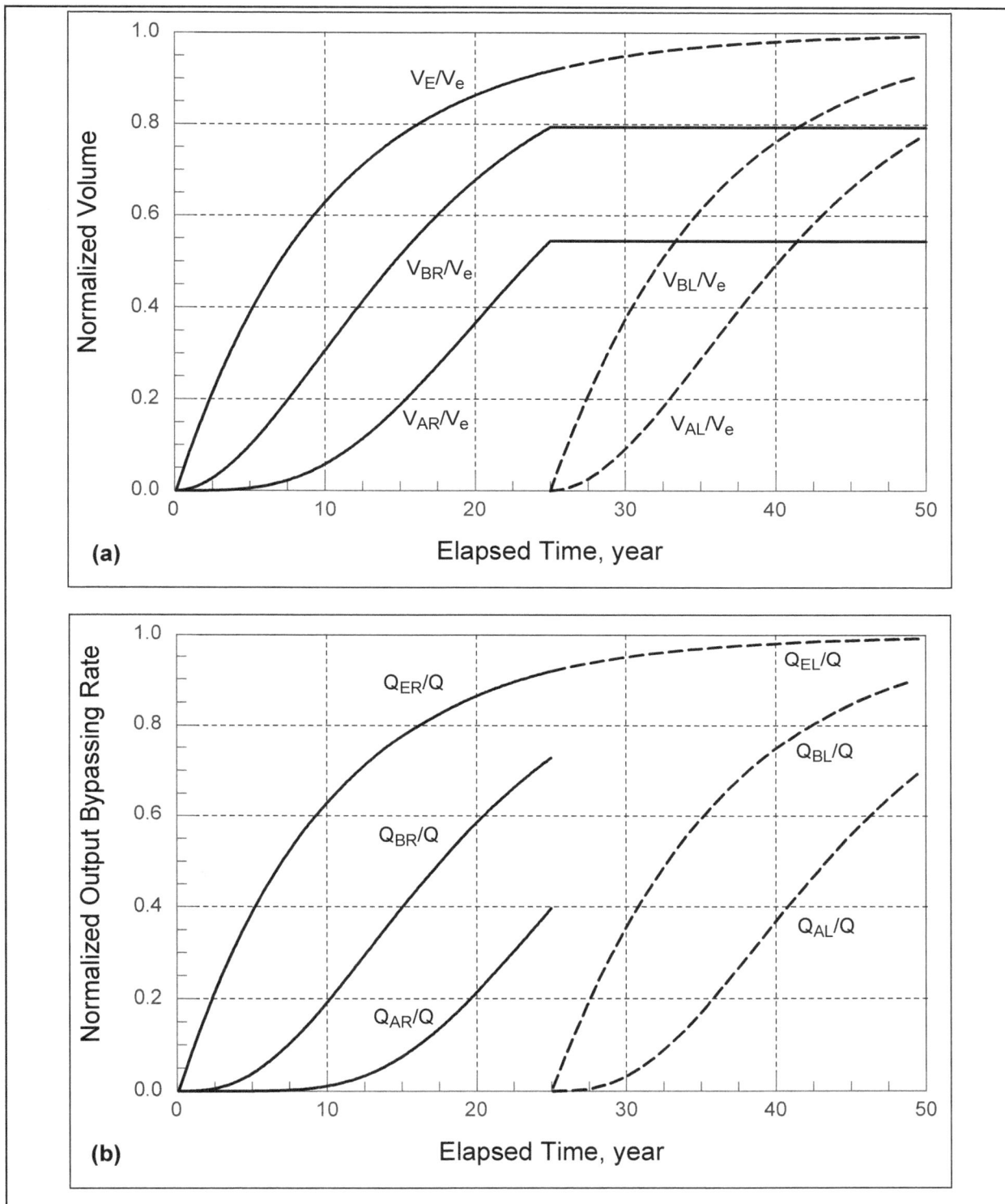

Figure V-6-48. Bypassing rates for simple bidirectional transport

(k) Conclusions. A reservoir model was introduced for describing changes in volume and bypassing rates of morphological components of ebb-tidal shoals. Required inputs for this aggregate model are compatible with the amount and quality of data typically available in engineering and science studies. The model requires estimates of the longshore transport rate, which may be the net or gross rate depending on the inlet configuration; equilibrium volume of the ebb shoal, bypassing bar, and attachment bar; and qualitative understanding of sediment pathways at the particular inlet. The reservoir model is robust in that solutions are bounded. The ratio of the input longshore transport rate and the equilibrium volume of the morphological feature is the main parameter governing volume change and bypassing rates. This parameter is directly related to the widely accepted Bruun and Gerritsen ratio. The reservoir model predicts a delay in sand bypassing to the downdrift beach according to the properties of the morphologic system and longshore transport rate.

- The reservoir method requires apportionment of material between the ebb shoal and the bypassing bar, although a simplified version of the model can combine these two sand bodies. A distinction between the shoal and the bypassing bar adds conceptual and quantitative resolution by allowing bypassing bars to develop and evolve according to the properties of the predominant transporting mechanism as either waves (longshore sediment transport) or tidal prism. With constant inputs, the analytic model takes about 1 sec to execute on a standard PC for 100 years of simulation time, allowing numerous runs to be made. The numerical model is also rapid, even if time-dependent rates are involved.

- The model as described here does not account for sediment exchange between the inlet channel and flood shoal. However, it can be readily extended both analytically and numerically to include these and similar interactions. The model appears capable of substantial generalization and incorporation of more detailed, processed-based data. In this regard, the reservoir model may serve as a source for preliminary design and provide a framework for generating questions about inlet morphology, sediment pathways, and the fundamental mechanisms of the collective behavior of sand bodies.

(3) Wave energy flux method (Jarrett 1977, 1991).

(a) Jarrett applied sediment budget analysis techniques together with estimates of wave energy flux (derived from wave refraction analysis) to infer longshore sand transport rates for the North Carolina coast (Jarrett 1977, 1991). Data required to apply the method include incident wave climatology, shoreline positions and/or beach profiles, inlet bathymetry, and engineering activities (dredging and placement volumes, beach-fill placement, and sand bypassing rates) for a common time period. Jarrett recommends that the best time period for analysis is one that is free from significant storm events. Storms have the potential to bias the sediment budget, whether it is to reverse the long-term sediment transport direction or cause short-term transport of sediment offshore. The procedure is as follows:

- Analyze basic data sets. Calculate volume change associated with the shoreline position, profile, inlet bathymetry, and engineering activity data sets. (see Sections (e) and (f)).

- Identify sediment budget cells. Based on the volume changes for each inlet and beach system, identify regions of the study area that appear to behave differently. Potential longshore transport rates will be determined for the boundaries of each of these cells. Jarrett recommends extending inlet cell boundaries several thousand feet up- and downdrift from the inlet to incorporate the effects of the inlet ebb tidal delta and tidal currents within the inlet cell. Natural bypassing across the inlet is left as an unknown that is determined through solving the sediment budget.

- Determine the longshore component of wave energy flux in the surf zone. The longshore component of wave energy flux, P_l, is proportional to the amount of sand being transported along the coast through a factor, β, as shown in Equation V-6-46. This quantity can be calculated by transforming incident waves over offshore bathymetry until wave breaking. Then P_l is related to the breaking wave height, depth at breaking, and breaking wave angle (see Part III-3). Longshore energy flux should be calculated for the entire study area both for left- and right-directed transport. Representative values of left- and right-directed P_l for each cell boundary should be determined.

$$Q = \beta P_\ell \qquad (V\text{-}6\text{-}46)$$

- Formulate equations between volume change and P_l. For each cell, equations can be written to solve for the relationship between the forcing (longshore energy flux, P_l), and the resulting sediment transport gradient (net volume change within the cell, considering all engineering activities within that cell). Jarrett relates the correlation as follows (notation has been adjusted to be consistent within this chapter):

$$\Delta V - P + R = \beta P_\ell \qquad (V\text{-}6\text{-}47)$$

where β is the unknown constant, and ΔV, P, and R are as defined previously.

- Solve for β in the vicinity of the inlets. At the inlet cells, there are three unknowns: bypassing to the updrift beach (during periods of reversal), bypassing to the downdrift beach, and the constant β. Three independent equations can be formulated considering the cell immediately updrift of the inlet, the cell downdrift of the inlet, and the inlet cell. After solving these three cells, a value for the parameter β is obtained which is then used to estimate sand transport between the other cells in the study area.

- Optimize the parameter β to best represent volume changes for all cells. Adjustments to the parameter β may be required to develop a balanced sediment budget for the entire region.

- The following example for the North Carolina coast is summarized from Jarrett (1991). Figure V-6-49 shows the study reach with the sediment budget cells identified in bold lettering.

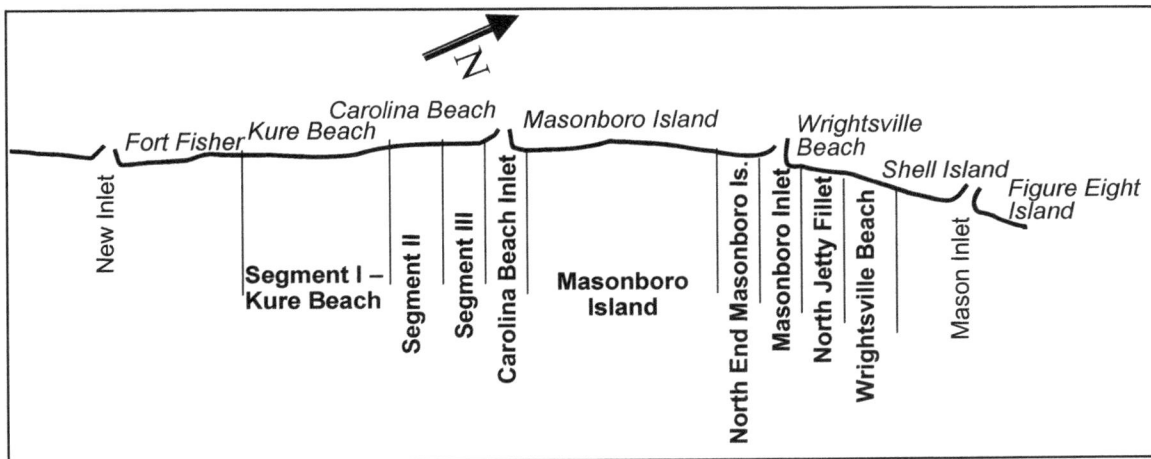

Figure V-6-49. Study reach for sediment budget (not to scale, adapted from Jarrett 1991)

Sediment Management at Inlets

- Segments of the study shoreline with similar trends, as determined from shoreline change data, were defined as sediment budget cells. Cell boundaries in the vicinity of inlets were located several thousand feet from the interior shoreline of the inlet. The inlet boundaries were set so that sediment transport within the inlet cell reflected the combined forcing of waves and tidal currents, with waves having been influenced by the ebb-tidal shoal. Within each shoreline cell, profile and shoreline change data were used to determine the rate of volume change. Sources of sediment, such as beach nourishment or sand bypassing operations, and sinks of sediment, such as inlets and capes, were determined from review of the engineering history, hydrographic surveys, and dredging records. The volume rate of material naturally bypassing each inlet in both the upcoast and downcoast directions were set as unknowns.

- A longshore energy flux analysis was conducted using wave information offshore of the study site, and transforming the waves over the offshore and nearshore bathymetry. Values of the longshore energy flux factor were related to both the left- and right-directed longshore transport rates using Equation V-6-46. These values are shown above each longshore transport rate arrow in Figure V-6-50. Once the transport potential was determined between the various cells, computations using Equation V-6-47 were used to solve for β. As calculations proceeded from cell-to-cell, some adjustments in the computed values of the longshore transport rate were required to develop a completely balanced sediment budget for the study area. The final sediment budget is shown in Figure V-6-50, with the calculated transport rates shown below each arrow.

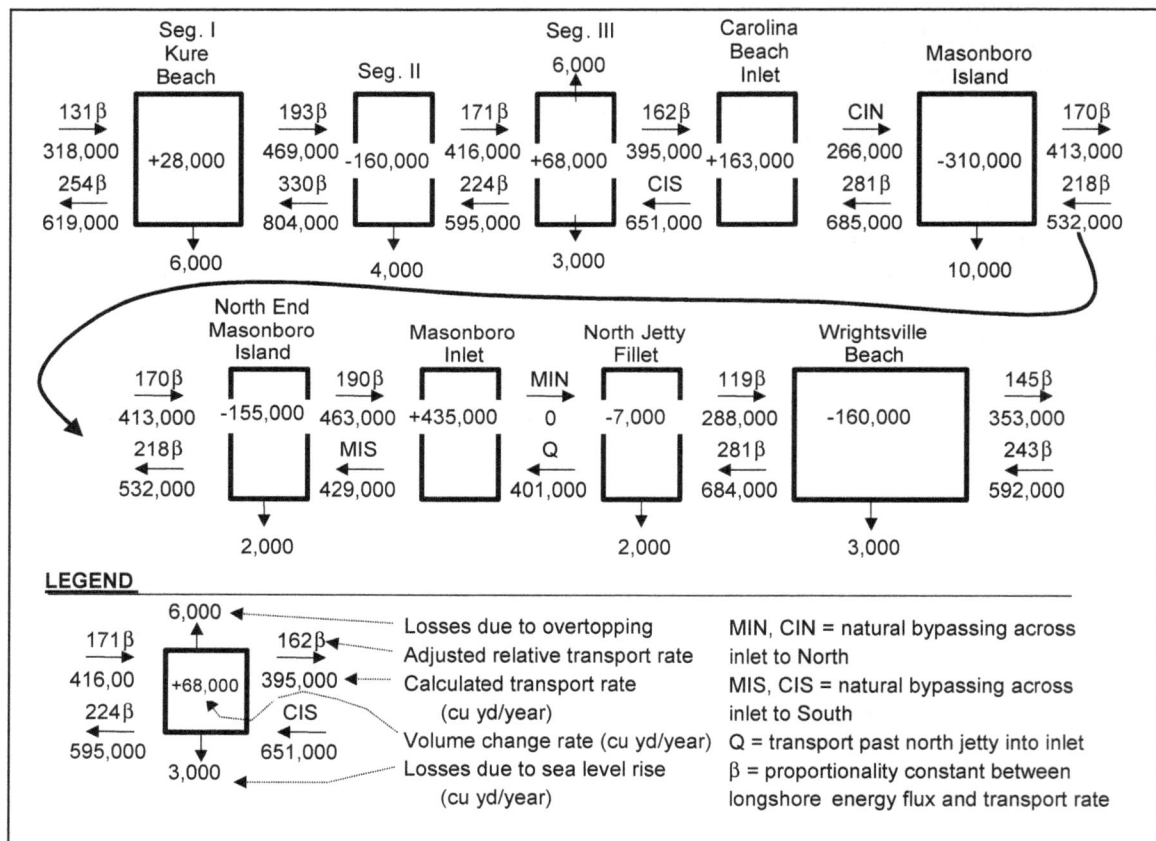

Figure V-6-50. Sediment budget from Kure Beach (south end of study) to Wrightsville Beach (north end of study), North Carolina, using the wave energy flux method (adapted from Jarrett 1991). Note that the Masonboro Island cell connects to the north end of Masonboro Island cell

(4) Bodge's method (1999).

(a) Bodge's approach is an extension of those described by Bruun (1966), Weggel (1981), as well as early case studies by Johnson (1959), Jarrett (1977), among others. Figure V-6-51 illustrates the sediment transport pathways at an idealized inlet for right- and left-directed incident transport, and mechanical transport (dredging and bypassing). Except as noted, all right-directed transport is positive-valued. All left-directed transport is negative-valued. Right- and left-directions are based upon an observer standing at the inlet, facing seaward. The study area boundaries are selected as locations outside of the inlet's direct influence on wave refraction and tidal currents.

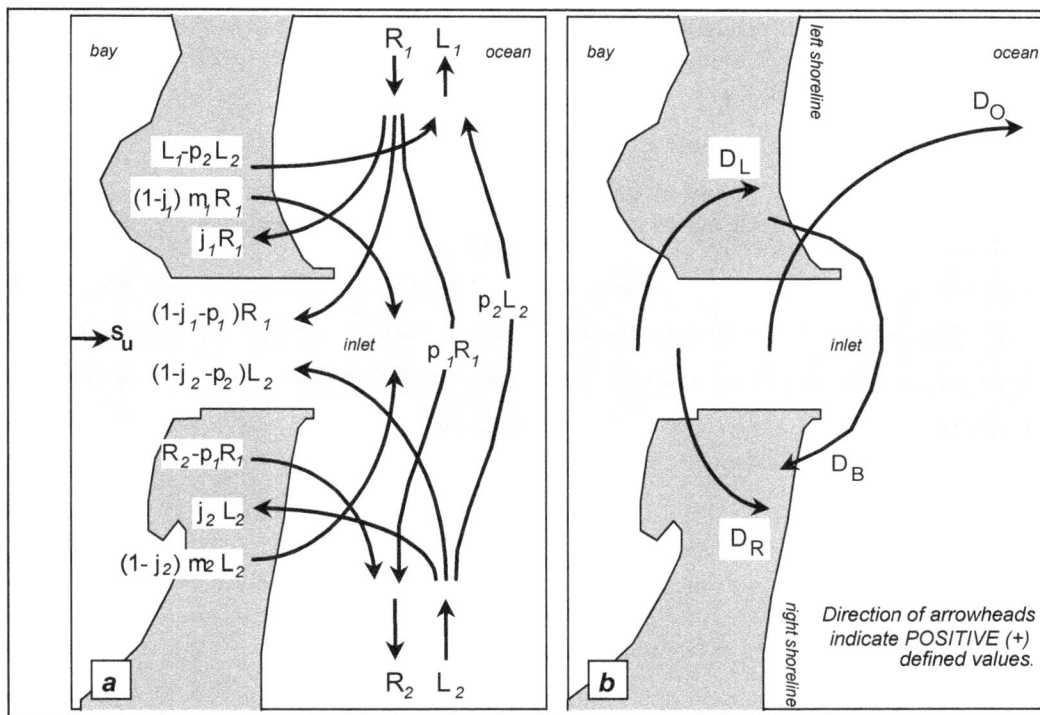

Figure V-6-51. Sediment transport pathways for Bodge method (1999)

(b) In Figure V-6-51a, each transport component is defined as a fraction or multiple of the right- and left-directed longshore transport rates at the boundaries. The subscript "1" refers to transport directed rightward from the left shore, and the subscript "2" refers to transport directed leftward from the right shore. In Figure V-6-51a and V-6-51b, the terms illustrated in the sediment transport pathways are as follows:

R, L = rightward- and leftward-directed incident transport values at the study area's boundaries

j_1, j_2 = fraction of incident transport (R or L) impounded by the inlet's jetties (j_1 = left jetty; j_2 = right jetty; 0.0 = transparent jetty; 1.0 = impermeable jetty)

p_1, p_2 = fraction of incident transport (R or L) naturally bypassed across the inlet (p_1 = from the left, p_2 = from the right; 0.0 = no bypassing; 1.0 = perfect bypassing)

m_1 = local inlet-induced transport from the left shoreline into the inlet (expressed as a fraction or multiple of the right-directed incident transport, R_1)

m_2 = local inlet-induced transport from the right shoreline into the inlet (expressed as a fraction or multiple of the left-directed incident transport, L_2)

S_u = transport of littoral material into the inlet from upland sources (positive value)

D_L, D_R = mechanical transfer of sand from the inlet to the left and right shorelines, respectively

D_B = mechanical transfer of sand from the left shoreline to the right shoreline (defined as positive from left to right; negative from right to left)

D_O = maintenance dredging and out-of-system disposal from the inlet (positive-valued; includes only material of littoral origin; includes deepwater (offshore) and upland disposal; excludes new work)

(c) The transport terms (R, L, Q, S_u, D_L, D_R, D_O) can be expressed as either volume quantities or volumetric rates, so long as the units of each term are consistent with one another. The ratio of left- to right-directed transport magnitude at the study area's boundaries is defined as:

$$r_1 = -L_1 / R_1 \text{ and } r_2 = -L_2 / R_2 \tag{V-6-48}$$

- By definition,

$$
\begin{array}{ll}
0 \le j_1 \le 1 & 0 \le j_2 \le 1 \\
0 \le p_1 \le 1 & 0 \le p_2 \le 1 \\
0 \le p_1 + j_1 \le 1 & 0 \le p_2 + j_2 \le 1 \\
m_1 \ge 0 & m_2 \ge 0
\end{array}
\tag{V-6-49}
$$

- The net volume changes of the left and right shorelines are, respectively,

$$\Delta V_L = (j_1 + j_1 m_1 - m_1) R_1 + L_1 - p_2 L_2 - D_B + D_L \tag{V-6-50}$$

$$\Delta V_R = (m_2 - j_2 m_2 - j_2) L_2 - R_2 + p_1 R_1 + D_B + D_R \tag{V-6-51}$$

where positive ΔV values imply net accretion and negative values imply net erosion. The gross volume of sand that enters the inlet, prior to maintenance dredging, is

$$\Delta V_G = (1 - j_1 - p_1 + m_1 - j_1 m_1) R_1 - (1 - j_2 - p_2 + m_2 - j_2 m_2) L_2 \tag{V-6-52}$$

- The inlet's net volume change after dredging, and neglecting upland/offshore input, is

$$
\begin{aligned}
\Delta V_N = \Delta V_G - D_L - D_R - D_O - S_U = & \left(1 - j_i - p_1 +_{,1} - j_1 m_1\right) \\
& R_1 - \left(1 - j_2 - p_2 + m_2 - j_2 m_2\right) L_2 - D_L - D_R - D_O - S_U
\end{aligned}
\tag{V-6-53}
$$

- Combining Equations V-6-50 through V-6-53 yields:

$$\Delta V_L = -(\Delta V_R + D_O + (\Delta V_N - S_u)) + \Delta R + \Delta L \tag{V-6-54}$$

$$\Delta V_R = -(\Delta V_L + D_O + (\Delta V_N - S_u)) + \Delta R + \Delta L \tag{V-6-55}$$

where $\Delta R = R_1 - R_2$ and $\Delta L = L_1 - L_2$.

- Physically, if the incident transport rates are identical on both sides of the inlet (that is, $\Delta R = \Delta L = 0$), Equations V-6-50 and V-6-55 demonstrate that the net volume change of an inlet-adjacent shoreline is the negative sum of the other shoreline's net volume change (e.g., impoundment); the volume of sand removed from the inlet by maintenance dredging and out-of-system disposal; and the net growth of the inlet shoal volumes (minus upland/offshore input). This simple and significant result states that an inlet's net volumetric effect to the downdrift shoreline is the sum of: the updrift impoundment; maintenance dredging and out-of-system disposal; and net shoal growth beyond that attributed to upland input. In this way, the global volumetric impact of the inlet to the downdrift shoreline (minus any impoundment fillet, or dead storage on the downdrift side) can be computed without reference to, or assumption of, measured downdrift shoreline changes, ambient longshore transport rates, or detailed mechanics of the inlet's transport pathways. While these data are ultimately useful, Equations (V-6-54) and (V-6-55) demonstrate that such data are not fundamentally required to assess the inlet's downdrift, volumetric impact, at least so long as differences in the ambient transport potential across the inlet are small or known.

- Equations V-6-50 through V-6-52 can be solved for p_1 and p_2 and combined to yield two coupled equations containing the volume change terms ΔV_L, ΔV_R, or ΔV_G. In practice, as previously noted, the net volume change of the downdrift beach is often most uncertain or suspect because of the effects of armoring or beach nourishment, or because the length of shoreline to consider is not known a priori. Accordingly, it is advantageous to remove the downdrift volume change term from the coupled equations, and to solve in terms of the volume changes of the updrift beach and the inlet.

- If the RIGHT shoreline is downdrift (or of less certain volume change), then from Equations V-6-50, V-6-51, and V-6-52:

$$p_1 = 1 - j_1 (1 + m_1) + m_1 - (L_2/R_1)(1 - j_2 (1 + m_2) + m_2 - p_2) - \Delta V_G/R_1 \qquad \text{(V-6-56a)}$$

$$p_2 = [(j_1 (1 + m_1) - m_1) R_1 + L_1 - \Delta V_L + D_L - D_B] / L_2 \qquad \text{(V-6-56b)}$$

and the corresponding, computed volume change of the right shoreline is:

$$\Delta V_R = (m_2 - j_2 m_2 - j_2) L_2 + p_1 R_1 - R_2 + D_R + D_B \qquad \text{(V-6-57)}$$

- If the LEFT shoreline is downdrift (or of less certain volume change), then from Equations V-6-50, V-6-51, and V-6-52:

$$p_1 = [(j_2 (1 + m_2) - m_2) L_2 + R_2 + \Delta V_R - D_R - D_B] / R_1 \qquad \text{(V-6-58a)}$$

$$p_2 = 1 - j_2 (1 + m_2) + m_2 - (R_1/L_2) (1 - j_1 (1 + m_1) + m_1 - p_1) + \Delta V_G/L_2 \qquad \text{(V-6-58b)}$$

and the corresponding, computed volume change of the left shoreline is:

$$\Delta V_L = (j_1 - m_1 + j_1 m_1) R_1 + L_1 - p_2 L_2 + D_L - D \qquad \text{(V-6-59)}$$

- To solve for the inlet sediment budget, a family of solutions is developed from either Equation V-6-56a,b or V-6-58a,b. Specifically, input values are identified for the updrift volume change (ΔV_L or ΔV_R) and the gross inlet volume change (ΔV_G) and for the dredging/bypassing quantities (D_O, D_L, D_R, D_B) and for upland/offshore inlet volume influx (S_U). A range of physically plausible values for the other transport parameters are additionally identified; i.e., the jetties' impermeability (j_1, j_2 (between 0 and 1)), the local inlet-induced transport (m_1, m_2), the tendency for natural bypassing (p_1 and p_2 (between 0 and 1)), and the incident transport components (R and L). In application, three incident transport components are required: R_1, R_2, and L_1 for Equation V-6-56, and L_1, L_2, and R_1 for Equation V-6-58.

- Candidate values for the various parameters, within their identified plausible ranges, are input to Equations V-6-56 or V-6-58. Those combinations of parameters that yield values of p_1 and p_2 within the allowed range of p_1 and p_2 (or, at least, values between 0 and 1), and which likewise satisfy

$$0 \le j_1 + p_1 \le 1 \quad \text{and} \quad 0 \le j_2 + p_2 \le 1 \tag{V-6-60}$$

  are retained as viable discrete solutions of the sediment budget. The set of all such viable, discrete solutions represents the sediment budget's family of solutions.

- The family of solutions can be conveniently plotted and inspected in the following way. For each discrete solution, values are computed for the incident (updrift) net transport Q, the natural net bypassing P, and the gross volume that shoals the inlet from the left and right shorelines, S_{LEFT} and S_{RIGHT}, respectively. If Equation V-6-56 is used, where the RIGHT shoreline is downdrift, then

$$Q = Q_1 = R_1 + L_1 \tag{V-6-61a}$$

  or if Equation V-6-58 is used, where the LEFT shoreline is downdrift, then

$$Q = Q_2 = R_2 + L_2 \tag{V-6-61b}$$

- The net natural bypassing and left- and right- gross inlet shoaling volumes are, respectively,

$$P = p_1 R_1 + p_2 L_2 \tag{V-6-62a}$$

$$S_{\text{LEFT}} = \Delta V_G - S_U - S_{\text{RIGHT}} = (1 - j_1 - p_1 + m_1 - j_1 m_1) R_1 \tag{V-6-62b}$$

$$S_{\text{RIGHT}} = \Delta V_G - S_U - S_{\text{LEFT}} = - (1 - j_2 - p_2 + m_2 - j_2 m_2) L_2 \tag{V-6-62c}$$

- The family of solutions can be narrowed by imposing additional constraints; e.g., requiring that the direction of any net natural bypassing, P, be coincident with the incident net transport rate, Q; or, that the jetties' impermeability values (j_1 and j_2) be similar to one another; or, that the shoaling from one side or the other be a minimum percentage of the inlet's total shoaling rate; etc. Or, families of solutions from different time periods can be overlaid, retaining only those subsets of solutions for which the incident transport, Q, solves the sediment budget for both time periods, etc.

- Methods to identify the sediment budget's parameters include the following. The volume change of the updrift shoreline (ΔV_L or ΔV_R) can be estimated from the cumulative volume change updrift of the inlet as in Figure V-6-18, where background changes are retained. The gross volume change of the inlet (ΔV_G) can be estimated from surveys and dredging histories, where care is taken to exclude changes due to new-work dredging, datum shifts, etc. The dredging and mechanical bypassing terms (D_R, D_L, D_O, D_B) are estimated from dredging records. The volume influx from upland or offshore sources (S_U) requires some insight as to the inlet's overall geology and/or the influence of, for example, episodic fluvial input, but in many cases can be neglected. A convenient method to identify ranges of values for the incident transport (R and L) is to consider the right- and left-directed transport potential from offshore (hindcast, etc.) wave data. The computed transport rates can be used directly (with ranges bound by some percentage or by standard deviation, etc.). It is additionally useful to compute the ratio of left- to right-directed transport, r, from Equation V-6-48. In solving the sediment budget equations, this ratio can be held constant, so that the value of L remains a computed, fixed fraction of specified values of R that will be considered. The values of R to be considered should be such that the net incident transport, $Q = R + L$, or $Q = (1-r)R$, are within a range of physically plausible values, perhaps defined by the range of values of Q for the inlet suggested by prior investigators.

- Plausible values for the jetties' impermeabilities (j_1 and j_2) are made from aerial photography, surveyed shoaling patterns, and physical inspection; or, in the limit, can be left to the default uncertainty range of 0 (permeable) to 1 (impermeable). Values for the inlet's net natural bypassing tendency (p_1 and p_2) are typically most uncertain and can be left to the default range of 0 (no bypassing) to 1 (full bypassing). In many cases, however, it might be accepted that an improved inlet is an imperfect bypassing system ($p < 1$) or a complete littoral barrier ($p=0$), etc.

- Values for the local, inlet-induced transport (m_1 and m_2) can be estimated from inspection of computed increases in the inlet-directed longshore transport potential near the inlet mouth. This can be discerned from wave refraction investigation whereby the longshore transport potential is computed from incipient breaking conditions along the shoreline developed for each of several representative offshore wave cases.

- The transport potential can be computed by the CERC formula with arbitrary coefficient. For example, an increase in the transport potential, R_1, for example from 0.5 to 0.7 (arbitrary units), induced by refraction across the inlet's ebb shoal, represents a 40 percent inlet-induced increase in local right-directed transport, suggesting a value of $m_1 = 0.4$. Additionally, the parameters m_1 and m_2 can be used to account for inlet-directed transport from the adjacent shorelines induced by tidal currents.

- Many other discrete solutions can, of course, be developed from the family of solutions presented in Figures V-6-53 and V-6-54. Those from the latter, narrowed family are more physically plausible, based upon general examination of the inlet. Although even this narrowed family appears broad, one will find that there are not extraordinarily significant differences between discrete solutions within the family, particularly after plotting the results as shown in Figure V-6-55. In particular, there are very modest differences between those solutions that define the central 50 percent of the narrowed family as a whole; i.e., the dark-shaded area in Figure V-6-54.

- The methodology described allows the coastal engineer a framework by which to:

- Investigate the volumetric and lineal extent of an inlet's impact upon adjacent shorelines.

- To develop a framework for a conceptual inlet sediment budget based upon ranges of physically plausible input values. The latter results in a family of solutions which bounds those solutions that can mathematically and reasonably solve an inlet sediment budget. This family of solutions inherently accounts for the underconstrained nature of an inlet sediment budget calculation; i.e., where there are more variables than known values. While it allows flexibility of solutions, it provides useful boundaries to the solution. These boundaries can be narrowed, or studied, to the degree that the investigator wishes to prescribe physical constraints to the inlet's transport patterns. The statistics of the values within the family of solutions can be examined to discern the occurrence with which various values of the inlet's transport parameters can solve the sediment budget. This can help to establish the degree to which a given discrete solution lies within the central or outer parts of the most likely (or at least, modal) solutions.

EXAMPLE PROBLEM V-6-13

To illustrate the method's application and utility, an example application at St. Lucie Inlet, Florida, is presented. St. Lucie Inlet is located on Florida's south-central east coast, on the Atlantic Ocean. The inlet was artificially opened in the 1920s, and jetties were later constructed on the north (left) and south (right) shorelines. The north jetty extends across most of the typical surf zone, but is low and permeable. The south jetty is sand-tight, but is very short, and sand is transported around its seaward end. The net incident drift is acknowledged to be from north to south, and is plausibly between 40,000 and 260,000 cu m/year (50,000 to 340,000 cu yd/year). The downdrift (south, or right) shoreline has experienced chronic erosion along a great distance (10's of km), and is it is generally acknowledged that the inlet exhibits some, but far less than perfect, natural sand bypassing.

Values of volume changes and typical dredging practices were identified over a period of 6 to 10 years in the late 1980s, and converted to equivalent annual rates (ATM 1992; USAED, Jacksonville, 1999). These included an updrift (left) shoreline change of ΔV_L = -16,000 cu m/year (net erosion); gross inlet shoaling of ΔV_G = 156,000 cu m/year; maintenance dredging and out-of-system disposal of D_O = 33,000 cu m/year; and maintenance dredging and placement to the downdrift (right) shoreline or nearshore of D_R = 60,000 cu m/year. Upland/offshore influx to the inlet shoals is assumed to be neglible; i.e., S_U = 0. From Equation V-6-53, the net inlet shoaling rate is ΔV_N = 63,000 cu m/year, representing the estimated net rate of accretion across the inlet's ebb and flood shoals and/or losses to the offshore.

Relevant results of a wave refraction and littoral transport analysis at the inlet (Browder 1996, conducted for USAED, Jacksonville, 1999) are illustrated in Figure V-6-52. Representative wave conditions from various offshore directions were refracted/diffracted to the point of incipient breaking, from which the longshore transport potential was computed. The results were weighted by each wave condition's hindcast occurrence and summed to develop an alongshore estimate of the right- and left-directed transport potential. The values in the figure are normalized by the maximum value of the computed transport potential. Source wave data were WIS Phase II hindcast, 1956-95 (Corson et al. 1982).

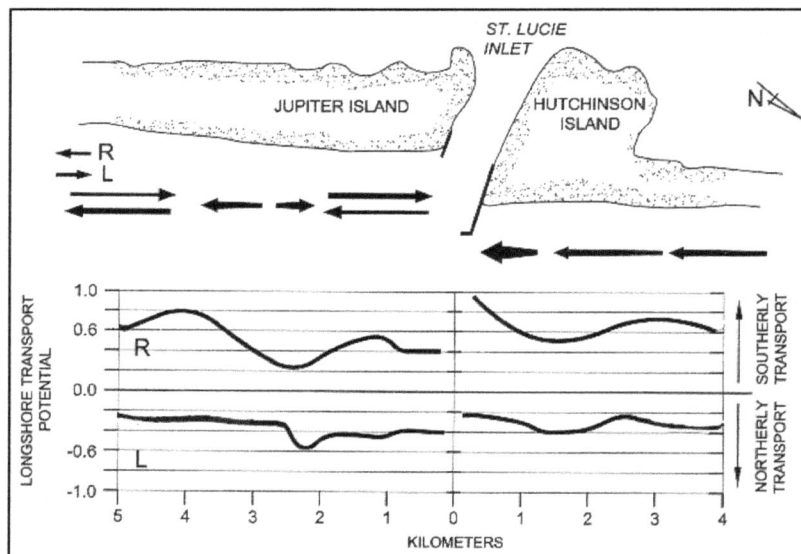

Figure V-6-52. Longshore transport potential in vicinity of St. Lucie Inlet, Florida (after Browder 1996)

(Continued)

EXAMPLE PROBLEM V-6-13 (Continued)

From Figure V-6-52, the ratio of left-to-right-directed transport at the updrift (left) boundary is about $r = -L/R = 0.45$. For an assumed range of net incident transport, $Q = 40,000$ to $260,000$ cu m/year, this implies a range of considered values for the incident, right-directed transport of

$$R_1 = Q/(1-r) = (40,000)/(1-0.45) \text{ to } (260,000)/(1-0.45) = 72,700 \text{ to } 472,700 \text{ cu m/year}$$

and corresponding incident, left-directed transport of

$$L_1 = -r\, R_1 = (-0.45)(72,700) \text{ to } (-0.45)(472,700) = 32,700 \text{ to } 212,700 \text{ cu m/year}$$

Left-directed transport at the downdrift (right) boundary is approximately 90 percent of that at the updrift (left) boundary; hence, it will be presumed that $L_2 = 0.9\, L_1$. Due particularly to shoal and reef features near the inlet mouth, the transport potential directed toward the inlet is computed to increase by about 30 to 50 percent in the immediate vicinity of the inlet. Local transport directed toward the inlet is likewise augmented by tidal (flood) currents, the contributions of which are not included in the figure. It is thus reasonably assumed that the local, inlet-induced increases in right- and left-directed transport, m_1 and m_2, are each in the range of at least 0.3 to 0.5, more or less.

For the sake of generality, no presumptions will be made regarding the jetties' impermeability values (j_1 and j_2) or the natural bypassing tendency (p_1 and p_2). These values, at least initially, will be allowed to range from 0 to 1.

As the right shoreline is downdrift (and of uncertain volume change), Equations V-6-56a and V-6-56b are solved for p_1 and p_2 for the ranges of values described; viz.,

$$p_1 = 1 - j_1\,(1 + m_1) + m_1 - (L_2/R_1)(1 - j_2(1+m_2) + m_2 - p_2) - (156000)/R_1$$

$$p_2 = [\,(j_1\,(1 + m_1) - m_1)\, R_1 + L_1 - (-16000) + 0 - 0\,]\,/\,L_2$$

where $0 \le j_1 \le 1$; $0 \le j_2 \le 1$; $0.3 \le m_1 \le 0.5$; $0.3 \le m_2 \le 0.5$; $72700 \le R \le 472,700$; and, $L_1 = 0.45\, R_1$; and $L_2 = L_1$. Only those combinations of values for which $0 \le p_1 \le 1$ and $0 \le p_2 \le 1$, and for which $0 \le p_1 + j_1 \le 1$ and $0 \le p_2 + j_2 \le 1$, are retained as viable solutions. For each of these, the net incident transport, shoaling volume from the left and right shorelines, and the net bypassing volume, are computed from Equations V-6-61a, V-6-62a, V-6-62b, and V-6-62c, respectively.

Figure V-6-53 illustrates the results of these computations; i.e., the general family of solutions. Any point within this family (the shaded area) can solve the sediment budget. The solutions lie along lines of constant net incident transport value (Q_1). The size of the solution family decreases as the local inlet-induced transport term (m_1, m_2) increases. This is indicated by the truncation of the family along its lower right-hand side as m_1 and m_2 increase from 0.3 to 0.5.

(Continued)

EXAMPLE PROBLEM V-6-13 (Continued)

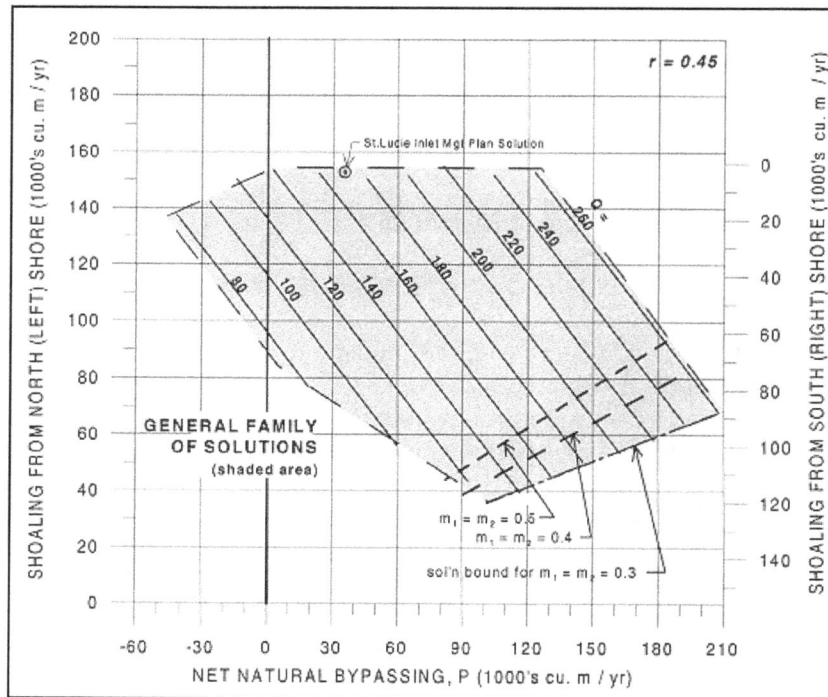

Figure V-6-53. General family of solutions for St. Lucie Inlet, Florida, sediment budget

From Figure V-6-53, it is noted that the minimum net incident transport (Q_1) is about 80,000 cu m/ year in order to solve the sediment budget. A numeric scan of the results demonstrates that the updrift jetty must be at least 25 percent impermeable but not more than 60 percent impermeable ($0.25 < j_1 < 0.6$) to solve the sediment budget; and that the modal impermeability value is about 46 percent ($j_1 = 0.46$). Values for the downdrift jetty's impermeability ranged from 0 to 1, though 90 percent of the values were less than 0.7. Likewise, natural bypassing of the right-directed transport must be less than 69 percent ($p_1 < 0.69$) to solve the sediment budget. A solution proposed in the St. Lucie Inlet Management Plan, prepared for the State of Florida, plots at the top edge of the family. While mathematically viable, it appears as an outlier relative to the bulk of the solution, and was ultimately not adopted by the state.

The family of solutions can be readily narrowed by imposing additional physical constraints, developed through observation of the inlet. For illustration, these shall include:

1. Direction of any net natural bypassing to coincide with the net drift ($P > 0$).

2. Net natural bypassing is not greater than two-thirds of the net incident rate ($P < 0.67 Q_1$).

3. Shoaling from the right comprises at least one-third of the total inlet shoaling ($S_{RIGHT} > 0.33 \Delta V_G$).

4. The downdrift, right jetty is not more than 70 percent impermeable ($j_2 \leq 0.7$).

(Continued)

EXAMPLE PROBLEM V-6-13 (Continued)

The resulting, narrowed family is plotted in Figure V-6-54. Requirements No. 1, 2, and 3, respectively, truncated the general solution's left, right, and top boundaries; while No. 4 had little effect in this case, as expected, given that 90 percent of j_2's values were naturally less than 0.7. The bounds of this narrowed family are identical for values of m_1 and m_2 from 0.3 to 0.5. The maximum net, incident transport rate must be less than 250,000 cu m/year in order to solve the sediment budget.

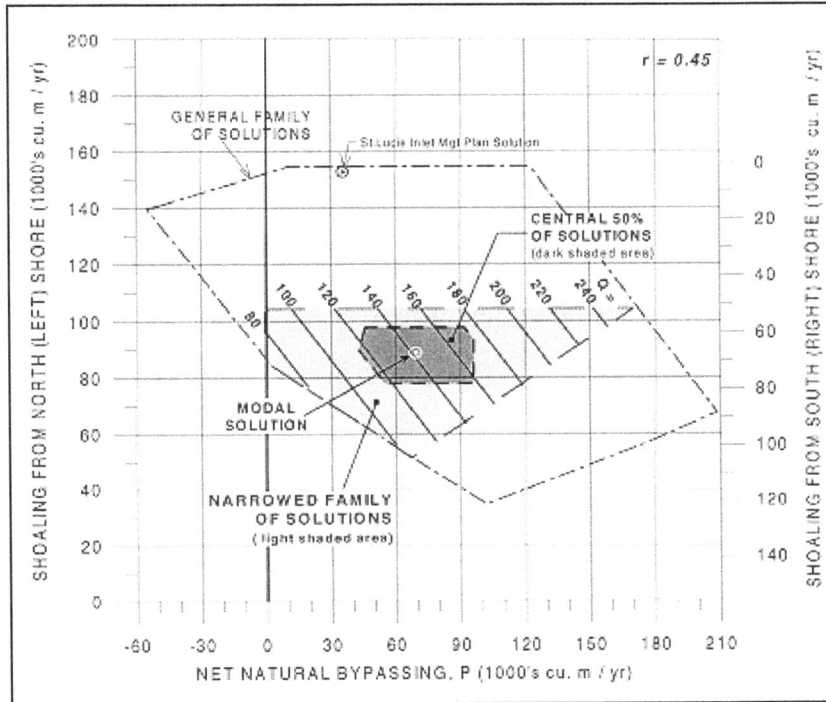

Figure V-6-54. Narrowed family of solutions, per requirements 1-4

A numeric scan of the narrowed family's results reveals the frequency with which values of the various transport parameters can solve the sediment budget for fixed increments of the input values j_1, j_2, and R_1 (or Q_1). For example, the modal (most frequent) value of shoaling from the left shore is 88,600 cu m/year, or 56 percent of the total gross shoaling rate, with half of all of the solutions falling within values of 44,000 cu m/year (28 percent of gross shoaling) to 96,000 cu m/year (62 percent). The modal value of the net incident transport rate is $Q_1 = 147,000$ cu m/year, with half of all of the solutions defined by values between about 120,000 and 170,000 cu m/year. The modal value of the updrift jetty impermeability is $j_1 = 0.47$, with half of all solutions described by $j_1 = 0.43$ to 0.53. The modal value of the downdrift jetty impermeability is $j_1 = 0.26$, with half of all solutions described by $j_1 = 0.13$ to 0.43. This suggests that the updrift jetty is about twice as impermeable as the downdrift jetty, a notion that is physically reasonable.

(Continued)

EXAMPLE PROBLEM V-6-13 (Continued)

The dark shaded area in Figure V-6-54 is a subset of solutions that represents the central 50 percent of the narrowed family. That is, it comprises those 50 percent of the solutions distributed about the inlet-parameter values that most frequently solve the sediment budget. Formally, it represents that half of the viable inlet solutions that occur most frequently. The modal solution in the middle indicates the narrowed solutions' weighted average value of computed net bypassing, P, and shoaling, S_{LEFT} and S_{RIGHT}.

As an example, a discrete solution is selected from the family and exploded to develop the corresponding sediment transport pathways. The modal solution is chosen as an illustration, where the most frequently occurring values of the transport parameters are selected as a discrete solution; viz., Q_1 = 142,000 cu m/year (R = 258,200 cu m/year, and L = -116,200 cu m/year); P = 69,400 cu m/year; S_{LEFT} = 88,600 cu m/year, and j_1 = 0.47. The family of solutions is scanned for these discrete conditions (or, a discrete solution can be computed directly for these values). For nominal values of m_1 = m_2 = 0.4, the parameters that correspond to these conditions are j_2 = 0.31, p_1 = 0.40, and p_2 = 0.32. For each of these values, and recalling that $L_2 = 0.9 L_1$, the transport pathways attendant to the inlet are computed with reference to the definition sketch, Figure V-6-51a,b. The results are plotted in Figure V-6-55.

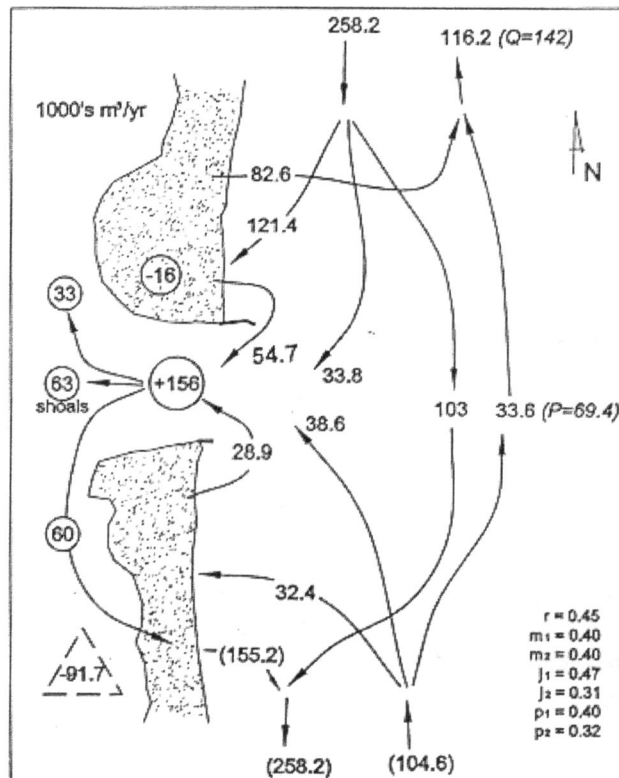

Figure V-6-55. Sediment transport rates (1000s cu m/year) computed for modal solution, St. Lucie Inlet, Florida (figure not to scale)

(Continued)

```
EXAMPLE PROBLEM V-6-13 (Concluded)


From Equation V-6-57, the volume change of the downdrift shoreline is also computed.  This requires
presumption of the rightward transport, $R_2$, directed away from the inlet on the downdrift side, not
otherwise needed up to this point.  In the present example, and by reference to Figure V-6-51a,b, it is
presumed that $R_1 = R_2 = 258,200$ cu m/year.  This suggests that the downdrift shoreline exhibits an
erosion potential of 91,700 cu m/year.  From Equation V-6-57, this is the sum of net inlet shoaling
(63,000 cu m/year) and out-of-system dredge disposal (33,000 cu m/year) minus net change of the
updrift shoreline (-15,900 cu m/year, computed), and accounting for a presumed decrease in the left-
directed transport potential across the inlet (11,600 cu m/year).

Measurements of net downdrift volume change vary widely, particularly because of fairly
frequent beach nourishment and nearshore (dredge) sand placement of uncertain success.  With sand
placement to the downdrift shoreline, the observed net volume change is between +67,000 cu m/year
to -139,000 cu m/year.  (These are the values to which to compare the computed net downdrift
volume change, above, $\Delta V_R = -91,700$ cu m/year.)  After subtracting sand placement, net downdrift
erosion is reported on the order of 160,000 to 367,000 cu m/year along 9 km south of the inlet
(USAED, Jacksonville, 1999).  The great uncertainty in these values is not atypical for downdrift
shorelines.

The modal solution depicted presumes a net incident transport rate of 142,000 cu m/year.  Prior
studies by the Corps estimate this value as about 175,000 cu m/year (USAED, Jacksonville, 1999).
The solution, suggests that the transport rates into the inlet from the north and south shorelines are
about 88,500 and 67,500 cu m/year, respectively.  Corresponding values estimated in USAED,
Jacksonville (1999) are about 103,000 and 56,000 cu m/year.
```

V-6-4. Engineering Approaches

   a. General considerations. Development of the inlet sediment budget allows estimation of the
degree to which an inlet has historically impacted the adjacent shorelines' littoral system, and is presently
bypassing sediment. Mitigation of the first approach may be possible by restoring sand to the affected
shores from the inlet's sediment (shoal) resources, although this may aggravate inlet impacts if done
improperly or may be otherwise limited by other concerns. Mitigation of the second approach requires
that the inlet be improved to decrease its littoral impact as a sediment sink. To the extent that these
mitigative approaches are limited, additional mitigation may be required in the form of beach
nourishment (or alternate shore protection methods) from sources external to the inlet. In developing
mitigative and/or inlet management strategies, the designer must be aware of social and environmental
constraints, as well as physical (site) constraints, that can ultimately limit these strategies' feasibility.
These include, for example, the presence of natural or historical resources that limit borrowing/sand
transfer from inlet shoals; upland property owners' ability to block sand transfer operations that are
perceived to remove sand from updrift beaches; questions as to which entity "owns" the sand resource to

EXAMPLE PROBLEM V-6-14

FIND:

The family of solutions that describes an improved inlet's sediment transport pathways.

GIVEN:

The apparent net transport direction across the inlet is from left-to-right. The net volume changes within the inlet's shoals, and along 6 km of the apparent, updrift shoreline, were estimated from survey data for a common period and averaged to annualized rates. Dredging and sand-bypass records are available for the same period. The volume change of the updrift (left) shoreline is -15,000 cu m/year (erosion). Prior to any dredging, the flood shoal accreted at a rate of 237,000 cu m/year and the ebb shoal accreted at a rate of 80,000 cu m/year. An average of about 55,000 cu m/year are dredged from the inlet and disposed of in deep water, offshore. Another 86,000 cu m/year are dredged from the flood shoals, on average, and placed upon the downdrift (right) shoreline. The inlet includes jetties on both shorelines, each of which appears to impound some, but not all, of the incident drift; $0.1 < j < 0.6$, for example. There is a bypassing bar across the inlet mouth that attaches to both shorelines within about 3 to 4 km of the entrance. Results of a wave refraction / transport rate analysis, based upon hindcast wave data, are shown in Table V-6-12. The values of the transport rate potential are normalized by the maximum value of the computed net transport rate. The actual magnitude of the average, net incident drift rate is uncertain, but is thought to be toward the right and on the order of 150,000 to 250,000 cu m/year.

Table V-6-12
Results from Wave Refraction and Transport Rate Analysis

Distance from Inlet (km)	Right-Directed Transport	Left-Directed Transport	Net Transport	Right-Directed Transport	Left-Directed Transport	Net Transport
	Left (updrift) shoreline			Right (downdrift) shoreline		
7	0.85	-0.28	0.57	0.84	-0.28	0.56
6	0.82	-0.28	0.54	0.85	-0.29	0.56
5	0.86	-0.30	0.56	0.82	-0.27	0.55
4	0.92	-0.20	0.72	0.70	-0.30	0.40
3	0.97	-0.11	0.86	0.59	-0.39	0.20
2	1.03	-0.08	0.95	0.33	-0.42	-0.09
1	1.03	-0.03	1.00	0.21	-0.41	-0.20
0.1	1.00	-0.00	1.00	0.15	-0.43	-0.28

SOLUTION:

Step 1. From Table V-6-11 of computed, potential transport rates, the right- and left-directed transport appear to be fairly stable and uniform at 6- to 7-km distance from the inlet. From the given values in the table and Equation V-6-48, the ratio of the right- and left-directed transport potential is about

$$r = - (-0.28) / (0.85) = 0.33$$

EXAMPLE PROBLEM V-6-14 (Continued)

Step 2. From the given data, the range of incident transport rate to be considered shall be $Q = 150$ to 250, where units of 1,000's of cubic meters per year are assumed throughout.

Step 3. The range of jetty impermeabilities to be considered shall be $0.1 < j_1 < 0.6$ and $0.1 < j_2 < 0.6$.

Step 4. From the table, the right-directed transport rate potential increases from about 0.85 to about 1.0 along the left shoreline adjacent to the inlet. This increase is assumed to reflect the local, inlet-induced transport associated with the term m_1. This represents a localized increase of 0.15 (normalized) units above the incident value of 0.85. Thus,

$$m_1 \approx 0.15 / 0.85 = 0.18$$

Likewise, from the table, the left-directed transport rate potential increases from about -0.29 to -0.42 along the right shoreline adjacent to the inlet. This increase is assumed to reflect the local, inlet-induced transport associated with the term m_2. This is a localized increase of -0.13 units above the incident value of -0.29. Thus,

$$m_2 \approx -0.13 / -0.29 = 0.45$$

Step 5. From the given data, $D_O = 55$, $D_R = 86$, $D_L = 0$, and $D_B = 0$, where units of 1000's of cu m/year are assumed. Riverine sediment input is assumed to be zero; $S_U = 0$.

Step 6. From the given data, the inlet's shoal complex accreted by $237 + 80 = 317$ (1,000's of cu m/year). The net change in inlet shoal volume, subsequent to maintenance dredging, is

$$\Delta V_N = 317 - 55 - 86 = 176 \text{ (1,000's of cu m/year)}$$

Step 7. Given the right-to-left net transport, the LEFT shoreline is the updrift shoreline, where $\Delta V_L = -15$ (1000's of cu m/year).

Step 8. For simplicity, require only that $0 < p_1 < 1$, and $0 < p_2 < 1$.

Step 9 - 10. Develop expressions for p_1 and p_2 for the values previously described. From Equations V-6-56a and V-6-56b), respectively,

$$p_1 = 1 + (0.33)(0.45 - j_2) - ((1-0.33)/Q)(0 + 86 + 55 + (-15) + 176 - 0)$$
$$= 1.15 - 0.33\,j_2 - 202.3/Q$$

$$p_2 = 1 + (1/0.33)[0.18 - j_1 + ((1-0.33)/Q)(-15 + 0 - 0)$$
$$= 1.55 - 3.03\,j_1 - 30.45/Q$$

(Continued)

EXAMPLE PROBLEM V-6-14 (Continued)

greater than 1; or, if the sum of $(p_1 + j_1)$ and/or $(p_2 + j_2)$ is less than zero or greater than 1, then the result is unreasonable and is excluded as a possible solution. For example, for the combination of values: $Q = 200, j_1 = 0.3,$ and $j_2 = 0.2,$

$$p_1 = 1.15 - 0.33\ (0.2) - 202.3/200 = 0.073$$

$$p_2 = 1.55 - 3.03\ (0.3) - 30.45/200 = 0.489$$

Compute the values of p_1 and p_2 for discrete values of Q between 150 and 250 (1000's of cu m/year), and for values of j_1 and j_2 between 0.1 and 0.6. If the the result yields a value of p_1 or p_2 that is less than 0 or

such that

$$p_1 + j_1 = (0.073 + 0.3) = 0.373$$

$$p_2 + j_2 = (0.489 + 0.2) = 0.689$$

This represents a potentially valid solution.

For another example, for the combination of values: $Q = 200, j_1 = 0.3,$ and $j_2 = 0.5,$

$$p_1 = 1.15 - 0.33\ (0.5) - 202.3/200 = -0.0265$$

$$p_2 = 1.55 - 3.03\ (0.3) - 30.45/200 = 0.489$$

which is an invalid solution because $p_1 < 0$.

Similarly, for the combination of values: $Q = 150, j_1 = 0.1,$ and $j_2 = 0.5,$

$$p_1 = 1.15 - 0.33\ (0.1) - 202.3/150 = -0.232$$

$$p_2 = 1.55 - 3.03\ (0.5) - 30.45/150 = -0.168$$

(Continued)

Sediment Management at Inlets

EXAMPLE PROBLEM V-6-14 (Continued)

which is an invalid solution because both p_1 and p_2 are less than zero. In fact, for this case of Q=150, no reasonable value of j_1 can yield a value of p_1 greater than zero; hence, the value of Q=150 is generally invalid and excluded from the family of solutions. An abbreviated list of additional example solutions is shown in the Table V-6-13.

Table V-6-13
Sediment Budget Example Solutions

Assumed Values			Computed Values				VALID Solution?
Q	j_1	j_2	p_1	p_2	$j_1 + p_1$	$j_2 + p_2$	
210	0.3	0.45	0.038	0.496	0.338	0.946	YES
210	0.3	0.5	0.022	0.496	0.322	0.996	YES
210	0.3	0.55	0.005	0.496	0.305	1.046	NO
210	0.35	0.1	0.154	0.345	0.504	0.445	YES
230	0.2	0.1	0.237	0.812	0.437	0.912	YES
230	0.2	0.15	0.221	0.812	0.421	0.962	YES
230	0.2	0.2	0.204	0.812	0.404	1.012	NO
230	0.25	0.1	0.237	0.660	0.487	0.760	YES
230	0.25	0.15	0.221	0.660	0.471	0.810	YES

Step 11. For each valid solution, compute total shoaling quantities from the left shoreline, S_{LEFT}, from the right shoreline, S_{RIGHT}; and the net natural bypassing quantities, P. Use Equations V-6-62a), V-6-62b, and V-6-62c, respectively. As an example, for the values listed on the last row of Table V-6-12:

$$S_{LEFT} = (0.18 + 1 - 0.25 - 0.221)(230 / (1-0.33)) = 243.4$$
$$S_{RIGHT} = (0.45 + 1 - 0.15 - 0.660)((0.33)(230)/(1-.33)) = 72.5$$
$$P = (230/(1-0.33))(0.221 - (0.33)(0.660)) = 1.1$$

each with units of 1,000's of cu m/year. Compute these values for all of the valid solutions and plot, as shown in Figure V-6-56.

(Continued)

EXAMPLE PROBLEM V-6-14 (Continued)

The shaded area in the graph represents the general family of solutions; i.e., all potentially valid solutions. The range of solutions can be narrowed by introducing additional physical restrictions. These might include, for example, requirements that

(1) the direction of net natural bypassing across the inlet be the same as the net incident transport (i.e., from left-to-right, or $P > 0$).

(2) The jetties' impermeabilities are relatively similar (e.g., j_1 and j_2 differ by no more than 50 percent).

(3) The tendency for natural bypassing of incident transport is similar for both shorelines (e.g., p_1 and p_2 differ by no more than 50 percent).

For these example assumptions, the family of solutions is narrowed to the smaller, cross-hatched area in Figure V-6-56

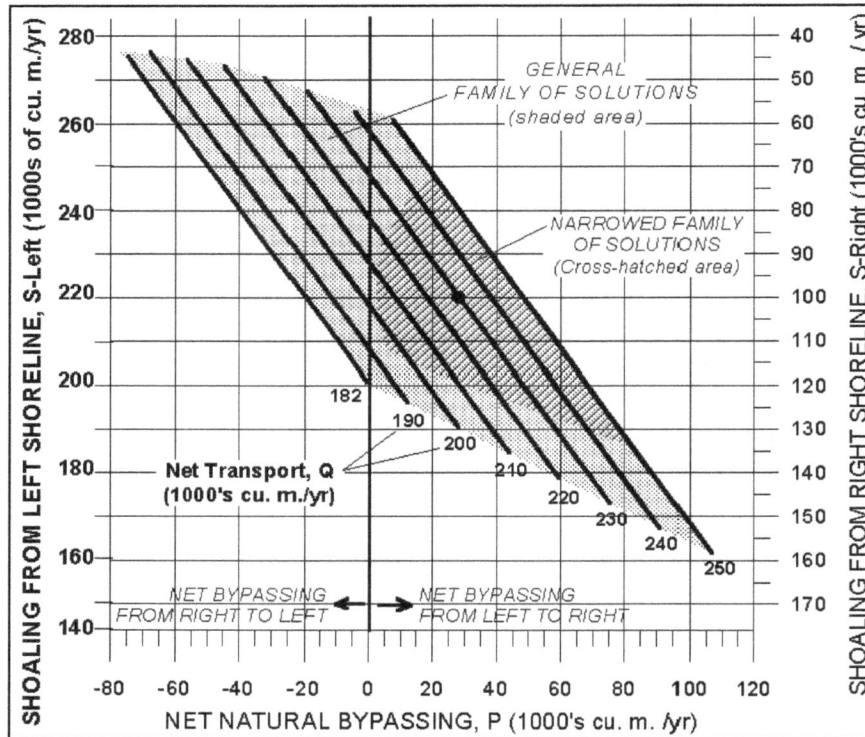

Figure V-6-56. Family of solutions for Example Problem V-6-14

(Continued)

EXAMPLE PROBLEM V-6-14 (Continued)

While any point within this family of solutions is potentially valid (i.e., can satisfy the inlet's sediment budget), it is useful to develop one or more discrete solutions. An example, discrete solution is developed using the values from the approximate center of the narrowed family of solutions (indicated by the bold dot in Figure V-6-56, see Figure V-6-57).

The discrete solution indicated in the figure approximately represents $Q = 230$, $S_{LEFT} = 216$, $S_{RIGHT} = 100$, and $P=28.6$. Scanning the computed family of solutions for values of j_1, j_2, p_1, and p_2 that result in these quantities yields $j_1 = 0.425$, $j_2 = 0.438$, $p_1 = 0.126$, and $p_2 = 0.130$. (While these values can be developed algebraically, it is simpler to scan the spreadsheets or other computations in order to identify the combination of input parameters that yield the desired, discrete solution.) For this solution, the components of the sediment transport pathways about the inlet are computed as follows:

Incident, right-directed transport, R (Equation V-6-48 and V-6-61a):
$R = (230) / (1-0.33) = 343.3$

Incident, left-directed transport, L (Equation V-6-48 and V-6-61a):
$L = (-0.33) (230) / (1-0.33) = -113.3$

Inlet-induced transport from left shoreline, $(m_1 R)$:	$m_1 R = (0.18) (343.3) = 61.8$
Inlet-induced transport from right shoreline, $(m_2 L)$:	$m_2 L = (0.45) (-113.3) = -51.0$
Transport impounded by left jetty, $(j_1 R)$:	$j_1 R = (0.425) (343.3) = 145.9$
Transport impounded by right jetty, $(j_2 L)$:	$j_2 L = (0.438) (-113.3) = -49.6$
Right-directed transport naturally bypassed, $(p_1 R)$:	$p_1 R = (0.126) (343.3) = 43.3$
Left-directed transport naturally bypassed, $(p_2 L)$:	$p_2 L = (0.130) (-113.3) = -14.7$

where units of 1000's of cu m/year are assumed throughout. The sediment transport pathways, including the given dredging practices, are illustrated in Figure V-6-57. In this example solution, the fraction of the net, incident transport that is naturally bypassed across the inlet is:

$$P / Q = 28.6 / 230 = 0.124 \quad (12.4 \text{ percent}).$$

For the method's assumption that the magnitudes of the left- and right-directed transport are equal at the updrift and downdrift boundaries of the study area, the predicted net volume change on the downdrift (right) side of the inlet is, from Equation V-6-55,

$$\Delta V_R = [(0.437)(0.33)-(0.45)(0.33)-1+0.126) (230/(1-0.33)] + 0 + 86 = -215.5 \text{ (1000's cu m/year)}$$

(Continued)

EXAMPLE PROBLEM V-6-14 (Concluded)

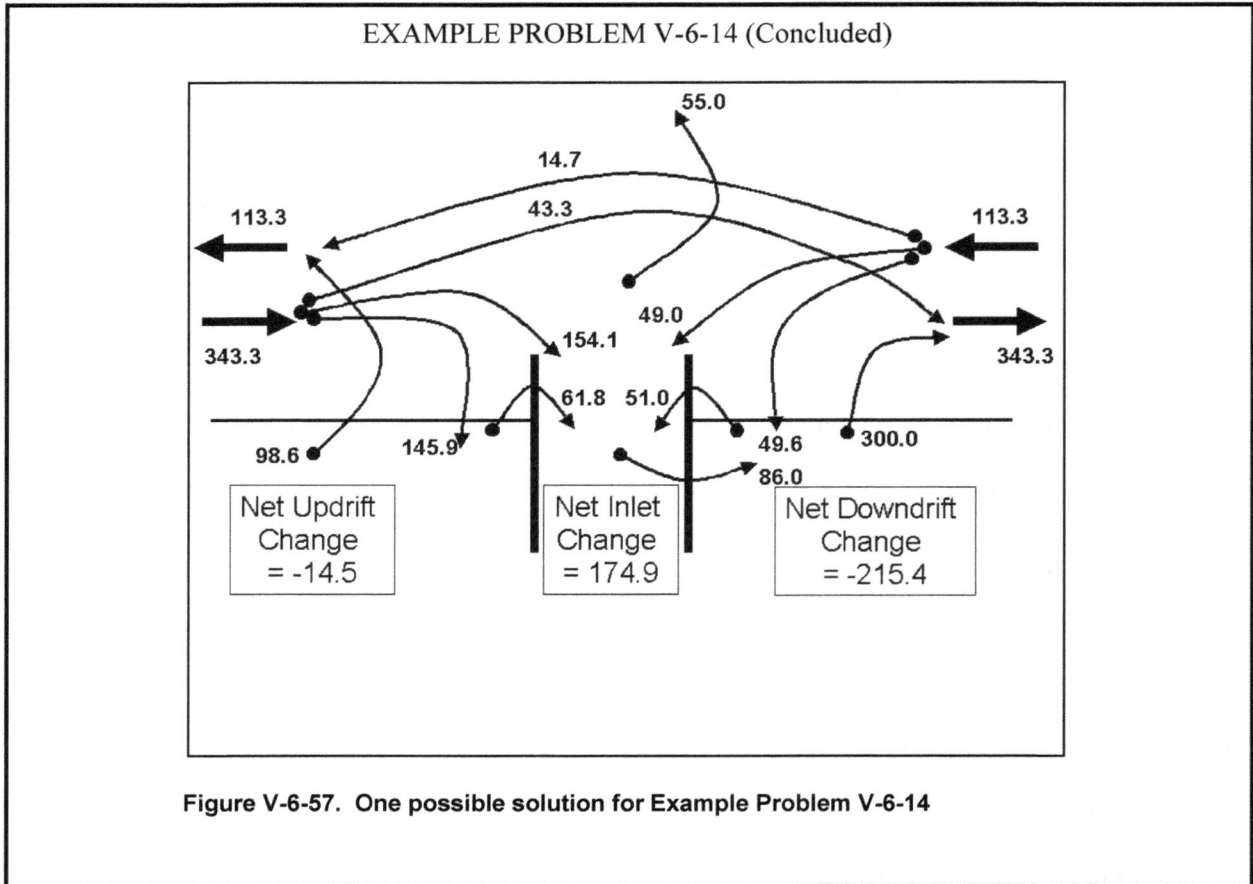

55.0

14.7

43.3

113.3   113.3

49.0

343.3   154.1   343.3

61.8 51.0

98.6   145.9   49.6   300.0
                                              86.0

| Net Updrift Change = -14.5 | Net Inlet Change = 174.9 | Net Downdrift Change = -215.4 |

Figure V-6-57. One possible solution for Example Problem V-6-14

be transferred; seasonal restrictions upon dredging activity imposed by marine turtle or bird nesting, storms, tourism, etc.; impacts to surfing and other recreation; inclusion of contaminated sediments; noise and aesthetic impact of sand transfer activity; limited upland access for sand-transfer operations (pipeline, maintenance, pumping stations, etc.); among others. The designer must also consider the ultimate (initial and annual) cost of implementing the proposed sand management strategy versus that of mitigating beach erosion by conventional nourishment from sand sources external to the inlet, assuming that authorization for the latter can be successfully sought and maintained.

b. Application of sediment budget to design. The sediment budget prepared for existing or historical conditions can be utilized to examine potential effects of modifications to the inlet system. The following example illustrates application of the Bodge method (Bodge 1999; see Part V-6-3) to estimate the optimum combination of jetty sand-tightening and mechanical bypassing. This application minimizes the total volume of sand that must be mechanically transferred in order to theoretically yield no net changes to the adjacent shorelines. The same approach can be used to examine the quantities and directions of mechanical sand transfer necessary for any given combination of jetty sand-tightening alternatives.

(1) In the limit, a jetty which is completely sand-tight ($j = 1$) will theoretically allow no transfer of sand past the jetty by either natural bypassing or local, inlet-induced transport; in which case, $p = 0$ and $m = 0$. If it is assumed that the degree to which the sand transfer past a jetty decreases linearly from its present state ($0 < j < 1$) to a completely sand-tightened state ($j = 1$), then the parameters p and m become:

$$p_{1'} = \left(\frac{p_1}{1 - j_1}\right)(1 - j_{1'}) = a\,(1 - j_{1'}) \tag{V-6-63}$$

$$p_{2'} = \left(\frac{p_2}{1 - j_2}\right)(1 - j_{2'}) = b\,(1 - j_{2'}) \tag{V-6-64}$$

$$m_{1'} = \left(\frac{m_1}{1 - j_1}\right)(1 - j_{1'}) = c\,(1 - j_{1'}) \tag{V-6-65}$$

$$m_{2'} = \left(\frac{m_2}{1 - j_2}\right)(1 - j_{2'}) = d\,(1 - j_{2'}) \tag{V-6-66}$$

where p_1', p_2', m_1', and m_2' are the natural bypassing coefficients and local, inlet-induced transport coefficients for the modified jetty-impoundment coefficients j_1' and j_2'. The nonprimed values of p, m, and j represent the existing (or nonmodified) condition, as determined from measured data and as described in the previous section.

(2) If it is required that there be no net volume change along the left and right shorelines, then from Equations (V-6-50), (V-6-51), and (V-6-63) through (V-6-66), bypass and dredging/beach disposal requirements are:

$$D_B' - D_L' = [Q/(1-r)]\,[(1+c)\,j_1' + (r(b-1)-c) - b\,r\,j_2'\,] \tag{V-6-67}$$

$$D_B' + D_R' = [Q/(1-r)]\,[a\,j_1' + (1-a+rd) - (r+rd)\,j_2'\,] \tag{V-6-68}$$

where, from Equations V-6-42 through V-6-45

$$a = p_1\,/\,(1-j_1) \tag{V-6-69a}$$

$$b = p_2\,/\,(1-j_2) \tag{V-6-69b}$$

$$c = m_1\,/\,(1-j_1) \tag{V-6-69c}$$

$$d = m_2\,/\,(1-j_2) \tag{V-6-69d}$$

(3) While this approach prescribes that there be no net volumetric change to each shoreline, there may be local changes (accretion and erosion) caused by redistribution of sand within each shoreline cell.

(4) If $D_B' - D_L' > 0$, then the net requirement is to transfer sand to the right; hence, any interim transfer to the left shoreline (D_L') would be superfluous. In that event, $D_L' = 0$. Likewise, if $D_R' + D_B' < 0$, then the net requirement is to transfer sand to the left; hence, any interim transfer to the right shoreline (D_R') would be superfluous. In that event, $D_R' = 0$. Finally, if $D_B' - D_L' < 0$ and $D_R' + D_B' > 0$, then both shorelines require placement of sand and, hence, any interim bypassing from one to the other (D_B') would be superflous. In that event, $D_B' = 0$. Concisely,

$$
\begin{array}{lll}
\text{For} & D_B' - D_L' \geq 0 & \text{then } D_L' = 0 \\
\text{For} & D_R' + D_B' \leq 0 & \text{then } D_R' = 0 \\
\text{For} & D_B' - D_L' \leq 0 \text{ and } D_R' + D_B' \geq 0 & \text{then } D_B' = 0
\end{array}
\tag{V-6-70}
$$

Sediment Management at Inlets

EXAMPLE PROBLEM V-6-15

FIND:

The requisite, total sand transfer quantities for zero net littoral impact to the adjacent shorelines for the inlet condition in Example Problem V-6-14 for the case of: (a) existing conditions, (b) complete and-tightening of both jetties, and (c) optimum sand-tightening to yield the theoretical, minimum sand transfer quantity.

GIVEN:

Same data as for Problem V-6-14. Assume existing conditions are represented by $j_1 = 0.425$, $j_2 = 0.438$, $m_1 = 0.18$, $m_2 = 0.45$, and $r = 0.33$ with an incident net drift value of $Q = 230,000$ cu m/year.

SOLUTION:

For the given values (see Example Problem V-6-14), $p_1 = 0.126$ and $p_2 = 0.130$. From Equation (V-6-69),

$$a = p_1 /(1-j_1) = (0.126) / (1-0.425) = 0.219$$
$$b = p_2 /(1-j_2) = (0.130) / (1-0.438) = 0.231$$
$$c = m_1 /(1-j_1) = (0.18) / (1-0.425) = 0.313$$
$$d = m_2 /(1-j_2) = (0.45) / (1-0.438) = 0.801$$

From Equations (V-6-67), (V-6-68) and (V-6-71), where quantities of 1000's of cu m/year are assumed,

$$D_B{}' - D_L{}' = A = [230/(1-0.33)] [(1+0.313) j_1{}' + (0.33 (0.231-1) - 0.313) - (0.231)(0.33) j_2{}']$$
$$= A = 450.7 j_1{}' - 26.2 j_2{}' - 194.6$$

$$D_B{}' + D_R{}' = B = [230/(1-0.33)] [0.219 j_1{}' + (1-0.219 + (0.33)(0.801)) - (0.33 + (0.33)(0.801) j_2{}']$$
$$= B = 75.2 j_1{}' - 204.0 j_2{}' + 358.9$$

(a) For the case of existing conditions (no jetty sand-tightening), $j_1{}' = 0.425$ and $j_2{}' = 0.438$:

$$D_B{}' - D_L{}' = A = (450.7) (0.425) - (26.2) (0.438) - 194.6 = -14.5$$

$$D_B{}' + D_R{}' = B = (75.2) (0.425) - (204.0) (0.438) + 358.9 = 301.5$$

From Table V-6-14, $D_B{}' = 0$. Thus, $D_L{}' = 14.5$ and $D_R{}' = 301.5$. That is, 14,500 cu m/year should be dredged from the inlet and placed upon the updrift (left) shoreline; and, 301,500 cu m/year should be dredged from the inlet and placed upon the downdrift (right) shoreline.

(Continued)

Table V-6-14
Method to Determine $D_R{}'$, $D_B{}'$, and $D_L{}'$

$A = D_B{}' - D_L{}'$	$B = D_R{}' + D_B{}'$	$D_L{}'$	$D_R{}'$	$D_B{}'$
+	+	0	B-A	A
-	+	-A	B	0
+	-	invalid solution		
-	-	B-A	0	B

EXAMPLE PROBLEM V-6-15 (Continued)

(b) For the case of total jetty sand-tightening, $j_1' = 1.0$ and $j_2' = 1.0$. Therefore,

$$D_B' - D_L' = A = (450.7)(1.0) - (26.2)(1.0) - 194.6 = 230$$

$$D_B' + D_R' = B = (75.2)(1.0) - (204.0)(1.0) + 358.9 = 230$$

From Table V-6-14, $D_L' = 0$. Thus, $D_B' = 230$ and $D_R' = 0$. That is, 230,000 cu m/year should be bypassed across the inlet from left-to-right. This is, as expected, equal to the net incident transport rate.

(c) Neglecting costs of jetty improvements, the optimum (theoretical) jetty sand-tightening alternative is that which minimizes the total quantity of material that must be mechanically transferred across the inlet, and/or from within the inlet to the adjacent shorelines. To investigate this, the dredging quantities $(D_B' - D_L')$ and $(D_B' + D_R')$ are solved for a variety of combinations of j_1' and j_2'. For each, the total requisite dredging quantity $(D_B' + D_L' + D_R')$ is recorded. An abbreviated list of example calculations is shown in Table V-6-15.

Table V-6-15
Example Calculations for Evaluating Dredging Requirements

j_1	j_2	$D_B' - D_L'$	$D_B' + D_R'$	D_B'	D_L'	D_R'	$D_B'+D_L'+D_R'$
0.40	0.6	-30.0	266.6	0.0	30.0	266.6	296.6
0.40	0.8	-35.3	225.8	0.0	35.3	225.8	261.1
0.40	1.0	-40.5	185.0	0.0	40.5	185.0	225.5
0.45	0.6	-7.5	270.3	0.0	7.5	270.3	277.8
0.45	0.8	-12.7	229.5	0.0	12.7	229.5	242.3
0.45	1.0	-18.0	188.7	0.0	18.0	188.7	206.7
0.50	0.6	15.0	274.1	15.0	0.0	259.1	274.1
0.50	0.8	9.8	233.3	9.8	0.0	223.5	233.3
0.50	1.0	4.6	192.5	4.6	0.0	188.0	192.5
0.60	0.6	60.1	281.6	60.1	0.0	221.5	281.6
0.60	0.8	54.9	240.8	54.9	0.0	186.0	240.8
0.60	1.0	49.6	200.0	49.6	0.0	150.4	200.0
0.70	0.6	105.2	289.1	105.2	0.0	184.0	289.1
0.70	0.8	99.9	248.3	99.9	0.0	148.4	248.3
0.70	1.0	94.7	207.5	94.7	0.0	112.9	207.5

(Continued)

EXAMPLE PROBLEM V-6-15 (Concluded)

The total requisite dredging quantity $(D_B' + D_L' + D_R')$ is the value in the last column of the table. This quantity is plotted as a function of the jetties' sand-tightening coefficients on the following graph.

From Figure V-6-58, the minimum dredging quantity is computed at or about $j_1' = 0.49$ and $j_2' = 1.0$. From Equations V-6-67, V-6-68 and V-6-71

$$D_B' - D_L' = A = (450.7)(0.49) - (26.2)(1.0) - 194.6 = 0$$

$$D_B' + D_R' = B = (75.2)(0.49) - (204.0)(1.0) + 358.9 = 191.7$$

This implies 100 percent sand-tightening of the right (downdrift) jetty, and minor sand-tightening of the left (updrift) jetty. There would be a theoretical requirement for the placement of 191,700 cu m/year from the inlet to the right shoreline.

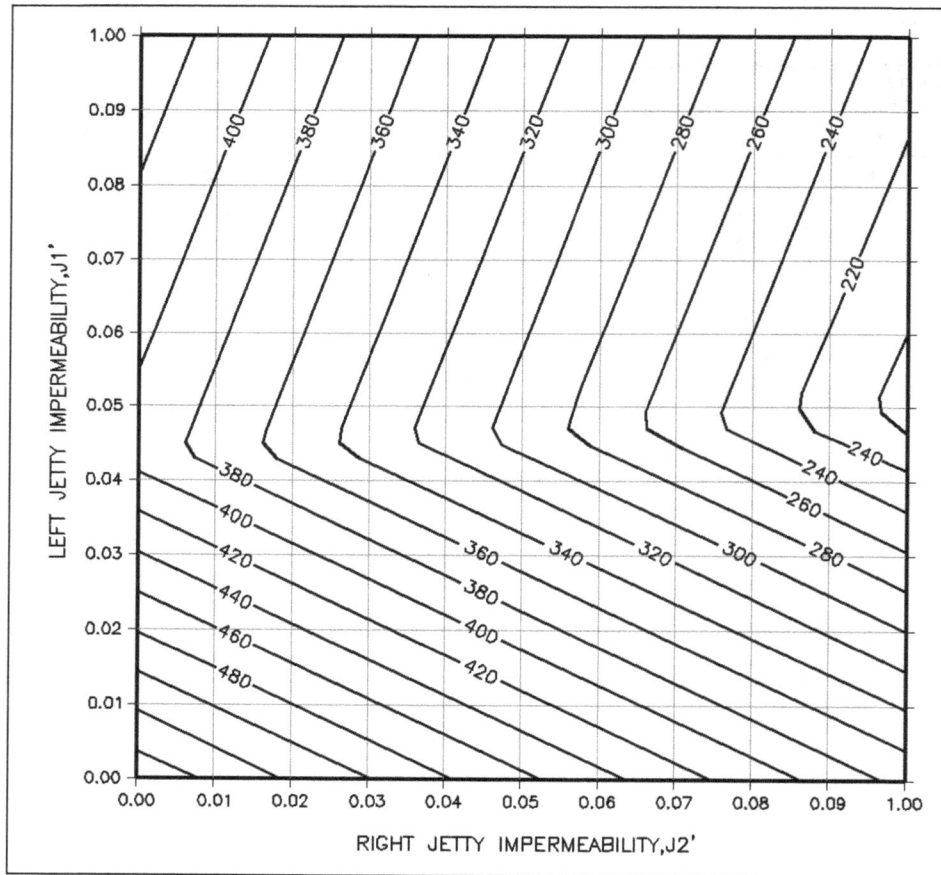

Figure V-6-58. Solution for Example Problem V-6-15

(Note that $D_B' - D_L' \geq 0$ and $D_R' + D_B' \leq 0$ is an invalid solution.) From Equation V-6-70, values of the sand transfer requirements D_L', D_R', and D_B', for a selected pair of jetty-tightening coefficients, j_1' and j_2', can be computed from Table V-6-14, where

$$A = D_B' - D_L' \quad \text{and} \quad B = D_R' + D_B' \quad\quad\quad\quad (V\text{-}6\text{-}71)$$

and where a positive (+) value implies a result that is greater than or equal to zero (≥ 0).

   c. *Principles of sand bypassing and backpassing.* Sand bypassing and backpassing refers to the mechanical transfer of littoral material between the inlet and/or adjacent shorelines to minimize the inlet's erosive impact to the adjacent beaches. Generally, *bypassing* is a transfer of sediment from the updrift shoreline or inlet shoals to the downdrift beach, and also represents a primary means by which to alleviate shoaling within the inlet. *Backpassing* is a transfer of sediment from the inlet shoals (or some portion of the shoreline) to the shoreline from whence the sediment came. The need for either bypassing or back-passing can be identified by the methods previously presented. The majority of sand bypassing/backpassing operations do not involve, nor require, fixed sand transfer plants; instead, they involve more intelligent dredge disposal practice and/or modification of the inlet's structures and channels.

   (1) For proper engineering design of an inlet improvement and bypassing/backpassing plan, it is necessary to:

   (a) Define the problem (e.g., through inspection and sediment budget approaches).

   (b) Determine the need, if any, for jetty improvements or other structure/channel modifications.

   (c) Define the requisite quantities and expected seasonal/annual variations thereof, if any.

   (d) Define the appropriate physical method(s) for the work.

   (e) Define the location(s) of sediment removal.

   (f) Define the appropriate location(s) for sediment placement.

The latter four requirements are, ultimately, interrelated.

   (2) General classes of bypass/backpass operations include:

   (a) Interior vs. exterior traps (i.e., temporary sediment storage within the inlet vs. sediment interception or storage within an updrift or downdrift jetty-fillet).

   (b) Periodic vs. continuous (i.e., periodic dredging vs. use of a dedicated plant).

   (c) Fixed vs. mobile plant (i.e., fixed plant vs. a conventional, trestle-mounted, or vehicle-mounted dredge plant).

   (3) Selection of the most appropriate approach is a function of the unique cirumstances of each inlet. In all cases, however, it is of paramount importance to maintain control of the sediment that enters the inlet system. Sediment can only be transferred if it can be recovered. Examples of transport pathways that, practically speaking, result in a permanent loss of sediment resources include: sediment that is deposited in a thin veneer across the seabed, lost to interior shoals which cannot be dredged for environmental reasons (e.g., high silt content, contaminants, sea grass beds, bird rookeries, etc.),

impounded along shorelines for which dredging cannot be undertaken, dumped offshore in deep water, or dumped nearshore in water depths or configurations that do not promote landward drift. Sand bypassing and backpassing can mitigate inlet impacts to adjacent shorelines only to the degree that sand can be practically recovered from the inlet and transferred. Accordingly, design of inlet and dredging practice that seeks to maximize the degree to which sediment can be intercepted or recovered is central to the development of an effective inlet/harbor sand management plan.

V-6-5. Engineering Methods

a. Jetty sand-tightening. As described in Part V-6-4(b), "Application of Sediment Budget to Design," an inlet's erosive impact to the adjacent shorelines can be reduced, and the requisite sand transfer quantities likewise reduced, with improvements to the terminal structures (jetties) that separate the shorelines and the inlet entrance. Example Problem V-6-15 illustrated the importance of sand-tightening the downdrift jetty in particular. Jetty sand-tightening involves raising the crest elevation, decreasing the porosity, and/or lengthening the structure. (See also Part V-6-3.) Common methods for sand-tightening include:

(1) Overbuilding the structure (i.e., adding armor and/or chinking rock to raise the crest elevation and/or decrease porosity.

(2) Injecting grout into the structure to decrease porosity.

(3) Capping the structure with grout/concrete/asphalt and/or armor units (to increase crest elevation and decrease porosity).

(4) Constructing sheet pile or other bulkhead adjacent to the structure, or placing impermeable barriers within the structure (to increase crest elevation and decrease porosity); and/or

(5) Lengthening the structure.

Specific implementation of these methods is described in Part VI.

b. Sand transfer plants. Selection of methods for sand transfer is described in EM 1110-2-1616, and in Richardson (1990) and Richardson and McNair (1981). The following is a brief overview of equipment and design considerations for facilitating sand transfer.

(1) Fixed systems. Fixed systems are essentially stationary dredging systems designed, built and operated for a specific location (Figure V-6-59). In this case, the sand mover (dredge pump, jet pump, etc.) operates without mobility. System components include the pump and pump motor; a housing to protect the pump and pump motor from waves, spray, and surge; a boom or brace that supports the intake(s); the dredge crater (created by the intake); a discharge pipeline; and booster pumps (depending upon the requisite diameter and length of the discharge pipeline and capacity of the pump). Fixed systems typically employ a conventional suction pump, jet pumps or eductors, submersible pumps, or dedicated dragline. Fluidizers, fixed to the intake or buried in the seabed, can also be employed to increase the sand intake capability. The pumps (or dragline) are generally stationed along the updrift jetty or a trestle immediately updrift of the inlet. Use of fixed systems typically implies the requirement for continuous or high-frequency sand transfer.

EM 1110-2-1100 (Part V)
31 Jul 03

(2) Mobile systems. Mobile systems include floating dredge plants, and shore-based plants mounted upon a vehicle. (Figure V-6-60). Use of floating plants typically allows for transfer of greater quantities of sediment and maximum mobility; however, it requires that the area to be dredged is reasonably protected from waves and can be accessed by the plant. Use of shore-based equipment typically involves transfer of smaller quantities, requires that beach access be available, and is limited to sand transfer from along the shore. At the same time, shore-based systems demand less requirement for protected waters and calm weather.

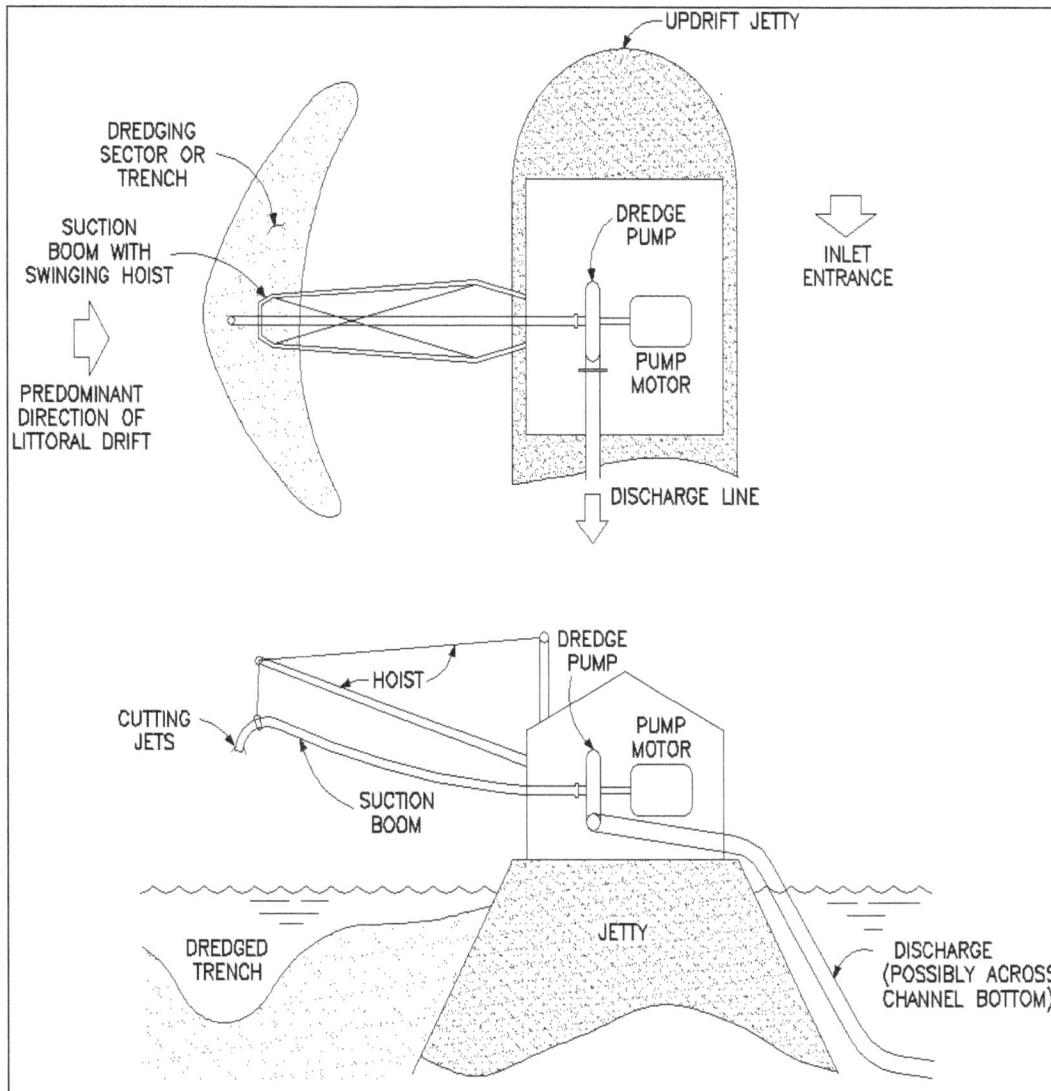

Figure V-6-59. Plan and section view of simple fixed bypassing plant (from EM 1110-2-1616)

(3) Semimobile systems. Semimobile systems include pumps or eductors that are deployed at one location in the inlet for some length of time, then moved to another location. Movement is by barge, truck, boat, etc. Components of mobile and semimobile systems include the dredging equipment (practically unlimited by type) and a temporary or dedicated pipeline or other means to discharge the sand. Use of mobile systems can be considered for either continuous or periodic sand transfer. The dredging equipment can be either contracted for periodic work or purchased (dedicated). The latter is usually considered when the requisite dredging frequency is high or the transfer quantity is modest.

V-6-132

Sediment Management at Inlets

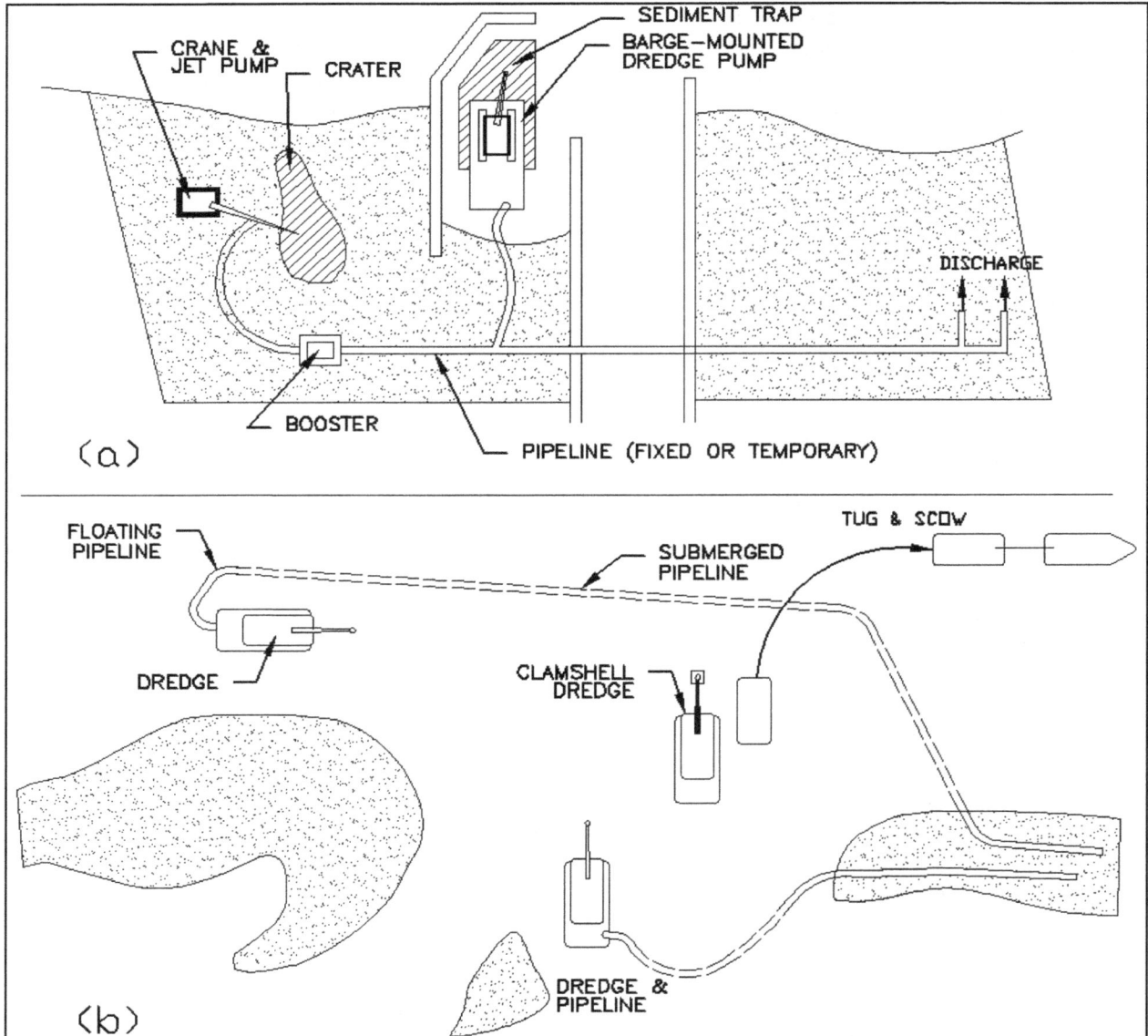

Figure V-6-60. Examples of (a) semimobile and (b) mobile bypassing systems

(4) Equipment for sediment extraction. Equipment most typically used to extract sediment for transfer includes the following systems.

(a) Dredges. Summaries of conventional dredge equipment and their use are presented in McKnight (1966); Bruun (1981); Herbich (1975, 2000); Herbich and Snider (1969); Richardson (1976, 1990); Huston (1970, 1986); and Turner (1984). Mechanical dredges include clamshell (or grapple) buckets, ladder buckets, dipper dredge, and dragline. The dredged material is placed in a hopper, barge or other storage area and is then transported autonomously to the disposal site. Hydraulic (suction) dredges include centrifugal pumps, trash pumps and other suction devices that intake and discharge a slurry mix of sand and water. The discharge is routed to a hopper or scow (from which it is transported to the disposal site) or through a pipeline to the disposal site. Conventional dredge equipment (mechanical or hydraulic) can be mounted on trucks, trailer, skids, or barges, and are available in a wide variety of sizes and capabilities. As such, they can be incorporated to either fixed or mobile, continuous or periodic, bypassing strategies. Clark (1983) presents a summary description of commercially available portable hydraulic dredges. Trade magazines (e.g., WODCON) offer monthly reviews of the dredging industry, equipment and technical consultation.

Hopper dredges recover sediment from the seabed through hydraulic means and store it in an onboard hopper for subsequent disposal. Most hopper dredges continuously move as they dredge, working along the channel axis or dredge area. Traditional, large hopper dredges are infrequently suitable for dredging small areas, and often mix littoral material from sandy deposits with silt and clay that is unsuitable for beach/nearshore disposal (Figure V-6-61). Shallow-draft hopper dredges feature loaded drafts of about 3.5 to 4.3 m (11 to 14 ft) with 45- to 60-m (150- to 200-ft) lengths and capacities of about 1,000 to 1,600 cu m (1,300 to 2,000 cu yd). Shallow-draft dredges are more maneuverable than traditional, large hopper dredges, and can work in wave heights of up to about 1.5 m (5 ft) (Bruun 1993).

(b) Jet pumps and eductors. Jet pumps (also known as eductors, ejectors, or injectors), are hydraulically powered pumps with no moving parts (Figure V-6-62). In a simple jet pump, a stream of high-velocity clear water from a supply pump (the motive flow) is forced through a reduction nozzle to create a high-velocity (low-pressure) jet. Upon exiting the nozzle, the jet entrains the surrounding fluid and forces the mixture through a mixing chamber and diffuser, then through the discharge line. When the nozzle/suction opening is buried in the sand, a sand-water (slurry) mixture is drawn into the discharge line. A conventional dredge pump (booster) downstream of the jet pump can be used to help move the slurry through the discharge pipe. Fluidizers are also typically employed near the suction intake to mobilize the seabed sediment and facilitate uptake.

- Typical jet pump capacities range from 75 to 300 cu m/hr (100 to 400 cu yd/hr), and feature suction intakes of 10 to 15 cm (4 to 6 in.), although larger pumps have been produced. Use of jet pumps may be ideal where there is a need for continuous or high-frequency bypassing of mostly modest quantities, littoral transport and/or sediment impoundment occurs over a limited (small) area, there are suitable locations for clear-water intake, the area is mostly free of debris, and the material to be moved is free of consolidated cohesive sediment. Clogging by debris is probably the greatest potential limitation of jet pump systems. Like conventional hydraulic (suction) dredges, jet pumps and their discharge pipelines also risk clogging by sediment, usually by excessive solids. Williams, Clausner, and Neilang (1994) and Clausner et al. (1994) provide additional guidance regarding use of jet pumps for sand bypassing.

- Jet pumps can be incorporated to either fixed or mobile sand bypass systems. Examples of the latter include jet pumps mounted upon a shore-based crane derrick, deployed from barges, and hauled about by lines and truck-mounted winches. Unlike jet pumps used in mobile systems, which are generally inherently capable of withdrawal from the seabed, jet pumps used in fixed systems should include a mechanism by which to raise the jet pump for purposes of maintenance and/or to remove debris that collects in the suction crater and clogs the intake.

(c) Submersible pumps. Submersible pumps are electrically or hydraulically driven pumps lowered directly into the material to be transferred (Figure V-6-63). In addition to the agitating action of the impeller, submersible pumps can be fitted with jetting rings and/or cutterheads to increase sand flow in compacted material. Most pumps are relatively small (order of 1 m) and weigh between 100 to 1,000 kg (220 to 2,200 lb). Because of their small size, submersible pumps can be deployed with a minimum of support equipment. Performance is comparable to jet pumps, with potential capacities of 40 to 320 cu m/hr (50 to 400 cu yd/hr) in fine to medium sand. Submersible pumps often require a booster pump for discharge distances of greater than about 600 m (2,000 ft). (EM 1110-2-1616). Submersible pumps offer the potential to pump material at higher solids contents than jet pumps or conventional dredges, and require neither a clear water intake, supply pump nor supply lines. On the other hand,

Figure V-6-61. Hopper dredge *McFarland* at Brunswick Harbor entrance, Georgia.
Lightly colored sandy sediment (right-side intake) is mixed with dark colored silt
and clay (left-side intake) during dredging

the degree to which debris cloggage and rapid wear of the impellers will impede their use is uncertain.
Wide prototype experience with submersible pumps for purposes of inlet sand bypassing is limited. Like
jet pumps and eductors, submersible pumps are potentially appropriate for either fixed or mobile bypassing systems. The risk of clogging the discharge line is fairly high with submersible pumps due to the
high solids-content slurries they are capable of producing. The Punaise is a submersible pump developed

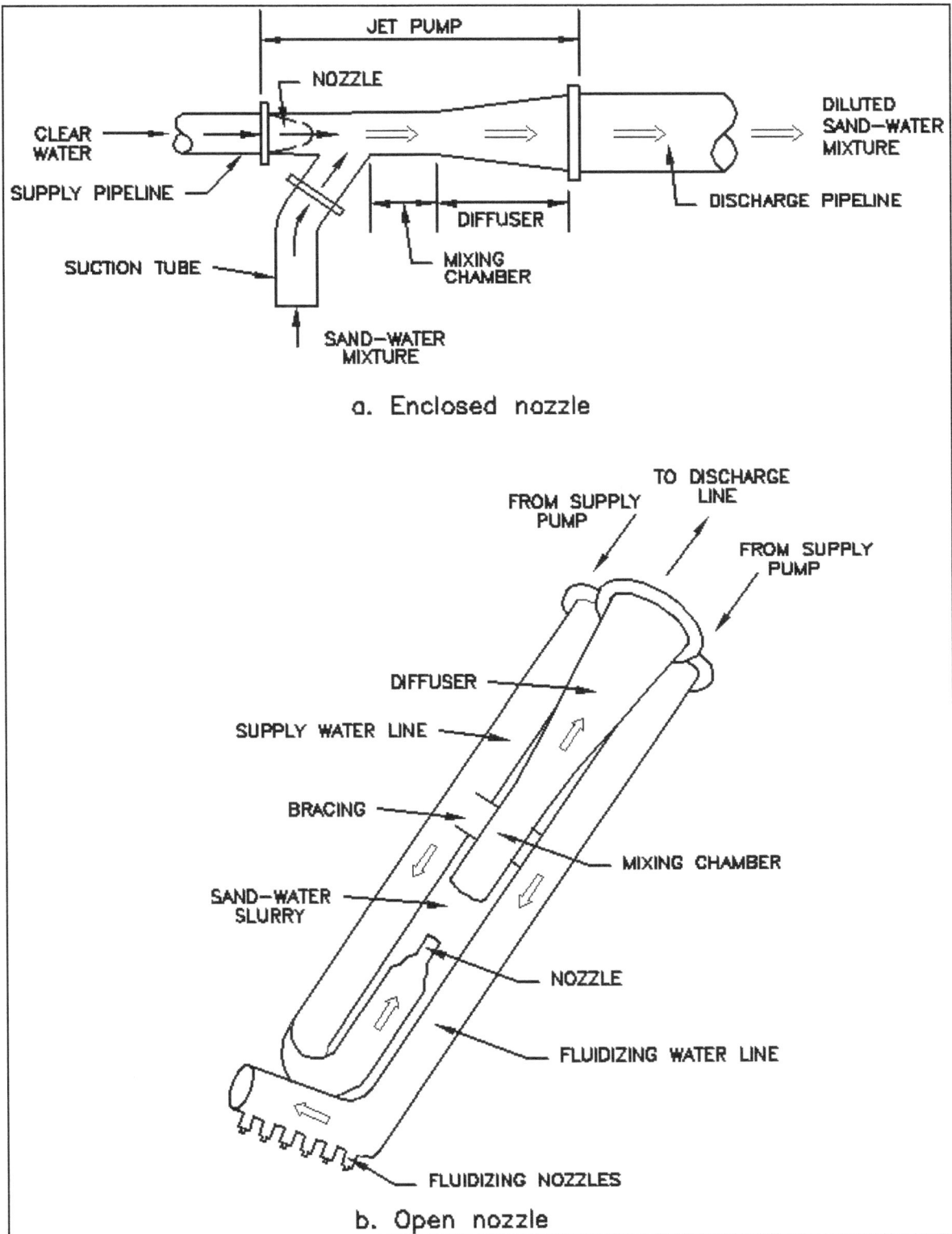

Figure V-6-62. Center-drive jet pumps (educators): (a) enclosed nozzle, (b) open nozzle (from EM 1110-2-1616)

Sediment Management at Inlets

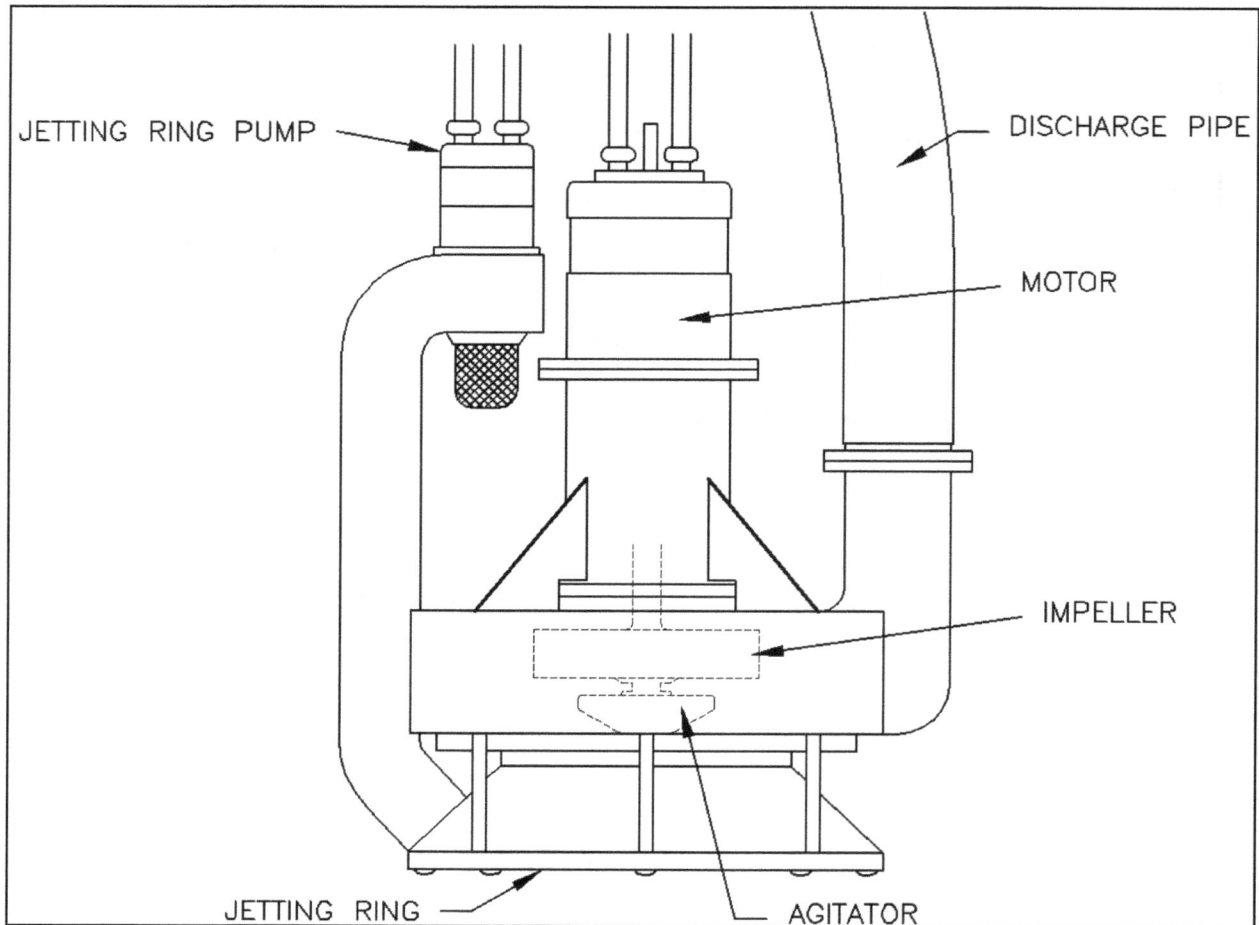

Figure V-6-63. Typical submersible pump (from EM 1110-2-1616)

by the Dutch for pit-dredging of the seabed (Figure V-6-64). Through the use of remotely operated flotation cells, it is designed to be both self-burying and self-emergent. The pump head digs a crater within the seabed and relies upon sediment to be gravity-driven down the crater slopes toward the pump. It is connected to shore or barge by a flexible pipeline (Brouwer, van Berk, and Visser 1992).

(d) Fluidizers. Fluidizers can be used in conjunction with any of the previously mentioned dredging systems to increase the mobility of the seabed sediments, and potentially increase the uptake and production rate of the bypass system (Figure V-6-65). Fluidizers inject clear water near or within the seabed in the vicinity of the suction intake. The fluidizer consists of one or more single jets directed at or around the intake; or, a manifold (a pipe with holes) that slopes downward toward the suction intake. In the former case, their use is intended to agitate, loosen and/or suspend the seabed sediments to facilitate their uptake. In the latter case, the clear-water flow from the fluidizer pipes is intended to mobilize the seabed sediments along each pipe and to induce a gravity flow toward the suction intake. In this way, fluidizers can increase the effective size of the suction crater. The practical benefits of fluidizers mounted directly at the suction intake is well established. On the other hand, long-term prototype experience with

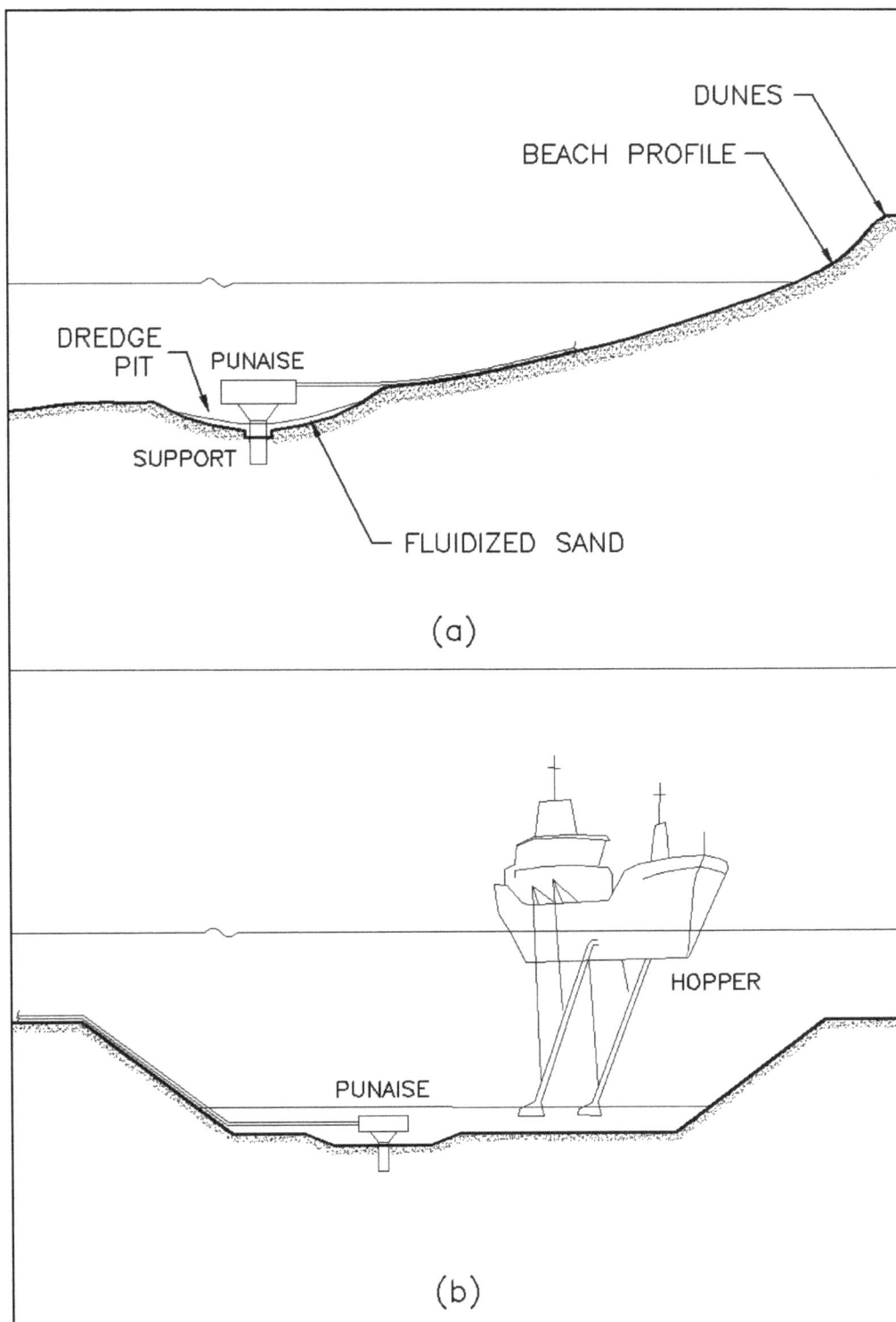

Figure V-6-64. Examples of the Punaise dredging system: (a) in the surf zone of a sandy coast, and (b) in a harbor with a trailing suction hopper dredge (after Brouwer, van Berk, and Visser 1992)

Sediment Management at Inlets

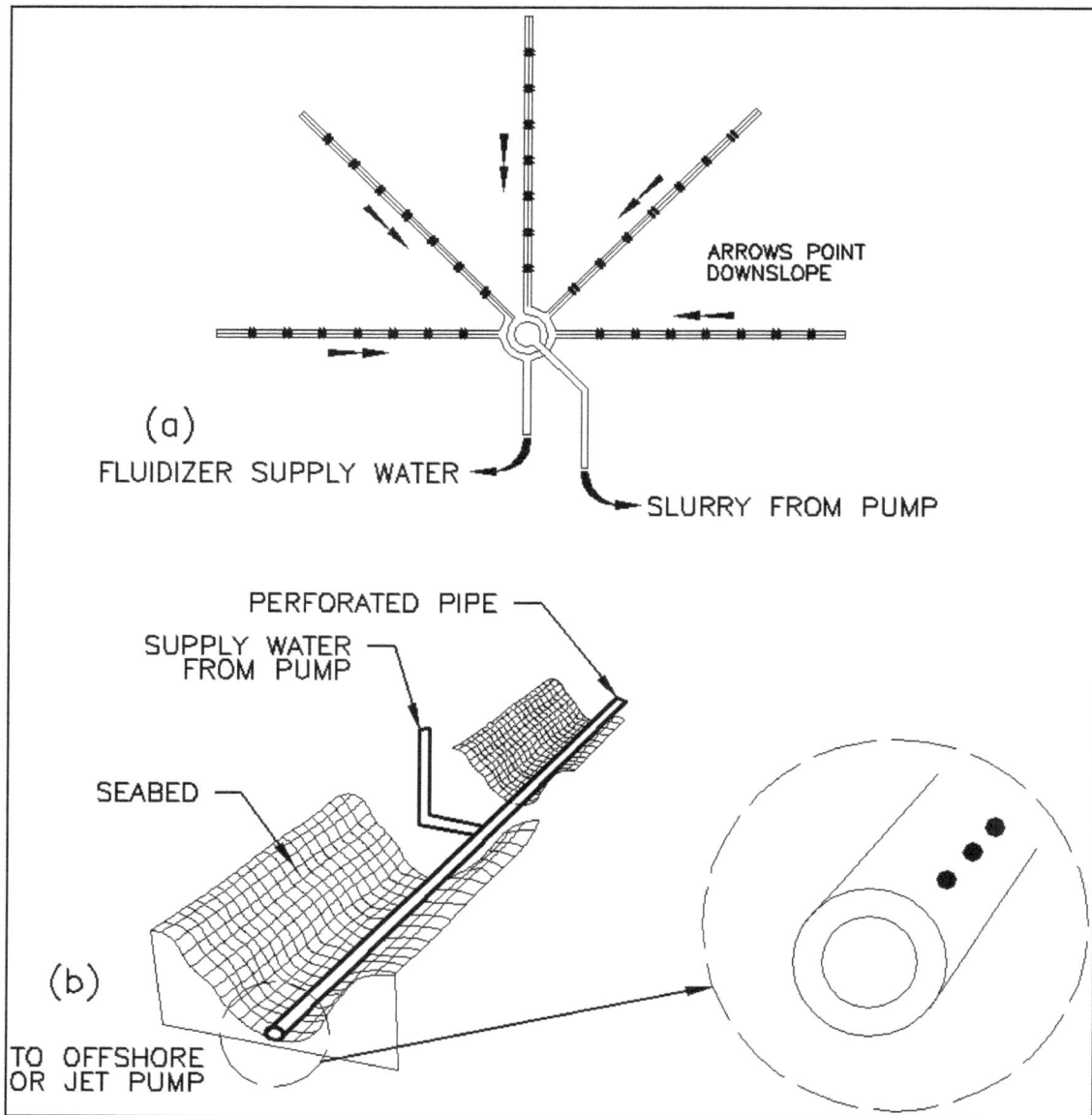

Figure V-6-65. Example of fluidizers: (a) multiple-pipe, manifold-type (planform view), and (b) single-pipe (seabed view). Adapted from EM 1110-2-1616)

manifold-type, seabed-imbedded fluidizers is limited; and there are problems regarding clogging of the fluidizer pipes (Bisher and West 1993). Practical guidance for the use of fluidizers for sand management is presented in Weisman, Lennon, and Clausner (1992, 1996). Methods to minimize clogging of manifold-type fluidizers can also be developed through reference to standard sewage treatment (aeration) equipment, where aerating manifolds, similar in dimension to fluidizers, consist of perforated pipes covered with elastomer membranes. Typical design values for fluidizer pipes include 1/8-in. holes at 2-in. spacing, horizontally opposed, with pipe slopes of about 1:100. To initiate and maintain fluidization, respectively, flow rates of 18 gpm/ft and 11 gpm are indicated (EM 1110-2-1616).

(5) Equipment for discharge. Equipment available for the transport and discharge of the sediment include the following classes.

(a) Hydraulic (pipeline) discharge. For modest transport distances (usually less than 8 km (5 miles), depending upon pump size and capacity), direct hydraulic transfer by fixed, temporary, floating, or submerged pipeline can be used. Sediment recovered by hydraulic means (dredge, jet pump, submersible pump, etc.) can be directly transferred from the intake to the discharge by pipeline, although booster pumps may be necessary en route. Sediment recovered by mechanical means, or otherwise stored in an upland stockpile, barge or scow, can be placed in a hopper, fluidized with clear water, and pumped through a pipeline to the discharge point.

(b) Mechanical, land-based discharge. Sediment recovered by hydraulic or mechanical means can be rehandled by truck transport, assuming that adequate roads and beach access locations are available. Because of its potential impact to roadways, traffic, safety, and noise pollution, this approach is generally suitable only for infrequent operations involving modest quanitites of material (e.g., less than 100,000 cu m (130,000 cu yd)); and typically, should be considered as a method of last resort.

(c) Hopper dredges, barges and scows. Hopper dredges, barges and scows are either closed-hull (also called fixed hull) or open-hull (also called split-hull, bottom-dump, doored, etc.). Closed-hull means that the hopper, barge, or scow can be loaded and emptied only from the top (Figure V-6-66a). Open-hull means that the hopper, barge, or scow can be opened from the bottom, by splitting the hull, opening trap doors on the bottom, or similar means, in order to drop the sediment to the seabed (Figure V-6-66b,c). Sediment recovered in closed-hull vessels must be pumped out or mechanically transferred (rehandled) for either nearshore or beach disposal. Sediment recovered in open-hull vessels can be dropped directly to the seafloor in water depths that accomodate the vessels' draft (bottom-dumping), or discharged to the beach or shallow waters by pump-out or mechanical transfer. Bottom dumping includes nearshore (berm) disposal, discussed in the following paragraphs. In pump-out systems, the sediment is fluidized by clear water and pumped from the vessel via pipeline. For beach disposal, the pipeline (and requisite boosters) are typically fixed, and the hopper/barge/scow ties up to the pipeline's seaward end to discharge its load (Figure V-6-66d). For open-coast pump-out, the pipeline is temporarily secured at the site. Alternately, for pump-out from protected waters, such as within an inlet entrance, harbor or bay, the pipeline might be permanently established (Figure V-6-66e).

(d) Offloaders and rainbow dredges. Offloaders are hydraulic (pump and pipeline) or mechanical (belt-conveyor) systems integral to the hopper, barge, or scow, that are capable of transferring the sediment to the nearshore or beach. Offloader systems vary widely by capacity and mechanism, and can often be customized for a particular application. Up-to-date information regarding such systems is available from the dredging industry, industry consultants and trade magazine (e.g., WODCON). Rainbow dredge discharge (also called over-the-bow discharge) is a hydraulic offloading system whereby a shallow-water hopper dredge offloads sediment by a bow-mounted pipeline that discharges a jet of slurry in an arc toward the shoreline (Bruun and Willekes 1992). (See Figure V-6-66f.) This theoretically enables placement in very shallow depths without need for pipeline or booster placement, but requires that the vessel can maneuver within 50 to 100 m (160 to 330 ft) of the desired placement area. The discharge distance can be theoretically extended with pontoon-floated pipeline deployed from the vessel's bow. To date, typically mild beach slopes combined with relatively deep drafts of existing U.S. hopper dredges greatly limited the application of this method in the United States.

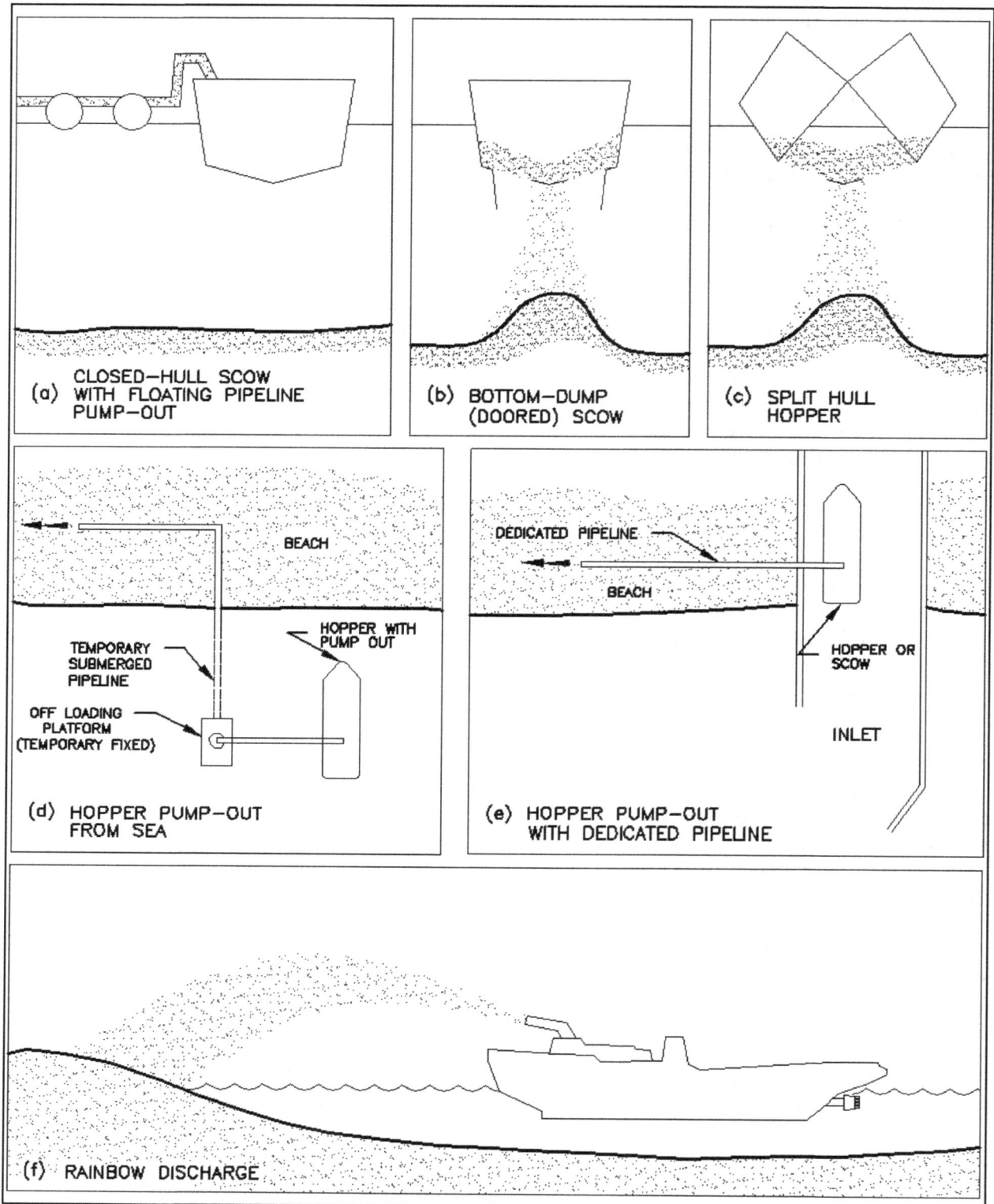

(a) CLOSED-HULL SCOW WITH FLOATING PIPELINE PUMP-OUT

(b) BOTTOM-DUMP (DOORED) SCOW

(c) SPLIT HULL HOPPER

(d) HOPPER PUMP-OUT FROM SEA

(e) HOPPER PUMP-OUT WITH DEDICATED PIPELINE

(f) RAINBOW DISCHARGE

Figure V-6-66. Examples of hopper dredge and scow discharge methods

(e) Hydrocyclones. A hydrocyclone separates coarse- and fine-grained particles within a slurry (Figure V-6-67). It has no moving parts, but instead, uses centrifugal force. The sediment-water slurry is pumped from the dredge or hopper into the hydrocyclone so as to create a vortex within. As the slurry spins within the hydrocyclone chamber, liquid and smaller-grained (lighter) particles migrate toward the inner, air core of the vortex and exit vertically upward as the overflow. Coarser-grained (heavier) particles crowd to the outer wall and exit at the bottom of the chamber as the underflow. Typical operating capacities of a single hydrocyclone are summarized in Table V-6-13. Potential applications of the hydrocyclone in inlet sediment management is to isolate coarser-grained, beach-compatible sediment from unsuitable sediments (silt and clay) dredged during maintenance or expansion of waterways. While widely applied in mining, their protoype use in coastal engineering is, thus far, mostly untested. In practice, their relatively modest capacities reasonably require the use of multiple hydrocyclones, connected in parallel. Limitations on operating pressures usually require use of a gravity-driven slurry supply (as opposed to direct connection to a dredge pump).

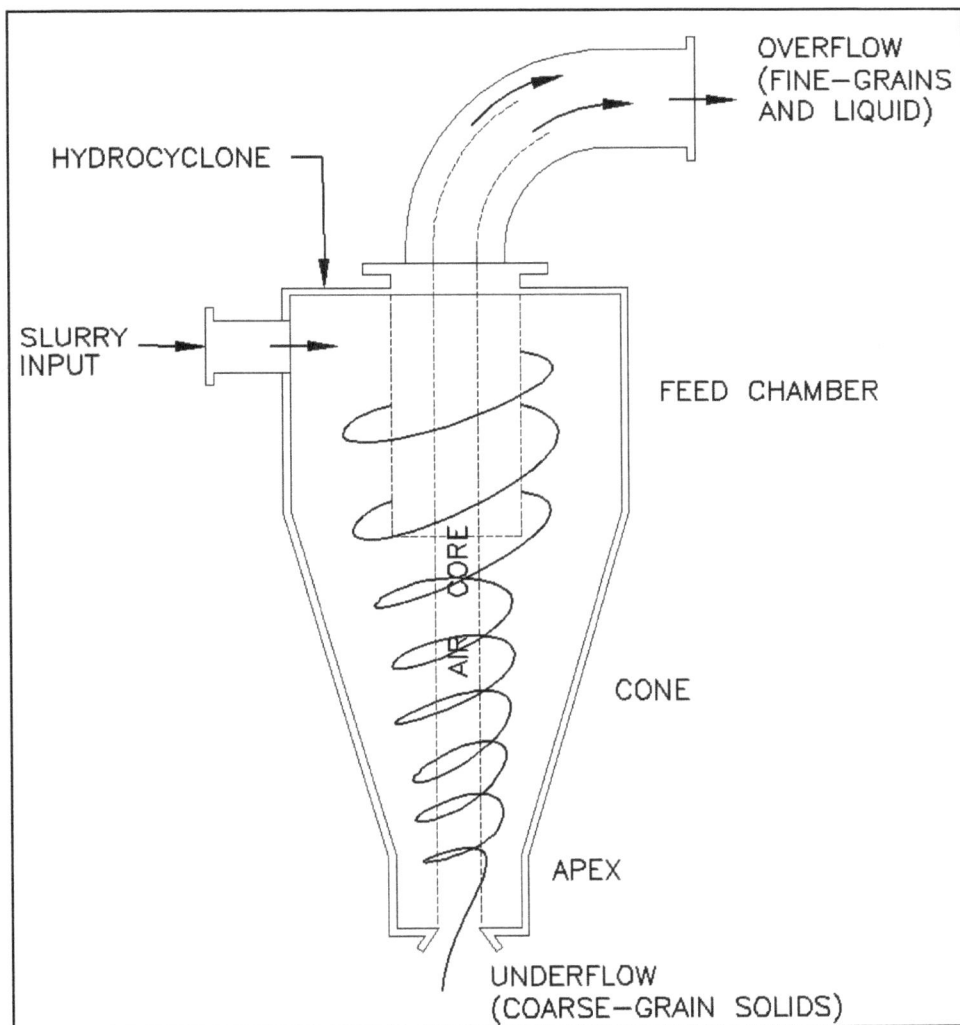

Figure V-6-67. Schematic of a typical hydrocyclone

Table V-6-16
Typical Hydrocyclone Capacities*

Size (in.)	Capacity (gpm)	Operating Pressure (lb/sq in.)	Separation
3	5 - 35	10 - 70	10 to 40 micron
4	20 - 90	10 - 60	10 to 40 micron
6	40 - 200	10 - 50	15 to 40 micron
8	90 - 300	5 - 40	20 to 44 micron
12	200 - 800	5 - 30	30 to 44 micron
18	300 - 1500	5 - 26	200 to 325 mesh
24	800 - 2400	5 - 25	150 to 250 mesh
30	800 - 3500	5 - 25	100 to 150 mesh

* Actual values depend upon suspended ratio and specific gravity of solids in slurry feed, among other factors. Data from Met Pro Refrax lined cyclones; Met Pro Supply, Inc., Bartow, Florida.

c. Sediment traps, deposition basins, and inlet sediment sources.

(1) Introduction. Sediment for bypassing and/or backpassing must be intercepted or temporarily stored to facilitate its extraction and subsequent transfer. Natural sediment traps include the ebb and flood tidal shoals, jetty-adjacent impoundment fillets, and inlet-interior spits. These traps can be augmented or created artificially through the construction of weir jetties, breakwaters, and deposition basins. Such artificial sand traps, if constructed properly and maintained frequently, localize the deposition of drift material and allow a more efficient dredging and/or bypassing operation to be planned. Deposition basins (or sediment traps) are located in regions where the wave climate is mild and the working environment is well suited for dredge or pumping operations; and, where the deposited drift is less likely to interfere with navigation (through shoaling or during dredging) or to be dispersed elsewhere within the inlet. Improperly constructed or maintained traps, on the other hand, can accelerate the inlet's sink effect upon the littoral transport. Sediment can be intercepted interior or exterior to the inlet. Interior traps include the flood shoals, channels, spits, and deposition basins within the jetties. Exterior traps include updrift impoundment fillets, the ebb tidal shoal; and, deposition basins, pumps, dredges, etc. constructed or operated on the outside (beach side) of the jetties. The decision of whether to use interior or exterior traps varies with each inlet; particularly, the size and morphology of the inlet entrance. The design capacity of a deposition basin or bypassing system is described in Part V-6-6.

(2) Weirs and weir jetties. A weir sand-bypassing system is typified by a low section of the updrift jetty (a weir jetty), or a low section of a similar structure built updrift of the updrift jetty (a sand weir or weir groin). A deposition basin, or sand trap, is located downdrift of the weir, from which sand is periodically excavated and then backpassed or bypassed. For wide inlets, the weir and deposition basin can be integral to the jetty and interior to the inlet (Figure V-6-68a). For narrower inlets, at which the deposition basin would interfere with navigation, the weir and deposition basin are located exterior to the inlet (Figure V-6-68b). The purpose of the weir is to: define a restricted area (the trap) into which sediment is transported and held until extracted by dredging or pumping, keep the trap area and sediment accumulation separated from the navigation channel, and ensure that the trap area is protected from waves to facilitate dredging and pumping. The weir also allows flood currents to enter the inlet during rising tide with subsequent channeling of the ebb current during falling tide. While this is intended to improve tidal flushing of the inlet channel, it likewise augments the inlet's sink effect upon the littoral system. That is, additional sediment is swept into the inlet from the adjacent shorelines during flood flow, and subsequently jetted seaward during ebb flow.

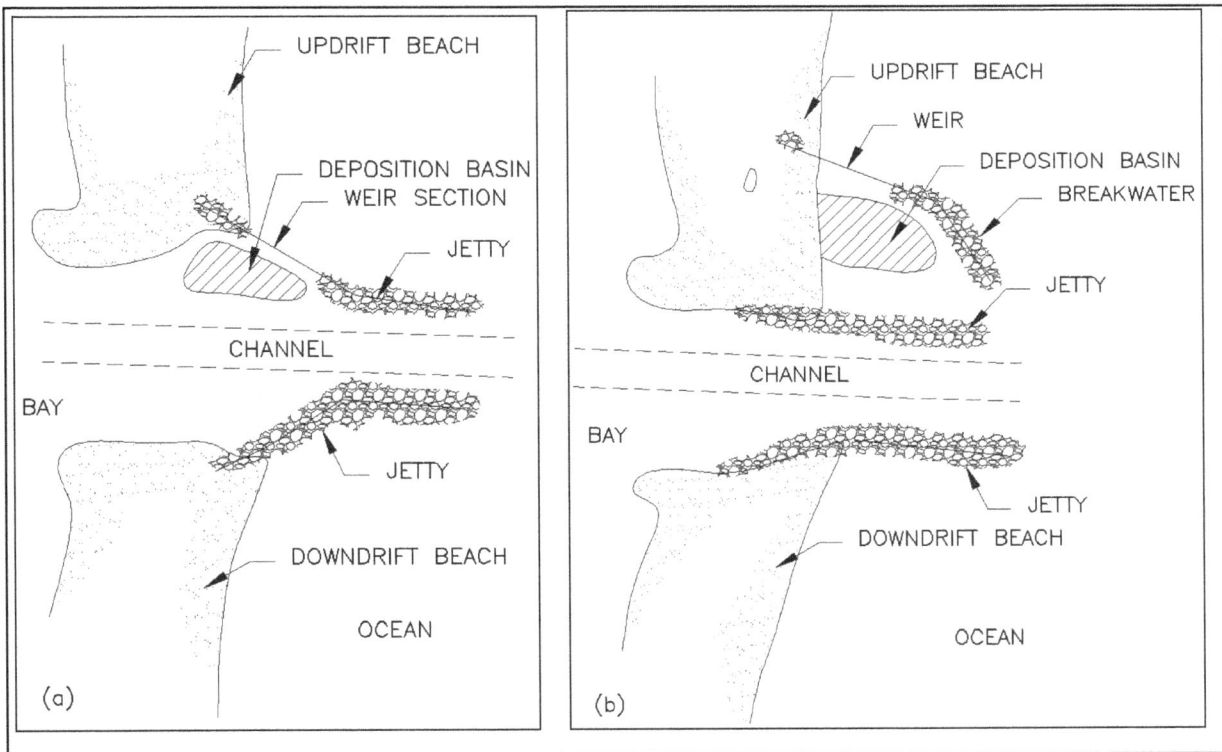

Figure V-6-68. Weir jetties, weir-groins, and weir deposition basins (a) interior to inlet, and (b) exterior to inlet

(a) Weggel (1981) and Seabergh (1983) describe important, specific design principles for weir jetty systems. In general, the weir is designed to allow a quantity of sediment ranging between the total incident sediment transport from the updrift direction and the net littoral drift to deposit in a semiprotected basin in which a dredge or other mechanical byapssing system can operate. Early weirs were constructed with vertical, smooth walls that often included adjustable-height panels. These materials are no longer recommended because of the associated wave reflection and the practical difficulty in adjusting the panels' height. Rubble-mound structures are preferred for weir sections.

(b) Weggel (1981) recommends that weir elevations be set near mean low water for tide ranges less than 0.6 m (2 ft); and, at mean tide level for tide ranges of 0.6 to 1.5 m (2 to 5 ft). Weir performance is poor for greater tide ranges, and weirs are not recommended for tide ranges greater than about 3.6 m (12 ft). Seabergh (1983) describes the effect of lower weir elevations as a function of an inlet's Keulegan "K" repletion coefficient. Weir sections lower than the adjacent beach induce inlet-directed transport, at all stages of the tide, because of the hydaulic gradient that is developed between the energetic, sandy beach outside of the weir and the less-energetic water inside of the weir. The design elevation of a weir should also consider the desired profile of the updrift beach, as the weir will more or less act as a template for that beach.

(c) The majority of sand transported over weirs occurs in the narrow region where the weir, beach, and waterline intersect (Weggel and Vitale 1981; Seabergh 1983). A much smaller portion of the drift moves over the weir as suspended material near the breaker zone. Current guidance is to establish the seaward end of the weir seaward of the normal breaker line (or offshore bars, if applicable), at low tide. This may, however, result in an exceptionally long weir that unintentionally accelerates inlet-directed transport from the updrfit beach via flood currents over the weir. The landward end of the weir is

Sediment Management at Inlets

established at the point where the updrift profile is to be approximately maintained. The landward weir profile should follow the updrift beach profile estimated from existing profiles measured far enough away so as not to be influenced by the inlet. Early weir lengths often extended longer than 350 m (1,000 ft) to account for the possibility of "sanding in" of shorter sectons during severe storms or elevated transport. Weggel (1981) reports instead that this has not occurred and, in fact, the shoreline of the updrfit beach has not typically extended beyond the landward section of the weir. Weir sections which are inappropriately long or low or close to the shoreline can induce anomalously high transport from the updrift beach toward the inlet resulting in net erosion of the updrift shoreline and accelerated shoaling rates and overflow of the deposition basin.

(d) Weir orientation with respect to the updrift shoreline and navigation channel is often determined by individual inlet characteristics and land availability. Weir trapping efficiency is generally independent of the weir orientation relative to the updrift shoreline. From model tests, Seabergh (1983) reported that updrift storage was greatest for weir angles, θ_w, of 90 deg (perpendicular to shore); became progressively less for angles of 60 and 45 deg to shore; and was least for an angle of 30 deg to shore. Likewise, offshore losses were least for 90 deg and greatest for 30 deg. The capability for the updrift storage to move back up the beach during drift reversals was greatest for the 30-deg angle and least for the 90-deg angle. Seabergh's (1983) results suggest that the ideal orientation is a weir section angled 30 to 60 deg from the shoreline, with the seaward ends of the jetties parallel to one another (Figure V-6-68a). It is recommended that the largest (widest and/or deepest) part of the deposition basin be landward of the weir-shoreline connection.

(e) Design of storage capacity is addressed in Part V-6-5, and in Weggel (1981). Likewise, methods to compute the flood tidal flow and wave transmission across the weir are specifically presented in Weggel (1981). Case histories and additional information on weir jetties are presented in Rayner and Magnuson (1966); Snetzer (1969); Magnuson (1967); Purpura (1974, 1977); Parker (1979); Weggel and Vitale (1981), among others. Coastal Tech (1997) describes measured and numerically-modeled flow over a low-weir section at St. Lucie Inlet, Florida.

(3) Sediment traps. Specific areas of an inlet system can be designated, designed, or dredged to serve as deposition basins (sediment traps or sand traps), with or without weir jetties. Examples are illustrated in Figure V-6-69. These "traps" serve as temporary repositories of sediment that enter the inlet. They should be located in areas where natural wave and current processes will readily transport sediment into the trap (but not out of it), in semiprotected waters suitable for periodic operation of a dredge or bypass pump, and so as not to impede navigation. Traps can be located on the downdrift side of a weir section (as described previously), or at other locations where chronic shoaling is observed (e.g., at the seaward ends of jetties, at interior spits, across the flood shoals, etc.), or at locations where shoaling is induced by the construction of breakwaters or other structures. Traps can be located within the inlet (interior traps) or outside the inlet jetties (exterior traps).

(a) Sediment traps can be an overdredged area (or dredge pit) into which sediment is deposited by waves and currents (Figure V-6-69a). Once sediment fills the trap to near the level of the adjacent seabed or channel, the trap must be dredged so as to preclude loss of additional sediment that enters the inlet. The seaward edge of flood shoals, and the interior seabed adjacent to the jetties shoreward ends, are typical locations of overdredged traps. A less typical example is an updrift, exterior trap (also called a nearshore borrow area), which is a pit dredged along the shoreline updrift of an inlet (Figure V-6-69b).

(b) Sediment traps can also be created by breakwaters located near the inlet entrance. Sediment is deposited in the lee of the breakwater and is removed by a dredge or pumps working in the protected waters of the breakwater (Figure V-6-68c).

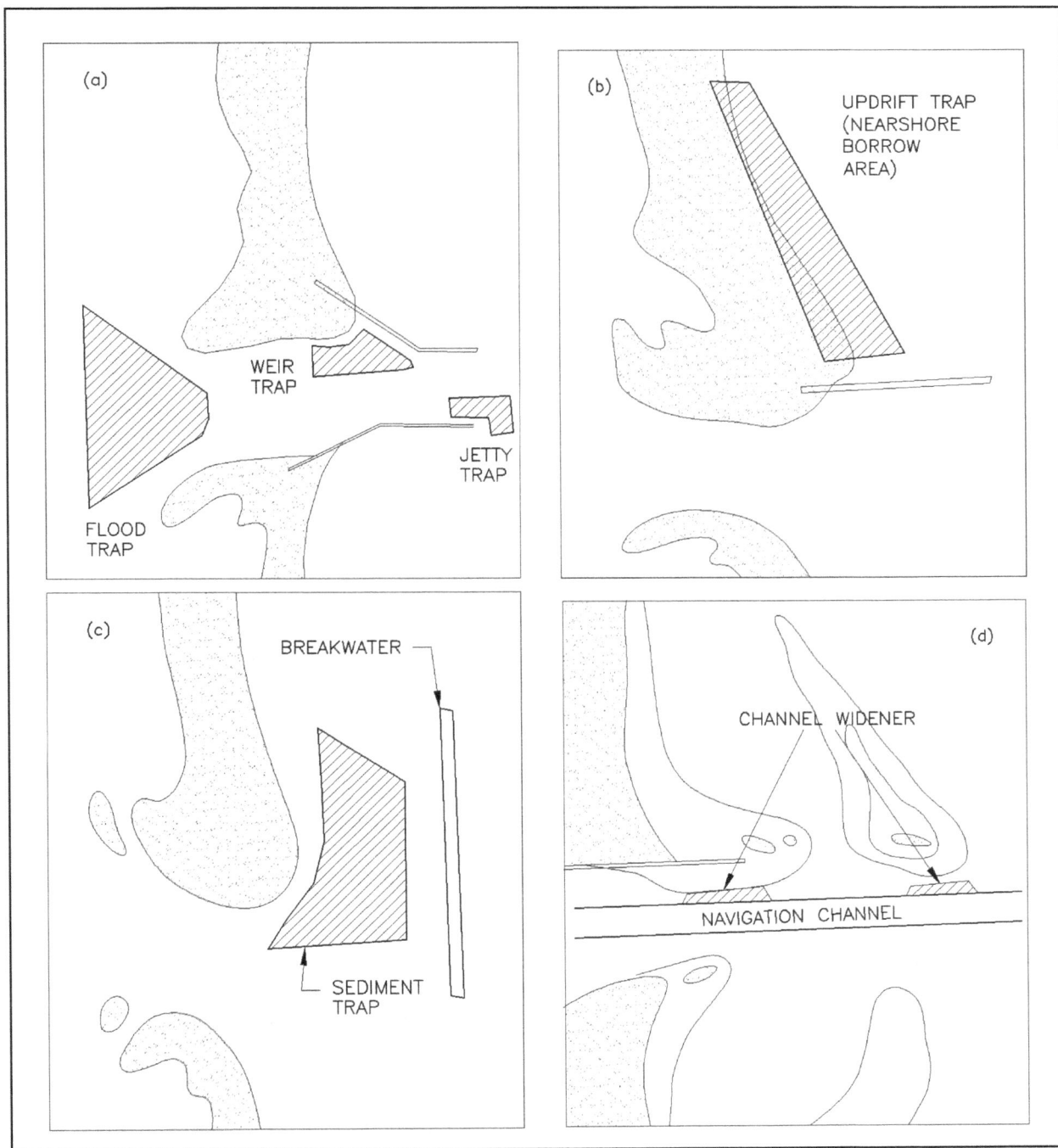

Figure V-6-69. Inlet sediment traps

(4) Channel wideners. The navigation channel can be widened along those areas where chronic shoaling by littoral drift is experienced (Figure V-6-69d). This variation of the sediment trap concept can also be termed advance maintenance, which presents possibilities for funding through Operations and Maintenance (O&M) monies. Channel wideners serve four purposes: they decrease the frequency of requisite maintenance dredging (which, in turn, reduces dredging costs by reducing the number of dredge

mobilizations); they improve the reliability of safe navigation; they provide a designated basin into which littoral sediment can deposit and be less likely to mix with non-beach-compatible sediments of the channel; and they increase the quantity of littoral sediments available for a given dredge job (thereby improving the economic viability of more expensive, beach disposal dredging practices). Methods to estimate the degree to which channel modifications (widening, deepening, etc.) will affect shoaling rates are approximate, at best. Some approaches are presented in O'Conner and Lean (1977); Lean (1980); Trawle (1981); and Galvin (1982). Application of sophisticated, three-dimensional numerical models which estimate sediment transport and shoaling rates from wave, tide and current effects are described in Vemulakonda et al. (1988).

(5) Flood shoals. Interior (flood) shoals present opportunities as a sand source for inlet-adjacent beach nourishment. Practically, however, their use may be limited by the shoal sediments' grain size (typically finer than the native beach sediments) and by the presence of environmentally sensitive resources such as seagrass beds, shellfish beds, or the presence of adjacent aquatic preserves in which dredge-related turbidity is restricted. As in the investigation of all sand borrow sources, core-borings of the flood shoals are critical to define the beach-compatibility and vertical extent of the sediment. Many flood shoals are thin, widespread deposits that overlay lagoonal clays and silts unsuitable for beach placement. The ability of conventional dredge equipment to excavate thin layers of material (i.e., less than 1.3 to 2 m (4 to 6 ft)) is limited, and may instead require the use of small, suction-dredge or jet-pump equipment. Extensive removal of interior shoals will likely increase the inlet's hydraulic efficiency and can be expected to increase the rate of interior shoaling. This may not be problematic so long as the geometry of the shoal borrow area is designed so as to act as a trap (in which the increased flux of sediment can be captured for bypassing/backpassing); and, is designed so as not to undercut or otherwise draw sand directly from the inlet-adjacent beaches. Increases of inlet hydraulic efficiency may also require examination of the hydrologic and biologic effects of increased encroachment of ocean (flood) waters to interior bays and aquifers.

(6) Ebb shoals. Inlet ebb shoals likewise present attractive, potential opportunities as sand sources for inlet-adjacent beach nourishment. Like the use of flood shoals, borrowing from the ebb shoal for these purposes can be viewed as mitigation for historical inlet impacts. Such retroactive sand bypassing/backpassing is particularly relevant for those cases in which inlet shoal volumes increased as a result of the inlet modifications' erosional effect upon the adjacent beaches.

(a) For those inlets where natural bypassing occurs quasicontinuously via transport along the ebb shoal, care must be taken so as not to overdredge, sever, or otherwise compromise the ebb shoal's ability to transport sand across the inlet's mouth. Improper borrowing would result in a short-term nourishment of the filled beach, followed by an acceleration of erosion, probably on both sides of the inlet, as the ebb shoal and bypassing bar recovers from the dredging event. In these instances, it may be prudent to limit ebb shoal dredging to the outer (seaward) edge of the plateau (Figure V-6-70a).

(b) For those inlets where natural bypassing occurs episodically through the migration of ebb shoals, spits, and baymouth bars, dredging and downdrift placement may simply speed the inlet's natural processes. This is a desirable outcome in those cases where the downdrift erosion stress is so high that the attendant development cannot withstand the time required for the inlet shoals to migrate naturally. The resulting dredge project usually acts to relocate the channel (which is otherwise compromised or forced downdrift by the migrating shoals), and thus results in at least an initial improvement of the inlet's navigability and hydraulic efficiency (Figure V-6-70b).

(c) The potential effect of ebb shoal bypassing to the wave refraction, current and associated littoral transport patterns along the inlet mouth and adjacent shorelines should be evaluated as part of the borrow

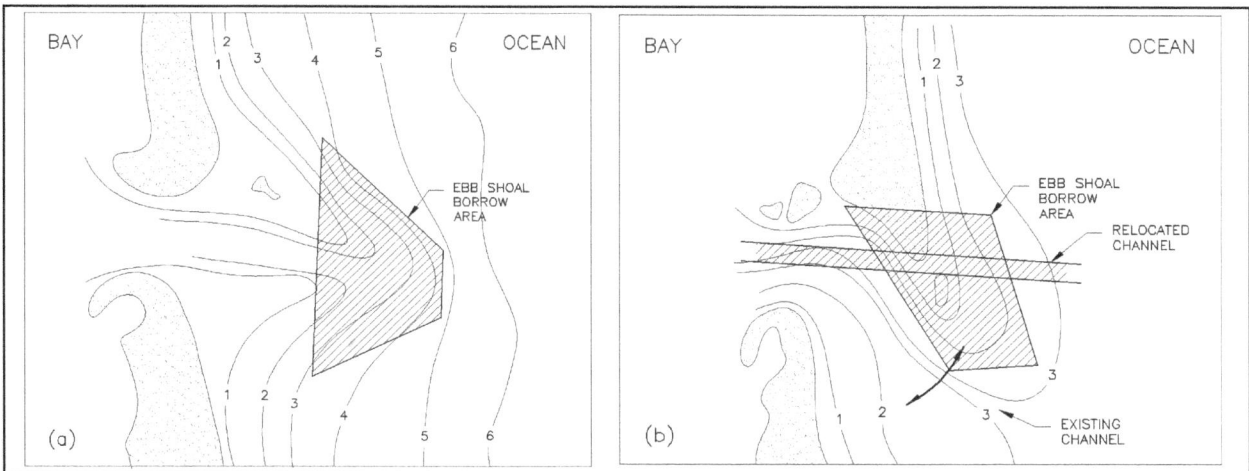

Figure V-6-70. Sediment borrowing from ebb shoal of an inlet

area design. Existing conditions should be compared to simulated post-borrow conditions. It is recommended that the wave height or wave energy density (H or H^2, respectively), be compared at each point across the refraction grid, for various incident wave conditions, to assess the degree to which the borrow activity may locally affect navigation. Likewise, the computed breaking wave energy density and littoral transport potential should be computed at each column of the grid along the adjacent shorelines, for various incident wave conditions, to discern possible changes in the local sediment transport patterns. The approach described in Part V, Chapter 6-2(d), method 4, is applicable in this regard. Like flood shoal borrowing, extensive removal of the ebb shoal may be expected to increase the trapping (shoaling) behavior and hydraulic efficiency of the inlet. Methods to estimate shoaling behavior are similar to those described for channel widening. Methods to examine inlet hydraulic efficiency are described in Part II- 6. Assessment of changes in tidal currents and associated sediment transport patterns is briefly described in Part II-7, and Part II-9.

(d) In the design of projects where ebb shoal dredging may result in channel relocation, it is essential to first understand the natural shoaling and channel-migration cycles of the inlet. Examinations of time-sequences of aerial photographs and bathymetric surveys, supplemented by anecdotal accounts, are valuable means by which to predict the degree to which the dredge project will advance, stagnate, or retard the inlet's natural patterns of sediment transport.

d. Bypass system capacity. Sizing of the requisite pumping equipment and geometry, sediment traps, and weir sections is a function of the gross bypassing requirements and the frequency with which bypassing will be undertaken. Sizing should also consider the potential effect of seasonal fluctuations and storms. Methods to determine the requisite bypass/backpass quantities are described in Part V-6-4 and Part V-6-5.

(1) Dredges, jet pumps, and other pumps. Specific guidance for the selection, layout and design of jet-pump systems is presented in Richardson and McNair (1981) and in EM 1110-2-1616. Introductions to the sizing and selection of conventional dredge equipment are presented in McKnight (1966); Bruun (1981); Herbich (1975, 2000); Herbich and Snider (1969); Richardson (1976, 1990); Huston (1970, 1986); and Turner (1984). Technical consultation is advised early in the design process. The Corps of Engineers' Marine Design Center, as well as industry consultants, are available in this regard.

(2) Excavation craters. Craters created by pumps or fixed dredge systems will remain empty until significant wave activity causes sufficient transport to fill them. Experience demonstrates that those craters located closest to shore fill fastest. For the purposes of intercepting the littoral drift, the apparent dominance of transport at the shoreline and in the swash zone is illustrated by the distribution of pumping hours from the Nerang River fixed bypassing plant in Queensland, Australia (Figure V-6-71). The theoretical yield volume V_C of a typical dredge crater, excluding use of fluidizers, can be estimated as

$$V_C = L_c d_c \left(b + \frac{d_c}{m} \right) + \pi \, d_c \left(\frac{b^2}{4} + \frac{f \, d_c}{2m} + \frac{d_c^2}{3m^2} \right) \qquad \text{(V-6-72)}$$

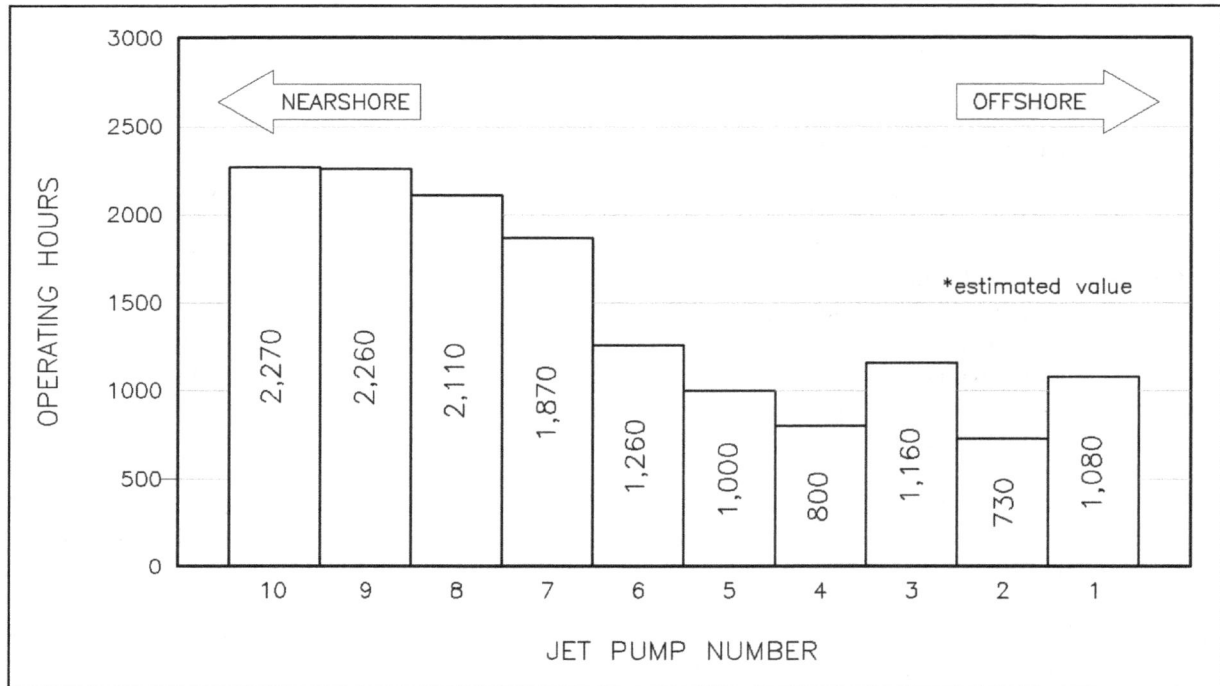

Figure V-6-71. Jet pump operating hours from May 1986 to February 1988; Nerang River bypassing plant; Queensland, Australia (data from Clausner 1988; figure adapted from Bodge 1993)

where L_c is the length (or arc-length) of the crater, d_c is the crater depth below the ambient seabed level, b is the width of the crater at the bottom, and m is the crater side slope as defined in Figure V-6-72. For medium to fine-grained beach sand, and for crater depths of about 1.5 to 2.75 m (5 to 9 ft), m is about 0.5, and b is between 6 and 9 times the diameter of the suction intake. For a circular crater (i.e., that made by a fixed pump), $L_c = 0$.

(3) Productivity. A single jet pump or eductor is typically capable of transferring between 50 and 250 cu m/hr when operating; or, between 50,000 and 150,000 cu m/year. However, average hourly and annual productivity is greatly influenced by mechanical/pipeline limitations, shoaling behavior at the crater, physical mobility of the intake(s), interference by marine debris, etc. considerations. A system's net productivity at any given site is therefore highly dependent upon the physical conditions specific to the site. Example considerations are described in the following paragraphs. See also Richardson and McNair (1981), EM 1110-2-1616, and Clausner et al. (1994).

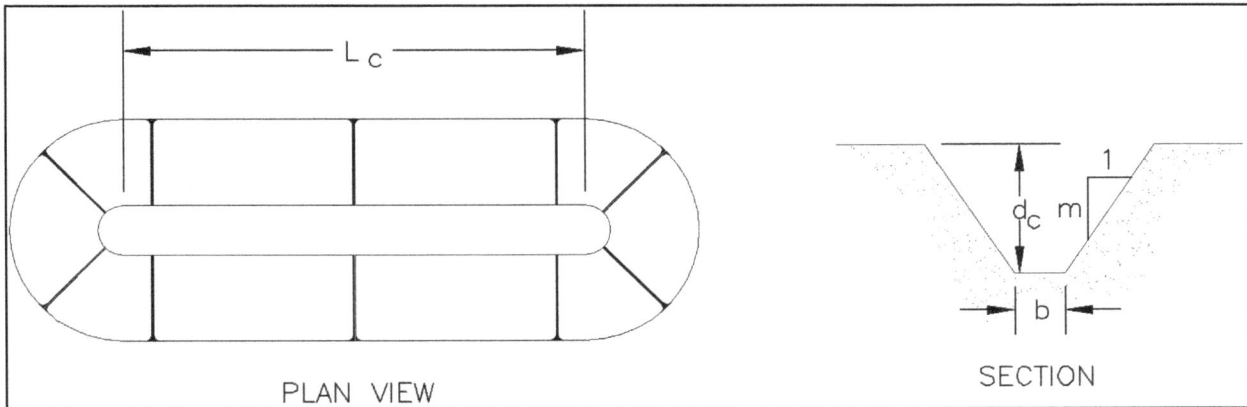

Figure V-6-72. Geometry of a dredge crater (definition sketch)

(a) A bypassing system's ultimate productivity is a function not only of the pumps' mechanical capacity, but also of the time required for the dredged crater(s) to refill. For example, Bodge (1993) observed that the bypass capacity of the fixed sand transfer plant at South Lake Worth Inlet, Florida, is more typically limited by the crater's shoaling rate than by the existing plant's capacity. The time required to excavate the crater, if full, is

$$T_E = V_C / (E - S) \qquad\qquad (V\text{-}6\text{-}73)$$

where S is the average volumetric shoaling rate of the crater (volume/time) during and after a given dredging event, and E is the net volumetric capacity of the pump (volume/time) after consideration of all mechanical limitations, shutdown times, etc.   The rate at which crater refills, S, will exceed the effective bypass capacity of the plant, E, if

$$S > \frac{1}{2} E \left[1 + \sqrt{1 - \frac{4V_C}{E T_B}} \right] \qquad\qquad (V\text{-}6\text{-}74)$$

where T_B is the elapsed time between the start of subsequent dredging events, For example, for daily operation, $T_B = 24$ hr. The (+) root implies that pumping continues throughout the interval T_B until the crater is completely emptied. The (-) root implies that pumping is discontinued the first time that the crater is completely emptied during the interval T_B. Undefined values of the radical (i.e., $4 V_C > E T_B$) indicate that the system is limited by the pump's volumetric capacity over the interval T_B, not by the crater volume.

(b) The distribution of crater shoaling rates can be approximated by a Rayleigh distribution centered about the site's mean shoaling rate – at least for the lower 90 percent of the distribution (Bodge 1993). (The Rayleigh distribution likely underestimates the occurrence of the largest 10 percent of the actual shoaling events.) The mean shoaling rate at the inlet can be approximated from existing bypass plant data (i.e., the average value of the volume bypassed divided by the time preceding each bypass event); or, approximated as the gross drift rate at the crater location minus the rate of natural (or other) bypassing around the crater's location. Values for the latter can be estimated from the inlet's sediment budget.

Sediment Management at Inlets

(c) The percentage of time for which Equation V-6-74 holds true, as derived from the distribution of the site's shoaling rates, indicates the percentage of time for which the plant's configuration is limited by mechanical capacity; i.e., that percentage of operating time for which the crater has completely refilled prior to the next start-up of the bypass plant. Alternately stated, the percentage of time for which Equation V-6-74 is false indicates the percentage of operating time for which the crater has not yet re-filled by the time the bypass plant is ready to start up again. For those times, increases in mechanical capacity of the plant are of no net value to overall productivity.

(d) Where sand transfer is limited by crater shoaling rates, net increases in inlet bypassing require that single pumps be moved during the day; or, that multiple pumps be used. Moving a pump implies that the pump assembly is to be physically lifted to another location; or, the pump's intake is mounted on a moveable support (e.g., along a trestle track, from a swinging or articulated boom, etc.) The intent is to increase the crater volume, V_C (precisely, so as to increase b, L_c and/or d_c in Equation V-6-72). It is also noted that the operation of fixed pumps tends to armor the crater's side slopes with coarser material over time, resulting in steeper slopes and smaller crater volumes.

(e) Sediment traps and channel wideners. For the typical case where sediment within a trap is not intended to feed the adjacent shorelines during transport reversals, the capacity of a sediment trap should be sized to accommodate all of the drift incident to that part of the inlet where the trap is located for the entire duration anticipated between bypass events. Accomodation for variations in the annual and/or seasonal transport rates must be made. At a minimum, this should be done by examing the statistical variation of the gross transport rate about the mean, as developed from hindcast wave data. Furthermore, it should not be assumed that deposition within the trap will occur uniformly. That is, accommodation must be made for varying vertical rates of deposition within the trap. Proper use of a sediment trap for effective inlet/harbor sediment management requires that the sediment trap be excavated at or before the time at which it reaches capacity. Accordingly, it is essential that the trap be sized large enough to accommodate periods of elevated transport and/or delays in bypassing operations.

(f) Figure V-6-73 illustrates an example of the temporal fluctuation of bypassing at a fixed sand transfer plant. The plant (South Lake Worth Inlet, Florida) operates whenever there is any sand within the crater and when waves are incident from the north. The record of the plant's bypassing operations is fairly well correlated with the southerly-directed longshore transport rate potential, where the latter was computed from hindcast wave data (Creed 1996).

(g) The ultimate design of a trap's capacity is determined by balancing the decreased cost of bypassing operations associated with larger traps and less frequent equipment mobilization, the increased initial cost of trap construction (if any), the availability of areas within the inlet area(s) in which to create a trap, the uncertainty as to shoaling rates, and in the case of channel wideners and advance nourishment, the decreased costs associated with requisite maintenance dredging to ensure safe navigation. Stochastic methods regarding the latter two issues are presented in Lund (1990).

(4) Weir jetty systems. Weggel (1981) describes the optimum weir-jetty system as one where only the net incident transport ($Q_{NET} = Q_R + Q_L$) enters the deposition basin for ultimate bypassing to the net downdrift beach (see Figure V-6-74). Neglecting local, inlet-induced transport, and assuming net transport directed to the right, the amount of sand carried to the weir from the updrift side is Q_R. An amount of sand equal to ($Q_R - Q_{NET}$) must be therefore retained in temporary (active) storage on the updrift beach to replace the material trapped by the downdrift jetty and inlet shoals during transport reversals (i.e., an amount equal to Q_L).

EXAMPLE PROBLEM V-6-16

FIND:
The potential net increase in bypassing productivity if a jet pump is changed from a fixed mount to a swinging-boom mount, and/or if the pump's operating schedule is revised.

GIVEN:
An inlet's small, existing 6-in. jet pump bypasses 25,000 cu m/year, operating for about 1,200 hr/year at a fixed location with maximum crater depth of d_c = 2.6 m below the seabed (limited by a clay layer). The plant is operated between 8 a.m. and 5 p.m., each day, as needed, but only when the crater is mostly full. At the beginning of one-quarter of those days for which bypassing is undertaken, the crater has been refilled to capacity overnight after having been mostly excavated the previous day. The proposed system would mount the jet pump at the end of a 5-m-long boom capable of swinging through a horizontal arc of 105 deg.

SOLUTION:
The plant's effective mechanical capacity is:

$$E = (25,000 \text{ cu m/year}) / (1,200 \text{ hr/year}) = 20.8 \text{ cu m/hr}$$

In SI units, the 6-in. pump diameter is 0.15 m. The crater's base width, b, is six to nine times this diameter, or 0.9 to 1.4 m; for example, if b = 1.2 m. Assume a crater side slope of 1 (vert) : 2 (horiz), or m = 0.5. From Equation V-6-72, and for d_c = 2.6 m and L_c = 0, the crater volume of the existing plant is:

$$V_C = \pi(2.6 \text{ m}) [(1.2 \text{ m})^2 / 4 + (1.2 \text{ m})(2.6 \text{ m}) / (2)(0.5) + (1/3)(2.6 \text{ m} / 0.5)^2] = 102 \text{ cu m}$$

For those events when the crater fills to capacity overnight (from 5 p.m. to 8 a.m., or 15 hr), the crater shoaling rate is at least 102 cu m / 15 hr = 6.8 cu m/hr. From the given data, this represents the 25th percentile of the shoaling rate for those days when bypassing is undertaken. From the Rayleigh distribution (Equation II-1-131), the probability, p, that a value S will exceed some value S_p is:

$$p \, (S > S_p) = \exp \, (-(S_p / S_m)^2)$$

where S_m is the mean value. Taking the natural logarithm of each side and rearranging yields:

$$S_m = [-S_p^2 / \ln \, (p(S_p))]^{1/2}$$

so that

$$S_m = [- (6.8 \text{ cu m/hr})^2 / \ln (0.25)]^{1/2} = 5.8 \text{ cu m/hr}$$

(Continued)

EXAMPLE PROBLEM V-6-16 (Continued)

From Equation V-6-74, assuming that bypassing is undertaken only when the crater is mostly full, the theoretical rate of shoaling which exceeds the existing pump system's capability, based upon no more than 9 hr of operation every 24 hr, is

$$S > (0.5)(20.8 \text{ cu m/hr}) \{ 1 - [1 - (4)(102 \text{ cu m}) / (24 \text{ hr})(20.8 \text{ cu m/hr})]^{1/2} \} = 6.0 \text{ cu m/hr}$$

From Equation V-6-73, this rate of shoaling would require $(102 \text{ cu m}) / (20.8 \text{ cu m/hr} - 6.0 \text{ cu m/hr})$ = 6.9 hr to excavate the crater. From the Rayleigh distribution, and based upon an assumed mean value of $S_m = 5.8$ cu m/hr, the probability of shoaling rates exceeding 6.0 cu m/hr is

$$p\ (S > 6.0 \text{ cu m/hr}) = \exp\ (-(6.0 / 5.8)^2) = 0.34 = 34 \text{ percent}$$

Therefore, for operations limited to 9 hr per day, the bypassing quantity could be theoretically increased for 34 percent of the bypassing events by increasing the crater capacity. Note that increases to the pump capacity, E, would not result in a net improvement.

Further, note that the positive (+) root of Equation V-6-74 yields 14.8 cu m/hr. From Equation V-6-73, that rate of shoaling would require $(102 \text{ cu m}) / (20.8 \text{ cu m/hr} - 14.8 \text{ cu m/hr})$ = 17 hr to excavate the crater. That value exceeds the 9-hr operating window of the plant. On the other hand, if operation beyond 9 hr per day was adopted, the probability of shoaling rates exceeding 14.8 cu m/hr is only

$$p\ (S > 14.8 \text{ cu m/hr}) = \exp\ (-(14.8 / 5.8)^2) = 0.001 = 0.1 \text{ percent}$$

Therefore, adopting bypassing operations beyond the existing 8 a.m. to 5 p.m. window would theoretically increase the bypassing quantity of the existing system without hardware improvements, but only for a small percentage of those days for which bypassing is undertaken.

The proposed hardware improvements (swing-boom) would create an arcuate crater shape of length

$$L_c = (60°)/(360°)\ 2\ \pi\ (5 \text{ m}) = 5.2 \text{ m}$$

From Equation V-6-72, the associated crater volume would be

$$V_C = (5.2 \text{ m})\ (2.6 \text{ m})\ (1.2 \text{ m} + (2.6 \text{ m}) / (0.5)) + 102 \text{ cu m} = 189 \text{ cu m}$$

(Continued)

EXAMPLE PROBLEM V-6-16 (Concluded)

From Equation V-6-74,

$S > (0.5)(20.8 \text{ cu m/hr}) \{ 1-[1-(4)(189 \text{ cu m/hr}) / (24 \text{ hr})(20.8 \text{ cu m/hr})]^{1/2} \} = (10.4 \text{ cu m/hr})(1- \sqrt{-0.5}\}$

for which the square root is undefined. Thus, with the proposed enlargement of the crater, the system's capacity would be theoretically limited by the pump capacity, E, instead of the crater volume, V_C.

While the crater volume for the proposed improvement is $(189 / 102) = 1.85$ times larger than the existing system, the improvements would not yield a 1.85-times increase in bypass capacity. This is because the existing system's bypass performance is limited in only 25 to 34 percent of the existing bypassing events. That is, only some fraction of the 85 percent increased crater capacity would be utilized for only 25 percent to 34 percent of the current bypassing events. Hence, the net increase in bypass quantity provided by the hardware improvements would be, at most, $(0.25 \text{ to } 0.34) \times (0.85) = 21$ to 29 percent. The decision to improve the plant's hardware or to periodically extend the operating hours (as needed), is a function of economic and other site-specific issues.

(a) Volumetric capacity requirements for the ideal system can be estimated from a mass curve. The mass curve is developed from a plot of the cumulative longshore transport rate versus time (Figure V-6-75). The active storage requirement of the updrift beach is the maximum vertical difference of the cumulative transport curve, as shown in the figure. The minimum storage requirement of the deposition basin is the vertical height of the curve's average slope calculated over the dredging interval of interest.

(b) Allowance on the updrift beach must be made for "dead" storage (Figure V-6-74). This is the impoundment fillet that forms immediately updrift of the weir from which sediment is not transported during reversals. The shape and volume of this dead storage can be approximated from analytic models such as Pelnard-Considère (See Part III-2-2) or numerical models such as GENESIS (See Part III-2-2). In all, the geometry of the weir section should be ideally designed so as to ensure that the requisite active storage will be developed in addition to the dead storage, and so as to ensure that the deposition basin leeward of the weir can be sufficiently sized.

  e. *Placement of material.* In order of priority, placement of bypassed/backpassed sediment should be: upon the beach and beach face; within the typical surf zone (so-called near-nearshore placement); and within the active depth of the beach profile (nearshore disposal). The alongshore placement should be outside of the inlet's direct influence; i.e., so as to minimze the potential for the sediment to return to the inlet and maximize its potential for return to the active littoral system external of the inlet.

(1) Alongshore location. The minimum distance from the inlet for sediment placement can be determined from refraction/transport analysis such as is described in Part V-6-2(c), method 4, in particular, and by Kana and Stevens (1992). Alternately, shoreline change models such as the one-line model of Perlin and Dean (1983) and GENESIS (Hanson and Kraus 1989; see also Part III-2-2, can be employed. Again, the principal objective is to define that distance at which the locally induced transport potential toward the inlet is diminished.

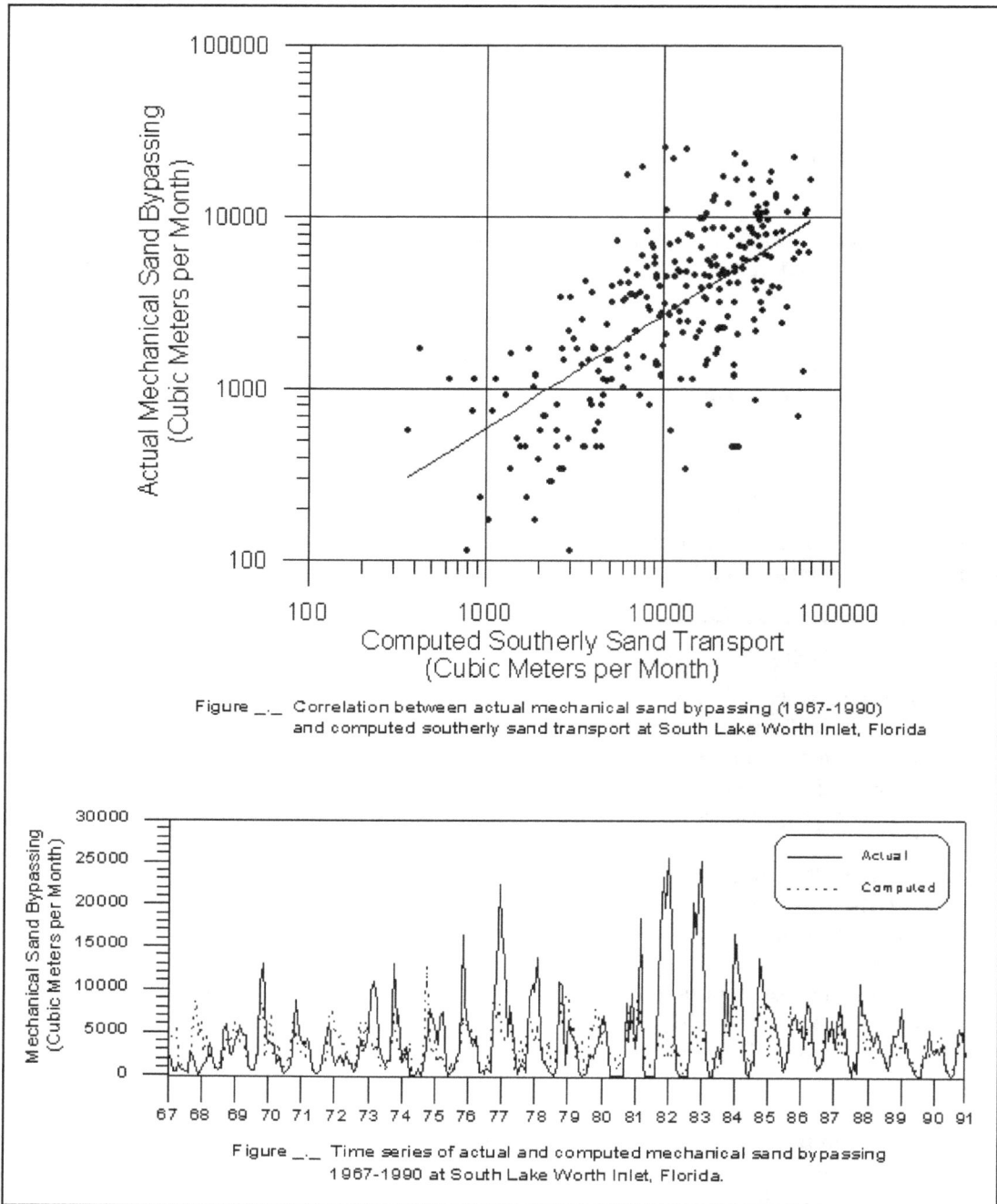

Figure _._ Correlation between actual mechanical sand bypassing (1967-1990) and computed southerly sand transport at South Lake Worth Inlet, Florida

Figure _._ Time series of actual and computed mechanical sand bypassing 1967-1990 at South Lake Worth Inlet, Florida.

Figure V-6-73. Measured sand bypass rates from the South Lake Worth Inlet, Florida, fixed transfer plant: (a) correlated with hindcast longshore transport potential, and (b) depicted as a time-series. "Actual" refers to monthly bypass volumes estimated from plant records. "Computed" refers to the southerly-directed transport potential as computed from hindcast WIS data for the same 24-year period of record, where the mean value of the computed transport potential was set equal to the mean value of the actual bypass volume over the period of record (Creed 1996)

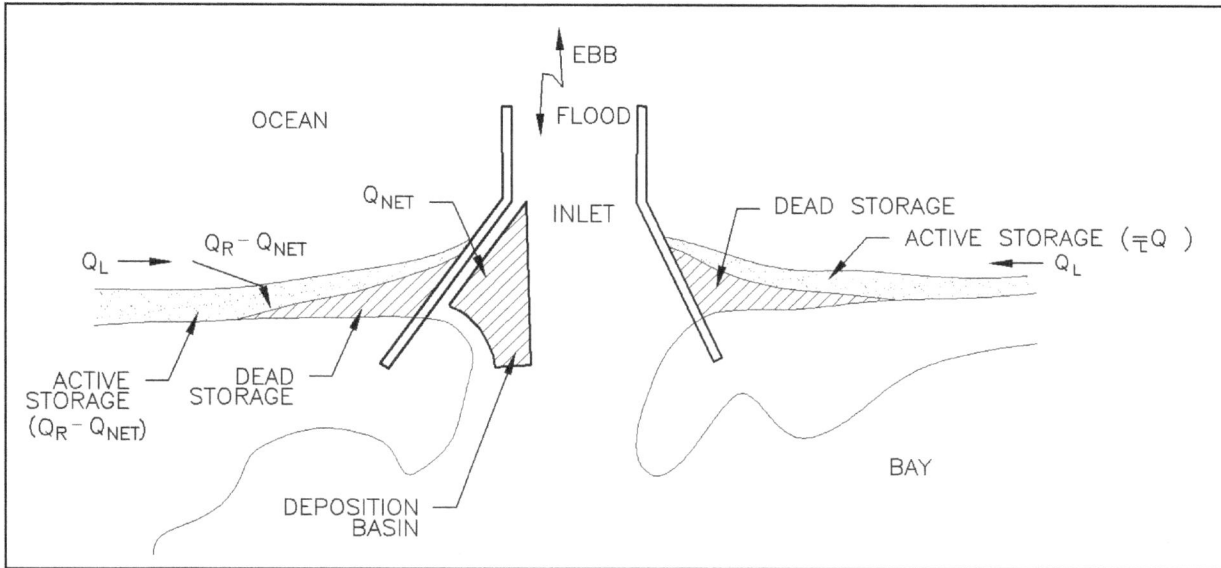

Figure V-6-74. Active and dead sediment storage on a beach adjacent to a weir jetty system (adapted from Weggel 1981)

(2) Nearshore disposal. When placement of the bypassed/backpassed sediment directly on the beach or within very shallow depths is not possible, nearshore disposal (also called nearshore berm disposal) may be considered. In inlet/harbor sediment management, the principal goal is to create an active berm from which the placed sediment will migrate shoreward and return to the shorelines' active littoral system. (A less preferable alternative is to create a stable berm which is intended to provide shore protection through diminution of wave energy (Zwamborn et al. 1970). The tangible benefits of this strategy are unproven; and, some field studies suggest that nearshore, submerged mounds and structures can induce beach erosion in their lee (USACE 1950; Browder 1995)).

(a) Hands and Allison (1991) present an analytic method by which to predict the seabed placement depth which separates active (shoreward-migrating) from stable berms. In this method, the nonexceedance probability of the maximum nearbed wave-orbital velocity, u_m, is computed and plotted for a given depth, using hindcast wave data for the site of interest. The probability distribution of such plots distinguish stable from active berms (see Figure V-6-76). Most briefly, Hands and Allison (1991) conclude that if the 75-percentile velocity far exceeds 40 cm/sec, or the 95-percentile velocity far exceeds 70 cm/sec, then sand berms should not be expected to remain stable, regardless of the depth or sand size. If the computed velocities are considerably less than these values, a stable sand berm is expected under all but unusual circumstances.

(b) From Equation II-1-22, the maximum nearbed horizontal wave orbital velocity is computed as

$$u_m = \pi \frac{H}{T} \frac{1}{\sinh(2\pi h/L)} \qquad (V\text{-}6\text{-}75)$$

where T is the (spectral peak) wave period, and H and L are the significant wave height and the wavelength, respectively, in the proposed water depth for berm disposal, h. In practice, nearshore or offshore wave data are transformed (refracted and shoaled) to some specified water depth, h; the velocity

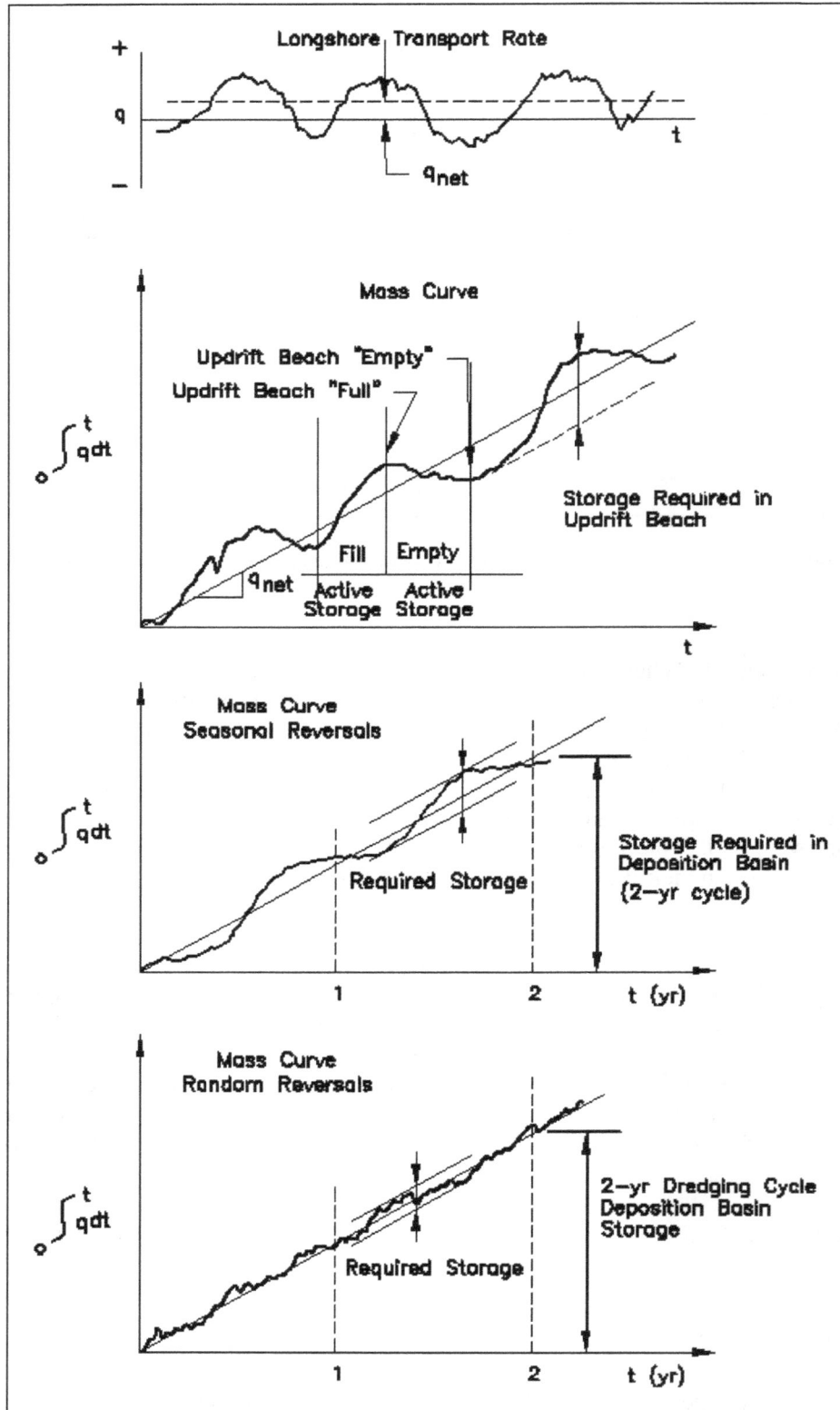

Figure V-6-75. Use of mass curve to predict volumetric capacities required for updrift beach and deposition basins (from Weggel 1981)

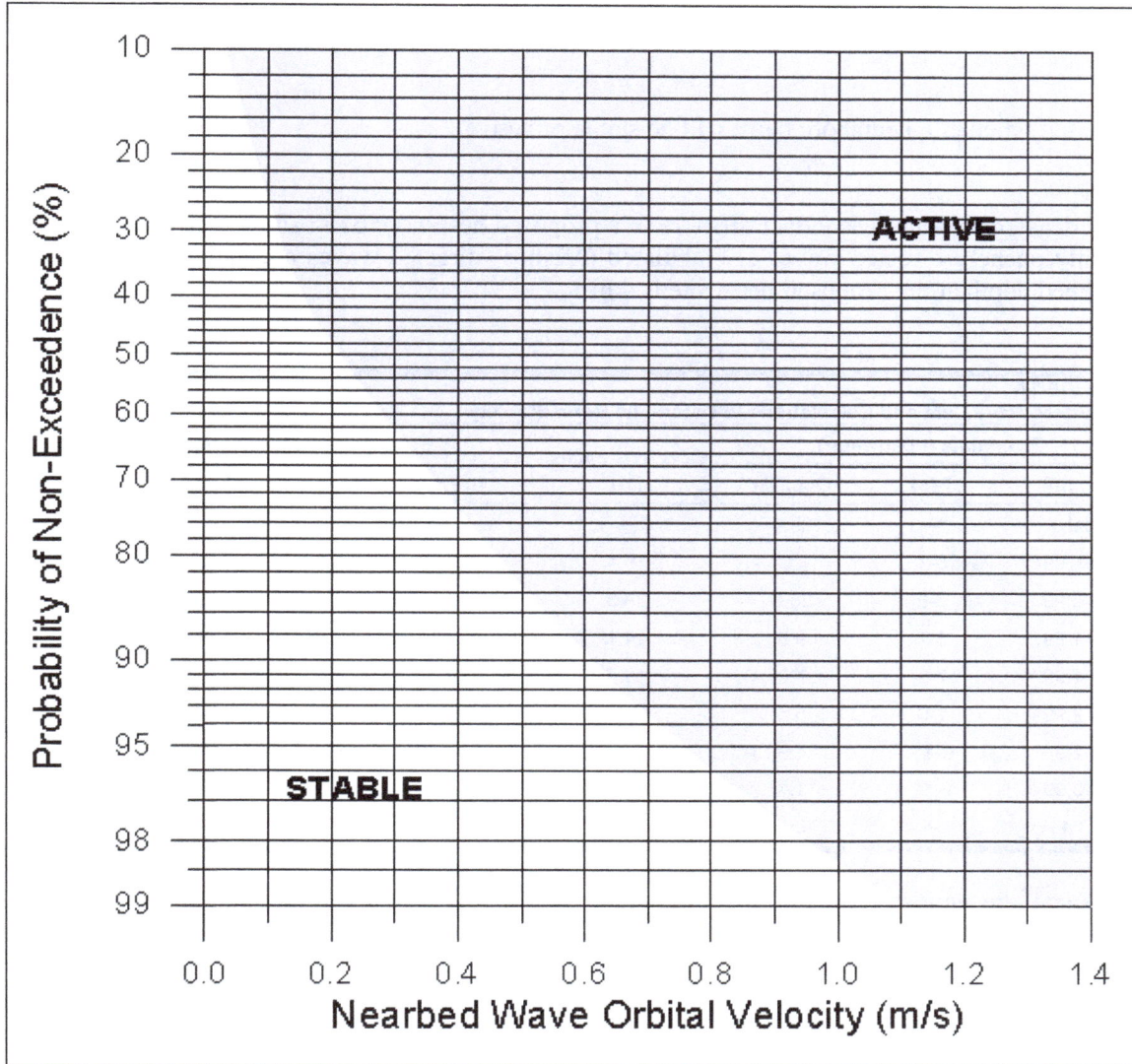

Figure V-6-76. Method to predict whether sediment placed in a nearshore berm will be active (move landward) or stable (after Hands and Allison 1991)

u_m is computed for each transformed wave condition, and the percent-occurrence of each velocity value noted. The velocity values are then ranked, and the percent of nonexceedence computed for each and plotted on a copy of Figure V-6-76. (The nonexceedence is simply the percent occurrence for which the computed velocity values are less than some given value.) If the plotted distribution falls mostly within the active half of the graph, then the berm can be expected to be active; i.e., to migrate shoreward. If not, then a shallower water depth, h, should be considered and the computation repeated, etc., until the distribution falls within the active half of the graph. The depth at which the distribution falls generally between the active and stable halves of the graph denotes the maximum seaward depth at which nearshore disposal can be undertaken so as to yield any detectable shoreward migration.

EXAMPLE PROBLEM V-6-17

FIND:
Predict whether a nearshore berm will be stable or active.

GIVEN:
Hindcast, deepwater wave data offshore of a proposed nearshore berm site are given in Table V-6-14. Only onshore-directed waves are tabulated (offshore-directed waves are included under calms). The seabed depth at the proposed berm site is 4 m.

Table V-6-17
Deepwater Wave Data and Calculations to Determine Berm Stability

Given data (deepwater)			Occ. exc. calms (%)	L_o (m)	water depth, h = 4.0 m		
H_o (m)	T (s)	Occ.(%)			L (m)	H (m)	u_m (m/s)
calm	--	32.4	--	--	--	--	--
0-0.99	5-7	13.1	19.4	56.2	34.8	0.35	0.23
0-0.99	7-9	25.9	38.3	99.9	48	0.29	0.21
1-1.99	5-7	9.5	14.1	56.2	34.8	1.05	0.7
1-1.99	7-9	7.2	10.7	99.9	48	0.87	0.62
2-2.99	7-9	5.5	8.1	99.9	48	1.44	1.03
2-2.99	9-11	4.9	7.2	156.1	61	1.24	0.92
3-3.99	9-11	1.1	1.6	156.1	61	1.73	1.28
>4	--	0.4	0.6	--	--	--	--

To emulate the empirical results of Hands and Allison (in which the probability of $u_m = 0$ is 0 percent), calm events are excluded. The occurrence of each wave condition is recomputed neglecting calms. For the first noncalm entry in the table:

Percent occurrence excluding calms = (13.1 percent) / (100 percent - 32.4 percent) = 19.4 percent

The wavelength at the reference (hindcast) site is computed for each wave condition. The median height and period values within the ranges given for wave condition are assumed. The wave data are given as deepwater values; hence, from Equation II-1-15, and for the first noncalm entry in the table:

$$L_o = 1.56 \, (T^2) = 1.56 \, (6.0)^2 = 56.2 \text{ m}$$

The wavelength, L, at the placement depth of h = 4.0 m, is computed from linear theory (see Example Problem Equation II-1-1). The corresponding wave height, H, is computed from Equation V-6-76. For the first noncalm entry in the table:

$$H = (0.5 \text{ m}) (34.8 \text{ m} / 56.2 \text{ m})^{3/4} = 0.35 \text{ m}$$

(Continued)

EXAMPLE PROBLEM V-6-17 (Concluded)

From Equation V-6-75, for the first noncalm entry in the table:

u_m = (3.14) (0.35 m / 6 sec) / [sinh {(2)(3.14)(4 m)/(34.8 m)}] = (0.183 m/sec) / (0.7862) = 0.23 m/sec

Rank the values of u_m in ascending order, and list the percent-occurrence for each (first two columns in TableV-6-15). Compute the cumulative occurrence for each event (last column in table), which is the same as the probability of nonexceedence. Plot the results atop Figure V-6-77. The plotted curve falls mostly well within the active (shaded) area, suggesting that the proposed berm is more likely to migrate landward than to remain stable.

Table V-6-18
Calculations of Maximum Nearbed Orbital Velocity

u_m, m/sec	Occ., percent	Cumulative Occ., percent
0.21	38.3	38.3
0.23	19.4	57.7
0.62	10.7	68.4
0.70	14.1	82.5
0.92	7.2	89.7
1.03	8.1	97.8
1.28	1.6	99.4

(c) In developing the empirical result illustrated in Figure V-6-76, Hands and Allison (1991) used the wave height transformation:

$$H = H * \left(\frac{L}{L*} \right)^{3/4}$$

(V-6-76)

where $H*$ and $L*$ are the wave height and length at some given water depth (such as the hindcast location), and where the standard dispersion relationship for linear waves was used to compute wavelength (Equation II-1-10).

(d) The propensity for shoreward migration increases with shallower water depths for disposal, and higher, better-defined berm elevations. A principal shortcoming of the Hands and Allison technique (Figure V-6-76) is that it does not consider the crest elevation or width of the berm. The more complicated method of Douglass (1995) demonstrates the importance of these two elements, particularly the former.

(e) Burke and Allison (1992) and Hands and Resio (1994) present additional guidelines on berm disposal geometries and depths, respectively. Douglass (1995) and Sheffner (1996) present analytic methods by which to predict postplacement migration of nearshore berms. Other Corps of Engineers' PC-based numerical methods to predict the fate of disposed dredged material include DIFID, DIFCD, DIFHD, ST-FATE and LT-FATE; described elsewhere. USACE (1950); Uda, Naito, and Kunda (1991); Johnson et al. (1994); Bodge (1994a); Foster, Mealy, and Delange (1996); Mesa (1996), among many others, describe case histories of nearshore berm stability and migration.

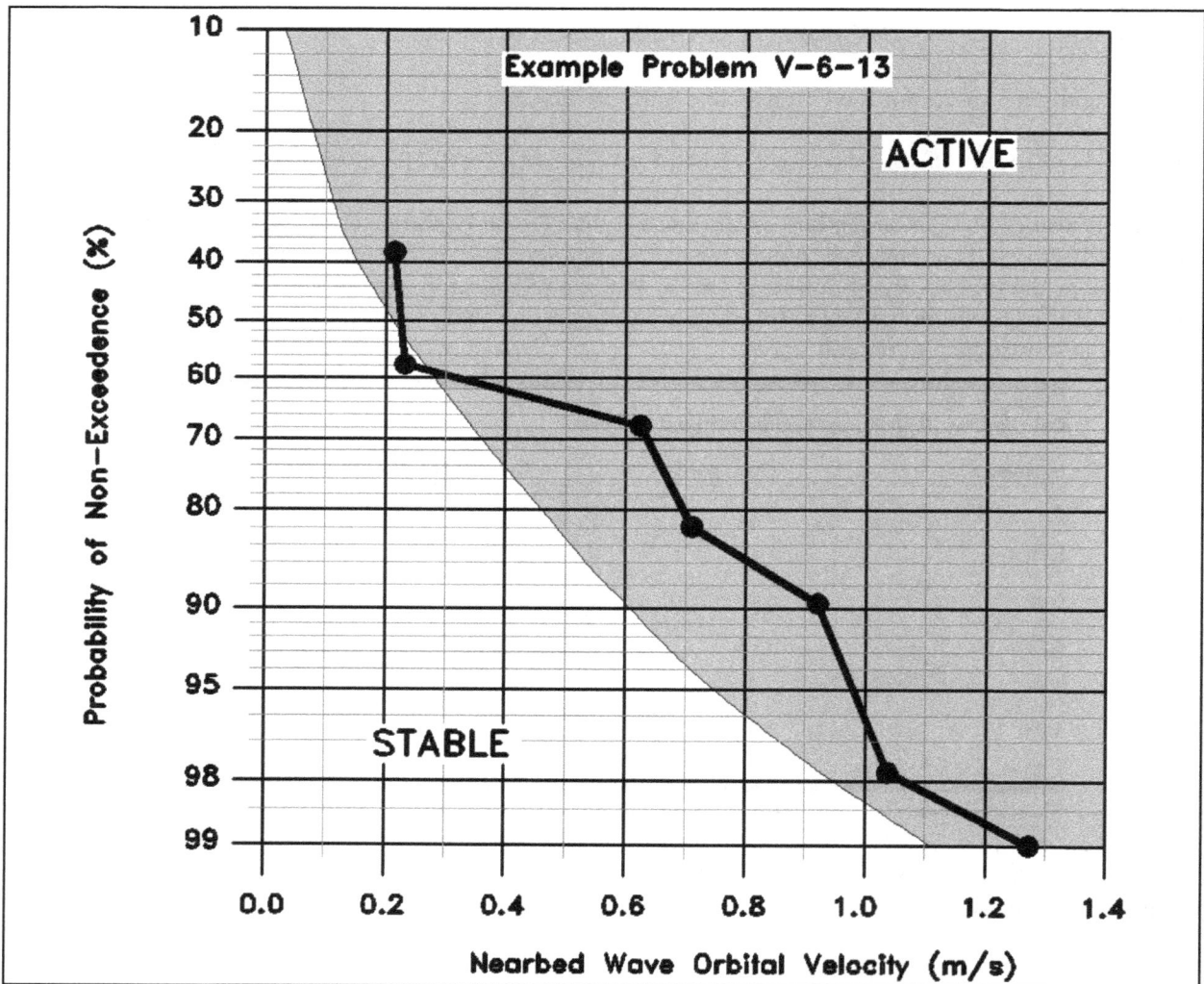

Figure V-6-77.   Results for Example Problem V-6-17

(f)   The profile zonation technique of Hallermeier (1981), and as described in the *Shore Protection Manual* (1984), generally overpredicts the water depths in which active berms are expected, and results in a wide buffer zone in which berm performance is uncertain (Hands and Allison 1991). The use of the Hallermeier Inner Limits and Outer Limits (HIL and HOL), and the annual seaward limiting depth of the littoral zone, are not recommended for design of nearshore berm disposal criteria.

(g)   McLellan and Kraus (1991) present an alternative method by which to predict whether a berm will be active or stable. The work is an extension of onshore/offshore transport predictors employed for beach profile response in the surf zone (see also, Kraus, Larson, and Kriebel 1991); and, as such, may be less appropriate for berms placed in deeper water toward the outer limit of the profiles' active depths. Nonetheless, the application of these transport predictors to berm depths may illuminate the relative degree to which shoreward transport may be anticipated. The simplest such predictor is the so-called "fall-velocity" or "Dean" parameter, N_o

$$N_O = \frac{H_o}{wT}$$

(V-6-77)

where H_o is the offshore wave height, w is the sediment's median fall-velocity, and T is the wave period. Concisely, for $N_o < 2.4$, shoreward movement is highly probable; for $N_o < 3.2$, shoreward movement is probable; for $N_o > 3.2$, seaward movement is probable; and for $N_o > 4.0$, seaward movement is highly probable. The utility of this predictor for nearshore berm stability is diminished, as the water depth of the berm placement or crest elevation is not explicitly specified.

f. Costs. Overviews of inlet sand bypassing costs are presented in Jones and Mehta (1977, 1980); EM 1110-2-1616; Bruun (1993), among others. Table V-6-19 presents a brief summary of costs for various inlet sand bypassing projects. It is noted that project costs will vary widely in the prototype as a function of the quantity and complexity of the work, and the potential to combine projects for which similar equipment can be utilized. By far, the greatest economies are to be gained where sediment management strategies (bypassing and backpassing) can be combined with requisite maintance dredging of the inlet/harbor, and can be combined between nearby project sites. Comparative economic analysis is required in those cases where the quantites of sediment bypass/backpass can be minimized by the construction of additional inlet structures, jetty improvements, or channel relocations. Specifically, the short-term (capital) costs of these works must be considered against the long-term reduction in costs associated with decreased dredging requirements. In all cases, the cost of improving inlet/harbor sediment management so as to minimize the erosive impacts to adjacent shorelines should be weighed against the cost of mitigating the erosion through separate beach nourishment projects, and/or the costs of accepting the increased storm damage, legal actions, and loss of recreation, revenue, and habitat should the erosion be left unabated.

V-6-6. Project Experience

a. Overview of existing prototype systems. The methods, mechanical equipment, and success of existing inlet sand management / bypassing systems vary widely, and are dictated by physical, environmental, and social considerations specific to each inlet site. Designers of potential sand management / bypassing systems are urged to consider the history and monitored performance of previous systems; and, whenever possible, to directly consult those engineers and operators that are intimately familiar with these systems' performance. Several prototype systems are summarized. Additional summary descriptions are presented in EM 1110-2-1616.

(1) Santa Cruz Harbor, California. Santa Cruz Harbor is located on the northern coast of Monterey Bay on the Pacific Ocean south of San Francisco (Figure V-6-78). During the project's design in the 1960s, the net longshore transport rate was believed to be less than 230,000 cu m/year, but subsequent studies in 1978 concluded that the rate is between 230,000 and 383,000 cu m/year (Moffatt and Nichol 1978). Within 2-1/2 years after the channel and jetties were constructed in 1962-63, the updrift (west) shoreline had impounded 400,000 cu m of sand, the channel experienced severe shoaling, and the downdrift beach had eroded. Annual channel dredging, usually by 30 cm (12-in.) hydraulic dredge, was begun in 1965 but could not maintain a clear channel against the significant and rapid shoaling experienced in winter. Between 1976-78, an experimental jet-pump system was tested using four mobile pumps and one fixed pump. The severe wave climate and rapid sand shoaling rates buried the jet pumps' supply and discharge lines. This hampered the pumps' mobility and also required frequent backflushing of the supply water lines to clear them of sand. Debris also presented severe problems to the jet pumps. Annual contract dredging was continued, as before, through the mid-1980s; and the channel continued to shoal and mostly close during winter.

Table V-6-19
Examples of Inlet Sand Bypassing Costs (U.S. Dollars in 1995)

Project	Volume (cu yd)	Frequency (yr)	Source	Placement	Method	UnitCost ($/cu yd)	Mob/Demob ($)	Total Unit Cost ($/cu yd)
Canaveral Harbor, FL	900,000	6	Updrift beach; -4 to -16 ft mlw)	Beach disposal: 0 to 2 miles downdrift	30" hyd. dredge	$4.20	$800K	$5.10
Canaveral Harbor, FL	200,000	1	Nav. Channel	Nearshore Berm; -20 ft mlw, 6 miles downdrift	Clamshell dredge and 2,500 cu yd dump scow	$0.00 (no added cost above offshore disposal)		$0.00
Canaveral Harbor, FL	< 200,000	1	Nav. Channel	Nearshore Berm; -10 to -16 ft mlw, 6 miles downdrift	Clamshell dredge and dump scow	< $0.25	$0 (inc. in basebid)	< $0.25
Masonboro Inlet, NC	696,000	4	Updrift Weir & Dep. Basin	Beach disposal: w/in 2- to 3- miles updrift and downdrift	27"- to 30"-pipeline dredge.	$2.76	$543K	$3.54
Carolina Beach Inlet, NC	517,000	3	Internal Sand Trap	Beach w/in 1 to 2 miles downdrift	Shallow-draft hopper; or pipeline dredge	$2.06	$310K	$2.66
Channel Is., CA	1.5 M	2	Int. Sand Trap	Beach w/in 2 miles downdrift	30"-pipeline dredge	$2.20	$750K	$2.70
Perdido Pass, AL	320,000	2-3	Updrift Weir & Dep. Basin	Beach w/in 1 mile downdrift	24"-pipeline dredge	$1.98	$132K	$2.39
E. Rockaway Inlet, NY	180,000	2	Nav. channel	Nearshore Berm; -16-ft depth, 1 mile downdrift	Hopper dredge (Atchafalaya)	$4.16	* $164K +	$5.07
Jones Inlet, NY	380,000	2	Nav. channel	Hempstead Beach; 1- to 2-miles away	Hopper dredge w/ pumpout and 5,000-ft pipeline	$7.56	$342K	$8.46
Indian River Inlet, DE	97,000	cont.	Updrift beach	Beach imm. north of inlet	Mobile crane and jet pump	$1.62	$1.7M**	$4.98**
Nerang River, Australia	600,000	cont.	Updrift beach	Beach imm. south of inlet	Fixed pier w/ 10 jet pumps	$1.20	**Aust.$ 6.3M (1986)	$3.24**
So. Lake Worth Inlet, FL	70,000	cont.	Updrift beach	Beach w/in 700 ft south of inlet	Fixed plant w/ suction dredge	$2.97	$30K per year maint.	$3.40

- Projects were contracted jointly.
- ** Initial Cost to Design and Construct Plant. Total Unit Cost assumes 30-year amortization of initial cost at 9 percent per year.

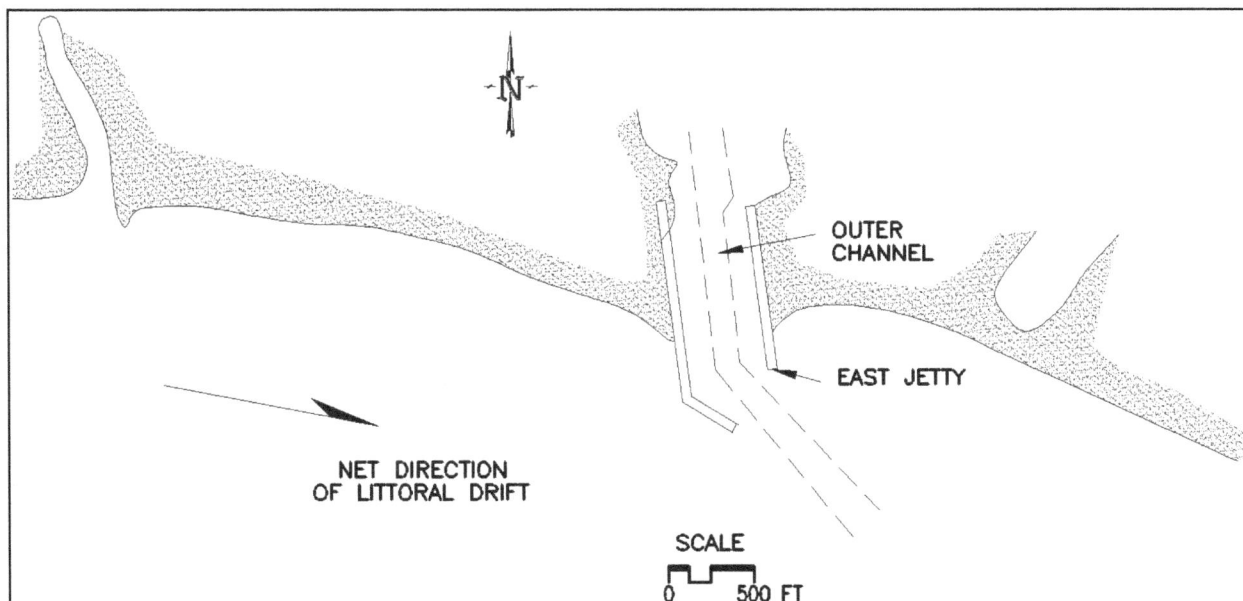

Figure V-6-78. Santa Cruz, California

In 1986, the Santa Cruz Port District obtained its own 40-cm (16-in.) hydraulic dredge. The $2.8 million purchase was 78 percent cost-shared by the Federal government. The dredge bypassed 176,000 cu m to the downdrift (east) beach in its first operational year (1986-87). Subsequently, the harbor has remained open almost continuously. The dredge operates primarily with a jet nozzle suction head and practices pothole dredging, wherein a series of discrete, deep craters is pumped (as opposed to continuous dredge cuts). Significant downtime has been reported due to debris blockage of the pump. Continuing maintenance and operational expenses have also proven to be a problem for the Port District (Walker and Lesnik 1990; EM 1110-2-1616).

(2) Santa Barbara Harbor, California. Santa Barbara Harbor is located on the Pacific Ocean northwest of Los Angeles (Figure V-6-79). The net and gross longshore transport rates are believed to be about 205,000 cu m/year and 282,000 cu m/year, respectively. The harbor's original 550-m-long offshore breakwater, constructed in 1927, was left detached from the coastline in the mistaken belief that this would allow littoral drift to pass through the harbor. The harbor immediately shoaled with sand and the downdrift (east) beach eroded. In 1930, the breakwater was extended 180 m in order to connect it to shore. Sand immediately impounded updrift of the new structure (advancing the beach over 300 m) while the downdrift beach retreated by over 120 m. Sand then bypassed the breakwater and impoundment fillet and shoaled the harbor. In 1935, 154,400 cu m of sand were hopper-dredged and placed in about 7-m water depth offshore of the downdrift beach, but surveys showed little shoreward movement of the sand. Beginning in 1938, hydraulic dredge and pipeline were employed. By 1952, an average of about 488,700 cu m of sand were bypassed every 2 years to the downdrift beach, representing about two-thirds of the area's estimated net littoral drift since the harbor's construction in 1927. In 1966, harbor dredging and sand bypass, by conventional hydraulic dredge and pipeline, became an annual operation. The quantity averages about 267,600 cu m/year or less (Bailard and Jenkins 1982; Walker and Lesnik 1990).

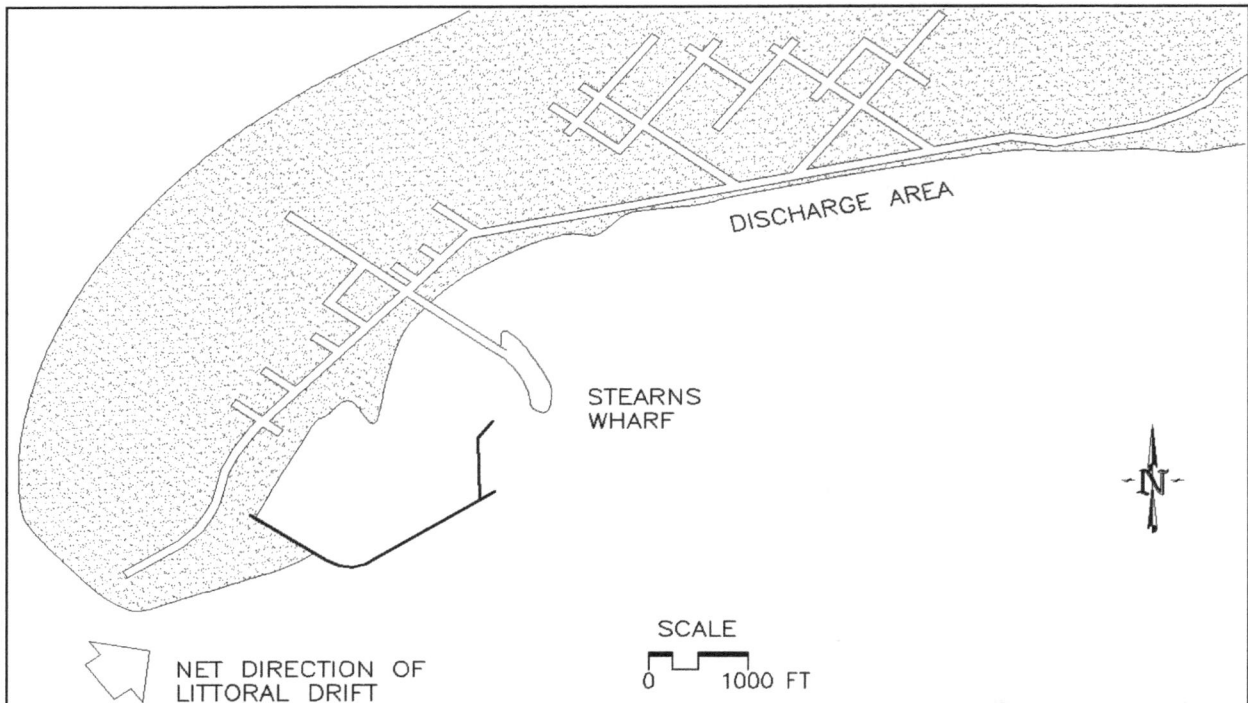

Figure V-6-79. Santa Barbara, California

(3) Port Hueneme, Channel Islands Harbor, and Ventura Marina, California. Port Hueneme is located about 70 km southeast of Santa Barbara Harbor on the Pacific Ocean (Figure V-6-80). The net longshore transport rate is estimated as 612,000 to 920,000 cu m/year. Port Hueneme's ocean entrance was dredged and stabilized by arrowhead jetties in 1938 at the head of the Hueneme submarine canyon. Despite placement of over 2 million cu m of dredged sand to the downdrift (east) shoreline during project construction, these beaches were completely cut off from the net littoral drift and retreated over 210 m within 20 years. Sand was impounded updrift of the inlet's jetties and diverted offshore into the submarine canyon. Coastal sand supply was simultaneously decreased by the construction of upriver dams and additionally by the interruption of littoral drift at Santa Barbara Harbor to the northwest. Between 1953 and 1960, hydraulic dredging of the updrift impoundment fillet and the navigation channel placed about 400,000 cu m to the downdrift beaches, primarily via submerged pipeline; but there remained a 16 million-cu m deficit. In bypassing the updrift impoundment fillet in 1953, the dredge cut into the shoreline and left a narrow barrier of sand seaward in order to provide a temporary wave shelter to the dredge plant. The narrow barrier was then dredged and bypassed at the end of the job.

(a) In 1960, Channel Islands Harbor was constructed about 1 mile northwest (updrift) of Port Hueneme entrance (Figure V-6-80). One intent of the project was to create a sand trap that would reduce losses to Hueneme Canyon and expedite mechanical bypassing to the downdrift beaches. The project consists of two jetties and an offshore, detached 700-m long breakwater, originally constructed in 9-m water depth, which shelter the harbor entrance. The littoral drift that deposits in the lee of the breakwater (i.e., in the trap) is hydraulically dredged by a conventional plant and pipeline and bypassed to the downdrift beach (principally south of Port Hueneme entrance). The equivalent, annual bypass rate has been about 990,000 cu m/year. To save on mobilization costs, bypassing typically takes place every other year (Herron and Harris 1967; Walker and Lesnik 1990).

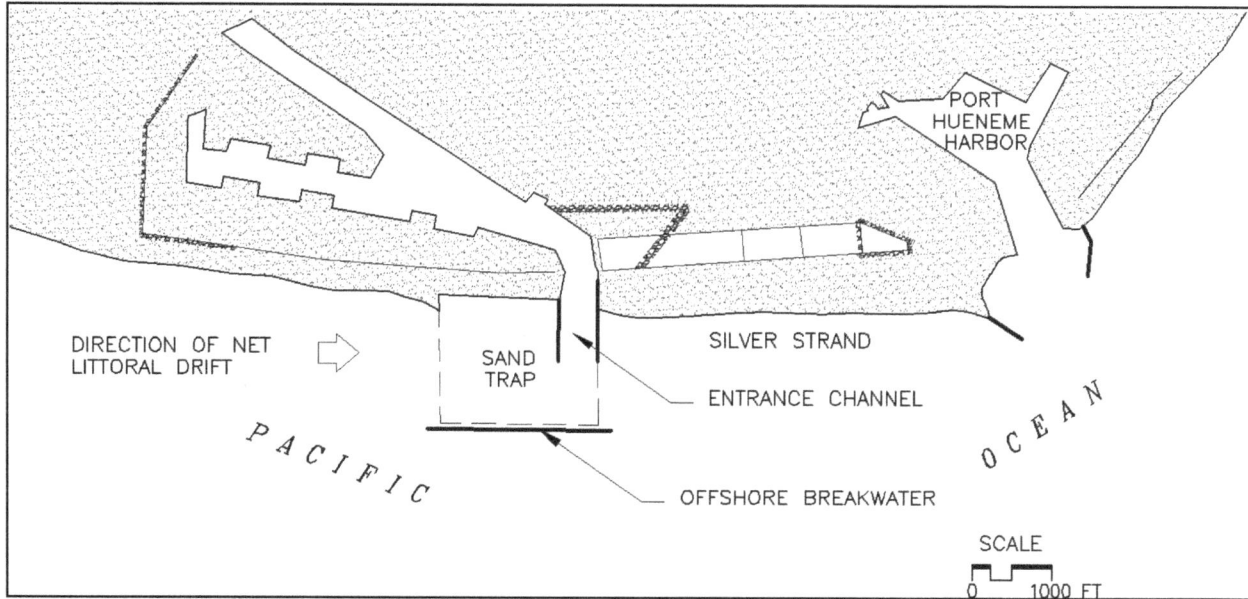

Figure V-6-80. Channel Islands and Port Hueneme Harbors, California

(b) Ventura Marina, located a few kilometers north of Channel Islands Harbor, was constructed by local authorities in 1962-63. Its design included features from the Channel Islands project (dual jetties, offshore breakwater, and leeward sand trap); but failed to recognize the area's high rate of littoral drift and did not plan for sand bypassing. In part because of the deficient length of the jetties and breakwater, sediment accumulates so rapidly that the entrance channel must be frequently dredged, often well before the sand trap is full. There are no measures to control the summer reversal in drift direction, and the resultant transport of sand around the south jetty and into the channel can be significant. While sand has been succesfully dredged from the project and placed upon the adjacent beaches, inlet navigation chronically suffers from rapid channel shoaling and wave inundation.

(4) Oceanside Harbor, California. Oceanside is located on the Pacific Ocean, about 50 km northwest of San Diego (Figure V-6-81). The estimated net littoral drift is small (75,000 to 190,000 cu m/year to the south) relative to the gross drift (on the order of 920,000 cu m/year). The Del Mar boat basin was constructed at the beginning of the 1940s and jetties were constructed in 1942. Shoaling of the entrance channel, updrift impoundment and downdrift erosion rapidly ensued. The north jetty was extended in 1958 and again in 1994. The Oceanside small craft harbor and its south jetty were constructed in the 1960s. Since 1957, the sand dredged from the harbor has been placed on downdrift beaches (about 230,000 cu m/year, on average); nonetheless, downdrift beach erosion continued to be a problem due to the inlet's historical impact to the littoral system and continuing losses associated with the shoreline's large gross transport rate. Mining of the impounded sand updrift (north) of, and between the two harbors, for purposes of bypassing and downdrift placement, was rejected by local interests. In 1982-83, an experimental sand bypass system was proposed consisting of the phased introduction and testing of fixed and mobile jet pumps. The latter, attached to a barge, would operate from locations along the north and south jetties, depending upon wave and sand transport conditions. A dedicated 35-cm pipeline and a shore-based booster station (used for bypassing from the north jetty location) would transport and discharge the sand to the downdrift shoreline.

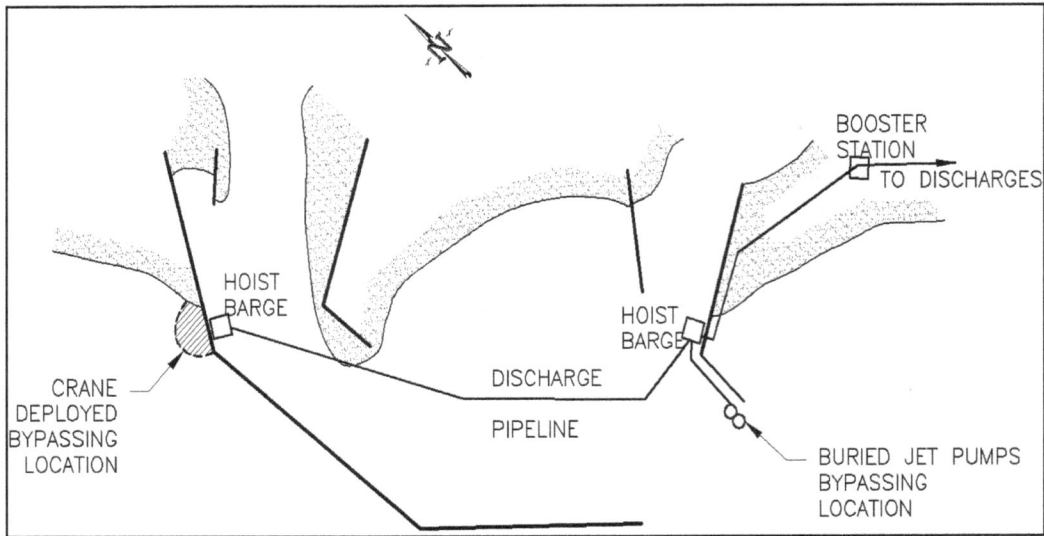

Figure V-6-81. Oceanside Harbor, California

(a) In 1989-90, the system's Phase I evaluation included 3-weeks' testing of a single crane-deployed jet pump operating at the north jetty fillet from upon the barge; and, 11 months' operation of two 4 in. x 4 in. x 6 in. Pekor jet pumps submerged in the entrance channel at the south jetty. Sand was pumped for 305 hr over this period, during which the production rate averaged 48 cu m/hr. The design value was about 150 cu m/hr. Principal difficulties involved continual clogging of the jet pumps (kelp root balls, rope, etc.) and the limited size and infilling rate of each pump's crater (about 4.6 m deep and 25-m diam).

(b) In 1991-92, Phase II evaluation included the addition of a single, pile-supported seabed fluidizer angled downward at about a 1:100 slope toward each of the two fixed jet pump's craters. The two fluidizers were 46 to 61 m in length and 20 to 25-cm diam. During a 13-month period, 81,000 cu m were bypassed with an average production rate of 73 cu m/hr. While the fluidizers appeared to improve productivity, the system exhibited difficulties from continual filling of the fluidizers with sand when the system was not operating. This required backflushing of the fluidizers prior to initiating sand transfer by the jet pumps. Productivity is predicted to improve if a separate fluidizer pump were provided, and if sand entry to the fluidizers could be minimized (Clausner, Patterson, and Rambo 1990; Walker and Lesnik 1990; Patterson, Bisher, and Brodeen 1991; Bottin 1992; Weisman, Lennon, and Clausner 1996).

(5) Indian River Inlet, Delaware. Indian River Inlet is located on the Atlantic Ocean approximately 16 km north of Ocean City, Maryland. The net littoral drift is estimated to be about 84,000 cu m/year to the north. The inlet was improved by dredging and the construction of jetties between 1938 and 1940. This resulted in immediate accretion of the updrift (south) beach and the inlet's ebb and flood shoals, and erosion of the downdrift (north) beach. In 1990, a dedicated, semimobile sand bypassing plant was installed at the inlet. The system consists of a jet pump deployed (suspended) by a 135-ton crawler crane with 37-m boom. The Genflo jet pump has a 6.4-cm nozzle with 15-cm mixing chamber, rated at 152 cu m/hr capacity. Discharge from the jet pump is through a 28-cm HDPE pipe to a booster pump, then through a dedicated pipeline across the inlet with discharge onto the beach at adjustable locations up to 460 m north of the inlet. Both the supply and booster pumps are stationed in a pump house adjacent to the south jetty. The system draws clear water from the inlet and powers the jet pump with a 340-hp supply pump supplying 415 ft of head and 2,500 gpm through a 25-cm supply line. The crane, jet pump, and crater geometry are similar to that shown in Figure V-6-60(a), and the jet pump is as shown in Figure V-6-62.

(a) The system is operated on a 4 day/week schedule with a three-person crew. To start bypassing from a fresh crater, the jet pump is jetted down to about -2.7 m (mlw) within the swash zone and allowed to create a small crater to provide a ready source of clear water to wash out the discharge line in the event of a potential line plug. The jet pump is then lowered to about -5.4 m (mlw), just above an existing clay layer, and kept at that depth for the remainder of the day. During calm surf or falling tide, the pump is raised and laterally moved about 4 m every 15 to 30 min to maximize productivity. Otherwise, little movement is required as wave action continually feeds sand to the crater. The craters produced by the operation are typically 5.5 m deep with 1:1.5 side slopes such that the beach surface diameter is approximately 15 m, and the nominal volume is about 300 cu m. The crane is able to excavate a trench about 3 diam long without moving.

(b) During the nonsummer (energetic) months, the jet pump operates from between 30 and 120 m south of the south jetty. This provides an area available for bypassing of approximately 3,700 cu m. Operation is limited by recreational activity, particularly during the summer months, and by potential nesting of endangered shore birds (piping plover) along the discharge area from March through August. The latter restricts sand discharge from within about 100 m of any observed nest. Productivity is limited by the amount of sand that is naturally transported to the 90-m-long stretch of beach utilized by the crane. Sheltering of this area by the inlet's ebb delta reduces the local rate of transport and the crater infilling rate during periods of low wave energy.

(c) In its first 11 months of operation, the system bypassed 85,500 cu m of sand, thus achieving its design objective. After gaining experience with the system, hourly productivity averaged 280 cu yd/hr. A remote production meter mounted in the crane's cab improved productivity by allowing the operator to monitor the effect of adjusting the pump position. Impacts to the south beach (narrowing) were limited to the area of bypass operation. Final cost of the system, including 610 m of discharge pipeline, was $1.7 million, with estimated annual O&M costs of $210,000/year (1990 dollars) (Rambo, Clausner, and Henry 1991; Clausner et al. 1991; Gebert, Watson, and Rambo 1992).

(6) *Rudee Inlet, Virginia.* Rudee Inlet, immediately south of Virginia Beach, was essentially non-navigable until 1952 when two short jetties were built and a channel was dredged. The channel immediately began to shoal with sand, and erosion occurred on the downdrift beaches. A fixed bypassing plant with a small capacity was installed in 1955 with little effect, and a floating pipeline dredge was added in 1956. The fixed plant was destroyed by a storm in 1962 and the inlet essentially closed, whereupon sand bypass resumed naturally. In 1968, the inlet was again improved with the construction of a jetty and a breakwater connected to the shore by a sand weir, similar to the geometry shown in Figure V-6-68(b). The weir jetty impoundment basin was never fully dredged initially, and the 25-cm dredge operations were hampered by wave action. From 1968 to 1972, sand bypassing was achieved by dredging sand from the channel and back bay and pumping it to the downdrift beaches. In 1972, 76,000 cu m of sand were removed from the impoundment basin. By 1975, the basin refilled with sand, and bypassing from the basin was repeated by the 25-cm dredge.

(a) Also in 1975, a semimobile jet pump system was added to the impoundment basin. The system consisted of two jet pumps (eductors) attached by flexible rubble hoses to fixed steel pipes. The steel pipes were connected to a pump house equipped with two centrifugal pumps having a combined nominal capacity of 115 cu m/hr. Discharge to the downdrift (north) beach was through a dedicated 20-cm steel pipe. During the system's first 6 months of operation, 60,400 cu m of sand were bypassed by the jet pumps and approximately 23,000 cu m were bypassed from the channel and impoundment basin by the floating dredge.

(b) Since late 1975, the system has been owned and operated by the city of Virginia Beach. Original estimates of pumping capacity were about 38 cu m/hr with effective pumping time of about 113 hr/month.

By 1980, only a single jet pump (eductor) was operated. At that time, it was moved within the impoundment basin by winching and redeploying via a steel cable operated from a truck (Richardson 1977; Dean et al. 1987).

(7) Masonboro Inlet, North Carolina. Masonboro Inlet is located near Wilmington, North Carolina, on the Atlantic Ocean (Figure V-6-82). Net littoral drift is southerly and in excess of 220,000 cu m/year. This natural inlet was first dredged in 1959 and the north (updrift) jetty was constructed in 1965-66. The jetty included a 305-m weir section built of concrete sheet piles with crest elevation at mean tide level. A deposition basin with a 283,000-cu m capacity was dredged along the interior of the weir. Because no south jetty was constructed, the area's strong reversal drift shoaled the inlet from the south and forced the navigation channel northward, undermining the north jetty and cutting through the deposition basin.

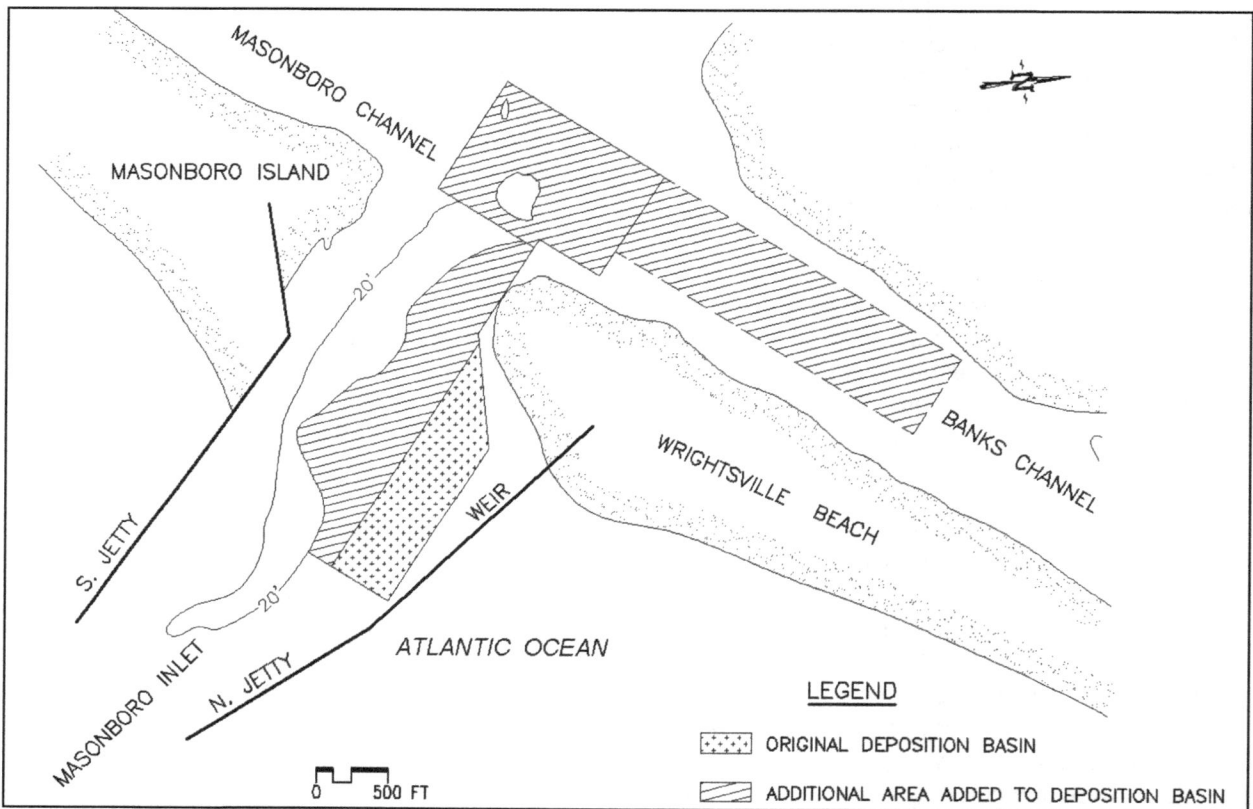

Figure V-6-82. Masonboro Inlet, North Carolina

(a) The south jetty was constructed in 1979-80, at which time 918,000 cu m were dredged from the inlet and placed on the north beach by pipeline. In 1981, dredging centered the channel between the two jetties. Since that time the inlet bathymetry has stabilized, and the need for a proposed training wall (separating the impoundment basin from the channel) has been eliminated. Additionally, the area of deposition expanded considerably to include the original basin, part of the inlet throat, and the inlet interior area. Transport over the weir is principally along the shoreline interface, and deposition is greatest at that area and landward thereof.

(b) Lack of available land along the updrift beach required that the weir-jetty be oriented at about 85 deg from the shoreline (i.e., almost perpendicularly), instead of at the recommended angle of about 60 deg. The latter would have provided greater deposition basin area along the shoreline interface of the weir, thus improving its capacity and reducing the proclivity of shoaling across the inlet throat and flood shoals. The expanded deposition basin area is typically dredged about every 4 years by contracted floating plant. The dredged material, on the order of 920,000 cu m per event, is bypassed to the south beach and/or backpassed to the north beach depending upon these beaches' conditions. The expansion of the deposition area has proven to be beneficial from the standpoints of both cost (reduced dredging frequency) and the ability to better capture sand within the inlet for purposes of bypassing and backpassing (Rayner and Magnuson 1966; Magnuson 1967; Vallianos 1973).

(8) Ponce de Leon Inlet, Florida. Ponce de Leon Inlet is located near Daytona Beach along east-central Florida's Atlantic coastline (Figure V-6-83). The net littoral drift was originally thought to be southerly-directed. Inlet improvements undertaken between 1968 and 1972 consisted of two jetties (the north jetty being a weir jetty), and the dredging of a navigation channel and an impoundment basin adjacent to the interior of the north weir. The weir was constructed of concrete panels that were to be adjusted in height between king-piles. This was never practicable as loose panels chattered and chipped between the piles, and/or the piles could not be aligned properly to allow adjustment. The impoundment basin adjacent to the weir has never been dredged. Instead, sand rapidly impounded against the south jetty and then created a large, chronic shoal within the inlet interior along the south jetty. The inlet's bypassing scheme was therefore altered. The entrance channel was dredged by hopper and the material placed in an offshore spoil area in about 6-m depth north of the inlet. The south shoal, adjacent to the jetty, was dredged by cutterhead and discharged via pipeline to the north beach. The continued growth of the south shoal forced the channel to migrate northward through the impoundment basin, rendering the basin ineffective and simultaneously acting to undermine the north jetty. The latter effect is exacerbated by the alignment of the north jetty, as the inlet's natural tidal and riverine flow is directed at and against the north jetty. The north jetty's weir section was closed in the 1980s (Partheniades and Purpura 1972; Parker 1979; Jones and Mehta 1980).

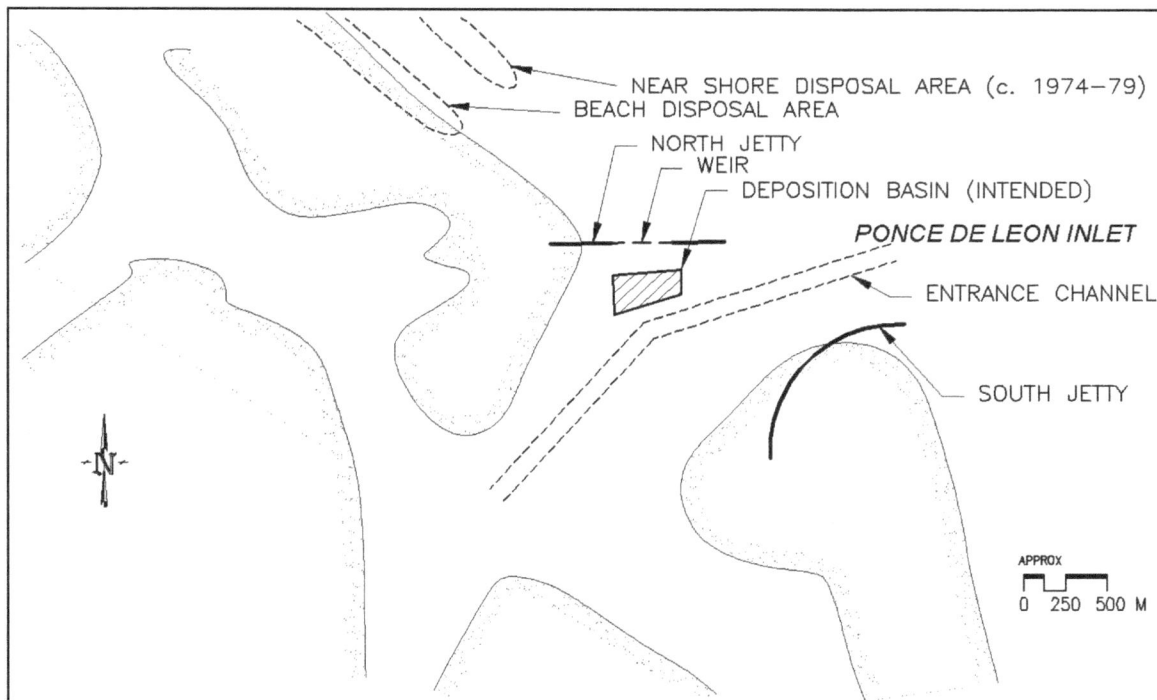

Figure V-6-83. Ponce de Leon Inlet, Florida

(9) Port Canaveral Entrance, Florida. The Canaveral Harbor Federal Navigation Project is located south of the Kennedy Space Center, at Cape Canaveral, along east-central Florida's Atlantic coastline. The net littoral drift is about 152,000 cu m/year to the south. The inlet was artificially created in 1950-52 and stabilized by two rock jetties. The inlet and harbor have always been hydraulically isolated from the interior waters. Tidal flow through the inlet is therefore minimal and there are no ebb or flood tidal shoals. It is estimated that the creation and maintenance dredging of the inlet have resulted in 6 million cu m of updrift impoundment (beyond historical conditions) plus deepwater disposal of another 6 million cu m of littoral material.

(a) Prior to inlet improvements and sand bypassing, sand shoaled the channel in four distinct areas at the seaward and landward ends of the inlet's short, low jetties (Figure V-6-84). The sand from these shoals (typically totaling about 150,000 cu m/year), mixed with up to another 500,000 cu m/year of silt and clay, were annually removed by hopper-dredging and disposed of in deep water, offshore. Beginning in 1992, barge-based clamshell dredging was required in lieu of hopper dredging in order to avoid impacts to marine turtles that loaf at the channel seabed. The mechanically-dredged material, usually placed in 900 to 1,300-cu m scows, was identified in terms of its sand content. Those scow loads containing suitably sandy material (generally less than 10 to 15 percent silt/clay) were placed in a nearshore disposal area about 11 km south (downdrift) of the inlet in 5.5 to 6.5-m water depth (msl). Between 1992 and 1997, about 500,000 cu m of sand were placed in the nearshore disposal area. Attempts to place the material in shallower depths (3.6 to 4.6 m depth) have, to date, been unsuccessful because of the danger and seabed impacts posed to the tugs that tow the scows.

(b) Sand transport over, through, and around the south (downdrift) jetty is estimated to have contributed up to about one-third of the inlet's shoaling and downdrift erosional effects. In 1993, the landward half of the jetty (about 145 m; or, to about -0.5 m, mlw) was temporarily sand-tightened by 1.8-m diam sand-filled geotextile tubes. This resulted in halting about 30,000 cu m/year (about one-third) of the sand transport through the south jetty. In 1995, the south jetty was permanently sand-tightened, raised, and lengthened by driving steel sheet pile immediately adjacent to, and seaward of, the original jetty, and then armoring the sheet pile with boulders. In 1998, the north jetty was sand-tightened by a sand-filled geotextile tube and monitored.

(c) Also in 1995, regular sand bypassing commenced whereby every 6 years, 690,000 cu m (or, 115,000 cu m/year) is dredged by a conventional hydraulic plant and discharged by temporary, submerged pipeline along 3.2 km of shoreline immediately south of the inlet. The dredge area is within the updrift impoundment fillet, along 2.6 km of shoreline immediately north of the inlet, between the mean high waterline and the -5 m-depth contour. The inaugural 1995 operation dredged material from between -1.2 and -5 m (mean sea level) and bypassed very fine sand relative to the native beach (overfill ratio > 2). Monitoring surveys indicate that the 1995 borrow area recovered at a rate equal to or greater than the proposed bypass rate, and that the grain size of the recovering sand is equal to, and sorts itself by depth similarly to, the predredged area. In the first 3 years subsequent to the 1995 bypassing operation and jetty tightening, shoaling of the inlet channel by littoral material appears to have been mostly stopped, and the downdrift beach began to recover toward its preinlet condition (Bodge 1994a, 1994b; Bodge and Hodgens 1997).

(10) South Lake Worth Inlet, Florida. South Lake Worth Inlet (a.k.a. Boynton Inlet) is located on southeast Florida's Atlantic coastline, about 16 km south of Palm Beach. It was opened artificially and stabilized by jetties in 1927 to provide increased flushing of Lake Worth. The net longshore transport rate is about 135,000 cu m/year to the south. By 1932, downdrift erosion prompted local property owners to construct seawalls and groins. By 1936, updrift impoundment saturated the north jetty and led to significant shoaling of the inlet and interior lake. In 1937, a fixed sand transfer plant was constructed atop the north jetty with a dedicated 365-m pipeline discharging sand to the south beach. Between 1937

and 1941, the 15-cm suction intake and 65-hp centrifugal pump bypassed about 55 cu yd/hr; or, 50,400 cu yd/year, on average. The plant was not operated from 1942 through 1945 due to wartime fuel shortages, and the inlet essentially closed. The plant was restarted in 1945 and upgraded in 1948 to a 25-cm suction intake mounted on a swinging boom with 9-m radius, a 20-cm 300-hp centrifugal pump, and 20-cm discharge pipe. The plant's bypassing capacity increased to 76 cu yd/hr. Overall bypassing was also increased by a floating hydraulic dredge that transferred sand from the interior (flood) shoals to the downdrift beach. In 1967, the north and south jetties were extended by 125 m and 20 m, respectively; and, the fixed sand transfer plant was shifted 36 m seaward and increased to a 25-cm pump with 400-hp motor. Portions of the north jetty were sealed in 1971.

(a) The present plant consists of a 30.5-cm suction intake and 25-cm discharge line driven by a 400-hp diesel engine rated to pump 4,000 gpm with up to 20 percent suspended solids. The plant is similar in appearance to that shown in Figure V-6-59. Productivity estimates vary from 95 to 122 cu m/hr. Palm Beach County operates the plant. Local (updrift) interests require that the plant be operated only during wave/transport events from the northeast. Productivity is limited by the crater's size (about 460 cu m) and its sand-infilling rate. The plant's annual bypassing rate averages about 50,000 cu m/year. Of the net incident littoral drift, it is estimated that about 45 percent is naturally bypassed across the ebb-delta plateau, 35 percent is bypassed by the fixed plant, 2 percent is bypassed by periodic hydraulic dredging of the interior (flood) sand trap. The remaining 18 percent is diverted from the littoral system by updrift impoundment and transport to the ebb shoal. It is predicted that the bypassing plant's capacity could be significantly improved by increasing the length of the boom upon which the suction intake is mounted, and by mounting the plant on a mobile platform atop the jetty or adding one or more jet pumps near the jetty's seaward end (Jones and Mehta 1977; Olsen Associates, Inc. 1990).

(b) A similarly-sized, electrically-powered plant operated from the north jetty of Lake Worth Inlet, to the north, from 1957 through the early 1980s. Between 1967 and 1978, the plant is estimated to have bypassed about 102,000 cu m/year (Jones and Mehta 1977).

(11) Boca Raton Inlet, Florida. Boca Raton Inlet is located about 25 km north of Fort Lauderdale on southeast Florida's Atlantic coastline (Figure V-6-84). The net longshore transport rate is estimated as about 124,000 cu m/year to the south. This natural inlet was improved by dredging in 1925-26. Jetties were constructed in 1930-31 and repaired in 1952. These works resulted in erosion of the downdrift (south) beach and promoted the growth of a large ebb shoal that threatened navigation through the inlet. In 1975, the city of Boca Raton undertook a 55-m extension of the north jetty and 165-m sand-tightening of the south jetty. In 1980, a 20-m weir was cut into the north (updrift) jetty to allow sand to enter the inlet throat to facilitate bypassing by the city-owned dredge.

(a) It is estimated that about 36,000 cu m/year enters the inlet through the weir and around the jetty from the north (updrift) shoreline, and that about 4,300 cu m/year enters the inlet from the south (downdrift) shoreline, for a total of 40,300 cu m/year. On average, the city's dedicated, floating hydraulic dredge bypasses this quantity, working mostly year-round, to the south shoreline. In addition, about 176,000 cu m of sand is removed by hydraulic, cutterhead pipeline dredge from the ebb shoal about once every 10 years, and is placed along 1.2 km of shoreline south of the inlet.

(b) Of the 124,000 cu m of net incident drift (approximate), it is estimated that about 30 percent is bypassed by the dedicated floating dredge, 50 percent is naturally bypassed across the ebb-tidal shoal, and 20 percent is diverted to the ebb shoal or offshore. The periodic (10-year) dredging of the ebb shoal is intended to capture at least a portion of the latter quantity for bypassing to the south (Jones and Mehta 1980; Coastal Planning and Engineering 1996; Olsen Associates, Inc. 1997).

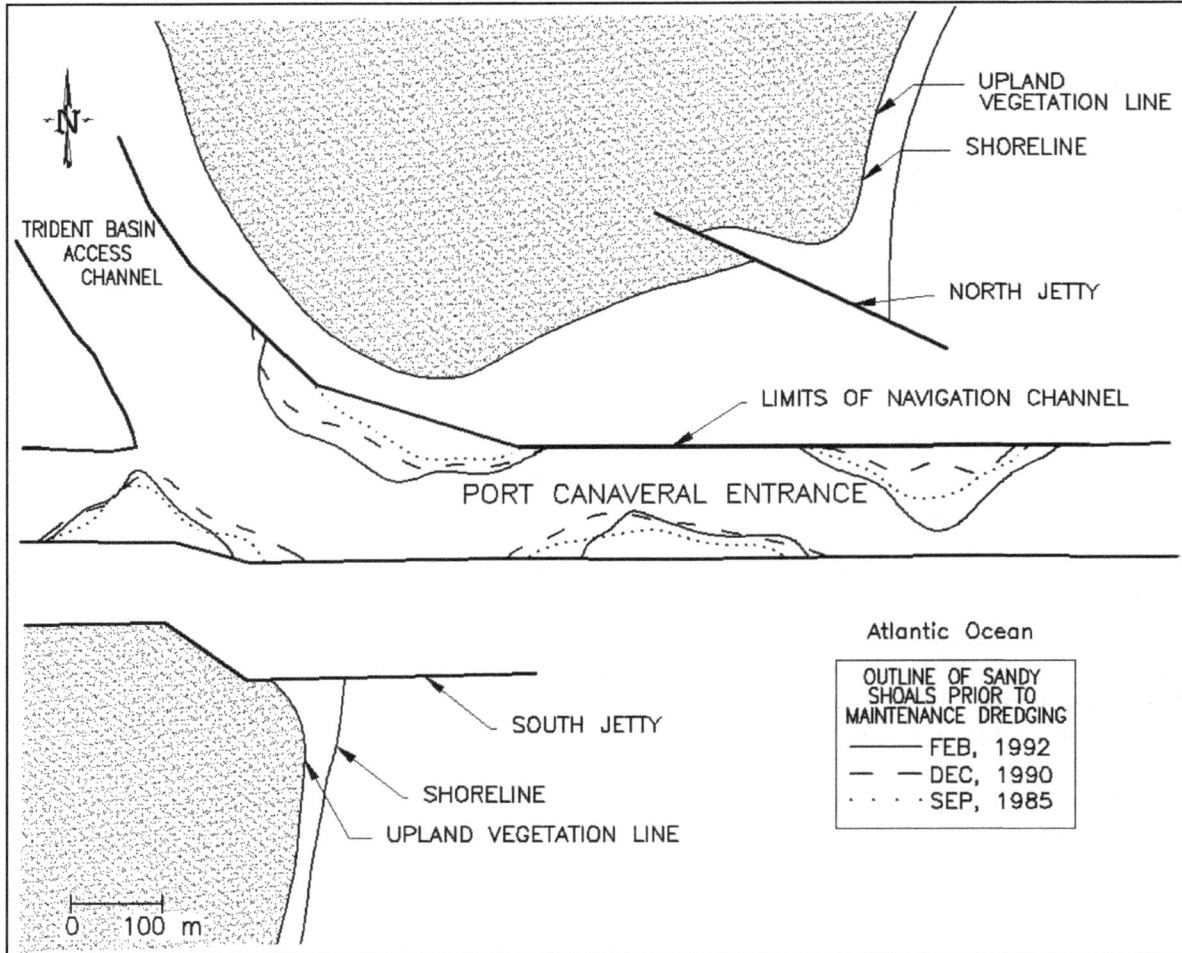

Figure V-6-84. Historical locations of sandy shoals within Port Canaveral Entrance, Florida (prior to implementation of inlet/jetty improvements). Adapted from Bodge 1994a

(12) Hillsboro Inlet, Florida. Hillsboro Inlet, about 9 km south of Boca Raton Inlet, or 16 km north of Ft. Lauderdale, is located on southeast Florida's Atlantic coastline. The net incident littoral drift is estimated as 92,000 cu m/year to the south. This inlet features a natural reef that forms its northern (updrift) boundary. This reef acts as a sand weir; and, in fact, the modern concept of an inlet weir originated at this inlet (Hodges 1955). Sand transported over this reef/weir and around the inlet's two jetties, amounting to about 74,700 cu m/year in total, is bypassed to the south (downdrift) shoreline via a dedicated 20-cm hydraulic dredge and pipeline owned and operated by the inlet's local authority. Of the incident net drift, it is estimated that about 14.5 percent is impounded updrift of the inlet, 64 percent is transported into the inlet (and bypassed by dredging), 18.5 percent is naturally bypassed across the ebb tidal shoal, and 3 percent is lost offshore. Of the net 74,700 cu m dredged and bypassed from the inlet each year, about 78 percent is estimated to originate from the updrift (north) shoreline, while the other 22 percent is sand that returns from the downdrift (south) shoreline into the inlet. Improvements to the south jetty and extending the discharge distance further downdrift of the inlet is predicted to improve sand management at the inlet (Jones and Mehta 1980; Coastal Planning and Engineering 1991; Olsen Associates, Inc. 1997).

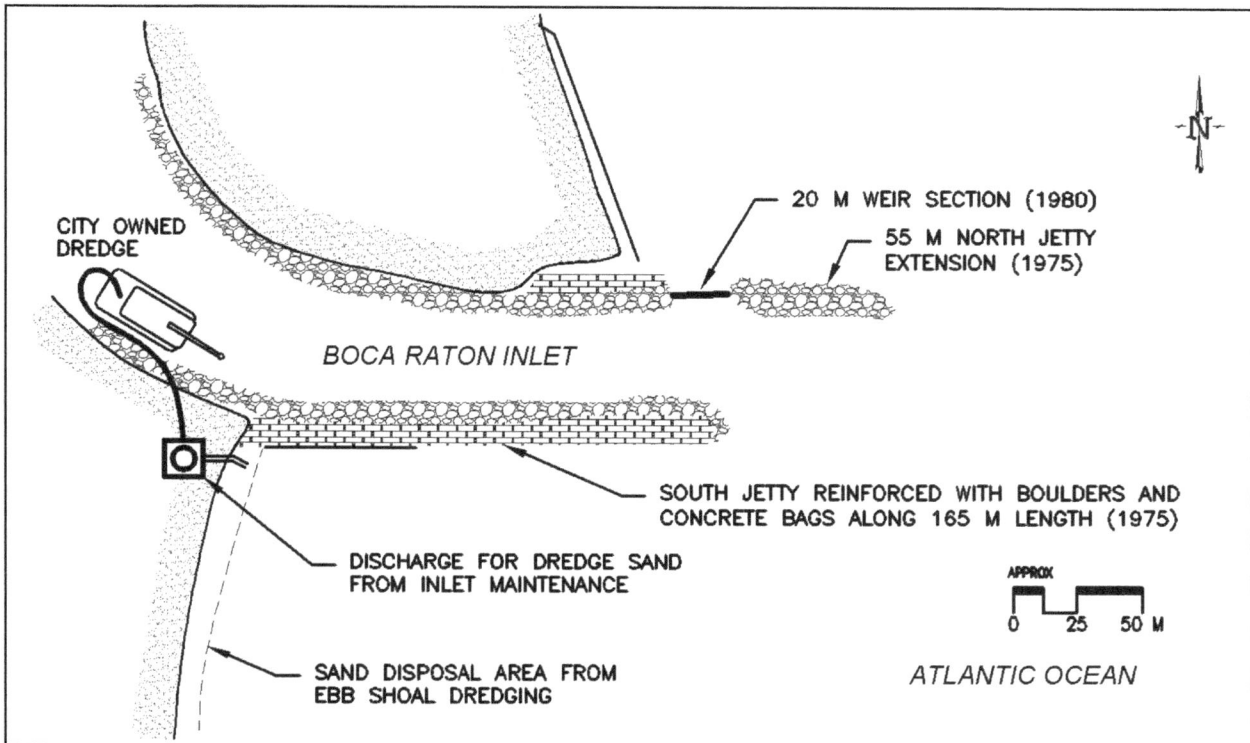

Figure V-6-85. Boca Raton Inlet, Florida

(13) Port Sanilac, Michigan. Port Sanilac is a small-craft harbor on Lake Huron, about 100 km north of Detroit. In the later 1970s, a portable, truck-mounted jet pump system was built by the U. S. Army Waterways Experiment Station (currently the U.S. Army Engineer Research and Development Center) and tested at the harbor for sand bypassing. A 12-m flatbed trailer was outfitted with a fuel tank, generator, air compressor, control room, and water supply and booster pumps with diesel engine drives. The jet pump water supply and slurry discharge lines were supported by foam floats designed to allow the jet pump to sink into the seabed while dredging. To raise the jet pump, air was pumped through the water supply line to a flotation unit attached to the jet pump. The jet pump was steered in the water by diverting water from the supply line through two nozzles attached to the sides of the pump. The system was designed to be driven onto an accretion fillet where the jet pump could be deployed and operated near the shoreline. Discharge from the pump was carried by pipeline along the harbor bottom to the downdrift beach. System technicians reported that the system worked extremely well for its designed purpose. However, resistance by updrift property owners often precluded bypassing operations from the accretion fillet. The system was used mainly as a rehandling device for material released by hopper dredges. For this purpose it was driven onto a barge and operated as a floating dredge. The system was surplused by the U.S. Army Engineer District, Detroit, in 1980, and acquired by the state of Michigan in July 1988 (EM 1110-2-1616).

(14) Viareggio Harbor and Marina di Carrara, Italy. The harbors of Viareggio and Marina di Carrara are located on the northwest coast of Italy on the Mediterranean Sea. The former is about 22 km south of the latter. The net littoral drift at Viareggio is about 200,000 cu m/year toward the north, while the net drift at Marina de Carrara is toward the south.

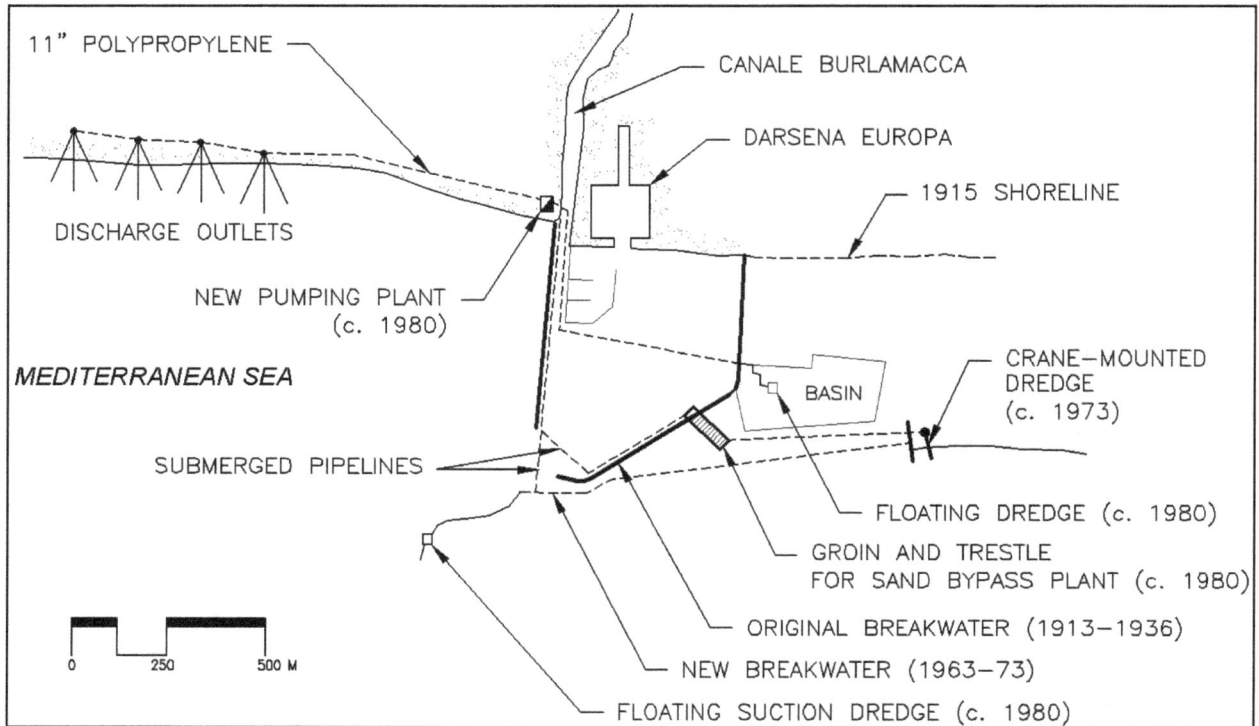

Figure V-6-86. Viareggio Harbor, Italy

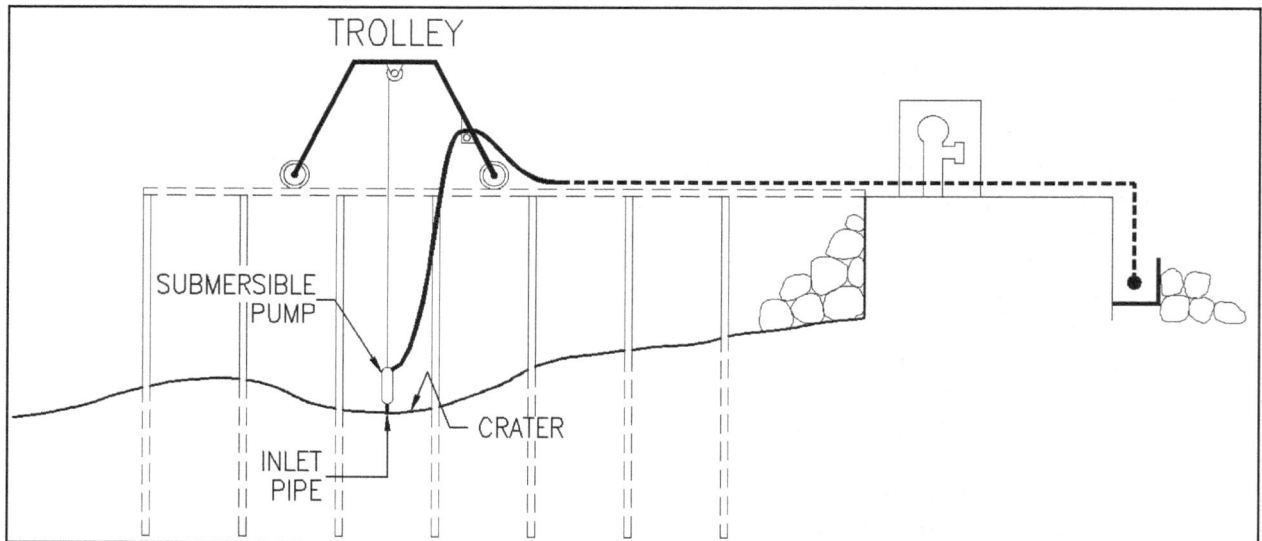

Figure V-6-87. Original sand bypassing plant at Viareggio Harbor, Italy

(a) Navigational improvements at Viareggio date from the Roman Empire. In 1913, an outer harbor was created by the construction of a shore-attached breakwater (Figure V-6-86). As a result, the updrift (south) beach accreted about 20 m/year, the harbor shoaled, and the downdrift beach eroded dramatically. In 1936, 220,000 cu m were dredged from the outer harbor and discharged to the downdrift (north) beach via 60 cm (24 in.) pipeline. This represented the first application of sand bypassing, and artificial beach nourishment, in Italy. In the early 1950s, a spur groin was built along the south breakwater in an

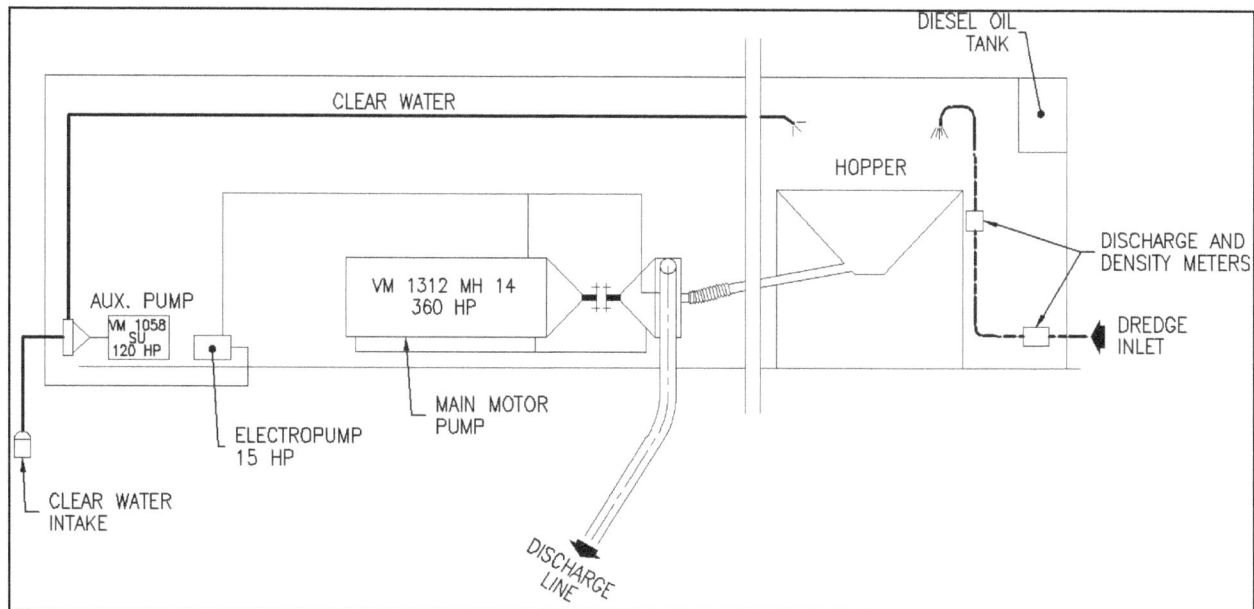

Figure V-6-88. Existing sand bypass pump system at Viareggio Harbor, Italy

attempt to arrest sand transport along the breakwater. A trestle-mounted sand bypassing plant was subsequently built atop this spur groin (Figure V-6-87). The plant consisted of an immersed pump body mounted upon a trolley that ran along tracks. The dredged material was discharged at the root of the north jetty via a 18-cm (7-in.) pipeline. The pump dredged a series of "holes" underneath the installation to serve as traps for the sand. These holes quickly filled with debris and seaweed, and interfered with the plant's intended operation. At the same time, the groin tended to push the sand offshore along a bypassing bar and reduced the rate at which sand reached the plant.

(b) In the early 1970s, a new breakwater was constructed seaward of the original breakwater, and a new basin was excavated into the beach that had been accumulated against the structure. The original sand bypassing plant, now within the uplands of the new breakwater, was relocated to the south. The trolley system was replaced by a crane that would allow sand to be removed from a larger area (i.e., greater potential crater size). The old works were transformed to a booster pump plant. The relocated, crane-mounted bypass plant was never put into operation, however, as the harbor's new authority suspended its operation and the new breakwater caused the updrift beach to rapidly overtake the new plant's location.

(c) In 1980, Viareggio Harbor's existing bypassing system was installed. It is based upon a flexible, mobile scheme to ensure bypassing productivity. Material can be dredged by floating suction dredges from the outer harbor, the new large basin, the updrift beach, or outside the harbor entrance. Suitable dredged material is pumped via submerged pipeline toward the principal pumping station located at the root of the north jetty. The new pumping plant was designed to receive variable sediment discharges and densities, transforming them into suitable slurries for beach discharge. Water-cooled diesel engines are used. An auxiliary plant is used to control the supply of material to the bypassing plant by pumping water into a hopper to compensate for periods when discharge from the dredge platforms is insufficient (i.e., when only one dredge is operating). (See Figure V-6-88). Recent improvements to the plant (combining the discharge from the two dredge platforms) allow up to 1,100 cu m/hr of slurry with 20 percent solids volume. From 1980 to 1985, 380,415 cu m of sand were bypassed over 3,623 operating hours, for an average sand discharge rate of 105 cu m/hr.

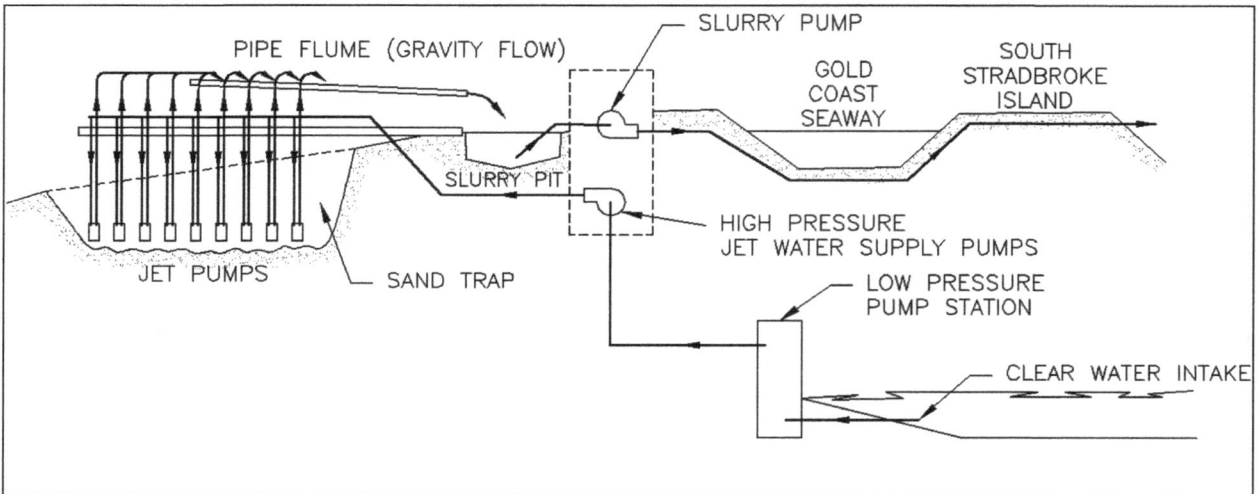

Figure V-6-89. Sand bypassing system at Nerang River entrance (schematic)

(d) Marina di Carrara is similar in geometry to Viareggio Harbor. A pilot bypassing plant was installed in 1965-66 using a floating dredge inside the harbor and five slurry booster pumps onshore along a 6-km-long 14-cm- (5.5-in.) discharge pipeline with multiple outlets. While the system's components reportedly worked well, net productivity was limited, and the material dredged from the harbor was mostly silt and fine sand that was poorly suited for beach placement.

(e) A second bypassing plant was installed in 1967-68. A fixed plant was installed updrift of the harbor entrance, as at Viareggio, but this time a mobile crane could operate on a large-diameter platform in order to reach a wide area far from the platform's foundation piles. After initial success in 1969, productivity declined as the dredged craters filled with debris, and the sand in-filling rate was dependent upon storm conditions. Operation was suspended when the platform collapsed after a ship collision (Fiorentino, Franco, and Noli 1985).

(15) Nerang River Entrance, Queensland, Australia. The Nerang River entrance is located on Queensland's Gold Coast, along Australia's central eastern coastline on the Pacific Ocean. The northerly longshore transport rate is large and dominant (580,000 cu m/year) compared to the southerly transport (84,000 cu m/year). In 1984-85, the inlet entrance, which naturally migrated northward at an average rate of over 36 m/year, was relocated southward by dredging a new navigation channel stabilized by dual rock jetties. In 1985-86, construction of a fixed sand bypassing plant was completed. The plant consists of a 490-m-long pier/trestle located 250 m updrift, and parallel to, the south jetty. Ten jet pumps are spaced approximately every 30 m along the outer 270 m of the pier. The 9-cm (3.5 in.) Genflo "Sand Bug" jet pumps, each rated at 103 cu m/hr (135 cu yd/hr), are attached to wide flange steel beams that slide down a second set of steel support beams attached to the pier's concrete piles. Stops on the beams prevent the pumps from penetrating below their design depth of -10 m (msl).

(a) Supply water to the jet pumps is drawn from the inlet interior via a 122-cm (4-ft) concrete pipe using two low-pressure pumps, each rated at 250 hp and 10,300 gpm (see Figure V-6-89). Water from these pumps flows through a 61-cm (2-ft), 700-m-long pipeline to the main pump house, where it feeds dual 450-hp supply pumps (10,200 gpm each). High pressure water from the supply pumps flows from the pump house through a 36-cm (14-in.) pipeline to the jet pumps. Solenoid-actuated valves control the flow of water to the 15-cm (6-in.) lines that feed each jet pump. Supply water can also be directed to fluidizers on each jet pump. These are used during installation and removal, and to improve transfer capacity when debris has collected around a jet pump. The jet pumps' slurry is discharged through 24-cm

(9-in.) pipes into an elevated pipe flume. The 58-cm- (23-in.-) diam flume, 370 m long and sloped at 2.5 deg, allows gravity flow of the slurry to a conical, 145 cu m-buffering hopper. The flume and hopper were intended to allow the incoming slurry to vary widely in solids content and volume. Make-up water for the flume and hopper, when necessary, can be supplied by clear water from a jet pump. A 950-hp (6,500 gpm) centrifugal booster pump transfers the slurry from the hopper, via pipeline under the inlet entrance, to three discharge points along the uninhabited, downdrift (north) shoreline.

(b) The system is computer controlled, allowing unattended bypassing operations at night to take advantage of lower electric rates. During the day, two full-time employees perform maintenance operations and program the system for the evening's operation as a function of the level of sand within each pump's crater. Should the percent solids fall below a preset value, the computer enables a new series of jet pumps to operate. If a major problem is sensed, the computer telephones an operator at home.

(c) In normal operation, the system bypasses sand at 333 cu m/hr (435 cu yd/hr) at 30.6 percent solids, for a discharge pipeline flow of 6,250 gpm. Under peak conditions, the system bypasses sand at 570 cu m/hr at 40 percent solids. The system was designed for an average yearly transport of about 500,000 cu m, a peak yearly transport of about 750,000 cu m, and maximum 5-day transport of about 100,000 cu m. During the first 22 months after commencing operation, the system bypassed approximately 1 million cu m of sand. The nearshore jet pumps bypass the great majority of the sand volume (see Figure V-6-71). The craters' infilling rate, and therefore overall productivity, is a function of incident wave energy.

(d) Principal problems with the system have included wear on the jet pump nozzles and debris within the craters. Nozzle replacement and periodic debris clearing requires use of a 20-ton crane to remove the jet pumps from their trestle support. In 1987-88, a large-diameter, 23-cm (9-in.) jet pump was introduced to dredge and bypass the debris that chronically collects at the bottom of the craters. This pump is intended for periodic operation at each of the jet pump's craters. Its large size requires the entire output from the supply pump. Experience has suggested that an alternate to the slurry's flume design, and variable (flexible) spacing of the jet pumps along the pier, would improve productivity (Polglase 1987; Pound and Witt 1987; Clausner 1988; Coughlan and Robinson 1990).

(16) Other projects. Numerous other sand bypassing systems have been attempted or incorporated at other tidal inlets that were not described. In the United States these include, among others, Fire Island Inlet, New York; Carolina Beach Inlet, South Carolina; Murrells Inlet, South Carolina; Little River Inlet, South Carolina; Carolina Beach Inlet, North Carolina; Cape Fear River Entrance, North Carolina; St. Marys River entrance, Florida; St. Johns River entrance, Florida; Sebastian Inlet, Florida; Jupiter Inlet, Florida; St. Lucie Inlet, Florida; Mexico Beach, Florida; East Pass, Florida; Perdido Pass, Alabama; and the Colorado River mouth, Texas.

b. *Lessons learned.*

(1) Inlet shoaling and its impact to navigation is intimately linked to beach response adjacent to inlets. Project planning that fails to recognize and take advantage of this link, that is, planning which isolates the objectives of navigation from those of shore protection, is less likely to develop an inlet sediment management program that is physically, economically, or environmentally optimal. Sand management at inlets is apt to be most successful when inlet improvements are viewed from a holistic approach that examines the inlet's sediment transport pathways, channel alignments and requirements, and adjacent beach processes as an integrated system. This implies that piecewise justification of proposed inlet / navigation project improvements (i.e., benefits based singularly upon navigation benefits or decreased dredging costs or oceanfront storm protection, etc.) should be discouraged in favor of system-wide benefits analysis.

(2) Failure to appreciate the direction, magnitude, and variability of a site's littoral drift, both gross and net, is central to many, if not most, problems of inlet shoaling and inlet-related beach erosion. Realistic evaluation of an inlet improvement project's potential performance requires that the site's transport portfolio be accurately understood. This includes an appreciation of how the littoral drift rates vary in the immediate vicinity of the inlet, particularly due to perturbations associated with the inlet's shoals and other nuances of the nearshore bathymetry. Development of an inlet sediment budget is an important step in determining requisite sand bypassing/backpassing requirements. Measurements of downdrift beach changes, by themselves, may not accurately reflect the magnitude of requisite bypassing. This is because shoreline and volume changes along downdrift beaches, especially those that have suffered chronic erosion, are usually obfuscated by shoreline armoring and beach nourishment projects; and, downdrift impacts may extend much further than expected in alongshore magnitudes that are difficult to perceive from periodic surveys. Beyond knowledge of the global, incident transport rates at an inlet site, understanding the specific transport pathways through and around an inlet, particularly at an improved inlet, is a prerequisite to identifying the most economically and physically effective system for sediment management. In many instances, project designers have failed to recognize the most obvious routes by which sand enters and is dispersed through the inlet and/or by which the adjacent shoreline erodes. Repeated visitation to the site during various wave regimes, discussions with local individuals knowledgeable of the inlet, and scrutiny of surveys that depict the patterns of sedimentation within the inlet are all elementary, but invaluable, means by which to preliminarily accomplish, or verify, this understanding.

(3) Historically, insufficient attention has been given to the height, length and porosity of downdrift jetties. Problems of both shoaling and beach erosion on the downdrift beaches can almost always be improved by sand-tightening the downdrift jetty. Sand-tightening of the updrift jetty, by improving crest elevation, length, and/or impermeability, may also be warranted unless transport through or past the jetty is desired for purposes of interior trapping. In improving an updrift jetty, however, it is important to ensure that updrift impoundment will not result along properties from which the sediment cannot be dredged (recovered) and will not result in an offshore diversion of the sediment.

(4) For sediment management systems that rely upon deposition traps, it is essential that the traps' capacities are large enough to stably contain the material between economically-optimal dredging events; the traps be dredged (downloaded) when they become full or near-full; the geometry and orientation of the traps be configured to intercept the inlet's specific transport pathways; and the traps be located outside of tidal- and wave-induced currents that will move material out of the traps. Poor performance of inlet weir systems, which must include one or more deposition basins (traps) to function properly, can almost always be attributed to a failure to achieve one or more of these four requirements. When a weir system is used, it must include a deposition basin of particularly large capacity, particularly at and landward of the shoreline/weir interface. Provisions must be likewise made to ensure reliable, safe dredging of the basin on a routine basis. In some cases (e.g., Ponce de Leon Inlet, Florida; East Pass, Florida), weirs have been constructed on the incorrect side of the inlet due to a lack of appreciation of the local sediment transport patterns that result from conditions particular to the inlet site.

(5) The productivity of bypassing systems that utilize fixed or semifixed pumps is most typically limited by the rate of crater infilling and debris problems. In most cases, the mechanical or hydraulic capacity of the pumps has not typically limited production; instead, productivity is often ultimately limited by the rate and reliability with which littoral material refills the pumps' dredged craters. Likewise, productivity decreases, and operating expense increases, with the amount of time and effort spent clearing debris from the pumps' craters. Success of these systems requires that that there are sufficiently large dredging areas which the pumps can access, and that these areas are within the zones of highest littoral drift, and that there is a means by which to limit (or remove) the influx of debris. Fluidizers mounted at the pumps' nozzles (intakes) agitate and suspend the seabed sediment. These are of

proven success in improving productivity. Fluidizers constructed across the seabed which are sloped toward the pumps' intake, intended to augment delivery of sediment to the pumps, may be of benefit in increasing the craters' effective size. However, such fluidizers have demonstrated some limitations due to sediment intrusion (blockage) and their use is still considered experimental. Where fluidizers, debris-clearing pumps, and other ancillary devices are used in conjunction with the bypassing pumps, experience demonstrates that there can be a significant net benefit in providing pumps and engines dedicated to these devices; i.e., separate from those used to supply and power the bypassing pumps. Reliable and economic means to lift, remove, and redeploy fixed pumps, for servicing and debris removal, should likewise be considered. Corrosion and wear of the pumps' elements should be given additional consideration beyond normal marine concerns, particularly those elements at and near the sediment intakes.

(6) At most inlets with long-term prototype sand bypassing experience, and at most new sites with dedicated plants, the systems have evolved toward, or have recommended for, increased mobility and flexibility. Because of the dynamic nature of the sediment transport pathways (and the bypassing systems' effects thereupon), and in view of earlier systems' chronic problems with unreliable infilling of dredge craters, modern systems are moving toward multiple pumping locations and/or pumps and dredges mounted on mobile devices. The former includes deployment of jet pumps or similar devices across a variety of areas within the inlet system, with the capability to bypass from one or more of these devices at different times of the year. The latter includes jet pumps or similar devices mounted on mobile cranes, trestles, barges or trucks; or, the use of a dedicated or contracted dredge to intercept sediment across a large (and potentially variable) area.

(7) Other problems encountered with fixed, semifixed, or other dedicated bypassing pumps have included operational limitations imposed by the presence of environmental resources, social (recreational) considerations, and/or upland property owners that protest the transfer of sand. An additional, frequent problem has been the physical reliability and dedication of the operation; i.e., lack of long-term funding, manpower, and/or commitment to operate the system. System interruptions are especially problematic for those plants that have difficulty starting up after extended layoffs or heavy sedimentation.

(8) Where littoral and nonlittoral material mix within an inlet, benefit can be derived by identifying structural means or dredging techniques to separate the material. This includes, for example, training structures that isolate littoral deposition from the influence of ambient silts and clays; or, selective use of smaller or mechanical dredges that can recover the littoral material (in lieu of hopper dredges that mix the material).

(9) The benefit of nearshore disposal of suitable, dredged material is maximized with increasingly shallow depths of placement. Poorly-controlled placement across large seabed areas, and/or in water depths predicted to result in stable berms, do not yield readily discernible benefits to the adjacent shorelines.

(10) The nature and grain sizes of the bypassed material and the adjacent beaches should be examined for compatibility. Sediment intercepted at deeper areas within the inlet system, or from potentially contaminated areas, may not be suitable for beach placement or may not fully represent a one-to-one transfer (bypass) of sediment across the inlet.

(11) As each tidal inlet is more or less unique unto itself, the most appropriate sediment management strategy at a given inlet is unique. A system that is successful at one inlet will not necessarily be viable or successful at another. Often, a combination of structural improvements, dredging equipment and changes in dredging practices will be required, rather than a single modification. Existing or historical practices may require significant revision or abandonment. Such improvements can often be the most difficult to

implement because of resistance to change and/or the perception that such changes are untested, more costly, or inconsistent with policy.

(12) Most generally, successful sediment management at inlets requires that the incident sediment be intercepted for purposes of subsequent handling. Uncontrolled deposition of sediment, or deposition within areas from which recovery is not feasible, is contrary to effective sediment management. When littoral material is transported across a large or dynamic area of the inlet, it is more likely to be mixed with unsuitable material, lost offshore, deposited in thin veneers that are not physically suited to dredging, and/or to interfere with navigation. There are also many areas within inlet systems from which dredging is severely limited. These include areas of environmental sensitivity (seagrass beds, shellfish areas, aquatic preserves, contaminated seabeds, etc.) and areas where upland property owners control or block local dredging activity. If deposition traps are not used to intercept the sediment within or before the inlet, then pumps or other dredging equipment must be positioned so as to reliably intercept, or to move and intercept, the material before it is lost to the inlet. This better ensures that the material will not contribute to navigation problems within the inlet, and that the sediment can be ultimately restored to the littoral system.

V-6-7. References

EM 1110-2-1616
EM 1110-2-1616, "Sand Bypassing System Selection"

EM 1110-2-1810
EM 1110-2-1810, "Coastal Geology"

Anders and Byrnes 1991
Anders, F. J., and Byrnes, M. R. 1991. "Accuracy of Shoreline Change Rates as Determined from Maps and Aerial Photographs," *Shore and Beach*, Vol 59, No. 1, pp 17-26.

Applied Technology Management, Inc. 1992
Applied Technology Management, Inc. (ATM). 1992. "St. Lucie Inlet Management Plan, Comprehensive Master Plan." (Report prepared for Board of County Commissioners, Martin County, Florida, 242 pp.)

Aubrey and Speer 1984
Aubrey, D. G., and Speer, P. E. 1984. "Updrift Migration of Tidal Inlets," *Journal of Geology*, Vol 92, pp 531-545.

Bailard and Jenkins 1982
Bailard, J. A., and Jenkins, S. A. 1982. "City of Carpenteria Beach Erosion and Pier Study," Report prepared for the City of Carpenteria, CA.

Barwis 1975
Barwis, J. H. 1975. "Catalog of Tidal Inlet Aerial Photography," U.S. Army Corps of Engineers, GITI Report 75-2, June 1975.

Barwis 1976
Barwis, J. H. 1976. "Annotated Bibliography on the Geologic, Hydraulic, and Engineering Aspects of Tidal Inlets," U.S. Army Corps of Engineers, GITI Report 4, Jan. 1976.

Berek and Dean 1982
Berek, E. P., and Dean, R. G. 1982. "Field Investigation of Longshore Transport Distribution." *Proceedings of the 18th International Conference on Coastal Engineering.* American Society of Civil Engineers, New York, pp 1620-39.

Birkemeier 1985
Birkemeier, W. A. 1985. "Field Data on Seaward Limit of Profile Change," *Journal of Waterway, Port, Coastal and Ocean Engineering*, ASCE, Vol 111, No. 3, pp 598-602.

Bisher and West 1993
Bisher, D. R., and West, F. W. 1993. "Jet Pumps and Fluidizers Working Together: The Oceanside Experimental Sand Bypass System," *Proceedings of the Beach Preservation Technology.* Florida Shore and Beach Preservation Association, Tallahassee, FL, pp 207-222.

Bodge 1993
Bodge, K. R. 1993. "Gross Transport Effects and Sand Management Strategy at Inlets," *Journal of Coastal Research,* Special Issue 18, Fall 1993, pp 111-124.

Bodge 1994a
Bodge, K. R. 1994a. "Performance of Nearshore Berm Disposal at Port Canaveral, Florida." *Proceedings of Dredging '94.* ASCE, New York.

Bodge 1994b
Bodge, K. R. 1994b. "The Extent of Inlet Impacts upon Adjacent Shorelines," *Proceedings of the 24th International Conference on Coastal Engineering*, ASCE, New York, pp 2943-57.

Bodge 1999
Bodge, K. R. 1999. "Inlet Impacts and Families of Solutions for Inlet Sediment Budgets." *Proceedings, Coastal Sediments '99.* ASCE, Reston, VA, pp 703-718.

Bodge, Creed, and Raichle 1996
Bodge, K. R., Creed, C. G., and Raichle, A. W. 1996. "Improving Input Wave Data for Use with Shoreline Change Models," Vol 122, No. 5, ASCE, New York.

Bodge and Hodgens 1997
Bodge, K. R., and Hodgens, E. 1997. "Recovery of a Nearshore Borrow Area for Inlet Sand Bypassing." *Proceedings of the Conference on Beach Preservation Technology '97.* Florida Shore and Beach Preservation Association, Tallahassee, FL.

Bottin 1992
Bottin, R. R. 1992. "Oceanside Harbor, California: Design for Harbor Improvements," Technical Report CERC-92-14, U.S. Army Engineer Waterways Experiment Station, Vicksburg, MS.

Bowen and Inman 1966
Bowen, A. J., and Inman, D. L. 1966. "Budget of Littoral Sand in the Vicinity of Point Arguello, California," Technical Memorandum No. 19, U.S. Army Coastal Engineering Research Center, 56 pp.

Brooke 1934
Brooke, M. M. 1934. "Shore Preservation in Florida," with discussion, *Shore and Beach*, Vol 2, No. 4, pp 151-154.

Brouwer, van Berk, and Visser 1992
Brouwer, J., van Berk, H., and Visser, K. G. 1992. "The Construction and Nearshore Use of the Punaise: A Flexible Submersible Dredging System," *Proceedings of the XIIIth World Dredging Congress*, Bombay, India, April 1992.

Browder 1995
Browder, A. E. 1995. "Wave Transmission and Current Patterns Associated with Narrow-Crested Submerged Breakwaters." *Proceedings of the Beach Preservation Technology '95*. Florida Shore and Beach Preservation Association, Tallahassee, FL, pp 348-364.

Browder 1996
Browder, A. E. 1996. "Wave Refraction and Littoral Transport Modeling of St. Lucie Inlet, FL." Olsen Associates, Inc., Jacksonville, FL, December, 1996.

Bruun 1962
Bruun, P. 1962. "Sea Level Rise as a Cause of Shore Erosion," *Journal of Waterways and Harbors Division*, ASCE, Vol 88, pp 117-130.

Bruun 1966
Bruun, P. 1966. "Tidal Inlets and Littoral Drift," *Stability of Coastal Inlets*. Vol 2, 193 pp. (Published by the author, copies available from Coastal Engineering Archives, Weil Hall, University of Florida, Gainesville, FL).

Bruun 1981
Bruun, P. 1981. *Port Engineering*. 3rd ed., Gulf Publishing Co., Houston, TX.

Bruun 1988
Bruun, P. 1988. "The Bruun Rule of Erosion by Sea-Level Rise: A Discussion of Large-Scale Two- and Three-Dimensional Usages," *Journal of Coastal Research*, Vol 4, pp 627-648.

Bruun 1993
Bruun, P. 1993. "An Update on Sand Bypassing Procedures and Prices," *Journal of Coastal Research*, Special Issue 18, Fall 1993, pp 277-284.

Bruun 1995
Bruun, P. 1995. "The Development of Downdrift Erosion," *Journal of Coastal Research*, Vol 11, No. 4, Fall 1995, pp 1242-57.

Bruun and Gerritsen 1959
Bruun, P., and Gerritsen, F. 1959. "Natural Bypassing of Sand at Coastal Inlets," *J. Waterways and Harbors Division* ASCE, Vol 85, No. 4, pp 75-107.

Bruun and Gerritsen 1960
Bruun, P., and Gerritsen, F. 1960. *Stability of Coastal Inlets*. North-Holland, Amsterdam, 140 pp.

Bruun, Mehta, and Johnsson 1978
Bruun, P., Mehta, A. J., and Johnsson, I. G. 1978. *Stability of Tidal Inlets*. Developments in Geotechnical Engineering, Vol 23, Elsevier, New York, 510 pp.

Bruun and Willikes 1992
Bruun, P., and Willikes, G. 1992. "Bypassing and Backpassing at Harbors, Navigation Channels, and Tidal Entrances: Use of Shallow-Water Draft Hopper Dredgers with Pump-Out Capabilities," *Journal of Coastal Research*, Vol 8, No. 4, pp 972-977.

Burke and Allison 1992
Burke, C. E., Allison, M. C. 1992. "Design Guidance for Nearshore Berms: Crest Length and End Slope Considerations," *Proceedings of the 1992 National Conference on Beach Preservation Technology*, Florida Shore and Beach Preservation Association, Tallahassee, FL, pp 180-193.

Byrnes and Hiland 1994
Byrnes, M. R., and Hiland, M. W. 1994. "Compilation and Analysis for Shoreline and Bathymetry Data (Appendix B)," In: N. C. Kraus, L. T. Gorman, and J. Pope (ed.) "Kings Bay Coastal and Estuarine Monitoring and Evaluation Program: Coastal Studies," Technical Report CERC-94-09, Coastal Engineering Research Center, Vicksburg, MS, B1-B90.

Caldwell 1951
Caldwell, J. M. 1951. "By-passing Sand at South Lake Worth Inlet, Florida." *Proceedings, First Conference on Coastal Engineering.* Long Beach, California, October 1950, J. W. Johnson, ed., Council on Wave Research, The Engineering Foundation, pp 320-325.

Caldwell 1966
Caldwell, J. M. 1966. "Coastal Processes and Beach Erosion," *Journal of the Society of Civil Engineers*, Vol 53, No. 2, pp 142-157.

Chasten 1992
Chasten, M. A. 1992. "Coastal Response to a Dual Jetty System at Little River Inlet, North and South Carolina," Miscellaneous Paper CERC-92-2, U.S. Army Engineer Waterways Experiment Station, Vicksburg, MS.

Chu, Lund, and Camfield 1987
Chu, Y.-H., Lund, R. B., and Camfield, F. E. 1987. "Sources of Coastal Engineering Information," TR-CERC-87-1, Coastal Engineering Research Center, U.S. Army Engineer Waterways Experiment Station, Vicksburg, MS.

Cialone and Stauble 1998
Cialone, M. A., and Stauble, D. K. 1998. "Historic Findings on Ebb Shoal Mining," *J. Coastal Res.*, Vol 14, No. 2, pp 537-563.

Clark 1983
Clark, G. R. 1983. "Survey of Portable Hydraulic Dredges," Technical Report HL-83-4, U.S. Army Engineer Waterways Experiment Station, Vicksburg, MS.

Clausner 1988
Clausner, J. E. 1988. "Jet Pump Sand Bypassing at the Nerang River Entrance, Queensland, Australia." *Proceedings of the Beach Preservation Technology '88.* Florida Shore and Beach Preservation Association, Tallahassee, FL, pp 345-55.

Clausner, Patterson, and Rambo 1990
Clausner, J. E., Patterson, D. R., and Rambo, A. T. 1990. "Fixed Sand Bypassing Plants -- An Update." *Proceedings of the 1990 National Conference on Beach Preservation Technology.* Florida Shore and Beach Preservation Association, Tallahassee, FL, pp 249-264.

Clausner, Geber, Rambo, and Watson 1991
Clausner, J. E., Gebert, J. A., Rambo, A. T., and Watson, K. D. 1991. "Sand Bypassing at Indian River Inlet, Delaware." *Proceedings of Coastal Sediments '91.* ASCE, New York, pp 1177-1190.

Clausner, Neilans, Welp, and Bishop 1994
Clausner, J. E., Neilans, P. J., Welp, T. L., and Bishop, D. D. 1994. "Controlled Tests of Eductors and Submersible Pumps," Technical Report DRP-94-2, Dredging Research Program, U.S. Army Engineer Waterways Experiment Station, Vicksburg, MS.

Cleary and Hosier 1988
Cleary, W. J., and Hosier, P. E. 1988. "Geomorphology and Shoreline History, Bald Head Island, NC," University of North Carolina, Wilmington, Wilmington, NC.

Coastal Planning and Engineering, Inc. 1991
Coastal Planning and Engineering, Inc. 1991. "Hillsboro Inlet, FL, Inlet Management Plan," Coastal Planning and Engineering, Inc., Boca Raton, FL.

Coastal Planning and Engineering, Inc. 1996
Coastal Planning and Engineering, Inc. 1996. "1995 Monitoring Report of the Boca Raton Inlet and Adjacent Beaches," Report submitted to the city of Boca Raton, FL, Coastal Planning and Engineering, Inc., Boca Raton, FL.

Coastal Tech 1997
Coastal Tech. 1997. "St. Lucie Inlet Hydrodynamic Monitoring and Modeling Study," Coastal Technology Corporation, Vero Beach, FL. Report prepared for Martin County, Stuart, FL.

Corson et al. 1982
Corson, W. D., Resio, D. T., Brooks, R. M., Ebersole, B. A., Jensen, R. E., Ragsdale, D. S., and Tracy, B. A. 1982. "Atlantic Coast Hindcast, Phase II Wave Information," WIS Report 6, Waterways Experiment Station, Vicksburg, MS.

Couglan and Robinson 1990
Couglan, P. M., and Robinson, D. A. 1990. "The Gold Seaway, Queensland, Austrailia," *Shore and Beach,* Vol 58, No. 1, ASBPA, January 1990, pp 9-16.

Creed 1996
Creed, C. G. 1996. "Modelling Inlet Sand Bypassing." *Proceedings, 25th International Conference on Coastal Engineering.* ASCE, New York.

Dean et al. 1987
Dean, R. G., Berek, E. P., Bodge, K. R., Gable, C. G. 1987. "NSTS Measurements of Total Longshore Transport." *Proceedings, Coastal Sediments '87.* ASCE, pp. 652-667.

Dean 1988
Dean, R. G. 1988. "Sediment Interaction at Modified Coastal Inlets: Processes and Policies," Lecture notes on coastal and estuarine studies. *Symposium on Hydrodynamics and Sediment Dynamics of Tidal Inlets*. D. G. Aubrey and L. Weishar, ed., Springer-Verlag, New York.

Dean 1993
Dean, R. G. 1993. "Terminal Structures at Ends of Littoral Systems," *Journal of Coastal Research*, Special Issue 18, Fall 1993, pp 195-210.

Dean and Perlin 1977
Dean, R. G., and Perlin, M. 1977. "Coastal Engineering Study of Ocean City Inlet, Maryland," *Coastal Sediments '77*. ASCE, pp 520-540.

Dean, Perlin, and Dally 1978
Dean, R. G., Perlin, M., and Dally, W. 1978. "A Coastal Engineering Study of Shoaling in Ocean City Inlet," Prepared for U.S. Army Engineer District, Baltimore, under contract with the University of Delaware, Newark, DE, 135 pp.

Dean and Work 1993
Dean, R. G., and Work, P. A. 1993. "Interaction of Navigational Entrances with Adjacent Shorelines," *Journal of Coastal Research*, Special Issue 18, Fall 1993, pp 91-110.

Dolan, Castens, Sonu, and Egense 1987
Dolan, T. J., Castens, P. G., Sonu, C. J., and Egense, A. K. 1987. "Review of Sediment Budget Methodology: Oceanside Littoral Cell, California," *Proceedings, Coastal Sediments '87*. ASCE, Reston, VA, pp 1289-1304.

Douglass 1987
Douglass, S. L. 1987. "Coastal Response to Navigation Structures at Murrells Inlet, South Carolina," Technical Report CERC-87-2, U.S. Army Engineer Waterways Experiment Station, Vicksburg, MS.

Douglass 1991
Douglass, S. L. 1991. "Simple Conceptual Explanation of Down-Drift Offset Inlets," *Journal of Waterway, Port, Coastal and Ocean Engineering*, March/April 1991, pp 44-59.

Douglass 1995
Douglass, S. L. 1995. "Estimating Landward Migration of Nearshore Constructed Sand Mounds," *Journal of Waterway, Port, Coastal and Ocean Engineering*, Vol 121, No. 5, ASCE, New York, pp 247-250.

Fenster and Dolan 1996
Fenster, M., and Dolan, R. 1996. "Assessing the Impact of Tidal Inlets on Adjacent Barrier Island Shorelines," *Journal of Coastal Research*, Vol 12, No. 1, Winter 1996, pp 294-310.

Fenton and McKee 1990
Fenton, J. D., and McKee, W. D. 1990. "On Calculating the Lengths of Water Waves," *Coastal Engineering*. Vol. 14, Amsterdam, The Netherlands, pp 499-513.

Fiorentino, Franco, and Noli 1985
Fiorentino, A., Franco, L., and Noli, A. 1985. "Sand Bypassing Plant at Viareggio, Italy." *Proceedings of the Australian Conference on Coastal and Ocean Engineering.* December 1985, Institute of Engineers, Australia.

FitzGerald 1984
FitzGerald, D. M. 1984. "Interactions Between the Ebb-Tidal Delta and the Landward Shoreline: Price Inlet, SC," *Journal of Sedimentary Petrology*, Vol 54, pp 1303-1318.

FitzGerald 1988
FitzGerald, D. M. 1988. "Shoreline Erosional-Depositional Processes Associated with Tidal Inlets," *Hydrodynamics and Sediment Dynamics of Tidal Inlets*, Vol 29, Lecture Notes on Coastal and Estuarine Studies, D. G. Aubrey and L. Weishar, ed., Springer-Verlag, New York, pp 186-225.

Foster, Healy, and De Lange 1996
Foster, G. A., Healy, T. R., and De Lange, W. P. 1996. "Presaging Beach Renourishment from a Nearshore Dredge Dump Mound, Mt. Maunganui Beach, New Zealand," *Journal of Coastal Research*, Vol 12, No. 2, pp 395-405.

Galvin 1982
Galvin, C. 1982. "Shoaling with Bypassing for Channels at Tidal Inlets," *Proceedings of the 18th International Conference on Coastal Engineering.* ASCE, New York, pp 1496-1513.

Gaudiano and Kana 2001
Gaudiano, D. J., and Kana, T. W. 2001. "Shoal Bypassing in Mixed Energy Inlets: Geomorphic Variables and Empirical Predictions for Nine South Carolina Inlets," *Journal of Coastal Research*, Vol 17, No. 2, pp 280-291.

Gebert, Watson, and Rambo 1992
Gebert, J. A., Watson, K. D., and Rambo, A. T. 1992. "57 Years of Coastal Engineering Practice at a Problem Inlet: Indian River Inlet, Delaware." *Proceedings of Coastal Engineering Practice '92.* ASCE, New York, pp 503-519.

Goda 1985
Goda, Yashimi. 1985. *Random Seas and Design of Maritime Structures.* University of Tokyo Press, Tokyo, Japan.

Gravens 1989
Gravens, M. B. 1989. "Bolsa Bay, California, Proposed Ocean Entrance System Study, Report 2: Comprehensive Shoreline Response Computer Simulation, Bolsa Bay, California," Miscellaneous Paper CERC-98-17, prepared for the State of California, State Land Commission, U.S. Army Engineer Waterways Experiment Station.

Hallermeier 1978
Hallermeier, R. J. 1978. "Uses for a Calculated Limit Depth to Beach Erosion." *Proceedings, 16th International Conference on Coastal Engineering.* ASCE, Hamburg, 1493-1512.

Hallermeier 1981
Hallermeier, R. J. 1981. "A Profile Zonation for Seasonal Sand Beaches from Wave Climate," *Coastal Engineering*, Vol 4, pp 253-277.

Hands and Allison 1991
Hands, E. B., and Allison, M. C. 1991. "Mound Migration in Deeper Waters and Methods of Categorizing Active and Stable Depths." *Proceedings of Coastal Sediments '91.* ASCE, New York, pp 1985-99.

Hands and Resio 1994
Hands, E. B., and Resio, D. T. 1994. "Empirical Guidance for Siting Berms to Promote Stability or Nourishment Benefits." *Proceeding of Dredging '94.* ASCE, New York.

Hansen and Knowles 1988
Hansen, M., and Knowles, S.C. 1988. "Ebb-tidal Response to Jetty Construction at Three South Carolina Inlets," *Hydrodynamics and Sediment Dynamics of Tidal Inlets,* Vol 29, D. G. Aubrey and L. Weishar, ed., Lecture Notes on Coastal and Estuarine Studies, Springer-Verlag, New York, pp 364-381.

Hanson and Kraus 1989
Hanson, H., and Kraus, N. C. 1989. "GENESIS: Generalized Numerical Modeling System for Simulating Shoreline Change, Report 1, Technical Reference Manual," Technical Report CERC-89-19, Coastal Engineering Research Center, U.S. Army Engineer Waterways Experiment Station, Vicksburg, MS.

Herbich 1975
Herbich, J. B. 1975. *Coastal and Deep Ocean Dredging.* Gulf Publishing Co., Houston, TX, 1975.

Herbich 2000
Herbich, J. B. 2000. *Handbook of Dredging Engineering.* 2nd ed., McGraw-Hill, New York.

Herbich and Snider 1969
Herbich, J. B., and Snider, R. H. 1969. "Bibliography on Dredging," Report 112-CDS, Texas A&M University, College Station, TX.

Herron and Harris 1967
Herron, W. J., and Harris, R. L. 1967. "Littoral Bypassing and Beach Restoration in the Vicinity of Point Hueneme, CA." *Proceedings of the 10th Conference on Coastal Engineering.* ASCE, New York, pp 651-75.

Hodges 1955
Hodges, T. K. 1955. "Sand Bypassing at Hillsboro Inlet, FL," *The Bulletin, Beach Erosion Board*, Vol 9, No. 2, Corps of Engineers, Washington, DC, April 1955.

Huston 1970
Huston, J. 1970. *Hydraulic Dredging, Theoretical and Applied.* Cornell Maritime Press, Inc., Cambridge, MD, ISBN 0-87033-142-6, 330 pp.

Huston 1986
Huston, J. 1986. *Hydraulic Dredging: Principles, Equipment, Procedures, Methods.* John Huston, Inc., Corpus Christi, TX, ISBN 0-9616260-0-3, 418 pp.

Inman 1991
Inman D. L. 1991. "Budget of Sediment and Prediction of the Future State of the Coast," State of the Coast Report, San Diego Region, Vol 1, Main Report, Chapter 9, 105 pp.

Inman and Frautschy 1966
Inman, D. L., and Frautschy, J. D. 1966. "Littoral Processes and the Development of Shorelines." *Proceedings, Coastal Engineering Specialty Conference.* ASCE, Reston, VA, 511-536.

Jarrett 1977
Jarrett, J. T. 1977. "Sediment Budget Analysis: Wrightsville Beach to Kore Beach, NC." *Proceedings of Coastal Sediments '77.* ASCE, New York, 1977.

Jarrett 1990
Jarrett, T. 1990. "Wilmington Harbor - Bald Head Island, Evaluation Report, Sec. 933, PL99-662," U.S. Army Engineer District, Wilmington, NC, June 1990, 44 pp plus appendices.

Jarrett 1991
Jarrett, J. T. 1991. "Coastal Sediment Budget Analysis Techniques." *Proceedings, Coastal Sediments '91.* ASCE, Reston, VA, pp 2223-2233.

Johnson 1959
Johnson, J. W. 1959. "The Supply and Loss of Sand to the Coast," *Journal of Waterways and Harbors Division,* Vol 85, No. WW3, ASCE, New York, September 1959, pp 227-251.

Johnson et al. 1994
Johnson, B. H., Scheffner, N. W., Teeter, A. M., Hands, E. B., and Moritz, H. R. 1994. "Analysis of Dredged Material Placed in Open Water." *Proceedings of the 22nd International Conference on Dredging and Dredged Material Placement, Dredging '94.* ASCE, New York.

Jones and Mehta 1977
Jones, C. P., and Mehta, A. J. 1977. "A Comparative Review of Sand Transfer Systems at Florida's Tidal Entrances." *Proceedings of Coastal Sediments ' 77.* ASCE, New York, pp 49-66.

Jones and Mehta 1980
Jones, C. P., and Mehta, A. J. 1980. "Inlet Sand Bypassing Systems in Florida," *Shore and Beach,* Vol 48, No. 1, January 1980, pp 25-33.

Kana 1995
Kana, T. W. 1995. "A Mesoscale Sediment Budget for Long Island, New York, *Marine Geology,* Vol 126, pp 87-110.

Kana and Stevens 1992
Kana, T. W., and Stevens, F. D. 1992. "Coastal Geomorphology and Sand Budgets Applied to Beach Nourishment." *Proceedings of Coastal Engineering Practice '92.* ASCE, New York, pp 29-44.

Komar 1976
Komar, P. D. 1976. *Beach Processes and Sedimentation.* Prentice-Hall, Inc., Englewood Cliffs, New Jersey, 429 pp.

Komar 1996
Komar, P. D. 1996. "The Budget of Littoral Sediments, Concepts and Applications," *Shore and Beach.* Vol 64, No. 3, pp 18-26.

Komar 1998
Komar, P. D. 1998. *Beach Processes and Sedimentation.* Prentice-Hall, Inc., Simon and Schuster, Upper Saddle River, NJ, pp 66-72.

Kraus 1989
Kraus, N. C. 1989. "Beach Change Modeling and the Coastal Planning Process." *Proceedings, Coastal Zone '89*, ASCE, pp 553-567.

Kraus 1999
Kraus, N. C. 1999. "Analytical Model to Spit Evolution at Inlets." *Proc, Coastal Sediments '99.* ASCE, Reston, VA, pp 1739-1754.

Kraus 2000
Kraus, N. C. 2000. "Reservoir Model of Ebb-Tidal Shoal Evolution and Sand Bypassing," *Journal of Waterway, Port, Coastal, and Ocean Engineering*, Vol 126, No. 6, pp 305-313.

Kraus, Larson, and Kriebel 1991
Kraus, N. C., Larson, M., and Kriebel, D. L. 1991. "Evaluation of Beach Erosion and Accretion Predictors," *Coastal Sediments '91*, ASCE, New York, pp 572-87.

Kraus, Larson, and Wise 1999
Kraus, N. C., Larson, M., and Wise, R. A. 1999. "Depth of Closure in Beach-Fill Design." *Proceedings, 12th National Conference on Beach Preservation.* Tallahassee, FL, pp 271-286.

Kraus and Rosati 1998a
Kraus, N. C., and Rosati, J. D. 1998a. "Estimation of Uncertainty in Coastal-Sediment Budgets at Inlets," Coastal Engineering Technical Note CETN-IV-16, U.S. Army Engineer Waterways Experiment Station, Vicksburg, MS.

Kraus and Rosati 1998b
Kraus, N. C., and Rosati, J. D. 1998b. "Interpretation of Shoreline-Position Data for Coastal Engineering Analysis," Coastal Engineering Technical Note CETN-II-39, U.S. Army Engineer Waterways Experiment Station, Vicksburg, MS.

Kraus and Rosati 1999
Kraus, N. C., and Rosati, J. D. 1999. "Estimating Uncertainty in Coastal Inlet Sediment Budgets." *Proceedings, 12th Annual National Conference on Beach Preservation Technology.* Florida Shore and Beach Preservation Association, Tallahassee, FL, pp 287-302.

Lean 1980
Lean, G. H. 1980. "Estimation for Maintenance Dredging for Navigation Channels." Hydraulics Research Station, Wallingford, Oxon, Crown Copyright, 73 pp.

Leatherman 1984
Leatherman, S. P. 1984. "Shoreline Evolution of North Assateague Island, Maryland, *Shore and Beach*, Vol 52, No. 3, pp 3-10.

Lund 1990
Lund, J. 1990. "Scheduling Maintenance Dredging in a Single Reach with Uncertainty," *Journal of Waterway, Port, Coastal and Ocean Engineering*, Vol 116, No. 2, ASCE, New York, pp 211-231.

Magnuson 1967
Magnuson, N. C. 1967. "Planning and Design of a Low-Weir Section Jetty (Masonboro Inlet, N.C.)," *Journal of Waterway and Harbors Division*, ASCE, New York, pp 27-40, 1967.

Mann 1993
Mann, D. W. 1993. "A Note on Littoral Budgets and Sand Management at Inlets," *Journal of Coastal Research*, Special Issue No. 18: 301-308.

McKnight 1966
McKnight, A. L. 1966. "Dredging - past - present and future." *Proceedings of the Coastal Engineering Specialty Conference, Dredging.* ASCE, New York, pp 727-747.

McLellan and Kraus 1991
McLellan, T. N., and Kraus, N. C. 1991. "Design Guidance for Nearshore Berm Construction." *Proceedings of Coastal Sediments '91.* ASCE, New York, pp 2000-2011.

Mehta, Dombrowski, and Devine 1996
Mehta, A. J., Dombrowski, M. R., and Devine, P. T. 1996. "Role of Waves in Inlet Ebb Delta Growth and Some Research Needs Related to Site Selection for Delta Mining," *J. Coastal Res.*, Vol SI 18, pp 121-136.

Meistrell 1972
Meistrell, F. J. 1972. "The Spit-Platform Concept: Laboratory Observation of Spit Development." In: *Spits and Bars.* M. L Schwartz, ed., Dowden, Hutchison, and Ross, Stroudsberg, PA, pp 225-283.

Mesa 1996
Mesa, C. "Nearshore Berm Performance at Newport Beach, CA." *Proceedings of the 25th International Conference on Coastal Engineering.* ASCE, New York, pp 4636-4649.

Migniot and Granboulan 1985
Migniot, C., and Granboulan J. 1985." Travaux de Genie Civil de Long des Cotes Cablonneuses," PIANC, S II-3, pp 47-68.

Moffatt and Nichol Engineers 1978
Moffatt and Nichol Engineers. 1978. "Santa Cruz Harbor Shoaling Study," Moffatt and Nichol Engineers, Prepared for U.S. Army Engineer District, San Francisco, June 1978, Long Beach, CA.

Moffatt and Nichol Engineers and URS Consultants, Inc. 1999
Moffatt and Nichol Engineers and URS Consultants, Inc. 1999. "Storm Damage Reduction Reformulation Study Inlet Dynamics-Existing Conditions," Prepared for U.S. Army Engineer District, New York.

Nersesian and Bocamazo 1992
Nersesian, G. K., and Bocamazo, L. M. 1992. "Design and Construction of Shinnecock Inlets, New York." *Proceedings, Coastal Engineering Practice '92.* ASCE, pp 554-570.

O'Conner and Lean 1977
O'Conner, B.A., and Lean, G. H. 1977. "Estimation of Siltation in Dredged Channels in Open Situations." *Proceedings of the 24th International Nav. Cong.* P.I.A.N.C., Section II, Subject 2, pp 163-177.

Oertel 1988
Oertel, G.F. 1988. "Processes of Sediment Exchange Between Tidal Inlets, Ebb Deltas, and Barrier Islands," *Hydrodynamics and Sediment Dynamics of Tidal Inlets 29.* Lecture notes on coastal and estuarine studies, D. G. Aubrey and L. Weishar, ed, Springer-Verlag, New York, pp 297-318.

Olsen 1977
Olsen, E. J. 1977. "A Study of the Effects of Inlet Stabilization at St. Mary's Entrance, Florida." *Proceedings, Coastal Sediments '77.* ASCE, pp 311-329.

Olsen Associates, Inc. 1989
Olsen Associates, Inc. 1989. "Feasibility Study of Beach Restoration at Bald Head Island, NC," Report prepared for the village of Bald Head Island, NC, Olsen Associates, Inc., Jacksonville, FL, April 1989, 112 pp plus appendices.

Olsen Associates, Inc. 1990
Olsen Associates, Inc. 1990. "South Lake Worth Inlet Sand Management Plan," Report prepared for Palm Beach County, Board of County Commissioners, 31 December 1990, 183 pp plus appendices, Olsen Associates, Inc., Jacksonville, FL.

Olsen Associates, Inc. 1995
Olsen Associates, Inc. 1995. "Sailfish Point Shoreline Stabilization Project, Analysis and Conceptual Design," Report prepared for Sailfish Point Property Owners' Association, November 1995, Olsen Associates, Inc., Jacksonville, FL.

Olsen Associates, Inc. 1997
Olsen Associates, Inc. 1997. "Feasibility Study of Structural Stabilization of Beach Fill in Broward County," Report prepared for Broward County, Department of Environmental Resource Protection, Ft. Lauderdale, FL, December 1997, Olsen Associates, Inc., Jacksonville, FL.

Parker 1979
Parker, N. E. 1979. "Weir Jetties - Their Continuing Evolution," *Shore and Beach,* Vol 47, No. 4, October 1979, pp 15-19.

Parson and Lillycrop 1998
Parson, L. E., and Lillycrop, W. J. 1998. "The SHOALS System: A Comprehensive Survey Tool," Coastal Engineering Technical Note CETN VI-31, U.S. Army Engineer Waterways Experiment Station, Vicksburg, MS.

Partheniades and Purpura 1972
Partheniades, E., and Purpura, J. A. 1972. "Coastline Changes Near a Tidal Inlet." *Proceedings of the 13th Conference on Coastal Engineering.* ASCE, New York, pp 843-864.

Patterson, Bisher, and Brodeen 1991
Patterson, D. R., Bisher, D. R., and Brodeen, M. R. 1991. "The Oceanside Experimental Sand Bypass System." *Proceedings of Coastal Sediments '91.* ASCE, New York, pp 1165-76.

Pelnard-Considére 1956
Pelnard-Considére, R. 1956. "Essai de Th'eorie de l'Evolution des Forms de Rivages en Plage de Sable et de Galets," *4th Journees de l'Hydralique,* les Energies de la Mer, Question III, Rapport No. 1, pp 289-298.

Penfield 1960
Penfield, W. C. 1960. "The Oldest Periodic Beach Nourishment Project," *Shore and Beach*, Vol 28, No. 1, April, pp 9-15.

Perlin and Dean 1983
Perlin, M., and Dean, R. G. 1983 "A Numerical Model to Simulate Sediment Transport in the Vicinity of Coastal Structures," Miscellaneous Report No. 83-10, Coastal Engineer Research Center, U.S. Army Engineer Waterways Experiment Station, Vicksburg, MS.

Polglase 1987
Polglase, R. H. 1987. "The Nerang River Entrance Sand Bypassing System." *Proceedings of the 8th Australian Conference on Coastal and Ocean Engineering.* November 1987.

Pope and Dean 1986
Pope, J., and Dean, J. L. 1986. "Development of Design Criteria for Segmented Breakwaters." *Proceedings, 20th Int'l Coastal Engr. Conference.* Taipei, Taiwan. ASCE, pp 2144-2158.

Pound and Witt 1987
Pound, M. D., and Witt, R. W. 1987. "Nerang River Entrance Sand Bypassing System." *Proceedings of the 8th Australian Conference on Coastal and Ocean Engineering.* November 1987, pp 222-226.

Purpura 1974
Purpura, J. A. 1974. "Performance of a Jetty-Weir Inlet Improvement Plan." *Proceedings of the Fourteenth International Conference on Coastal Engineering.* ASCE, New York, pp 1470-1490.

Purpura 1977
Purpura, J. A. 1977. "Performance of a Jetty-Weir Inlet Improvement Plan." *Proceedings of Coastal Sediments '77.* ASCE, New York, pp 330-349.

Rambo, Clausner, and Henry 1991
Rambo, A., Clausner, J., and Henry, R. 1991. "Sand Bypass Plant: Indian River Inlet, Delaware." *Proceedings of the Conference on Beach Preservation Technology '91.* Florida Shore and Beach Preservation Association, Tallahassee, FL.

Rayner and Magnuson 1966
Rayner, A. C., and Magnuson, N. C. 1966. "Stabilization of Masonboro Inlet," *Shore and Beach*, Vol 34, No. 2, October 1966, pp 36-41.

Richardson 1976
Richardson, T. W. 1976. "Beach Nourishment Techniques, Report 1," Technical Report H-76-13, U.S. Army Engineer Waterways Experiment Station, Vicksburg, MS.

Richardson 1977
Richardson, T. W. 1977. "Systems for Bypassing Sand at Coastal Inlets." *Proceedings of the Fifth Symposium of the Waterway, Port, Coastal, and Ocean Division.* ASCE, New York, November 1977, pp 67-84.

Richardson 1990
Richardson, T. W. 1990. "Sand Bypassing," *Handbook on Coastal and Ocean Engineering.* J. B. Herbich, ed., Gulf Publishing Company, Houston, TX.

Richardson and McNair 1981
Richardson, T. W., and McNair, E. C. 1981. "A Guide to the Planning and Hydraulic Design of Jet Pump Remedial Sand Bypassing Systems," Instruction Report HL-81-1, Hydraulics Laboratory, U.S. Army Engineer Waterways Experiment Station, Vicksburg, MS.

Rosati and Ebersole 1996
Rosati, J. D., and Ebersole, B. A. 1996. "Littoral Impact of Ocean City Inlet, Maryland, USA." *Proc, 25th Coastal Eng. Conf.* ASCE, Reston, VA, pp 2779-2792.

Sargent 1988
Sargent, F. E. 1988. "Case Histories of Corps Breakwater and Jetty Structures, Report 2, South Atlantic Division," Technical Report REMR-CO-3, U.S. Army Engineer Waterways Experiment Station, Vicksburg, MS.

Savage 1990
Savage, R. 1990. "A Comparison of Mean High Water Shoreline Change and Bluff Movement." *Proceedings of the 1990 National Conference on Beach Preservation Technology.* Florida Shore and Beach Preservation Association, Tallahassee, FL, pp 324-338.

Schmidt and Schwichtenberg 2000
Schmidt, D. E., and Schwichtenberg, B. R. 2000. "Regional Sediment Management in Action, Multi-Purpose Project At Post Everglades." *Proceedings, Nat'l Conf. on Beach Preservation Technology.* Florida Shore and Beach Preservation Association, pp 17-24.

Scheffner 1996
Scheffner, N. W. 1996. "Systematic Analysis of Long-Term Fate of Disposed Dredged Material," *Journal of Waterway, Port, Coastal and Ocean Engineering*, Vol 122, No. 3, ASCE, New York, pp 127-133.

Schneider 1980
Schneider, C. 1980. "The Littoral Environment Observation (LEO) Data Collection Program," CETA 80, U.S. Army Engineer Waterways Experiment Station, Vicksburg, MS.

Seabergh 1983
Seabergh, W. C. 1983. "Weir Jetty Performance: Hydraulic and Sedimentary Considerations," Technical Report HL-83-5, U.S. Army Engineer Waterways Experiment Station, Vicksburg, MS.

Shore Protection Manual 1984
Shore protection manual. (1984). 4th ed., 2 Vol, U.S. Army Engineer Waterways Experiment Station, U.S. Government Printing Office, Washington, DC.

Snetzer 1969
Snetzer, R. E. 1969. "Jetty-Weir Systems at Inlets in the Mobile Engineer District," *Shore and Beach*, Vol 37, No. 1, April 1969, pp 28-32.

Stauble 1997
Stauble, D. K. 1997. "Appendix A: Ebb and Flood Shoal Evolution," In: "Ocean City, Maryland, and Vicinity Water Resources Study, Draft Integrated Feasibility Report and Environmental Impact Statement," U.S. Army Engineer District, Baltimore, Baltimore, MD.

Stauble et al. 1988
Stauble, D. K., Da Costa, S. L., Monroe, K. L., and Bhogal, V. K. 1988. "Inlet Flood Tidal Delta Development Through Sediment Transport Processes," *Hydrodynamics and Sediment Dynamics of Tidal Inlets*. Vol 29, Lecture Notes on Coastal and Estuarine Studies, D. G. Aubry and L. Weishar, ed, Springer-Verlag, New York, pp 319-47.

Taney 1961a
Taney, N. E. 1961a. "Geomophology of the South Shore of Long Island, New York," Technical Memorandum 128, U.S. Army Corps of Engineers, Beach Erosion Board, Washington. DC, 97 pp.

Taney 1961b
Taney, N. E. 1961b. "Littoral Materials of the South Shore of Long Island, NewYork," Technical Memorandum 129, U.S. Army Corps of Engineers, Beach Erosion Board, Washington. DC, 97 pp.

Terich and Komar 1974
Terich, T. A., and Komar, P. D. 1974. "Bayocean Spit, Oregon: History of Development and Erosional Destruction," *Shore and Beach*, Vol 42, No. 2, pp 3-10.

Trawle 1981
Trawle, M. J. 1981. "Effects of Depth on Dredging Frequency," Technical Report H-78-5, Report 2, U.S. Army Engineer Waterways Experiment Station, Vicksburg, MS.

Turner 1984
Turner, T. M. 1984. *Fundamentals of Hydraulic Dredging.* Cornell Maritime Press, Centerville, MD.

Uda, Naito, and Kanda 1991
Uda, T., Naito, K., and Kanda, Y. 1991. "Field Experiment on Sand Bypass off the Iioka Coast (Japan)," *Coastal Engineering in Japan*, Vol 34, No. 2, pp 205-221.

Underwood and Hiland 1995
Underwood, S. G., and Hiland, M. W. 1995. "Historical Development of Ocean City Inlet Ebb Shoal and Its Effect on Northern Assateague Island," Report prepared for U.S. Army Engineer Waterways Experiment Station, Coastal Engineering Research Center, Vicksburg, MS.

U.S. Army Corps of Engineers 1950
U.S. Army Corps of Engineers. 1950. "Test of Nourishment of the Shore by Offshore Deposition of Sand, Long Branch, NJ," U.S. Army Corps of Engineers, Beach Erosion Board, Technical Memorandum No. 17, June 1950, 32 pp.

U.S. Army Corps of Engineers 1957
U.S. Army Corps of Engineers. 1957. "Shore of New Jersey from Sandy Hook to Barnegat Inlet, Beach Erosion Control Study," Letter from Secretary of Army, House Document No. 361, 84th Congress, 2nd Session, U.S. Government Printing Office, Washington, DC.

U.S. Army Corps of Engineers 1958
U.S. Army Corps of Engineers. 1958. Shore of New Jersey from Sandy Hook to Bernegat Inlet, Beach Erosion Control Study. Letter from the Secretary of the Army, House Document No. 362, 85th Congress, 2nd Session, U.S. Government Printing Office, Washington, DC.

U.S. Army Engineer District, Jacksonville, 1973
U.S. Army Engineer District, Jacksonville. 1973. "Survey Review Report on St. Lucie Inlet," U.S. Army Engineer District, Jacksonville, Jacksonville, FL.

U.S. Army Engineer District, New York 1987
U.S. Army Engineer District, New York. 1987. "General Design Memorandum – Shinnecock Inlet, Long Island, new York," 2 VOL, revised 1988, New York.

U.S. Army Engineer Waterways Experiment Station 1995
U.S. Army Engineer Waterways Experiment Station. 1995. "Kings Bay Coastal and Estuarine Physical Monitoring and Evaluation Program: Coastal Studies," Technical Report CERC-94-9, Coastal Engineering Research Center, U.S. Army Engineer Waterways Experiment Station, Vicksburg, MS, January 1995.

U.S. Army Engineer District, Jacksonville, 1999
U.S. Army Engineer District, Jacksonville. 1999. "St. Lucie Inlet, FL, Navigation Project, General Design Memorandum," U.S. Army Engineer District, Jacksonville, FL. Rev., March 1999.

Vallianos 1973
Vallianos, L. 1973. "A Recent History of Masonboro Inlet, North Carolina." *Proceedings of the 2nd International Estuarine Research Conference.* ASCE, Columbia, SC.

Vemulakonda 1988
Vemulakonda, S. 1988. "Kings Bay Coastal Processes Numerical Model," Technical Report CERC-88-3, Coastal Engineering Research Center, U.S. Army Engineer Waterways Experiment Station, Vicksburg, MS.

Walker and Lesnik 1990
Walker, J. R., and Lesnik, J. R. 1990. "Impacts of Ocean Entrances on Beaches in Southern California." *Proceedings of the 1990 National Conference on Beach Preservation Technology.* FSBPA, Tallahassee, FL, pp 203-217.

Walther and Douglas 1993
Walther, M. P., and Douglas, B. D. 1993. "Ebb Shoal Borrow Area Recovery," *Journal of Coastal Research*, Vol SI 17, pp 211-223.

Walton and Adams 1976
Walton, T. L., and Adams, W. D. 1976. "Capacity of Inlet Outer Bars to Store Sand." *Proceedings of the 15th Coastal Engineering Conference.* ASCE, New York, pp 1919-1937.

Weggel 1981
Weggel, J. R. 1981. "Weir Sand-Bypassing Systems," Special Report No. 8, Coastal Engineering Research Center, U.S. Army Engineer Waterways Experiment Station, Vicksburg, MS.

Weggel and Vitale 1981
Weggel, J. R., and Vitale, P. 1981. "Sand Transport over Weir Jetties and Low Groins," Symposium on Coastal Physical Modeling, University of Delaware, Newark, DE, pp 163-197.

Weishar and Fields 1985
Weishar, L. L., and Fields, M. L. 1985. "Annotated Bibliography of Sediment Transport Occurring over Ebb-Tidal Deltas," Miscellaneous Paper CERC-85-11, Coastal Engineering Research Center, U.S. Army Engineer Waterways Experiment Station, Vicksburg, MS.

Weisman, Lennon, and Clausner 1992
Weisman, R. N., Lennon, G. P., and Clausner, J. E. 1992. "A Design Manual for Coastal Fluidization Systems," *Proceeding of Coastal Engineering Practice '92*. ASCE, New York, pp 862-878.

Weisman, Lennon, and Clausner 1996
Weisman, R. N., Lennon, G. P., and Clausner, J. E. 1996. "A Guide to the Planning and Hydraulic Design of Fluidizer Systems for Sand Management in the Coastal Environment," Technical Report DRP-96, Dredging Research Program, U.S. Army Engineer Waterways Experiment Station, Vicksburg, MS.

Wiegel 1954
Wiegel, R. L. 1954. *Gravity Waves, Tables of Functions*. University of California, Council on Wave Research, The Engineering Foundation, Berkeley, CA.

Williams, Clausner, and Neilans 1994
Williams, G. L., Clausner, J. E., and Neilans, P. J. 1994. "Improved Eductors for Sand Bypassing," Program Technical Report DRP-94-6, U.S. Army Engineer Waterways Experiment Station, Dredging Research November 1994, 36 pp.

Wise 1998
Wise, R. A. 1998. "Depth of Closure in Beach-Fill Design," Coastal Engineering Technical Note, CETN II-40, U.S. Army Engineer Waterways Experiment Station, Vicksburg, MS.

Zwamborn, Fromme, and Fitzpatrick 1970
Zwamborn, J. A., Fromme, C. A. W., and Fitzpatrick, J. B. 1970. "Underwater Mound for the Protection of Durban's Beaches." *Proceedings of the 12th Coastal Engineering Conference*. ASCE, New York, pp 975-994.

V-6-7. Definition of Symbols

α	Wave angle relative to the grid at the reference point [deg]
α_b	Angle of the breaking wave crests relative to the shoreline [deg]
β	Shoreline angle relative to the grid at the alongshore column of interest [deg]
ε	Longshore diffusivity (Equation V-6-6) [length^2/time]
κ	Ratio of local breaking wave height to local depth [dimensionless]
b	Width of a dredge crater at the bottom [length]
B	Beach berm height [length]
C	Wave celerity [length/time]
C_g	Wave group velocity [length/time]
D_A	Active shoreline depth [length]
D_B	Mechanical transfer of sand from the left shoreline to the right shoreline [length^3/time]
d_c	Dredge crater depth below the ambient seabed level [length]
D_C	Profile closure depth [length]
D_L, D_R	Mechanical transfer of sand from the inlet to the left and right shoreline [length^3/time]
D_O	Maintenance dredging and out-of-system disposal from the inlet [length^3/time]
E	Net volumetric capacity of a dredge pump [length^3/time]
$erfc()$	Error function complement
g	Gravitational acceleration [length/time^2]
h_*	Depth of active sediment transport (depth of closure) [length]
h	Water depth [length]
H	Wave height [length]
H_b	Breaking wave height [length]
H_O	Offshore wave height [length]
j_1, j_2	Fraction of incident transport impounded by the inlet's jetties [length^3/time]
K	Empirical coefficient from the longshore transport Equation III-2-5 (of order 1) [dimensionless]
L	Wavelength [length]

L	Leftward-directed incident transport values at the study area's boundaries [length^3/time]
L_c	Length (or arc-length) of a dredge crater [length]
L_c	Cross-shore distance from datum to the long-term depth of closure [length]
m	Dredge crater side slope (Figure V-6-72)
m_1, m_2	Local inlet-induced transport from the left and right shoreline into the inlet [dimensionless]
M_{tot}	Average annual littoral brought to the inlet [length^3]
n	Sediment porosity (about 0.4) [dimensionless]
N_O	Fall-velocity or Dean parameter (Equation V-6-77) [dimensionless]
p	Describes the degree to which the inlet's sand removal alters each of the inlet's two adjacent shorelines (Equation V-6-8)
P	Tidal prism [length^3]
P	Beach fill or dredged placement [length^3]
P	Natural net bypassing [length^3/time]
p_1, p_2	Fraction of incident transport naturally bypassed across the inlet [length^3/time]
P_l	Longshore component of wave energy flux
Q	Longshore transport rate [length^3/time]
Q	Incident (updrift) net transport [length^3/time]
Q_R	Annual longshore transport to the right (looking seaward) [length^3/time]
Q_{NET}	Net annual longshore transport [length^3/time]
Q_{GROSS}	Gross annual longshore transport [length^3/time]
Q_L	Annual longshore transport to the left (looking seaward) [length^3/time]
Q_{Aout}	Bypassing rate of the attachment [length^3/time]
Q_{Bout}	Bypassing rate of the bar [length^3/time]
Q_{Eout}	Rate of sand bypassing the ebb shoal [length^3/time]
R	Material removed from a sediment budget cell [length^3]
R	Rightward-directed incident transport values at the study area's boundaries [length^3/time]
s	Ratio of sediment to water specific gravity (about 2.65) [dimensionless]
S	Average volumetric shoaling rate of a dredge crater [length^3/time]

S	Sea level change [length]
S_u	Transport of littoral material into the inlet from upland sources [length^3/time]
T	Wave period [time]
T_B	Elapsed time between the start of subsequent dredging events
t_c	Crossover time [time]
T_E	Time required to excavate a dredge crater (Equation V-6-73)
t_f	Time at which the structure becomes filled to capacity and begins to bypass [time]
t'	Lag time that delays development of the bar
t''	Lag time that delays development of the attachment
u_m	Maximum nearbed horizontal wave orbital velocity [length/time]
V_A	Volume of attachment bar [length^3]
V_B	Bypassing bar volume [length^3]
V_c	Theoretical yield volume of a dredge crater [length^3/time]
V_E	Volume of sand in the shoal [length^3]
∇V	Sediment volume change [length^3]
w	Sediment's median fall velocity [length/time]
x	Distance downdrift of structure [length]
y	Shoreline recession [length]
Y	Structure length [length]
Δy_{st}	Long-term beach loss because of an increase S in relative sea level

V-6-8. Acknowledgments

Authors of Chapter V-6, "Sediment Management at Inlets and Harbors:"

Kevin R. Bodge, Ph.D., Olsen Associates, Jacksonville, Florida.
Julie D. Rosati, Coastal and Hydraulics Laboratory (CHL), U.S. Army Engineer Research and Development Center, Vicksburg, Mississippi.

Reviewers:

Lyndell Z. Hales, Ph.D., CHL
Nicholas C. Kraus, Ph.D., CHL
Andrew Morang, Ph.D., CHL
Gregory L. Williams, U.S. Army Engineer District, Wilmington, North Carolina

Chapter 7
COASTAL ENGINEERING FOR
ENVIRONMENTAL ENHANCEMENT

EM 1110-2-1100
(Part V)
31 July 2003

Table of Contents

List of Figures

Page

Part V-7-1
Coastal Engineering for Environmental Enhancement

V-7-1. Introduction

The purpose of this chapter is to provide an overview of coastal habitats and information resources on the creation and restoration of coastal habitats. For the purpose of this report, coastal habitats include both marine nearshore habitats and estuarine (both brackish and freshwater tidal) habitats. Many projects are or can be designed to restore, create, enhance, or protect critical coastal environments and the natural resources (fisheries, wildlife, etc.) that depend on them. Innovative uses of new and traditional coastal engineering technology have enabled scientists, engineers, and resource managers to rehabilitate degraded coastal habitats, create new habitats, better identify habitat needs and opportunities, and manage environmental impacts from development projects. Opportunities to provide environmental enhancements or mitigation in traditional coastal engineering projects (those that do not include habitat creation/restoration or protection as the primary objective) are also presented in this chapter, along with a discussion of potential environmental constraints and issues that may affect coastal engineering projects in general.

V-7-2. Coastal Habitat Projects

a. Individual habitat restoration, creation, or protection projects in the coastal zone provide identifiable benefits onsite. However, in the context of an ecosystem, watershed or landscape, they provide a continuum of benefits which may not be realized at the outset by all project participants. Recent Federal directives and other agency initiatives in restoring coastal habitats have encouraged the implementation of ecosystem-level planning in the design phase of projects (Thom 1997).

b. One example of ecosystem evaluation incorporates the concept of landscape ecology. Landscape ecology focuses on the interaction of three characteristics: (a) structure (the spatial relationship between ecosystem elements), (b) function (the interaction among these elements), and (c) change (the alteration in structure and function over time) (Forman and Gordon 1986). Several components make up natural landscapes and they include: (a) Matrix (the dominant landscape type, in coastal situations this may be water), (b) Patch (a nonlinear surface area differing in appearances from its surroundings, such as a coral reef surrounded by water), (c) Corridor (a narrow strip of habitat that differs from the matrix on either side, such as a band of seagrass), and (d) Node (an intersection of corridors). Individual patches collectively form a heterogeneous mosaic. Organisms within an individual patch (e.g., fish residing in a sub-tropical seagrass bed) may migrate between or among adjacent habitat patches using corridors (e.g., mangrove forests, coral reefs), depending on seasonal, diel, or environmental influences. Effective management of entire ecosystems entails improvements or enhancements to multiple habitat types, with an understanding of the role of habitats. An example of this would be to consider both adjacent and distant habitat types that Pacific salmon are heavily dependent on. That is, near coastal areas and estuaries for juvenile rearing, open ocean areas where they mature and spawning areas located as much as 500 miles upriver in freshwater. Recognition of the interactions between coastal, freshwater, and terrestrial systems, and an understanding of processes occurring in nearby watersheds such as agricultural or industrial activity, are key to planning habitat projects on an ecosystem scale.

c. Several other approaches are valid when considering restoration in the large scale. Examples include the analysis of limiting factors on keystone species. Other restoration efforts have focused on comparison of the historical ecosystem with the current regime to determine what critical elements have been lost. No one single methodology is right for every situation but it is important to evaluate individual restoration efforts in the larger context.

d. Recently, scientists and managers have developed recommendations for improving the state-of-the-art in habitat restoration (Pastorok et al. 1997). These recommendations include refinements in developing goals and objectives, consideration of spatial and temporal scales in design parameters, flexibility and adaptive management in project planning and design, and the importance of establishing long-term monitoring programs to document structural and functional attributes of habitats being restored. Additionally, the concept of ecosystem management is becoming the more common, especially on Federal lands. The primary objective of managing an ecosystem is to maintain its integrity of function, diversity, and structure.

V-7-3a. Habitat Trade-offs: Issues Associated with Compensatory Mitigation

a. Within the last two decades, habitat restoration and creation have increasingly been used as requirements to mitigate for damage to natural resources. This is required in order to compensate for habitat losses as specified in Section 404(b)(1) of the Clean Water Act of 1972 and many state and local regulations. When most of these laws and regulations were passed, wetland and aquatic habitat creation and restoration were still developing technologies. By the mid-1980's, mitigation for human-induced wetland losses became standard practice in the United States.

b. Mitigation is often described as three general types: avoidance, minimization, and compensatory. Under Section 404 of the Clean Water Act, these actions are sequenced in such a way that avoidance is preferred, impacts that cannot be avoided are minimized to the greatest extent practicable, and finally a determination of appropriate levels of compensatory mitigation are determined based on the analysis of lost functions and values. One useful approach to evaluating coastal habitats for mitigation needs has been developed by the U.S. Army Engineer Research and Development Center (ERDC), formerly the U.S. Army Engineer Waterways Experiment Station (Ray 1994).

c. Mitigation may occur at or near the affected habitat *(onsite mitigation)*. It is also considered acceptable to mitigate for damages by creating or restoring habitat elsewhere *(off-site mitigation)*, especially when on-site mitigation would be adversely affected by the surrounding development. *In-kind* mitigation is accomplished by restoring/creating the same habitat type (e.g., mitigate for intertidal salt marsh lost by marina construction by creation of salt marsh on nearby dredged material deposit). *Out-of-kind* mitigation involves compensating for the loss of a particular habitat type by replacement with a different habitat type (e.g., creating a salt marsh to mitigate a seagrass bed destroyed by construction of a ferry terminal). The various combinations of mitigation strategies include *onsite in-kind, onsite out-of-kind, off-site in-kind, and off-site out-of-kind*. Preservation of valuable habitats is sometimes used as compensatory mitigation. Yet another alternative involves the use of mitigation banks, large habitat creation projects that developers have the option of contributing funds to in lieu of actually constructing a mitigation wetland. Though many have criticized mitigation banking as a way to skip avoidance and minimization of adverse impacts, in practice, mitigation banking is probably a superior alternative to haphazard construction of small, poorly designed wetland creation projects that are unlikely to be monitored or maintained. In all cases, avoidance and minimization of project impacts are considered preferable to compensatory mitigation.

d. Compensatory mitigation has been criticized and deemed largely unsuccessful in coastal habitats (Race 1985, Zedler 1996a). Restoration of lost ecological functions is difficult to achieve in created wetlands, particularly those that are small and/or isolated and affected by surrounding land use. Even when vastly more habitat area is created than was lost, it may be insufficient to provide functional equivalency to tidal wetlands lost (Zedler 1996b). In recent years, there has been considerable research on measurement and assessment of functional equivalency in restored and created coastal habitats. The results suggest that even in the case of the most well-designed and carefully executed projects, restoration of certain ecological functions may not occur for decades (Simenstad and Thom 1996).

V-7-3b. Habitat Restoration: Issues and Initiatives Beyond Compensatory Mitigation

a. National policy concerning the protection, restoration, conservation, and management of ecological resources encourages initiatives to go beyond traditional compensatory mitigation concepts and develop environmental projects, including coastal habitat improvements, based purely on environmental benefits. Accordingly, ecosystem restoration has become one of the primary missions of the Civil Works program of the U.S. Army Corps of Engineers. The purpose of Civil Works Ecosystem restoration activities is to restore significant ecosystem function, structure, and dynamic processes that have been degraded. Ecosystem restoration efforts will involve a comprehensive examination of the problems contributing to the system degradation, and the development of alternative means for their solution. The intent of restoration is to partially or fully reestablish the attributes of a naturally functioning and self-regulating system. Study and project authorities through which the Corps can examine ecosystem restoration needs and opportunities are found in Congressionally authorized studies, pursued under General Investigations, i.e., new start reconnaissance and feasibility studies for single-purpose ecosystem restoration or multiple purpose projects that include ecosystem restoration as a primary purpose. Other authorities through which the Corps can participate in the study, design, and implementation of ecosystem restoration and protection projects include: (1) Section 1135, Project Modifications for Improvement of the Environment (Water Resources Development Act (WRDA) of 1986, as amended); (2) Section 206, Aquatic Ecosystem Restoration (WRDA) 1996); (3) Section 204, Beneficial Uses of Dredged Material (WRDA 1992, as amended); and (4) dredging of contaminated sediments under Section 312 of WRDA 1990, as amended. All of these authorities can be used to restore coastal habitats and resources are usually appropriated each year under the specific authority to accomplish such work in the Corps' portion of the Water and Energy budget.

b. Additional opportunities for ecosystem restoration and protection may also be pursued through existing project authorities for the management of operating projects, e.g., through water control changes, or as part of natural resources management.

c. All of these authorities, which vary in their particularities related to specific applications of Corps interest, local cost-sharing, multiple agency participation, etc., are potential tools for developing coastal engineering/environmental projects. Additionally, other Federal agencies provide grants to local jurisdictions or citizen groups to plan, design, and/or construct habitat restoration projects. Examples of alternative means of acquiring and restoring coastal habitats include the National Coastal Wetlands Conservation Grant Program, and North American Wetlands Conservation Grants administered by the U.S. Fish and Wildlife Service; the National Fish and Wildlife Foundation Challenge Grants administered by the National Fish and Wildlife Foundation; and the Wetlands Reserve Program administered by the Natural Resources Conservation Service. Partnering with other agencies is also an effective way of leveraging limited assets in an area that usually has more needs identified than resources to accomplish them.

V-7-4. Ecosystem Function and Biodiversity in Coastal Habitat Projects

a. Coastal habitats are part of a connected ecosystem of freshwater, estuarine, terrestrial, and marine habitats. A primary consideration of habitat restoration is what functions a particular habitat provides and how those functions relate to functions provided by other parts of the ecosystem. Additionally, it is important to determine what natural physical processes form and change natural habitats and whether these natural processes still exist in an ecosystem. A primary goal of habitat restoration and creation should always be to maintain or recreate natural physical processes because this will help maintain habitats over a much longer time scale.

b. Coastal habitats also contribute substantially to biological diversity, simply defined as the number of species within a habitat, bioregion, or worldwide. Concern over protecting against loss of species in coastal (and terrestrial) habitats is increasing, and environmentally responsible coastal habitat projects should be planned and implemented with maintenance of biological diversity as a key concern. The primary cause of species loss in many areas is the destruction of critical habitat for one or more life-history stages of an organism. In coastal areas, these may include barrier beaches, maritime forests, salt marshes, tidal wetlands, seagrass meadows, and coral reefs. Pollution and loss of water quality may further stress populations, and may contribute to a reduction in genetic diversity due to differential mortality and local extinctions.

c. In certain cases, preservation of biodiversity also may be a primary goal of a habitat restoration, creation, or preservation project, and many projects have attempted to restore or rehabilitate critical habitats in areas that have undergone extensive habitat loss or degradation.

V-7-5. Defining Success and Project Maintenance and Monitoring

a. A key component of designing and implementing habitat restoration or creation projects is to define how and when your project will be considered a success. This requires identifying a project's goals, objectives, and performance standards. Performance standards should be specifically stated in terms such as composition and density of a plant community or number of fish utilizing a site. While such standards can often be difficult to develop, they will provide a clear direction for what parameters must be monitored and how long monitoring will need to be conducted.

b. The most often cited reason for failure of habitat restoration projects is failure to properly monitor site development and implement corrective action as needed. The National Research Council (NRC 1992) has identified a need to develop a systematic approach to improving the state of the art in habitat restoration. A primary area for improvement is in the development and implementation of monitoring programs. A well-designed monitoring program allows project managers to make crucial changes or mid-course corrections to projects, ensuring long-term success. Monitoring data can be used by project managers to demonstrate the ability of the project to meet stated goals and objectives. Monitoring also allows others to learn from previous projects, and avoid pitfalls in future restoration efforts.

c. Recent guidance on monitoring aquatic and marine habitat restoration projects (Thom and Wellman 1996) outlines the components of a monitoring program. A monitoring program should be designed during the planning phase as a direct result of the project objectives and performance standards. Failure to do so may result in the inability to evaluate project performance relative to the stated standards.

d. Baseline data collection helps in setting clear, realistic objectives, provides site-specific information, and guides the development of the monitoring plan. This is considered the initial phase of the monitoring program, and provides pre-project conditions against which to evaluate project performance. During construction, monitoring is carried out to ensure that project design criteria are followed, and to assess any off-site damage that may occur during construction. Upon project completion, performance is assessed and management or modifications to the project can be carried out, as necessary, to achieve the desired project objectives.

e. Reference site selection is often critical to the development of a restoration monitoring program. Reference sites are used as models upon which to base a project design. They provide a target for the development and evaluation of performance standards. Reference sites also provide a control, useful in assessing the degree of natural ecosystem fluctuations. Pre- and post-project conditions can be assessed in the absence of reference sites; however, project performance can only be determined relative to reference conditions (Thom and Wellman 1996).

V-7-6. Adaptive Management

Two of the key elements of success in habitat restoration/creation projects are: (a) clear, technically sound objectives, and (b) the flexibility and capability to deal with unforeseen problems or physical changes. These elements represent the foundation of an adaptive approach to ecosystem management. In recent years, adaptive management plans have been specifically recommended for ecosystem management programs by Federal and State Governments, and other entities (Thom and Wellman 1996). An adaptive management approach literally involves "learning by doing" - a sequential reassessment of system states and dynamic relationships should be an integral part of a well-designed monitoring program (Walters and Holling 1990). Since our knowledge of ecosystems is often incomplete, project managers must rely on continuous assessment and data collection to guide modifications intended to optimize restoration projects. Data collection, comparison with carefully chosen reference sites, and experimentation, where feasible, should be used to indicate the need for adjustments or modifications to the system (and in some cases reevaluation of original goals and project objectives). This can be instrumental in avoiding the pitfalls that typically result from inflexible project designs and represents a "safe-fail" approach, in contrast to the "fail-safe" approach of traditional civil and coastal engineering projects (Pastorok et al. 1997). However, in order for adaptive management to work, or even be utilized at all, there must be a clear mechanism for the monitoring results to be evaluated and a funding source to come back and provide necessary maintenance or construct actual changes to the project.

V-7-7. Design Considerations and Information Sources for Habitat Projects

a. Underwater projects.

(1) Artificial reefs. Artificial reefs have been used to enhance fishing productivity, in both artisanal and highly developed fisheries, for centuries. The National Fishing Enhancement Act of 1984 authorizes states and other government entities to develop and responsibly manage artificial reef programs in coastal waters of the United States. Since the passage of the Act, reef construction in U.S. coastal waters has increased dramatically. There are now over 300 artificial reefs in U.S. waters, most in nearshore waters of the Southern Atlantic and Gulf coasts.

(a) Artificial reefs enhance marine habitat by providing structurally complex substrate and food resources (in the form of encrusting and epiphytic invertebrates) in topographically homogeneous areas of the ocean floor. The new substrate is quickly colonized by epiphytic algae, sponges, bryozoans, and hydroids. In tropical waters, corals are among the initial colonizers. Ultimately, small fish, crustaceans, and larger predatory fish take up residence. Size, vertical relief, structural complexity and location relative to source areas are the primary factors that determine community composition of artificial reefs. Most artificial reefs are constructed to attract and support populations of recreationally or commercially important finfish; however, reefs have also been constructed to specifically target other resource species such as giant kelp, lobsters, and corals.

(b) Engineering and design concerns in artificial reef construction/deployment focus on the following (Bohnsack and Sutherland 1985):

- Material composition.

- Surface texture.

- Shape, height, and profile of the reef.

- Reef size, spatial scale, and dispersion.

(c) Proper placement of artificial reef structures is of critical importance. Most reported failures of artificial reefs are due to improper siting. If deployed too close to an existing natural reef, they may provide little or no enhancement, and may reduce habitat function of the natural reef.

(d) Artificial reefs may be placed at any depth; most reefs in U.S. waters have been deployed at depths of 10-110 m (30-360 ft). Substrate type must also be considered. Hard substrate must be located at or near the surface in order to prevent the artificial reefs from sinking. Natural sedimentation rates in the placement area should be low enough to prevent burial of structures over time. In high-energy coastal waters, reef structures must often be mechanically stabilized to prevent shifting or relocation by waves and storm surges. Water quality parameters in the vicinity of the reef should be assessed prior to deployment. Chronic low oxygen conditions (hypoxia) and rapid changes in salinity, water clarity, temperature, and nutrient concentration will negatively impact reef fish and invertebrate communities.

(e) A variety of materials have been used to construct artificial reefs. Some of the earliest recorded artificial reefs, deployed by Japanese fishermen in the early 18th century, consisted of simple wooden structures weighted with rocks. Bamboo racks (payaos) have been used extensively in artificial reef programs in the Philippines; in recent years these have been replaced with modern concrete structures. Recycled materials are often used to construct artificial reefs. Discarded automobile tires are particularly common due to their low cost and durability. Fly-ash composites and fiber-reinforced plastics have been used. Wooden street cars, automobiles, barges, and ships have been deployed at artificial reef sites in the United States and elsewhere. Concrete structures of various sizes and configurations are also commonly deployed. Quarry rock and rubble derived from demolition activities, such as channel maintenance, have been used extensively in reef construction. Detailed reviews of artificial reef construction and design are provided by Seaman and Sprague (1991), and Sheehy and Vik (1992).

(f) The effectiveness of artificial reefs as a fishery management tool has been questioned by coastal resource managers. It is recognized that artificial reefs can dramatically enhance fish harvests by concentrating fish around discrete, identifiable structures. However, it is often argued that the reefs merely serve to redistribute existing resources from natural sites to artificial sites, and do not actually increase fish production (Bohnsack 1989). Artificial reefs should be considered habitat creation rather than restoration of a naturally occurring habitat type.

(2) Oyster reefs. Harvests of the American oyster (*Crassostrea virginica*) along the U.S. Atlantic coast have declined dramatically during the latter half of the 20th century. In Chesapeake Bay, which historically supported the largest oyster fishery along the Atlantic coast, the principal cause of the decline was overharvesting of the resource (Wennersten 1981). During the early 20th century, yearly harvests declined rapidly as the fishery became more technologically advanced and efficient. Since the early 1960's, oyster diseases and predation, along with reduced environmental quality, have further contributed to the decline. Currently, the oyster fishery in Chesapeake Bay is threatened with economic extinction.

(a) Oysters are filter-feeding bivalves (Phylum Mollusca, Class Bivalvia). They reproduce by shedding eggs and sperm directly into the water column. Oysters are hermaphroditic; they may be either male or female, and change sex at various times during their life cycle. Larval oysters, or veligers, begin to grow shells at approximately two weeks of age and sink to the bottom. If veligers are not able to settle upon a suitable hard substrate (culch), they die. Upon cementing themselves to hard substrate (rock, shell, another oyster) the juvenile oysters (spat) remain sessile for life, filtering the surrounding water, and forming structurally complex reefs. Because they are stationary for life, oysters are susceptible to a wide variety of predators, including gastropod oyster drills, whelks, crabs, boring sponges, and starfish. Most invertebrate oyster predators thrive in salinity above 25 psu. In recent years, oyster populations in the mid-Atlantic have been severely impacted by disease microorganisms. While not harmful to man, these diseases have seriously reduced oyster harvests in Chesapeake Bay and other mid-Atlantic estuaries.

(b) The primary means of increasing oyster populations is by providing additional hard substrate for veligers to settle on. Typically old oyster shells are used, but in some coastal areas this material has become scarce and alternatives such as clamshells, concrete, rubble, or fly-ash (not recommended) composites are used. In recent years, coastal engineers have created experimental oyster reefs by depositing dredged material in areas historically known to support oyster populations, and capping the dredged material mounds with a layer of oyster shell (Earhart, Clarke, and Shipley 1988). This technique holds considerable promise as a tool for enhancement of shellfish resources and as a viable alternative to traditional dredged material disposal options in shallow coastal waters.

(c) Restoring *ecological functions* attributed to oyster reefs, including water-filtration capacity and sediment stabilization, is more likely to provide a self-sustaining ecosystem. Oyster reefs provide structurally complex refuge for a variety of fish and invertebrate species, including many of commercial and recreational significance. The restoration technique is the same; provision of additional hard substrate to increase survivorship of newly recruiting juvenile oysters. However, in the case of restoring oyster reefs as *habitat*, careful attention is given to the structural characteristics of the shell deposits and orientation/spacing of the mounds in order to derive maximum use of the resource by target finfish and macrocrustacean species.

(3) Coral reefs. Coral reefs are very productive ecosystems and occur worldwide along shallow tropical coastlines. The majority of coral reefs occur between 22.5°N and 22.5°S latitude in the tropical Western Pacific and Indian Oceans, and the Caribbean Sea. However, warm ocean currents such as the Gulf Stream in the western Atlantic and the Kuroshio Current in the western Pacific allow limited growth of coral reefs as far north as 34°N latitude (Maragos 1992). Coral assemblages found in waters of the continental United States (South Florida) are dominated by four species: the elkhorn coral (*Acropora palmata*), staghorn coral (*Acropora cervicornia*), common star coral (*Montastraea annularis*), and large star coral (*Montastraea cavernosa*) (Jaap 1984).

(a) Coral reefs are classified according to geomorphic attributes. The most common type of coral reef is the *fringing reef*. Fringing reefs occur slightly below low tide level, occasionally extending into the intertidal zone, and are situated parallel to the shoreline. The proximity of fringing coral reefs to shore increases their susceptibility to human-induced environmental degradation. *Patch reefs* are isolated coral reefs situated shoreward of larger, offshore reefs. Offshore *barrier reefs* are linear in configuration, arising from an offshore shelf, and separated from the mainland by a lagoon. The Great Barrier Reef system of Australia is a well-known example of this reef type. *Atolls* are circular or semicircular reefs that are situated atop subsiding sea floor platforms. They maintain their position above the water column by vertical accretion as the platform subsides.

(b) Coral reefs are the product of a symbiotic association of corals (a colonial invertebrate) and the microscopic vegetative stage of dinoflagellates (zooxanthellae) living within the tissue of the coral animal. Zooxanthellae are responsible for the bulk of primary production within the reef system. The coral animal and zooxanthellae remove calcium ions from the surrounding seawater and incorporate them into the coral skeleton, forming the reef. A variety of invertebrates and fish utilize the structurally complex reef habitat as a predation refuge and nursery. Because corals grow in oligotrophic waters, nutrients such as nitrogen and phosphorous are recycled within the system. Gross primary production is high due to the presence of the zooxanthellae, but net primary productivity is low due to high rates of respiration within the reef community. Coral reefs function as a sink for nutrients originating from outside the system.

(c) Maragos (1992) lists a number of "functional values" attributed to coral reefs. These include:

- Provision of a food source.

- Shoreline protection.

- Sand and mineral extraction.

- Habitat for rare species.

- Tourism and recreation.

- Scientific, medicinal, and educational resources.

(d) Coral reef ecosystems are fragile and have undergone considerable losses in recent years. Sources of environmental degradation in coral reef systems include mechanical destruction due to channel dredging, vessel groundings, anchors, dynamite fishing, collection for the aquarium industry and trampling by recreational divers and snorkelers.

(e) Coral reef restoration is still very much an experimental technology, and only a few case studies have been documented (Maragos 1992). Coral reef restoration typically occurs in two phases. First, before any physical reconstruction can take place, water quality conditions in the vicinity of the degraded reef must closely approximate that of the system before degradation occurred. Optimal salinity, temperature, water clarity, and hydrological conditions must be met prior to any attempt to mechanically repair or enhance the reef structure. This may be accomplished by cessation or diversion of sewage discharges and runoff, and reestablishment of natural flow regimes in the area being considered for restoration. When the environmental parameters listed above are conducive to reef development, then mechanical repair or restoration of reef structures can be implemented. In the case of physical damage due to ship grounding, it is imperative to salvage as many living corals and sponges as possible. The living surfaces of hard corals die rapidly if allowed to sit on the bare seafloor. Rubble associated with the damaged reef is typically removed and discarded offsite in deep water; occasionally it is feasible to reconstruct portions of the reef structure using cements or epoxies. If the underlying limestone reef platform is damaged, it is often necessary to reconstruct the substrate using limestone boulders. When the underlying reef platform has been reconstructed and the site cleared of debris and rubble, surviving corals and sponges are transplanted using a variety of techniques, including stainless steel wires, nylon cable ties, and epoxy/cement (Maragos 1992).

(4) Live bottom and worm rock reefs. Live bottom habitats are common along all of the coastlines of the United States. They are particularly prevalent along the western coast of Florida and along the North and South Atlantic Coast. Live bottom consists of hard subtidal habitats colonized by sessile and mobile invertebrates, including sponges, hydroids, bryozoans, crustaceans, echinoderms, mollusks, and polychaetes. Typically, these organisms live directly upon submerged rock or fossil reefs. Macroalgal communities (e.g., kelp beds) may dominate in some hard bottom areas. Hard bottom in the surf zone is dominated by low-relief boring and encrusting organisms. As depth increases with distance from shore, species richness and vertical relief increase. Live bottom habitats are known to attract fish and mobile invertebrates, which use the vertical relief as a predation refuge, and prey upon the various organisms that comprise the live bottom community.

EM 1110-2-1100
31 Jul 03

In general, nutrient concentrations in the vicinity of live bottom habitats are low, and net primary productivity is also low, due to respiration by resident organisms.

(a) Worm rock reefs are a specific type of hard bottom habitat found along the southeastern Florida Coast (Zale and Merrifield 1989). These reefs are composed of the tubes of the polychaete *Phragmatopoma lapidosa*. The worms construct elaborate, complex aggregations of tubes by cementing sand grains. These structures may extend for hundreds of kilometers. A diverse assemblage of motile and sessile invertebrates, including many other species of polychaetes and crustaceans, are found in association with the worm reefs.

(b) Live bottom habitats are particularly susceptible to changes in sedimentation rates associated with storms, dredging, and artificial beach nourishment. Invertebrate faunas can adapt to periodic burial due to natural processes, but chronic or persistent burial resulting from construction or coastal engineering activities will destroy live bottom. Disturbances due to commercial fishing, such as trawling, are detrimental to live bottom communities. Vessel groundings and anchor scars may also cause permanent physical damage to live bottom habitat.

(c) There are few documented instances of live bottom restoration other than coral reefs (Maragos 1992) and kelp beds (Schiel and Foster 1992). However, many of the general guidelines for coral reef restoration can be applied to restoration of live bottom. Assuming appropriate hydrodynamic and water-quality regimes, restoration involves removal of debris and mechanically rebuilding damaged structures. In the case of worm rock, fragments may be removed from nearby source areas and transplanted onto damaged reefs to speed recolonization.

(5) Seagrasses. Seagrasses are submerged, flowering marine angiosperms, of which approximately 35 species are known worldwide. Seagrass beds occur mainly in low-energy subtidal and intertidal habitats along the Atlantic, Pacific, and Gulf coasts of the United States, with species composition and areal extent varying greatly along each of these coasts. Along the Atlantic coast, eelgrass (*Zostera marina*) beds occur from the Canadian Maritime Provinces south to Albemarle-Pamlico Sound in North Carolina. Along the Pacific coast, eelgrass beds occur from Mexico to Alaska in bays and inlets. Turtle grass (*Thalassia testudinum*) dominates along the east coast of Florida, the Florida Keys, and the Gulf of Mexico, along with shoalgrass (*Halodule wrightii*) and manatee grass (*Syringodium filiforme*). In deeper waters off the Florida continental shelf, two other species (*Halophila engelmanni* and *H. decipiens*) occur. Seagrass communities of the U.S. Pacific coast are comprised of three species of surfgrass (*Phyllospadix scouri*, *P. torreyi*, and *P. serulatus*) and two species of eelgrass (*Zostera marina* and *Z. japonica*).

(a) Seagrass beds are critical nursery areas for many recreational and commercial fishery species, including bay scallop (*Argopecten irradians*), summer flounder (*Paralichthys dentatus*), and blue crab (*Callinectes sapidus*). On the Pacific coast, eelgrass meadows provide nursery habitat for dungeness crab (*Cancer magister*), english sole (*Parophrys vetulus*) and starry flounder (*Platichthys stellatus*). Juveniles of these and other fishery species are afforded refuge from predators and benefit from abundant food within the complex seagrass canopy. Eelgrass beds are also important as spawning substrate for bait fish species such as herring (*Clupea pallasi*).

(b) Critical environmental parameters for seagrass beds include wave energy, salinity, temperature, water clarity, and nutrient concentrations. Depth and water clarity exert the primary controls over seagrass zonation and the degree of colonization by epiphytes. Redistribution of sediments by waves and storm surges can severely impact seagrass beds. Diseases can have a catastrophic effect on seagrass communities. During the 1930's, widespread infection by the slime mold *Labryinthula macrocystis* decimated Atlantic coast eelgrass populations (Short, Muehlstein, and Porter 1987). Many coastal areas have not yet recovered from this "wasting disease."

Coastal Engineering for Environmental Enhancement

V-7-9

(c) Seagrass beds are susceptible to an array of human-induced degradations. Dredge and fill operations associated with navigation channel maintenance have taken a toll. Deterioration of water quality conditions associated with human population in coastal areas remains a primary cause of seagrass bed degradation. Physical damage to seagrass beds may be caused by recreational boating in shallow waters. This type of chronic disturbance is common in populated areas and is persistent; turtle grass beds in Florida Bay may require upwards of a decade to recover from propeller scarification (Zieman 1976).

(d) There has been considerable interest and effort expended to devise effective methods of creating or restoring seagrass beds, primarily by transplantation. Most efforts have failed, usually because site conditions are not suitable. The parameters of the transplant site must closely match those of the donor, or reference site, if restoration is to be successful.

(e) The earliest recorded eelgrass transplant efforts along the U.S. Atlantic coast were documented by Addy (1947). The first successful transplantation of turtle grass was reported by Kelly, Fuss, and Hall (1971) in Boca Ciega Bay, Florida. Attempts to reestablish subtropical seagrasses on dredged material deposits in Port St. Joe, Florida, are described by Phillips (1980).

(f) Thorhaug and Austin (1976) discuss types of coastal engineering projects that could benefit from transplantation of seagrass beds. These include (a) dredging of canals where both sides can be replanted, (b) stabilization of the shallow subtidal and intertidal portions of dredged material islands, and (c) other miscellaneous dredge and fill impact projects (road, bridge construction, marinas, etc.). Transplantation of turtle grass, manatee grass, and shoalgrass beds in Biscayne Bay, Florida, and other locations in the Caribbean are documented by Thorhaug (1985, 1986). Darovec et al. (1975) described transplant techniques for seagrasses along Florida's west coast. Lewis (1987) reviews seagrass restoration in the southeast United States, and discusses reasons for failures, as well as success, using well-documented case studies in south Florida.

(g) Recent efforts to reestablish eelgrass in lower Chesapeake Bay are documented by Moore and Orth (1982). Thom (1990) reviewed eelgrass transplanting projects in the Pacific Northwest and Wyllie-Echeverria, Olson, and Hershman (1994) documented the state of seagrass science in policy in the Pacific Northwest.

(h) Planting techniques, along with cost and labor estimates for establishment of eelgrass, shoalgrass, manatee grass, and turtle grass on dredged material and other unvegetated substrates are documented by Fonseca, Kenworthy, and Thayer (1987). Fonseca (1994) reviews all aspects of seagrass restoration, including planting guidelines and monitoring programs for the Gulf of Mexico; however, this information is applicable to seagrass restoration in general.

(i) A variety of transplant methods have been used to restore seagrasses, including broadcast seeding, seed tapes, stapling of individual plants, and use of "peat pots" or sediment plugs containing whole plants. The latter method appears to be most successful (Fonseca 1994). Fertilizer applications have been tried in some instances, although performance has been inconclusive (Fonseca 1994). Careful attention must be paid to spacing of individual planting units in order to achieve site coalescence. Subtropical seagrass beds in Florida Bay and the eastern Gulf of Mexico have achieved coalescence in as little as 9 months, or as long as 3-4 years, depending on planting distance between individual units. In high-energy areas, beds may never fully coalesce.

(j) Careful monitoring is critical to the success of any seagrass restoration project. Performance indicators should include survival rates of planting units, areal coverage, and number of shoots per planting unit. Fonseca (1994) recommends quarterly monitoring intervals during the first year following planting and semi-annual intervals for the next 2 years. Areal coverage should be monitored in successive years.

Assessment of ecological functions (e.g., nutrient cycling, primary production, utilization by benthic invertebrate and fishery species), should also be conducted, and more recent studies have focused on restoration of specific functions of seagrass beds, including the ability of seagrasses to modify their surrounding hydrodynamic environment (Fonseca and Fisher 1986) and their potential for utilization as a nursery by fishes and invertebrates (Homziak, Fonseca, and Kenworthy 1982, Fonseca et al. 1990).

(6) Use of dredged material for creating shallow habitat. In the face of increasing restrictions on open-water disposal of dredged material and limited capacity in existing disposal facilities, coastal engineers have proposed placing dredged material in nearshore waters. In highly developed bay and harbors where shallow intertidal or subtidal habitat is limited, the placement of clean dredged material can create or restore important nearshore habitat. Shallow coastal waters are often home to seagrass beds, macroalgae beds, and other communities.

b. Projects at the land-sea interface.

(1) Mud/sand flats. Intertidal mud and sand flats are a conspicuous coastal habitat type present along all coasts of North America. They are most abundant and expansive in high tidal range areas such as Puget Sound and the New England coast. A variety of fish and invertebrate species, many of commercial and recreational importance, depend on intertidal and shallow subtidal unvegetated marine habitat, particularly during early life stages. Along the U.S. Atlantic coast these include the sandworm (*Nereis virens*) and bloodworm (*Glycera dibranchiata*). These two species represent an important bait industry in the northeast (Wilson and Ruff 1988). Bivalves, which occupy mud and sand flats along the Atlantic coast, include the hard clam (*Mercenaria mercenaria*) and softshell clam (*Mya arenaria*); both are harvested commercially. On the west coast, several commercially important species are associated with mud and sand flats, including the Pacific razor clam (*Saliqua patula*), littleneck clam (*Protothaca saminea*), pismo clam (*Tivela stultorum*), Pacific oyster (*Crassostrea gigas*) and dungeness crab (*Cancer magister*).

(a) Mud and sand flats are also very productive for benthic and epibenthic invertebrates and algae. They are an important source of nutrients to the entire coastal ecosystem. In virtually all estuarine and coastal areas, mud and sand flats are important forage sites for a myriad of resident and migratory waterfowl as well as wading bird species, which feast on the abundance of invertebrate prey items (worms, small crustaceans, bivalves) available at low tide.

(b) Restoration and creation of unvegetated intertidal habitats has not received the level of attention given to restoration/creation of vegetated intertidal habitats, such as salt marshes and mangrove forests. However, deposition of fine dredged material in shallow coastal waters may inadvertently result in the creation of intertidal mud and sandflats. Many such artificial habitats were created prior to the implementation of the National Environmental Protection Act (NEPA) and therefore, are not well-documented. A study by the U.S. Army Engineer Division, New England and ERDC compared benthic invertebrate community dynamics at a recently constructed mudflat with a nearby natural mudflat at Jonesport, Maine (Ray et al. 1994). A diverse infaunal assemblage was present 2 years after construction and included commercially important species such as sandworms and softshell clams. As with most shallow-water habitat creation and restoration projects, habitat tradeoffs must be considered. In certain geographic areas (e.g., coastal New England) creation of intertidal mud or sand flats may represent an attractive beneficial use alternative to conventional dredged material disposal. Mud and sand flats can also be restored or created by the removal of fill material that may have been placed in the nearshore zone to create uplands for development. When industrial areas adjacent to the water are abandoned, it provides a perfect opportunity to remove fill and recreate aquatic habitat. Another technique used in highly modified shorelines (such as ports) in the Pacific Northwest includes the creation of intertidal benches that are surrogates for once prevelent mudflats. Long linear stretches of riprap bankline are altered to accommodate a rock crib that is lined with filter fabric. These cribs are filled with fine-grained material (50 percent sand, 25 percent silt, and 25 percent clay) and become sites for benthic recruitment and algae attachment. These intertidal benches are also placed at specific tidal elevations for epibenthic production as well. As juvenile fish migrate downstream to the ocean, they follow the shoreline through the developed port areas and can feed during their migration.

(2) Intertidal salt and brackish marshes. Intertidal marshes are found in all coastal areas of the United States, except for the southern half of Florida, where they are replaced by mangroves; and certain areas of the Western Gulf of Mexico, where they are replaced by wind-tidal flats. They are a conspicuous landscape feature along the gently sloping Atlantic coastal plain, from New England to east-central Florida, in association with drowned river-valley estuaries and back-barrier lagoons. Intertidal marshes are ubiquitous in the lower Mississippi Delta and back-barrier systems of the Gulf of Mexico, ranging from west-central Florida to Southwest Texas. Intertidal marshes are found along the Pacific coast from Baja, Mexico to Alaska, but are less extensive than those along the Atlantic and Gulf coasts. Large Pacific coast estuaries (e.g., San Francisco Bay and Puget Sound) historically supported large areas of intertidal marsh, but have experienced dramatic losses. In the past, the losses were primarily due to the building of dikes and reclamation of the marshes as farmland, but more recently have primarily been due to dredge and fill activities.

(a) Intertidal marshes occur along the entire estuarine salinity gradient from tidal freshwater (<0.5 psu) to polyhaline (>20 psu) conditions (Figure V-7-1). Ecological functions attributed to intertidal wetlands include shoreline stabilization, storage of surface waters, maintenance of surface water and groundwater quality, and the provision of nursery habitat for estuarine-dependent finfish and shellfish species.

(b) Vegetation communities in intertidal wetlands are dominated by grasses (Poaceae), rushes (Juncaceae), or a combination of the two. Variability in environmental factors (e.g., nutrient availability, duration and depth of intertidal flooding, and pore water salinity) limits plant species diversity in intertidal salt and brackish marshes. The spatial extent of the major zones of intertidal marsh vegetation is largely determined by elevation and its effect on the tidal flooding regime.

(c) Intertidal marshes provide habitat for a variety of organisms, including many commercially important fish species. Examples of marsh-dependent fish and crustaceans along the Atlantic and Gulf coasts include blue crab, brown, white and pink shrimp (*Penaeus* spp.), red drum (*Sciaenops ocellatus*), and spotted sea trout (*Cynoscion ocellatus*). Early life stages of these organisms are afforded refuge from predators and benefit from abundant prey resources in shallow tidal marsh habitats. Prey species include various killifishes (*Fundulus* spp.) and caridean shrimp (*Palaemonetes* spp.) commonly encountered on the vegetated intertidal marsh surface and in shallow subtidal creeks and pools. Characteristic marsh invertebrates include fiddler crabs (*Uca* spp.), amphipod and isopod crustaceans, terrestrial insects and arachnids, various bivalve and gastropod mollusks, and annelids. Wading birds, such as egrets and herons, prey upon resident fishes and invertebrates of intertidal marshes. Many other birds, both arboreal and aquatic, feed and nest in upper intertidal marsh habitats. A variety of mammals, including deer, fox, raccoon, and otter, use intertidal marshes for foraging, breeding, and refuge.

(d) The restoration and creation of intertidal salt and brackish marshes have received much attention in coastal engineering. This is likely due to the considerable acreage of salt marsh that has been lost along U.S. coastlines, recognition of the functions provided by intertidal marshes, and the relative ease in which salt marsh vegetation can be propagated upon dredged material. Restoration of tidal marsh environments may involve removal or breaching of dikes and berms, or installation of culverts under roads to reestablish the natural tidal prism. Hydrologic restoration of tidal marshes has occurred in New England (Sinicrope et al. 1990), the mid-Atlantic (Shisler 1990), central Florida (Rey et al. 1990), central and southern California (Niesen and Josselyn 1981), and the Pacific Northwest (Frenkel and Morlan 1991). Reestablishing tidal conduits increases accessibility of previously impounded or restricted wetlands to estuarine-dependent finfish and wildlife. Invasive plant species such as common reed (*Phragmites australis*), which often predominate in hydrologically altered wetlands, may be controlled via reestablishment of historical tidal regimes (Roman, Niering, and Warren 1984).

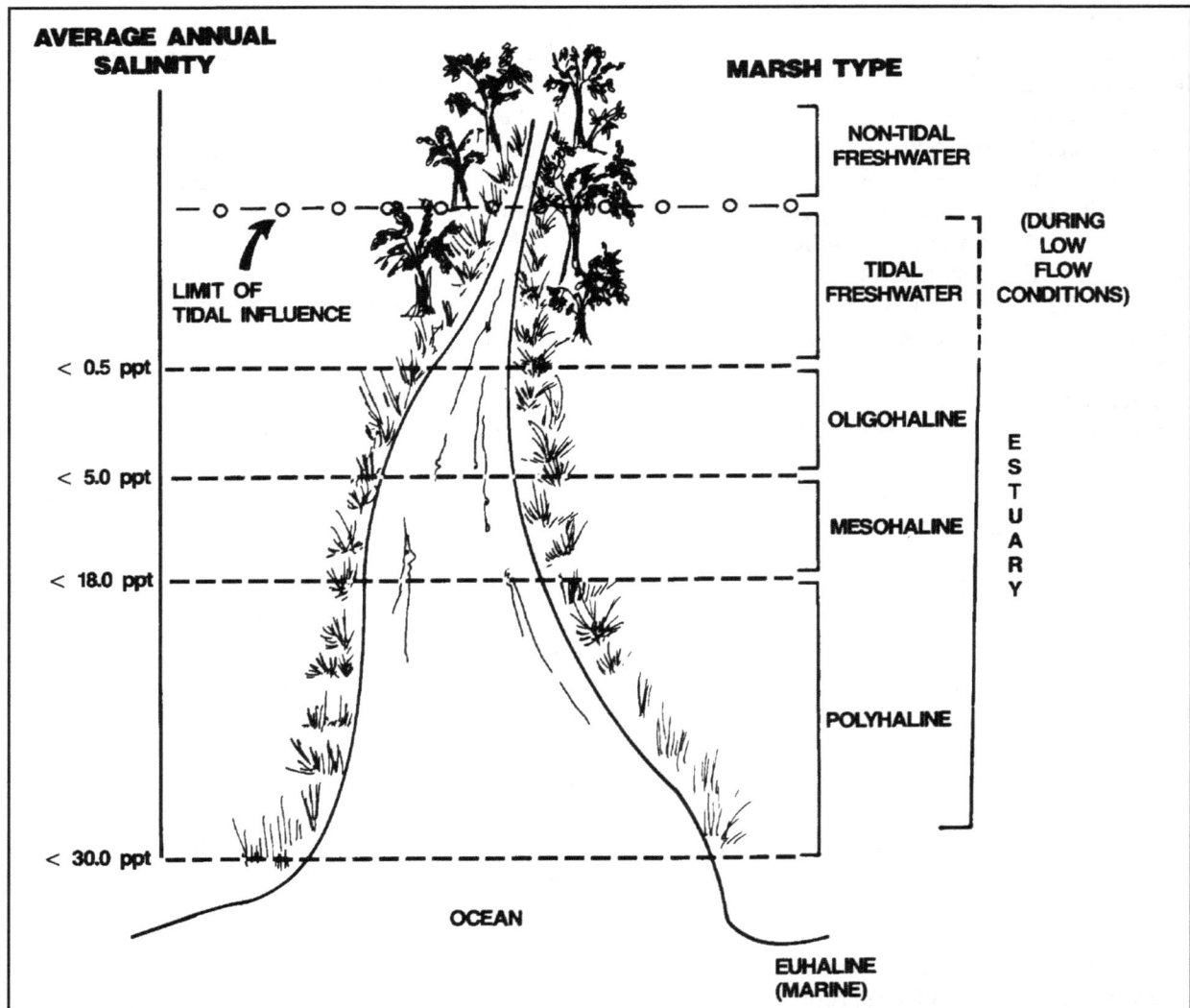

Figure V-7-1. Distribution of tidal marsh type along an estuarine salinity gradient

(e) Marsh *creation* is often a component of the restoration process, especially in projects involving the removal of fill and/or regrading of adjacent uplands to intertidal elevations. However, it is important to recognize that marshes can often be created in upland or shallow subtidal areas that have not historically supported intertidal vegetation.

(f) Techniques for establishing *Spartina alterniflora* marshes on dredged material deposits along the south Atlantic coast were pioneered by researchers at North Carolina State University in the late 1960's to early 1970's (Seneca 1974, Seneca et al. 1976). The objective of these early studies was to provide stabilization of shorelines and dredged materials, and to recoup some of the losses to coastal habitats that had occurred as a result of human population growth in coastal areas. The U.S. Army Corps of Engineers Dredged Material Research Program (DMRP) pioneered large-scale salt marsh establishment on all three U.S. coastlines in the 1970's (Barko et al. 1977, Smith 1978). Tidal marshes established under the DMRP were monitored from 1974-1987 (Landin, Webb, and Knutson 1989). Parameters studied include plant propagation success, shoreline stabilization properties, and utilization by fish and wildlife species. Successive research has focused on refining the techniques developed by the DMRP, and in recent years, increased attention has focused on replication of ecological function in created or restored intertidal marshes.

(g) Tidal marsh creation/restoration, with emphasis on planting techniques, is discussed by Broome, Seneca, and Woodhouse (1988). Darovec et al. (1975) provide guidance on planting salt marsh vegetation, including information on handling of seed stock, transplant units, planting intervals, and elevation requirements, with emphasis on the Florida coast. Roberts (1991) evaluated the habitat value of 22 created coastal marshes in northern and central Florida, concluding that the basic habitat requirements of fish, bird, and mammal species that use natural marshes in this region were being met by the majority of the man-made habitats. Marsh shape and size were critical features determining the degree of use by fish and wildlife species. Minello, Zimmerman, and Medina (1994) determined that geomorphic features such as tidal creek edges strongly influenced the abundance and distribution of fishes and invertebrates on the surfaces of intertidal marshes created from dredged materials in Galveston Bay, Texas.

(h) Salt marsh creation/restoration has been conducted at a number of sites in central and southern California during the last two decades. There has been considerable debate on the benefits of marsh creation/restoration as mitigation for destruction of natural wetlands along the west coast and elsewhere (Race 1985, Zedler 1996a). Josselyn and Bucholz (1982) provide a detailed overview of marsh restoration projects in the San Francisco Bay estuary, including case histories and monitoring studies. Zedler (1988) reviews salt marsh restoration in Southern California, and discusses the importance of hydrologic concerns, and the value of experimentation in planning and monitoring marsh restoration projects. Several "lessons" from restoration efforts in California are outlined, including the importance of planning for the maintenance of rare or endangered plant and animal species in tidal marsh restoration projects, and a recommendation against the use of offsite mitigation for development projects that impact natural wetlands. Recently, the San Francisco District has taken dike breaching to large scale. The Sanoma Baylands project incorporated both levee breaching of an old salt evaporation pond and beneficial uses of dredged material. The salt pond had been in use for almost 100 years and as a result the original marsh elevation had greatly subsided. In order to offset the subsidence, dredged material was spread out over the 117-ha (289-acre) project area to reestablish tidal marsh elevations. Two other larger scale dike breaches are planned for Hamilton Army Airfield (a 364-ha (900-acre) abandoned facility) and Napa Marsh. Similar projects also have taken place in the Pacific Northwest. Trestel Bay, a project in the Portland District, restored over 202 ha (500 acres) by a dike breach in five separate locations. The ability of created/restored tidal wetlands to perform the functions attributed to natural tidal wetlands is addressed by Simensted and Thom (1996) in their study of a restored brackish intertidal wetland in the Puyallup River, Washington. These authors contend that only a few of 16 functions investigated displayed a tendency toward equivalency with natural tidal wetlands in the Pacific Northwest in the first several years subsequent to construction. Natural variability among reference sites was cited as an impediment to assessing the degree of functional equivalence between restored and natural marshes.

(3) Tidal freshwater wetlands. Tidal freshwater wetlands occur along the upper reaches of rivers characterized by moderate to strong tidal influence. They are most extensive along the Atlantic coast between Georgia and New England, especially in the mid-Atlantic/Chesapeake Bay region and along the coastal rivers of South Carolina and Georgia. Tidal freshwater wetlands are also common in the Pacific Northwest, where considerable freshwater influence and high-amplitude tides prevail (Odum et al. 1984). The vegetation community of tidal freshwater marshes is diverse, in comparison to salt and brackish marshes (Figure V-7-2). Rarely does any one species dominate, and notable changes in plant community structure can occur within a single growing season. Freshwater tidal marshes can be dominated by emergent herbaceous species, shrubs, or trees (particularly in the Pacific Northwest), or a combination of all three. As with salt and brackish marshes located downstream, tidal freshwater marshes support populations of both resident and migratory fishes, many of which have recreational or commercial significance, including salmon and trout (*Oncorhynchus sp.*), striped bass (*Morone saxatilis*), yellow perch (*Perca flavescens*), and black bass (*Micropterus* spp.). Wildlife, including mammals, reptiles, and amphibians reside or forage in tidal freshwater marshes. Many arboreal and aquatic bird species are temporary or year-round residents of tidal freshwater marshes.

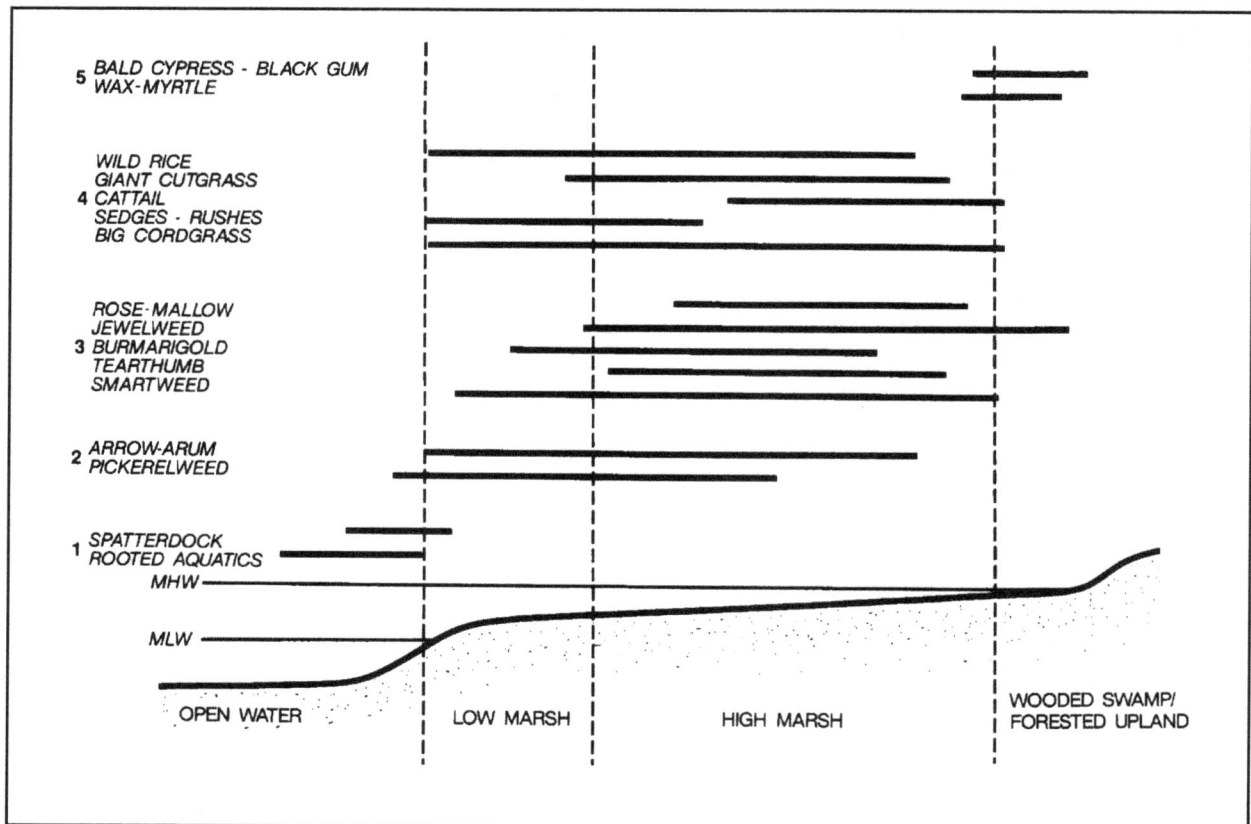

Figure V-7-2. Characteristic profile of a mid-Atlantic tidal freshwater marsh

(4) Tidal freshwater marsh restoration. Restoration and creation of tidal freshwater marshes is technically similar to that of salt and brackish marshes. One of the earliest of the Corps' DMRP wetland creation efforts was the Windmill Point Marsh project conducted in the James River, Virginia (Lunz et al. 1978; Landin, Webb, and Knutson 1989). In 1974, fine-grained dredged material from the James River was used to construct an 8-ha (20-acre) island surrounded by a temporary sand dike. Upon completion of the island, the dike was breached to allow natural formation of tidal channels. Vegetation colonization occurred within one growing season without planting, attesting to the value of seed banks in tidal freshwater sediments. Information collected on fisheries use and benthic invertebrate communities at Windmill Point represents some of the most comprehensive data available on faunal utilization of tidal freshwater habitats. Ultimately, much of the original island site was lost due to erosion and subsidence; however, a spatially complex system of intertidal marsh and shallow subtidal habitat persists in providing nursery habitat for resident and migratory fish and wildlife. Several smaller tidal freshwater marsh restoration projects have also been conducted in upper Chesapeake Bay (Garbisch and Coleman 1978).

(a) Another DMRP project involving the creation/restoration of tidal freshwater wetlands is the Miller Sands Island habitat development project in the lower Columbia River, near Astoria, Oregon (Clairain et al. 1978). This large island/wetland complex was also constructed in 1974 and monitored extensively to document vegetation and soils development, and utilization by fisheries and wildlife (Landin, Webb, and Knutson 1989). This represents one of the few published efforts to date documenting faunal utilization of tidal freshwater wetlands in the Pacific northwest.

(b) On the west coast, a common method of restoring tidal freshwater wetlands is to breach dikes or levees constructed for farmland creation in the past. The Sacramento District breached dikes at Cache Slough to create approximately 8.9 ha (22 acres) of freshwater marsh habitat (Stevens and Rejmankova 1995). The Seattle District restored over 162 ha (400 acres) of freshwater wetlands to tidal influences in a WRDA 1986, Section 1135 project. The Deepwater Slough project used explosives experts from the Army's 168th Division to detonate charges to create some of the breaches. Forested tidal freshwater wetlands are relatively rare habitats and should be considered an important habitat to restore. Unfortunately, it may take 25-100 years, as they are not yet mature.

(5) Mangroves. Mangroves are woody trees and shrubs of the family Rhizophoracea, and represent a tropical/subtropical analog to herbaceaous intertidal vegetation of temperate regions. In the United States, mangroves occur primarily in southern Florida, especially along the southwest coast, and in scattered locations in Louisiana, Texas, and Hawaii.

(a) Like salt marshes, mangrove forests provide critical nursery and foraging habitat for resident and transient fish populations, many of which are recreationally and commercially significant, including red drum (*Sciaenops occelatus*), tarpon (*Megalops atlantica)*, and Snook (*Centropomus undecimalis*). A variety of wildlife and avifauna, including American alligator (*Alligator mississippiensis*), American crocodile (*Crocodylus acutus*), Roseate spoonbill (*Ajaia ajaja*), white ibis (*Eudocimus albis*) and several species of herons and egrets use mangroves as refuge and breeding habitats. Important tropic linkages have been established between mangrove forests and adjacent ecosystems, such as seagrass beds and coral reefs. Mangroves, like intertidal marshes, provide shoreline stabilization. Mangroves intercept and retain nutrients moving downstream and maintain water quality via filtration of tidal surface waters.

(b) Loss and degradation of mangrove forests has resulted from hydrologic alteration, industrial and agricultural land use, dredge and fill activities, and, in some areas, direct harvest for wood products (Cintron-Molero 1992).

(c) Mangrove restoration. Mangrove restoration is typically accomplished by transplanting individual propagules (seedlings) along unvegetated intertidal shorelines. Important factors to consider in attempting to transplant mangroves include plant size and source of donor plantings, salinity, shoreline energy, and tidal flooding depth (elevation). The latter factor is of considerable importance, particularly when transplantation is being conducted on dredged material deposits, which may settle over time. Commonly cited reasons for failure of mangrove restoration efforts include excess wind/wave energy at the transplant site, improper hydrologic regime (inadequate tidal flushing), and failure to replace planting units lost to mortality (Cintron-Molero 1992).

(d) The earliest documented efforts to restore mangroves date back to the early 1970's in Florida. There are anecdotal reports of earlier attempts to transplant mangroves for soil stabilization in the early 1900's (Pulver 1976). Teas (1977) describes the life history of various mangrove species in Florida, with implications for their restoration. Pulver (1976) describes transplant techniques for red, black, and white mangroves in southwest Florida. Darovec et al. (1975) provide detailed planting guidelines for mangroves in South Florida, including information on elevation requirements, planting unit height, age and spacing of planting units, soil types, and fertilization. Lewis (1982) discusses a variety of mangrove restoration projects from Florida and the U.S. Virgin Islands, with recommendations for improving the success of mangrove projects. A more recent assessment of the state of the art in mangrove restoration is provided by Cintron-Molero (1992) with examples from Puerto Rico and other Caribbean locales.

(6) Rocky intertidal shores. Rocky intertidal shorelines are found worldwide, primarily in high-energy littoral environments. Biotic assemblages of rocky intertidal shores include macroalgae, particularly brown algae (*Fucus* spp.); various mollusks, including limpets, mussels, and gastropods; and barnacles. The complex interactions among these organisms along rocky shores of the east and west coasts of North America have been the subject of numerous experimental studies in the last several decades, forming the basis for

much of our understanding of the structure and function of shallow marine communities (Connel 1972, Paine and Levin 1981). Shallow rocky habitats along the U.S. west coast are used as spawning sites for juvenile fishes, including some commercially important species such as Pacific herring (*Clupea harengus pallasi*). Shorebirds forage extensively along rocky shorelines and in tidepools. Marine mammals such as sea lions, otters, and seals breed and reside along rocky shores.

   c. *Projects in coastal uplands.*

   (1) Coastal dunes. Coastal dunes are highly dynamic sand deposits located landward of beaches. They occur along all coasts of the United States, including the Great Lakes. They are especially common along the Atlantic and Gulf barrier island shores, and also dominate in some Pacific coast areas (California, Oregon). Dunes supply sand to beaches during erosive storm events, and act as buffers to wave energy. Removal of dunes from coastal areas can result in significant economic loss from damage to homes, businesses, and natural areas (Woodhouse 1978).

   (a) In general, dunes and other transient lands such as spits are highly desired by developers, because of their proximity to the shoreline. This practice should be strongly discouraged as the forces causing the dynamic nature of the land masses do not cease once houses are built on a spit, etc. Engineering structures such as steel sheet piles and stone revetments are a temporary solution and frequently cause harm to downdrift properties.

   (b) Dunes may be vegetated, and, therefore, relatively stable, or they may be naturally unvegetated. Three types of vegetated dunes are recognized:

   •   *Foredune Ridges* are linear, low-amplitude sand ridges oriented parallel to the beach.
       These are common along the Atlantic and Gulf coasts.

   •   *Parabolic Dunes* are sparsely vegetated, U-shaped dunes, open to the direction of prevailing winds.
       These generally form behind foredunes and are common along the lower shoreline of Lake Michigan,
       Cape Cod, and northwest Florida.

   •   *Precipitation Dunes* are blown sand deposits which intrude onto adjacent forested uplands. These
       commonly occur along the U.S. Pacific coast.

   (c) Unvegetated dunes are the most dynamic and also vary considerably in morphology. *Transverse Dunes* are sand deposits that lie perpendicular to the prevailing winds. *Longitudinal Dunes* are oriented parallel to prevailing winds. *Transverse Dunes* are rapidly migrating dunes that move landward in response to prevailing winds. Transverse dunes may be straight or sinuous in shape and may persist for up to 1 km (0.6 mile) in length.

   (d) Wave climate and local wind regimes are the primary physical factors responsible for dune establishment. Littoral processes (longshore drift) and storm-associated overwash events provide the sand; dune morphology is determined by wind climate (the direction of prevailing winds relative to the orientation of the beach).

   (e) Characteristic perennial grass species that grow on dunes are instrumental in maintaining dune integrity and stabilization function. Sea oats (*Unionicola paniculata*) are the dominant dune grass along most of the south Atlantic and Gulf coasts. This and other dune-building species are able to withstand high salinity, wind, evaporation, and periodic burial by drifting sand. Other dune-building plants include American beach grass (*Ammophila breviligulata*), bitter panic grass (*Panicum amarum*), saltmeadow cordgrass (*Spartina patens*), and seacoast bluestem (*Schizachyrium maritinum*). European beach grass

(*A. arenaria*) has been widely introduced, especially in the Pacific Northwest, and has displaced native American dunegrass (*Elymus molli*) throughout much of this region.

(f) Dunes are fragile coastal habitats and are subject to destruction by trampling, vehicular traffic, construction, and livestock grazing. Once disturbed, the dunes become unstable, and sand is distributed via wind and wave action either into adjacent waters, or inland.

(2) Coastal dune restoration. In recognition of the potential importance of dunes as shoreline stabilizers, considerable research has been conducted to develop restoration techniques. Historically, dunes have been stabilized with vegetative plantings in many coastal areas; however, this frequently led to a loss of dune processes (longshore movement, etc.), and may even eliminate unvegetated habitat used by native wildlife such as snowy plovers on the west coast. Early settlers on Cape Cod, Massachusetts, planted American beach grass to stabilize dunes. European beach grass was introduced to the San Francisco Bay area at the turn of the century for dune and shoreline stabilization purposes. During the 1930's, dune restoration projects were completed by the Civilian Conservation Corps in the Pacific Northwest and along the Outer Banks of North Carolina. Dune creation/restoration has also been conducted in Europe, Israel, Australia, and South Africa (Dahl et al. 1975). Recent efforts have sought to improve the state of the art of dune restoration with regard to appropriate selection of native plant species, improved planting techniques, removal of nonnative invasive species, and the use of new technology. Much of the research to develop effective methods of dune building was conducted at the U.S. Army Engineer Waterways Experiment Station's Coastal Engineering Research Center (CERC) in the mid-1960's. These techniques have been applied to dune creation and restoration in coastal areas throughout the United States, and elsewhere (Gage 1970, Darovec et al. 1975). Knutson (1977) provides a general overview of dune restoration, including a discussion of specific applications and recommendations for transplant species. Woodhouse (1978) reviews the ecology of dune plants and provides detailed guidelines for planting, fertilization, and maintenance of artificial dunes, on a region-specific basis.

(a) Typically, dunes are constructed mechanically using earth-moving equipment, or sand fences are used to trap sand in desired areas. Transplant units can be obtained from nursery stocks, or removed from nearby intact dune systems (this may be illegal in some areas). The most commonly used species in dune transplantation are American beach grass, European beach grass, sea oats, and bitter panic grass. Plantings are done by hand or with a strawberry or tobacco planter. Irrigation can be used to stabilize unconsolidated sand prior to planting. Fertilization is usually necessary to establish plant growth in these nutrient-poor systems. Typically, nitrogen, phosphorous, and potassium fertilizers are broadcast by hand or applied using a mechanical spreader.

(b) Once the primary species are established, additional species will eventually grow due to natural colonization processes. Vegetative planting serves two functions: (a) it stabilizes sand deposits, and (b) it provides a baffle that encourages deposition of wind-blown sand on the dune surface. The ability of the various transplant species to perform the two functions varies widely among geographic regions.

(c) Removal of nonnative vegetation is another technique for dune restoration, particularly on the west coast. Methods include manual pulling/removal of plants, mowing, or grubbing with equipment. Often these treatments need to be repeated over several seasons to ensure removal of all or most of the non-native plants. Pickart, Miller, and Duebendorfer (1998) and Pickart et al. (1998) removed yellow bush lupine from northern California dune systems and investigated how the nutrient input from the lupine influenced colonization of the dunes by both nonnative and native plant species.

(d) Sand fences (longitudinal wooden or fabric fence erected along the beach face to encourage deposition of sand) can be used to initially create dunes in the absence of vegetation (Figure V-7-3). This is typically done in an area where dunes do not currently exist, and must be established prior to vegetative stabilization. Sand fences may be constructed in successive "lifts" in order to encourage the accretion of large

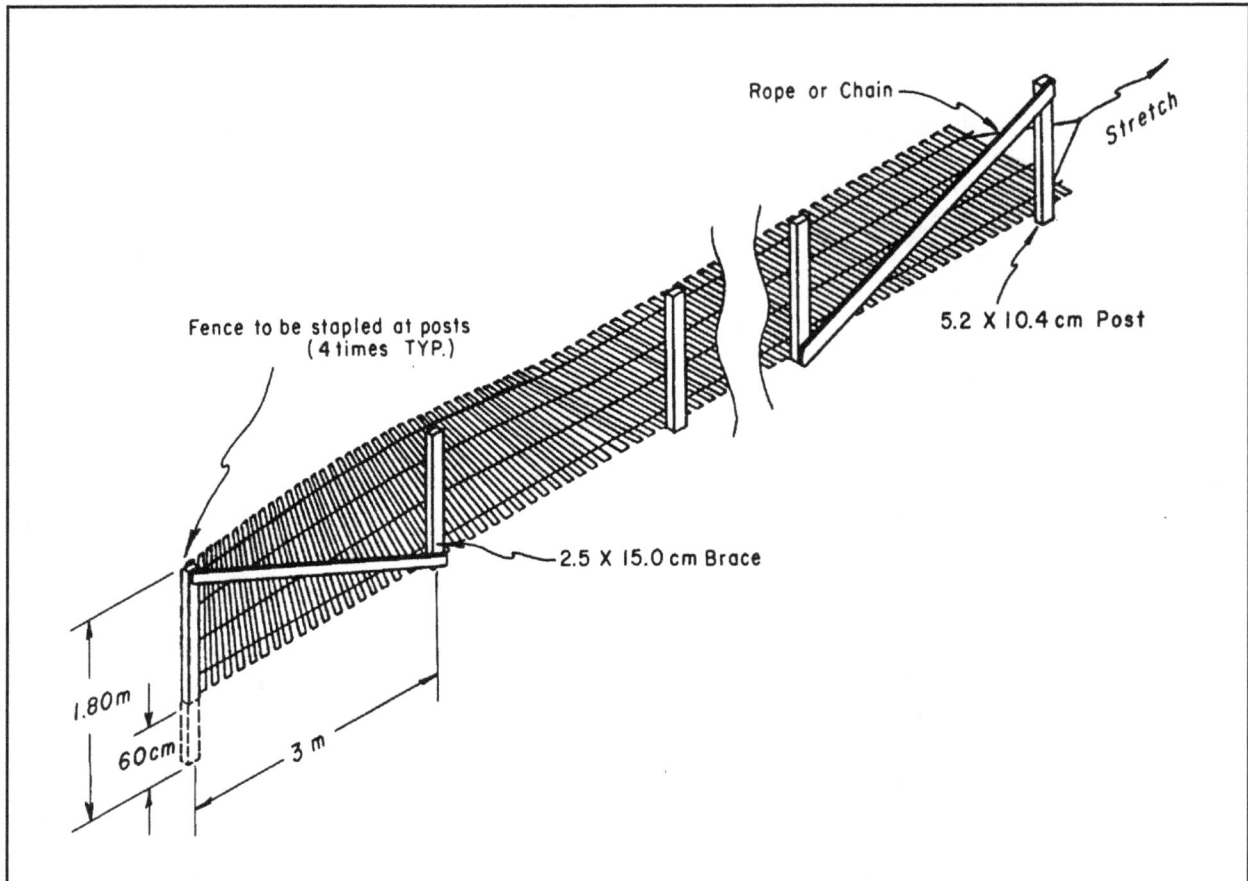

Figure V-7-3. Diagram of a sand fence

dunes (Woodhouse 1978). Most constructed dunes are linear. However, in some cases, it may be advantageous to attempt to replicate the sinuous contours often observed in natural dune systems. Sand fences are advantageous because they trap sand at a rapid rate. Once established, dune-building plants can trap sand at a rate comparable to that of sand fences (Knutson 1980).

(e) Although benefits to reestablishing or stabilizing coastal dunes are considerable, environmental issues need to be considered (Knutson and Finkelstein 1987). Dune construction may interfere with natural geomorphological processes, such as barrier island migration and salt marsh development. Overwash provides the sediment necessary for the maintenance of salt marshes, and periodically provides new substrate for the formation of new marshes. Large-scale dune establishment along barrier islands could potentially interfere with natural cycles of marsh burial and reestablishment associated with overwash events.

(f) Another environmental concern involves changes in natural plant communities as a result of artificial dune construction. Dunes provide protection from salt spray, flooding, and wind/wave energy. A reduction in environmental stress may induce rapid and unwanted changes in vegetation communities behind the primary dunes. Dense growth of shrubs is the most commonly encountered result, with a resulting change in microclimate. This may or may not be considered ecologically desirable, depending on the relative abundance and perceived importance of this habitat type in a specific region. Many coastal wildlife species (arboreal birds, small mammals) will readily inhabit coastal shrub thickets. However, most colonial waterbirds (gulls, terns) prefer to nest in sparsely vegetated dunes.

(3) Maritime forests. Maritime forests are evergreen tree and shrub systems that occur in a narrow discontinuous band along barrier islands and low-lying mainland areas. In the United States, they are most common from North Carolina to Florida. Maritime forests support a distinct fauna and flora adapted to a unique set of environmental parameters, including high soil salinities resulting from periodic storm-induced seawater inundation, wind, and limited fresh water. Canopy height is often restricted from the effects of near-continuous salt spray. Bellis (1995) described the following maritime forest sub-community types based on dominant vegetation and hydrologic patterns:

- Maritime Shrub

- Maritime Evergreen Forest

- Maritime Deciduous Forest

- Coastal Fringe Evergreen Forest

- Coastal Fringe Sandhill

- Maritime Swamp Forest

- Maritime Shrub Swamp

- Interdune Pond

(a) Common plants of eastern maritime forests include wax myrtle (*Myrica cerifera*), live oak (*Quercus virginiana*), and loblolly pine (*Pinus taeda*); on the west coast common species include shore pine (*Pinus contorta*) and Sitka spruce (*Picea sitchensis*). Exposure to salt aerosols is the primary factor that determines species composition and canopy height. As distance from the ocean increases, the floristic composition of maritime forests more closely resembles a typical mainland forest community. Maritime forests provide habitat for wildlife, and stabilize barrier island soils. A diverse assemblage of terrestrial and semi-terrestrial invertebrates inhabit maritime forests, but these communities have not been well-studied. Wildlife species richness can be high, especially on large barrier island systems or on islands that are periodically connected to the mainland and provide dispersal corridors.

(b) Maritime forests have been lost or decimated by timber harvesting and livestock grazing since colonial times. They are subject to development impacts and are becoming increasingly fragmented. Recent recognition of maritime forests as a rare and unusual habitat type has led to various protection initiatives along the U.S. Atlantic coast. As of yet, there have been no documented attempts to restore or create maritime forests in areas where significant acreage has been lost. Certain maritime forest species, such as wax myrtle, are known to rapidly recolonize protected areas behind man-made dunes, suggesting the potential for large-scale creation or restoration of these unique habitats.

(4) Bird/wildlife islands. Terrestrial islands and coastal uplands are often created using dredged material specifically to provide nesting and refuge habitats for birds and other wildlife. Many of the Corps DMRP habitat development sites included bird/wildlife islands as part of the overall habitat mosaic created (Buckley and McCaffrey 1978, Chaney et al. 1978, Soots and Landin 1978). Dredged-material islands are used extensively as rookeries by various birds, including terns, gulls, pelicans, skimmers, stilts, willets, and oystercatchers. Maximum utilization by a diverse bird and wildlife assemblage is attained when habitat heterogeneity is maximized. Creation of terrestrial islands will require careful consideration of the environmental consequences. It is important to focus on the system as a whole; not benefitting one species over another. This becomes apparent during the last few years when Sand Island created in the lower Columbia estuary from dredged material for nesting terns required extensive modification to discourage birds

from using the project. Pit tag analysis (these are small markers placed on young fish) at the nesting site of the terns revealed that they had taken an enormous amount of juvenile salmon as they migrated down the Columbia River to the ocean. Several species of the salmon are listed as threatened or endangered under the Endangered Species Act and it seems that the nesting terns were one of many factors that have led to salmon decline.

(a) Dunes, vegetated swales, ponds, and mud/sandflats are all important components of a well-functioning bird/wildlife island. Tree and shrub species rapidly colonize created dredged material islands, providing rookeries for species that nest in canopies. Small mammals, such as mice, shrews, and voles will readily colonize these islands, especially if nearby upland habitats are available to provide dispersal corridors. Deer, fox, raccoon, and other species are typical residents of created upland habitats in coastal areas.

(b) Studies along the North Carolina coast (Parnell, DuMond, and Needham 1978; Parnell, DuMond, and McCrimmon 1986) suggest that the presence or absence of retaining dikes is an important determinant of plant colonization rates and subsequent utilization by wildlife. The earlier constructed wildlife islands in North Carolina (prior to the mid-1970's) were mostly undiked, while islands constructed after 1975-76 were diked. Diked islands were not used as extensively by colonial waterbirds compared to the undiked islands. Terrestrial plant species colonize diked islands more rapidly than undiked islands; species which prefer dense vegetative cover (some small mammals, arboreal birds) benefit at these sites; however, most colonial waterbirds prefer to nest along sparsely vegetated beachfront, which is available mainly at older, undiked islands. Nest sites located behind dikes are subject to flooding following heavy rain. Although many previously undiked islands have been diked at a later date for stability and longevity, this has decreased available nesting habitat for colonial waterbirds.

V-7-8. Environmental Features of Traditional Coastal Engineering Projects

The majority of coastal engineering projects will no doubt be required to provide habitat mitigation or restoration as a result of regulatory requirements and there are many opportunities and considerations for habitat protection in the planning and implementation phase of traditional coastal engineering projects. Consideration of sound environmental design criteria and potential environmental benefits can be invaluable in generating public and agency approval in high-profile or controversial coastal engineering projects.

a. Example project types.

(1) Beach nourishment. The dynamic nature of the littoral environment ensures that beaches will remain an ephemeral resource in many areas. Wave, wind, and tidal action combine to continually erode or accrete beach landscapes; these processes occur on a variety of temporal and spatial scales. In areas experiencing relative sea level rise, the net result is gradual loss of beach environment. Given the penchant for man to live, work, and recreate near the beach, maintenance of some steady state is considered economically desirable. Thus, considerable effort has been expended to stabilize and maintain beaches and counteract the effects of beach erosion.

(a) Artificial renourishment of beaches is considered to be the most cost-effective and environmentally desirable method of maintaining beaches in the short term (NRC 1995); however, recreating natural beach processes is likely to be much more successful ecologically, over longer time scales. The most appropriate sources of artificial beach fill are generally (a) dynamic accreting shorelines, such as those located around inlets, (b) channel maintenance dredging, and (c) offshore sand deposits at a depth of greater than about 15 m. These sand reserves are generally located far enough from shore such that they do not affect littoral processes and, in most cases, do not represent critical habitats (Hobson 1981). Sand from upland or riverine sources may also be used as beach fill; however, this material may vary considerably in quality and the costs of transporting large volumes over land or downstream may be prohibitive (NRC 1995).

(b) Many of the beaches in the Great Lakes are composed of a relatively thin layer of sand over glacial or lacustrine clay. When the sand supply is interrupted, the clay is exposed and eroded, thus allowing (larger) higher energy waves to impact the bluff. The Detroit District has observed that beach nourishment fills that contain a coarse fraction (gravel to cobble-sized material) would lag the underlying clay, reducing the lake bed downcutting of the clays.

(c) Excessive siltation and increased turbidity associated with the sand mining process can cause serious environmental impacts to marine organisms (Auld and Schubel 1978, Snyder 1976). Siltation and burial of benthic organisms and reef/hard bottom habitat is an issue of concern, because the increase in turbidity affects both filter-feeding organisms and fishes. Larval and juvenile fish, in particular, are especially sensitive to dredging-induced turbidity as their gills may become clogged or abraded by floating particulates. Feeding ability of larval and juvenile fishes is decreased due to a reduction in available light. Sand mining activities must be timed properly so that they do not coincide with recruitment or migration of larval and juvenile marine organisms, especially at inlets, where these organisms are often concentrated.

(d) Substrate factors, especially grain size, are of critical importance in planning beach nourishment. It is important to match grain size of the donor sites with that of the beach site being nourished (James 1975). Analysis of grain size using standard sediment testing sieves is preferred. Generally, a single beach fill will not maintain a beach indefinitely; replenishment must be conducted over the long term in a gradual series of nourishment events. The amount of fill used will vary depending on location and severity of loss; however, a rule of thumb is to provide enough material to offset losses occurring naturally. Often, it is desirable to place excess material at the upper end of an eroding beach, allowing natural littoral processes (longshore drift) to naturally redistribute material over time.

(e) A variety of marine organisms are potentially affected by placing fill on intertidal or subtidal portions of the beachface. In general, these organisms are able to persist in the dynamic beach environment because they have adapted to conditions such as high wind and wave energy and periodic burial. The ability of most benthic invertebrates to survive a fill event depends on their ability to burrow up into the newly deposited substrate. Therefore, substrate composition and depth are the major factors determining survival rates of beach invertebrates subjected to instantaneous burial (Culter and Mahadevan 1982, Nelson 1985). Although benthic invertebrate communities on beaches are generally able to recover rapidly from fill events, the time between mortality and recovery represents a decrease in available prey resources for wildlife (primarily shore birds) which utilize the lower intertidal zone as foraging habitat.

(f) Sea turtles may be affected by beach nourishment activities (Nelson 1988; Nelson, Mauck, and Fletemeyer 1987). Spring through late summer is the primary sea turtle nesting period along the U.S. southeast and Gulf coasts. Although nourishment programs can potentially result in greater available nesting habitat for sea turtles, most of the construction activity associated with beach nourishment is likely to deter nesting females. Physical properties of the newly deposited fill (grain size, density, shear strength, color, temperature, moisture content) have been demonstrated to influence incubation, and hatching success (Nelson, Mauck, and Fletemeyer 1987). Various nesting relocation programs have been attempted to minimize the impacts associated with beach nourishment during nesting seasons.

(2) Shoreline structures. Hard structures used in beach stabilization include breakwaters, groins, revetments, seawalls, and bulkheads. These are costly to build compared to beach filling alone, but are often necessary along developed shorelines. In most cases, "hard" structures are used in combination with an artificial beach nourishment program to maintain their effectiveness in reducing beach loss in coastal communities. Environmental impacts associated with breakwaters, groins and sills are largely related to the

interruption of natural littoral drift processes and the necessity of periodically replacing or bypassing sand. The environmental impacts of shore protection structures are not limited to the beach being stabilized. When longshore drift is interrupted by stabilization structures, the effect is manifested downdrift. Animals which rely on the beach or surf zone for all or part of their lives are directly impacted by both soft and hard beach stabilization techniques (e.g., sea turtles, mole crabs and other invertebrates, grunion). Nesting and feeding shorebirds are deterred by noise and construction. Construction of shoreline stabilization structures typically involves excavation, installation of piles, backfilling, and material transport. These activities can induce temporary conditions of high turbidity in nearby waters, which may result in increased mortality to larval and juvenile finfish and shellfish. Increased turbidity may impact sensitive habitats nearby, such as coral reefs and seagrass beds. The presence of shoreline structures may alter natural patterns of circulation, resulting in altered flushing rates and scour/deposition patterns.

(a) Breakwaters and sills. Offshore breakwaters are linear structures placed parallel to the beach. Their purpose is to diminish wave energy and reduce beach erosion, while still providing for longshore transport of sand along the beach. Both single and multiple nearshore breakwater systems have been employed. In some cases, breakwaters are connected to the shoreline via a jetty. Shore-connected breakwaters may interrupt longshore transport processes and require increased maintenance (e.g., sand-bypassing) in order to maintain the integrity of downdrift shorelines (NRC 1995).

- Submerged offshore sills provide a similar function to breakwaters; they diminish some wave energy and thus reduce erosive forces acting upon the beach. Sills interrupt movement of sediments offshore, resulting in a "perched" beach; however, the characteristics of a sill which reduce loss of sand during erosive episodes also function as an impediment to sand being deposited on the beach during periods of accretion. A sill provides some environmental and aesthetic advantages relative to breakwaters, because the sill is rarely located above the high-tide mark.

- Breakwaters and sills function by modifying the nearshore wave environment. This may result in undesirable habitat changes for species which have adapted to the high-energy surf zone. However, species adapted to low-energy sheltered environments may colonize the area in the shadow of the breakwater/sill. In addition, the addition of hard substrate provides habitat for fouling organisms, and shelter for fish and motile invertebrates. Exposed portions of detached breakwaters may be used by colonial seabirds. In some areas, these structures may represent the only form of hard-bottom habitat available, resulting in an increase in local biodiversity.

(b) Groins. Groins are oriented perpendicular to the beach and extend into the surf zone. They are most commonly constructed of rubble (quarrystone) but may also be made from timber, steel, or concrete. They are typically constructed in series, commonly referred to as a groin field, along a beach. The purpose of a series of groins is to prevent sand from being transported off the beach via longshore drift. However, the presence of groins can lead to increased erosion downdrift due to interruption of longshore drift patterns. An advantage of groins over parallel nearshore structures is that wave energy is not affected by perpendicular structures, although the prevalence of rip currents is increased. Subtidal benthic invertebrate communities are not likely to be affected, because species which persist in the surf zone are adapted to dynamic patterns of scour and deposition. Sea turtles and shore birds are not likely to be deterred by the presence of groins, although initial construction and maintenance activities must be timed so as to avoid critical nesting periods for these special concern species. As with other rubble-mound protective structures, the submerged rock surfaces can function effectively as fish habitat, resulting from the provision of shelter and the abundance of epifaunal invertebrates which colonize exposed rock surfaces.

(c) Seawalls and revetments. Some seawalls are vertical structures (bulkheads) intended to protect developed shorelines or fill from wave erosion. They should only be used in situations where reflected wave energy can be tolerated. This generally eliminates bodies of water where the reflected wave energy may interfere with or impact on harbors, marinas, or other developed shore areas. A revetment is protective armour placed on an existing bank and therefore are sloped. They are typically employed to absorb the direct impact of waves more effectively than a vertical seawall. Although revetments displace intertidal beach habitat, a potential environmental benefit may be realized by the colonization of the submerged portions of the structures (particularly rock revetments) by fouling and motile epifaunal invertebrates. Juvenile finfish and shellfish may use the submerged rocks as a feeding area and to avoid predators. In time, revetments are undermined or flanked, resulting in failure. Both seawalls and revetments eliminate the supply of literal material from the protected bluff.

(3) Navigation, ports, and marinas.

(a) Construction and maintenance channels. Channel construction and deepening to facilitate passage of vessels may result in a variety of potential impacts to shallow coastal habitats. The severity and extent of dredging-related impacts depends on the type of dredging equipment used, and the susceptibility of nearby habitat types and aquatic biota. *Mechanical dredging* involves the use of dragline/bucket dredges, or clamshell dredges. *Hydraulic dredging* involves the use of suction pumps and pipelines, which may or may not be equipped with rotating cutterheads. Mechanical dredging operations generate localized, pulsed, conditions of high turbidity as the bucket or clamshell apparatus is deployed. Hydraulic dredging may generate continuous conditions of high turbidity, and the area affected may be large, relative to that of mechanical dredging operations, especially if mobile equipment (e.g., hopper dredge) is used (McClellan et al. 1989, Raymond 1984).

- Both mechanical and hydraulic dredging may result in direct and indirect impacts to a variety of aquatic fauna. Benthic invertebrate communities are subject to direct mortality and burial (Hirsch, DiSalvo, and Peddicord 1978); however, these organisms usually recover from disturbance in a relatively short time. Direct impacts and disturbance are usually confined to within a few hundred meters of the dredging equipment while in operation. Initially, invertebrate species diversity may increase in dredged channels, as opportunistic species rapidly colonize the denuded or disturbed area. Elevated turbidity associated with dredging and channel construction activities has adverse secondary impacts on a variety of organisms, including filter-feeding bivalves, and fishes; the larval and juvenile forms of many species are susceptible to mortality induced by gill-clogging and abrasion. If underwater blasting is necessary to remove rocks, this may result in fish kills due to rupturing of swim bladders.

- Dissolved oxygen levels may be substantially decreased during dredging operation, due to the resuspension of oxidizable particulates (Brown and Clark 1968). However, dissolved oxygen levels may improve considerably following cessation of dredging due to the removal of organic sediments from the system. Deepening of navigation channels may result in increased saltwater intrusion into inland waterways, potentially altering the distribution and abundance of freshwater, estuarine and marine organisms.

- Channel construction in tropical and sub-tropical locales can directly impact valuable seagrass and coral reef communities. Recovery of damaged seagrass or reef communities can take decades, and in some cases they may never recover. These sensitive habitats are also subject to indirect impacts resulting from increased turbidity, as they depend on adequate light penetration for survival. Floating silt curtains can be used to partially alleviate the effects of dredging on critical habitats and organisms, however, careful attention must be given to critical or special-concern habitats, seasonal migration patterns of non-resident fishery species, and spawning seasons of resident species in the specification of optimal windows for conducting maintenance dredging.

(b) Inlets and jetties. Inlets are dynamic coastal features primarily associated with barrier islands. Most inlets shift continuously in response to short and long-term physical factors, such as storms and near-shore sediment distribution patterns. Coastal engineers are faced with the difficult challenge of stabilizing inlets in order to provide for safe and efficient navigation. Accumulation of sediments in inlets necessitates frequent maintenance dredging in order to maintain navigation conditions. Sand stored in inlets may reduce the volume available for downdrift beaches.

- Inlet dredging can cause burial or direct mortality of benthic organisms and chronic turbidity. However, many of the organisms adapted to life in inlets are minimally impacted by maintenance dredging. Recolonization by benthic organisms generally occurs within a short time following disturbance. Dredging windows (seasonal restrictions) are used to reduce the level of impacts to migratory organisms such as sea turtles and marine mammals. Avoidance of dredging during the known spawning seasons of marine and estuarine finfish and shellfish is critical to reducing impacts to the delicate early life stages (eggs, larvae, juveniles) of these organisms.

- Jetties are used to protect and maintain inlets, marinas, and port facilities. They serve multiple purposes, including protecting vessels entering a port/marina from dangerous waves and controlling the movement of sand into dredged channels and harbor entrances. Like groins, jetties will induce accretion of sand on the updrift side, with consequent erosion of sand on the downdrift shoreline. Environmental considerations for jetties are similar to those outlined previously for groin systems. These are primarily associated with the interruption of natural longshore sediment transport. In addition, scour on the downdrift side of a jetty may adversely affect colonization by benthic invertebrates. A major environmental concern associated with jetty construction involves disruption of natural transport patterns of ichthyoplankton (fish eggs and larvae). Eggs and larvae which are naturally transported into estuaries via longshore drift may become entrained into harbors via disruption of normal hydrodynamics. The inherently poor water quality and increased turbidity in harbors may result in significant mortality of ichthyoplankton.

(c) Piers and docks. A common environmental problem in urban coastal areas is habitat and water quality degradation in abandoned pier and dock basins (Hawkins et al. 1992). These degraded habitats, abundant in industrialized areas, may date back to colonial times in large U.S. cities (e.g., New York, Boston, Philadelphia) and even earlier in European cities. Many older docks were abandoned in the latter half of the 20th century as larger seagoing vessels were unable to access them, and newer, modern port facilities were established. Most abandoned docks are polluted from decades or centuries of unregulated dumping and discharge from vessels and shore facilities. Poor tidal circulation in docks, in part due to the prevalence of numerous dead-end canals and deep basins, combined with the prevalence of organic matter in the sediments, results in near continuous hypoxia or anoxia. Benthic invertebrate communities of docks and basins are species depauparate, and dominated by a few opportunistic taxa, mostly annelid worms or insect larvae. Bacterial mats and noxious algal blooms are prevalent. Certain blue-green algae and dinoflagellates associated with these blooms produce toxins known to cause significant mortality in fishes, and may cause sickness in humans. Heavy metals and other industrial contaminants are also often found in high concentrations in abandoned basin sediments.

- A variety of technologies have been applied to the problem of dock and pier restoration (Hawkins et al. 1992). The first step is to identify and curtail discharges of industrial and human wastes. Following cessation of discharge, there are several possible means of improving water and habitat quality. Mechanical aeration of anoxic basin sediments has been used to promote mixing and oxygenation. Hydraulic pumps and paddles are used to promote mixing in enclosed areas. Bivalve mollusks have been transplanted in dock/pier areas as biological filters to improve water quality. Dead-end canals and deep basins can be filled in or recontoured to improve circulation. Decaying piers and pilings are removed, primarily for aesthetic purposes. However, this activity may reduce overall habitat quality because epifauna associated with piers and pilings provide a food resource for fish.

- The complex structure of the piers and pilings provide a predation refuge for juvenile fishes. Some commercially significant species along the U.S. Atlantic coast (e.g., striped bass) may be found in high concentrations in and among piers in urban reefs which can provide habitat diversity, along with the visual appeal of an unobstructed waterfront.

(d) Boat basins and marinas. Construction of marinas, or basins for small boats measuring less than 31 m (102 ft) constitutes a significant coastal engineering issue worldwide. The environmental effects of marina construction and maintenance are many, and sound construction practice and management are necessary to prevent serious environmental degradation. Critical shallow-water habitats, which may be present in the vicinity of marinas, include intertidal marshes and mudflats, seagrass beds, shellfish beds, and, in tropical locations, mangroves and coral reefs. Pollution (from both point and non-point sources) is a potentially serious environmental problem associated with marinas. Construction of boat-launching facilities and parking lots may cause in chronic turbidity in the vicinity due to increased runoff. Discharge of fuels and engine oils, garbage, paints, and other waste materials directly impacts water quality. Noise pollution associated with the day-to-day activities of the marina will affect use of the area by wading birds and other wildlife. Wakes caused by the constant movement of small vessels in and out of the area lead to increased shoreline erosion and may disrupt life-cycle larval and juvenile fishes and invertebrates. Construction of a basin increases the residence time of water in the area, inhibiting the natural pollution abatement function attributed to tidal flushing. Use of breakwaters in the vicinity of the marina may further increase residence time of water in the area.

- Mitigation of direct impacts resulting from marine construction is often attempted by creation/restoration of shallow, vegetated habitats elsewhere. However, the success rate of these projects is unacceptably low in many cases (e.g., seagrass projects). The U.S. Army Corps of Engineers provides detailed guidelines for marina construction in order to minimize environmental impacts (USACE 1993). These include construction of the marina to avoid dead-end canals, thereby reducing the likelihood of creating stagnant pools of water. Square or rectangular basins are not recommended, and it is imperative that basins not be constructed such that they are deeper than the associated access channels.

- Wide access channels are recommended, and gradual slopes from channel to harbor are best. Floating, rather than fixed breakwaters situated at the marina entrance are preferred for maintaining adequate flushing. Giannio and Wang (1974) provide a number of recommendations to reduce adverse environmental effects in small boat basins. These include implementation of designs which promote tidal flushing and control water quality, use of structures which encourage colonization by fouling communities as a food source for fishery species, use of sloping, rather than vertical sidewalls for channels where feasible, and provision to beneficially use material removed by dredging of the marina (e.g., habitat development projects).

(e) Port facilities. Many of the environmental concerns associated with the construction and maintenance of small boat basins are applicable to larger commercial and industrial port facilities, albeit at a much greater scale. Elimination of significant acreage of critical spawning and nursery habitats, such as intertidal marsh and seagrass beds is one major problem, and numerous attempts have been made to mitigate for the detrimental effects of port construction and expansion on critical shallow coastal habitats. The environmental effects of shoreline protective structures such as jetties on early life stages of finfish and invertebrates should be addressed in port facility projects. Point and non-point source pollution effects are a primary concern in port construction/expansion projects, especially considering the variety and magnitude of hazardous materials which continuously pass through ports in the United States and elsewhere.

(f) *Confined disposal facilities.* Confined disposal facilities (CDFs) are structures designed to store dredged materials. Approximately 30 percent of total material dredged from U.S. waters is placed in confined aquatic disposal facilities, the remaining 70 percent being disposed of at unconfined disposal areas or used in beach nourishment or beneficial-use programs (USACE 1987). In recent years, environmental issues associated with open-water disposal of dredged material has increased the demand for confined disposal. This demand is expected to increase in the future (Averett, Palermo, and Wade 1988).

- A typical confined disposal facility is a diked containment area. The design is such that solids can be retained while carrier water is released from the containment facility (Palermo, Montgomery, and Poindexter 1978). Dikes are constructed in order to form a confined area. Dredged materials are hydraulically pumped in the form of a slurry into the containment area. After solids have settled out of suspension, clarified water is released from the facility through a weir. Filtration and or chemical treatment of effluent may be necessary in some cases in order to meet water quality criteria for suspended solids concentrations (Schroeder 1983). The projected functional life span of a confined disposal area ranges from years to decades.

- Environmental considerations in the construction of confined disposal facilities are varied. Critical habitats such as intertidal marsh, seagrass, mangroves, and coral reefs should be avoided in the site selection phase. Near-shore or on-shore CDF's are more likely to displace critical shallow water habitats than CDF's which are constructed in deeper, open waters. Nursery grounds and migratory pathways for fishes and marine vertebrates should be identified and avoided. Ground-water impacts are of primary importance. Leachates may eventually work their way into aquifers and even in uncontaminated dredged materials may contain elevated levels of chloride, potassium, sodium, calcium, iron, and manganese, especially when marine or estuarine sediments are deposited over a freshwater aquifer. Contaminated dredged materials may leach a variety of metals and organic contaminants (e.g., PCB's, pesticides). These materials may also be present in effluent water discharged from the CDF. Engineering considerations to reduce groundwater impacts include site location, topography and slope, underlying stratigraphy, subsurface hydrology, climatological factors, soil properties, and the degree of use of the aquifer by humans. Aquifer recharge areas should be avoided. In some cases, either natural (clay) or artificial liners are necessary to prevent introduction of potentially contaminated leachate into sensitive aquifers.

V-7-9. Environmental Issues To Be Considered for All Projects

Certain critical habitats and individual species may be of particular importance in coastal engineering projects, regardless of whether or not the project is intended to provide environmental benefits. Threatened and endangered (T&E) species of fish, birds, mammals, and reptiles are high-profile issues, and merit special consideration and accommodations during the reconnaissance, planning, implementation, and monitoring phases of projects. The following are specific examples of T& E species considerations or habitats of special concern encountered in the coastal zone.

a. In the water.

(1) *Sea turtles.* Maintenance dredging of navigable waterways can potentially induce mortality of threatened or endangered sea turtles. Relatively little is known about the distribution and abundance of sea turtles in navigable waterways and harbors, but there have been documented cases of sea turtle mortality during hopper dredging in channels of the southeastern United States (Dickerson et al. 1995). The three species which are at risk along the South Atlantic and Gulf coasts are the loggerhead (*Caretta caretta*), green (*Chelonias mydas*), and Kemp's ridley (*Lepidochelys kempi*). Since 1980, an observer program has been in effect during dredging operations in the southeast U.S., in accordance with the Endangered Species Act of

1973. Observation and documentation of sea turtle mortality is essential to develop technical and management solutions to dredging-related impacts. Mortality has decreased substantially since inception of the observer program. This may be due to improved management and operations, but may also reflect long-term fluctuations in turtle abundance in southeast U.S. Waters (Dickerson et al. 1995). Management and operations alternatives which can help reduce mortality include seasonal restriction on dredging, relocation of individual live turtles, and improvement to dredging equipment which minimizes turtle entrainment. Comprehensive surveys have been conducted in several channel/harbor areas where sea turtle mortality due to maintenance dredging has occurred (Butler, Nelson, and Henwood 1987, Dickerson et al. 1995, Van Dolah and Maier 1993). Results of these and similar studies are critical in specifying seasonal dredging windows, when sea turtles are least likely to be present and dredging may be conducted with minimal risk.

(2) Cetaceans. Whales, dolphins, and porpoises are large mammals which have adapted completely to the marine environment. They spend their entire lives at sea, although a few species are known to migrate into freshwater rivers and lakes. Some 80 species of cetaceans are recognized worldwide; many of these are in serious decline due to harvesting. Cetaceans are divided into two functional groups based on feeding morphology. The Mysticeti are the ten species of baleen whales, which feed by filtering water through a bony sieve (the baleen). The Odontoceti (toothed cetaceans) include some 70 species of whales, porpoises and dolphins.

(a) Many cetaceans have a large home range and undergo extensive seasonal movements. Dolphins and porpoises do not have set migration patterns, but larger cetaceans do. Cetaceans usually reproduce in warmer tropical waters, but food availability is greater in temperate waters, and extensive feeding journeys are undertaken.

(b) Because many cetacean populations are in decline, it is important to consider potential detrimental effects of coastal engineering projects on these animals. In general, whales are not likely to be encountered in shallow nearshore waters, where most coastal engineering projects are likely to occur. Occasionally, however, an individual will become disoriented and enter channels and harbor areas, or become stranded along a beach. Dolphins and porpoises are common in shallow coastal waters, and may be seen riding the bow waves of supertankers and other large commercial vessels entering and leaving harbors. These animals are strong and graceful swimmers, but nonetheless, they are occasionally killed by large commercial vessels.

(c) Dredging operations in channel harbor areas are potentially threatening to small cetaceans. The turbidity plume associated with maintenance dredging interferes with feeding, since these animals are visual predators. Dolphins and porpoises are known to frequent highly polluted estuaries and harbors; thus, poor water quality and high contaminant levels can pose serious threats to their health. In recent years, there have been increasing numbers of mass dolphin strandings along the southeast U.S. coast. It is not known to what extent chronic pollution has contributed to this phenomenon.

(3) Fish and invertebrates. A number of marine and estuarine fish species, many of commercial or recreational significance, have undergone population declines in the latter half of the 20th century, primarily due to overharvesting and loss of shallow-water spawning and nursery habitat. Some species, such as striped bass on the U.S. Atlantic coast and red drum on the Gulf coast, have rebounded dramatically in the last decade, as a result of wise management practices, including a moratorium on commercial harvests in some areas. Many other species continue to decline, and some, such as the shortnose sturgeon (*Acipenser brevirostrum*) along the Atlantic coast, and multiple species of Pacific salmon, are listed as federally threatened or endangered.

(a) Coastal engineering projects should consider potential effects on fish and the aquatic food web. Construction activities, especially those that include dredging and the resultant increase in turbidity, should be timed so as not to interfere with critical spawning or migration times. Spawning and nursery areas are known for most species; these should be identified and avoided if possible.

(b) A number of marine and estuarine invertebrate species have also undergone significant declines in the latter half of the 20th century. American oyster populations have been completely decimated in many coastal areas, particularly along the Atlantic coast. Current harvests are but a fraction of what was historically taken in once-productive estuaries such as Chesapeake and Narragansett Bays. Blue crab harvests have declined in the mid-Atlantic region in recent years, likely a result of overfishing. Improved management practices may provide some relief, but destruction of habitat remains a critical issue. Other invertebrate communities are valued not for their commercial significance but for aesthetics and recreational value (e.g., coral reef communities). Coastal engineering projects should seek to avoid or minimize impacts on invertebrate populations. Construction activities that involve dredging are most detrimental. Many crustaceans (shrimp, blue crabs) burrow and overwinter in channels and are likely to be encountered during maintenance dredging. Sessile invertebrates (oysters, mussels, clams) are subject to burial and subsequent mortality from dredging. Even if burial does not occur, increased turbidity associated with nearby dredging activity is detrimental to these filter-feeding invertebrates. Corals and other hard-bottom invertebrate communities are subject to direct mechanical damage and burial effects from construction projects that include dredging. Projects that result in periodic or chronic conditions of poor water quality (including turbidity, elevated nutrient levels, low dissolved oxygen) are detrimental to both sessile and motile invertebrate communities.

a. At the shore/water interface.

(1) Sea turtles. A primary concern related to beach nourishment along the southeast Atlantic and Gulf coasts is the nesting activity of sea turtles. Several species of threatened and endangered sea turtles commonly nest on beaches in the southeast United States, including the loggerhead, leatherback, and green turtles (Nelson 1988; Nelson, Mauck, and Fletemeyer 1987). Nesting season occurs from late spring through summer, although there is considerable variation within and between regions. Sea turtles ascend the beach at night, lay eggs in a nest cavity in the zone between MHW and the top of the primary dune, and return to the ocean. Beach nourishment and restoration activities can disrupt the nesting patterns of sea turtles, but effective beach nourishment can also provide additional nesting habitat in areas where beaches have eroded. The issue of beach nourishment vs. turtle nesting has sparked contentious debate in heavily populated coastal areas of the southeast, particularly in Florida.

(a) Specific activities that interfere with turtle nesting include direct burial of existing nest cavities with new fill and placement of barriers along the beach (e.g., pipelines), which deter or impede migration of nesting females. Bright lights, noise, and vehicles disrupt nesting activity of female turtles, and may interfere with the orientation of hatchlings as they migrate towards the ocean. Disturbance may cause a nesting female to return to the water without digging a nest cavity, or she may abandon a cavity without depositing eggs. Minimization of construction activities and illumination along beaches at night is recommended. In some cases, turtle nests have been successfully relocated to undisturbed areas of beach.

(b) Post-construction effects of beach nourishment may also impact sea turtle nesting success. If the topography of the resulting beach differs from that found in nature (e.g., formation of false scarps), nesting females may be deterred. Compacted sands typical of newly filled beaches may inhibit nest excavation by sea turtles. This can be mediated to some extent by tilling, and by mandating the use of wide-tracked vehicles during construction. Differences in grain size, color, shear resistance, density and gas exchange potential of newly deposited sand vs. natural sand may negatively affect hatching success and/or hatchling sex ratios. New fill material can be placed seaward of the primary nesting area and graded to a gentle slope to facilitate access by nesting turtles. Timing of nourishment activities should be outside of the nesting season, when possible. Appropriate timing, minimization of light and noise disturbance, and an effort to properly replicate natural conditions can potentially reduce impacts to nesting turtles, and many beach nourishment projects have actually increased suitable sea turtle nesting habitat in the southeast United States.

(2) Marine mammals. Coastal engineering projects can potentially affect marine mammals that use coastal habitats as breeding areas or rookeries. Several marine mammals are commonly found near populated shorelines in the United States and worldwide. Many of these species are threatened or endangered. Most marine mammals frequent temperate waters, although a few species are known to exist in tropical or subtropical waters. Seals, sea lions, and walruses (Pinnipedia) are amphibious mammals that rely on isolated, protected shorelines to mate and bear young. Three pinniped groups are recognized and include the walrus (Odobenidae), the eared seals (Otaridae), and true seals (Phocidae). Eared seals include the northern fur seal and sea lion and are distinguished by the presence of small but visible external ears and the ability to rotate the hind flippers for increased mobility on land. The true seals (Phocidae) include the ringed, harbor, and elephant seals. All pinnipeds are strong swimmers and effective divers. Direct mortality to these animals due to construction activity is unlikely in most cases, but destruction of breeding habitat may be a significant concern in some areas.

(a) River otters (*Lutra canadensis*) are common in shallow coastal habitats, including intertidal wetlands. Unlike pinnipeds, which have evolved specifically for life in the marine environment, otters are highly mobile on land and water. Destruction of shallow coastal habitats in populated areas is likely the primary threat to this species. The sea otter (*Enhydra lutris*) was once common along the entire Pacific coast of North America, but was hunted nearly to extinction during the 19th century. Populations have recovered during the last half-century, but destruction of breeding and refuge habitats remains a potential threat to this species.

(b) Mammals that belong to the order Sirena include the three species of manatee (Trichechidae), and dugongs (Dugongidae). These large, slow-moving vegetarians are highly vulnerable due to their gentle, gregarious nature; the Caribbean sea cow and the Stellers sea cow (a cold water species) are now extinct. Populations of the manatee (*Trichechus manatus*) in southern Florida and the Caribbean are seriously threatened by coastal development. Manatees are frequently injured or killed in collisions with recreational or commercial vessels, and many bear tell-tale propeller scars from such encounters. The largest concentration of manatees occurs in sub-tropical waters of south and central Florida, and they are known to occasionally migrate as far as North Carolina and Virginia, presumably in search of food. Manatees are listed as Federally endangered in U.S. waters and many areas have implemented restrictions on use of motorized vessels in canals and other areas frequented by these animals. Coastal engineering projects that may potentially impact manatee populations include dredge and fill projects that result in destruction of critical habitat (e.g., seagrass beds), and any activity that may result in an increase in boat traffic.

(3) Sea, shore, and wading birds. Aquatic birds rely on shallow coastal habitats such as mudflats, intertidal wetlands, and both natural and man-made islands as critical breeding grounds and forage sites. True *seabirds* are those species that spend most of their time at sea, primarily foraging on surface waters by diving or skimming for food, but rely on shore habitats for breeding. These birds have evolved various adaptations to an exclusively marine existence, notably the presence of a salt gland, which regulates blood salinity. True seabirds include gulls, terns, pelicans, albatrosses, cormorants, and the frigate birds and tropic birds of warm regions. Many of these species, notably the albatross, are wide-ranging travelers, and fly vast distances over open ocean in search of food. Others, such as gulls and pelicans, spend most of their time near the coast.

(a) Waterfowl, including diving and dabbling ducks, geese, and swans are common in saline and freshwater tidal marshes and shallow tidal embayments. These habitats provide abundant food and refuge for these birds, especially during winter. Seeds and rhizomes of emergent marsh vegetation are the preferred food of most waterfowl species.

(b) *Shorebirds* forage along intertidal margins and *wading birds* forage primarily on invertebrate prey in shallow waters or on sand/mudflats. During fall migration, many shorebirds switch to a diet of seeds of tidal marsh plants, such as wild rice. Shorebirds include sandpipers, rails, and soras; these birds are found on intertidal flats and in both saline and fresh tidal marshes. Wading birds include herons, egrets, ibises, and bitterns. These are commonly encountered on intertidal flats and in marshes. Wading birds feed on small fish and invertebrates captured in tidal shallows.

(c) Coastal engineering projects that may potentially affect sea, shore, and wading birds are those that result in habitat destruction, alteration, and disturbance. Construction activities should be timed so as not to occur during critical nesting seasons. Beach nourishment activities can deter shorebirds from feeding and nesting; however, if carefully timed, beach nourishment may potentially improve shorebird habitat via increased beach width. Dredge and fill projects which result in destruction of seagrass or marsh habitat will negatively impact waterfowl, and wading/shore birds that use these habitats. Chronic noise and human activity will deter seabirds, shorebirds, and wading birds from using an area. Long-term pollution impacts such as eutrophication and contaminant input can have both immediate and long-term effects on the viability of sea, shore-, and wading bird populations in coastal areas.

(4) Fish and invertebrates. Many important fishery species rely on coastal wetlands as nurseries and as feeding and refuge areas (Boesch and Turner 1984, Rozas and Hackney 1983, Weinstein 1979). Along the Atlantic and Gulf coasts of the United States, species such as blue crab, summer flounder, red drum, and penaeid shrimp (*Penaeus* spp.) depend on shallow tidal marsh creeks, which provide refuge from predators and an abundant food source in the form of zooplankton and benthic invertebrates. These estuarine-dependent fishery species grow rapidly during their temporary stay in coastal wetlands before migrating back to deeper coastal waters. Adults of these and other transient species periodically enter tidal wetlands in search of food. Loss of valuable coastal wetland habitat is a primary concern for marine fisheries managers, and habitat destruction is considered as serious a threat to the integrity of many coastal fisheries as is overharvesting.

(a) In the Pacific northwest, all of the Pacific salmonids (trout and salmon) are known to directly utilize estuarine tidal wetlands as critical nursery habitat for feeding and avoiding predators (Shreffler, Simenstad, and Thom 1992), as well as migratory corridors along the shorelines. Other species, such as dungeness crab and various flatfish, rely on the extensive tidal flats and shallows as a primary nursery habitat. Bivalve mollusks, including the pismo clam (*Tivela stultorum*), littleneck clam (*Protothaca staminea*), and Pacific razor clam (*Siliqua patula*) are found in shallow intertidal sand/mudflats. These species are the basis for important commercial and recreational fisheries.

(b) The California grunion, which spawns in burrows dug in saturated sand, is directly threatened by coastal habitat destruction throughout its range along the central and southern California coast (Figure V-7-4). This species is harvested by recreational fishermen throughout most of its range. Disturbance resulting from an increase in human activity along the coastline is a potential threat to this unusual fish (Fritzsche, Chamberlain, and Fisher 1985).

e. Critical areas.

(1) Many marine habitats and other areas of special concern are subject to degradation from coastal engineering activities. These include, but are not restricted to, intertidal wetlands, shallow-water nursery habitats, and cultural/archeological resources. In many cases, the actual project "footprint" may not encroach upon the habitat in question; however, both short- and long-term impacts may affect critical areas. For example, a dredge and fill project may create a turbidity plume that could reduce light penetration in a nearby seagrass bed. Beach nourishment projects may affect nearby offshore areas, including coral reefs and other hard-bottom habitats by anchoring dredges or placement of pipelines. Prolonged beach nourishment or

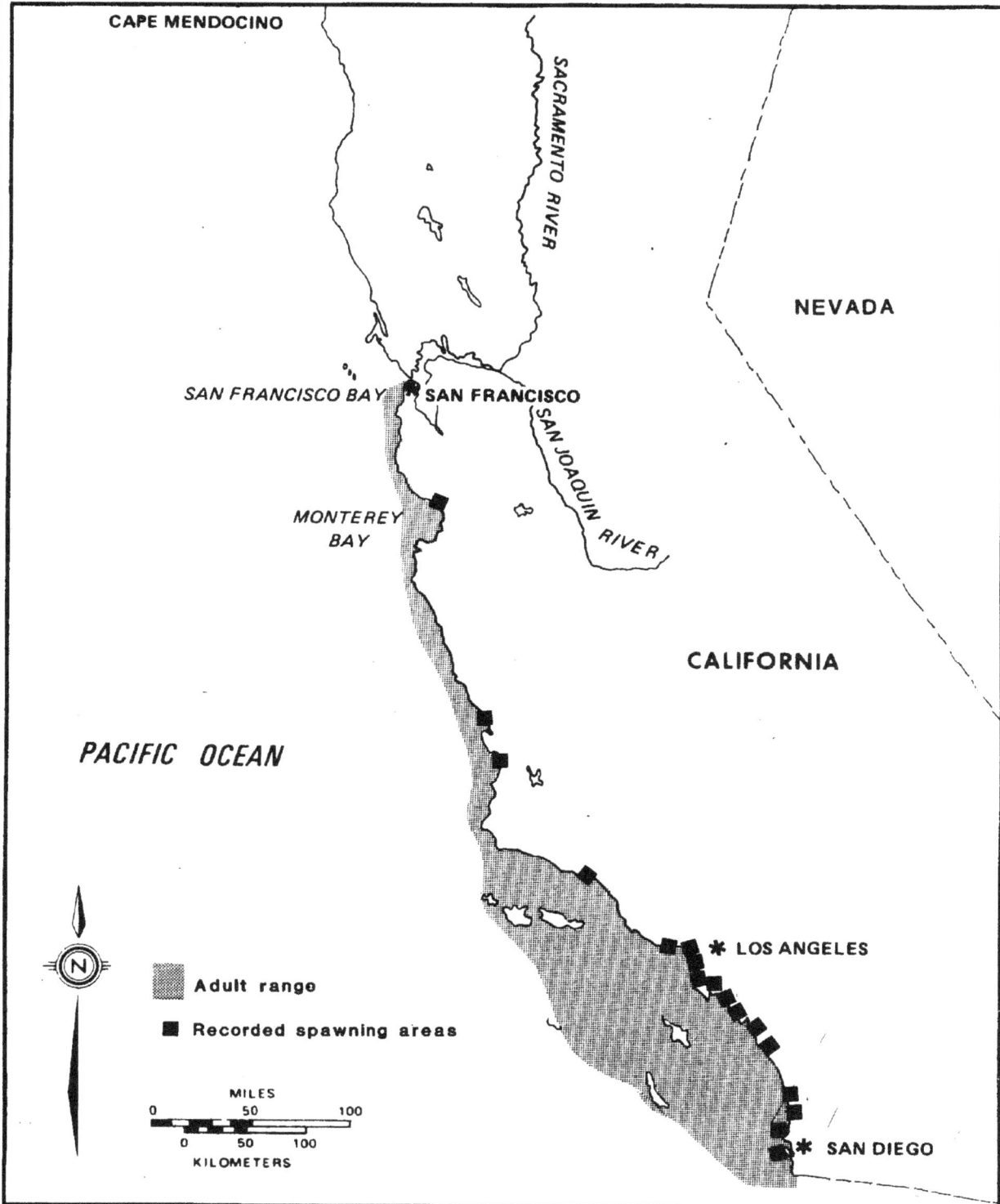

Figure V-7-4. Distribution of the California grunion along the southwest U.S. coast

Coastal Engineering for Environmental Enhancement

excessive erosion of recently nourished beaches may result in chronic conditions of high turbidity or smothering in offshore habitats. Marine construction may result in chronic nutrient or contaminant loading, which could affect biota in adjacent tidal wetlands or other nursery areas. The timing of coastal engineering projects should consider known spawning and migration patterns of fishes and invertebrates. Many commercially important fishery species enter shallow coastal waters during spring and summer as larvae or juveniles, and are especially susceptible to environmental disturbances associated with coastal engineering projects. Similarly, beach nourishment and shoreline stabilization projects should not be conducted during the nesting season for shorebirds or sea turtles (typically late spring through summer).

(2) Sites of known or potential cultural or archaeological importance are frequently encountered in coastal areas. Cultural and archaeological resources provide valuable physical evidence (e.g., structures, sites, artifacts) of past human history. This information can be used to assess historic and prehistoric use of an area for habitation, fishing, agriculture, and navigation. Native Americans and early European colonists made extensive use of coastal areas in North America, and archaeological remains associated with Indian middens and early colonial settlements can often be encountered in coastal habitat or engineering projects. Construction can potentially disrupt or destroy valuable cultural or archaeological remains. Offshore dredging projects may encounter historic shipwrecks. Pre-construction site assessments using divers, side-scan sonar, and magnetometers will identify potential cultural resources prior to construction or dredging. The potential of a project site to support such artifacts should be assessed prior to initiation of projects in coastal areas. The National Historic Preservation Act of 1966 directs the Federal Government to preserve, restore, and maintain historic or cultural sites; coastal engineering and construction activities associated with Federal navigation or shore protection projects must coordinate with Federal, state, or local historic preservation authorities.

(3) Avoidance or minimization of adverse effects to critical areas and habitats need not be viewed as an impediment to coastal engineering projects. Rather, these issues can be viewed as opportunities to apply innovative technology to environmental problem-solving in the coastal zone. Multi-disciplinary partnerships between scientists, resource managers, and engineers, and the continued development of comprehensive, regional management plans in coastal areas are necessary to ensure the success of both habitat restoration projects and traditional shoreline protection projects. Institutional and regulatory authorities are increasingly recognizing the need for adaptive management, design flexibility, and applications of new engineering technology in coastal areas; future efforts will surely benefit from these opportunities, and result in improved management of coastal resources.

V-7-10. Information Sources

a. An important part of any coastal habitat project is doing the proper research. First steps usually involve talking with local resource agencies that are familiar with the area. Sometimes a watershed analysis or restoration plan for the area has already identified specific needs. Corps libraries have a wealth of information on historic conditions, hydraulic analyses, and coastal project descriptions.

b. Several scientific societies are partially or wholly dedicated to habitat restoration issues. Some of these are:

- Society for Ecological Restoration (SER). The SER hosts conferences, provides proceedings, and has a journal. They maintain a Web site at http://ser.org/.

- National Estuarine Research Federation (ERF). The ERF hosts conferences, provides proceedings , and has a journal. They maintain a Web site at http://www.erf.org/.

- Society for Conservation Biology (CB). The CB hosts conferences, provides proceedings, and has a journal. They maintain a Web site at http://conbio.rice.edu/scb/.

- Society of Wetland Scientist (SWS). The SWS hosts conferences, provides proceedings, and has a journal. They maintain a Web site at http://www.sws.org.

In addition, the Corps offers some good information through its Web sites:

- Institute for Water Resources, http://www.wrsc.usace.army.mil.iwr.

- U.S. Army Engineer Research and Development Center http://www.erdc.usace.army.mil.

- U.S. Army Corps of Engineers Headquarters policy homepage, http://www.usace.army.mil.

V-7-11. References

EM 1110-2-5027
EM 1110-2-5027, "Confined Disposal of Dredged Material"

EM 1110-2-1206
EM 1110-2-1206, "Environmental Engineering for Small Boat Basins"

Addy 1947
Addy, C. E. 1947. "Eelgrass Planting Guide," *Maryland Conservationist*, Vol 24, pp 16-17.

Auld and Schubel 1978
Auld, A. H., and Schubel, J. R. 1978. "Effects of Suspended Sediment on Fish Eggs and Larvae: A Laboratory Assessment," *Estuarine and Coastal Marine Science*, Vol 6, pp 153-64.

Averett, Palermo, and Wade 1988
Averett, D. E., Palermo, M. R., and Wade, R. 1988. "Verification of Procedures for Designing Dredged Material Containment Areas for Solids Retention," Technical Report D-88-2, U.S. Army Engineer Waterways Experiment Station, Vicksburg, MS.

Barko et al. 1977
Barko, J. W., Smart, R. M., Lee, C. R., Landin, M. C., Sturgis, T. C., and Gordon, R. N. 1977. "Establishment and Growth of Selected Freshwater and Coastal Marsh Plants in Relation to Characteristics of Dredged Sediments," Technical Report D-77-2, U.S. Army Engineer Waterways Experiment Station, Vicksburg, MS.

Bellis 1995
Bellis, V. J. 1995. "Ecology of Maritime Forests of the Southern Atlantic Coast: A Community Profile," Biological Report 30, U.S. Department of the Interior, National Biological Service, Washington, DC.

Boesch and Turner 1984
Boesch, D. F., and Turner, R. E. 1984. "Dependence of Fishery Species on Salt Marshes: The Role of Food and Refuge," *Estuaries*, Vol 7, pp 460-68.

Bohnsack 1989

Bohnsack, J. A. 1989. "Are High Densities of Fishes at Artificial Reefs the Result of Habitat Limitation or Behavioral Preference?" *Bulletin of Marine Science*, Vol 44, pp 631-645.

Bohnsack and Sutherland 1985

Bohnsack, J. A., and Sutherland, S. P. 1985. "Artificial Reef Research: A Review with Recommendations for Future Priorities," *Bulletin of Marine Science*, Vol 37, pp 11-39.

Broome, Seneca, and Woodhouse 1988

Broome, S. W., Seneca, E. D., and Woodhouse, W. W., Jr. 1988. "Tidal Salt Marsh Restoration," *Aquatic Botany*, Vol 32, pp 1-22.

Brown and Clark 1968

Brown, C. L., and Clark, R. 1968. "Observations on Dredging and Dissolved Oxygen in a Tidal Waterway," *Water Resources Research*, Vol 4, pp 1381-1384.

Buckley and McCaffrey 1978

Buckley, F. G., and McCaffrey, C. A. 1978. "Use of Dredged Material Islands by Colonial Seabirds and Wading Birds in New Jersey," Technical Report D-78-1, U.S. Army Engineer Waterways Experiment Station, Vicksburg, MS.

Butler, Nelson, and Henwood 1987

Butler, R., Nelson, W. A., and Henwood, T. A. 1987. "A Trawl Survey Method for Estimating Loggerhead Turtle, *Caretta caretta*, Abundance in Five Eastern Florida Channels and Inlets," *Fishery Bulletin*, Vol 85, pp 447-453.

Chaney et al. 1978

Chaney, A. H., Chapman, B. R., Karges, J. P., Nelson, D. A., Schmidt, R. R., and Thebeau, L. C. 1978. "Use of Dredged Material Islands by Colonial Seabirds and Wading Birds in Texas," Technical Report D-78-8, U.S. Army Engineer Waterways Experiment Station, Vicksburg, MS.

Cintron-Molero 1992

Cintron-Molero, G. 1992. "Restoring Mangrove Systems." *Restoring the Nation's Marine Environment*. G. W. Thayer, ed., Maryland Sea Grant, College Park, MD, pp 223-278.

Clairain et al. 1978

Clairain, E. J., Jr., Cole, R. A., Diaz, R. J., Ford, A. W., Huffman, R. T., Hunt, L. J., and Wells, B. R. 1978. "Habitat Development Field Investigations, Miller Sands Marsh and Upland Habitat Development Site, Columbia River, Oregon. Summary Report," Technical Report D-77-38, U.S. Army Engineer Waterways Experiment Station, Vicksburg, MS.

Connel 1972

Connel, J. H. 1972. "Community Interactions on Marine Rocky Intertidal Shores," *Annual review of Ecology and Systematics*, Vol 3, pp 169-192.

Culter and Mahadevan 1982

Culter, J. K., and Mahadevan, S. 1982. "Long-Term Effects of Beach Nourishment on the Benthic Fauna of Panama City Beach, Florida," Miscellaneous Report No. 82-2, Coastal Engineering Research Center, U.S. Army Engineer Research and Development Center, Vicksburg, MS.

Dahl et al. 1975
Dahl, B. E., Fall, B. A., Lohse, A., and Appan, S. G. 1975. "Construction and Stabilization of Coastal Foredunes with Vegetation: Padre Island Texas," Miscellaneous Paper No. No. 9-75, Coastal Engineering Research Center, U.S. Army Engineer Research and Development Center, Vicksburg, MS.

Darovec et al. 1975
Darovec, J. E., Jr., Carleton, J. M, Pulver, T. R., Moffler, M. D., Smith, G. B., Whitfield, W. K., Jr., Willis, C. A., Steidinger, K. A., and Joyce, E. A., Jr. 1975. "Techniques for Coastal Restoration and Fishery Enhancement in Florida," Florida Marine Research Publication No. 15, Bureau of Marine Research, Florida Department of Natural Resources.

Dickerson et al. 1995
Dickerson, D. D., Reine, K. J., Nelson, D. A., and Dickerson, C. E., Jr. 1995. "Assessment of Sea Turtle Abundance in Six South Atlantic U.S. Channels," Miscellaneous Paper EL-95-5, U.S. Army Engineer Waterways Experiment Station, Vicksburg, MS.

Earhart, Clarke, and Shipley 1988
Earhart, G., Clarke, D., and Shipley, J. 1988. "Beneficial Uses of Dredged Material in Shallow Coastal Waters; Chesapeake Bay Demonstrations," Environmental Effects of Dredging Information Exchange Bulletin D-88-6, U.S. Army Engineer Waterways Experiment Station, Vicksburg, MS.

Fonseca 1994
Fonseca, M. S. 1994. "A Guide to Planting Seagrasses in the Gulf of Mexico," Texas A&M University Sea Grant College Program, TAMU-SG-94-601.

Fonseca and Fisher 1986
Fonseca, M. S., and Fisher, J. S. 1986. "A Comparison of Canopy Friction and Sediment Movement Between Four Species of Seagrass with Reference to Their Ecology and Restoration," *Marine Ecology Progress Series*, Vol 29, pp 15-22.

Fonseca, Kenworthy, and Thayer 1987
Fonseca, M. S., Kenworthy, W. J., and Thayer, G. W. 1987. "Transplanting of the Seagrasses *Halodule wrightii*, *Syringodium filiforme*, and *Thalassia testudinum* for Sediment Stabilization and Habitat Development in the Southeast Region of the United States," Technical Report EL-87-8, U.S. Army Engineer Waterways Experiment Station, Vicksburg, MS.

Fonseca et al. 1990
Fonseca, M. S., Kenworthy, W. J., Colby, D. R., Rittmaster, K. A., and Thayer, G. W. 1990. "Comparisons of Fauna Among Natural and Transplanted Eelgrass *Zostera marina* Meadows: Criteria for Mitigation," *Marine Ecology Progress Series*, Vol 55, pp 251-264.

Forman and Gordon 1986
Forman, R. T., and Gordon, M. 1986. "Landscape Ecology," John Wiley and Sons.

Frenkel and Morlan 1991
Frenkel, R. E., and Morlan, J. C. 1991. "Can We Restore Our Salt Marshes? Lessons from the Salmon River, Oregon," *The Northwest Environmental Journal*, Vol 7, pp 119-135.

Fritzsche, Chamberlain, and Fisher 1985
Fritzsche, R. A., Chamberlain, R. H., and Fisher, R. A. 1985. "Species Profiles: Life Histories and Environmental Requirements of Coastal Fishes and Invertebrates (Pacific southwest) - California Grunion," U.S. Fish and Wildlife Service Biological Report 82 (11.28), Technical Report EL-82-4, U.S. Army Engineer Waterways Experiment Station, Vicksburg, MS.

Gage 1970
Gage, B. O. 1970. "Experimental Dunes of the Texas Coast," Miscellaneous Paper No. 1-70, U.S. Army Engineer Waterways Experiment Station, Coastal Engineering Research Center, Vicksburg, MS.

Garbisch and Colemen 1978
Garbisch, E. W., Jr., and Coleman, L. B. 1978. "Tidal Freshwater Marsh Establishment in Upper Chesapeake Bay." *Freshwater Wetlands: Ecological Processes and Management Potential.* R.E. Good, D.F. Whigham, and R.L Simpson, eds., Academic Press, New York, pp 285-298.

Giannio and Wang 1974
Giannio, S. P., and Wang, H. 1974. "Engineering Considerations for Marinas in Tidal Marshes," Report No. DEL-SG-9-74, College of Marine Studies, University of Delaware, Newark, DE.

Hawkins et al. 1992
Hawkins, S. J., Allen, J. R., Russell, G., White, K. N., Conlan, K., Hendry, K., and Jones, H. D. 1992. "Restoring and Managing Disused Docks in Inner City Areas," *Restoring the Nation's Marine Environment.* G. W. Thayer, ed., Maryland Sea Grant, College Park, MD, pp 473-582

Hirsch, DiSalvo, and Peddicord 1978
Hirsch, N. D., DiSalvo, L. H., and Peddicord, R. 1978. "Effects of Dredging and Disposal on Aquatic Organisms," Technical Report DS-78-5, U.S. Army Engineer Waterways Experiment Station, Vicksburg, MS.

Hobson 1981
Hobson, R. D. 1981. "Beach Nourishment Techniques; Report 3: Typical U.S. Beach Nourishment Projects Using Offshore Sand Deposits," Technical Report H-76-13, U.S. Army Engineer Waterways Experiment Station, Coastal Engineering Research Center, Vicksburg, MS.

Homziak, Fonseca, and Kenworthy 1982
Homziak, J., Fonseca, M. S., and Kenworthy, W. J. 1982. "Macrobenthic Community Structure in a Transplanted Eelgrass Meadow," *Marine Ecology Progress Series*, Vol 9, pp 211-221.

Jaap 1984
Jaap, W. C. 1984. "The Ecology of South Florida Coral Reefs: A Community Profile," U.S. Fish and Wildlife Service, Division of Biological Services, Washington, DC, FWS/OBS-82/32.

James 1975
James, W. R. 1975. "Techniques in Evaluating Suitability of Borrow Material for Beach Nourishment," Technical Manual 60, U.S. Army Engineer Waterways Experiment Station, Coastal Engineering Research Center, Vicksburg, MS.

Josselyn and Buzholz 1982
Josselyn, M., and Bucholz, J. 1982. "Summary of Past Wetland Restoration Projects in California." *Wetland Restoration and Enhancement in California, Proceedings of a Conference, February 1982, California State University, Hayward.* M. Josselyn, ed., California Sea Grant Report T-CSGCP-007, Tiburon Center for Environmental Studies, Tiburon, CA, 1-10.

Kelly, Fuss, and Hall 1971
Kelly, J. A., Jr., Fuss, C. M., Jr., and Hall, J. R. 1971. "The Transplanting and Survival of Turtle Grass, *Thalassia testudinum*, in Boca Ciega Bay, Florida," *Fishery Bulletin*, Vol 69, pp 273-280.

Knutson 1977
Knutson, P. L. 1977. "Planting Guidelines for Dune Creation and Stabilization," Coastal Engineering Technical Aid No. 77-4, U.S. Army Engineer Waterways Experiment Station, Coastal Engineering Research Center, Vicksburg, MS.

Knutson 1980
Knutson, P. L. 1980. "Experimental Dune Restoration and Stabilization, Nauset Beach, Cape Cod, Massachusetts," Technical Paper No. 80-5, U.S. Army Engineer Waterways Experiment Station, Coastal Engineering Research Center, Vicksburg, MS.

Knutson and Finkelstein 1987
Knutson, P. L., and Finkelstein, K. 1987. "Environmental Considerations for Dune-Stabilization Projects," Technical Report EL-87-2, U.S. Army Engineer Waterways Experiment Station, Vicksburg, MS.

Landin, Webb, and Knutson 1989
Landin, M. C., Webb, J. W., and Knutson, P. L. 1989. "Long-Term Monitoring of Eleven Corps of Engineers Habitat Development Field Sites Built of Dredged Material, 1974-1987," Technical Report D-89-1, U.S. Army Engineer Waterways Experiment Station, Vicksburg, MS.

Lewis 1982
Lewis, R. R., III. 1982. "Mangrove Forests." *Creation and Restoration of Coastal Plant Communities.* R.R. Lewis, III, ed., CRC Press, Inc., Boca Raton, FL, pp 153-172.

Lewis 1987
Lewis, R. R., III. 1987. "The Restoration and Creation of Seagrass Meadows in the Southeast United States." *Proceedings of the Symposium on Subtropical-Tropical Seagrasses of the Southeastern United States.* M. J. Durako, R. C. Phillips, and R. R. Lewis, III, eds., Research Publication No. 42, Bureau of Marine Research, Florida Department of Natural Resources, pp 153-173.

Lunz et al. 1978
Lunz, J. D., Zeigler, T. W., Huffman, R. T., Diaz, R. J., Clairain, E. J., Jr., and Hunt, L. J. 1978. "Habitat Development Field Investigations, Windmill Point Marsh Development Site, James River, Virginia. Summary Report," Technical Report D-77-23, U.S. Army Engineer Waterways Experiment Station, Vicksburg, MS.

Maragos 1992
Maragos, J. E. 1992. "Restoring Coral Reefs with Emphasis on Pacific Reefs." *Restoring the Nation's Marine Environment.* G. W. Thayer, ed., Maryland Sea Grant, College Park, MD, pp 141-222.

McClellan et al. 1989
McClellan, T. N., Havis, R. N., Hayes, D. F., and Raymond, G. L. 1989. "Field Studies of Sediment Resuspension by Selected Dredges," Technical Report HL-89-9, U.S. Army Engineer Waterways Experiment Station, Vicksburg, MS.

Minello, Zimmerman, and Medina 1994
Minello, T. J., Zimmerman, R. J., and Medina, R. 1994. "The Importance of Edge for Natant Macrofauna in a Created Salt Marsh," *Wetlands*, Vol 14, pp 184-198.

Moore and Orth 1982
Moore, K. A., and Orth, R. J. 1982. "Transplantation of *Zostera marina* L. into Recently Denuded Areas." *The Biology and Propagation of* Zostera marina, *Eelgrass, in the Chesapeake Bay, Virginia*. R.J. Orth and K.A. Moore, eds., Special Report No. 265 in applied Marine Science and Ocean Engineering, Virginia Institute of Marine Science, Gloucester Point, VA, pp 92-148.

NRC 1992
National Research Council. 1992. *Restoration of Aquatic Ecosystems: Science, Technology, and Public Policy.* National Academy Press, Washington, DC.

NRC 1995
National Research Council. 1995. *Beach Nourishment and Protection.* National Academy Press, Washington, DC.

Nelson 1988
Nelson, D. A. 1988. "Life History and Environmental Requirements of Loggerhead Turtles," U.S. Fish and Wildlife Service Biological Report 88(23), Technical Report EL-86-2, U.S. Army Engineer Waterways Experiment Station, Vicksburg, MS.

Nelson, Mauck, and Fletemeyer 1987
Nelson, D. A., Mauck, K., and Fletemeyer, J. 1987. "Physical Effects of Beach Nourishment on Sea Turtle Nesting, Delray Beach, Florida," Technical Report EL-87-15, U.S. Army Engineer Waterways Experiment Station, Vicksburg, MS.

Nelson 1985
Nelson, W. G. 1985. "Guidelines for Beach Restoration Projects, Part 1: Biological Guidelines," Report No. 76, Florida Sea Grant College Program, Gainesville, FL.

Niesen and Josselyn 1981
Niesen, T., and Josselyn, M. 1981. "The Hayward Regional Shoreline Marsh Restoration: Biological Succession During the First Year Following Dike Removal," Technical Report No. 1, Tiburon Center for Environmental Studies, Tiburon, CA.

Odum et al. 1984
Odum, W. E., Smith, T. J., III., Hoover, J. K., and McIvor, C. C. 1984. "The Ecology of Tidal Freshwater Marshes of the United States East Coast: A Community Profile," U.S. Fish and Wildlife Service, Division of Biological Services, Washington, DC, FWS/OBS-83/17.

Paine and Levin 1981
Paine, R. T., and Levin, S. A. 1981. "Intertidal Landscapes: Disturbance and the Dynamics of Pattern," *Ecological Monographs*, Vol 51, pp 145-178.

Palermo, Montgomery and Poindexter 1978
Palermo, M. R., Montgomery, R. L., and Poindexter, M. 1978. "Guidelines for Designing, Operating, and Managing Dredged Material Containment Areas," Technical Report DS-78-10, U.S. Army Engineer Waterways Experiment Station, Vicksburg, MS.

Parnell, DuMond, and McCrimmon 1986
Parnell, J. F., DuMond, D. M., and McCrimmon, D. A. 1986. "Colonial Waterbird Habitats and Nesting Populations in North Carolina Estuaries: 1983 Survey," Technical Report D-86-3, U.S. Army Engineer Waterways Experiment Station, Vicksburg, MS.

Parnell, DuMond, and Needham 1978
Parnell, J. F., DuMond, D. M., and Needham, R. N. 1978. "A Comparison of Plant Succession and Bird Utilization on Diked and Undiked Dredged Material Islands in North Carolina Estuaries," Technical Report D-78-9, U.S. Army Engineer Waterways Experiment Station, Vicksburg, MS.

Pastorok et al. 1997
Pastorok, R. A., MacDonald, A., Sampson, J. R., Wilber, P., Yozzo, D. J., and Titre, J. P. 1997. "An Ecological Decision Framework for Environmental Restoration Projects," *Ecological Engineering*, Vol 9, pp 89-107.

Phillips 1980
Phillips, R. C. 1980. "Planting Guidelines for Seagrasses," Coastal Engineering Technical Aid 80-2, U.S. Army Engineer Waterways Experiment Station, Coastal Engineering Research Center, Vicksburg, MS.

Pickart, Miller, and Duebendorfer 1998
Pickart, A. J., Miller, L. M., and Duebendorfer, T. E. 1998. "Yellow Bush Lupine Invasion in Northern California Coastal Dunes; I: Ecological Impacts and Manual Restoration Techniques," *Restoration Ecology*, Vol 6, No. 1, pp 59-68.

Pickart et al. 1998
Pickart, A. J., Theiss, K. C., Stauffer, H. B., and Olsen, G. T. 1998. "Yellow Bush Lupine Invasion in Northern California Coastal Dunes; II: Mechanical Restoration Techniques," *Restoration Ecology*, Vol 6, No. 1, pp 69-74.

Pulver 1976
Pulver, T. R. 1976. "Transplant Techniques for Sapling Mangrove Trees, *Rhizophora mangle*, *Laguncularia racemosa*, and *Avicennia germinans*, in Florida," Florida Marine Research Publication No. 22, Bureau of Marine Research, Florida Department of Natural Resources.

Race 1985
Race, M. S. 1985. "Critique of Present Wetlands Mitigation Policies in the United States Based on Analysis of Past Restoration Projects in San Francisco Bay," *Environmental Management*, Vol 9, pp 71-82.

Ray 1994
Ray, G. L. 1994. "A Conceptual Framework for the Evaluation of Coastal Habitats," Technical Report EL-94-33, U.S. Army Engineer Waterways Experiment Station, Vicksburg, MS.

Ray et al. 1994
Ray, G. L., Clarke, D., Wilber, P., and Fredette, T. J. 1994. "Ecological Evaluation of Mud Flat Habitats on the Coast of Maine Constructed of Dredged Material," Environmental Effects of Dredging Information Exchange Bulletin D-94-3, U.S. Army Engineer Waterways Experiment Station, Vicksburg, MS.

Raymond 1984
Raymond, G. L. 1984. "Techniques to Reduce the Sediment Resuspension Caused by Dredging," Miscellaneous Paper HL-84-3, U.S. Army Engineer Waterways Experiment Station, Vicksburg, MS.

Rey et al. 1990
Rey, J. R., Shaffer, J., Tremain, D., Crossman, R. A., and Kain, T. 1990. "Effects of Re-establishing Tidal Connections in Two Impounded Subtropical Marshes on Fishes and Physical Conditions," *Wetlands*, Vol 10, pp 27-45.

Roberts 1991
Roberts, T. H. 1991. "Habitat Value of Man-Made Coastal Marshes in Florida," Technical Report WRP-RE-2, U.S. Army Engineer Waterways Experiment Station, Vicksburg, MS.

Roman, Niering, and Warren 1984
Roman, C. T., Niering, W. A., and Warren, R. S. 1984. "Salt Marsh Vegetation Change in Response to Tidal Restriction," *Environmental Management*, Vol 8, pp 141-150.

Rozas and Hackney 1983
Rozas, L. P., and Hackney, C. T. 1983. "The Importance of Oligohaline Estuarine Wetland Habitats to Fisheries Resources," *Wetlands*, Vol 3, pp 77-89.

Schiel and Foster 1992
Schiel, D. R., and Foster, M. S. 1992. "Restoring Kelp Forests." *Restoring the Nation's Marine Environment.* G.W. Thayer, ed., Maryland Sea Grant, College Park, MD, pp 279-342.

Schroeder 1983
Schroeder, P. R. 1983. "Chemical Clarification Methods for Confined Dredged Material Disposal," Technical Report D-83-2, U.S. Army Engineer Waterways Experiment Station, Vicksburg, MS.

Seaman and Sprague 1991
Seaman, W., Jr., and Sprague, L. M. 1991. *Artificial Habitats for Marine and Freshwater Fisheries*, Academic Press, San Diego, CA.

Seneca 1974
Seneca, E. D. 1974. "Stabilization of Coastal Dredge Spoil with *Spartina alterniflora*." *Ecology of Halophytes.* R. J. Reimold and W. H. Queen, eds., Academic Press, New York, pp 525-529.

Seneca et al. 1976
Seneca, E. D., Broome, S. W., Woodhouse, W. W., Jr., Cammen, L. M., and Lyon, J. T., III. 1976. "Establishing *Spartina alterniflora* marsh in North Carolina," *Environmental Conservation*, Vol 3, pp 185-188.

Sheehy and Vik 1992
Sheehy, D. J. and Vik, S. F. 1992. "Developing Prefabricated Reefs: An Ecological and Engineering Approach." *Restoring the Nation's Marine Environment*. G. W. Thayer, ed., Maryland Sea Grant, College Park, MD, pp 543-582.

Shisler 1990
Shisler, J. K. 1990. "Creation and Restoration of the Coastal Wetlands of the Northeastern United States." *Wetland Creation and Restoration: the Status of the Science*. J. A. Kusler and M. E. Kentula, eds., Island Press, Washington, DC, pp 143-170.

Short, Muehlstein, and Porter 1987
Short, F. T., Muehlstein, L. K., and Porter, D. 1987. "Eelgrass Wasting Disease: Cause and Recurrence of a Marine Epidemic," *Biological Bulletin*, Vol 173, pp 557-562.

Shreffler, Simenstad, and Thom 1992
Shreffler, D. K., Simenstad, C. A., and Thom, R. M. 1992. "Foraging by Juvenile Salmon in a Restored Estuarine Wetland," *Estuaries*, Vol 15, pp 204-213.

Simenstad and Thom 1996
Simenstad, C. A., and Thom, R. M. 1996. "Functional Equivalency Trajectories of the Restored Gog-Le-Hi-Te Estuarine Wetland," *Ecological Applications*, Vol 6, pp 38-56.

Sinicrope et al. 1990
Sinicrope, T. L., Hine, P. G., Warren, R. S., and Neiring, W. A. 1990. "Restoration of an Impounded Salt Marsh in New England," *Estuaries*, Vol 13, pp 25-30.

Smith 1978
Smith, H. K. 1978. "An Introduction to Habitat Development on Dredged Material," Technical Report DS-78-19, U.S. Army Engineer Waterways Experiment Station, Vicksburg, MS.

Snyder 1976
Snyder, G. R. 1976. "Effects of Dredging on Aquatic Organisms with Special Application to Areas Adjacent to the Northeastern Pacific Ocean," *Marine Fisheries Review*, Vol 38, pp 34-38.

Soots and Landin 1978
Soots, R. F, Jr., and Landin, M. C. 1978. "Development and Management of Avian Habitat on Dredged Material Islands," Technical Report D-78-18, U.S. Army Engineer Waterways Experiment Station, Vicksburg, MS.

Stevens and Rejmankova 1995
Stevens, M. L., and Rejmankova, E. 1995. "Cache Slough/Yolo Bypass Ecosystem Monitoring Study to Determine Wetland Mitigation Success." Technical Report WRP-RE-11, U.S. Army Engineer Waterways Experiment Station, Vicksburg, MS.

Teas 1977
Teas, H. J. 1977. "Ecology and Restoration of Mangrove Shorelines in Florida," *Environmental Conservation*, Vol 4, pp 51-58.

Thom 1990
Thom, R. M. 1990. "A Review of Eelgrass (*Zostera marina* L.) Transplanting Projects in the Pacific Northwest," *The Northwest Environmental Journal*, Vol 6, pp 121-137.

Thom 1997
Thom, R. M. 1997. "System-Development Matrix for Adaptive Management of Coastal Ecosystem Restoration Projects," *Ecological Engineering*, Vol 8, pp 219-232.

Thom and Wellman 1996
Thom, R. M., and Wellman, K. F. 1996. "Planning Aquatic Ecosystem Restoration Monitoring Program," IWR Report 96-R-23, U.S. Army Corps of Engineers Institute for Water Resources, Alexandria, VA.

Thorhaug 1985
Thorhaug, A. 1985. "Large-Scale Seagrass Restoration in a Damaged Estuary," *Marine Pollution Bulletin*, Vol 16, pp 55-62.

Thorhaug 1986
Thorhaug, A. 1986. "Review of Seagrass Restoration Efforts," *Ambio*, Vol 15, pp 110-117.

Thorhaug and Austin 1976
Thorhaug, A., and Austin, C. B. 1976. "Restoration of Seagrasses with Economic Analysis," *Environmental Conservation*, Vol 3, pp 259-268.

Van Dolah and Maier 1993
Van Dolah, R. F., and Maier, P. P. 1993. "The Distribution of Loggerhead Turtles (*Caretta caretta*) in the Entrance Channel of Charleston Harbor, South Carolina, U.S.A.," *Journal of Coastal Research*, Vol 9, pp 1004-1012.

Walters and Holling 1990
Walters, C. J., and Holling, C. S. 1990. "Large-Scale Management Experiments and Learning by Doing," *Ecology*, Vol 71, pp 2060-2068.

Weinstein 1979
Weinstein, M. P. 1979. "Shallow Marsh Habitats as Primary Nurseries for Fishes and Shellfish, Cape Fear River, North Carolina," *Fishery Bulletin*, Vol 77, pp 339-357.

Wennersten 1981
Wennersten, J. R. 1981. *The Oyster Wars of Chesapeake Bay.* Tidewater Publishers, Inc., Centreville, MD.

Wilson and Ruff 1988
Wilson, W. H., and Ruff, R. E. 1988. "Species Profiles: Life Histories and Environmental Requirements of Coastal Fishes and Invertebrates (North Atlantic) - Sandworm and Bloodworm," U.S. Fish and Wildlife Service Biological Report 82 (11.80), U.S. Army Engineer Waterways Experiment Station, Technical Report EL-82-4, Vicksburg, MS.

Woodhouse 1978
Woodhouse, W. W., Jr. 1978. "Dune Building and Stabilization with Vegetation," Special Report No. 3., U.S. Army Engineer Waterways Experiment Station, Coastal Engineering Research Center, Vicksburg, MS.

Wyllie-Echeverria, Olson, and Hershman 1994
Wyllie-Echeverria, S., Olson, A. M., and Hershman, M. J. 1994. "Seagrass Science and Policy in the Pacific Northwest," Proceedings of a Seminar Series SMA 94-1), EPA 910/R-94-004.

Zale and Merrifield 1989
Zale, A. V., and Merrifield, S. G. 1989. "Species Profiles: Life Histories and Environmental Requirements of Coastal Fishes and Invertebrates (south Florida) - Reef-Building Tube Worm," U.S. Fish and Wildlife Service Biological Report 82 (11.115), U.S. Army Engineer Waterways Experiment Station, Technical Report TR EL-82-4, Vicksburg, MS.

Zedler 1988
Zedler, J. B. 1988. "Salt Marsh Restoration: Lessons for California," *Rehabilitating Damaged Ecosystems*. 2nd ed., J. Cairns, Jr., ed., CRC Press, Inc., Boca Raton, FL, pp 75-96.

Zedler 1996a
Zedler, J. B. 1996a. "Coastal Mitigation in Southern California: The Need for a Regional Restoration Strategy," *Ecological Applications*, Vol 6, pp 84-93.

Zedler 1996b
Zedler, J. B. 1996b. "Tidal Wetland Restoration: A Scientific Perspective and Southern California Focus," Report T-038, California Sea Grant College, University of California, La Jolla, CA.

Zieman 1976
Zieman, J. C. 1976. "The Ecological Effects of Physical Damage from Motorboats on Turtle Grass Beds in Southern Florida," *Aquatic Botany*, Vol 2, pp 127-139.

V-7-9. Acknowledgments

Authors of Chapter V-7, "Coastal Engineering for Environmental Enhancement:"

David J. Yozzo, Ph.D., Barry A. Vittor & Associates, Kingston, New York

Jack E. Davis, Ph.D., Coastal and Hydraulics Laboratory (CHL), U.S. Army Engineer Research and Development Center, Vicksburg, Mississippi

Patrick T. Cagney, U.S. Army Engineer District, Seattle, Seattle, Washington

www.ingramcontent.com/pod-product-compliance
Lightning Source LLC
Chambersburg PA
CBHW081346190326
41458CB00018B/6092